MEMOIRS

ON
THE LATE WAR
IN
NORTH AMERICA
BETWEEN
FRANCE AND ENGLAND

by

Pierre Pouchot

• Revised Edition •

Translated by

Michael Cardy

Edited and Annotated by

Brian Leigh Dunnigan

Old Fort Niagara Association, Inc.

Youngstown, New York

Published by:
OLD FORT NIAGARA PUBLICATIONS
Old Fort Niagara Association, Inc.
Fort Niagara State Park
P.O. BOX 169
Youngstown, New York 14174-0169

Memoirs on the Late War in North America Between France and England
by Pierre Pouchot;
translated by Michael Cardy; edited and annotated by Brian Leigh Dunnigan.

Revised Edition – 2004

First Printing, October 2004

Cataloging-in-Publication Data (provided by Old Fort Niagara Association, Inc., based on Library of Congress practice).

Pouchot, Pierre 1712-1769
Memoirs on the late war in North America between France and England /
Pierre Pouchot (revised edition); translated by Michael Cardy ; edited and annotated by Brian Leigh Dunnigan.
640 p. : ill. ; 26 cm.

Translation of: *Mémoires sur la dernière guerre de l'Amérique septentrionale entre la France et l'Angleterre. Suivis d'Observations, dont plusieurs sont relatives au théatre actuel de la guerre, & de nouveaux détails sur les mœurs & les usages des Sauvages, avec des cartes topographiques* (3 vols.; Yverdon, Switzerland, 1781).

Includes bibliographical references (pp. 596-617) and index.

1. Pouchot, Pierre, 1712-1769--Personal narratives. **2.** United States--History--French and Indian War, 1755-1763--Campaigns. **3.** Fort Niagara (N.Y.)--Siege, 1759. **4.** Fort Levis (N.Y.)--Siege, 1760. **5.** Niagara River Region (N.Y. and Ont.)--History--18th Century. **6.** Indians of North America--Social life and customs---18th Century.
I. Cardy, Michael, tr. **II.** Dunnigan, Brian Leigh, ed. **III.** Title
IV. Title: Pouchot's Memoirs

E199.P8913 1994 973.2
ISBN: 0-941967-24-7 (Case Bound)
Printed in the United States of America on acid-free paper.

Printing & Binding: Thomson-Shore, Inc., Dexter, Michigan.

Table Of Contents

...(continued next page)...

Table of Contents – Vol. III: Topographical Observations . . . continued

Maps, Plans & Other Illustrations:

Preface

It is thus a solemn duty to assemble the facts, or to safeguard them against the sponge of oblivion....By publishing his memoirs, we pay off a debt of his fellow citizens & by giving him here the proper tribute of our praise, we fulfill our duty....(Mémoires, I, xi-xii)

When the Old Fort Niagara Association undertook the publication of Pierre Pouchot's *Mémoires,* in 1994, it was hoped that Michael Cardy's new translation would make this important work more readily accessible to both academic and non-academic scholars of the French and Indian War. That goal appears to have been achieved, and the Old Fort Niagara edition of Pouchot's commentary has been cited in many of the fine new studies on the conflict that have appeared over the past ten years. The 1994 edition of Pouchot is now out of print, but there has been a continuing demand for it. Thus, we have decided to produce a revised edition.

Tinkering with a completed project seems to be a basic human impulse, and a new edition of Pouchot provides the opportunity to make a few improvements and to correct some annoying errors. It has also been possible to incorporate a number of recent works on the French and Indian War into the notes and bibliography. The most substantial change will be found in "Appendix D," where we have added a journal of Indian activities at Fort Niagara during the summer of 1757. This was almost certainly kept by Pouchot, and it

The closest thing to a likeness of Pierre Pouchot is this fanciful rendering of the surrender of Fort Niagara. It was inset on an engraving entitled "Conquests in the Glorious 1759" published in the "London Gazette Extraordinary." *Courtesy, National Archives of Canada (C7223).*

supplements his narrative for a period that is dealt with in cursory fashion in his *Mémoires.*

One thing that has not changed in ten years is the motivation behind our efforts. The purpose of this new translation of Pierre Pouchot's *Mémoires sur la dernière guerre* is to make an important work on the Seven Years War in America more readily available to both scholars and enthusiasts, especially those who do not read French. The rarity of both the original edition of 1781 and the 1866 Franklin Hough translation has limited the use of the information provided by Pouchot's rich narrative. Hough's nineteenth century translation is, moreover, dated and in many cases imprecise or inaccurate, particularly when dealing with military terminology and events. It is hoped that a fresh, annotated translation will open new and broader vistas on the French conduct of the war in North America. At the very least, this edition will make it possible, as Franklin Hough had intended with his own, for a much larger number of libraries and students to have a copy of Pouchot's *Mémoires* on their shelves.

Additional research involves additional thanks. To the names of those who assisted with preparation of the 1994 edition we would like to add those of Robert Andrews, Carl Benn, Andrew Gallup, Tim Todish, Lee Vedder, and the late Dean Ulrich. Particular thanks, for a second time, to René Chartrand, André Gousse, Don Graves, Paul Huey, and Nicholas Westbrook. And, to Claire Beausom Dunnigan, who was only just learning to talk in 1994, thanks for your help in editing copy for this second effort.

Brian Leigh Dunnigan
Grass Lake, Michigan

Michael Cardy
Swansea, Wales

Harry M. DeBan,
Chairman and Publisher
Old Fort Niagara Publications
Youngstown, New York

For

Bernice
&
Candice

Elles entrent en effet dans les grands conseils
de la nation, & y influent beaucoup.

and for

the Captain

Ses généraux le regréterent,
& sa perte fut
pleurée par sa famille...

This publication was fostered, in part, by the generosity, support & artistry of
Robert Griffing and Paramount Press
who dedicated proceeds from "Departure at Daybreak" to the original project.

INTRODUCTION

The Seven Years War, or the "French and Indian War" as its colonial equivalent has long been popularly known in the United States, left many important milestones in the history of North America. Two well-established colonial powers resolved, by a protracted war, the question of dominance over the northern half of the continent. New France became an English possession, but one with a culturally durable Francophone population that would profoundly affect the development of modern Canada. The original English colonies, relieved of the threat presented by a rival colonial power, were freed to seek their own course of development. Within two decades the global empire that fell to England in 1763 would be shaken by the loss of much of North America to the emergent United States.

On a narrower plane, the American war witnessed the first significant commitment of regular military forces to both colonies. Large numbers of English and French officers, trained in Europe and, in many cases, veterans of its wars, were introduced to an alien environment possessing extremes of climate, geography and warfare unfamiliar to their experience. American service brought these professional soldiers into contact with two distinct native populations with whose members they served or fought: European colonists and American Indians. French and English colonists had inhabited North America for several generations by 1755, and they had developed behavior, both social and military, far enough removed from their origins to make them seem quite foreign to officers from the mother countries. The other population, the original peoples of North America, provided an experience with a culture and a manner of making war that was as fascinating and frustrating to the professional soldiers as it was unfamiliar.

While many European officers recorded their impressions of the war and the peoples of America, only a relative few of such narratives found their way into print during the eighteenth century. All but one present the English side of the conflict. The reason is easily understood. The war was one of victory for England, and a number of histories were soon published in that country. John Entick's *The General History of the Late War* (1763), John Knox's *An Historical*

Journal of the Campaigns in North America (1764) and Thomas Mante's *The History of the Late War in North-America and the Islands of the West-Indies* (1772) familiarized English readers with the events and the settings of the North American conflict. The French, having sustained staggering losses at sea and in their colonial possessions, were far less inclined to celebrate the events of 1754-1760.

There was, in fact, only one contemporary French work that addressed the subject, and it did not appear until twenty-one years after the capitulation of New France and more than a decade subsequent to the death of its author. The three small volumes, entitled *Mémoires sur la dernière guerre de l'Amérique septentrionale,* were published in 1781 in Yverdon, a town on Lac de Neuchâtel in the west of Switzerland, not far from the French border.[1] Its author was identified as "M. Pouchot", a former captain in the Béarn Regiment and a Chevalier of the Royal and Military Order of St. Louis. Pouchot's work was termed *mémoires,* a genre akin to the modern autobiography but more properly a personal history of the author's participation in great events. It was a form of literature well suited for vindicating to the French public the conduct of both the author and of the military force with which he served in what had proved to be a lost cause.[2]

The publication of Pouchot's work so long after the events it described was likely due to a number of factors. The author did not seriously begin preparation of his manuscript until 1768 or 1769, and it was still far from complete when he was killed in action in the latter year. It is probable that the draft, which was part of the estate bequeathed to one of his brothers, languished until sometime in the next decade. Another reason for the delay is that much of the subject matter of the *Mémoires* is highly critical of the manner in which the war had been waged in the French colony. Pouchot's text contains

---

1 Pierre Pouchot, *Mémoires sur la dernière guerre de l'Amérique septentrionale entre la France et l'Angleterre. Suivis d'Observations, dont plusieurs sont relatives au théatre actuel de la guerre, & de nouveaux détails sur les mœurs & les usages des Sauvages, avec des cartes topographiques* (3 vols.; Yverdon, Switzerland, 1781).

2 An analysis of the form and the intent of Pouchot's *Mémoires* may be found in Michael Cardy, "The Memoirs of Pierre Pouchot: A Soldier's View of a Doomed Campaign", *War, Literature, and the Arts,* IV (Spring 1992), pp. 1-23.

Cardy notes that the word "autobiography" (*autobiographie*) did not exist in either French or English when Pouchot's *Mémoires* were published in 1781. Modelled on the word "biography", a coinage of the mid-eighteenth century, the term "autobiography" appeared around 1800.

repeated indictments of the corruption of colonial officials. Publication of his book outside France and twenty years after the fact supports the belief that its contents remained sufficiently controversial for there to have been concern about official reaction. Finally, the outbreak of a new American conflict, followed by a French declaration of war against England in 1778 and the commitment of naval and military forces to support their rebellious colonists, gave new relevance to Pouchot's narrative of the events that had helped shape the contemporary situation.

Pouchot's *Mémoires* is thus the only comprehensive French account of the war in America published during the eighteenth century. The quantity of the original printing is unknown, and there is no indication that further editions were produced. By the beginning of the nineteenth century the book was recognized as a historical source, but the fact that it had been published in French made it largely inaccessible to English-speaking North American historians. This situation began to change with the publication, in 1858, of a translation of that part of Pouchot's second volume recounting the siege of Fort Niagara.[3]

This tantalizing bit of Pouchot's colorful and often intensively detailed narrative was followed, in 1866, by a two-volume English edition of the complete work.[4] Translated and edited by Franklin B. Hough, a New York antiquarian and local historian, it included Pouchot's original text as well as the footnotes provided by his 1781 editor. Hough added a number of his own notes to clarify or identify events, places and terms. He also included a brief introduction outlining his purpose. Hough hoped, first, "to contribute a valuable addition to the literature of an interesting period of American history." He observed that, although the work "has been in the hands of historians engaged upon the special period to which it relates, it is scarcely known in our public libraries, or in private collections of American history." In order to make Pouchot's text more available, Hough translated from a copy of the 1781 edition owned by the library of Harvard University.[5]

---

[3] E.B. O'Callaghan (ed.), *Documents Relative to the Colonial History of the State of New York* (Hereafter cited as *DRCHSNY;* 15 vols.; Albany, 1856-77), X, pp. 977-92.

[4] Pierre Pouchot, *Memoir Upon the Late War in North America Between the French and English, 1755-60.* Translated and edited by Franklin B. Hough. (2 vols.; Roxbury, MA, 1866).

[5] *Ibid.,* pp. iii-iv.

It is in a similar spirit that this edition has been prepared. The realization of Franklin Hough's goal was limited by the fact that only two hundred copies of his translation were printed. These have today become rare books. And, despite the fact that many French documents have been published in both languages in the years since 1866, there nonetheless remains a startling lack of French primary source material available to the historian or enthusiast fluent only in English. A translation of Pouchot's narrative provides an accessible French counterpart to the English histories of the conduct of the Seven Years War in America.

In addition to its value as a French account of the war, the *Mémoires* contain a fascinating personal chronicle of the events in which Pouchot participated. There are also invaluable first-hand observations on the country and its Indian inhabitants. In several cases, most notably for events at Niagara and the 1760 campaign on the St. Lawrence River, no other source provides as much detail from the French perspective.

The military career and services of Captain Pouchot are presented in some detail in the pages of his *Mémoires,* and the information provided by the author and his 1781 editor has been augmented in the notes to this edition. A summary and evaluation of Pouchot's career is, nonetheless, in order here. Much of the personal information is based upon details from his *Mémoires* and the official record of his military promotions. Further particulars may be found in the *Dictionary of Canadian Biography* and in the sources cited below.[6]

Pouchot was born on April 8, 1712 in Grenoble, France where, the *Mémoires* imply, his father was a merchant. Pouchot's Christian name appears in few places. Among them is an entry in the parish records of the church of Notre Dame de Montréal that identifies him as "Pierre."[7] He usually signed simply "Pouchot" and, occasionally, as

---

6 Pouchot's promotions within the Béarn Regiment are detailed in a document entitled "Contrôle des Capitaines, 1746-1763; Béarn", Service historique de l'Armée, Paris (hereafter cited as SHA), Y^b 121. Copy in the National Archives of Canada. The most current biographical summary is Peter N. Moogk, "Pierre Pouchot", *Dictionary of Canadian Biography* (Hereafter cited as *DCB*; 12 vols. and index; Toronto, 1966-1991), III, pp. 534-37. Another discussion of his career is Brian Leigh Dunnigan's, "Pierre Pouchot: A French Soldier Views America," in Ian Steele and Nancy L. Rhoden (eds.), *The Human Tradition in Colonial History* (Wilmington, DE, 1999), pp. 197-216.

7 *Ibid.,* p. 537.

"Pouchot de Maupas" in the parish register and in his 1759 correspondence with English Major General Jeffery Amherst.[8]

Virtually nothing is known of Pouchot's physical appearance. No portrait has ever been identified. Even written descriptions are almost nonexistent. The only one of any substance comes from an English observer who saw Pouchot arrive at Albany in August 1759. The "late Governor of Niagara," he wrote, "is a soldierly-looking man."[9]

The early death of Pouchot's father and a litigious reaction by Pouchot and his siblings to his mother's remarriage apparently propelled him towards a military life. In 1733, at the age of 21, he entered the French Army as a volunteer. Although the introduction to the *Mémoires* states that Pouchot volunteered in the capacity of engineer, it is just as likely that he began service in the Béarn Regiment of infantry in which he received a commission as lieutenant *en second* on May 1, 1734. Although subaltern commissions were not actually purchased in the French Army of that period, it was not uncommon for money to facilitate the approval of a new junior officer by both the Ministry of War and the colonel of the regiment. Pouchot's bourgeois background and his ultimate purchase of a company suggest that either he or his family possessed sufficient financial resources to launch the young man on an officer's career in an age of extreme social stratification. His background and the manner by which he entered the army identify Pouchot as a *roturier,* or commoner. It has been suggested that men of his origin might, by the 1750s, have comprised as much as one-third to one-half of the French officer corps.[10]

It is evident from every stage of his career that Pierre Pouchot had an inclination towards engineering. The *Mémoires* suggest that he might have gained some experience even before joining the army. These talents were utilized by his superiors in Europe and America, and the frequency of Pouchot's independent commands in New France was a direct result of his technical expertise. He appears, nonetheless, to have been essentially self-taught, having studied "the

8 Pouchot to Amherst, Aug. 14, 1759, Great Britain, Public Record Office, Jeffery Amherst Papers, MSS WO 34/10 (hereafter cited as "Amherst Papers" by volume).

9 Quoted in Frank H. Severance, *An Old Frontier of France* (2 vols.; New York, 1917), II, p. 360. The original source is not given but was probably either a New York or Boston newspaper.

10 Lee Kennett, *The French Armies in the Seven Years' War: A Study in Military Organization and Administration* (Durham, NC, 1967), pp. 55-58.

great art of Vauban and Cohorn amid the tumult of the camps."[11]

Despite his competence, Pierre Pouchot was never admitted to the Royal Corps of Engineers, and it was as an infantry officer that he gained promotion. The commencement of his service coincided with the War of the Polish Succession, and he first served with French armies in northern Italy. With the approach of peace, however, came a setback. On January 7, 1737, Pouchot was given the rank of ensign, a reduction in grade that he perhaps found necessary to accept in order to remain in the army. His advancement resumed during service on Corsica when he was promoted lieutenant on October 28, 1739. Pouchot's status in the Béarn Regiment further improved in May 1742 with his appointment as lieutenant of the colonel's company, the senior company of the regiment.

As in all ages, it was wartime that brought the most rapid promotion. The outbreak of the War of the Austrian Succession found Pouchot serving north of the Alps. On September 27, 1743 he was named assistant major of his regiment, a staff appointment to the chief administrative officer.[12] Advancement to captain came on August 2, 1745, but the rank was without a company. This development seems to support the contention, made in the introduction to the *Mémoires,* that the promotion was for his service at the siege of Tournai and, perhaps, for reconnoitering the road across the Tirol in 1744. At the same time, Pouchot was awarded the coveted cross of the Royal and Military Order of St. Louis, further indication of favor.

His next goal was command of a company, and this step would mark the final crossing of the line between the junior *officiers subalternes* and the *officiers supérieurs.* It also entailed purchase of the company and the financial burden of maintaining it.[13] Pouchot took four more years to reach that important objective, obtaining his own company on September 22, 1749. His ambition had perhaps become more easily attainable in 1746 when the Béarn Regiment was authorized to form a second battalion, thereby doubling the number of its companies.[14]

---

11 Pouchot, *Mémoires,* I, pp. xiv-xv. Citations are to the pages of the 1781 edition. These are identified in the text of this edition by page numbers in square brackets. Their position indicates the end of that page of the 1781 edition.

12 Kennett, p. 56.

13 *Ibid.,* pp. 55-56.

14 Maurice Sautai, *Les Débuts de la guerre de la succession d'Autriche. Tome premier. L'entrée en Bavière et en Haute-Austriche* (Paris, 1909), p. 358.

It was with the second battalion of Béarn that Pouchot sailed for New France in 1755 and the adventures related in his *Mémoires.* He was, by that time, the fourth most senior fusilier captain of his battalion.[15] This was a grade beyond which it proved impossible to advance, despite outstanding service in America and the recommendations of his superiors. Pouchot's disappointment and bitterness were obvious. He "could not cross the barrier that separates subordinate functions from upper ranks, an obstacle," his editor continued, "which intrigue & wiles overcome without difficulty."[16]

Pouchot had, in fact, been actively soliciting a lieutenant colonelcy prior to his departure for New France. The Marquis de Paulmy, later to serve as Minister of War, had even promised during an interview in Paris sometime before April 1755 "to be favorable" to his ambitions.[17] Recommended for promotion to lieutenant colonel after the capture of Oswego in 1756, Pouchot received instead the first of four pensions that, by 1759, would total 800 *livres* per year. Although he described this sum as "modest," it was equivalent to more than eight months pay (at 95 *livres* per month) for an infantry captain.[18] Pouchot's disappointment at not receiving the recommended promotion, tinged with apprehension that "my time is passing ... for I am 46 years of age," was apparent in a letter to the Minister of War.[19] Although his 1781 editor claimed that Pouchot had again been recommended for the rank of lieutenant colonel by the Marquis de Montcalm following the battle at Carillon (or Ticonderoga) in 1758, no record of this has been found.[20] Thereafter, although Pouchot

15 "Contrôle des Capitaines .... Béarn", SHA, Y^b 121. See also "Officiers du régiment de Béarn", *Le bulletin des recherches historiques,* Vol. LI, No. 10 (October 1945), pp. 356-57. Senior to Pouchot were the captain of the grenadier company and three fusilier captains. Seven of the more junior company commanders had received their captaincies in 1746, the year in which the second battalion was formed.

16 Pouchot, *Mémoires,* I, p. xxiv.

17 Pouchot to Paulmy, Nov. 3, 1757, *DRCHSNY,* X, p. 668. The letter is also printed in Appendix A of this volume.

18 Pouchot, *Mémoires,* I, p. xxiv. For citations on Pouchot's pensions see note 119. The figure for a captain's pay is from Kennett, p. 96n.

19 Pouchot to Paulmy, Nov. 3, 1757, *DRCHSNY,* X, p. 668.

20 Pouchot, *Mémoires,* I, p. xxiv. No other source corroborates this claim, and his editor might have confused the 1756 recommendation with that supposedly made in 1758.

cited his achievements at every opportunity to influential officers in Paris, his efforts to obtain further promotion were to no avail.[21]

The conclusion of Pouchot's American service and his return to France as a paroled prisoner of war in the winter of 1761 further dimmed his hopes for advancement. Unable to serve actively until formally exchanged, he joined his family in Grenoble. His prospects must have appeared completely hopeless when the Béarn Regiment was disbanded on November 25, 1762.[22] It is likely that Pouchot had not even been exchanged by that time. The war against England and her allies, with all its opportunities for enhancing a military career, ended soon after.

These substantive blows struck Pouchot only nine months after the emotional trauma of thinking his honor questioned when he had been summoned to Paris in connection with the *affaire du Canada*. Although Pouchot quickly cleared his name, and, indeed, might have been sought merely to provide evidence against the dozens of former officers and officials of New France implicated in the scandal, the incident added to his disillusionment. "The loss of my patrimony is the sole fruit remaining to me of my hard work and my service," he complained of the episode. Promised rewards for his cooperation, Pouchot was again overlooked. His editor attributed it to the fact that, "At Court they drink incessantly from the waters of forgetfulness."[23]

It is in the context of this period of his life that Pouchot's two final important actions must be viewed. The first was the composition of his *Mémoires,* clearly with the intention of publication. The other was his resumption of military service on Corsica in 1768 or 1769 at the age of 57. The latter decision was as likely made for financial reasons

---

21 See, for example, Pouchot to Belle-Isle, April 14, 1758, *DRCHSNY,* X, pp. 694-95. This letter is also included in Appendix A of this volume. When Bougainville went to France in the autumn of 1758 he carried a *mémoire* from Pouchot to be presented at court. *RAPQ,* 1923-24, p. 7. The testimonial of the Marquis de Vaudreuil, obtained on May 6, 1761 and printed on pp. xxix-xxxii of the *Mémoires,* was also clearly for the purpose of obtaining preferment.

Ironically, Belle-Isle, who was named Minister of War on March 3, 1758, was in the forefront of those advocating that senior commissions be given only to members of the nobility. In March 1758 he wrote that "the King desires henceforth that the regiments which are most used in the war be commanded only by those of quality." Belle-Isle to Cadillac, March 18, 1758, quoted in Kennett, p. 59. In at least one instance this resulted in the replacement of a lieutenant colonel of common birth.

22 Louis Susane, *Histoire de l'infanterie française* (5 vols.; Paris, 1876), V, p. 308.

23 Pouchot, *Mémoires,* I, pp. xix-xxiv.

as from a desire to serve. There was probably also the glimmer of hope that his talents would yet be properly recognized through advancement in rank. Although Pouchot obtained a staff appointment as an engineer, however, it resulted only in his death, on May 8, 1769, in a country considered by many Frenchmen to be as wild and as foreign as New France.

The *Mémoires,* as pointed out by translator Michael Cardy, would prove to be Pouchot's most enduring monument.[24] There is no doubt that they were intended to justify his conduct and illuminate his many and important contributions to the brave but futile defense of New France. The narrative also justifies the role of the *troupes de terre,* the eight infantry battalions and supporting units of the regular Army sent to assist in the defense of the colony. Seldom do these soldiers or their officers appear in an unfavorable light. Pouchot is less laudatory of the officers and men of the *troupes de la marine,* the colonial regulars administered by the Marine Department, or of their local auxiliaries, the Canadian militia and allied Indian warriors. But it is for the colonial administrators, particularly the financial personnel, and those military officers with commercial interests that Pouchot reserves his greatest criticism. These were the men who not only, in his opinion, fatally hindered the defense of New France but who had sullied his personal reputation. He takes every opportunity to detail their peculations and to document their contributions to the loss of the colony. The *Mémoires* present and sharply contrast the venal actions of Intendant Bigot and his associates with the selfless and professional motives of Pouchot and his brother officers of the *troupes de terre.* The same theme may be found in the writings of most of the Army officers who served in New France.

Despite the justificatory tone of much of Pouchot's writing, he maintains throughout the *Mémoires* a strong element of objectivity. Officers of the Marine or Indian warriors who performed well or faithfully are given due recognition. Honorable actions by the enemy are acknowledged. The adjustments in military tactics necessary to fight a war in the wilderness of North America are reported with interest and military pragmatism. Pouchot's observations on the colonial and Indian peoples, although very much from his own perspective, are less subjective and judgmental than those of most of his contemporaries. He maintains, for example, a much healthier

---

24 Cardy, p. 19.

respect for the military intelligence provided by Indians and Canadians than do his fellow officers, such as Louis-Antoine de Bougainville.[25] Perhaps this was a result of his frequent independent front-line assignments, usually in areas where the support of the Iroquois and other nations was critical to success or even survival.

The author's prominence in the pages of his *Mémoires* is, of course, an inevitable feature of the literary genre. This is, after all, a personal history and one composed by a man at a time of his life when other sources of career satisfaction had virtually disappeared. He therefore had every reason to report those situations where he believed his actions had influenced important events. Pouchot records instances of personal counsel to senior officers and participation in critical decisions that, although they may be accurately presented, cannot be corroborated from other sources.[26] He likewise magnifies his importance in other situations.[27] Although Pouchot participated in all the events he describes, and very likely deserved greater recognition, his contributions frequently did not receive mention in the reports of his superiors.

One particular aspect of Pouchot's writing style might prejudice the modern reader about his veracity, or at least his modesty. Pouchot's consistent reference to himself in the third person, the style employed by such classical military authors as Julius Caesar, seems affected today. It is likely, however, that he intended this technique as further

25 Bougainville's opinions are freely expressed throughout his journals. In one instance he baldly states that "All the Indian reports are false, they do not go where one wishes them to, but where they want to go, that is to say where there is no danger." Edward P. Hamilton (trans. and ed.), *Adventure in the Wilderness: The American Journals of Louis Antoine de Bougainville, 1756-1760* (Norman, OK, 1964). p. 130. Bougainville's journals are printed in their original French in *Rapport de l'Archiviste de la Province de Québec pour 1923-1924* (Hereafter cited following initial reference as *RAPQ* with years of report; Québec, 1924), pp. 202-393.

26 These are most apparent in his conversations with the Marquis de Montcalm at critical points in the siege of Oswego in 1756 (Pouchot, *Mémoires,* I, pp. 74-75), the defense of Carillon in 1758 (I, pp. 148-49) and preparations for the 1759 defense of Fort Niagara (II, pp. 3-4).

27 Such as his responsibility for fortifying the camp at Frontenac in 1755 (Pouchot, *Mémoires,* I, p. 48), direction of the siege of Oswego in 1756 (I, p. 72), command of the *troupes de terre* detachments during the abortive campaign against the Mohawk Valley in 1758 (I, p. 133), and the selection of Isle aux Noix as the primary defensive position for the defense of Montréal (I, p. 177).

evidence of his impartiality and objectivity.[28] Unsupported claims, exaggerations and style aside, Pouchot is highly dependable when recording events in which he took part. The additional details provided in the notes to this edition usually support his claims and opinions and show him, for the most part, to have been an honest and forthright observer.

If the composition of the *Mémoires* gave Pierre Pouchot the opportunity to highlight just claims for greater recognition, then it also presented the temptation to employ hindsight to overlook or alter some of his original opinions. The *Mémoires* are silent, for example, concerning his estrangement from the Marquis de Montcalm and other senior officers of the *troupes de terre* because of his optimism about Governor Vaudreuil's plans to resume the offensive in the Ohio Valley during 1759. Instead, the *Mémoires* claim that Pouchot expressed a fatal reluctance to undertake the mission to Niagara "because he could foresee what must result from the shortage of resources in the country."[29] While his naive acceptance of Iroquois promises to warn him of any threat caused Pouchot to divide his considerable forces, and no doubt hastened the fall of Fort Niagara, his account of the decision to send reinforcements to the Ohio is vaguely presented.[30] In light of the loss of Niagara, Pouchot's original positions on both these critical issues of the 1759 campaign were best de-emphasized.

Regardless of such flaws, perceived or real, in Pouchot's reporting of events, there is no doubt that he was one of the most capable officers of the *troupes de terre* to serve in America. He deserved, and received, extraordinary confidence and trust from his superiors. This was most apparent in the frequency and importance of his independent commands which often carried great responsibility. Aside from relatively

---

28 Cardy, p. 16.

29 Pouchot, *Mémoires,* II, pp. 3-4.

30 As late as June 1, 1759 Pouchot stated, in a letter to Captain François Le Marchand de Lignery printed in the *Mémoires,* that he believed he had "at least two months of security." In the following paragraph, however, he says that he was suddenly "absolutely convinced" that the Iroquois, upon whom he was depending for this reassuring intelligence, "had taken up arms against us." Pouchot, *Mémoires,* II, pp. 32-33. Three weeks later, a letter from Pouchot expressed concern over nothing more than the temporary unruliness of the Seneca and no real anticipation of the English attack on his post that would come only nine days later. Pouchot to Lévis, June 27, 1759, in H.R. Casgrain (ed.), *Collection des manuscrits du Maréchal de Lévis* (12 vols.; Montréal and Québec, 1889-95), X, pp. 121-22. A translation of the letter is printed in "Appendix A" of this volume.

brief periods in 1755, 1756 and 1758, Pouchot did not serve with his own company of the Béarn Regiment. Instead, he was absent on engineering duties, leading detachments (termed *piquets*) or commanding the important posts of Fort Niagara and Fort Lévis, especially when they were threatened.

Pouchot did, as he claimed, enjoy the personal confidence of his general, the Marquis de Montcalm. In this respect, his position was somewhat similar to the officers of Montcalm's own staff and gave him access and influence far above his station as a fusilier captain. His letters were accorded the same priority as those of regional commanders, such as Colonel François-Charles de Bourlamaque, or the commandants of the *troupes de terre* battalions.[31] There appears to have been personal warmth as well, enough for the marquis to refer to Pouchot as his "friend" in private correspondence with his second-in-command.[32] This particular reference was made not long before the time, in March 1759, when Pouchot recorded one of his unsubstantiated confidential discussions with the general.[33]

The captain was certainly well-regarded by Montcalm, at least until 1759. Pouchot initially drew his attention in June 1756 for his work at Niagara, soon after the general's arrival in New France.[34] The two men first met (at least in America) during the siege of Oswego. Pouchot quickly earned Montcalm's respect for his engineering skill and his efficiency in cutting the waste of stores at Niagara. Soon thereafter, Pouchot's capable dealings with the Indians further enhanced the general's opinion of him.[35]

Pouchot found himself in the unique position of also enjoying the favor of Governor General Vaudreuil. Given the tense relationship between colonial military personnel and the officers of the *troupes de terre,* and especially between Vaudreuil and Montcalm, this was no

---

31 Montcalm to Lévis, Sept. 9, 1757, Casgrain, *Collection,* VI, p. 51.

32 Montcalm to Lévis, Jan. 17, 1759, *Ibid.,* p. 153.

33 Just prior to his departure for Niagara, Pouchot claims to have discussed his mission with Montcalm and to have expressed the fatalistic belief that it "appears that we shall see one another next in England." Pouchot, *Mémoires,* II, p. 4.

34 Montcalm to Argenson, June 12, 1756, *DRCHSNY,* X, p. 414.

35 Montcalm journal, Feb. 9, 1757, Casgrain, *Collection,* VII, p. 151; Montcalm to Bourlamaque, May 23, 1757, *Ibid.,* V, p. 152.

small achievement. Pouchot became known to the governor soon after arriving in the colony in 1755 because of his "engineering talents." He was personally selected by Vaudreuil to reconstruct Fort Niagara.[36] The captain did not disappoint, and the first phase of the Niagara work gained Vaudreuil's praise to the Minister of Marine for Pouchot's "zeal and activity" and a recommendation to the Minister of War for "such favors from the King as he will deserve."[37] The siege of Oswego raised Pouchot's stock even more. By that time the governor considered Pouchot "one of the best officers I know among the troops of the line." Vaudreuil again recommended him to the Minister of War and made sure that Pouchot was included in the list of *grâces* or "favors" requested for participants in the siege of Oswego.[38] The governor's good opinion of Pouchot seems never to have wavered. He considered the captain "experienced and intelligent" and, following the repatriation of both men to France, provided a glowing letter of recommendation.[39]

It was his influence with the governor that gained Pouchot a unique distinction - command of Fort Niagara, one of the frontier posts with profitable commercial opportunities that was traditionally assigned to an officer of the Marine. This practice had continued during the first winter of Vaudreuil's administration when Pouchot labored on Niagara's fortifications but remained subordinate to colonial Captain François Lefebvre Duplessis Faber.[40] In September 1756, however, Pouchot was given the appointment despite the protests of colonial officers with whom, Montcalm noted, Governor Vaudreuil had "made considerable difficulty in granting that command to an officer of the troops of the Line."[41] Ultimately, the complaints were too

---

36 Vaudreuil to Machault, Sept. 25, 1755, *DRCHSNY*, X, pp. 325-26.

37 Vaudreuil to Machault, Feb. 2, 1756, *Ibid.*, p. 391; Vaudreuil to Argenson, June 8, 1756, *Ibid.*, p. 411.

38 Vaudreuil to Argenson, Aug. 20, 1756, *Ibid.*, p. 474; "Mémoire - Grâces proposées ....", Aug. 30, 1756 in Antoine Roy (ed.), "Les Lettres de Doreil", *Rapport de l'Archiviste de la Province de Québec pour 1944-1945* (Québec, 1945), pp. 82-83.

39 "Summary of the Plan of ... Operations for ... 1759", April 1, 1759, *DRCHSNY*, X, p. 953; Vaudreuil Testimonial, May 6, 1761 in Pouchot, *Mémoires*, I, pp. xxix-xxxii.

40 Doreil to Argenson, Oct. 28, 1755, *DRCHSNY*, X, p. 371.

41 Montcalm to Argenson, April 24, 1757, *Ibid.*, p. 548.

much. In the autumn of 1757, soon after he reported completion of the Niagara fortifications, Pouchot was replaced by a captain of the Marine. Both Bougainville and Montcalm echoed Pouchot's opinion that his relief was a result of jealousy among the colonial officers. The marquis even reported that Vaudreuil regretted the decision and "has frequently told me that he wished him still there."[42]

The source of Vaudreuil's regret was the discovery of another talent in Captain Pouchot. He had a way with the Indians and had done much to maintain the loyalties of those who resorted to Niagara. This ability was rare enough in Marine officers and even more so among those of the *troupes de terre.* Fort Niagara was a particularly sensitive post because of its location in the territory of the Seneca, westernmost of the Six Nations of the Iroquois. Iroquois support, or at least neutrality, was critical to both the French and the English, and agents for both sides energetically pursued those goals. Pouchot's ability was immediately recognized. He soon had the distinction of receiving a complimentary name from the Iroquois - *Sategariouaen,* "the center of good dealings."[43] Montcalm believed that Pouchot was "wonderfully liked by the Indians," and saw tangible evidence that he was positively influencing the Seneca.[44] The governor was equally impressed with Pouchot's achievements and later praised his "talent for dealing with the Indians & enticing them for the good of the service."[45] Pouchot's departure from Niagara was, according to Montcalm, much regretted and resulted in a marked decrease in the number of Iroquois visiting the post.[46]

The opinion held of Pouchot by the Iroquois is recorded primarily in his own *Mémoires* and is thus not entirely objective. It is obvious, however, that he was respected by many of them, as evidenced by his name in their language. He also seems to have forged strong bonds

---

42 Hamilton, *Adventure,* p. 201; Montcalm to Paulmy, April 10, 1758, *DRCHSNY,* X, pp. 693-94. Pouchot's account is in *Mémoires,* I, pp. 114-18.

43 Pouchot. *Mémoires,* II, p. 10. Pouchot was given his Iroquois name during the winter of 1757. Bougainville and Montcalm both recorded it as early as March 5. Hamilton, *Adventure,* p. 90; Montcalm journal, March 5, 1757, Casgrain, *Collection,* VII, p. 169.

44 Montcalm to Argenson, April 24, 1757, *DRCHSNY,* X, p. 548; Montcalm to Paulmy, July 11, 1757, *Ibid.,* p. 575.

45 Testimonial of Vaudreuil, May 6, 1761, in Pouchot, *Mémoires,* I, p. xxx.

46 Montcalm journal, May 15, 1758, Casgrain, *Collection,* VII, p. 350.

with some of their chiefs, notably Kaendaé, a Seneca who remained with him in Fort Niagara throughout the 1759 siege. Pouchot's account of the reaction of two Iroquois to his departure in 1757 may be indicative of his status. One protested that "Our father [Vaudreuil] no longer loves us then, since he is taking from us a chief whom we all love."[47] The timing was unfortunate, for the coming year would prove to be pivotal to the Iroquois position. When Pouchot returned to Niagara in 1759, after an absence of seventeen months, it would be too late. The Six Nations had committed themselves to support the English.

Inevitably, Captain Pouchot's favor in both camps was bound to involve him in the widening rift between Governor Vaudreuil and the Marquis de Montcalm. It was ironic that his estrangement should have been from his own general of the *troupes de terre* because Montcalm thought Pouchot had become too close to the colonial governor. At the root of the problem was the issue of the defense of Niagara and the western margins of the colony.

The difficulty began with discussion between Montcalm and Vaudreuil about reinstating Pouchot as commandant of Niagara in the wake of the disastrous English raid on Fort Frontenac. By early September 1758 they had agreed that Pouchot should proceed to Niagara to improve and, if necessary, defend the post. Within the next month, however, Vaudreuil changed his mind.[48] This delayed Pouchot's return to Niagara until the following spring, although his assignment had again been agreed upon by January 1759.[49] It was during the preparations for Pouchot's departure that Montcalm's attitude toward him suddenly and radically changed.

Montcalm and Vaudreuil had been in disagreement since the fall of 1758 about the best strategy for defending New France. Montcalm wished to concentrate his forces and maintain a defensive posture to preserve the heart of the colony in anticipation of a European peace settlement. Vaudreuil's intention was to maintain efforts in more

---

47 Pouchot, *Mémoires,* I, p. 115.

48 Memoir on Lake Ontario, Sept. 12, 1758, *DRCHSNY,* X, pp. 870, 872; Montcalm to Belle-Isle, Oct. 27, 1758, *Ibid.,* p. 860. Montcalm attributed the delay to Vaudreuil's unwillingness to dismiss a Marine officer. Montcalm to Belle-Isle, April 12, 1759, *Ibid.,* p. 961.

49 Montcalm to Lévis, Jan. 30, 1759, Casgrain, *Collection,* VI, pp. 156-57.

distant parts of the colony, especially the Ohio Valley.[50] Pouchot was surely drawn into the dispute because of his experience at Niagara, his knowledge of conditions there and his imminent return to the post.

Pouchot's instructions for his Niagara duty originated with Vaudreuil, and it was the governor who ordered him to reassume command and defend the place.[51] At some time after January Pouchot must have expressed too much support or optimism for the governor's plans. Vaudreuil, in fact, later referred to Pouchot as the "best disposed officer," presumably in regards to his strategy.[52] On March 23, a few days before Pouchot's departure from Montréal, Montcalm began to express negative opinions about the captain, his mission and his relationship with the governor. Vaudreuil, he wrote, had issued "bombastic" orders giving Pouchot the status of a "tutelary god" on Lake Ontario. This was, he believed, because "France never had as much confidence in Marshal de Saxe as the Marquis de Vaudreuil has in Pouchot."[53] Rather than being gratified that one of his officers of the *troupes de terre* had gained such favor with the governor and that his long-sought appointment of Pouchot to Niagara was finally a reality, Montcalm had become more concerned that the captain had gone over to his rival's camp. Thereafter, Pouchot's former "friend," as well as other officers of the general's staff, had little good to say about either the captain or the Niagara "adventure."

Montcalm's annoyance increased as dispatches from Niagara brought Pouchot's optimistic appraisals of the security of his position and the good intentions of the Six Nations. Montcalm became ever more convinced that the captain had fallen under the spell of the governor's quixotic desire to hold the Ohio, or Belle-Rivière. He believed

---

50 See George F.G. Stanley, *New France: The Last Phase, 1744-1760* (Toronto, 1968), pp. 220-21.The dispute is summarized in greater detail in Lawrence Henry Gipson, *The British Empire Before the American Revolution, Volume VII: The Great War for the Empire, The Victorious Years, 1758-1760* (New York, 1946), pp. 381-84. A series of *mémoires* expressing the opinions of the two French leaders is in *DRCHSNY*, X, pp. 868-77.

To Stanley's and Gipson's histories of the Seven Years' War in American must be added the most recent comprehensive work on the subject: Fred Anderson, *Crucible of War: The Seven Years' War and the Fate of Empire in British North America, 1754-1766* (New York, 2000). He discusses the differing strategic visions of the two leaders on pp. 345-47.

51 Memoir, March 21, 1759, Casgrain, *Collection*, IV, p. 147.

52 "Summary of the Plan of ... Operation for ... 1759", April 1, 1759, *DRCHSNY*, X, p. 953.

53 Montcalm to Bourlamaque, March 23, 1759, Casgrain, *Collection*, V, p. 302; Montcalm to Bourlamaque, March 31, 1759, *Ibid.*, p. 306.

Pouchot had been "praised in the cabinet of the Marquis de Vaudreuil" and "imbued with the false principle that he had nothing to fear there and that two thousand men could be sent to the Belle-Riviere." Pouchot, in his view, had been "seduced" by those who had a personal interest in the preservation of a French presence in the Ohio Valley.[54] In July, when informed by Pouchot that the English had commenced a siege of Niagara, Montcalm sarcastically recalled the captain's "Canadian arguments" and noted his "great regrets, but too late."[55]

It was, in fact, Pouchot's actions prior to the 1759 siege of Fort Niagara that drew the only real criticism of his wartime performance from his brother officers. Like Montcalm, they believed that he had been falsely optimistic, buoyed by the flattery of Vaudreuil but, more important, deceived by the Six Nations.[56] On June 1, exercising the latitude permitted by Vaudreuil's instructions, Pouchot sent a large part of his garrison to support offensive action against the English posts of Pittsburgh and Fort Ligonier. He did so only after having become convinced, largely by assurances from the Iroquois, that it would be some time before the English would be able to approach Niagara. The Iroquois promised to warn him of enemy movements. Pouchot's miscalculation accelerated the loss of Niagara and made him a prisoner of war for the first time.

Pouchot had no opportunity to reconcile with Montcalm because the general was killed in September 1759 while the captain was a prisoner in New York. The defeat at Niagara does not seem to have had any effect on Pouchot's employment during the 1760 campaign, however, although it is apparent from his actions and his writings that he was far more circumspect in his dealings with the Iroquois. He retained the respect of Governor Vaudreuil as well as that of General François de Lévis, Montcalm's successor. Pouchot's important assignment to defend Fort Lévis was based on the same rationale as his Niagara command - he "combined within himself the qualities of a good Engineer, and had acquired experience in managing the Indian Nations."[57]

---

54 Casgrain, *Collection*, VII, p. 532.

55 Montcalm to Lévis, July 16, 1759, *Ibid.*, VI, p. 190; Montcalm's journal, July 16, 1759, *Ibid.*, VII, p. 576.

56 Lévis journal, June 30, 1759, *Ibid.*, I, p. 182.

57 Vaudreuil to Berryer, June 24, 1760, *DRCHSNY*, X, p. 1090.

Pouchot's ability to gain the esteem of the leaders of both the colony and the *troupes de terre* seems to have applied as well to relations with junior officers of both forces. Aside from criticism of his blind trust in the Iroquois during 1759, it is nearly impossible to find a negative opinion about the man or his performance. Army engineer Captain Jean-Nicolas Desandrouins thought him an "energetic officer." Colonial official Louis de Courville was highly complimentary of Pouchot in his own *mémoire* of the war, characterizing him as "active and enterprising," as well as "brave, full of spirit and skill."[58] Aside from their jealousy over his appointment to Niagara, command of which was "usually conferrd [sic] on a Canadian," he seems to have worked well with most officers of the Marine.[59]

Far less is known of Pouchot's reputation among the soldiers he commanded. That Pouchot was a competent disciplinarian is evident throughout the *Mémoires*, and no one implies that he was deficient in this respect. His many special assignments permitted little time with his own company of the Béarn Regiment. During one of these brief periods, at Montréal in the winter of 1757-1758, the captain found himself dealing with the ultimate consequence of eighteenth century military discipline. After investigating the murder of one of his men, nicknamed "Lyonnois", by a comrade known as "Sans-Chagrain" (sic), Pouchot's actions led to a sentence, probably of execution.[60]

Pouchot was also wise enough to maintain morale by encouraging activities among his men. When the intense cold in February 1757 made work on the Niagara fortifications impossible, he allowed the soldiers to perform a play composed by one of them, a comedy entitled *Le Vieillard dupé*.[61] Nor was he averse to set an example by working

58 Abbé Charles-Nicolas Gabriel, *Le Maréchal de camp Desandrouins, 1729-1792* (Verdun, 1887), pp. 314-15; Louis de Courville, *Mémoires sur le Canada depuis 1749 jusqu'à 1760* (Québec, 1838), pp. 125, 149.

59 See the comments about him in the examination of Michel Marest de La Chauvignerie, Jr., Oct. 26, 1757, in Sylvester K. Stevens and Donald H. Kent (eds.), *Wilderness Chronicles of Northwestern Pennsylvania* (Harrisburg, 1941), p. 117. La Chauvignerie noted that Pouchot was "esteemed a very good engineer" and for that reason had been given command of Fort Niagara.

60 Montcalm to Lévis, Dec. 16, 1757, Casgrain, *Collection*, VI, p. 89; Montcalm to Lévis, Feb. 3, 1758, *Ibid.*, p. 125.

61 Montcalm journal, March 5, 1758, *Ibid.*, VII, p. 169; Gabriel de Maurès de Malartic and Paul Gaffarel (eds.), *Journal des campagnes au Canada de 1755 à 1760 par le comte de Maurès de Malartic* (Paris, 1890), p. 99.

alongside his men on swampy road construction in 1756. And his kindness during his second New York captivity to "J.C.B.", a colonial artilleryman who had been his secretary at Fort Lévis, further attests to the fact that the captain was a firm but fair leader.[62]

Favorable opinions were also expressed by Pouchot's English adversaries who met him either in battle or during his time as a prisoner of war. Major Herman Munster of the Royal Americans considered him "a very gallant man."[63] Another Royal American, Captain Walter Rutherford, best expressed the impression made by Captain Pouchot. Having breakfasted with him after delivering Brigadier Prideaux's summons to surrender Niagara, Rutherford reported, "I believe him to be a good soldier and a man of spirit."[64]

The frequent mention of Pouchot by English officers raises an interesting aside about his place in the war in America. Except for the siege of Louisbourg, few French officers were captured until the large-scale capitulations at Québec and Montréal, and prisoners taken on these three occasions were held outside the English colonies. Pouchot was twice a prisoner of war at New York. In both instances he was the senior officer of the prisoners and, thus, their spokesman. As a result, he became the *troupes de terre* officer best-known to his adversaries.

Pouchot was also familiar to the English for his achievements. Captain "Pauchot" had been identified through intelligence reports as Niagara's commandant as early as 1757, but it was his defense of that post and Fort Lévis that made his name common among military men and civilians alike in the northern English colonies.[65] He was personally acquainted with many regular and colonial officers. During the last two years of the conflict, Pouchot dined with senior English leaders such as Sir William Johnson, General Jeffery Amherst, engineer Colonel James Montresor, and many others. When Amherst's army encountered Pouchot at Fort Lévis, in 1760, regular and provincial officers, such as Connecticut provincial Colonel Nathan

---

62 Pouchot, *Mémoires,* I, p. 82; Sylvester K. Stevens, Donald H. Kent, and Emma Edith Woods (eds.), *Travels in New France by J.C.B.* (Harrisburg, 1941), pp. 129-30.

63 Munster to Haldimand, Aug. 1, 1759, Frederick Haldimand Papers, Great Britain, British Library, Add. MSS 21728.

64 Rutherford to Haldimand, July 14, 1759, *Ibid.,* Add. MSS 21728.

65 Examination of Michael La Chauvignerie, Jr., Oct. 26, 1757, in Stevens and Kent, p. 117.

Whiting, were well aware that they were again facing "Monsieur Pisheaue who Commanded Niagra Last year."[66]

It was as an engineer, however, that Pouchot was best known to many of his contemporaries, and, as noted previously, he owed his most important postings to that reputation. Pouchot was acknowledged by the major of his own Béarn Regiment to be a "good engineer", no doubt because of his earlier service in Europe.[67] The fact that he "was in possession of engineering talents" came to the attention of Governor Vaudreuil during the summer of 1755 as the *troupes de terre* units were deployed along the frontiers of New France.[68] Both temporary and permanent fortifications were needed in the colony, but the troops newly arrived from Europe had lost all three of their engineers when the ship of the line *Alcide* was taken by the English.[69] Pouchot's abilities were therefore at a premium in a country where, until the spring of 1756, he was the most experienced military engineer. Vaudreuil offered him the opportunity to refortify Niagara and, by June 1756, Pouchot's performance had convinced the governor that the captain "perfectly understands all the departments of engineering."[70]

Montcalm, too, was pleased with Pouchot's work at Niagara.[71] His repute spread within the colony. Louis de Courville remembered him as the officer "most capable of all to fortify a place." Even colonial Cadet Michel de La Chauvignerie knew that Pouchot was "esteemed a very good Engineer."[72] But Vaudreuil's justification for giving

---

66 Whiting to Whiting, Aug. 24, 1760, in Lemuel Aiken Welles (ed.), "Letters of Col. Nathan Whiting, Written From Camp During the French and Indian War", *Papers of the New Haven Colony Historical Society, Vol. VI* (New Haven, 1900), p. 146.

Pouchot recorded his own celebrity. When Amherst's boats ran the guns of Fort Lévis on August 18, 1760, a number of English officers "wished him good day as they passed by, while others shouted out to him to let them through as they were his friends." Pouchot, *Mémoires,* II, p. 265.

67 Malartic and Gaffarel, p. 27.

68 Vaudreuil to Machault, Sept. 25, 1755, *DRCHSNY,* X, p. 325.

69 Report of the Army, Oct. 15, 1755, *Ibid.,* p. 357; Doreil to Argenson, Oct. 28, 1755, *Ibid.,* p. 370.

70 Vaudreuil to Argenson, June 8, 1756, *Ibid.,* p. 411.

71 Montcalm to Argenson, June 12, 1756, *Ibid.,* p. 414.

72 Courville, p. 93; Examination of Michael La Chauvignerie, Jr., Oct. 26, 1757, in Stevens and Kent, p. 117.

Pouchot command of Niagara derived largely from the need for his engineering skills. Once that work had been completed, in 1757, he was relieved by an officer of the Marine.[73]

Pouchot's practical experience encompassed the attack and the defense of fortified places as well as their construction. He had the opportunity to participate in only one siege of an enemy post. At Oswego, in 1756, accident again placed his skills in demand when the chief engineer was killed. Captain Desandrouins, the next-ranking officer of the Royal Engineer Corps whom Pouchot superseded in matters of practical siegecraft, credited him for the French success and added that "he did not disappoint those who confided in him."[74]

Pouchot received the most attention for his defense of fortified places. In this he combined the experience of an engineer with that of an infantry officer. During the planning for two hopeless sieges - Fort Niagara and Fort Lévis - Pouchot was the logical choice for command in an army that could not afford to sacrifice one of its few regular engineers. Montcalm stated the obvious when preparing to send Pouchot to Niagara - he was "assuredly ... better qualified to defend that place than any Colonial Captain."[75] Governor Vaudreuil believed him "qualified to complete and even to defend the works, if attacked." Combined with his demonstrated abilities for dealing with the Indians, Pouchot's skills made him indisputably the best man to hold the western fortresses of New France.[76]

Pouchot, esteemed by most of his contemporaries as a practical military engineer, was himself brutally frank when evaluating the talents of others working in his field. He had little use for the Marine Department engineers whom he found in the colony. Lieutenant Gaspard-Joseph Chaussegros de Léry was termed "a so-called engineer," and Pouchot believed Captain François Le Mercier to be more interested in personal gain than in the good of the service.[77] He also had slight regard for decisions made by fellow infantry officers of the

---

73 Montcalm to Paulmy, Nov. 4, 1757, *DRCHSNY,* X, p. 670.

74 Letter from Desandrouins, Aug. 28, 1756, *Ibid.,* p. 466.

75 Montcalm to Belle-Isle, Oct. 27, 1758, *Ibid.,* p. 860.

76 Vaudreuil's Observations, Sept. 12, 1758, *Ibid.,* p. 872; Vaudreuil to Berryer, June 24, 1760, *Ibid.,* p. 1090.

77 Pouchot, *Mémoires,* I, pp. 17-18.

troupes de terre when they presumed to dabble in the duties of an engineer. He attributed the selection of the island site of Fort Lévis to Captain Jean de Lapause de Margon of the Guienne Regiment and is critical of the position at several points of his narrative.[78]

Captain Pouchot had greater regard for the officers of the Royal Corps of Engineers who began to arrive in the colony in 1756. Even they were not immune from his criticism, however. Although he was himself unlikely to aspire to a commission in that prestigious corps at such a late stage of his career, Pouchot took every opportunity to emphasize the breadth of his practical experience in comparison with that of the new arrivals. Captain Desandrouins, for example, was characterized as inexperienced at Oswego in 1756, an evaluation he did not dispute.[79]

If the *Mémoires* represent Pouchot's most enduring legacy, he also left some tangible evidence of his engineering prowess. His activities encompassed the broad range of duties expected of an eighteenth century military engineer. Most of his fortifications were merely temporary field works, such as the retrenchments constructed at Frontenac in 1755 and at Pointe au Baril and Isle aux Galots in 1759. These have disappeared altogether. The "Redan de Pouchot", built in the late summer of 1758 to anchor the northern end of the Carillon lines, was only slightly more substantial. Some of his duties were mundane, such as the frustrating repairs to the road between La Prairie and St. Jean in the late summer of 1756. Other projects entailed eleventh-hour repairs and alterations to Fort Niagara and Fort Lévis on the eve of sieges at both places.

Pouchot's most substantial and enduring physical monument is, without doubt, Fort Niagara. His design for the place was derived from the traditions of the great French engineer Vauban, and it effectively converted a stockaded post with a stone citadel building into the strongest fortification west of Québec. Properly termed a "hornwork", Pouchot's design took advantage of the position, which was likely to be attacked only from one direction, and placed the strongest defenses on the land side of a peninsula. Montcalm considered the reconstructed post a good and regular fortification. Vaudreuil was even more enthusiastic when he informed the Minister of War

78 *Ibid.*, II, pp. 136-37; III, pp. 90-94.

79 *Ibid.*, I, p. 72. Desandrouins admitted that he had not previously conducted a siege. Gabriel, p. 49.

that the "fort which was abandoned, and beyond making the smallest resistance, is now a place of considerable importance, in consequence of the regularity, solidity and utility of its works."[80] Although superficially altered by later English and American garrisons, the trace and basic earthen fabric of Pouchot's walls remain intact today, preserved as part of Old Fort Niagara.[81]

In addition to the fortifications, Pouchot directed the construction of at least a dozen major buildings of wood or masonry within the expanded Fort Niagara. The design of one of these, a massive arched stone powder magazine, certainly required his engineering skills. The magazine stands today, the only building constructed by Pouchot known to survive. There is also the likelihood that he supervised the alterations to Fort Niagara's "machicolated house", converting the structure built in 1726-1727 into an officers' quarters and his own residence. That building also exists, known popularly today as the "French Castle." [82]

Other evidence of Pouchot's skill includes at least four plans of fortifications. Two were published as plates in his *Mémoires.* One of these is of Fort Niagara as it appeared in the winter of 1755-1756. The other depicts the siege of Fort Lévis. Two additional manuscript plans of Fort Niagara are also known. They represent the fort in October 1756 and again in the autumn of 1757.[83]

Closely related to Pouchot's talents as an engineer were his interests in topography, geography and cartography and their influence on the conduct of the war in New France. All three disciplines were part of the normal responsibilities of a military engineer, and, once again,

---

80 Montcalm to Argenson, June 12, 1756, *DRCHSNY,* X, p. 414; Vaudreuil to Argenson, June 8, 1756, *Ibid.,* p. 411. Further information on the improvement to Fort Niagara may be found in Brian Leigh Dunnigan, *Siege, 1759: The Campaign Against Niagara* (Revised edition, Youngstown, NY, 1996), pp. 15-16.

81 For a summary of the development of Fort Niagara see Brian Leigh Dunnigan and Patricia Kay Scott, *Old Fort Niagara in Four Centuries: A History of Its Development* (Youngstown, NY, 1991).

82 *Ibid.,* pp. 26-27. Regarding alterations to the "Castle", see Brian Leigh Dunnigan, *Glorious Old Relic: The French Castle and Old Fort Niagara* (Youngstown, NY, 1987), pp. 28-30.

83 Pouchot's "Plan de Niagara et des Fortifications faites en 1755 et 1756" is in the French Archives with a copy in the National Archives of Canada (NMC-26647). The later plan is entitled "Niagara Cette Place a été Commencée le 14 Janvier 1756 et Finie le 12 Octobre 1757 avec 80 Travailleurs partours." The original is in the British Library, Add. MSS 15332.H.

Pouchot's prior service, particularly that on Corsica and his 1744 survey of the road through the Tirol, provided valuable experience. Cartography occupied him during much of his North American service. This resulted in the collection of the information found in the first half of Volume III of the *Mémoires* as well as in the preparation of his detailed map of the French and English frontiers.

Pouchot's concern for gathering information about the nature of the North American countryside probably began on his first trip up the St. Lawrence in 1755. It intensified once he had been assigned to Niagara, a place that guarded French interests in the poorly known country of the Seneca and faced several possible avenues for English offensives. By February 1756 he was able to forward a memoir outlining a plan for disrupting English operations at Oswego.[84] This was certainly based on intelligence provided by the Indians and colonial officers familiar with the Iroquois country. Pouchot's collection of geographical data increased following his appointment to command Niagara in the fall of 1756. He later reported that he had "applied myself ... throughout the whole of my mission, to be in a position to render an account of the state of the country as regards the frontier of the English."[85] Information was forwarded to headquarters with his dispatches from Niagara. Two of Pouchot's "geographical observations" survive, the earliest of them sent to Montcalm in April 1757.[86]

Pouchot's return to Montréal late in 1757 provided an opportunity to apply his accumulated knowledge to the preparation of a map of the English and French frontiers. This had been completed by the end of February 1758. By that time, Montcalm wrote, Pouchot and the Chevalier de Lévis were being "consulted as geographers" by Governor Vaudreuil, a task that occupied them through the long Montréal winter.[87] Both officers apparently depicted Vaudreuil's plan for the boundaries of the colony, a further sign of the governor's favor of

---

84 Pouchot, *Mémoires,* I, p. 66.

85 Pouchot to Paulmy, Nov. 3, 1757, *DRCHSNY,* X, p. 668.

86 "Observations géographiques ...", April 11, 1757, Casgrain, *Collection,* VII, p. 197; "General Observations on the Frontier of Canada", April 14, 1758, *DRCHSNY,* X, p. 695. Both are printed in Appendix A of this edition.

87 Montcalm to Bourlamaque, Feb. 22, 1758, Casgrain, *Collection,* V, p. 200; Montcalm to Bourlamaque, March 30, 1758, *Ibid.,* p. 220.

Pouchot but also a harbinger of suspicions about Pouchot's personal loyalties that would later mar relations with his general. It is also likely that Vaudreuil's interests in map-making lay in his own preparations for a 1758 offensive in the Mohawk Valley, a scheme he kept secret from Montcalm until the very last moment.

Both Lévis and Pouchot drew maps, and Pouchot's was immediately dispatched by Governor Vaudreuil to the Minister of Marine. Montcalm intended to send another copy to the Ministry of War. He forwarded it early in May 1758, but, as late as July, it had not yet arrived in Paris. Montcalm had intended multiple copies of Pouchot's map to serve as duplicates to that drawn by Lévis in the event the latter's was intercepted.[88] It is possible that the copies of Pouchot's map sent by Montcalm fell prey to English vessels and that only the version sent to the Minister of Marine actually reached France.

Pouchot's map survives as plate III of his *Mémoires.* Entitled "Carte des Frontieres Françoises et Angloises dans le Canada depuis Montreal jusques au Fort Duquesne", it is clearly a version of the work completed at Montréal during the winter of 1758. Much of the data about the country west of Oswego must have been Pouchot's own, collected while at Niagara. Considering his close contact with Lévis during the preparation of his own map, it seems likely that some exchange of information occurred, especially for the eastern regions with which Pouchot was probably less familiar. Lévis based his work on "information carefully collected from several Colonial officers and intelligent Indians" and "conversations with recent English prisoners."[89] Pouchot later travelled the Mohawk and Hudson waterways, and these voyages no doubt influenced the final version of the map published with his *Mémoires.* It is also clear, from information in Volume III, that Pouchot consulted the works of American geogra-

---

88 Montcalm to Paulmy, Feb. 23, 1758, *DRCHSNY,* X, p. 690; Montcalm to Paulmy, April 18, 1758, *Ibid.,* p. 698. The map compiled by Lévis might, in fact, survive. Number 444 in the collection of Louis C. Karpinski's photostats of "Manuscript Maps Prior to 1800...." is described as "[Anonymous map including the northern parts of the Ohio, the Susquehanna, the Hudson, and Connecticut Rivers through part of Lake Erie, Lake Ontario and the St. Lawrence River to Montréal]." The manuscript original appears to be in SHA, 7B59. Like Pouchot's map, it focuses on the frontier between the English and French colonies. Inset plans depict Fort Herkimer, the fort at Canajoharie, Fort Hunter, Fort Johnson, and Fort Edward. The details and nomenclature of these plans correspond closely with a c.1757 French account of the "Topography of the Country between Oswego and Albany" in *DRCHSNY,* X, p. 676.

89 Montcalm to Paulmy, April 18, 1758, *Ibid.,* p. 698.

pher Lewis Evans, particularly in regard to his descriptions of the mountain ranges of North America.[90] The printed version of Pouchot's map must have been engraved from a copy of the original but also, according to the editor, "benefitted from additional care in the drafting."[91] Utilized with the "Topographical Observations" in Volume III of the *Mémoires,* Pouchot's map provides a strong sense of the information available to the officers who conducted the defense of New France.

It was a logical progression from Pouchot's concern for the terrain to his interest in the nature and habits of its original inhabitants. Published histories and journals of travel in New France and other parts of North America provided, whenever possible, a description of the Indians. These people were both essential to the conduct of warfare and trade in the wilderness and exotic in the eyes of European observers - and readers. Earlier French writers, notably Father Pierre-François-Xavier de Charlevoix, had described the culture and appearance of the Indians as part of their description of the country and its flora and fauna.[92] Charlevoix had been read by the Marquis de Montcalm and by other officers of the *troupes de terre* who came to New France.[93]

While Pouchot had almost certainly also read Charlevoix, his observations on the Indians are fresh and straightforward. His own discussion begins without preamble, perhaps indicative of incompletely organized material. Although his editor used Charlevoix as a source for a number of footnotes, Pouchot seems to owe no particular debt to the priest for his own narrative. Instead, he had every opportunity to either observe or hear of many of the practices he describes. Most of Pouchot's comments relate to the Iroquois or the Loups (Delaware) and Mississauga who regularly visited Fort Niagara. Many western nations were also to be seen at the post, however, particularly during the campaigns of 1757 and 1759.

Like his geography, Pierre Pouchot's description of the native peoples is quite objective, although he did not always understand the

---

90 Pouchot, *Mémoires,* III, p. 190. For information on Evans and his works see note 972.

91 Pouchot, *Mémoires,* III, p. 3.

92 See Pierre-François-Xavier de Charlevoix, *Histoire et description générale de la Nouvelle France, avec le Journal historique d'un voyage fait par ordre du roi dans l'Amérique septentrionale* (6 vols.; Paris: 1744).

93 Montcalm to the Marquise de St. Véran, June 16, 1756, "Montcalm's Correspondence", *Report on the Public Archives for 1929* (Ottawa, 1930), p. 44.

cultural contexts of their behavior, and his perspective was inevitably that of a European military man. His commentary is, nonetheless, relatively non-judgmental for its time, with only occasional comparisons to the European practices familiar to his own experience. Pouchot seems to have been genuinely interested in the ways of the people with whom he interacted so frequently at Niagara and Fort Lévis. He had somewhat less opportunity to become familiar with their life in the village and home than in warfare and diplomacy, but he was a good observer and recorded much useful and interesting information about his Native American allies and enemies. If he is to be criticized, it is for his sweeping generalizations that indicate a lack of understanding or an inability to recognize the rich cultural differences between the Indian nations of North America. "Apart from one or two minor differences, in customs or in character," he states at one point, "the Indians of this continent are all the same."[94] Other comments and observations found throughout all three volumes indicate that he was well aware that this was not the case.

The fortunes of war and the independence of action permitted by his considerable talents and responsibilities placed Captain Pierre Pouchot in a unique position to observe the Seven Years War in North America, particularly events on the Lake Ontario frontier. His keen powers of observation and his interest in the land and peoples of America allowed him to accurately record events in which he participated and encouraged him to gather information about other theaters of the war. Most fortunately, the impulse to justify his own conduct and that of his comrades and to indict those whom he considered to have betrayed his King and the people of New France motivated him to prepare his work for publication.

In 1858 E.B. O'Callaghan termed Pouchot the "French Mante," a comparison to one of the best known English authors on the war. Pierre Pouchot might be more properly compared to John Knox because of the personal nature and regional emphasis of his narrative. His work possesses equal or greater significance to that of either English writer because it is the only contemporary published account of the conflict from the perspective of the defenders of New France.

---

94 Pouchot, *Mémoires,* III, p. 362. For more on Pouchot's observations of the habits of Native Americans see Michael Cardy, "Life Styles of Eighteenth-Century North American Indians as Perceived by Two French Officers" in Colin Smethurst (ed.), *Romantic Geographies* (Glasgow, 1996), pp. 199-214.

Pierre Pouchot: A Biographical Addendum

The revised edition of the *Mémoires* was ready to go to press when a fortunate combination of technology and the international interest in genealogy provided further detail on the Pouchot family. While most of it corroborates or complements information given by Pouchot or his editor, there are conflicts as well. We believed this material to be worthy of inclusion, but with layout and indexing complete, it was necessary to present it as an addendum.

The family information was obtained from a web site entitled:

"Base de données généalogiques MICHAL comprenant les familles Michal, Lereboullet, Marc et des familles alliées ou apparentées" (www.chez.com/cgvvr/michal/dat2.htm).

The relevant record is found under the name of Élisabeth Durand, Pouchot's mother. The information – primarily the names and dates of birth and death of Pouchot and his siblings – appears to be drawn largely from the records of the church of St. Hugues in Grenoble.

Pierre Pouchot was the son of Léonard Pouchot and Élisabeth Durand, who wed at St. Hugues on April 17, 1703. Their union produced ten sons and two daughters between 1705 and 1720. Pouchot's editor noted – perhaps wryly – that Leonard was unable to enrich his numerous heirs or "make useless sybarites of them," something that seems unlikely given his trade as a *marchand gantier* — a glover or glove merchant.

Pierre Pouchot was the couple's sixth child, and the date of his birth in this genealogical record is at odds with that in his entry in the *Dictionary of Canadian Biography* (DCB)and in the "Tribute" in the *Mémoires* (I, xiii). We have used the date from the DCB, April 8, 1712, while the records of St. Hugues show both his birth and baptism as having taken place on December 10, 1710. It seems likely that the latter is correct and that Pierre's birth date might have been confused with that of his next younger brother, François, who was born on April 9, 1712, according to the St. Hugues records. It should be noted that Franklin Hough, in his 1866 translation, mistakenly believed Captain Pouchot's given name to have been François. Since Pouchot's name is not revealed in the *Mémoirs,* and the "Tribute" says he was born in 1712, this would have been a logical assumption if Hough had access to the records of St. Hugues.

The premature death of Léonard Pouchot occurred on August 5, 1720. Although the "tribute" to Pouchot states that his mother "hastened" to remarry (I, xiii), she did not do so until January 21, 1730, when she wed Pierre Jullien, a minor government official. Pierre Pouchot was nineteen years of age at the time. Although the new marriage apparently did not result in further children, it estranged Élisabeth from her children and sparked prolonged litigation, presumably over their inheritance.

The list of Pierre Pouchot's siblings permits some speculation as to which of his nine brothers was the heir who presumably made the *Mémoires* available for publication. He is identified only as "having spent much time in financial posts" (I, xxvii-xxviii). One brother died before 1721, while three others entered the priesthood. The occupations of three others are not recorded in the genealogical entry. The last two, Leonard (b. 1707) and Paul (1713-1781), were *avocats au parlement,* or barristers of the pre-revolutionary court of justice. Paul was specifically associated with that of Grenoble. Any of the five likely possibilities, had they survived, would have ranged in age from 62 to 72 years by the time the censor approved the *Mémoires* for publication on December 20, 1780.

It is thus impossible to say which brother was the heir who had the foresight to preserve and publish the *Mémoires.* Lacking further information, Paul seems the most likely possibility given his profession and the fact that he is known to have been living at the time of publication and probably resided in Grenoble.

TRANSLATOR'S NOTES

In this volume, I have been responsible for translating Pierre Pouchot's *Mémoires* and almost all other material concerning him. I have not consulted the 1866 Franklin Hough translation of Pouchot, preferring to do the work afresh from the point of view and in the language of the late twentieth century.

Pierre Pouchot's *Mémoires* represent his most enduring monument. In translating them, I have attempted to keep two principal aims constantly in view: to render the text into clear modern English; and to avoid betrayal of either the sense or the spirit of the original. At certain moments the sense has proven elusive. Pouchot evokes, often wonderfully well, the North American environment in which his campaigns were fought, but occasionally he deals with topics, such as his encounters with the native peoples or the details of contemporary fortification, which may be as remote from modern experience as they were strange and noteworthy to his eighteenth century readers. Grasping his real meaning can therefore sometimes present problems, and those problems are certainly not eased by the evident haste with which he composed the text. Moreover, no discourse can ever be entirely impartial or objective, even when the writer or speaker wishes it to be so. As a soldier fighting in a campaign that became increasingly desperate, Pouchot was involved in exceptional circumstances, and his translator must grapple not just with the overt details of a life lived *in extremis* but also with the assumptions and value judgments which, explicitly or implicitly, sustain his text.

To know what Pouchot had in mind at any moment in time, the translator must acquire some knowledge of the way his mind worked. Translation involves not just a voyage into Pouchot's life in North America but also into his head and into the collective mentality of his peer group, the officer class of the French Army in North America during the late 1750s. To understand this is to encounter, however imperfectly, the spirit of his writings and to make contact with the personality who composed them.

The engagement with Pouchot's memoirs has already led me, in my article cited in the bibliography of this volume, to reflect on the nature of military autobiography, especially when it is written from the point of view of the losing side. While the world to which Pouchot belonged may be largely lost to us, some very significant traces of it remain in his *Mémoires.* To confront that world through the medium of translation is to venture into unknown territory and to become progressively familiar with its topography.

I would be remiss if I did not conclude with acknowledging here my gratitude to Brian Dunnigan, who has shared the exploration of Pouchot's world with me and who has so often helped me to focus my mind on a given passage and to crystallize my ideas as to its meaning. Our collaboration has been a very happy one indeed; and, in consequence, the journey, in its spatial, temporal and psychological dimensions, has been throughout one of the utmost fascination.

Michael Cardy
Swansea, Wales

EDITORIAL NOTES

Pouchot's *Mémoires* were originally produced as three volumes, and that organization has been maintained here in a single tome. The first volume includes introductory material and an account of the conduct of the war from 1754 through the close of 1758. The second covers the last two years of the conflict in New France and contains detailed accounts of Pouchot's defense of Niagara and of Fort Lévis. Volume III is divided between his topographical observations and comments on the "manners and customs" of the Indians of North America. We have also decided to supplement this edition with three appendices containing related material. The first of these presents translations of the known letters and brief *mémoires* written by Pouchot and preserved in sources other than his published work. The other two are translations of journals by French officers relating to the 1759 defense of Fort Niagara.

The rationale for Michael Cardy's translation of Pouchot's text is provided in his notes. The flavor of the original has been retained as much as possible through the preservation of Pouchot's mis-spellings or phonetically rendered names of persons and places and his erratic usage of accents on proper names. The French names for Indian nations or major places on the French-English frontier (i.e Sonnontoin for Seneca; Chouegen for Oswego) have been used here. Place names in the English colonies have been translated directly unless they were phonetically mangled in Pouchot's version. Further evidence of the eighteenth century production of the text will be found in the retention of the frequently used ampersand and the inconsistent manner in which numbers are rendered.

A few minor changes have been made in the text. Pouchot identifies quotations either with quotation marks or by italicizing the word or sentence. Direct quotations, names or terms have been standardized by placing them within quotation marks. Words foreign to Pouchot - other than English - have been italicized as is usually the case in the original. Words, other than place names, left in the original French have also been italicized. Another change is in Pouchot's word for the Indians. His usual term is *sauvage* or "savage," although he also occasionally uses the word *Indien.* All such references have been rendered as "Indian" in this translation.

For readers in a position to compare this translation with a copy of the original *Mémoires,* page citations for the 1781 edition have been included in the body of the text. These are shown in square brackets with their position indicating the end of that page in the appropriate volume of the original.

The large number of notes accompanying this edition have been included for the purpose of clarifying and complementing Pouchot's narrative. He is, in a number of cases, incomplete, obscure, imprecise, or simply inaccurate in his treatment of an event. These faults probably resulted from the haste in which he prepared his manuscript, lapses of memory, unfamiliarity with the details of events in which he did not participate, and even carelessness caused by intimate familiarity with his own material. The goal in preparing the notes to this edition has been to identify or clarify and, in many cases, to expand upon interesting elements of the original text. Every effort has been made to provide a complete identification and a secondary biographical reference for all individuals mentioned. Likewise, primary sources have been provided for citations, and appropriate references to secondary works have been included in the commentary about major battles or events.

The 1781 edition included a substantial number of footnotes prepared by the editor or, in some cases, based on marginal information in Pouchot's own manuscript or his incomplete notes. These have been identified as "Note in original" with the page number of the 1781 edition and the original letter designation of the eighteenth century footnote. The text of the 1781 notes has been italicized to further differentiate it. Modern additions to the eighteenth century footnotes are set in regular type.

This edition reproduces the three maps and plans included in the 1781 edition of the *Mémoires*. Two, "Plan Du Fort de Niagara...." and "Plan Des attaques du Fort Levis...." are from an original copy in the William L. Clements Library. It proved impossible to reproduce Pouchot's detailed "Carte des Frontieres Françoises, et Angloises" from the 1781 edition. The version included here is a faithful copy made in the nineteenth century for E.B. O'Callaghan's *Documents Relative to the Colonial History of the State of New York*. It differs from the 1781 engraving primarily by the addition of a numerical key for points along the St. Lawrence River.

The two other known manuscript plans by Pouchot have also been reproduced here. In addition, we have included a contemporary English plan of the siege of Niagara and a German engraving of the region depicted in Pouchot's map of the French and English frontiers.

Brian Leigh Dunnigan
Youngstown, New York

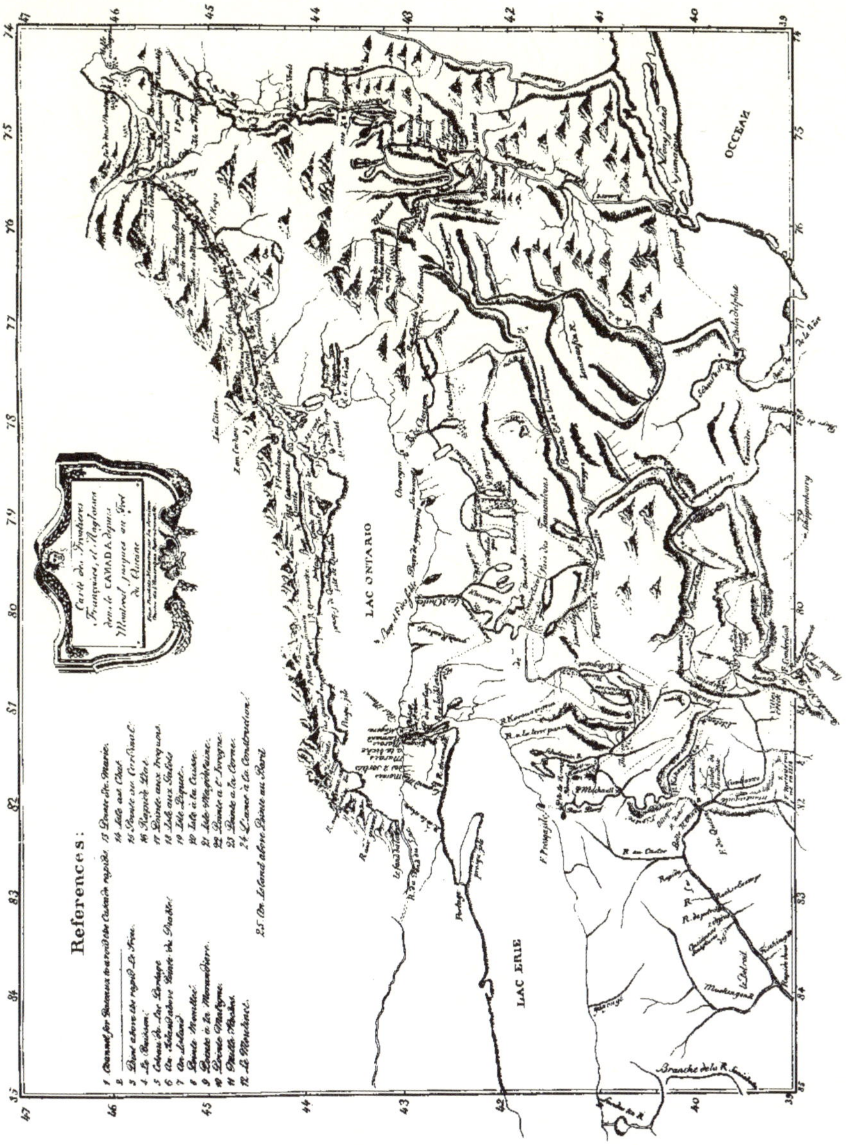

The original of Pouchot's *Carte des Frontieres...*was probably drawn during the winter of 1757-1758. It was printed as "Plate 3" of the *Mémoires*. This version was carefully copied and engraved in 1858 for Volume X of O'Callaghan's *Documents Relative to the Colonial History of the State of New York.*

MÉMOIRES
SUR
LA DERNIERE GUERRE
DE
L'AMÉRIQUE SEPTENTRIONALE,
ENTRE
LA FRANCE ET L'ANGLETERRE.

Suivis d'Obſervations, dont pluſieurs ſont relatives au théatre actuel de la guerre, & de nouveaux détails ſur les mœurs & les uſages des Sauvages, avec des cartes topographiques.

Par M. POUCHOT, Chevalier de l'Ordre Royal & Militaire de St. Louis, ancien Capitaine au Régiment de Béarn, Commandant des forts de Niagara & de Lévis, en Canada.

TOME PREMIER.

YVERDON.

M.DCC.LXXXI.

MEMOIRS
ON
THE LATE WAR
IN
NORTH AMERICA
BETWEEN
FRANCE AND ENGLAND

Followed by observations, of which several are
pertinent to the current theater of the war,
and fresh details on the customs and habits of the Indians,
together with topographical maps.

By M. POUCHOT,
Knight of the Royal & Military Order of St. Louis,
formerly Captain in the Béarn Regiment,
Commandant of the Forts of Niagara & Lévis,
in Canada.

VOLUME I

YVERDON
MDCCLXXXI

❀ Publisher's Preface ❀

One would seek in vain, amidst that mass of periodical publications, the archives of falsehood & ignorance, for materials with which to write history; only personal memoirs can provide us with them. In publishing today those of M. Pouchot on the late war in North America, we believe we have fulfilled that important objective [v] at a timely moment, that of the revolution which is breaking the shackles of the continent & changing the political system of Europe.[95]

Several works relative to the war have long been in print in England.[96] Their lack of objectivity rendered desirable the publication of more accurate accounts, capable of transmitting to posterity those brave deeds which so strikingly upheld the honor of the French nation in the New World, even in the midst of setbacks. The latter were so unremitting that fortune, by assailing us with them from every quarter, seemed to give the lie to her own character. [vi]

If national prejudices appear on occasion to animate M. Pouchot, we venture to assert that they have never been sufficiently strong to impel him to betray the interests of truth, whether by altering or concealing it. The truth is to be found imprinted throughout his narration & is expressed with a simplicity that too often degenerates into a quite marked negligence of style. Everyone will readily forgive this blemish in an officer who was more concerned with deeds than with eloquence. In any case, the language of the military is not always that of the public tribune.

The parts of this work whose [vii] sense was unclear, or obscure, are almost the only ones which we have thought fit to correct. We have also allowed ourselves a few minor omissions. Perhaps the author would have

---

95 The American War for Independence, with France an active ally of the former American colonists, was nearing its climax when Pouchot's book went to press.

96 Among the English histories of the war, all of which were in print before Pouchot's *Mémoires*, are John Entick, *The General History of the Late War Containing It's Rise, Progress, and Event in Europe, Asia, Africa, and America* (5 vols.; London, 1763); John Knox, *An Historical Journal of the Campaigns in North America for the Years 1757, 1758, 1759, and 1760* (2 vols.; London, 1764). Edited by Arthur G. Doughty and reprinted in 3 vols., (Toronto, 1914-1916); and Thomas Mante, *The History of the Late War in North-America and the Islands of the West-Indies* (London, 1772). John Dobson, *Chronological Annals of the War; From Its Beginning to the Present Time* (London, 1763) is another contemporary English work that presents the events of the conflict in a day-by-day format.

No similar French histories of the war in America had been published prior to Pouchot's *Mémoires.*

made even more himself, if he had had the time to revise his memoirs, as it was also necessary to change the order of the different sections.

In fact, a number of readers would have been irritated to find the historical narrative placed after excellent geographical observations which serve to illuminate them & to which we have added a few remarks on the Appalachian Mountains and [viii] Niagara Falls. We have thus quite properly placed these observations after the memoirs & before the precious details on the customs and habits of the Indians. This last text arises out of relations which our author's various commands obliged him to establish with the principal nations of the North American continent. [ix]

A Tribute To M. Pouchot

A man of letters becomes identified with his works: their quality is the measure of his fame, and their existence is alone sufficient for his reputation. By contrast, the man who devotes himself to the service of his country, more concerned to shed blood for her than to perpetuate the memory of his deeds, leaves to posterity the task of consecrating them. It is thus [xi] a solemn duty to assemble the facts, or to safeguard them against the sponge of oblivion, especially when they had as their theater a remote country and so run the risk of being buried there. Such are those of the intrepid defenders of Canada, among whom M. Pouchot occupies a distinguished position.

By publishing his memoirs, we pay off a debt of his fellow citizens & by giving him here the proper tribute of our praise, we fulfill our duty. Truth will have no reproach to make us: it is not to afflict her that we borrow her language. Simple [xii] & precise, it rarely becomes that of falsehood, which always requires a surface glitter.

M. Pouchot was born in Grenoble in 1712, of a virtuous father who had not sought in commerce the dangerous means of flaunting scandalous luxury and enriching his heirs in order to make useless sybarites of them. Through his premature death he left his widow the fatal freedom to marry again and she hastened to take advantage of it. Maternal feelings losing from day to day their power in the arms of a new husband, the voice of self-interest became stronger & plunged her [xiii] children by her first husband into the deep abyss of litigation from which they have not yet extricated themselves.

The author of these memoirs, in order to avoid sinking his talents into that abyss, as did so many others less prudent but greedier, joined the army in 1733 as a volunteer engineer, a state to which nature had called him. She did not lose her rights when, in the next year, he passed into the Béarn Regiment.[97] Far from thinking that his new posting freed him from diligent application, & that he could live in uneasy idleness, he continued to study the great art of Vauban[98] [xiv] & Cohorn,[99] amid the tumult of the camps.

His knowledge soon came to be appreciated. It did not escape the attention of M. de Maillebois,[100] who set M. Pouchot on the course to which his talents called him. This general posted him to work, under M. Bourcet,[101] on the earthworks of Borgo-Forte, in the

---

97 The Béarn Regiment, "72nd" in numerical order of the regular line infantry of the French Army, was formed September 3, 1684 and disbanded on November 25, 1762 during the reorganization of the army. Lucien Mouillard, *Les Régiments sous Louis XV* (Paris, 1882), "Béarn"; Susane, V, p. 308.

Pouchot's commission as lieutenant *en second* in the Béarn Regiment was dated May 1, 1734. His record of promotions makes no mention of a year spent as a volunteer engineer. He might, in fact, have volunteered in the Béarn Regiment in anticipation of a commission. "Contrôle des Capitaines ... Béarn", SHA, Y^b 121. Despite his engineering talents, Pouchot was never admitted to the prestigious Royal Corps of Engineers. Anne Blanchard, *Dictionnaire des ingénieurs militaires* (Montpellier, 1981).

98 Sébastien Le Prestre de Vauban (1633-1707), premier military engineer of seventeenth century France whose systems for the attack and defense of fortified places formed the basis of eighteenth century siegecraft. For a summary of his life and the significance of his work see Christopher Duffy, *The Fortress in the Age of Vauban and Frederick the Great* (London, 1985), pp. 71-97. Another biographical summary is in the modern edition of Vauban's *A Manual of Siegecraft and Fortification* (Ann Arbor, 1968), translated and edited from the 1740 edition by George A. Rothrock.

99 Menno van Cochoorn (1641-1704), Dutch military engineer. A contemporary of Vauban, he developed his own systems of attack and defense and had an influence on siege warfare second only to his French rival. See Duffy, pp. 63-71.

100 Jean-Baptiste-François Desmarets, Marquis de Maillebois (1682-1762). Appointed lieutenant general in 1731, he temporarily commanded French forces in Italy in 1733 and served in a subordinate capacity in that theater during 1734-1735. Maillebois quickly subdued Corsica in 1739. Pouchot came to his attention during these campaigns. Maillebois was appointed marshal in 1741 and fought in Germany and Italy during the War of the Austrian Succession. *Biographie universelle (Michaud) ancienne et moderne .... Nouvelle édition* (45 vols.; Paris, 1854-1865), XXVI, pp. 124-25. See also Jean B.D. Courcelles, Chevalier de Courcelles, *Dictionnaire historique et biographique des généraux français depuis le onzième siècle jusqu'en 1823* (8 vols.; Paris, 1823), VII, pp. 350-59.

101 Pierre-Joseph de Bourcet (1700-1780) served in Italy in 1733-1735 as an officer of engineers. He left the service but reentered it in 1741 as an artillerist and engineer, later commanding those branches in Germany in 1756. Bourcet gained a reputation as an expert on mountain warfare during service on the Alpine frontier of northwestern Italy during the war of the Austrian Succession. *Biographie universelle,* V, pp. 290-91; Spencer Wilkinson, *The Defence of Piedmont, 1742-1748* (Oxford, 1927), p. 59.

Sarraglio, and on those of Ferrara on Mount Baldo.[102] The war in Corsica provided our author with other opportunities to satisfy that need, irresistible and ever-recurring, of strong personalities, that of being useful. He was employed in fortifying positions & laying out [xv] roads in the interior of that unhappy island which, despite its poverty, could not shake off the yoke of inflexible tyrants, the Genoans, who were republicans & despots at the same time.[103]

We shall not follow M. Pouchot in all the campaigns in Italy, Flanders and Germany, in which he shared the glory with all the French troops. In 1744 he was entrusted by the court to reconnoitre the road over the Tirol, & drew up a map of it to which he appended an instructive memoir.[104] In the second to last year of the war, he retrenched the camp at Tournai, which was under the command of

---

[102] Pouchot's earliest field service was against the Austrians in northern Italy during the War of the Polish Succession (1733-1738). French and allied forces invaded Lombardy from the Dauphiné in 1733. Maneuvering during 1734 centered on the Po River south of Mantua, and the armies clashed at Parma and Guastalla. In 1735 the Austrians withdrew northward to Trento leaving a garrison in Mantua which the French blockaded until a truce ended active fighting in November. See Duffy, p. 101 and John L. Sutton, *The King's Honor and the King's Cardinal: The War of the Polish Succession* (Lexington, KY, 1980), pp. 88-111, 162-75, 187-90.

The Béarn Regiment served in Italy from 1733 to 1736, seeing action at the Battle of Parma on June 29, 1734 where its colonel was killed. Susane, V, p. 308. Pouchot's engineering activities occurred during 1735. Borgoforte is on the north bank of the Po in the Seraglio, the low plain between the river and Mantua. The Austrian fortress was about 10 miles farther north. Ferrara is situated to the east between the Po di Volano and the Po Morto and was occupied after the Austrian withdrawal.

[103] Corsica was in a virtually continuous state of revolt after 1729 as the islanders struggled to free themselves from the rule of the Republic of Genoa. In 1737 the Genoans requested French aid in subduing the island. French troops arrived in 1738, and the Corsicans had laid down their arms by November 1739. Thadd E. Hall, *France and the Eighteenth Century Corsican Question* (New York, 1971), pp. 34-44.

Pouchot's regiment served on Corsica from 1739 to 1741. Susane, V, p. 308. The French intervention provided him with a much-needed opportunity for field service. Pouchot's military career had suffered a setback when, as the War of the Polish Succession neared its end, he found it necessary to accept a lower grade, probably so as to remain in the army. On January 7, 1737 the second lieutenant had been reduced to ensign, the most junior commissioned rank. Active campaigning on Corsica seems to have been what was needed for promotion. On October 28, 1739 Pouchot was appointed lieutenant. "Contrôle des Capitaines ... Béarn", SHA, Y^b 121.

[104] It is unfortunate that few details are provided of Pouchot's activities during the War of the Austrian Succession (1740-1748). Based on the movements of the Béarn Regiment, his service was extensive.

...(continued next page)...

M. de Villemur.[105] [xvi] Through this service our brave officer obtained a special payment, his commission as Captain and the Cross of St. Louis.[106] For him, these were just rewards & not favors, the latter being a term in the vocabulary of the minister and in truth very

---

In 1741 a French army joined that of Bavaria in the capacity of "volunteers" in the Bavarian invasion of Austrian Bohemia. By the end of the year the army had taken Prague but was blockaded by the Austrians for most of 1742. The Béarn Regiment was then in Flanders following its return from Corsica. The 1743 campaign found the regiment in Germany with the army of the Duc de Noailles facing Hanoverian, Dutch and English troops under King George II. Béarn fought at the Battle of Dettingen on June 27, 1743. For the regiment's service see Susane, V, p. 308. A summary of operations of the War of the Austrian Succession is in R. Ernest Dupuy and Trevor N. Dupuy, *The Encyclopedia of Military History* (New York, 1970), pp. 630-37. The opening campaigns against the Austrians are detailed in Sautai, *Les Débuts de la guerre de la succession d'Austriche.*

The new war brought Pouchot further advancement. He became lieutenant of the colonel's company in May 1742, making him the senior lieutenant of the Béarn Regiment. Then, on September 27, 1743 Pouchot received the regimental staff appointment of assistant major. "Contrôle des Capitaines ... Béarn", SHA, Y^b 121.

France finally declared war in April 1744. The Béarn Regiment was then part of an army that lay siege to Freiburg in southwestern Germany. The Austrian garrison surrendered at the end of November. Susane, V, p. 308; Duffy, p. 106.

Pouchot claims to have spent part of 1744 on reconnaissance in the "Tirol." It has not been possible to confirm this mission, although the *mémoire* he prepared might well exist. The Tirol, lying between Austria and northeastern Italy, lay deep in enemy territory, and a survey there by a French officer was a hazardous undertaking. The road from Innsbruck to Verona via the Brenner Pass was one of a very limited number of natural routes through the Alps and was critical to Austrian operations in Italy. Wilkinson, p. 2. Freiburg, where Béarn spent much of 1744, was about as close to the Tirol as Pouchot's regiment would get during the war.

105 Pouchot returned to Flanders for the 1745 campaign. Tournai, in the Austrian Netherlands (modern Belgium), was besieged by a French army under Marshal Saxe (note 1006) in April 1745. The fortress capitulated on June 20. Saxe, meanwhile, turned back an English and allied relief force under the Duke of Cumberland at Fontenoy on May 10. For details of the campaign and battle see Francis Henry Skrine, *Fontenoy and Great Britain's Share in the War of the Austrian Succession, 1741-1748* (Edinburgh, 1906), pp. 138-224.

Retrenching or fortifying the camp of the besiegers was common practice. The Béarn Regiment was not among the units engaged at Fontenoy and would thus seem to have been occupied in the siege of Tournai. *Ibid.*, pp. 371-72. Pouchot's superior, Villemur, had been a general and divisional commander in the Franco-Spanish army that invaded Piedmont in 1744. He served again in Italy in 1747 with the rank of lieutenant general. Wilkinson, pp. 122, 302, 312.

A contemporary English source for the War of the Austrian Succession is Richard Rolt, *An Impartial Representation of the Conduct of the Several Powers of Europe Engaged in the Late General War ....* (4 vols.; London, 1749). Duffy, pp. 101-11, deals with the war from the point of view of the influence of fortifications.

106 A number of biographical references give the date of Pouchot's promotion to captain as 1749. That is indeed the year in which he received a company (September 22). He had,

...(continued next page)...

improper in origin, but unfortunately it is often possible these days to justify its use.[107]

Since peace had not been established on firm foundations, the ambition of England soon sought to destroy them. In order to forestall or stop its initiatives, France sent to Canada several battalions. Among them was that of Béarn, **[xvii]** to which M. Pouchot was attached. The memoirs which we are publishing spare us from entering into any detail about the brilliant exploits & important service of this officer. The defense of Niagara required of him very great resourcefulness which his genius never failed to supply him with. He always faced up to his needs, which increased in proportion to the superiority of the enemy forces. They did not overwhelm him at Fort Lévis, in the ashes of which his grave ought to have been dug after his death in order to raise a monument worthy of his courage. **[xviii]**

If he then escaped from the fetters of his country's enemies, it was only to be exposed, back home, to the keen shafts of calumny.

---

however, been promoted to the rank of captain, without a company, on August 2, 1745 as implied here. "Contrôle des Capitaines ... Béarn", SHA, Y^b 121. See also "Officiers du régiment de Béarn", pp. 356-57. It is possible that the promotion came as a reward for his service in Flanders that spring. Realization of the goal of having his own company was probably made easier when, in 1746, a second battalion was raised for the Béarn Regiment. Sautai, *Les Débuts,* p. 358. It was apparently in the second battalion that Pouchot received his company and thus found himself assigned to New France six years later.

The tribute to Pouchot suggests that he also received the Cross of St. Louis in 1745 in recognition of his services. This decoration was established on April 5, 1683. For background on the coveted order, see Aegidius Fauteux, *Les Chevaliers de Saint-Louis en Canada* (Montréal, 1940). This award might have provided additional social distinction. By royal edict in 1750, officers received into the Royal and Military Order of St. Louis were to be elevated to the nobility. The language of the edict provided many exceptions, however, and, as will be seen below, officers of common origin were unlikely to be fully accepted in their new status by members of the old nobility. René Chartrand, *Canadian Military Heritage: Volume I, 1000-1754* (Montréal, 1993), p. 154.

The Béarn Regiment remained in Flanders for the 1746 campaign and participated in the French victory at Rocoux on October 11. It was then moved south for a renewed French invasion of Piedmont through the Maritime Alps of northwestern Italy. The regiment fought in the bloody action at the Colle dell' Assietta, west of Turin, on July 19, 1747. Another officer present at this French defeat was Pouchot's future commanding general, the Marquis de Montcalm. Susane, V, p. 308; Duffy, pp. 104, 270. This and the other Alpine campaigns of the War of the Austrian Succession are discussed in Wilkinson, *The Defence of Piedmont.*

107 "Favors" or *grâces,* in the form of promotions, decorations or pensions, were solicited of the King by most of his officers and were granted in large numbers after most military actions. The ease with which these were obtained no doubt accounts for the negative connotation here. See Kennett, pp. 63-64.

M. Berryer, Minister of Marine,[108] recalling that he had once been Lieutenant of Police, thought it necessary to apply that civil inquisition, so appropriate for keeping in check a vile and corrupted populace, but too often fatal to the peace of honest people, in order to discover those responsible for so many abuses and prevarications which had caused the loss of Canada.[109] As a result, he hired an informer, a sort of parasite, with the aim of identifying all those whose [xix] conduct might be suspect through their connections with the general supplier. He found at his table M. Pouchot, who had come to speak to him about the provisioning of Fort Lévis.[110] This was quite enough for M. Kervisian (that was the name of the informer) to accuse this brave officer upon his return to France.[111]

108 Nicolas-René Berryer, Comte de La Ferrière (d.1762), Minister of Marine from November 1, 1758 to October 13, 1761. Berryer had been appointed Lieutenant of Police in 1747 and was deeply involved in the intrigues of Mme. de Pompadour. *Biographie universelle,* IV, pp. 115-16. The general knowledge that Berryer's appointment as Minister of Marine was a result of court intrigues is discussed by Louis-Antoine de Bougainville (note 283) in his journals. Bougainville, who had traveled to France late in 1758 to "beg" for reinforcements, soon learned of the new minister's (or Pompadour's) attitude towards the colony. Beset by enemies in Europe, Berryer reportedly told Bougainville that "one did not try to save the stables when the house was on fire." Bougainville, according to the editor of his journals, responded: "At least, Monsieur, no one will say you talk like a horse." Hamilton, *Adventure,* pp. 322-23.

109 The *affaire du Canada* erupted in the wake of the loss of the colony and the return to France of most of its officials. A decree of December 12, 1761 ordered that action be taken against "the authors of the monopolies, abuses, troubles and double-dealings which have been committed in Canada." A total of 55 officials or military officers of the colony were accused. Not all were arrested or tried by the judges of the Paris Châtelet. The proceedings had been completed and verdicts issued by December 10, 1763. Severance, II, pp. 411-22; Guy Frégault, *Canada: The War of the Conquest* (Toronto, 1969), pp. 33-34. Many details about the accused may be found in Pierre-Georges Roy, *Bigot et sa bande et l'Affaire du Canada* (Lévis, Québec, 1950).

110 Joseph-Michel Cadet (1719-1781) was purveyor to the French forces in Canada. *DCB,* IV, pp. 123-28. Pouchot's efforts to obtain supplies for the defenders of Fort Lévis were made in the first half of March 1760. The *Mémoires* imply that he received little support from Cadet or the intendant, Bigot (note 116). Pouchot departed Montréal for Fort Lévis on March 17.

111 The informer "Kervisian" was Charles-François Pichot (or Pichon) de Querdisien Trémais (1724-1784), a commissary of the Marine. Pouchot's rendering of his name was not too far off as it was alternately spelled "Kerdisien." The bitter references to this man are somewhat ironic, although understandable in light of Pouchot's perceived implication in the scandal. Querdisien Trémais was charged with documenting "all the abuses which have crept into any parts whatsoever of the service in the colony," a goal that would otherwise seem to have been welcome to Pouchot considering his many criticisms of the officials of New France. Querdisien Trémais was dispatched to the colony by the Minister of Marine early in 1759. His reports resulted in the implication of a number of the principals in the *affaire du Canada. DCB,* IV, pp. 632-33.

At that point M. Berryer had left the ministry. His successor,[112] without having recourse to informers, nonetheless wished to take advantage of their testimony in such unhappy circumstances, where truth had as its adversaries guilty men who sought to multiply their accomplices in order to [xx] avoid exemplary punishment. A number of *lettres de cachet* were sent & the order was given to take the accused to the Bastille. M. Pouchot was resting in the bosom of his family from the fatigues of the war when he learned to his surprise that he was to be arrested. He did not wait for his arrest, but set out at once for the Court.[113]

Having introduced himself to the Minister, he spoke to him in these terms: "I have just arrived from Canada, where I have risked my life a thousand times in the interests of my country, whose enemies have offered me positions, [xxi] money & a very favorable settlement. I have rejected their offers.[114] The loss of my patrimony is the sole fruit remaining to me of my hard work and my service. What is wanted of me? What am I accused of?"

The thunder of authority ceases to rumble when the voice of innocence manages to make itself heard. They replied to M. Pouchot only by praising him, & he was told that they needed the testimony of a man like him in order to convict the guilty. "I realize now," he replied, "that a vile informer, an infamous [xxii] spy, whom you have shamefully promoted to the rank of an officer, is at the heart of this

---

112 Étienne-François Choiseul, Duc de Choiseul, Comte de Stainville (1719-1785) succeeded Berryer as Minister of Marine on October 13, 1761. He held the post until April 1, 1766. *Biographie universelle,* VIII, pp. 182-86.

113 A *lettre de cachet* was a document with contents sealed under the privy seal. These were of two types. The first was an official summons; the other was an order for exile or imprisonment. Skrine, p. 360n. Pouchot appears to have received the first kind.

Pouchot probably learned of his impending arrest when inquiries concerning his whereabouts were made in Beziers, Grenoble and Metz late in 1761 and early in 1762 by Antoine-Raymond-Jean-Gabriel de Sartine (1729-1801), Lieutenant General of Police. Sartine learned that Pouchot had come to Paris, probably, as stated in his *Mémoires,* to forestall formal charges. The authorities had not discovered his residence in the city as of February 1. Sartine to Accaron, Feb. 1, 1762, F. et L. Ravaisson-Mollien (eds.), *Archives de la Bastille: documents inedits* (19 vols.; Paris, 1866-1904), XVIII, p. 281. A summary of Sartine's career is in Étienne Taillemite, *Dictionnaire des marins français* (n.p., 1982), p. 308.

114 Nowhere else in his *Mémoires* does Pouchot record bribes or offers by the English. Nor has any evidence of such offers been found in other sources.

process.[115] What affinity can I have with M. Bigot[116] & his accomplices? Does one subpoena a witness who refuses to testify?"

M. Pouchot having promised to return whenever it was required, not only was his freedom granted him but he was assured that he would soon enjoy the rewards due his service.[117] He received none. At Court they drink incessantly from the waters of the river of forgetfulness. To forestall its effects, there is only one **[xxiii]** opportunity. Thanks to the informer, M. Pouchot was prevented from seizing it.

After the Carrillon [sic] affair, M. de Montcalm[118] had requested for him a lieutenant-colonel's commission. The minister granted no

---

115 Pouchot's anger was directed at the manner in which information on the corruption in New France had been gathered. Querdisien Trémais continued his investigations in France in 1761-1762 and was rewarded with a substantial pension. In December 1762 he was sent to Saint-Domingue (Haiti) with the appointments of Subdelegate General of the Intendant, Financial Commissary of Cap-François and First Councillor of the *Conseils Supérieurs* of Cap-François and Port-au-Prince. He spent the rest of his career in that colony and died there in 1784. *DCB,* IV, pp. 632-33.

116 François Bigot (1703-1778). As Intendant of New France from 1748 until 1760, Bigot was, in effect, the chief administrator of the colony, directly concerned with finance, supplies, housing, hospitals, timber, shipbuilding, and the populace in general. Bigot's position provided many opportunities for personal gain, and he was one of the most prominent figures implicated in the *affaire du Canada.* A particularly balanced account of Bigot's career, which places his actions in the context of contemporary French government, may be found in *DCB,* IV, pp. 59-71. Pouchot, like many officers of the Army (*troupes de terre*), was extremely critical of the pervasive corruption he saw as a major cause of the military disaster and the misery of the populace of New France. Pouchot's attitude was not improved by his own brush with the *affaire du Canada.*

117 Pouchot apparently never testified before the Paris Châtelet regarding the *affaire du Canada.*

118 Louis-Joseph de Montcalm, Marquis de Montcalm (1712-1759) was appointed major general (*maréchal de camp*) on March 11, 1756 and arrived at Québec May 13. Montcalm replaced General Jean-Armand Dieskau (note 184) as commander of the *troupes de terre* battalions in New France. He served as field commander until his death on September 14, 1759 from wounds received the previous day on the Plains of Abraham. Montcalm was promoted lieutenant general on October 28, 1758. *DCB,* III, pp. 458-69. The "Carillon affair" of July 8, 1758, in which Montcalm repulsed a vastly superior English force, was one of his most dramatic successes. Carillon was the French name for Ticonderoga, New York.

Many of Montcalm's papers have been published in Casgrain, *Collection des manuscrits du Maréchal de Lévis.* See particularly VI, *Lettres du marquis de Montcalm au chevalier de Lévis* and VII, *Journal du marquis de Montcalm durant ses campagnes au Canada de 1756 à 1760.* Much of his official correspondence, translated into English, is in *DRCHSNY,* X. Another useful translation, including many personal letters to his wife, mother and others, is "Montcalm's Correspondence," *Report on the Public Archives for 1929,* pp. 31-108.

more than a modest pension.[119] A man who had distinguished himself in that memorable combat and who had subsequently defended with as much valor as intelligence Fort Niagara and Fort Lévis, thus could not cross the barrier that separates subordinate functions from upper ranks, an obstacle which intrigue & wiles overcome without difficulty.[120]

Having obtained neither favors nor [xxiv] promotion & possessing no more than the satisfaction of being judged honest, M. Pouchot returned to Grenoble. He could imagine no other way of avenging himself on the ingratitude of his country than further service & eagerly grasped the opportunity presented by the Corsican troubles, which were caused in the first instance by tyranny and subsequently fomented by the habit of excessive freedom & the fear of being shackled once again.[121]

---

119 No recommendation for promotion has been found in the aftermath of the Carillon action. A lieutenant colonel's commission was requested for Pouchot in a *mémoire* of "Grâces proposées pour les troupes Employées au siège de Chouaguen" dated August 30, 1756. This was for his work at Niagara and service as an engineer during the siege of Oswego. *RAPQ*, 1944-45, pp. 82-83. The subsequent list of *grâces* of May 1757 included only a pension of 200 *livres*. Pouchot would have received this disappointing news at Niagara in August 1757. "État général des grâces", May 1757, Casgrain, *Collection*, III, p. 73; "Relation de Poulariès Envoyée à Mr le Marquis de Montcalm", Aug. 8, 1757, *Rapport de l'Archiviste de la Province de Québec pour 1931-1932* (Québec, 1932), p. 60.

When Bougainville returned from France in May 1759, he brought with him a list of further promotions, pensions and "gratifications." Pouchot received two pensions of 200 *livres* each in addition to "another one that he had before." *Ibid.*, May 12, 1759, p. 88. The additional pensions were probably for his service at Carillon in 1758 and/or his completion of the works at Niagara in the autumn of 1757.

It appears that one other small mark of favor was to follow. The list of *grâces* of February 12, 1760 includes "an augmentation" of 200 *livres* to the pension of 200 *livres* from the royal treasurer, probably the one granted in 1757. "Grâces particulières", Feb. 12, 1760, Casgrain, *Collection*, III, p. 217. It is possible that this was granted in recognition of Pouchot's spirited defense of Fort Niagara in 1759. The list of February 10, 1761 contained no further rewards. *Ibid.*, pp. 226-31.

120 The matter of Pouchot's social status appears at this point. He was certainly a *roturier*, a member of the bourgeoisie who could afford the financial burden of maintaining a company but was unlikely to advance to higher rank. His inclination towards engineering and the fact that he entered the army as a volunteer would seem to support this supposition. See Kennett, pp. 57-59.

121 Two other factors certainly influenced Pouchot's decision to return to Grenoble in 1762. First, he might not have been free to resume active service. Although his surrender at Fort Lévis in August 1760 meant that his status as a prisoner of war was dictated by Article LIV rather than by Articles I-III of the capitulation of Montréal, he still had to be

...(continued next page)...

Three months before his departure for that island, he undertook to compose his memoirs at the request of his friends. That brief space of time did not permit him to compose them [xxv] with care & was only really enough for him to arrange the material. They are none the less fascinating and useful for that, since the author combines knowledge of his profession with the courage to tell the truth, which is so often reduced to silence by fortune, of which cowardice is one of the principal attributes.

Once arrived in Corsica, M. Pouchot was employed according to his inclinations but on this occasion they were to prove fatal to him.[122] Having received the order to advance with a detachment of fifty men to reconnoitre a position, he was [xxvi] abandoned by them & killed, on 8 May 1769, by a number of Corsicans hidden in the bush. Thus death, which he had defied so many times, sought to lie in wait out of

---

exchanged before he could fight again in the conflict that would end in 1763. "Articles of Capitulation", Sept. 8, 1760, *DRCHSNY*, X, pp. 1107-20. Pouchot had not yet been exchanged when he returned to France on March 8, 1761. Bonneau to Choiseul, March 9, 1761. *Collection de manuscrits contenant lettres, mémoires, et autres documents historiques relatifs à la Nouvelle-France, recueillis aux Archives de la Province de Québec, ou copiés à l'étranger* (4 vols.; Québec, 1883-1885), IV p. 307. It is not known if his exchange was completed before the end of hostilities. Second, Béarn was one of the regiments disbanded during the reorganization of the army in 1762. Pouchot thus became a captain without a regiment. See Mouillard, "Béarn" and Susane, V, p. 308.

Peacetime was always difficult for officers without influence, and it seems likely that Pouchot had nowhere to serve until Corsica again presented an opportunity. Following their intervention in 1739, the French had remained concerned about the instability of Genoa's island possession. Corsica was seen as critical to the protection of French trade with the Levant as well as a possible venue for the English to gain a foothold in the northern Mediterranean. This, and the importance of the alliance with Genoa to French interests on the Italian mainland, resulted in a series of agreements for the maintenance of French garrisons at strategic ports on the island. The growing strength of an independence movement under the leadership of Pasquale Paoli doomed any hope of the Genoese reestablishing control. After years of uncertainty about the island's future, they ceded the island to the French in 1768. Hall, *passim* and pp. 172-75.

During the summer of 1768 the French troops already on Corsica were reinforced by 16 infantry battalions and two legions in addition to artillery and engineers. The Corsicans determined to resist, and hostilities commenced on July 30-31 when French troops began to occupy parts of the island. The poorly planned campaign of 1768, culminating in the defeat at Borgo on October 5-9, was humiliating for the French. In January 1769 a new commander was appointed. *Ibid.*, pp. 191-99. He was Lieutenant General Noel Jourda, Comte de Vaux (1705-1788). *Biographie universelle*, XLIII, pp. 49-50; Courcelles, VII, pp. 30-34.

122 The "Tribute" does not reveal whether Pouchot returned to Corsica with the French contingent of 1768 or with Vaux's substantial reinforcements in 1769. Although the period of preparation for the first expedition would have given him three months to organize his *Mémoires*, it is also possible that Vaux's appointment provided the personal connection and

...(continued next page)...

sight, & through a kind of vengeance, resorted to the hands of vile assassins in order to deliver the final blows.[123]

His generals missed him & his death was mourned by his family, which was still numerous. Among all its members, who were equally distinguished by their virtue & their service, he had chosen as his heir, before his departure, one of his [xxvii] brothers who, having spent much time in financial posts, never breathed their contagious air & proved himself deserving of public esteem. [xxviii]

❁ Testimonial Of ❁ M. Le Marquis De Vaudreuil

We, Pierre de Rigaud, Marquis de Vaudreuil,[124] Great Cross of the Royal & Military Order of St. Louis, formerly Governor and Lieutenant General for the King of all New France:

Certify that M. Pouchot, Captain in the Béarn Regiment, conducted himself in the different commands with which we entrusted him at Forts Niagara & Lévis - either [xxix] for their construction with

---

influence necessary for Pouchot's resumption of active military service. The two men were very likely acquainted since Vaux's earlier campaigns in Europe had closely paralleled those of Pouchot: Italy in 1734, Corsica in 1738-1739, Bohemia and Germany in 1743-44, and Flanders (including the siege of Tournai and the Battle of Fontenoy) in 1745-1747. Courcelles, VII, pp. 30-32. Vaux, who had offered to serve on Corsica as early as March 1768, was appointed in January 1769 and arrived on the island during the first week of March. Hall, pp. 198-99. This would have provided considerably less time for Pouchot's literary preparations.

Pouchot's "inclinations" included road-building, a major activity of French troops on Corsica in 1768-1769 and the same duty that had occupied him during service on the island thirty years previously. *Ibid.*, pp. 184, 190, 199.

123 The French opened the campaign of 1769 on May 3, advancing south from the base of Cap Corse at the northern extremity of the island. The army moved in three columns with Vaux in the center. The left quickly seized Borgo near the east coast. Vaux pushed back the Corsicans under Paoli in heavy fighting and, by May 7, had placed his troops between the towns of Canavaggio and Lento overlooking the Golo River. Hall, p. 200-01.

Pouchot must have marched with Vaux's column. On May 7 or 8 the general ordered him to reconnoitre the road from Lento to Borgo, probably to improve communications with the left column. On the 8th, Pouchot and his escort of grenadiers from the Royal-Italien Regiment were attacked by inhabitants of the village of Vignale, and the engineer was killed in the skirmish. Rinatu Coti (ed.), *Ponte Novu: Journal d'un officier français an 1769* ([Corsica], 1986), p. 33. The next day Paoli attacked Vaux to contest the crossing of the Golo, and the French won the decisive victory of Ponte Novu. Hall, p. 201.

124 Pierre de Rigaud de Vaudreuil de Cavagnial, Marquis de Vaudreuil (1698-1778). Destined to be the last Governor General of New France, Vaudreuil was Canadian by birth.

...(continued next page)...

which he had been charged or their defense - with all good sense, zeal, intelligence, & economy that one is entitled to expect from a first-rate officer. This includes his talent for dealing with the Indians & enticing them for the good of the service into the sieges that unhappy circumstances compelled him to sustain & in which he upheld the honor of the King's arms with rare valor, which enabled him to acquire a well-deserved reputation. 1st. At Niagara, he held out for 19 days in open trenches with 450 regulars & militiamen of whom 117 were killed or wounded, against a force of about 5000 English and Indians, of whom the two leading generals, Prideaux & Johnston were killed.[125] 2nd. At Fort Lévis [xxx] he sustained the siege with 250 regulars & militiamen, of whom sixty were killed or wounded. Among the latter was his artillery officer. For eleven days he held out with this small force against the army of General Amherst, who had eleven thousand regular troops & Indians, together with overwhelming artillery. It should be noted that, although vigorously attacked by the enemy on land & by water, on 22 August, he caused a 22-gun brigantine, with 300 men on board to run aground & surrender. He was also responsible for the grounding of two others, one of 18 guns, & the other of 10 12-pounders. He conducted himself with equal distinction in the other operations in which we employed him & in other actions [xxxi] in which he served, as at Chouegen,[126] where he directed the siege, and also in the Carrillon [sic] action, upon which we reported to the Minister at the time.[127] By virtue of the above, we have signed the present certificate & apposed the seal of our arms. Paris, 6 May 1761.[128]

Signed VAUDREUIL [xxxii]

After gaining much experience in the western reaches of the colony and as Governor of Louisiana from 1743 to 1753, he was appointed to New France on January 1, 1755. His commission gave him command of all military forces in the colony, a responsibility that ultimately brought him into conflict with the *troupes de terre* commanders, Dieskau and Montcalm. Vaudreuil arrived at Québec on June 23, 1755 and held his office until forced to surrender to the English on September 8, 1760. He too was implicated in the *affaire du Canada* but exonerated. *DCB,* IV, pp. 662-74.

125 John Prideaux (1718-1759), lieutenant colonel (1758) of the 55th Regiment of Foot and brigadier general. Prideaux had been appointed in May 1759 to command the expedition against Fort Niagara. *Dictionary of National Biography* (Hereafter cited as *DNB*; 22 vols.; London, 1908), XVI, pp. 356-57. Prideaux's second in command was a provincial officer, John Johnston (1719-1759), colonel of the New York Regiment. A resident of Perth Amboy, New Jersey, he had served in the 1758 campaign with troops from his own

...(continued next page)...

Introduction

In spite of a century & a half of occupation, France had hardly been able to derive any advantage from that huge country in North America known as Canada. The colony they had founded there was still, so to speak, in its infancy when it passed under a foreign yoke. It could doubtless have emerged from this state of feebleness, or rather prostration, and have become one day very useful to the mother country, if only France had [xxxiii] known more about it & had not been so often deceived by those who ought to have provided accurate information. People in France had such false ideas of the country that they considered its usefulness to lie entirely in the fur trade & failed to distinguish clearly between the colonials and the Indians.[129] In fact, ignorance [xxxiv] & blindness were carried to such a degree that some people were even delighted at its loss.

England, in order to prevent its rival from opening its eyes as to the advantages of Canada, began to plan an invasion even in peacetime. The court in London had long repented using up its strength & wealth in vainly expressing its hatred of the French by giving aid to their enemies. Immediately after signing the Treaty of Aix-la Chapelle,[130]

---

province. "Journal of Peter Pond", *Wisconsin Historical Collections* (31 vols.; Madison, 1855-1931), XVIII, p. 321n. See also Severance, II, p. 298.

Both officers were killed in the trenches before Fort Niagara on the evening of July 20, 1759. Dunnigan, *Siege*, pp. 69-71.

126 The name used by the French to describe the site where the Oswego River enters Lake Ontario. The English knew the place as "Oswego". It is today Oswego, New York.

127 Although Vaudreuil wrote to the Minister of Marine on August 4, 1758 about the Carillon action, his letter focuses on his disagreements with Montcalm and makes no mention of Pouchot. *DRCHSNY*, X, pp. 779-83.

128 It appears from the date of Vaudreuil's letter that Pouchot went to Paris immediately after his return to France on March 8 to solicit a recommendation from the former governor general.

129 Note in original, p. xxxiv (a): *In France people even supposed that Canadians were facially different & that they had even stranger customs. Although neighbors of Canada, the residents of New England in the late war regarded the inhabitants of that country as half-savage because they were persuaded that the French married Indian wives.*

130 Signed October 18, 1748, the Treaty of Aix-la-Chapelle concluded the War of the Austrian Succession (1740-1748). Skrine, pp. 348-49.

she conceived the notion of seizing the French colonies. Canada, or New France, which had to be considered as their most solid bulwark [xxxv] thus became the object of her initiatives. The first step was to extend as far as the St. Lawrence River the settlements she was hastening to found in the region of Acadia.[131] Those which were envisaged, whether in the area of Hudson's Bay, or beyond the Appalachians towards the banks of the Ohio, or on the shores of Lakes Ontario & Erie, were absolutely limitless. The intended consequence of this plan was that what would have remained to France would have been useless to her since entry to the territory would have been blocked.

The Council at St. James[132] fully expected complaints & claims on the part of France, [xxxvi] which proposed the appointment of commissioners from each nation to decide upon the boundaries of the respective colonies. This proposal was accepted & it was decided as a first step that there should be no changes to the territories in question. Scarcely had this agreement been signed than England unscrupulously violated it. Her commissioners could never agree with those of France since they had secret orders to raise difficulties & to agree upon nothing. However, Louis XV wished for peace at any [xxxvii] price. His ministers believed they could secure its continuance on conditions that England would not have rejected under any other circumstances. But she was by now resolved to use her overwhelming naval strength to destroy French commercial interests which, through their vigor, awakened her hatred & aroused her jealousy.

If English aspirations had been fully understood, no one would have been surprised when the English minister only replied to the advantageous proposals made by the Court at Versailles by making unacceptable demands. Despite this, he pretended to negotiate & to desire [xxxviii] peace, "but he had no other aim," as a judicious statesman observes, "than to concentrate all his initiatives & to inspire in France a sense of security which would prevent her from preparing for war by dispersing her forces."[133]

---

131 The French colony in what is today Nova Scotia that had been given up to the English in 1713 by the Treaty of Utrecht.

132 The British court, or the Lords of Trade.

133 Note in original, p. xxxix (a): Droit public de l'Europe, *Vol. III, p. 194.*

The referenced work has not been further identified. It would appear to have been a text concerning international European law.

A number of philosophers, or rather men who imagine they deserve the name, every time they can demonstrate the wrong-headedness of princes & can ridicule it, have obstinately declined to perceive the true cause of this war. They have persistently claimed **[xxxix]** that France only exposed itself to so many setbacks & shed so much blood in order to retain a precarious hold on "frozen territory," a few "savage regions" or "useless deserts." Such was the language of cynical ignorance. These are the terms that pride & presumption would dearly like to consecrate as the oracles of sublime reason.[134]

Some persons, rather better informed, have certainly not reproached France in this manner, but they have accused her commissioners of incompetence & ill will and her ministers of ambition & lack of a sincere **[xl]** desire for peace. The author of the memoirs that follow seems to have shared some of these prejudices. We believe it proper to warn his readers about this. They will find in the public documents of France innumerable & incontestable proofs that the pretensions of England were not the cause, but the pretext, of the late war. **[xli]**

---

134 French intellectuals were not disappointed by the loss of New France. François-Marie Arouet de Voltaire (1694-1778) was in the forefront of those depreciating the value of the colony. The reference here is probably to his *Précis de Louis XV* (1768): "We have lost in one day ... fifteen hundred leagues of ground. These ... being frozen deserts, are perhaps not really a loss. Canada has cost much and brought in very little." Quoted in Stanley, p. 269.

In *Candide,* published in 1759, Voltaire remarked of England and France in a similar vein that, "You know these two nations are at war for a few acres of snow in Canada, and that they are spending more on this fine war than all Canada is worth." Voltaire, *Candide and Other Writings* (New York, 1956), p. 167.

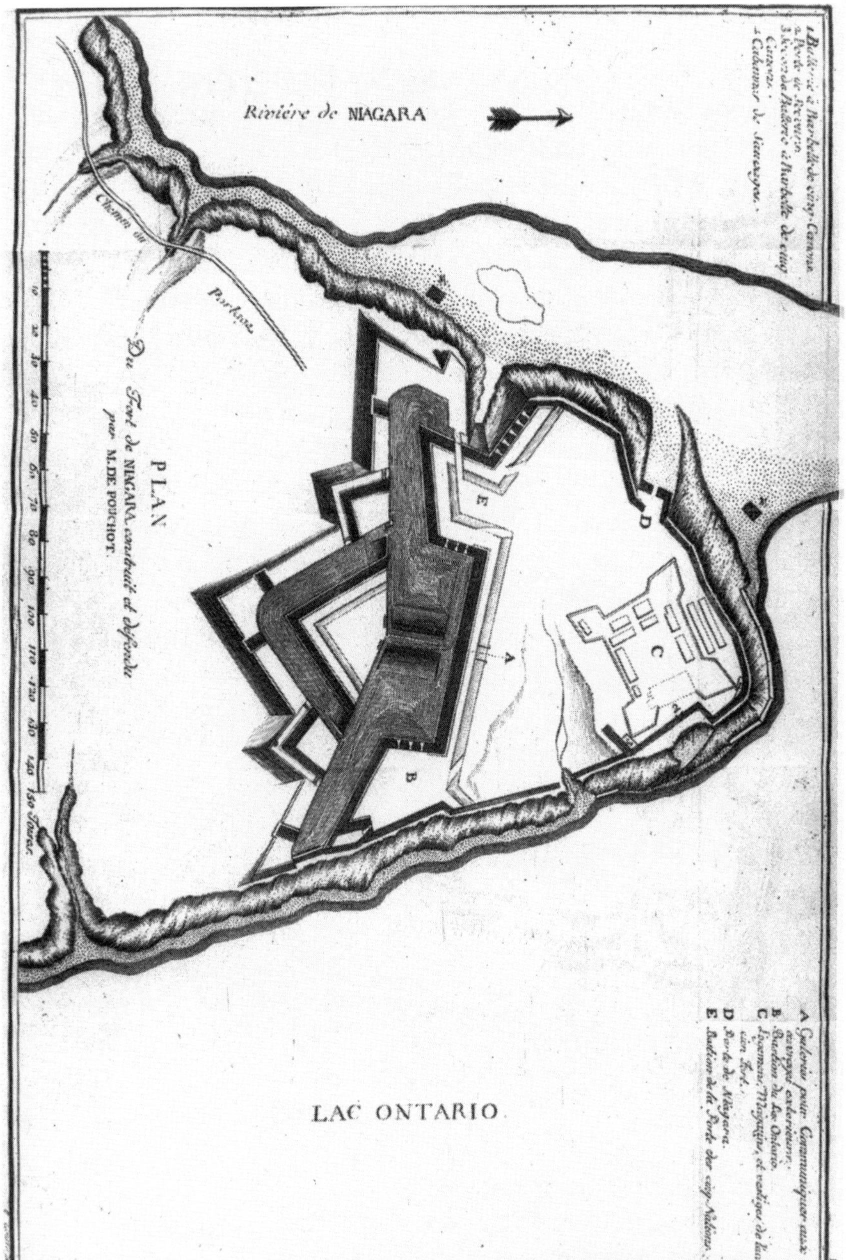

"Plate 1" of the *Mémoires* is a plan of Fort Niagara. It appears to represent the fortifications as laid out in the autumn of 1755 or as partially completed by the early summer of 1756. *Courtesy, William L. Clements Library, University of Michigan.*

Memoirs On The Late War In North America, Between France & England

Before entering into a few details about the war which led to the loss of Canada, it is necessary to speak briefly about the claims of the two powerful nations which established themselves in a continent to which they originally had no right [1] & could acquire none except in respect of the territories they settled. We shall not speak of the rights of the natural owners of the country, which these great powers entirely discount, even though the natives find it very strange that others should fight for a country where the author of life has, in their view, created them, where they have always lived & of which the bones of their ancestors have had possession from the beginning of time. They are unwilling to recognize any foreigner as their master, just as they have none amongst themselves.

Only through force have these foreigners been able to establish and maintain their settlements. The latter, by meeting at several points, have become the cause of conflict. The English, having penetrated to the upper reaches of the Hudson River, [2] encountered the settlements of the French in the region of Lake Champlain. The colonies of Virginia and Pennsylvania became prosperous and sought to extend their trading contacts with the Indians beyond the Appalachians. They interfered with our trading posts, and this caused complaints in the two courts. Each side sent military detachments ahead in order to secure the places which were occupied. Jealousy soon led to armed conflict.

Let us begin with the claims of Acadia, or Nova Scotia, without entering into details prior to the Treaty of Utrecht.[135] In 1714, the English sent a regiment to Annapolis, or Port Royal,[136] in order to take possession of Acadia & made no other attempt to settle in the rest of the peninsula, of which at least three [3] quarters were occupied by the French & their Indian allies. By the peace treaty of 1748 it was

135 Signed April 11, 1713, the Treaty of Utrecht ended the participation of most combatants in the War of the Spanish Succession (1701-1714). Among other concessions, France ceded Newfoundland and Acadia to England. DuPuy and DuPuy, p. 625.

136 Annapolis Royal, Nova Scotia.

agreed that the boundaries of the English & French possessions should be determined in a friendly manner by a joint commission.[137] In 1749, the English settled at Chiboucton and founded Halifax.[138] Everything had remained calm until the war of 1740 in which the Acadians, the French of Acadia, supported by the Canadians, made several attacks on the English, without it appearing that the latter claimed to consider the Acadians as their subjects.[139]

During the arguments of the commissioners, the English built the fort of Beau-bassin[140] & the French that of Beau-séjour.[141] The

---

137 The Treaty of Aix-la-Chapelle referred the boundary question to an international commission. The French members were Étienne de Silhouette and the Marquis de La Galissonière (note 149), former acting Governor General of New France. The English representatives were William Mildmay and William Shirley (note 255), Governor of Massachusetts. Shirley was recalled in 1752. The commission began discussions in the summer of 1750 and met sporadically until December 1755. Stanley, pp. 124-27.

The deliberations of these commissioners, with supporting historical documentation, were published in several versions including *Mémoires des Commissaires du roi et de ceux de Sa Majesté britannique, sur les possessions & les droits respectifs des deux couronnes en Amérique; avec les actes publics & piéces justificatives ....* (4 vols.; Paris, 1755-1757).

138 The British landed at Chebucto Bay in June 1749 and founded a settlement named for the Earl of Halifax. Stanley, pp. 65-68.

139 During the War of the Austrian Succession (1740-1748), commonly known as King George's War in the English colonies of America where military action occurred from 1744 to 1748, the Acadians were in a difficult position. Technically subjects of King George II, their cultural ties were with France, and they were pressured by the government of New France to participate in the fight against the English. Their dilemma is discussed in Lawrence Henry Gipson, *The British Empire Before the American Revolution, Volume V: Zones of International Friction: The Great Lakes Frontier, Canada, the West Indies, India, 1748-1754* (New York, 1942), pp. 167-206. See also N.E.S. Griffiths, "The Acadians", *DCB,* IV, pp. xvii-xxi.

140 Fort Lawrence, established in 1750 near the site of the settlement of Beaubassin on the south bank of the Missaguash River. The village had been abandoned and burned by its Acadian inhabitants prior to the arrival of English troops. Fort Lawrence guarded the isthmus of Chignecto and defined the frontier between Nova Scotia and the French colonies. It was located near modern Amherst, Nova Scotia. Stanley, pp. 72-75.

141 Fort Beauséjour, established in 1751 on the Missaguash River opposite Fort Lawrence. The post was located near modern Sackville, New Brunswick. Stanley, pp. 74-75. Its ruins are preserved and interpreted in Fort Beauséjour National Historic Site.

142 At least 6,000 Acadians were forcibly removed from their land and dispersed during 1755. Thousands more fled to New France and Isle Royale (Cape Breton Island). The expulsion of the Acadians and the English rationale for this act are discussed in Lawrence Henry Gipson, *The British Empire Before the American Revolution, Volume VI: The Great War for the Empire, The Years of Defeat, 1754-1757* (New York, 1949), pp. 243-344.

English called upon the Acadians to swear an oath of loyalty. When they [4] refused, the English removed them, with the exception of those who withdrew to Canada, and dispersed them among their different colonies.[142] Although the English extended their claims to include the St. Lawrence, it nonetheless appears from the construction of Beau-séjour[143] that they desired to possess only the Acadian peninsula. M. de la Jonquiere, commandant of Canada,[144] & Colonel Laurence[145] for the English side, were charged with determining the boundaries. They could only be established by force or through determined goodwill on the part of the two nations. As a result the two commissioners did nothing but quarrel. M. Laurence was the shrewder of the two & M. de la Jonquiere, a man of the sea, was the blunter. But the latter became angry very easily, said harsh things to his opponent and even challenged him to several duels.[146] These [5] disputes spilled over to the commandants of the frontier posts and this caused the death of a commandant of Beau-bassin who was killed by one of our Indians, supposedly at the instigation of one of our Sulpician missionaries, who was attached to the Indian nations in this territory. This man was later caught by the English, who left him to die in a prison cell, a rather understandable reward for his excessive zeal.[147] This act did not lead to any open break in relations, although the treaty of 1748 had certainly not ended the war.

---

143 Pouchot presumably means Beaubassin - the English Fort Lawrence.

144 Jacques-Pierre de Taffanel de La Jonquière, Marquis de La Jonquière (1685-1752). Although appointed Governor General of New France in 1746, wartime captivity prevented him from assuming his office until August 14, 1749. He governed the colony until his death at Québec on March 17, 1752. *DCB,* III, pp. 609-12.

145 Charles Lawrence (c.1709-1760), lieutenant colonel (1750) of the 40th Regiment of Foot and Governor of Nova Scotia, 1756-1760. It was Lawrence who carried out the expulsion of the Acadians. *DCB,* III, pp. 361-66.

146 No record of these challenges has been found.

147 Edward How (c.1702-1750), an officer of the Nova Scotia militia, was shot and killed on October 4, 1750 after meeting with French representatives on the Missaguash River near Fort Lawrence. See *DCB,* III, pp. 297-98. The murderer was an Indian, but, although it was never proven, the English blamed Abbé Jean-Louis Le Loutre (1709-1772), missionary to the Micmac Indians since 1738. Le Loutre's position and instructions from the French government during the War of the Austrian Succession had made him a military leader of his parishioners, and he was a major contributor to tensions between the French and English of Acadia. The abbé was indeed captured by the British on September 15, 1755 while on a voyage to France. Although he was confined by them for eight years, however, he did not die in prison. *DCB,* IV, pp. 453-58.

In the meantime, the English having desired to station troops in des Mines village, which claimed to be French, the governor of Canada sent in wintertime a detachment to remove them. They arrived in the middle of the night preceding Twelfth Night &, as they knew [6] the locality well, they split up in order to attack all the houses at the same time. It was close to midnight. The commandant, who had dined with his officers and had gone outside to satisfy some need, saw a great deal of snow falling and exclaimed: "My God! It's beautiful weather for a Canadian." Nothing could have been closer to the truth. Our soldiers in hiding heard him, let out a yell and rushed into the houses together, where they took 300 men prisoner, of whom the most wide-awake were doubtless the officers.[148]

M. de la Jonquiere succeeded M. de la Galissoniere,[149] one of the most able men to command in these territories because of his talents, but in addition to not staying long enough, he had made no progress in the negotiations, in which neither of the two powers was prepared to make concessions. He was relieved by M. du Quesne,[150] [7] who was entrusted with the same negotiations. Continual minor skirmishes in Acadia continued to make negotiations necessary, which increasingly embittered relations between the two countries.

On his arrival, M. du Quesne fell in love with an agreeable lady & became friendly with her family & friends. Care was taken, as was the

---

148 Pouchot was confused here, perhaps because the events in Acadia had occurred well before his arrival in America. He describes the attack on Grand Pré, situated on the south side of the Minas Basin (near Wolfville, Nova Scotia). On the night of February 11-12, 1747 some 250 Canadians, led by Captain Nicolas-Antoine Coulon de Villiers (1708-1750; *DCB*, III, pp. 149-50), surprised a force of 500 New England troops under Lieutenant Colonel Arthur Noble (*DCB*, III, pp. 483-84). Noble and about 70 of his men were killed and the rest paroled by the French. For detailed accounts of this action see Francis Parkman, *A Half-Century of Conflict* (New York, 1962), pp. 369-79 and Stanley, pp. 23-25.

149 Roland-Michel Barrin de La Galissonière, Marquis de La Galissonière (1693-1756) served as Commandant General of New France in the absence of the appointed governor general, Jacques-Pierre de Taffanel de La Jonquière, from 1747 until 1749. *DCB*, III, pp. 26-32.

150 Ange Duquesne de Menneville, Marquis Duquesne (c.1700-1778), Governor General of New France, 1752-1755. *DCB*, IV, pp. 255-58.

151 No mention of this alleged affair has been found in biographies of Duquesne. See *DCB*, IV, pp. 255-58 and Charles W. Dahlinger, "The Marquis Duquesne, Sieur de Menneville, Founder of the City of Pittsburgh", *Western Pennsylvania Historical Magazine*, XV (1932), pp. 2-33, 121-51, 219-62.

...(continued next page)...

custom, to send her husband to one of the most profitable posts in the upper country.[151] M. Bigot moved from the intendance of Louisbourg to that of Canada at about the same time.[152] He also fell for Madame Péan, the wife of the adjutant of the citadel of Québec, & took a great interest in the family.[153] The intendant was responsible for everything to do with finance, the food supply which was let out to tender, & the supply of [8] trade goods. For these two leaders to be useful to one another, they had to see eye to eye and this their two companies managed to achieve. Since their appointments were for barely three or five years, these gentlemen normally tried to make their fortune & that of their friends, or those who had helped them during the time. Whether the plan of a settlement on the Ohio was hatched by them in order to create an opportunity highly favorable to their interests, or whether the court decided upon it because of the advantages to be derived from it, the plan was set in motion in the winter of 1753-1754.[154] [9]

Duquesne was either very discreet or Pouchot, who arrived in New France just as Duquesne was leaving, confused this affair with the one between Intendant Bigot and Mme. Péan (note 153). Pouchot's language suggests that he might have been referring to her. Another possibility is Marie-Marguerite Pennisseaut, the wife of Louis Pennisseaut, another member of Bigot's commercial circle. Mme. Pennisseaut became mistress to Péan and later to General Lévis (note 284). *DCB,* IV, pp. 621-22.

152 Bigot's appointment to New France occurred in 1748. *DCB,* IV, p. 63.

153 Pouchot's comment reflects common knowledge of the most blatant illicit relationship of New France. Michel-Jean-Hugues Péan (1723-1782), captain (1750) in the *compagnies franches de la marine* (hereafter referred to as "the Marine") and adjutant of Québec, was closely associated with Bigot and his commercial activities and was one of those later tried in the *affaire du Canada. DCB,* IV, pp. 614-17. His wife, born Angélique Renaud d'Avène Des Méloizes (1722-1792), became known as "the Pompadour" of Bigot. Their affair continued until the fall of New France. *DCB,* IV, pp. 659-60.

154 Note in original, p. 9 (a): *M. Pouchot presented the facts in a false light. The motives of which he speaks may have determined the choice of the Governor of Canada as to the officer who was to command on the Ohio, without committing him to found a settlement. His predecessor, M. de la Jonquiere, had planned it before M. du Quesne and the latter hastened to put it into effect in order to forestall the designs of the English, who wanted to cut communications between Louisiana and Canada. They were, moreover, making great preparations to attack the French in that area, under the pretext of helping the Indians, whom they had stirred up against them.*

Concern over growing English influence in the Ohio Valley indeed predated Duquesne's arrival. In 1749 La Galissonière sent an expedition through the region to demonstrate French sovereignty. Command was given to Pierre-Joseph Céloron de Blainville (1693-1759), captain (1738) in the Marine. *DCB,* III, pp. 99-101. Céloron portaged from Lake Erie to the Allegheny River, descended the Ohio to the Miami River and returned to Lake

...(continued next page)...

They equipped[155] & provisioned 7 to 800 Canadians under the orders of M. Marin.[156] Several officers from the colony first went off to establish a post at the Niagara portage.[157] **[10]** In the spring, large quantities of food and equipment, tools & trade goods were despatched. They assembled in that territory everything that could be found in the way of merchandise of every kind, even velvets, silk cloths, women's damask shoes, silk stockings, and so on, & also quantities of Spanish wine. These goods were donated by the groups of which we have spoken & bought at the king's expense. It can be assumed that there was very little haggling over the price asked.

These goods took a long time to travel from the Niagara portage and Presqu' Isle to the Ohio because of the lack of horses & vehicles.

Erie by way of the Maumee River. Along the way he buried a number of lead plates claiming the territory for King Louis XV. La Galissonière urged his successor to establish posts to help protect communications with Louisiana. Before his death in 1752, Governor La Jonquière was unable to do more than encourage French traders to become more active in the region, so it was left to Duquesne to initiate the construction of military posts. Donald H. Kent, *The French Invasion of Western Pennsylvania, 1753* (Harrisburg, 1954), pp. 6-12. Preparations for the Ohio expedition were completed during the winter of 1752-1753.

For French and English rivalry over the Ohio, see also Lawrence Henry Gipson, *The British Empire Before the American Revolution, Volume IV: Zones of International Friction: North America South of the Great Lakes Region, 1748-1754* (New York, 1939), pp. 186-312.

155 Note in original, p. 10 (a): *Every time that Canadian troops went on campaign, they were supplied with a coat* [capot], *two shirts, a pair of breeches, mittens, blanket & a pair of sealskin shoes* [moccasins] *each month. In addition, the officers were given a bottle of wine a day, two jars of brandy a month, a ham or a sheep, hunting shot & powder.*

For information on the organization and equipment of the militia of New France, see Francis Back and René Chartrand, "Canadian Militia, 1750-1760", *Military Collector & Historian,* XXXVI, No. 1 (Spring, 1984), pp. 18-21 and Chartrand, *Canadian Military Heritage,* pp. 154-56.

156 Paul Marin de La Malgue (1692-1753), captain (1748) in the Marine and an experienced officer who had seen much duty at the posts of the western Great Lakes and in the Acadian and New York campaigns of King George's War. Marin died October 29, 1753 at Fort de la Rivière au Boeuf (Waterford, Pennsylvania), a post he had recently established. *DCB,* III, pp. 431-32.

157 Pouchot refers to the establishment of an advance depot since a permanent post on the portage predated the expedition. Known as Fort Little Niagara, it had been constructed at the head of the portage (Niagara Falls, New York) in 1751. Severance, I, pp. 374-80.

In the autumn of 1752, the commandant of Fort Niagara received orders to improve the portage road around the falls and to prepare for the passage of an army of 2,000 men. The advance detachment of this expeditionary force departed Montréal in February 1753 under the command of Charles Deschamps de Boishébert et de Raffetot (1727-1797; *DCB,* IV, pp. 212-15), lieutenant (1748) in the Marine. Boishébert was to enlarge the warehouses on the Niagara portage, build pirogues and proceed to the Tchatacoüen (Chautauqua) portage to establish a fortified storehouse. Kent, pp. 17-19.

In consequence, nearly 400 men perished, either from scurvy or from the excessive strain of carrying these goods on their backs.[158] During this period [11] the officers drank the Spanish wine and they all removed from the stores everything that took their fancy, especially the velvets, &c. which were certainly not intended as trade goods for the Indians. As a result, the provisions which arrived at the post called "du Quesne"[159] were in very small quantities and were diminished even further by waste & damage suffered on the trail. The officers and the militiamen who returned to Canada were thus very well supplied & all this was covered by bills of sale.

At this point Chevalier Péan, whom the intendant was quite happy to separate from his wife, was charged with leading an expedition with 400 militiamen to Détroit, in the same general area. He was well supplied with all kinds of provisions and goods as [12] presents for the Indians on the pretext of seducing them to our side.[160]

Such a mission was futile, since that area had for long been settled by the French, who enjoyed very close contacts with the Indians of the

---

158 The tremendous effort of moving men, boats, supplies, and equipment across the Niagara and Presqu' Isle portages and its effect on the men of the expedition is graphically detailed in Kent, *The French Invasion* and in Severance, II, pp. 1-33.

Throughout his narrative, Pouchot consistently renders the name of the important post of Presqu' Isle without capitalization as simply *la presqu'isle* or "the peninsula". We have capitalized the name whenever he refers specifically to the fort at the site of Erie, Pennsylvania on the south shore of Lake Erie. The spacious bay, sheltered by a long peninsula that gave the place its name, marked the beginning of one of several portages to the Ohio River. It was selected as the route to be used by Marin's expedition, and construction of a fort commenced on May 15, 1753. The post became known as Fort de la Presqu' Isle. See Max Schoenfeld, *Fort de la Presqu' Ile* (Erie, PA, 1989) and Charles Morse Stotz, *Outposts of the War for Empire: The French and English in Western Pennsylvania: Their Armies, Their Forts, Their People, 1749-1764* (Pittsburgh, 1985), pp. 72-73. Stotz's work is particularly useful for its reproduction of both contemporary plans and his own architectural reconstructions of the forts discussed.

159 Fort Duquesne, at the site of modern Pittsburgh, guarded the confluence of the Allegheny and Monongahela Rivers and was last in the chain of four French posts linking Lake Erie with the Ohio River. Construction began after the French had expelled a small English force from the site on April 16-18, 1754. Stotz, pp. 80-87.

160 Péan's trip to the West was ordered in May 1754. He was to proceed from Montréal to the Ohio by way of the Chautauqua portage. From the Forks of the Ohio he was to go to the Great Miami, remove any English traders and continue to Detroit and Michilimackinac, returning to Montréal by the Ottawa River. His voyage was abbreviated, and Péan proceeded to Detroit along the south shore of Lake Erie. His mission is summarized in Severance, II, pp. 47-51.

...(continued next page)...

region.[161] There were French officers in all the posts capable of undertaking this task. But they were getting rid of a husband & also a considerable quantity of merchandise for the company's benefit. Péan returned to Canada in triumph after this splendid feat.

• 1754 •

M. de Contre-coeur[162] remained commandant at Fort du Quesne which M. Mercier,[163] an artillery officer, designed & had built. They left M. de Villiers,[164] M. Jumonville[165] and several other officers at the post.

In the summer they received information that [13] an English detachment had passed through the area of the Fourche de la Manenguelée[166] & was moving towards the Ohio to set up a post there. The officers' council at Fort Duquesne decided to send out

---

Like Presqu' Isle, Pouchot renders the name of the post at Detroit, Michigan as a geographical term and without capitalization (*détroit*). We have capitalized the name when it refers to the fortified village at the site of Detroit, Michigan.

161 French interests were not so firmly established among the tribes of the West as Pouchot implies. Penetration of the Ohio by English traders remained a concern, even after the destruction of a pro-English village at Pickawillany, Ohio by French-Indian raiders in June 1752. See Stanley, pp. 45-49.

162 Claude-Pierre Pécaudy de Contrecoeur (1705-1775), captain (1748) in the Marine, to whom command of the Ohio expedition had devolved in December 1753 following the death of Marin and the resignation of Captain Jacques Legardeur de Saint-Pierre (note 270). Contrecoeur had led the first confrontation with English military forces on April 16-18, 1754 when he seized their incomplete fort at the Forks of the Ohio. He commanded Fort Duquesne until 1756. *DCB,* IV, pp. 617-18.

163 François-Marc-Antoine Le Mercier (1722-c.1798), captain (1753) in the colonial artillery company of *canoniers-bombardiers* and commandant of the artillery of New France. He served as engineer for the Ohio expedition. Later accused in the *affaire du Canada,* he was acquitted. Le Mercier was disliked by Pouchot and other regular officers, including Montcalm, who expressed a poor opinion of both his technical ability and his ethics. *DCB,* IV, pp. 458-61.

164 Louis Coulon de Villiers (1710-1757), captain (1753) in the Marine. Sent from Canada with reinforcements for Fort Duquesne, he did not arrive at the post until June 26, 1754. *DCB,* III, pp. 148-49.

165 Joseph Coulon de Villiers de Jumonville (1718-1754), second ensign (1741) in the Marine. *DCB,* III, pp. 150-51. He was the younger brother (or half-brother) of Louis Coulon de Villiers. J. Martin West (ed.), *War for Empire in Western Pennsylvania* (Ligonier, PA, 1993), p. 23. The colonial artilleryman "J.C.B." maintained that the two were brothers. Stevens, *Travels in New France,* p. 106.

Jumonville with a detachment of thirty armed men to oblige them to go back and he was provided with a letter so informing their commander. The English officer, warned by allied Indians of the arrival of this detachment, lay in wait for him in a kind of ambush. Jumonville, who did not consider himself to be in the stronger position, tried to show the letter he was bearing. The English, not wishing to become involved in negotiations, opened fire on the detachment, killing Jumonville & several others, and took the remainder prisoner.[167] As soon as the news [14] arrived at Fort du Quesne, Villiers, deeply upset over the death of his brother, demanded to go and avenge it after the manner of the Indians. They held a council of war, of which Mercier was the leading light.[168] They arrived at a decision in writing, the terms of which stated that, without seeking to infringe the Treaty of Utrecht, Villiers would set out with 300 men in [15] search of the English who, 500 in number, were beginning to construct a fort at a

---

166 The confluence of the Monongahela and Youghiogheny rivers at the site of McKeesport, Pennsylvania. The English detachment, opening a road towards Fort Duquesne, had reached the "Great Crossing of the Youghiogheny" by May 17. Gipson, VI, pp. 28-29.

167 Note in original, p. 14 (a): *Could M. Pouchot really have intended to insinuate here that M. de Jumonville had aggressive intentions? The limited size of his force does not permit one to suppose so. Even the Indians did not think so, for they had such a high regard for the character of this officer that, as he was being murdered, they threw themselves between the French & the English in order to protect him. It is painful to recall that the celebrated General Washington was in command of M. de Jumonville's murderers on that occasion. He was indeed acting on specific orders of his court, but he could have obeyed them in a less distasteful manner.*

The English commander was, of course, George Washington (1732-1799), then a lieutenant colonel of Virginia militia. Richard L. Blanco (ed.), *The American Revolution, 1775-1783: An Encyclopedia* (2 vols.; New York, 1993), II, pp. 1733-49; Allen Johnson and Dumas Malone (eds.), *Dictionary of American Biography* (Hereafter cited as *DAB*; 10 vols and supplements; New York, 1927-1936 and subsequent editions), X, pp. 509-27.

Washington's attack on Jumonville's detachment took place at dawn on May 28, 1754. The ensign and nine of his companions were killed. One Frenchman was wounded and 21 were captured while only one escaped to Fort Duquesne. Washington lost just one man killed and two or three wounded. Gipson, VI, pp. 30-32. The editor's uneasy apology for Washington's actions was due to his popularity in 1781 as the military leader of France's new ally, the United States of America.

168 The French reaction was not quite so swift as Pouchot states. The elder Coulon de Villiers did not arrive at Fort Duquesne until June 26. There he found a force of 500 men preparing to march against Washington under the command of Le Mercier. In a council of war the next day, Villiers successfully argued that he should command the expedition. The French set out from Fort Duquesne on June 28, one month after Jumonville's death. Gipson, VI, pp. 35-37; *DCB*, III, pp. 148-49.

place they had baptized "Of Necessity."[169] The French surrounded this fort, positioning themselves behind trees & a small abatis built by the English. They had begun a ditch which was already knee-deep. As they were all packed together, the volleys of the men who are good marksmen had soon felled a large number of them. On seeing themselves overwhelmed by this murderous fire, the English asked to surrender. They were taken prisoner on parole, on condition they would immediately release their own prisoners & that we would keep two of their officers as hostages.[170] M. de Villiers also forced them to hand over a statement as to the manner in which Jumonville had been murdered against the **[16]** rules of war.[171] Moreover, they had no

---

169 Washington's Fort Necessity was located at the "Great Meadows" near Farmington, Pennsylvania. Stotz, pp. 95-97. See also J.C. Harrington, *New Light on Washington's Fort Necessity: A Report on the Archaeological Explorations at Fort Necessity National Battlefield Site* (Richmond, VA, 1957). The site, with a reconstruction of the fort, is today preserved and interpreted by the National Park Service.

170 Following a nine-hour action on July 3, 1754, Washington surrendered Fort Necessity to de Villiers whose troops took possession of the post on the 4th. Gipson, VI, pp. 38-42.

The hostages were both officers of the Virginia Regiment. One was Robert Stobo (1726-1770), a captain (1754) who remained a prisoner until making a daring escape from Québec in May 1759. *DCB*, III, pp. 600-02. See also Robert C. Alberts, *The Most Extraordinary Adventures of Major Robert Stobo* (Boston, 1965).

Stobo's fellow hostage, Jacob Van Braam (b.1729), had been appointed captain only on May 8. He was a Dutch immigrant to Virginia who had served as an interpreter on Washington's 1753 journey to confront the French on the Allegheny. He would remain a prisoner in Canada until 1760. See *The Journal of Major George Washington* (Williamsburg, VA, 1959), p. 33 and Van Braam to Amherst, Nov. 9, 1759, Amherst Papers/78. Following his release, Van Braam was commissioned captain in the 60th Regiment of Foot in 1761, went on half-pay in 1765 and returned to active service in 1775 with the third battalion of the regiment. He advanced to major in 1777 and disappears from the *Army List* after 1779. See also Alberts, pp. 22, 334.

171 Note in original, p. 17 (a): *This statement never existed. M. Washington was still in command at the fort when he surrendered to M. de Villiers. The account of the latter does not quite conform with that of M. Pouchot. See the supporting documents of* mem. de la France, no. IX.

The documents were printed in [Jacob Nicolas Moreau], *Mémoire contenant le précis des faits avec leurs piéces justificatives, pour servir de résponse aux Observations envoyées par les ministres d'Angleterre dans le cours de l'Europe* (Paris, 1756). "No. IX" of the appended *Piéces justificatives* includes Villiers' journal with a copy of the articles of capitulation on pp. 153-55.

The statement concerning Jumonville's death was contained in article VI of the capitulation which termed the act an "assassination" of "an officer charged with a summons." Washington, relying on a translation by Captain Van Braam who imprecisely rendered the phrase "the defeat of M. Jumonville," signed the document. The articles of capitulation are also printed in Mante, pp. 12-13.

choice because they had had great difficulty in feeding such a large body of men in their fort.

The French returned triumphantly to their post & had no more problems for the rest of the campaign. Several officers returned to Canada, among them Mercier who was relieved by Lery, a so-called engineer.[172] Mercier & Péan were sent to France to present an account of the glorious & interesting events of their campaign. It is not difficult to guess whether they made a point of stressing to the Court the need to [17] support these useful outposts. They spoke above all in their own interest, since they foresaw that the costs involved would all benefit their company.[173]

As soon as the English learned of the events in that area of America, they decided to send to the continent in the winter of 1754-1755, the regiments of Peperel, Syrley, Halquet, & Dumbard in order to reinforce their colonies. The first two were intended for Chouegen & the two others went to Virginia in order to march on the Ohio.[174]

---

172 Gaspard-Joseph (or Joseph-Gaspard) Chaussegros de Léry (1721-1797), lieutenant (1751) in the Marine and former assistant engineer of New France. He resigned that post in 1749, but various governors and military officers continued to draw upon his experience. De Léry's father and grandfather had been engineers, and his work received favorable mention from many officers. *DCB,* IV, pp. 145-47.

In 1755 de Léry, then at Niagara, was informed that Pouchot had been dispatched to contrive, in cooperation with him, a plan to strengthen the place. Vaudreuil to de Léry, Sept. 23, 1755 in Pierre-Georges Roy (ed.), *Inventaires des papiers de Léry* (3 vols.; Québec, 1939-40), III, p. 192. Vaudreuil further reported that he had ordered de Léry to construct temporary entrenchments at Niagara. Pouchot, upon his arrival, was to "turn all his works to the best advantage possible." Vaudreuil to Machault, Sept. 25, 1755, *DRCHSNY,* X, p. 326. The captain substituted his own plan for whatever de Léry had begun. Pouchot's low opinion probably originated with this encounter and reflects his lack of regard for a colonial engineer.

173 Le Mercier and Péan departed for France in October 1754 and returned with Governor Vaudreuil and the *troupes de terre* reinforcements in the spring of 1755. *DCB,* III, p. 459.

174 The English regiments assigned to service in America in 1754 included two from Ireland that were to be recruited to full strength in the colonies: the 44th (Sir Peter Halkett's) and 48th (Thomas Dunbar's). Two more were to be raised in America: the 50th (William Shirley's) and 51st (Sir William Pepperrell's). The 44th and 48th were sent to Virginia and the 50th and 51st massed at Oswego on Lake Ontario. Gipson, VI, p. 58. A brief but useful introduction to the English forces committed to North America during the war is C.P. Stacey, "The British Forces in North America During the Seven Years' War", *DCB,* III, pp. xxiv-xxx. For a much more comprehensive and very useful study of the role of British regulars in the North American war see Stephen Brumwell, *Redcoats: The British Soldier and War in the Americas, 1755-1763* (Cambridge, 2002).

• 1755 •

When France learned of the departure of these regiments for America, it was decided also to send there the regiments of La Reine, Artois, Bourgogne, Languedoc, Guienne, & Béarn, all of which travelled to Brest at the [18] beginning of April 1755.[175] They found a squadron of 22 warships ready to welcome aboard the second battalions of these regiments that were all destined for America.[176] There follows the detail of this squadron, the finest[177] that ever set sail from our ports during the late war.[178]

---

175 The Guienne Regiment was the first to embark on April 4. Pouchot's Béarn Regiment arrived at Brest on April 7 and boarded *Opiniâtre* and *Léopard* the following day. La Pause, "Mémoire", *RAPQ*, 1931-32, p. 3.

176 The six battalions represented the first direct involvement of the Army, or *troupes de terre*, in the defense of New France since the 1660s. The battalions of Artois and Bourgogne were ordered to the fortress of Louisbourg on Isle Royale (Cape Breton Island). The others went to Québec. See René Chartrand, *The French Soldier in Colonial America* (Ottawa, 1984), pp. 28-36 for a summary of their service.

Infantry regiments of the *troupes de terre* at this time were composed of one or more battalions. For the organization of the French army see Kennett, *The French Armies in the Seven Years' War*. W.J. Eccles, "The French Forces in North America During the Seven Years' War", *DCB*, III, pp. xv-xxiii, offers a concise evaluation of the *troupes de terre* and the other military resources of New France. French military manuals, arms and equipment and their impact on military efficiency are discussed in J.A. Houlding, *French Arms Drill of the 18th Century* (Bloomfield, Ont., 1988).

177 Note in original, p. 19 (a): *That of Marshal Conflans was larger.*

In November 1759 Admiral Hubert de Brienne, Comte de Conflans (1690-1777; Taillemite, pp. 68-69) led 21 ships of the line and several frigates against Admiral Edward Hawke's English squadron. Conflans was defeated in a stormy action at Quiberon Bay on November 20, and French naval power was neutralized for the remainder of the Seven Years' War. Lawrence Henry Gipson, *The British Empire Before the American Revolution, Volume VIII: The Great War for the Empire, The Culmination, 1760-1763* (New York, 1954), pp. 17-24.

178 Pouchot divides the vessels into categories of fully armed ships of the line, frigates and ships of the line armed *en flûte*. These last had been partially disarmed to provide space for transporting troops. The vessels were organized into several groups. A covering squadron under Macnémara was to escort the troop convoy out of European waters. This included the ships of the line *Formidable, Palmier, Héros, Eveillé, Inflexible,* and *Aigle* and the frigates *Ametisthe, Fleur de lys* and *Héroïne*. Most of the others, including only three fully armed ships of the line and two frigates (*Sirenne* and *Comète*) under the command of Dubois de La Motte, were to continue to America. The frigates *Diane* and *Fidéle* sailed separately.

For an account of both the diplomatic and operational details see Gipson, VI, pp. 99-126. A French history of the operation is Jacques Aman, *Une campagne navale méconnue à*

...(continued next page)...

ARMED WARSHIPS

The *Formidable,* 80 guns & a crew of 900 men. On board, M. de Macnemara, the squadron admiral and fleet commander,[179] M. de Kersaint, the flag captain[180] & M. Duchaffault, second captain.[181]

The *Entreprenant,* 74 guns & a crew of 700 men, **[19]** commanded by M. Dubois de la Mothe, squadron admiral,[182] & M. de Sauzay, second captain.[183] Destined to lead the troops to Canada, the vessel had on board M. de Vaudreuil, governor of Canada, M. Dieskau, *maréchal de camp,* commander of the French troops,[184] & M. Daureil,[185] paymaster general of the *troupes de terre.*

---

la veille de la guerre de sept ans: L'escadre de Brest en 1755 (Vincennes, 1986). For an overview of the French navy see James Pritchard, *Louis XV's Navy, 1748-1762: A Study of Organization and Administration* (Kingston, Ont., 1987).

Governor Vaudreuil's journal of the voyage is in *DRCHSNY,* X, pp. 297-99. His list of Dubois de La Motte's squadron gives the normal full armament of the ships armed *en flûte* and differs slightly from Pouchot in their ratings. La Pause's "Mémoire et observations sur mon voyage en Canada" provides another journal and an enumeration of the ships' crews and the soldiers and officers carried aboard each transport. *RAPQ,* 1931-32, pp. 3-9.

179 Jean-Baptiste Macnémara (1690-1756), *lieutenant général* (1752). Although born in France, he was of Irish extraction. Taillemite, pp. 222-23.

180 Guy-François de Coëtnempren, Comte de Kersaint (1703-1759), *capitaine de vaisseau* (1747). Taillemite, pp. 179-80.

181 Louis-Charles, Comte Du Chaffault (1708-1794), *capitaine de vaisseau* (1754). Taillemite, p. 95.

182 Emmanuel-Auguste de Cahideuc, Comte Dubois de La Motte (1683-1764), rear admiral (1751). *DCB,* III, pp. 92-93.

183 Vicomte de Sanzay (d.1762), *capitaine de vaisseau* (1748). G. Lacour-Gayet, *La Marine militaire de la France sous le règne de Louis XV* (Paris, 1910), p. 528.

184 Jean-Armand Dieskau, Baron de Dieskau (1701-1767), *maréchal de camp* (major general) in the French Army, appointed on March 1, 1755 to command the *troupes de terre* battalions sent to New France. He was subordinate in authority to the governor general. *DCB,* III, pp. 185-86. Dieskau was lieutenant colonel of the Bentheim Regiment, one of the German infantry units of the French Army. La Pause, "Mémoire", *RAPQ,* 1931-32, p. 5.

185 André (Jean-Baptiste) Doreil, appointed financial commissary of wars (roughly deputy commissary general) for New France in 1755. He was recalled to France in 1758. *DCB,* III, pp. 187-89. Pouchot gives his title as *commissaire ordonateur.* A collection of his papers have been published in *RAPQ,* 1944-45.

The *Palmier,* 74 guns and 750 men, M. le chevalier de Beaufremont captain[186] and d'Orvilliers second captain.[187]

The *Héros,* 74 guns, 750 men, M. de Montlouet admiral[188] and M. de Kermaban, second captain.[189]

The *Bizarre,* 64 guns & 500 men, M. de Salvert, admiral[190] and de Marolles second captain[191] destined for Louisbourg.

The *Alcide,* 64 guns & **[20]** 500 men, M. Hocquart captain[192] and M. de Paruveau second captain.[193] Had on board M. de Rostaing, colonel, deputy commander of the French troops[194] & two engineers.[195]

The *Eveillé,* 64 guns & 500 men, M. de Fontes captain.[196]

The *Inflexible,* 64 guns & 500 men, M. de Guoëbriant.[197]

The *Aigle,* 54 guns, M. de Cousage.[198]

---

186 Joseph de Bauffrement, Prince de Listenois (1714-1781), *capitaine de vaisseau* (1742). Taillemite, p.27.

187 Louis Guillouet, Comte d'Orvilliers (1710-1792), *capitaine de vaisseau* (1754). Taillemite, p. 256.

188 Rémi-Claude Bullion, Marquis de Montlouet (d. 1772), *chef d'escadre* (1754). Lacour-Gayet, p. 165; Aman, p. 197.

189 Sébastien-André or Célestin de Kermabon, *capitaine de vaisseau* (1754). Aman, p. 196.

190 Admiral Antoine-Alexis Perier de Salvert (1691-1757). He had been appointed to succeed Duquesnel as governor of Louisbourg in 1744 but was unable to reach his post. In 1745 he had commanded a relief fleet for Louisbourg, but the place fell before his arrival. Taillemite, p. 265. See also *DCB,* IV, pp. 247, 254, 588.

191 Chevalier de Marolles (d.1770), *capitaine de vaisseau* (1754). Lacour-Gayet, p. 533.

192 Toussaint Hocquart de Blincourt (d.1772), *capitaine de vaisseau* (1746). Lacour-Gayet, p. 497; Aman, p. 195.

193 Chevalier de Parcevaux (fl.1718-1764), *capitaine de vaisseau* (1751). Lacour-Gayet, p. 497.

194 In the instructions to Dieskau, this officer is identified as the Sieur de Rostaing, assigned to serve as second in command with the rank of colonel of infantry. *DRCHSNY,* X, p. 286. La Pause says he held the rank of captain in the Regiment du Roy. "Mémoire", *RAPQ,* 1931-32, p. 5. Rostaing was killed when the English captured *Alcide* on June 8, 1755. Report on the Staff of the Army, *DRCHSNY,* X, p. 357.

195 Dieskau's staff was decimated by the loss of *Alcide.* The captives included not two but all three engineers assigned to the army. They are identified in Doreil's Report of

...(continued next page)...

FRIGATES

The *Ametisthe*, 30 guns & 220 men, M. Dubos captain.[199]
The *Fleur de lys*, 30 guns, 220 men, M. le chevalier de Marinniere.[200]
The *Sirene*, 30 guns, 220 men, M. de Tourville.[201] **[21]**
The *Héroïne*, 30 guns, 220 men, M. de Bory.[202]
The *Comette*, 30 guns, 200 men, M. de Ruis.[203]
The *Diane*, 30 guns, M. de l'Aiguille.[204]
The *Fidéle*, 30 guns, M. de la Jonquiere.[205]

October 15, 1755 as Lieutenant Dumoulin and cadets Geoffroy and Aguitton. One of Dieskau's two aides-de-camp, Captain Godart d'Helincourt, and one of his two commissaries, M. Dubois de Crancé, were also taken. *DRCHSNY*, X, p. 357. This misfortune created the demand for Pouchot's engineering talents. Doreil wrote to Argenson, Minister of War, on October 28, 1755, that it was "necessary ... to order out the same number of Engineers that you had sent and were taken on board *l'Alcide*. They are urgently required. In this colony there are only some Naval [Marine] Engineers, who are unable to attack or defend places." *Ibid.*, X, p. 370.

196 N... Darot de Fontais, *capitaine de vaisseau* (1743). Aman, p. 195.

197 Chevalier de Guébriant Budes (d.1760), *capitaine de vaisseau* (1746). Lacour-Gayet, p. 523.

198 Henri-François, Comte de Cousage de La Rochefoucauld (d.1784), *capitaine de vaisseau* (1747). Taillemite, p. 73. Aman, p. 196, renders the name as La Rochefoucauld-Cousage.

199 Charles-Jacques Du Bot de Lohan (d.1758), *lieutenant de vaisseau* (1746). Lacour-Gayet, p. 499; Aman, p. 194.

200 Gilles-René Marin de Marnières (d.1771), *lieutenant de vaisseau* (1748). Aman, p. 196; Lacour-Gayet, p. 182.

201 Chevalier de Tourville (d.1758), *lieutenant de vaisseau* (1749). Lacour-Gayet, p. 533.

202 Gabriel de Bory (1720-1801), *lieutenant de vaisseau* (1751). Taillemite, p. 40; Aman, p. 193.

203 Jacques, Chevalier de Ruis (fl.1734-1764). Lacour-Gayet, p. 536. He was appointed *aide-major* in 1747 according to Aman, p. 197.

204 Michel-Joseph Froger de l'Eguille (d.1772), *capitaine de vaisseau* (1752). Taillemite, p. 127.

205 Clément Taffanel de La Jonquière (1706-1795), *lieutenant de vaisseau* (1746). This officer was a nephew of the former Governor General of New France (note 144). Taillemite, p. 191.

VESSELS ARMED *EN FLUTE* FOR TRANSPORTING THE *TROUPES DE TERRE*

The *Defenseur,* 74 guns reduced to 24, M. de Beaussier captain,[206] 9 companies of the Artois Regiment, embarked April 12.[207]

The *Dauphin Royal,* 74 guns reduced to 24, M. de Montalais captain,[208] 9 companies of the Bourgogne Regiment.

The *Algonquin,* 74 guns reduced to 24, M. de Villeleon captain,[209] 9 companies of the La Reine Regiment embarked on the 14th. [22]

The *Espérance,* 70 guns reduced to 24, M. de Bouville captain,[210] a company of grenadiers, 3 *compagnies factionnaires*[211] of the Artois Regiment, & 3 of the Bourgogne Regiment.

The *Actif,* 64 guns reduced to 22, M. de Caumont captain,[212] 9 companies of the Languedoc Regiment, embarked on the 6th.

The *Illustre,* 64 guns reduced to 22, M. de Choiseul,[213] 9 companies of the Guienne Regiment, embarked on the 4th.

The *Opiniâtre,* 64 guns reduced to 22, M. de Moleans captain,[214] 9 companies of the Béarn Regiment, embarked on the 8th.[215]

---

206 Louis-Joseph Beaussier de Lisle (1701-1765), captain in the Navy and port captain (1749). He later commanded *Entreprenant* during the 1758 siege of Louisbourg. *DCB,* III, pp. 54-55.

207 The French Army battalion of 1755 was composed of 12 companies of fusiliers and one of grenadiers. See Chartrand, *The French Soldier,* p. 30. Pouchot accounts for all 13 companies of each battalion. No more than nine companies could be accommodated aboard each transport. La Pause confirms the embarkation dates given here. *RAPQ,* 1931-32, p. 3.

208 Jean-René Séraphin du Tertre de Montalais (d.1759), *captaine de vaisseau (1751).* Lacour-Gayet, p. 526; Aman, p. 196.

209 La Villéon (d.1762), *capitaine de vaisseau* (1751). Lacour-Gayet, p. 497; Aman, p. 196.

210 Jubert de Bouville, Vicomte Bouville (fl.1721-1764), *capitaine de vaisseau* (1751). Lacour-Gayet, p. 538; Aman, p. 193.

211 *Compagnies factionnaires* were fusilier companies as opposed to the grenadiers. See Donald E. Graves, *French Military Terminology, 1670-1815* (St. John, N.B., 1979), p. 33. The grenadiers of the Bourgogne Regiment were also aboard *Espérance.* The grenadier companies, in all cases, sailed with the smaller detachments of their respective battalions.

212 Tancrède de Caumont, Chevalier de Caumont (fl.1724-1764), *capitaine de vaisseau* (1751). Lacour-Gayet, p. 528; Aman, p. 194.

213 Antoine-Nicolas, Marquis de Choiseul-Praslin (d.1760), *capitaine de vaisseau* (1751). Lacour-Gayet, p. 531; Aman, p. 194.

The *Lis,* 64 guns reduced to 22, M. de Lorgerie captain,[216] 4 companies of the La Reine Regiment & 4 of the Languedoc Regiment.

The *Léopard,* 60 guns reduced to 22, M. de Chifreville captain,[217] 4 companies of the Guienne Regiment & 4 of the Béarn. [23]

The *Apollon,* 60 guns, M. de Gomain,[218] 4 companies, destined to serve as a hospital ship.[219]

The *Aquillon,* likewise [60 guns], M. de Rigaudieres,[220] 4 companies.

We must not forget to say that Mercier & Péan came on board with M. dc Vaudreuil. They had been consulted at Versailles about the war as though they were oracles, & then assigned to M. Dieskau as men to whose advice he would absolutely have to defer in order to succeed.

---

214 Joseph-René de Moléans (d.1786), *capitaine de vaisseau* (1754). Lacour-Gayet, p. 258; Aman, p. 196.

215 The battalion commandant and *aide-major,* 18 captains and lieutenants of the fusilier companies and one ensign attached to the first company of fusiliers were assigned to *Opiniâtre. Léopard* was to carry the captain, lieutenant and *sous-lieutenant* of the grenadier company, six captains and lieutenants of the fusilier companies and one ensign attached to the second fusilier company. "État des officiers ... embarqués", H.R. Casgrain (ed.), *Extraits des archives des ministères de la Marine et de la Guerre à Paris* (Québec, 1890), pp. 254-58. The embarkation return for *Opiniâtre,* however, places both ensigns aboard her and lists the officers by name. Pouchot is not among them and thus sailed aboard *Léopard.* SHA, A[1], 3404, no. 111.

216 Louis-François-Nicolas de Lorgeril, *dit* l'aîné [the first-born; the eldest brother] (fl.1725-1762), *capitaine de vaisseau* (1754). Lacour-Gayet, p. 501; Aman, p. 196:

217 De Chiffrevas (d.1759), *capitaine de vaisseau* (1754). Lacour-Gayet, p. 501; Aman, p. 194.

218 Yves Gomain (d.1756), *lieutenant de port* (1746). Aman, p. 195. Gomain died in March 1756 while commanding *Léopard* on another voyage to Québec. *DRCHSNY,* X, p. 402.

219 Although Pouchot seems to erroneously imply that eight additional companies sailed aboard *Apollon* and *Aquillon,* he might have meant that the vessels each had the capacity to transport four companies. Vaudreuil lists "hospitals, several passengers, &c." aboard the two ships. *DRCHSNY,* X, p. 299. In a letter to the Minister of War of April 5, 1755, Doreil shows medical personnel aboard *Apollon. RAPQ,* 1944-45, p. 4. La Pause lists six surgeons on each of the two vessels and describes quartering arrangements for officers and men. *RAPQ,* 1931-32, pp. 4, 6.

220 N... Froger de La Rigaudière (d.1756), *lieutenant de vaisseau* (1747). Lacour-Gayet, p. 497; Aman, p. 195.

Upon their arrival the regiments were reviewed by M. de Crémille, inspector, who made up the companies that were to cross at the expense of men from the same regiment.[221] These battalions were handed over to the Marine. This change from land to marine command upset the troops without them knowing quite why. Subsequent events [24] will show that they were much worse off than if they had passed into the service of an absolutely foreign prince, either because of different traditions or the way they were treated.[222]

The fleet remained at anchor until 3 May owing to contrary winds. On that day they blew from the N. N. E. and around eight o'clock the general signalled to raise sail. At 10:30 the voyage began.

---

221 Louis-Hyacinthe Boyer de Crémilles (1700-1768), a lieutenant general identified in Dieskau's instructions as inspector general of infantry. *Biographie universelle,* IX, p. 464.

Only the second battalion of each regiment was assigned to Canada. The complete regiments marched to Brest where men were drafted from the first battalion of each to bring the second up to full strength. A battalion, in 1755, was to number 31 officers and 525 men. Chartrand, *The French Soldier,* p. 30.

Crémilles had been instructed to complete the companies with volunteers from the first battalions but, if this was not possible, to draw on militia from two regiments in Brittany. This proved to be unnecessary as a more than sufficient number volunteered. There had been rumors of a lack of enthusiasm for colonial service in the Béarn Regiment. Crémilles reported, however, that "Without exception all answered the roll call, and there was no other precaution to be taken, but to prevent the absolute annihilation of the battalion which is not going by the total absorption of its soldiers." He noted, in addition, that although "the men so far embarked are fine and good, one must acknowledge the 2nd battalion of Béarn surpasses the others by their fine appearance and the height of the men, as well as by the way they are clothed and equipped." Quoted in Maurice Sautai, *Montcalm at the Battle of Carillon* (Ticonderoga, NY, 1941), pp. 19-22.

The embarkation return for nine fusilier companies of Béarn aboard *Opiniâtre* shows them to have been at full strength with two sergeants, two corporals, three *anspessades* (lance corporals), one drummer, and 31 soldiers each. SHA, A[1], 3404, no. 111. The organization of an infantry battalion as of February 1756 is given in Casgrain, *Collection,* III, pp. 28-30. Although specific to the battalions of La Sarre and Royal Roussillon sent to Canada that spring, it differs only in the addition of one sergeant.

The depleted first battalions marched to garrison posts in France. That of Béarn went to "Xaintes" (Saintes, in the modern French department of Charente-Maritime) for coastal defense. La Pause, "Mémoire", *RAPQ,* 1931-32, p. 5; Susane, V, p. 308.

222 Note in original, p. 25 (a): *A way of apprehending or envisaging things which is false as well as being contrary to state interests & the glory of the nation. Happily dire prejudices such as these are disappearing more and more every day.*

Pouchot refers to the fact that the *troupes de terre* battalions, and their general, were placed under the orders of the Governor General of New France, an official of the Department of the Marine. The relationship between the Army and Marine forces inevitably caused friction and complicated defense of the colony. See Eccles, "The French Forces ...", *DCB,* III, pp. xv-xxiii.

...(continued next page)...

The *Formidable* and the *Entreprenant* having stationed themselves at the head of the squadron, by noon the fleet was out of the Brest Channel[223] with a fine fresh wind blowing from the N. 1/4 N. E. We made full sail until six in the evening, when the squadron formed up **[25]** in three columns line ahead. M. de Macnemara & the warships to windward, M. de Salvert in the center & M. Dubois de la Mothe to leeward.

On the 5th, at 45d. 47" latitude & 11d. 21" longitude from Paris after noon, the general signalled to proceed full sail without any fixed order. Towards evening, there appeared to the West three vessels one of which left us at nightfall and the two others kept us in sight following in our wake. They were judged to be two English frigates.

On the 7th, at one o'clock in the afternoon, the *Entreprenant* fired a cannon shot & signalled to the ships armed *en flûte* to follow it. The squadron of M. de Macnemara remained hove to. We advanced in the direction W. N. W. & 1/4 N. W. At four in the afternoon we lost sight of the squadron. It also appeared that one of the frigates following us had **[26]** lost one of its topmasts & we never saw them again from that day on. On the 25th, a light breeze from the E. S. E. & mist. At six in the morning, the general hove to with the whole squadron. The winds veered to the S. S. W. blowing hard. We hove to W. N. W. at 46d. 9" latitude, 46d. 9" longitude from Paris. We were 47 leagues[224] from the Grand Banks. We saw icebergs about a quarter of a league in circumference.

Morale was probably not improved when the battalions sent to New France were issued clothing by the Marine that differed in style and facing colors from their own distinctive regimentals. Marine clothing was provided to the *troupes de terre* until 1758. Chartrand, *The French Soldier,* pp. 32-35; Jack L. Summers and René Chartrand, *Military Uniforms in Canada, 1665-1970* (Ottawa, 1981), pp. 20-21; Eugène Lelièpvre and René Chartrand, "Béarn Regiment, New France, 1755-1757", *Military Collector & Historian,* XXXIX, No. 3 (Fall, 1987), pp. 126-27. A *mémoire* of items to be furnished to the *troupes de terre* in Canada, dated March 22, 1754, is in "Les 'papiers' La Pause", *Rapport de l'Archiviste de la Province de Québec pour 1933-1934,* (Québec, 1934), p. 225.

223 The channel is identified in the original as *du goulet,* literally "narrow entrance". This was the popular name of the channel connecting Brest to the sea.

224 Pouchot consistently uses the *lieue* or league to describe distances between places. Unfortunately, in most cases, he only estimates, a fact that confuses many of the topographic descriptions in the latter part of his work. A modern French league is 4 kilometers or 2.59 miles. The league, as used in New France, was equivalent to 3.0519 miles. See J. Roy Reasonover, *Land Measures: French, Spanish and English Land Measures of the United States with Units, English Equivalents, Reduction and Conversion Factors* (Houston, TX, 1946), p. 33.

On the 26th, the winds from the S. S. W. were variable through the west, and to the N. N. W. the squadron became separated. Each vessel sought to advance in formation with those it could encounter. The mists became so thick that from the post of the steersman one could sometimes not see the mainmast. That day was spent among ice floes taller than the rigging and from 1/4 to 1/2 a league in **[27]** circumference. The squadron did not meet up again until Québec.

The English whose squadrons were on patrol between the Grand Banks and Newfoundland came upon some of our vessels in fog clearings. The fate of the *Alcide* & the *Lis* is well known.[225] The *Entreprenant* would also have been captured if it had not lost its way in the fog. It did not enter the river until it reached the Isle aux Coudres[226] for fear of hitting the sandbar, which English vessels of 110 cannons later managed to cross. The *Opiniâtre* & the *Algonquin* were the first to arrive. They dropped anchor at Québec on 19 June and the others were not long in arriving one after the other.[227]

M. du Quesne reviewed the regiments upon their arrival & reported to M. de Vaudreuil and M. Dieskau about the state in which he was leaving the affairs of the country to them. He assured them **[28]** that everything was calm and that the English who, it was said, were threatening the installation of Fort du Quesne, could not cross the mountains with sizeable forces. How could he know this for sure?[228]

Although the country lacked everything, whether war material or victuals or trade goods, it was exhausted by previous campaigns. Nevertheless, they found it strange that such a considerable force should have been sent to a colony that was well able to defend itself without aid. Several officers whose curiosity impelled them to visit the King's magazines were very surprised to find only 300 poor trade

---

225 *Alcide* (64 guns), *Lys* (*en flûte,* 22) and *Dauphin Royale* (*en flûte,* 24) were attacked off Newfoundland by an English squadron on June 8. *Alcide* and *Lys* were taken, and the four companies each of the Languedoc and La Reine regiments aboard the latter were prevented from reaching New France. *Dauphin Royale* escaped, arriving at Louisbourg on June 11. Gipson, VI, pp. 110-13.

226 Isle aux Coudres lies off the north shore of the St. Lawrence and the mouth of the Rivière Gouffre at Baie St. Paul, Québec.

227 *Léopard,* with Pouchot aboard, arrived at Québec on June 23. The troops disembarked the next day, and the battalion of Béarn was reviewed on June 25. Malartic and Gaffarel, p. 7.

228 Duquesne's July 6, 1755 *mémoire* to Vaudreuil gives no such impression but, rather, focuses on the military condition of the colony. *DRCHSNY,* X, pp. 300-02.

muskets, a barrel & a half of ball & very little gunpowder in a country that had been threatened with war for four years & where preparations had been made [29] for offensive operations. It may be inferred from this that everyone was thinking of his own private interest rather than of the good of the service.

M. Dieskau began to treat the troops in the German manner. He was no longer willing to speak to common officers and only received representations from corps commanders whom he never consulted - a very great drawback in a small army. He listened only to the advice of Péan & Mercier. They alone were his eyes & ears and he no longer deferred to the opinions & feelings of M. de Vaudreuil who, being head of the colony & a native of it, necessarily had a more intimate knowledge of it than he did. The two officers in question, wholly inept except so far as their own interests were concerned, were boastful like all Canadians. They had always imagined that one Canadian could put ten Englishmen to flight, a prejudice that unexpected success [30] had rooted more and more deeply in them. They persuaded him to take certain measures, the consequences of which will be seen later. Even before all his operations which were directed solely by those persons, M. Dieskau discovered the secret of displeasing the governor of the country and the troops.[229]

The French regiments spent the rest of the month of June at Québec & left in bateaux only during the early days of July to go to Mont-Réal where they remained until the 19th of the same month.[230]

During this period, the officers became acquainted with their colleagues of the *compagnies franches de la marine* with whom they were to serve through seniority of commission, according to a regulation especially drawn up for this purpose. These officers only rose from the rank of captain in order to make up the general staff. [31] Those who had been awarded the Cross of St. Louis were as highly esteemed as lieutenant generals & *cordons bleus* in France. Those who had commanded armies, of which the largest were 300 men, were respected in

---

229 No other French officers record Dieskau's "German manner". As for Péan and Le Mercier, Duquesne, piqued by Vaudreuil's disregard of his advice, reported that he had "persuaded" Dieskau to employ the two officers "for the arrangement and prompt execution" of the movement of his forces against Oswego. Duquesne to Machault, July 15, 1755, *DRCHSNY*, X, p. 306.

230 The Béarn Regiment left Québec for Montréal in two divisions on June 30 and July 1. Pouchot probably travelled with the second division comprising the last five companies. The divisions arrived at Montréal on July 7 and 9 respectively. Malartic and Gaffarel, pp. 8, 12.

the colony like field marshals in France.[231] A captain of the King's vessels who arrived in the country was regarded as a divinity & habitually behaved with a great deal of arrogance.

The Canadian officers, although courageous, knew hardly anything of their profession. The Major of Québec, Ramsay,[232] did not even know how to give orders. They hardly knew their own troops, who were always billeted with the habitants. Even if they were [32] together in their postings, they all thought of their own petty interests, which created conflicts between officers & soldiers. Isolation & their limited means for enforcing discipline rendered the latter insubordinate. Often they were not entirely in the wrong. The most admired quality of both an officer & a soldier in that country was to be reputed to have a handsome leg.[233] Since no operation was undertaken without Indians, it was they who were in charge, whether on the march or in the attack.[234] With more knowledge of the profession & more discipline, it would have been possible to make these companies excellent [33] light infantry, as soldiers and officers alike were highly

---

231 Note in original, p. 32 (a): *All of this is highly exaggerated.*

Pouchot uses the term *l'Etat Major,* translated here as "general staff". In a *troupes de terre* infantry regiment of the period, these men were the field officers above company grade. Graves, p. 32. In the context here, Pouchot means the colonial garrison appointments such as the governors of important towns, their town majors, and other positions. Chartrand, *Canadian Military Heritage,* pp. 107-08.

The *compagnies franches de la marine* were independent companies of regular soldiers maintained by the Department of the Marine (or Navy) to garrison colonial posts. The first such companies were sent to New France in 1683, and the organization would continue until the English conquest. At the time of Pouchot's arrival, the *compagnies franches* in New France numbered 30 companies, each of 50 men and four officers. Another 24 companies were assigned to Isle Royale. Thirty-six more were stationed in the colony of Louisiana, and some of these saw action in the Ohio country. See Chartrand, *The French Soldier,* pp. 9-14, 16-19.

A *cordon bleu* was a Knight of the Order of the Holy Ghost.

232 Jean-Baptiste-Nicolas-Roch de Ramezay (1708-1777), an experienced captain (1734) of the Marine who was appointed town major of Québec in 1749 and King's lieutenant in 1758. It was Ramezay who would prematurely surrender the city to the English on September 19, 1759. *DCB,* IV, pp. 650-53.

233 The phrase in the original is *pour avoir une bonne jambe,* and the expression might relate to the contemporary belief that "well-turned calves" were a desirable attribute of recruits for the French Army. See Kennett, p. 112n. It is also possible that the expression *bonnes jambes* was a nickname for Marine troops or Canadian militia. Pouchot used it to describe a detachment he sent to reconnoiter the British position before Niagara on July 6, 1759. "Siège de Niagara", *Rapport de l'Archiviste de la Province de Québec pour 1933-1934* (Québec, 1934), p. 92.

234 Note in original, p. 33 (a): *A multiplicity of facts give the lie to the author here. He was too often blinded by military prejudice.*

courageous. Their way of making war suited them well for this role. It would have thus been very useful to employ in the legions those officers & troops who were trained in France & who were dismissed upon their arrival.[235]

These troops were extremely ill armed, because they sent to the country all the rejects from the King's magazines. The same was true of the artillery; the guns were all rust-worn.[236]

The news soon arrived of the capture of Beau-séjour, which was attacked by six thousand English. The garrison was sent back to Canada, on condition that they did not see service for six months.[237] Vercors, the commander of the post, was court-martialled because the

---

235 Enlisted men of the *compagnies franches* were recruited in France, but, by 1755, most of the officers were Canadian-born. Chartrand, *The French Soldier,* p. 10. The nature of the companies of colonial regulars and their enlisted personnel is discussed in Gilles Proulx, *The Garrison of Québec from 1748 to 1759* (Ottawa, 1991). A more comprehensive summary of the organization, origin and service of these troops may be found in Chartrand, *Canadian Military Heritage,* pp. 83-156.

The comment about Pouchot's prejudices, made by his 1781 editor, appears to be apt. Inter-service rivalry and the superior attitude of the European regulars colored their opinions of colonial forces. Montcalm expressed similar sentiments when he described the *compagnies franches* as "a troop knowing neither discipline nor subordination. Within six months I would make grenadiers of them, and now I would carefully abstain from placing as much dependence on them as the unfortunate M. de Dieskaw did, by having given too much ear to the confident talk of Canadians, who believe themselves, in all respects, the first nation in the world." Montcalm to Argenson, Aug. 28, 1756, *DRCHSNY,* X, p. 463.

236 Pouchot's negative appraisal might be based on the fact that the Marine Department contracted independently for its small arms. By the 1740s Marine troops were receiving muskets similar to the Army's Model 1728, but Chartrand does not indicate that the *compagnies franches* were poorly armed. Artillery for the colonies was also of Marine rather than Army pattern and was almost always constructed of iron, thus the comment about rust. Chartrand, *The French Soldier,* pp. 19-21; 27-28. Eccles, "The French Forces", *DCB,* III, p. xvi briefly discusses the generally poor quality of much of the French equipment.

237 News of the fall of Beauséjour reached the Béarn Regiment at Montréal on July 12. Malartic and Gaffarel, p. 12.

English operations against Fort Beauséjour had commenced on June 3, 1755. The commandant, over the protests of some of his garrison, requested a truce on June 16, and the place was given up to the English on the 18th. Fort Gaspereau, located on the north side of the isthmus at the mouth of the Gaspereau River (Baie Verte, N.B.), was incapable of defense and surrendered soon after. Stanley, pp. 108-17. Beauséjour was renamed Fort Cumberland and Gaspereau, Fort Monckton.

Useful accounts of the siege are J. Clarence Webster (ed.), *Journals of Beauséjour. Diary of John Thomas; Journal of Louis de Courville* (Sackville, N.B., 1937) and J. Clarence Webster (ed.) and Alice Webster (trans.), *The Siege of Beauséjour in 1755: A Journal of the Attack on Beauséjour Written by Jacau de Fiedmont* (St. John, N.B., 1936). These reproduce plans and views of both French posts. A detailed map of the environs of Beauséjour is printed in Mante, p. 17 with the articles of capitulation on pp. 18-19.

artillery officer had protested against the surrender.[238] One cannot [34] say that the type of fort constructed in those regions is impregnable, given that there is no hope of rapid relief.[239]

Once the regiments of Guienne & Béarn had been equipped for the campaign, they set out for Frontenac on July 19th.[240] They went aboard bateaux loaded with victuals for that post, and arrived at Lachine where M. Péan came to distribute their rations to them. There were a few arguments because they were not being given the quantities specified for the troops who, [35] if they ran out of provisions, would not have found any in uninhabited country. One officer, with whom he had a fairly lengthy squabble because his

---

238 Beauséjour's commandant was Louis Du Pont Duchambon de Vergor (1713-after 1775), captain (1750) in the Marine. He was another officer with close connections to Bigot and owed his posting to the intendant. Vergor was acquitted by the court martial held in October 1757. He later had the misfortune to command the guard at Anse au Foulon above Québec when Wolfe's army landed there, precipitating the Battle of the Plains of Abraham. *DCB,* IV, pp. 249-51.

The commandant of Fort Gaspereau, Benjamin Rouer de Villeray (1701-1760) was a co-defendant with Vergor. He was a captain (1751) in the Marine and was also acquitted. *DCB,* III, pp. 570-71. The court martial included officers of the *troupes de terre,* who agreed that the surrender had much to do with the fact that the Acadians had "forced the Commandant to capitulate to save their lives" and that Gaspereau was indefensible. Montcalm to Paulmy, Nov. 4, 1757, *DRCHSNY,* X, p. 671.

Louis-Thomas Jacau de Fiedmont (c.1723-1788), lieutenant (1753) in the colonial *cannoniers-bombardiers* at the time of the capture of Fort Beauséjour, was the post artillery officer and engineer. *DCB,* IV, pp. 382-83. He was accorded a high compliment by the ever-critical Bougainville who called him "a very honest and very exact man, *rara avis in terris et presertim in America septentrionale* [a rare bird in the land and especially so in North America]." Hamilton, *Adventure,* p. 93.

239 Note in original, p. 35 (a): *M. Pouchot ought to have added here that the English general, Monckton, in order to occupy Forts Beau-séjour & Gaspareaux, abused the sense of security inspired in M. Vercors by the formal promise of the court of London that no operations or troop movements would be undertaken in Acadia before the decision of the respective commissioners to determine its boundaries.*

Robert Monckton (1726-1782) was lieutenant colonel (1752) of the 47th Regiment of Foot. Monckton was later Lieutenant Governor of Nova Scotia and Wolfe's second in command in the attack on Québec in 1759. *DCB,* IV, pp. 540-42. Boundary negotiations were, in fact, still proceeding sporadically, and England and France would not formally declare war until 1756.

240 Fort Frontenac, the French post at the head of the St. Lawrence River, had been established in 1673. Later abandoned and then reoccupied in 1695, it had been maintained since that time. Fort Frontenac was located at the site of Kingston, Ontario. For a history of the post and significant documents relating to it see Richard A. Preston and Leopold LaMontagne, *Royal Fort Frontenac* (Toronto, 1958).

...(continued next page)...

company lacked 80 or 100 pounds of bread or pork, reached an agreement with Péan at last and the latter kindly lent him his shoulders in order to have the receipt signed on his back, thus getting rid of him that much sooner.[241]

The troops' instructions to proceed to Frontenac commanded them to march with caution & that, in the event that the enemy had already seized the fort, they should install themselves in the advantageous positions in the area of the fort of La Présentation.[242] From that one can judge how little M. du Quesne knew about the enemy's movements when he claimed that we had nothing to fear from them.

On 3 August the regiments [36] arrived at Frontenac.[243] On August 1st, having halted at the Petit Détroit,[244] they were ap-

---

Campaign equipment issued to the two battalions between July 13 and 15 included, for each man, a coat of "Cadiz cloth", a woolen blanket, two cotton shirts, a pair of breeches, *mitasses* (Indian-style leggings), thread, needles, a pair of shoes (described as *tannés* and probably meaning mocassins) each month, and a pound of tobacco. A tent was issued for every four officers. Two tents per battalion were provided for servants and four for each company of soldiers (39 non-commissioned officers and men). One cooking pot was given to each officer, and four to each company. Malartic and Gaffarel, p. 13. Pouchot's list of the campaign items normally provided to Canadian troops (note 155) is very similar.

241 While the ration of the French soldier varied throughout the war, the official daily issue was one and one-half pounds of bread, one-half pound of salt pork and one-half pound of peas. Edward P. Hamilton, *The French Army in America* (Ottawa, 1967), p. 13. Fifteen days worth of these rations, plus two quarts of wine per man for the trip, were issued to the first division of the Béarn Regiment at Lachine on July 18. The battalion marched from Montréal in three divisions: the first (grenadier and four fusilier companies) on July 18, the second (four fusilier companies) on July 19 and the third (four fusilier companies) on the 20th. Pouchot seems to have accompanied the second division, reaching Frontenac on August 2. The first detachments of Guienne did not arrive until August 9 and the last on the 11th. Malartic to Argenson, Oct. 6, 1755, *DRCHSNY*, X, pp. 348-50; Malartic and Gaffarel, pp. 13-17.

A daily account of the journey of the first division was kept by Anne-Joseph-Hippolyte de Maurès de Malartic (1730-1800; *DCB*, IV, pp. 530-31), captain (1746) and *aide-major* or adjutant (1749) of the Béarn Regiment.

242 Located at the site of Ogdensburg, New York, Fort de la Présentation was established in 1749 to protect the mission of Abbé Picquet (note 710). The place was also referred to as Chouegatchi, Oswegatchie and La Galette. See Persis Yates Boyesen, "The French and Indian Settlement at Ogdensburg", *The Quarterly*, XXXV, No. 1 (January 1990), pp. 4-10. Also Paul Fortier, "La Galette versus Cataraqui: The Forgotten Rivalry for Dominance on the Upper St. Lawrence, 1673-1796", unpublished paper, 1979. Copy in the Old Fort Niagara Association library.

243 The third division of the Béarn Regiment arrived on this date. Malartic and Gaffarel, p. 18.

244 Tar Island Narrows between Tar Island, Ontario and the north shore of the St. Lawrence River.

proached by Indians bearing scalps. They gave the troops the first news of the action which had taken place at Fort du Quesne on 13 July.[245] We shall give here the account related by the Canadian officers who were there & following the order of battle adopted by the English. M. de Contre-Coeur had learned through the Indians about the advance of a large body of English who had left Fort Comberlant[246] & were opening up a trail as they advanced. He formed a detachment of 200 Canadians & soldiers [37] of the colony under the orders of Captain de Beaujeu and Captain Dumas & several other officers. Some of the latter commanded Indians from the upper country & others from our own settlements, 500 in number.[247] This detachment expected to find the enemy still some distance away and intended to try, through an ambush or some other action, to slow down their advance rather than preventing them from reaching Fort du Quesne. The officers had been informed that the enemy had a very superior force. On the contrary, the English were relying on the size of their force and intended to set up a post in the belief that they scarcely need do more than show themselves & they would be able to take the fort in twenty-four hours. However, they were advancing with the greatest caution. They arrived [38] about three leagues from Fort du Quesne and, having crossed a small river, they halted near the house of a blacksmith named Frazer, a German who had settled there in order

245 Braddock's defeat had occurred on July 9.

246 Note in original, p. 37 (a): *General Braddock, who was in command of these troops, had arrived there on 10 May 1753, & the rest of his army was there by the 17th after an extremely hard march.*

Fort Cumberland, located on the site of Cumberland, Maryland, had been established in 1755 at the junction of the north branch of the Potomac River and Wills Creek. It was intended as the base for Braddock's march to Fort Duquesne. Stotz, pp. 91-94.

The date for Braddock's arrival at Fort Cumberland is in error by two years. The English general reached the post on May 10, 1755. Edward Braddock (1695-1755) was colonel (1753) of the 14th Regiment of Foot and had been named major general and commander of English troops in America on March 29, 1754. *DNB*, II, pp. 1061-63. See also Lee McCardell, *Ill-Starred General: Braddock of the Coldstream Guards* (Pittsburgh, 1958).

247 Daniel-Hyacinthe-Marie Liénard de Beaujeu (1711-1755), captain (1749) in the Marine, had been appointed to succeed Captain Contrecoeur (note 162) as commandant on the Ohio. He arrived at Fort Duquesne late in June but remained subordinate to Contrecoeur during the preparations to oppose Braddock. *DCB*, III, pp. 400-02.

Jean-Daniel Dumas (1721-1794) was a captain (1747) in the Marine. *DCB*, IV, pp. 242-43.

The combined French-Indian force was reported to have numbered 72 colonial regulars, 146 militia and 637 warriors. Journal of Operations, *DRCHSNY*, X, p. 337.

to ply his trade with the Indians, [but] which he had abandoned as soon as the French settled on the Ohio.[248]

Around 11 in the morning, the English began to advance along a track cut into a hillock. In the English party there were twenty mounted men in the lead, then ten carpenters, two companies of grenadiers of Halket, the seven companies of the same regiment, six independent companies of Virginia, three on the right & three on the left, the regiment of Dumbard and his grenadiers in the rear guard. Finally the workmen & twenty horses completed this column under the command of General Braddock. The **[39]** artillery was in the center and the baggage train and provisions brought up the rear. All of this equipment was covered by the troops who were drawn up by company, in alternate order.[249]

The horsemen who had arrived at the top of the gully saw the French, who were advancing to proceed down it, and wheeled round on their advance guard who were only about a musket shot away from them.

The French, for their part, threw themselves behind trees as soon as they saw the English & began to fire their muskets. The Indians advanced left & right along the sides of the gully. They were at first shaken by the musket & artillery fire of the English column since they were not accustomed to hear such loud firing. However, seeing **[40]** that the French were holding firm & that the gunfire was doing little damage, they took up their positions at the base of each tree with their customary shrieks.

The English were not really expecting this encounter. However, they bore up well, covering their forward positions & their flanks. Seeing that we held as much ground as they did, they began to advance standing up & firing aggressively. Their officers, all mounted and sword in hand, urged them on. After the death of M. de Beaujeu, who

---

248 John Frazier (or Fraser) was a Pennsylvania Indian trader who had established a trading house on Turtle Creek, a tributary on the north bank of the Monongahela. *Journal of Major George Washington,* p. 33.

249 Gipson, VI, pp. 92-97 gives a summary of the action. A more detailed account is Paul E. Kopperman, *Braddock at the Monongahela* (Pittsburgh, 1977). The movements of Braddock's army are shown in "Six Plans of the Different Dispositions of the English Army Under the Command of the Late General Braddock in North America ... by an Officer" (London, 1758). A copy is in the William L. Clements Library, Ann Arbor, Michigan.

was killed at the first volleys, M. Dumas[250] took over command of the French, or rather everyone continued to do [41] his best in the position he found himself in.

Soon after the English abandoned two cannon and withdrew onto the rear of the column, which was still trying to advance in order to attack. But they lost their cannon one after the other which were shattered by musket fire during the space of five hours. The Indians took this withdrawal as a retreat and broke through the column with their tomahawks. The French broke through too, whereupon the column collapsed & was massacred.

We pursued the English, who threw themselves into the river where many of them were killed as they attempted to swim across. They could not be pursued any further because the Indians had remained behind to pillage and drink. We counted on the [42] battlefield 600 killed and wounded, about 400 on the trail and 300 along a little river. Their total losses were estimated at 1270,[251] since they abandoned their wounded, almost all of whom perished in the woods. Of 160 officers, only six survived. We captured two 12-pounder cannon, 4 6-pounders made of brass, 4 howitzers, 12 cohorn mortars, their ammunition & victuals, about a hundred covered wagons, the paymaster's cash and all the officers' effects. Since the latter were well equipped, the Indians and the Canadians had very good pickings.[252]

This action was the most vigorous & the most glorious in which the Indians were involved [43] and a part of the glory can be attributed to

---

250 Note in original, p. 41 (a): *After the peace, he was promoted brigadier & appointed governor of the isles de France & de Bourbon.*

Dumas' later postings were to the French Indian Ocean possessions known today respectively as Mauritius and Réunion.

251 Note in original, p. 43 (a): *Out of a total of 2000 fighting men before the engagement, as was confirmed in the papers of General Braddock, who lost his life after having five horses shot from under him.*

Returns of Braddock's army made at Will's Creek on June 8 list a total of 1,864 regulars and seamen and 497 provincials. Stanley Pargellis (ed.), *Military Affairs in North America, 1748-1765* (New York, 1936), pp. 86-91. Of 86 officers and 1,373 men actually engaged, the English lost 977 killed and wounded. Gipson, VI, p. 96.

252 A French return of ordnance recovered on the field lists four brass 12-pounders, four brass six-pounders, four brass eight-inch howitzers, three "grenade mortars" (probably coehorns of 4 2/5-inch caliber), and a considerable quantity of equipment. *DRCHSNY*, X, p. 311. Another report in *Ibid.*, p. 304 says "six brass twelves and sixes." Captain Thomas Ord's return of July 18, 1755 admits the loss of four 12-pounders, two six-pounders, four eight-inch howitzers, and three coehorn mortars. Pargellis, *Military Affairs*, pp. 96-97.

the accuracy of their fire. Our losses were only 11 dead & 29 wounded.[253]

If, on terrain without real problems, such a disaster could happen to brave & well-disciplined troops, through an inability to direct fire & ignorance of the nature of the enemy they were engaging, then this provides a good lesson that these two aspects of warfare should receive close attention. This victory, which was won on 9 July, left all our positions in the upper country out of danger for that campaigning season & caused cancellation of the projected general invasion of Canada. According to the plan[254] which had been devised between M. Shirley[255] & M. Lawrence, the Governor of Acadia, and previously sent for approval to the Court in London, it was agreed: "1. That Colonel Mockton [sic] [44] would be entrusted with attacking without delay the French forts in the area of Acadia. Orders were immediately sent to him to undertake this expedition, the outcome of which we have already recounted.

"2. It was agreed that M. Johnson[256] with an army corps of about 4000 troops raised in the northern colonies would make a surprise attack on Fort Fréderic & would capture it.[257]

---

253 The French reported three officers and 25 soldiers and Indians killed with about 25 wounded. Journal of Operations, *DRCHSNY,* X, p. 338. Adjutant Malartic recorded heavier losses of 3 officers and 40 soldiers and Canadians killed and 100 wounded. Malartic and Gaffarel, p. 18.

254 A "Sketch for the Operations in North America. Novr 16: 1754", in Pargellis, *Military Affairs,* pp. 45-48 summarizes the English plan of campaign. Braddock's secret instructions, dated November 25, 1754, are in *DRCHSNY,* VI, pp. 920-22.

255 William Shirley (1694-1771), Governor of Massachusetts and colonel of the 50th Regiment of Foot. Appointed major general in 1755, he assumed command of English forces in America upon the death of Braddock. *DNB,* XVIII, pp. 142-43; *DAB,* IX, pp. 120-22. A selection of Shirley's papers have been printed in Charles Henry Lincoln (ed.), *Correspondence of William Shirley, 1731-1760* (2 vols.; New York, 1912).

256 William Johnson (c.1715-1774), named by Braddock in April 1755 to command the expedition against Fort St. Frédéric on Lake Champlain. Johnson held a provincial commission as major general. *DCB,* IV, pp. 394-98. See also his papers published in James Sullivan and Milton W. Hamilton (eds.), *The Papers of Sir William Johnson* (Hereafter cited as *SWJP;* 14 vols.; Albany, 1921-65).

257 The goal of Johnson's force of New York and New England provincials was Fort St. Frédéric at Crown Point, New York. The stone fort and redoubt had been constructed in 1734-1737 as part of a French effort to secure and settle the Champlain Valley. For histories of the post see Guy Coolidge, *The French Occupation of the Champlain Valley from 1609 to 1759* (Harrison Hill, NY, 1979), pp. 143-313 and Pierre-George Roy, *Hommes et choses du Fort Saint-Frédéric* (Montréal, 1946).

"3. That M. Shirley, with his regiment & that of Peperel, would undertake an attack on Fort Niagara; that he would be given a sufficient number of bateaux to transport his troops & artillery on Lake Ontario & that the garrison at Oswego would be reinforced, so as to provide support if necessary & to facilitate his retreat if he was pursued.[258] [45]

"4. In addition to the attack on Fort Fréderic, Colonel Johnson was entrusted with a very important mission: to persuade the Iroquois to go to war. He was given speeches already in draft & 2000 pounds which were to be used to buy presents for them.[259]

"5. There remains the expedition that General Braddock was keeping for himself. It was decided that he would set out on 20 April to go to Frederic-Town[260] & that from there he would reach the mountains during the first days of May in order to be well prepared to finish in the month of June all of the things he intended to do on the Ohio, or the Belle-Riviere.[261] [46]

---

258 Shirley concentrated his army at Oswego during the summer of 1755. The post was deep in the country of the Six Nations of the Iroquois and tenuously connected by Oneida Lake and the Oneida Portage (Rome, New York) to the Mohawk River and the colony of New York. The English had, since 1727, maintained a fortified stone trading house on the west bank of the Oswego River. See Maj. W.H. Bertsch, "The Defenses of Oswego", *Proceedings of the New York State Historical Association,* XIII, 1934, pp. 108-27. A recent discussion of the English establishment of the post is Ronald E. Mohar, "Oswego's First Permanent Settlement: The Connection Between Oswego & Niagara", unpublished manuscript, 1993. Copy in the Old Fort Niagara Association library.

Shirley's army included the newly raised 50th (Shirley's) and 51st (Pepperrell's) as well as the New Jersey Regiment, totalling about 2,400 men. See Gipson, VI, pp. 127-61 for Shirley's campaign. A journal kept by an officer of the expedition is printed in William Hervey, *Journals of the Hon. William Hervey* (Bury St. Edmunds, England, 1906), pp. 1-17. A recent discussion of Shirley's American redcoats and the hardships they endured on the shores of Lake Ontario may be found in Peter Way, "Soldiers of Misfortune: New England Regulars and the Fall of Oswego, 1755-1756", *The Massachusetts Historical Review,* III (2001), pp. 50-88.

259 After nearly a century of hostile relations with the French and, during the late seventeenth century an alliance of necessity with the English, the Iroquois had, after 1701, assumed a neutral posture between the competing European powers. Successful involvement of the Iroquois in the war against the French would have been a major coup, but their full support was not obtained until the winter of 1758-1759. See Richard Aquila, *The Iroquois Restoration: Iroquois Diplomacy on the Colonial Frontier, 1701-1754* (Detroit, 1983) for a history of Iroquois neutrality.

260 Frederick, Maryland where Braddock arrived May 21.

261 Note in original, p. 46 (a): *We have derived all these details from the French* mémoire justificatif, *based on the papers of General Braddock. We believe that they are necessary for a clear understanding of the events of the campaign.*

...(continued next page)...

The regiments of Shirley & Peperel, militiamen of New York and New Jersey had arrived at Chouegen at the end of June as a consequence of the plan mentioned above and from this position they threatened to take Frontenac or Niagara.[262] The bad weather & diseases which they contracted prevented them from carrying out their intentions. During the campaign they busied themselves making a retrenched camp around Chouegen & built Fort Ontario on the other side of the river. They also set about building vessels to form a navy on the lakes.[263]

The regiments of Guienne & Béarn for their part built a retrenchment below Frontenac to cover their camp & protect the [47] fort. As there were no engineers among them, M. Pouchot, a captain in the Béarn, undertook this work which, while simple, derived all its advantages from its site. The Canadian officers, who had never seen anything like it, spread the news in the colony that it was an impregnable fortress.[264]

Piéces justificatives XI-XII of [Moreau], *Mémoire,* pp. 160-206, include various instructions and letters to Braddock. A copy of the English plan was taken with Braddock's papers at the Monongahela. The French perception of this useful intelligence is recorded in Dieskau to Doreil, Aug. 16, 1755, *DRCHSNY,* X, p. 312.

La Belle-Rivière was the French name for the Ohio.

262 Situated at the mouth of the Niagara River near the southwest corner of Lake Ontario, Fort Niagara (Youngstown, New York) controlled the portage around Niagara Falls and thus the route from New France to the interior of the continent and the colony of Louisiana. A stone house and stockade had been constructed on the site in 1726-1727 and improved during the 1740s. The most comprehensive account of Niagara during its French occupation is Severance, *An Old Frontier of France.*

263 Fort Ontario was constructed in 1755 on the east side of the Oswego River opposite the 1727 fort. English shipbuilding activities were intended to counter the small armed vessels maintained by the French on Lake Ontario. See Bertsch, pp. 108-27 and Gipson, VI, pp. 149-50, 156-58. The ships constructed by both sides are identified and discussed in Robert Malcomson, *Warships of the Great Lakes, 1754-1834* (London, 2001).

264 Pouchot, at this point, introduces us to his North American service. Throughout the narrative, he consistently refers to himself in the third person.

For the duration of the war, Pouchot would regularly act in the capacity of engineer. No other source explicitly confirms his claim to have fortified the camp at Frontenac, although Pouchot again took credit for it in his letter to Marshal Belle-Isle of April 14, 1758. See *DRCHSNY,* X, pp. 694-95 and Appendix A of this work. It is likely, for several reasons, that he did direct the work: his experience at Tournai qualified him for the task; the shortage of engineers caused by the loss of *Alcide* put his talents at a premium; and his subsequent assignment to Niagara attests to his reputation. Adjutant Malartic, while not identifying Pouchot as the designer of the retrenchments, referred to the captain on October 1 as a "good engineer." Malartic and Gaffarel, p. 27.

...(continued next page)...

M. Dieskau left in the month of August with the battalions of La Reine & Languedoc, 1200 Canadians & 300 Indians and made for Fort Fréderic. He marched out of the latter fort in the early days of September, doubtless in order to take up a position at the head of Lac St. Sacrement, which has since been renamed Lake George by the English. He crossed the Bay[265] with his detachment. Johnson had just arrived there with a corps of 5 to 6 thousand militiamen & had stationed himself at the edge of the lake.[266] On receiving the news of the **[48]** French advance, he created around his army a sort of retrenchment with his bateaux.

Persuaded by Mercier & Péan of the advantage that the Canadians & the Indians had over the English & confirmed in this opinion by the action at Fort du Quesne, M. Dieskau advanced with confidence, sure that he would beat these troops. He left the two regiments to guard his bateaux, considering them little suited for the war, & only took with him two *piquets*[267] & the two companies of grenadiers.

The retrenchments laid out at Fort Frontenac on August 20 were begun on the 22nd. They included "a grand redoubt at the head of the camp of the regiment of Guyenne, two at that of Béarn, joined to the one of Guyenne by a curtain and an epaulement on the bank of the river." The lines were completed by September 8 but ordered reduced in extent during October to be more practical for the small winter garrison. Malartic to Argenson, Oct. 6, 1755, *DRCHSNY*, X, pp. 350-51; Malartic and Gaffarel, p. 21. The redoubts, connected by pickets, are also described in La Pause's "Mémoire", *RAPQ*, 1931-32, p. 14. Parts of these retrenchments were later used for cover by the English when they besieged the post in 1758. They are shown in the "Plan of Fort Frontenac" printed in Mary Ann Rocque, *A Set of Plans and Forts in North America, Reduced from Actual Surveys* (London, 1763).

Captain La Pause made further and less complimentary observations about the Frontenac retrenchments in 1756. He found the faces of the redoubts "badly defended, having been badly placed." *RAPQ*, 1933-34, pp. 84-87. A translation is in Preston and LaMontagne, pp. 250-53.

265 South Bay, the extreme southern end of Lake Champlain near Whitehall, New York. Dieskau had intended to attack the new fort at the Hudson River end of the portage (Fort Edward, NY) but was dissuaded by his Indian allies.

266 Johnson had arrived at the south end of Lake George on August 28 with 1,500 soldiers and 40 Indians. He was joined by 200 more warriors on the 29th. Johnson to Shirley, Sept. 1, 1755, *SWJP*, I, pp. 892-94. Johnson's entire army numbered some 3,000 provincials from New York, New Hampshire, Massachusetts, and Connecticut as well as 500 Mohawks, but his men were strung out along the portage road from the Hudson to Lake George. Stanley, p. 101. By September 8 Johnson had about 2,200 troops in camp. Peter Wraxall to Lieutenant Governor Delancey, Sept. 10, 1755, *DRCHSNY*, VI, pp. 1003-04.

267 Pouchot uses the term *piquets* to describe detachments of picked men drawn from *troupes de terre* battalions for detached service. The word is usually translated as "platoons"

...(continued next page)...

While advancing in the direction of the lake, the Indians & the Canadians met a unit of 300 men, who had left Fort Lydius in order to join up with Johnson. It was utterly routed within sight of the retrenchment.[268] M. Dieskau, doubtless wishing to take advantage of the confusion that this action [49] must have inspired in the English, decided to attack the English immediately. The Indians pointed out to him, though in vain, that they needed to draw breath, & to have time to regroup, as well as the Canadians, so that they might advance together. Dieskau insisted on not delaying the attack on the retrenchment with the two *piquets* & the grenadiers. The Indians shouted at him: "Father, you must have lost your mind. Wait for us."

With his troops, he presented himself before the enemy, who greeted him with extremely heavy artillery fire. This small force

---

in *DRCHSNY.* This does not convey the proper meaning. Nor were *piquets* "light infantry", as translated by Hamilton in *Adventure,* or "pickets" in the sense of advance guards posted before a camp as defined by Captain George Smith, *An Universal Military Dictionary* (London, 1779; reprinted Ottawa, 1979), p. 202.

The *piquet* was, instead, a common provisional unit in the French Army of the period. The infantry manual of 1750 detailed the manner of forming a *piquet* within a battalion. It was to be composed of men selected from the fusilier companies and was to comprise a captain, lieutenant, two sergeants, one drummer and 47 men. When the battalion was drawn up in line, the *piquet* was to form on its left while the grenadier company took its position on the right. The *piquet* thus served as a second élite company for special duty, and in this sense it was roughly equivalent in status to the light infantry company later added to English regiments. The *piquet* was not, however, necessarily trained in light infantry tactics. See *Ordonnance du Roy sur le Maniement des armes de l'Infanterie Françoise & Étranger Du 7 May 1750* (Arras, 1750), pp. 4-7.

The *piquets* mentioned throughout Pouchot's *Mémoires* correspond in strength to the 50-man detachments specified in the 1750 manual with the single exception of the larger formations (three officers and 67 men) organized for Lévis' abortive 1758 expedition (note 404). We have retained the original French word, as did many English officers of the period, although they usually spelled it "picquet".

268 Pouchot's Fort "Lydius" was established on the site of a trading house owned by Colonel John Henry Lydius. It marked the beginning of the portage from the Hudson to Lake George. A regular fortification begun on the site in August 1755 was first named Fort Lyman for provincial officer Phineas Lyman. It was soon re-christened Fort Edward after the Duke of York. The post was located at the site of Fort Edward, New York. See Gipson, VI, pp. 140, 166-67.

The Battle of Lake George, fought on September 8, 1755, actually comprised three separate engagements. The attack described here was the morning action against a detachment of about 1,000 soldiers and 200 Indians sent from Johnson's camp towards Fort Edward. Pouchot also describes the assault on Johnson's hastily fortified camp. After the French were repulsed, they were attacked late in the day by another English force moving north from Fort Edward, perhaps the source of Pouchot's confusion here. Accounts of the day's fighting are in Gipson, VI, pp. 169-74 and Stanley, pp. 100-04. Johnson's version is in his letter to Shirley of September 9, 1755, *SWJP,* IX, pp. 228-34.

sustained the fire with great courage, however, losing 90 men.[269] M. Dieskau [50] fell wounded from a musket shot. The Canadians & their officers who could rush to help him, did their best to support him but seeing that they were in no state to rush the English who were very numerous & in a very advantageous position, they withdrew. M. de St. Pierre, an elderly Canadian officer of the highest reputation, was killed in the first attack.[270] M. Dieskau, wounded in the thigh, had remained on the battlefield and was captured. An English volunteer who reached him first saw him place his hand over his waistcoat pocket and, believing he was reaching for a pistol, shot him in the belly with his musket. The English also captured his aide-de-camp. [51]

It is quite probable that if M. Dieskau had taken his two battalions with him, which were alone capable of standing firm in such an attack, & had been satisfied to receive support from the Indians & the Canadians, whom he ought to have used to guard the bateaux, he would have succeeded in this engagement. It has since been learned that the English militiamen had little heart for battle and that only with difficulty were they urged to fight.[271]

---

269 Note in original, p. 50 (a): *Some English writers have claimed that on that day the French had 7 to 800 men killed, & Johnson only 200, including those from the detachment on the previous day under the command of Colonel Williams.*

Doreil reported total French casualties of six officers and 101 *troupes de terre,* colonial regulars, militia, and Indians killed and five officers and 125 men wounded in addition to Dieskau and his adjutant who were left on the field. Of the total of 237 killed and wounded, 78 were men of the *troupes de terre* who had stood in the open against Johnson's artillery. Doreil to Argenson, Oct. 20, 1755, *DRCHSNY,* X, pp. 360-61.

On September 11 the English reported 120 soldiers killed, 80 wounded and 62 missing with another 38 Indians killed and 12 wounded during the morning engagement and the assault on the camp. Two more men were killed, 11 wounded and 5 missing in the evening attack on the French made by New Hampshire and New York troops from Fort Edward. *DRCHSNY,* VI, pp. 1006-07. Others died later of their wounds. Among those killed in the morning action was Ephraim Williams (1714-1755), colonel of the 3rd Massachusetts Regiment. *DAB,* X, pp. 259-60.

270 Jacques Legardeur de Saint-Pierre (1701-1755), captain (1748) in the Marine. He had served with the Ohio expedition of 1753-1754. It was Saint-Pierre who received the summons from the Governor of Virginia delivered by George Washington in 1753. *DCB,* III, pp. 374-76. For an in-depth study of Saint-Pierre that does much to explain the complicated mix of military and business affairs of colonial officers, see Joseph L. Peyser, (trans. and ed.), *Jacques Legardeur de Saint-Pierre: Officer, Gentleman, Entrepreneur* (East Lansing, 1997).

271 Note in original, p. 52 (a): *M. Pouchot ought to have added that, despite his advantage, not only did Johnson not dare to pursue the French, but after their retreat he did not even attack Fort Saint-Fréderic, but withdrew to Albany.*

...(continued next page)...

However, this episode inspired in the Indians the very highest opinion of the bravery of the French troops, through the firmness they showed ever after when in the line. [52] They always asked to fight with those Frenchmen who had no intelligence & got themselves killed out in the open.

M. Dieskau was taken to New York where a capable English surgeon dressed his serious wounds. His convalescence was lengthy. For some reason or other, he drew upon himself the ill will of the English who held him in very close custody. He perhaps owed this treatment to his aide-de-camp whose indiscretion had greatly annoyed the English. They sent the latter back to France during the winter and there he had the opportunity to recount the very real abuses he had witnessed during his brief stay in Canada. This man was the son of a peasant from the Dauphiné and had risen from being a teacher to the rank of lieutenant in the Royal Suédois Regiment. [53] To reward him, the Court sent him back to Canada during the next campaigning season as commissary of war. In this position he somehow managed, throughout the war, to make himself hated by the English troops & little loved by the French. Upon his return to France, his ability as an informer, which constituted his sole merit, earned him fresh favors from the court.[272]

After this setback, the French withdrew to Carillon. M. de Montreuil, an *aide-major général* holding a lieutenant colonel's com-

---

Dieskau's actions and his decisions about the disposition of his considerable force were also widely criticized. Much of this is reflected in the French documents printed in *DRCHSNY,* X. Johnson did not withdraw to Albany but began construction of Fort William Henry near the site of the battle.

272 Dieskau's aide was Benoît-François Bernier (1720-1799) who was held prisoner in England until 1757. In June 1758 he was appointed to replace Doreil as financial commissary of wars for New France. Bernier was the son of a cloth merchant and had entered the Royal-Suédois (Royal Swedish) regiment in 1746. During 1759-1760 Bernier, probably because of fluency in English, served as liaison officer with the enemy on such matters as the capitulation of Québec and the repatriation of French troops to Europe. Following his return to France, he received the Cross of St. Louis, a substantial pension and an appointment as financial commissary at Dunkirk. *DCB,* IV, pp. 57-58.

Whether or not Bernier was rewarded for his service as a commissary or as an informer, presumably in regards to the *affaire du Canada,* Pouchot's allegations reflect his generally negative attitude toward administrative and financial personnel. Doreil considered Bernier to have had "talent and intelligence", and Montcalm thought him "qualified to acquit himself well of the duties of his station." *DRCHSNY,* X, pp. 765, 959.

273 Pierre-André Gohin, Comte de Montreuil (1722-1793), assistant chief of staff to the Army troops assigned to New France and second in command to Dieskau after the death of Rostaing. *DCB,* IV, pp. 307-08.

mission,[273] found himself in command of the French. M. de Vaudreuil ordered a fort to be constructed and this work was done under the supervision of M. Lobiniere, an engineer from the colony.[274] The French troops withdrew to Canada as soon as winter arrived. A detachment of 400 men commanded by M. [54] d'Hebecourt, a captain in the La Reine Regiment,[275] was left on the frontier.

The news of the retrenchments at Frontenac had spread, as has been mentioned, and this decided M. de Vaudreuil to write M. Pouchot, captain in the Béarn Regiment, requesting him to proceed to Niagara and to restore the fortifications there to the highest possible standard. As a result, he left with the Guienne Regiment which was being sent to cover the outpost. The English at Chouegen were a threat to it and could easily have taken it. About sixty Canadians or soldiers from the colony made up the entire garrison of this post which was constructed of stakes all of which were rotten & useless for defensive purposes.[276]

---

274 Michel Chartier de Lotbinière, Marquis de Lotbinière (1723-1798), lieutenant (1753) in the Marine and engineer. *DCB*, IV, pp. 143-45. His new fort was that of Carillon, today known as Ticonderoga. Lotbinière's assessment of the post is in his letter to the minister of November 2, 1756, *DRCHSNY*, X, pp. 493-96.

Lotbinière's fort drew much criticism from officers of the *troupes de terre*. Montcalm reported it "full of defects, [and] serves to enrich the Colonial Engineer, one of M. de Vaudreuil's relations." *Ibid.*, p. 491. Doreil was even more caustic. It was "worth nothing," he wrote, and an "ignoramus constructed it ... whose fortune 'twas desirous to make and who has made it." Doreil to Belle Isle, July 31, 1758, *Ibid.*, p. 769. Army engineer Captain Pontleroy thought the place "little susceptible of defence" and that he could take it with "only six mortars and two cannon." Memoir on Fort Carillon, 1758, *Ibid.*, p. 720. Bougainville sarcastically dismissed Lotbinière as "the Vauban of Canada." Hamilton, *Adventure*, p. 55.

Pouchot does not mention that Lotbinière was assisted by two *troupes de terre* officers, Captain François-Joseph, Chevalier de Germain (fl.1736-1760) of La Reine and Captain Armand de Joannès (fl.1747-1769) of Languedoc. Montcalm to Argenson, June 12, 1756, *DRCHSNY*, X, p. 414. Germain seems to have had some engineering experience and occasionally served in that capacity in Canada. "Officiers du régiment de la Reine", *Le bulletin des recherches historiques*, LI, No. 11 (November, 1945), p. 387. For Joannès see "Officiers du régiment de Languedoc", *Le bulletin des recherches historiques*, LI, No. 8 (August 1945), p. 287.

275 Louis-Philippe Le Dossu d'Hébécourt (fl.1741-1760), captain (1745) in the La Reine Regiment. "Officiers du régiment de la Reine", p. 388.

276 The tiny garrison was, however, supported by a larger detachment. Sometime before July 24 Captain Louis Coulon de Villiers (note 164) had been sent from Fort Duquesne to Niagara with 200 men to establish a camp of observation. By September Villiers' force numbered 500 "marines, Canadians and Indians." Vaudreuil to Machault, July 24, 1755,

...(continued next page)...

For the good of the service, therefore, M. Pouchot decided to undertake this mission.[277] The Guienne Regiment set out on 5 October. The [55] journey lasted until the 28th because of the bad weather.[278] On his arrival, M. Pouchot attempted to make the fort less of a joke & sent his plan of fortification to M. de Vaudreuil.[279] The Guienne Regiment left on 16 November to go down to Canada & M. Du Plessis, first captain of the colony, arrived at Niagara at the same period to take command there.[280] Three *piquets* from the Guienne Regiment & the same number of Canadians from the colony re-

---

and Vaudreuil to Machault, Sept. 25, 1755, *DRCHSNY,* X, pp. 307, 326. Pouchot knew that he would find Villiers and his men at Niagara. Malartic and Gaffarel, p. 25.

As to the potential for the fort itself to resist an assault, Vaudreuil reported that, "should the English once attack it, 'tis theirs." The stockade was "so dilapidated, that 'tis impossible to put a peg in it without causing it to crumble," and the garrison was only "thirty men without any muskets." Vaudreuil to Machault, July 24, 1755, *Ibid.,* p. 307.

277 Pouchot's seemingly presumptuous implication that the governor requested rather than ordered him to take the Niagara assignment appears to be accurate. Vaudreuil later wrote that the captain "was so good as to take on himself the direction of the fortifications I proposed constructing at Niagara." Vaudreuil to Argenson, June 8, 1756, *DRCHSNY,* X, p. 411.

278 This frustrating voyage is described in La Pause's "Mémoire", *RAPQ,* 1931-32, pp. 15-17.

279 Vaudreuil had ordered Pouchot to "draw the plan and sketch of the other works necessary to place Niagara in the best posture of defence." Vaudreuil to Machault, Sept. 25, 1755, *DRCHSNY,* X, p. 326. The plan of Fort Niagara printed in Pouchot's *Mémoires* as "Plan Du Fort de NIAGARA construit et défendu par M. de Pouchot" is probably based on the draft sent to Vaudreuil in the autumn of 1755. Pouchot, *Mémoires,* I, plate II. It is reproduced in this edition. Pouchot's plan shows the old stockade, his earliest improvements to it and the trace of the new fortifications. See also Dunnigan and Scott, pp. 22-23.

280 La Pause says Guienne sailed on the 14th. "Mémoire", *RAPQ,* 1931-32, p. 18.

The arrival of a *troupes de terre* detachment at Niagara created a dilemma. The commandant was then François-Clément Boucher de la Perrière (d.1759), lieutenant (1748) in the Marine. *Ibid.,* p. 17; Pierre-Georges Roy, "Les Commandants du Fort Niagara", *Le bulletin des recherches historiques,* LIV, No. 6 (June 1948), pp. 176-77; "Tableau ... des Officiers de la Marine", Douglas Brymner, *Report on Canadian Archives, 1886* (Ottawa, 1887), p. clxxxii.

La Perrière was a junior Marine officer, and his continuation in command would have been intolerable to Pouchot and the captains of the Guienne *piquets.* On the other hand, command of the frontier posts, with their lucrative trading concessions, was a jealously guarded perquisite of colonial officers. Doreil had learned from Governor Vaudreuil that "officers belonging to our troops (*troupes de terre*) are not to pretend to such appointments, at least for this year." Doreil to Argenson, Oct. 28, 1755, *DRCHSNY,* X, p. 371. Thus, even Pouchot was excluded from command.

...(continued next page)...

mained behind.[281] It was immediately necessary to construct houses in the Canadian style as lodgings for these troops. These houses are sheds made of round oak logs dovetailed at the corners. In this country of forests such houses are soon built. The chimney is in the middle and there are windows and a roof covered with planks. Four posts are placed in the shape of a truncated pyramid, **[56]** which is open at the bottom on all four sides to the height of three feet. The top is a sort of daub and wattle arrangement. You take reeds or long hay or straw and roll it in a mixture of heavy earth. Then you twist it up and down from one strut to another of the horizontal studs & plaster over the whole structure, including the gaps in the walls. This type of construction can be used for armies in Europe, if the country is wooded, for housing an army comfortably, because these houses can be made as large as required & also because every soldier can contribute to the work as soon as someone has outlined what has to be done.

Throughout the winter we worked on the new fort as assiduously as possible. The new engineer was obstructed by the officers of his detachment, **[57]** who, considering him as ignorant as they were themselves, laughed disdainfully when they saw work being started that, in their view, could not be finished in four or five years with double the troops. In this they were to be proved wrong.[282]

The new commandant, François Lefebvre Duplessis Faber (1689-1762), was a very senior captain (1727) of the Marine. *DCB,* III, pp. 372-73. He passed Fort Frontenac on his way to Niagara on November 3. Malartic and Gaffarel, p. 34. Although Duplessis was appointed Town Major of Montréal on March 17, 1756, he was at Niagara as late as August 13. *DRCHSNY,* X, pp. 438-39. He probably remained in command until relieved by Pouchot on October 12, 1756.

281 Other sources confirm that four *piquets* of the Guienne Regiment, totalling 200 men were left at Niagara and that, with Marine troops and militia, the garrison numbered 300. Doreil to Argenson, Oct. 20, 1755; Doreil to Argenson, Oct. 28, 1755; Vaudreuil to Machault, Feb. 2, 1756, all in *DRCHSNY,* X, pp. 361, 371, 391. La Pause says that the four "last companies" (*dernières compagnies*) of the battalion remained at Niagara. *RAPQ,* 1931-32, p. 18. Malartic agrees. Malartic and Gaffarel, p. 62.

282 Pouchot seems to imply that the obstructive officers were of the Guienne Regiment rather than of the Marine.

Concern about the "disagreeableness" of leaving officers of Army *piquets* under the command of "juniors" led to the policy of assigning only younger captains to the frontier posts for the winter of 1755-1756. One of the officers at Niagara with Pouchot was probably the Sieur de Saint-Vincent (d.1758) who had been promoted captain only in 1755. Doreil to Argenson, Oct. 28, 1755, *DRCHSNY,* X, pp. 371-73; "Les Officiers du régiment de Guyenne", *Le bulletin des recherches historiques,* LI, No. 5 (May 1945), p. 194.

...(continued next page)...

• 1756 •

The defeat of M. Dieskau was perhaps a stroke of good fortune for Canada, because the court, confident in the country's forces, might have neglected it & we would have been utterly incapable of resisting enemy operations. M. de Vaudreuil made representations to the court, which decided to send over in the spring of 1756 M. de Montcalm, *marechal de camp,*[283] M. de Lévis, brigadier,[284] M. de Bourlamaque, [58] colonel,[285] two engineers, Descombles & Des Androins,[286] & the two battalions of La Sarre & Royal-Roussillon, with provisions, ammunition & trade goods.[287]

While the names of most of the officers of Guienne who spent the winter of 1755-1756 at Niagara are not recorded, it is possible to identify the companies and their captains, most of whom probably stayed with their men. A series of issuing lists for the regiment in 1755-1756 makes it clear that the four "last companies" left at Niagara were (from the most senior) those of the Sieur de La Bretèche (captain July 17, 1754), Louis Preissac de Bonneau (Aug. 14, 1754), Saint-Vincent (Feb. 12, 1755), and Jean de Carecy du Bousquet (April 3, 1755). *RAPQ,* 1933-34, pp. 69, 70, 71-72, 74; "Les Officiers du régiment de Guyenne", pp. 190, 194. All four were junior to Pouchot.

283 Montcalm (note 118), holding the rank of major general (*marechal de camp*), arrived at Québec on May 13, 1756. Pouchot does not mention another officer who traveled with him, Louis-Antoine de Bougainville (1729-1811). Bougainville was a newly promoted captain (February 1756) in the dragoon Regiment d'Apchon when he was appointed aide-de-camp to Montcalm and ordered to New France. *DCB,* V, pp. 102-06. Bougainville notes, in his detailed journal, that he arrived at Québec on May 12. Hamilton, *Adventure,* p. 3.

284 Note in original, p. 58 (a): *Later Marquis de Lévis, Knight of the king's orders, lieutenant general of his armies, governor of the province of Artois, &c.*

François (François-Gaston) de Lévis, Duc de Lévis (1719-1787), second in command to Montcalm. He arrived at Québec May 31, 1756. *DCB,* IV, pp. 477-82. The 12 volumes of Casgrain's *Collection* include *Journal des campagnes du Chevalier de Lévis en Canada de 1756 à 1760* (Vol. I) and *Lettres du Chevalier de Lévis concernant la guerre du Canada (1756-1760)* (Vol. II).

285 Note in original, p. 59 (a): *At the peace, field marshal, subsequently governor of Guadeloupe, where he died.*

François-Charles de Bourlamaque (1716-1764), captain in the Dauphin Regiment, had been commissioned colonel of infantry in Canada (March 11, 1756) and third in command to Montcalm. He arrived at Québec May 15, 1756. *DCB,* III, pp. 84-87.

286 Jean-Claude-Henri de Lombard de Combles (1719-1756), captain (1751) in the Royal Engineer Corps. *DCB,* III, pp. 405-06. His assistant, Jean-Nicolas Desandrouins (1729-1792), was a second-captain (1755) in the same corps. *DCB,* IV, pp. 211-12. An account of Desandrouins' Canadian service, based heavily on his own writings, is Gabriel, *Le Maréchal de camp Desandrouins, 1729-1792.* De Combles and Desandrouins were the first officers of the Royal Engineer Corps to reach New France. They replaced the men lost on *Alcide.*

Before recounting the actions of this campaign, we shall give an idea of the administration of the King's stores which fell into a ruinous state & led to the trial of the intendant among other things. It must first be understood that the storekeeper reported directly to the intendant on his activity. Thus one can consider the stores as belonging to a merchant whose agent is obliged to report to him.[288]

When commodities arrived in a store, it was necessary to check the **[59]** lists in the presence of the commandant. An inventory was drawn up showing anything missing & anything damaged. This inventory was subsequently sent back to the intendant signed by the commandant & the storekeeper.

Everything that was issued from the stores, whether food or equipment due to the troops, was issued to them on the order of the commandant in each individual case. This order always began in these terms: "I request the storekeeper," &c., which shows how polite the Marine service was. Whenever the commandant desired that an issue of food, ammunition or equipment be made to the Indians, if they were going on a war party or had come to talk business & even when nations whose friendship we wished to gain came to **[60]** negotiate, it was made on his orders. In the event of a shortage of given commodities, he could have them set aside or even stop them being issued in order to be able to use them for the purpose of the service.

At several posts, if the commodities which were necessary for the service were lacking, the commandant could have a supply bought by the storekeeper on his authority, a document which was signed by the governor of the territory and covered by the intendant.

Since all types of commodities were in the stores only to be issued, all officers, soldiers & militiamen were authorized to buy both

---

287 The second battalions of the Royal Roussillon and La Sarre regiments arrived at Québec during May 1756. Chartrand, *The French Soldier,* pp. 31-32.

288 An account of the activities of Intendant Bigot and his associates and their effect on the colony may be found in Stanley, pp. 201-12. A very detailed contemporary critique of the administration of the outposts of the Canadian colony, with recommendations for reform, may be found in Joseph L. Peyser (trans. and ed.), *On the Eve of the Conquest: The Chevalier de Raymond's Critique of New France in 1754* (East Lansing, 1997).

Many officers of the *troupes de terre* were openly critical of the administration of colonial affairs and of the King's property. Soon after Montcalm's arrival, Montreuil, veteran of a year in New France, wrote: "It is incredible to what a degree luxury prevails in this country, and to what an extent the King is robbed in consequence of the bad administration of affairs. All the French who arrive here are shocked at the waste that is made. The Governor and Intendant are too easy and too remiss in a country where greater strictness is required than in any other. There is no police." Montreuil to [?], June 12, 1756, *DRCHSNY,* X, p. 419.

necessities and luxuries, either with ready cash or with promissory notes issued to cover their labor or their pay. As it was advantageous to the suppliers of the commodities and to the **[61]** storekeepers to issue them, they were not sparing with them as long as they lasted. If one adds to that the consumption of commodities sold to the Indians as trade goods, it will easily be understood that these stores were very soon exhausted.

As long as provisions were a monopoly, no goods could be brought in, which led to highly inaccurate estimates of consumption. The more goods supplied, the higher the profits of the suppliers. They acquired provisions from all sources, good or bad, provided there were sufficient quantities to fill the stores. Subsequently, a chief supplier was appointed, who had commodities brought over from France, which ensured better quality goods.[289] However, the greater the quantity of goods sent to him, the higher his profits. He permitted goods to be brought in, which only added to his profits, as he was considered to have delivered **[62]** them in their entirety. The price of all commodities rose to an exorbitant level because of English gains and, since he had to continue his functions as supplier, he acquired them where he could but at the highest price. Although he was responsible for victuals, his vessels nonetheless brought over from France a variety of trade goods, which he sold to the King at the price then current in Canada. Since the cost of supplying victuals was controlled by the King & there was thus no profit to be gained from selling to the garrisons, he falsified statements of consumption in order to indemnify himself against the price disparity. For example, if he had supplied a bottle of brandy at 3 *livres,* he accounted for the consumption of 4 bottles in order to recoup his purchase price, & similarly with other items. This resulted in a prodigious inflation of the true state of consumption, since the price of a bottle **[63]** of brandy rose to 100 *écus.* It is easy to guess from that what happened to other items.

Private individuals who had profitably traded with one another in goods bought from the stores subsequently sold them back at the going price to the same chief supplier as soon as the latter received orders from the commandant to purchase them, since the goods in question were imperatively needed by the service.

---

289 Pouchot probably refers to Jacques-Michel Bréard (1711-1775), comptroller of the Marine at Québec. He was a partner with Bigot and the Bordeaux firm of David Gradis et Fils. *DCB,* IV, pp. 90-92. Bréard, Bigot and Abraham Gradis formed the *Sociéte du Canada* for the purpose of importing goods to the colony. Stanley, pp. 202-03.

One can easily imagine the way in which this speculation continued to increase as the English closed in & shortages developed because private stores had been captured. On the grounds that more vessels might arrive from France in the summer, the intendant only made purchases at the end of winter, when everything was in short supply and when those who had goods to **[64]** trade were only prepared to let them go at a very advantageous price. Since they refused to sell before the market peaked, they were able to maximize their profits. The King believed he had 60 thousand *livres* of goods in his stores, when in fact he had only 10 thousand *livres* worth, for which he had paid premium rates. He lost altogether on any surplus, since it was sold back to the French & Canadians, those who had made the original purchase enjoying huge profits and covering their tracks with a doctored balance sheet.

As a result, as soon as the price of the goods had risen, French officers & soldiers who, according to the arrangement made in France, should have been able to find everything they needed at a fixed price, were unable to obtain anything in the stores & were **[65]** to buy at current prices those very articles that had left France destined for them.

Such is the source of the disorders which ensued in that administrative sector. Let us now return to the military operations. In the early days of February, M. Pouchot sent from Niagara a memoir about Chouegen, in which he demonstrated how the English might be disrupted at that fort & their operations delayed.[290] On the 25th, there arrived at Niagara a deputation from the Five Nations, made up of Goyogoins & Sonnontoins,[291] 112 persons in all, including women & children. These Indians, who had taken up the hatchet against us, had arrived to speak of peace. "Their words were sent to the general."[292] **[66]**

---

290 This document has not been found.

291 Pouchot consistently refers to the Iroquois as the "Five Nations", even though, since the early part of the eighteenth century, the original members of the League (the Mohawk, Oneida, Onondaga, Cayuga, and Seneca) had been joined by a sixth nation, the Tuscarora. For essays on each of the six Nations see Bruce G. Trigger (ed.), *Handbook of North American Indians, Volume 15: Northeast* (Washington, 1978), pp. 418-524.

The French names for the Iroquois nations, as used by Pouchot, are: *Agnier* or *Anier* (Mohawk), *Onnoyotte* (Oneida), *Onontague* (Onondaga), *Tascarorin* (Tuscarora), *Goyogoin* (Cayuga), and *Sonnontoin* (Seneca). Useful identifications of French names for native peoples are to be found in "Glossary of Indian Tribal Names", *DCB*, III, pp. xxi-xlii and "Glossary of Native Peoples", *DCB*, IV, pp. l-lvii. Pouchot's original spellings have been retained for all Indian nations.

In the month of March, the artillery captured in the Braddock action arrived at Niagara.[293] At the end of the same month, a party of French, Canadians & Indians, 300 men in all, under the command of M. de Lery, which had set out from Montréal, arrived at the portage of Chouegen where the English had begun construction of two forts, Stenvox & Brueil.[294] They attacked the latter of these, which was the smaller of the two, & captured it. Only sixty or so men were taken prisoner, but the English sustained a much greater loss of supplies destined for Chouegen,[295] which made it impossible for them to build up reserves [67] in that sector as early as they had originally planned.[296]

---

292 Note in original, p. 66 (a): *A manner of speaking used in the country with respect to the Indians.*

The "general" was, in this case, Governor General Vaudreuil.

293 Vaudreuil intended Braddock's guns to remain at Niagara to strengthen the negligible armament of that post. Vaudreuil to Machault, Sept. 25, 1755, *DRCHSNY*, X, p. 326. After a visit to Niagara in 1757, an Indian or half-breed named Job Chilloway reported that the fort's guns included the brass field pieces taken from Braddock, but that they were to be sent to Fort Frontenac that summer. Shippen to Burd, Jan. 20, 1758, quoted in Severance, II, p. 206. This decision was probably made after Niagara had been well-armed with artillery taken at Oswego in 1756. Most, if not all, of Braddock's guns were later recaptured by Bradstreet at Fort Frontenac. Doreil to Belle-Isle, Aug. 31, 1758, *DRCHSNY*, X, p. 821.

294 Pouchot is premature in identifying the first of these post as Fort "Stenvox" (Stanwix) which was not constructed on the Oneida Carry (Rome, New York) until 1758. He actually refers to Fort Williams, established in November 1755, a short distance downstream from the Fort Stanwix site on the Mohawk River end of the carry. Fort "Brueil" is Fort Bull rendered phonetically. It was built in November 1755 on Wood Creek at the west end of the carry. See Gilbert Hagerty, *Massacre at Fort Bull: The de Léry Expedition Against the Oneida Carry, 1756* (Providence, RI, 1971), p. 24. Another summary of the forts of the Oneida Carry is in John F. Luzader, "Fort Stanwix: Construction and Military History, 1758 to 1777", *Fort Stanwix: History, Historic Furnishing, and Historic Structures Report* (Washington, 1976), pp. 3-6.

Pouchot's phonetic rendering of Stanwix as "Stenvox" and, later in Volume III, "Sténix" approximates the English pronunciation of "Stennix".

295 Note in original, p. 67 (a): *The explosion of the powder magazine, which occurred a few moments after the capture of the fort, prevented the French from using the supplies and munitions they found there.*

296 On March 27, 1756 Lieutenant Joseph-Gaspard Chaussegros de Léry (note 172) descended on the Oneida Carry with a mixed force of 362 Canadians, *troupes de terre* and Indians and destroyed Fort Bull. Some 62 English soldiers, bateau men and women were killed or captured, and large quantities of supplies destined for Oswego were destroyed. See Hagerty, pp. 57-66 for details of the expedition. De Léry's journal of the expedition has been published in French in *Rapport de l'Archiviste de la Province de Québec pour 1926-1927* (Québec, 1927). Another contemporary French account is in *DRCHSNY*, X, pp. 403-05.

On May 7, two parties of Indians, Sauteurs & Mississakes,[297] left Niagara, the first of 25 men, & the second of 21. They headed to Chouegen. On the 15th, they returned with 12 scalps & three English prisoners who were ship's carpenters working on vessels beneath the fort.[298]

These early successes decided the Five Nations in our favor & encouraged the Indians from the upper country, who came in bands one after the other to Niagara to form war parties. We were obliged to feed them & equip them for war. This appearance of a successful campaign was counterbalanced by less agreeable perspectives. On 30 May M. Pouchot wrote to M. de Vaudreuil from Niagara in these terms: "The whole question of victualling **[68]** must be very closely watched. Please note, Monsieur, that only 40 quarts[299] of flour arrived at Niagara as our entire supply. We have been compelled to give food

---

297 Two related but distinct peoples, the Sauteurs (Ojibwa or Chippewa) were from the northern Lake Huron and Lake Superior regions of Michigan and Ontario while the Mississakes (Mississauga) occupied that part of modern Ontario between Lakes Ontario, Erie and Huron. "Glossary", *DCB,* IV, pp. liv-lv. See also Helen Hornbeck Tanner (ed.), *Atlas of Great Lakes Indian History* (Norman, OK, 1987), pp. 40-41. Also Robert E. Ritzenthaler, "Southwestern Chippewa" and E.S. Rogers, "Southeastern Ojibwa" in Trigger, pp. 743-59, 760-71.

298 The "Journal of Occurrences in Canada" to June 1756 claims that the raiders attacked a party of 40 men, killed 14 carpenters and took three prisoners. *DRCHSNY,* X, p. 402.

The raiders from Niagara struck on May 12, according to the senior English naval officer at Oswego. He reported eight of his carpenters killed and scalped and four more taken prisoner. Broadley to [?], May 20, 1756, in W.L. Grant, "The Capture of Oswego in 1756", *Proceedings of the New York State Historical Association,* XIII, 1914, p. 355.

A unique personal account of the incident was published in 1757 following the escape from Niagara of one of the prisoners. William Phelps gave the date of his capture as May 11, 1756 and named his two fellow captives as Charles Carter and James Flanagan. The fourth prisoner, Lewis Dunning, was wounded in the skirmish and later killed by the Indians when he could not keep up. They arrived at Niagara in four days. Carter and Flanagan were sent to Montréal, but Phelps was adopted by one of his captors and lived with the Indians until they turned him over to the commandant of Niagara (probably Duplessis) on September 28, 1756. Phelps and another prisoner, Peter Luney, remained at the fort until they were able to effect their escape on June 19, 1757. The pair made their way to New York in 31 days. *New York Mercury,* July 25, 1757.

299 Pouchot commonly uses the word *quart,* literally "quarter", in reference to quantities of provisions. This is probably an abbreviation or corruption of *quartaut,* a unit of liquid measure equivalent to 67.05 liters. Graves, p. 48.

Quartaut is also defined as a "quarter cask", and Pouchot likely refers not to a specific weight or measure of provisions but to a container of a standard size. In enumerating supplies for Montcalm's 1756 campaign, La Pause lists *sacs, quards* and *barrils,* the first and

...(continued next page)...

& equipment to the Indians. We haggle over bread with both French & Indians, which is of such dangerous consequence that only pure chance currently prevents us either from starving to death or abandoning this post." These shortages continued as long as the war lasted. After giving to M. de Vaudreuil an account of his work, M. Pouchot continued thus: "Try, sir, to compel those responsible for provisioning to submit to you punctual and faithful accounts & to ensure that what they send is in good condition. Make those in charge of transport accountable, **[69]** for everything arrives here spoiled. One of the greatest economies that could be made in this country would be to avoid these problems. Little attention is paid to them, but nothing could be more true than that none of the absolutely essential supplies have arrived here, & very few others."

During May, M. de Montcalm sent M. de Lévis with the battalions of La Reine, Languedoc & Royal-Roussillon, a corps of colonial regulars and one of Canadians to set up an observation post at Carillon.

M. de Montcalm, accompanied by M. de Bourlamaque & two engineers, went up to Frontenac with the regiments of La Sarre, Guienne and Béarn, of whom 1500 **[70]** men had spent the winter at that fort.[300] He set to work immediately preparing for the siege of Chouegen. First of all, he stationed a corps of 500 Canadians & Indians, under the command of M. de Villiers,[301] on the peninsula of the Bay of Niaouré[302] in order to observe the enemy's movements. Then, around June 17, he sent the Béarn Regiment to Niagara, where

---

last definitely references to containers. His *quards* vary in weight according to the commodity: 150 pounds (tobacco), 170 (flour), 200 (rice), and 300 (brandy). "Memoire", *RAPQ,* 1931-32, p. 27. Bougainville gives the standard credit for a "quarter" of flour as 180 pounds. Hamilton, *Adventure,* p. 177. If Pouchot's *quards* were of standard sizes, Niagara received only between 6,800 and 7,200 pounds of flour.

300 "1500" is a misprint. A detachment of 150 men of the Béarn Regiment had been left at Frontenac during the winter of 1755-1756. Doreil to Argenson, Oct. 28, 1756, *DRCHSNY,* X, p. 371.

301 Captain Louis Coulon de Villiers (note 164). A "Journal de l'expédition de M. de Villiers" is printed in Casgrain, *Collection,* XI, pp. 65-70. The size of Villier's force is given as 700 in "Journal of the Siege of Oswego", *DRCHSNY,* X, p. 440.

302 The Bay of Niaouré is the area composed of Henderson Bay, Black River Bay and Chaumont Bay in the vicinity of Sackets Harbor, New York. The peninsula on which the camp of observation was established is Six Town Point or Association Island at the northern tip of Stony Point near Henderson, New York.

it arrived on the 22nd. Thanks to M. Pouchot's energy, work on that place was almost finished by July 22 and so it set out to return to Frontenac.[303]

On August 5, M. de Montcalm left with the La Sarre & Guienne regiments to camp at the Bay of Niaouré, & on the 7th the Béarn Regiment set out with the artillery.[304] M. de Rigaud[305] and M. de Villiers left to take up positions on a bay about a quarter of a league from Chouegen with 500 Indians & Canadians. On the 9th, M. de Montcalm [71], accompanied by the La Sarre & Guienne regiments, set out to join up with them.[306]

Béarn, having left on the 10th with the artillery, arrived on thc 12th at the camp beneath Chouegen. The day before, the engineer M. Descombles was killed by one of his escorting Indians who, having taken up a position near the fort to observe if any English were leaving it, made a mistake over his uniform.[307]

---

303 Ten companies of the Béarn Regiment departed Frontenac for Niagara as early as June 14. Delayed by stormy weather, their vessels only reached the Niagara River on the 22nd. During the next month the troops largely completed the fortifications begun by Pouchot the previous autumn. The soldiers of Béarn sailed from Niagara on July 23 aboard four vessels to return to Frontenac leaving a *piquet* at Niagara. Malartic and Gaffarel, pp. 59-67. A partial translation of Malartic's journal for this period is in Severance, II, pp. 145-50. Pouchot left Niagara with his regiment.

304 Montcalm left Frontenac on August 4. "Journal", *DRCHSNY,* X, p. 441.

305 François-Pierre Rigaud de Vaudreuil (1703-1779), captain (1724) in the Marine. He was Governor of Trois Rivières and the brother of the governor general. Rigaud had been captured aboard *Alcide* in 1755 and had only returned to New France in May 1756. *DCB,* IV, pp. 660-62.

306 Pouchot is confused here. The Canadian and Indian force first moved to Ance aux Cabanes, today Sunset Bay just west of Nine Mile Point in the town of Scriba, New York and within ten miles of Oswego. Following Montcalm's arrival from Niaouré Bay on August 10, the army moved forward to land, undetected, at Baldwin's Bay about one mile east of Fort Ontario. See "Journal", *DRCHSNY,* X, pp. 441-42. The operations of the siege of Oswego may also be followed in Bertsch, pp. 117-21, which includes a sketch of siege operations superimposed on a modern map, and in Grant, "The Capture of Oswego in 1756", pp. 338-67.

307 The Indian is later identified by Pouchot as Ochik (note 336). He is recorded as Aouschik, a Nipissing chief, in an account of an Indian council held at Montréal in April 1757 (*DRCHSNY,* X, p. 560) and alternately as Hotchig in *DCB,* III, p. 406. A detailed account of de Combles' death and the grief expressed by Ochik is given in a letter from Desandrouins of August 28, 1756, *DRCHSNY,* X, pp. 465-66.

...(continued next page)...

This event made rather a difference to M. de Montcalm, because there only remained one young engineer, who was capable but who was new to the profession, never having been on campaign. Therefore the general asked M. Pouchot, who had rejoined his regiment, to take over this responsibility. Pouchot accepted and was given formal orders to undertake these functions.[308] He immediately went to reconnoitre Fort Ontario. On his return, M. de Montcalm showed him letters from the fort [72] that had just been intercepted.[309] Although there was not a single fascine in the camp, but only a hundred or so gabions, M. Pouchot urged the general to begin trenching that very evening in order to surprise the enemy. Orders were given to work with the utmost diligence on the gabions & the saucissons with all available men.[310]

Malartic says the engineer was taken for an English soldier because of his "vest with red facings" (*sa veste à parements rouges*). Malartic and Gaffarel, p. 71. The uniform established for engineers on December 8, 1755 was a blue coat with red cuffs, collar and lining, red waistcoat and red breeches. Thus de Combles could easily have been mistaken for one of the New Jersey provincials of the English garrison who wore uniforms of the same color combination. At the siege of Fort William Henry, in 1757, Desandrouins put on a white infantry coat "so as not to be taken for an Englishman because of my blue coat." See Francis Back and René Chartrand, "French Engineers, New France, 1750-1763", *Military Collector & Historian,* XXXVIII, No. 1 (Spring 1986), pp. 26-27. It is also possible that de Combles had discarded his coat, exposing his red sleeved waistcoat and breeches.

308 Pouchot refers here to the lack of field engineering experience of Captain Jean-Nicolas Desandrouins. Although he had seen infantry service during the War of the Austrian Succession, Desandrouins had been admitted to the engineer corps only in 1752 (note 286). To Pouchot, a veteran of campaigns in Italy, Corsica, Germany, and Flanders, he must have seemed very green, and Desandrouins frankly stated that he had not previously directed a siege. Gabriel, p. 49.

Pouchot gives the impression that he was ordered to conduct the siege, and Vaudreuil says the same in his testimonial at the beginning of Pouchot's *Mémoires.* Several French accounts confirm that Pouchot was ordered to "act as engineer" during the siege. See *DRCHSNY,* X, pp. 442, 459, 468, 916. Lévis (Casgrain, *Collections,* I, p. 63), La Pause (*RAPQ,* 1931-32, p. 32) and Desandrouins (Gabriel, p. 49) all clearly indicate that Pouchot was ordered to assist the young engineer. It would seem that Pouchot was ordered to act as "an" engineer and took over immediate direction of siege operations while Desandrouins remained chief engineer.

309 Malartic records only that Desandrouins and the artillery officers reconnoitered Fort Ontario at 2:00 p.m. on August 12. Malartic and Gaffarel, pp. 71-72.

The intercepted letters had been sent from the English garrison about 4:00 a.m. on the 12th. They were captured by some Nipissings and delivered to the French six hours later. "My hopes were increased," wrote Montcalm, "by the manner in which the Commandant expressed himself." The English officer had been unaware until the 11th that he was about to be besieged. Montcalm to Argenson, *DRCHSNY,* X, p. 462.

At 11:30 on the night of the 12th, M. Pouchot opened up a parallel trench 60 geometric feet from Fort Ontario, about 50 *toises*[311] long and formed of abatis of very large trees which almost all had to be cut down. However, the work was completed by daybreak.[312] On the 13th, they worked at mounting batteries & at 3 in the afternoon, the enemy was surprised to find us so close to them. [73] They abandoned the fort, which might have held out for a considerable time, as the artillery was very advantageously sited. They withdrew in the greatest disorder to the old fort on the other side of the river.[313]

As soon as they were in position there, they subjected the abandoned fort to very heavy fire, believing that they were greatly inconveniencing us. That day, M. de Bourlamaque was slightly wounded in the cheek. During the night we busied ourselves digging a communi-

---

310 Several basic items of eighteenth century siegecraft are mentioned in this paragraph. "Fascines" were tightly bound bundles of sticks used to support the earthen walls of trenches and batteries. Those for siege work were usually from six to 18 inches in diameter and about ten feet in length. "Saucissons", sometimes called "double fascines", were longer, generally 16 feet. "Gabions" were round, wicker-work baskets or forms, without top or bottom, that could be set up along the line of a trench and rapidly filled with earth to establish a wall. They varied in size but were usually three feet in diameter and from three to six feet in height. See the definitions in John Muller, *A Treatise Containing the Elementary Part of Fortification, Regular and Irregular* (London, 1746; reprinted Ottawa, 1968) and Smith, *Dictionary.*

Much of the Béarn Regiment was employed constructing fascines and gabions on August 12. Malartic and Gaffarel, p. 72.

311 Most of Pouchot's measurements for siegework and fortification are given in *toises,* archaic French fathoms or six French feet. One *toise* is equivalent to slightly under 77 inches (76.71) or two yards, five inches. Hamilton in *Adventure,* p. 65, renders a *toise* as 2.1315 yards. Graves, p. 53 gives the metric equivalent as 1.94904 meters. We have retained Pouchot's use of *toise* throughout this translation.

312 La Pause gives another reason for the delay. During his reconnaissance on the evening of the 12th, Pouchot strayed or became lost. He was therefore unable to lay out his trench before 11:30 p.m. The workmen began to dig at midnight. "Mémoire", *RAPQ,* 1931-32, p. 32.

313 The French directed their attack against Fort Ontario on the east bank of the river. Two other fortifications stood on the west bank: "Old" Fort Oswego, established in 1727, and a smaller post, "New" Fort Oswego or Fort George (referred to by some as "Fort Rascal"), built during the year preceding the French attack. Bertsch, pp. 108-27. The British defenses are detailed in a "Plan of Oswego and the Forts shewing the state of that Garrison as it was on the 28th May 1756", Archer Butler Hulbert (ed.), *The Crown Collection of Photographs of American Maps* (3 series, 12 vols.; Cleveland, 1907), II/1, pp. 16-17. A frank evaluation of the defenses by English engineer Patrick Mackellar is in Pargellis, *Military Affairs,* pp. 190-93. A copy of de Combles' original plan of attack on the Oswego forts is printed in Severance, II, p. 182 and a reconstruction of the actual lines is in Bertsch, p. 120.

cation trench along the moat to reach the river's edge, where we mounted a barbette battery for 20 cannon, although only 8 were mounted. This battery, under the command of M. Mercier, would have been sited just as favorably on the flanks of the fort, & would have avoided the disadvantage of having to get to it without cover, a fact which nearly caused the failure of the operation.[314] [74] The ground was soaked by very heavy rain during the night & since no precautions were taken to build platforms, the wheels of the gun-carriages sank into the ground, thus making it very difficult to service the cannon. There was no cover & ammunition was in short supply. As a result we were compelled to place men at intervals, one to carry the powder, another the ball, a third the wadding, for each shot that was fired in a cannonade of more than 30 guns. One can just imagine how effective this arrangement was. M. Pouchot had warned M. de Montcalm of this disadvantage. The latter replied that he was obliged to defer to the opinions of the artillery officer. M. Pouchot tried to obviate the difficulty by means of a communication trench with the battery.

However, M. Pouchot judged [75] that the enemy could not hold out for more than 24 hours, given their poor position which left them completely exposed in their trenches, if a corps of our men crossed to the other side of the river. As a result, M. de Montcalm ordered M. de Rigaud to cross with the Canadians & the Indians, which he did at daybreak by using the ford upstream, without being seen by the enemy.[315]

As soon as it was light, the enemy began a very heavy fire on our battery which, although well operated, could use only four cannon.

---

314 Guns mounted *en barbet* fired over the parapet and not through embrasures. The crews were thus more exposed, but their cannon had a wider field of fire. Desandrouins commented on the dangerous position of the French gunners and noted that the English were "slaughtering us with their artillery." *DRCHSNY,* X, p. 466.

The effort required to establish this battery was immense. The 20 cannon "were conveyed, during the night, in men's arms, a labor which occupied the whole army." "Journal", *Ibid.,* p. 442.

315 The English gunners in "Old" Fort Oswego were also badly exposed to the French battery, placed on higher ground. "We took them *de revers;* they were firing their cannon almost in the opposite direction, and were exposed down to their feet," exulted Desandrouins in a letter of August 28, 1756. *DRCHSNY,* X, p. 466.

In this paragraph Pouchot records the first of several personal conversations with Montcalm that are unconfirmed by any other source. He also claims credit for the idea of sending Rigaud's Canadians and Indians to the west bank of the river, a crucial event of the siege.

M. de Montcalm wanted to send M. de Bougainville at 9 a.m. to summon the enemy to surrender, but M. Pouchot persuaded the general not to do this for fear of encouraging them by allowing them a breathing-space. A **[76]** half an hour later they sounded the chamade on the drum.[316] The garrison, composed of the Shirley and Peperel regiments, the Jersey militia & the employees of the fort, 2400 men in all, surrendered. Colonel Mercier had been killed around eight a.m. and the English were deeply upset by his death.[317]

In the fort, we found one hundred twenty pieces of artillery, ammunition, tools, copious quantities of food & trade goods and 9 war vessels, one of which, of 18 guns, had just been completed. The others had been laid down & all their rigging was in the stores.[318] All this equipment, which was assembled with care, was not re-allocated quite so punctiliously to the places where the King's service required it. A proportion of the flour & the pork was sent to Frontenac & to Niagara; [77] those two posts owed their subsistence to these

---

316 The "chamade" was a beating played when a besieged commander wished to parley or discuss terms. "In that case one or more drums mount the rampart, and beat what the military call a *chamade.*" Smith, *Dictionary,* p. 232. A French officer reported that "when we heard the *rappel* [an alternate term for *chamade*] beat and saw the White flag waving from the tower of their crenellated building; cries of *Vive le Roi* communicated the tidings to the trench." Letter, Aug. 22, 1756. *DRCHSNY,* X, p. 456. Desandrouins was puzzled that the enemy surrendered so soon and attributed it to the fact that "the English are, it must be confessed, a people that understand war very imperfectly." *Ibid.,* p. 466.

317 James F. Mercer (d.1756), lieutenant colonel (1754) of the 51st Regiment of Foot, was killed on the morning of August 14. Command devolved to John Littlehales (fl.1754-1765), lieutenant colonel (1755) of the 50th Regiment of Foot. Following a council with his officers, Littlehales asked for terms. For the service of both officers see the *Army List.* Littlehales never held another active command, remaining on half-pay until his name disappeared from the *Army List* after 1764.

Littlehales later reported that he had lost only 12 or 14 men killed in action. Gipson, VI, pp. 199-200. Pouchot exaggerates the number of prisoners. A detailed French list enumerates 1,742 English in the forts, including 152 dead. Casgrain, *Collection,* VII, pp. 103-07. Other French accounts in *DRCHSNY,* X generally agree on a total of nearly 1,700 captives with as many as 150 English killed for only about 30 French casualties. Some of the English were killed by the Indians following the capitulation. See *Ibid.,* pp. 443, 456, 460, 462, 468, 484.

Two versions of the articles of capitulation are in *DRCHSNY,* X, pp. 444, 474-75.

318 A return of captured equipment and supplies, "exclusive of the pillage," lists 121 pieces of artillery ranging from 18-pounders to "patereros" or light swivel guns. The artillery presumably includes the armament of the vessels, seven of which are listed in addition to 200 bateaux. *DRCHSNY,* X, p. 485. A slightly different account is on p. 444. A series of useful documents concerning English shipbuilding efforts at Oswego, mostly letters from Captain Housman Broadley, are in Grant, "The Capture of Oswego in 1756", pp. 348-67.

provisions for nearly two years. The artillery, which was initially taken to Frontenac was used there & at Niagara.[319] Several cannon were transported to Mont-Réal. The tools, which were abundant & of high quality, found their way into the hands of the artillery commander, M. Mercier, & were never seen again. The rigging, which had been set aside pending the creation of a naval force, was taken away by the government agents & never used by the military. All the provisions were immediately distributed to the Canadian officers & the employees. There was a huge quantity of tea. They only left for the King's use things which were difficult to carry away. A single example will be sufficient to prove the point. M. Pouchot, who was one [78] of the first to enter the fort, found a very fine *ceine*[320] & asked one of the commissaries to have it kept and sent to Niagara, where he was to take command & where it would have been useful in a fort that risked being short of food. He was promised it with great feeling. Despite that, it disappeared along with everything else. In 1758 M. Pouchot was very surprised to see it at Carillon, being used for fishing & the property of the King, to whom the agent in question had sold it for 12 or 1500 *livres.*

It would have been fortunate for the King if the tools had been sold back to him. At least they would have been of better use to him than those which M. Mercier, the artillery commander, supplied to the King to [79] his own advantage. The axes were only used to destroy the fort of Oswego, the fort of Chouegen & another called *au boeuf,* about a musket shot away from the latter. This operation lasted from the 15th to the 20th, when the army went back to Mont-Réal.[321]

---

319 The captured artillery was an important addition to the defenses of Fort Niagara which previously had only six cannon, probably the six heavier guns taken from Braddock. A total of 24 guns "of the largest calibre" from Oswego had been sent to Niagara by September. Abstract of Dispatches, Sept. 12, 1756, *DRCHSNY,* X, p. 481.

Captured artillery and ammunition bolstered the defense of Canada throughout the war. "Were it not for the ammunition furnished me successively by the Beautiful River [Braddock's Defeat], Chouagouin and Fort George [William Henry], I should not have had enough either for attack or defence," wrote Vaudreuil to Massiac on November 1, 1758. *Ibid.,* p. 863.

320 Note in original, p. 79 (a): *A sort of net.*
The English word is "seine".

321 By the first two forts, Pouchot means Ontario and Oswego. Fort *au boeuf* was Fort George or "New" Fort Oswego. La Pause provides a particularly detailed account of the demolition of the English works. *RAPQ,* 1931-32, pp. 34-35. He and Bougainville say that the army departed on the 21st as does Malartic. Hamilton, *Adventure,* p. 28; Malartic and Gaffarel, p. 77.

The capture of Chouegen had a very great influence on all the Indian nations, because the English claimed to be decidedly superior to us &, through their boasting about their power & courage, intended to make the Indians believe that we could not resist them. The latter noted how easily we had taken this fort where there were as many defenders as there were attackers & observed that the very heavy artillery fire, the like of which they had never heard before, did not trouble the French troops. One might say **[80]** that, since that operation, they increased their loyalty to & friendship for the French, whom they generally like more than the English because of their easy-going attitude towards life & their cheerfulness. But the principal reason for their conduct derives from the fact that they understand very well the advantage of adhering to the stronger side, for, although some of them are genuinely fond of us, they only like Europeans in relation to their own self-interest.

When the news of the siege of Chouegen was brought to Corlac,[322] the English sent a corps of troops to try to relieve it but as soon as they learned at the portage that the fort had surrendered they filled the little Woodskever River with abatis & concentrated on strengthening the defenses of Fort **[81]** Stenix. They feared that we might break through to their territory by taking it.[323]

Since M. de Montcalm wanted to take the troops he had brought with him to Carillon as soon as possible, he ordered M. Pouchot to improve the quality of the track from La Prairie in order to speed his march. As a result M. Pouchot went there on the 28th with the Béarn battalion. It was necessary to cross a plain about a league and a half long covered in water to a depth of 2 to 3 feet. As soon as he saw it, he estimated that he would be able to use the stumps of cedar & pine that had been burned & cut down in that area. He had the soldiers of the regiment uproot them and for 3 days they worked with very great

322 The common French name for Schenectady, New York.

323 "Woodskeever" is a phonetic rendering for Wood Creek which flows into the eastern end of Oneida Lake and provides the connection between that body of water and the Oneida Carry at Rome, New York. Pouchot again uses the name of the 1758 post on the Mohawk River end of the carry. "Fort Stenix" (Stanwix) here is actually Fort Williams. Following the destruction of Fort Bull, the English built a new post on Wood Creek and two more, Fort Craven and Fort Newport, along the portage road. Brigadier Daniel Webb ordered these and Fort Williams abandoned and destroyed in August 1756 after the loss of Oswego, thus exposing the Mohawk Valley to French incursions. See Hagerty, pp. 67-73, Luzader, pp. 5-6 and Gipson, VI, p. 195.

intensity from morning till nightfall in water nearly up to their thighs. M. Pouchot worked alongside them. He thus made a roadway and, once it was covered with [82] reeds cut from the plain, it allowed the army and its equipment to pass without getting wet.[324]

The Carillon campaign consisted of the formation of a number of parties of Canadians & Indians, one after the other, and always to our advantage. We also worked vigorously at the defenses of Fort Carillon &, in the month of November, the army went back to Canada to take up winter quarters.[325]

On September 21, M. Pouchot left with a *piquet* of the Béarn Regiment to take over command of Fort Niagara & work at completing the defenses. Captain de Blot & a *piquet* from the Guienne and Captain de la Ferté & a *piquet* from La Sarre,[326] were detailed with him for this posting, where M. de Montcalm had at last persuaded M. de Vaudreuil to allow a Frenchman to command, to the great [83] annoyance of the colonial officers & perhaps of the intendant

---

324 Pouchot's narrative makes an awkward transition at this point. From a discussion of events at the Oneida Carry, he switches to the rapid movement of Montcalm's army from Lake Ontario to Lake Champlain. Pouchot went with his regiment to La Prairie, across the St. Lawrence River from Montréal.

Improvements to the road from La Prairie to St. Jean on the Richelieu River allowed troops to move more expeditiously to Lake Champlain. Montcalm considered it "a road in our rear, of the greatest consequence for our communications." Montcalm to Argenson, Sept. 26, 1756, *DRCHSNY,* X, p. 489. Bougainville reported that 200 men of the Béarn Regiment repaired the road between August 27 and September 4. Hamilton, *Adventure,* p. 29. Malartic recorded the Béarn Regiment at La Prairie on August 28 when 150 men were ordered to work under Pouchot. The detachment labored through the 31st when the regiment concentrated at St. Jean on the Richelieu. Malartic and Gaffarel, pp. 79-81.

Despite the hard work of Pouchot's party, La Pause says that the effort was in vain, and the road remained "absolutely impracticable." *RAPQ,* 1931-32, p. 35.

325 Pouchot continued to Carillon with the Béarn Regiment. Adjutant Malartic describes the work done at that fort but does not mention either Pouchot's presence or his departure for Niagara on September 21. Malartic and Gaffarel, pp. 80-83.

326 François-Louis, Chevalier de Blau (fl.1727-1760), captain (1747) in the Guienne Regiment. "Les Officiers du régiment de Guyenne", pp. 189-90.

Le Chevalier de La Ferté de Mun (fl.1748-1759), captain (1748) in the La Sarre Regiment. "Officiers du régiment de la Sarre", *Le bulletin des recherches historiques,* LI, No. 7 (July 1945), p. 255. La Ferté was later wounded on the Plains of Abraham and captured on September 13, 1759. Knox, III, p. 143.

The total strength of Pouchot's *piquets* was 150 soldiers, but this probably includes the men of Béarn left at Niagara in July 1756 (note 303). Montcalm to Argenson, April 24, 1757, *DRCHSNY,* X, p. 552. The garrison also included 100 Marine troops. Casgrain, *Collection,* VII, p. 82.

himself.[327] The latter considered M. Pouchot very unlikely to accept the views of his company, whose activities he wished to conceal.

This officer left with M. de Lignery, a colonial captain,[328] who was going to relieve M. Dumas, the commandant at Fort du Quesne. M. de Contre-Coeur, considered inadequate for a difficult command, had already been relieved. By contrast, the company found M. Dumas far too shrewd for it.[329]

M. Pouchot took pains to have himself supplied at Frontenac with everything that might be necessary for his fort during the winter. On October 4 he set out for Niagara in the barks with his troops and arrived there on the 12th.

In October the Loups[330] came to establish an alliance with the **[84]** French & they were followed by the Iroquois. All the nations which up till that time had embraced the English side began to stir & to take up the hatchet against them. Thanks to the excellent treatment they received from M. Pouchot, they made a firm decision to go to war. He was ably assisted by M. Chabert, a Canadian officer in command of the little fort at the portage.[331] He enjoyed the very highest esteem among the Iroquois who considered him a member of their nation. He

327 Montcalm commented to Argenson on April 24, 1757 that the governor general "has made considerable difficulty in granting that command to an officer of the troops of the Line." Vaudreuil was, nonetheless, satisfied since Pouchot was "wonderfully liked by the Indians." *DRCHSNY*, X, p. 548.

328 Francois-Marie Le Marchand de Lignery (1703-1759), captain (1751) in the Marine. *DCB*, III, pp. 378-79.

329 We have translated *la société* as "the company". This is a reference to Bigot's trading interests.

330 While at Niagara, Pouchot had frequent contact with the people called by the French the Loups (literally, "Wolves"). This nation was known to the English as the Delaware, an Algonkian-speaking group originally from the middle and lower Delaware River valley of Pennsylvania. By 1756 their territory was primarily in Pennsylvania west of the Susquehanna River and in eastern Ohio. "Glossary of Indian Tribal Names", *DCB*, III, p. xxxiv and Tanner, pp. 40-41. See also Ives Goddard, "Delaware", in Trigger, pp. 213-39.

331 Daniel-Marie Chabert de Joncaire de Clausonne (1714-1771), ensign (1748) and lieutenant (1757) in the Marine. He helped construct Fort Little Niagara above the falls in 1751 and obtained the concession for that post and the Niagara Portage. *DCB*, IV, pp. 137-38. Daniel and his elder brother, Philippe-Thomas (note 517), were important French agents among the Iroquois. Pouchot distinguishes between them by referring to Philippe as "Joncaire".

...(continued next page)...

was equally known to the other peoples to whom he was constantly of service & whose languages he spoke. It was at that time that work on the fort resumed & a start was made to some proper buildings.[332]

When M. de Montcalm returned to Canada, he was bearing an order to withdraw from the officers [85] all the rations, except the two jugs of brandy a month, granted to the marines as part of their emoluments. They were reduced to two pounds of bread & a half pound of pork per officer. They say that M. de Montcalm had requested the order when in France because he believed this ration to be excessive.[333] It was not long before he regretted the measure. He would have done better to withdraw part of the equipment which was of hardly any use to the officers. The initial expense would not have been very great, if it had been economically channeled towards cases of need by higher-ranking French officers. This lack of forethought arose from ignorance of the country and left the officers and the

---

Following his return to France in 1761, Daniel was confined to the Bastille in connection with the *affaire du Canada.* He was found guilty in 1763 and released with a warning. *Ibid.*, pp. 137-38. In the same year he published *Mémoire pour Daniel de Joncaire-Chabert, ci-devant commandant au petit fort de Niagara* (Paris, [1763]) justifying his conduct. An extensive, though dated, biography is in Severance, II, pp. 375-447. Pouchot's assessment of Chabert's character lends support to the latter's claims of innocence.

332 Pouchot refers to additional barracks, storehouses and other structures begun at Fort Niagara late in 1756. Although he describes them as *bâtimens civils,* he does not mean "civilian" structures. See Dunnigan and Scott, p. 24-27.

333 Pouchot was on his way to Niagara when Montcalm received Governor Vaudreuil's orders to reduce the officers' allowances. Writing to the Minister of War, the general reported that "The Marquis de Vaudreuil and M. Bigot have communicated to me an order they have received from the Keeper of the Seals, whereby all the allowances granted to the officers for subsistence, even the wine, is to be cut off in the next campaign, and they are to be reduced, the same as the soldier, to the simple ration, leaving them at liberty to provide themselves with the surplus as they think proper. This article interests equally the officers of the Line [*troupes de terre*] and of the Marine." Pouchot picked up on Montcalm's subsequent comment that "It is possible that they received too much last year, and that too much is to be deducted for the next." Montcalm to Argenson, Sept. 26, 1756, *DRCHSNY,* X, p. 489.

Vaudreuil's order was followed on October 3 by instructions to reduce the ration in general because of "the lack of provisions in the storehouses, this year's poor harvest, the feeble resources of the colony, and the uncertainty of help from France." Although Montcalm was ordered to make the reductions immediately among his troops at Carillon, the general feared that it would "expose our poverty and could even make it known to the enemy." He therefore deferred action until November 1. Bougainville noted that the French troops had never drawn more than a pound and one-half of bread and one-half pound of salt pork and that the surplus was given to them in money. Hamilton, *Adventure,* pp. 63-64.

French troops at the mercy of the monopoly company & the chief supplier. By bringing in all provisions on their [86] own account, they set the prices for the officers, & raised them at will, as we shall see later. The booty captured from the English, although quite considerable, was of very little use to the service for the reasons we have indicated above. As a result, the stores contained few supplies & the posts even less and so it was necessary to request large quantities of provisions from France. M. de Montcalm also asked for troop reinforcements because the English, after the siege of Mahon, sent to America Scottish troops & 4 battalions of the Royal American regiment which they raised for service in their colonies.[334]

In the course of the winter, several war parties were sent out from Mont-Réal to Fort George,[335] where a number of prisoners & scalps [87] were taken. The Indian Ochik, who had the misfortune to kill our engineer, distinguished himself in these raids in order to obtain his pardon.[336] In the course of one year, he killed more than 33 Englishmen. In the month of March, a party of 3 to 400 men commanded by M. Rigaud marched to Fort George in an attempt to take it by surprise. When his presence was discovered, he contented himself with burning a sawmill together with their bateaux.[337] On the enemy side, Volf, a German officer, came and burned a bark on the stocks

---

334 The English garrison of Fort St. Philip at Port Mahon on the Mediterranean island of Minorca capitulated to the French on June 28, 1756. The invasion of Minorca brought an English declaration of war on May 18, followed by one from France on June 9 after two years of undeclared conflict in North America and the western Atlantic. Gipson, VI, pp. 398-417. The *Gentleman's Magazine* for May 1756 says the English declaration was signed at Whitehall on May 17.

Escalation of the conflict in North America caused the English to augment their forces. The Scottish troops mentioned here were of the 42nd Regiment of Foot, ordered to America with the 35th Foot early in 1756. It was further decided to raise a four-battalion regiment in the American colonies. At first numbered the 62nd, it became the 60th (Royal American) Regiment of Foot following the disbanding of Shirley's and Pepperel's regiments as a result of their capture at Oswego. The decision to provide these reinforcements was made in January 1756, long before the French invasion of Minorca. See Stanley M. Pargellis, *Lord Loudon in North America* (New Haven, CT, 1933), pp. 47-49, 61-66.

335 Pouchot, like most other French officers, consistently refers to Fort William Henry as "Fort George". Fort William Henry was constructed by the English in 1755 at the south end of Lake George following the Battle of Lake George. It was located on the site of the modern village of Lake George, New York. Two plans of the 1755 fort are printed in Rocque and in Gipson, VII, p. 78. Fort William Henry has been reconstructed on its original site and is operated as a tourist attraction.

336 This warrior is elsewhere identified as Hotchig or Aouschik, a Nipissing chief (note 307).

beneath Fort Niagara, which also sent out several raiding parties during the winter.[338] Since M. Pouchot had won over all the Iroquois nations & the Loups of upper Pennsylvania, **[88]** they always arrived at the fort with several scalps or some prisoners.

At the end of the fall, 13 Englishmen & Catabas who had set out from Virginia arrived in the vicinity of Fort du Quesne where they took the scalps of three Chaouanons.[339] M. de Lignery had them

---

337 Note in original, p. 88 (a): *A total of 150 of them, together with four brigantines of 10 to 14 guns & two 50-oar galleys and in addition a large quantity of construction lumber.*

Rigaud's force was much larger than stated by Pouchot, numbering some 1,500 Army and Marine troops, militia, and Indians. The French lost the element of surprise but invested Fort William Henry from March 19 to 23, 1757, destroying a number of sailing vessels and boats and nearly burning the fort itself. Gipson, VII, pp. 67-69. Three French accounts are in *DRCHSNY,* X, pp. 542-46. Another (*Ibid.,* pp. 571-72) claims only three sloops destroyed because the fourth would not catch fire. Yet another account, recorded by Bougainville, is in Hamilton, *Adventure,* pp. 95-99. Two journals of Rigaud's expedition are in Casgrain, *Collection,* XI, pp. 71-86.

338 Pouchot was confused at this point. There was no English raid on Fort Niagara in 1757. "Volf" (Wolf or Wolff) was a French officer with Rigaud's force. An account of the expedition reported that, on March 22, "Mr. Volff, a partisan officer of the regular troops, took it upon himself to burn the sloop" beneath the walls of Fort William Henry. *DRCHSNY,* X, p. 572.

Bougainville identifies Wolf as a half-pay lieutenant of the Bentheim Regiment who had accompanied the *troupes de terre* to New France. Hamilton, *Adventure,* p. 97. Montcalm said he was "formerly sergeant in Bentheim, and now a seconded officer, attached to the land forces." Montcalm to Paulmy, *DRCHSNY,* X, p. 693.

Bentheim, numbered 46th in the French line, was a German regiment. Most foreign (German, Irish, Swiss) units in the French service were identified by the name of their colonel. In the case of the 46th this was, from 1751 to 1759, the Comte de Bentheim. Mouillard, "Bentheim". Wolf appears in numerous sources, frequently serving as a messenger to the English or leading raiding parties. He is described variously as a "volunteer", "officer of partisans" or officer of regular troops and was one of two "officers of partisans" authorized for the staff of General Dieskau. *DRCHSNY,* X, p. 286. Wolf and his fellow partisan officer, Lieutenant Carpentier, are so listed in the "Report on the Staff of the Army" of October 15, 1755. *Ibid.,* p. 357. Wolf very likely owed his appointment to the fact that Dieskau was lieutenant colonel of the Bentheim Regiment.

339 The first named group were the Siouian-speaking Catawbas of the Carolinas. They were almost universally supportive of the English. "Glossary of Indian Tribal Names", *DCB,* III, p. xxxiv. See also Frederick Webb Hodges (ed.), *Handbook of American Indians North of Mexico.* (2 vols.; New York, 1959), I, pp. 213-16.

Chaouanons were the Shawnee, an Algonkian-speaking tribe. "Glossary of Indian Tribal Names", *DCB,* III, p. xl. In 1756 the Shawnee were centered in the upper Ohio Valley from western Pennsylvania to northern Kentucky. Tanner, pp. 40-41. See also Charles Callender, "Shawnee", in Trigger, pp. 622-35.

The English-Catawba raid against the Shawnee was reported in Montcalm to Argenson, April 24, 1757, *DRCHSNY,* X, p. 548 and recorded on March 8, 1757 in Malartic and Gaffarel, p. 99.

pursued by a party of Frenchmen & Indians, who killed 3 of them & brought back 2 prisoners. Rocheblave, a colonial cadet,[340] who was travelling from Fort Cumberland, came across two of them & took them prisoner. One of them was English and had his hair cut in the Indian manner; it was the custom of their partisans. They could not accuse the French of adopting this disguise, which they believed made them appear more terrifying.

• 1757 •

In January 1757, the Iroquois & the Loups, or Chaouanons, of Theogen,[341] met [89] in large numbers at Niagara for a great council. The Loups gave back to the Iroquois the *Machicote,*[342] which the latter

---

340 Philippe-François Rastel de Rocheblave (1727-1802), then a cadet in the Marine and appointed second ensign in 1757. "Tableau ... des Officiers de la Marine", Brymner, 1886, p. clxxxiv. He would later figure in the American Revolution in the Illinois country. See E. Fabre Surveyer, "Philippe Francois de Rastel de Rocheblave" in Percy J. Robinson, *Toronto During the French Regime* (Toronto, 1933), pp. 233-42. There is a brief entry on his later service in Mark M. Boatner III, *Encyclopedia of the American Revolution* (New York, 1966), p. 918.

Vaudreuil received news from the Ohio in April 1757 that Rocheblave had returned with one prisoner from the Potomac River, three days' journey from Fort Cumberland, and that Lieutenant Duverger de St. Blin (note 558) had taken two scalps and one prisoner near Fort Cumberland. Vaudreuil to Moras, July 12, 1757, *DRCHSNY,* X, p. 581.

341 Pouchot here seems to associate or confuse the Loups of Theogen with the Chaouanons (Shawnee). The village of "Theaggen" is shown on Pouchot's map at the junction of the Chemung and Susquehanna rivers. This town, more commonly known as Tioga, was near modern Athens, Pennsylvania, and was occupied by the Munsee, a people related to the Delaware or Loups. Pouchot usually identifies the "Loups of Theogen" as a distinct group. See "Glossary of Indian Tribal Names", *DCB,* III, pp. xxxiv; Tanner, pp. 40-41; Trigger, p. 236.

342 Note in original, p. 90 (a): *A skirt indicating that the Iroquois considered them as women & would not permit them to engage in war.*

The Iroquois had referred to the Delaware as "women" since the late 1600s, a term that was not originally meant as an insult but rather as a designation of an important tributary group of the Six Nations. The establishment of Iroquois control over the tribes of Pennsylvania and their lands after 1701, particularly in the period 1728-1746, is discussed in Aquila, pp. 156-204. The Delaware bitterly resented this and turned against the Iroquois during the war. *Ibid.,* pp. 239-42.

Bougainville recorded news of the Niagara council and the growing support for the French by the Delaware. He was told that they had even threatened the Iroquois with war. Bougainville, in classical terms, likened the *machicoté* and the control it represented to "the day of the Caudine Forks [321 B.C.] which the Romans never forgot." Hamilton, *Adventure,* p. 90.

had imposed on them when they had left their maritime territory to settle on the Iroquois frontiers. They informed the Iroquois that henceforth they wished to be men & to make war against the English, of whom they had killed or captured 3 to 400 in the previous campaign. They also expressed their wish to become allies of the French, whom they had not known before, & advised the Iroquois to grasp the hand of their father, the Frenchman, as they themselves were resolved to do.

In April, we were obliged to send [90] 43 men & a sergeant back from Presqu' Isle to Niagara because there were no supplies left at that post. They travelled on foot.

There also arrived at Niagara some Renards, or Outagumis, who lived among the Loups & the Sonnontoins. Since their war with the French, they had not frequented the vicinity of our forts.[343] A total of 30 of them arrived to assure the commandant that they also wished to grasp the French by the hand and that they had sent their young warriors on a raid with the Loups. It was learned from them that the English were building many bateaux on the Susquehanna in the vicinity of Fort Schamokin.[344]

These Indians added that Johnson had sent wampum belts through an *ânier*[345] & an Onontague to the Sonnontoins & the Loups, in order to bury the hatchet of their father, [91] and that they could no longer go out for their most pressing needs, for fear of losing their scalp. Despite that, the nations replied that their decision was made & that their father, the Frenchman, had shown them the direction they

343 The French "Renard" is usually translated literally as "Fox". These central Algonkian-speaking people, residing in Wisconsin near Green Bay and on the Fox-Wisconsin river trade route, are also known as the Mesquakie. They had hostile relations with the French from 1712 until a series of wars ended in 1732. Most of the Renards or Outagamis became associated with the Sauks and moved farther west to the Mississippi River. "Glossary of Indian Tribal Names", *DCB,* III, pp. xxxiv-xxxv and Tanner, pp. 40-41. See also Charles Callender, "Fox", in Trigger, pp. 636-47. For a very thorough study of the conflict between the French and the Renards, see R. David Edmunds and Joseph L. Peyser, *The Fox Wars: The Mesquakie Challenge to New France* (Norman, OK, 1993).

Tanner, p. 41, shows the Fox or Mesquakie village on the upper Allegheny River.

344 Fort Augusta or Schamokin guarded the confluence of the west and north branches of the Susquehanna at the site of Sunbury, Pennsylvania. Stotz, pp. 101-05.

345 Pouchot here uses the word *ânier* - "donkey driver" in French - in reference to a Mohawk. On occasion he refers to the Mohawk River as *la rivière des âniers* and alternately spells the word *agnier.* The origin of this name appears to come not from French but rather from a corruption of Huron and other Indian names for the Mohawk. Trigger, p. 479.

should take. As a consequence, from all the posts, from Frontenac to as far as Fort du Quesne, parties of Indians & some Frenchmen left in waves to lay waste the frontiers of New York, New Jersey, Pennsylvania, Maryland & Virginia. They always brought back scalps & prisoners. This forced the settlers along the length of the frontier to abandon their homesteads & to construct small forts at all rivers mouths.

On May 6, an Iroquois & an Englishman who had settled [92] among them arrived at Niagara. They told M. Pouchot that the King of France had been killed & that the King of Prussia had captured the Queen of Hungary, although no vessel had yet arrived from France. The commandant found this news so extraordinary that he thought it his duty to inform M. de Vaudreuil. It was confirmed that the King had been wounded and that the Queen of Poland had been detained by the King of Prussia.[346]

The commanders of all the forts of the upper country were sending Indians of different nations to Mont-Réal to join the army, but the smallpox prevalent in some places made many of them return home because they were deeply afraid of the disease. Nonetheless, 1900 of them were recruited, including those resident in our territory. There had never [93] been such a large number assembled in Canada.

---

346 An attempt had indeed been made on the life of Louis XV on January 5, 1757. Robert-François Damiens, a mildly retarded and highly disturbed man, stepped from a crowd and slipped between two guards as the King was descending a stairway in Versailles. He plunged a small knife into Louis XV's right side. The wound, feared at first to be mortal, proved to be slight. Damiens was apprehended, tortured and publicly executed in a particularly painful manner on March 28. Olivier Bernier, *Louis the Beloved: The Life of Louis XV* (Garden City, NY, 1984), pp. 185-88. Damiens and the incident are discussed in detail in Dale K. Van Kley, *The Damiens Affair and the Unraveling of the Ancien Régime, 1750-1770* (Princeton, 1984).

As for the Queen of Poland, King Frederick of Prussia had invaded Saxony at the end of August 1756. The King of Poland withdrew from Dresden, leaving his Queen there. The Prussians occupied the city on September 10. Dupuy and Dupuy, p. 668. See also Entick, II, p. 74.

News of both events reached Niagara very quickly and was probably based on items from Pennsylvania newspapers. Accounts of the attack on King Louis XV were carried in four consecutive issues of the *Pennsylvania Gazette* between March 17 and April 14, 1757. Bougainville did not record the attack on the King until June 10. The news had just arrived in letters from Europe. This was at least two days after the receipt of dispatches from Niagara dated May 23. Hamilton, *Adventure*, p. 113. Montcalm, who might have first learned of the incident from Pouchot, was careful not to mention the assassination attempt to the Indians of Canada. "It caused us all to shudder with horror, and these barbarians so ferocious in war, so humane in their lodges, might waver in their esteem for us, seeing us capable of producing such monsters," he wrote to Paulmy on July 11, 1757. *DRCHSNY*, X, p. 575.

At that period, General Loudon[347] had arrived from England. He went to Philadelphia to hold a great council and to persuade the Quakers to provide him with subsidies, which they refused, saying that they were not permitted to make war.

The Missisakes who were to travel to Mont-Réal, 90 of them in all, had gathered at Taronto, which was garrisoned by only about ten men although there was a substantial stock of trade goods. The trade of this fort belonged to M. Varrin.[348] The Indians conceived a plan to slaughter the small garrison and pillage the fort. M. de Noyelle,[349] who commanded there, was warned by a Frenchman who lived among them that they were only waiting for news from Détroit before putting their plan into effect. He sent a [94] canoe & two men to Niagara, 30 leagues away, to ask for help. M. Pouchot immediately sent M. de la Ferté, a captain in the La Sarre Regiment, & M. de Pinsun, an officer of the Béarn,[350] with 61 men. Each of them had a *pierrier* on board his bateau.[351] They arrived at Taronto at 4 in the afternoon the next day

---

347 John Campbell, Fourth Earl of Loudon (1705-1782), colonel (1749) of the 30th Regiment of Foot and major general. Loudon had been named colonel-in-chief of the newly raised 60th Regiment of Foot in 1755 and, on March 20, 1756, commander-in-chief in North America. He arrived at New York in July 1756. Loudon would be replaced by Jeffery Amherst in 1758. *DNB,* III, pp. 828-29. For an analysis of his American service see Pargellis, *Lord Loudon in North America.* Loudon's papers are preserved in the Huntington Library, San Marino, California.

348 Fort Rouillé, the tiny post at Toronto, had been constructed in 1750-1751 just west of the entrance to the bay of the modern city. An account of the history of the post is in Robinson, pp. 93-140. Pouchot notes that the trade at Fort Rouillé was the concession of Jean-Victor Varin de La Marre (1699-between 1780 and 1786), commissary and comptroller of the Marine and subdelegate of the intendant in New France. *DCB,* IV, pp. 749-51.

349 Charles-Joseph de Noyelles (1719-1767), lieutenant (1756) in the Marine. Fauteux, *Les Chevaliers,* p. 187. He was commandant of Fort Rouillé from 1754 to 1757. Robinson, p. 128.

350 Jacques de Pinsens (fl.1744-1760), lieutenant (1752) in the Béarn Regiment. He was later mortally wounded in the fierce action at Ste. Foy on April 28, 1760. "Officiers du régiment de Béarn", p. 356.

351 A swivel gun but, more specifically, an archaic iron breech-loading weapon known commonly as a *pierrier* or "perrier" and often called by the English a "paterero". It could throw a small ball or a handful of shot. These weapons were very common at the frontier posts of New France. Fort Niagara had five of them, with 13 loading chambers, in 1749. Return, Oct. 20, 1749, *DRCHSNY,* X, p. 196. Three original examples have been excavated from the ruins of Fort St. Frédéric. See Joseph Thatcher, *Cast for War: A History of Muzzle-Loading Artillery* (Albany, 1983).

and found the Indians in their huts by the fort. As they passed by them they saluted their encampment with their artillery & their muskets. But they only fired in the air, as M. Pouchot had recommended in his orders to them. They then summoned the Missisakes to a council. The Indians were much astonished by this occurrence & also at having been discovered in the fort. They admitted everything and confessed their stupidity. They added that **[95]** evil rumors had circulated among them; that we had brought troops over from France to kill them, that we had made peace with their enemies, the Têtes Plates,[352] that the English had beaten us and that we were hiding the fact from them, & a host of other reasons of this kind. But the only true reason was that they realized their superior numbers & sensed that they might obtain a great deal of brandy for nothing. Without foresight for the consequences, they had dreamed up their plan.

From this plot, one can judge how much confidence one may place in these peoples. That nation was nonetheless the most devoted to the French and served them with the utmost competence & fidelity in the war. The Missisakes are more dangerous than the Iroquois because they live solely by hunting. They set out immediately for Mont-Réal and conducted themselves well **[96]** in the Fort George campaign. The nation repudiated those warriors & was very upset by the event.

At the beginning of this year, the Cherakis,[353] the Têtes Plates, & the Catabas asked to make peace with the nations affiliated to the French, promising to strike at the English, if they were left alone, which some of them did. The Chicachas[354] did not wish to enter into this agreement.

It would be too long & boring to give the details of all of the raids which were made.[355] Around this same month, we had three Cana-

---

352 Literally "Flat Heads", a name applied by the French to a number of southern tribes, particularly the Siouian Catawbas and the Muskogean peoples such as the Chickasaws. "Glossary of Indian Tribal Names", *DCB,* III, p. xxxiv; Hodges, I, pp. 213-16, 260-62.

Elsewhere in his *Mémoires,* Pouchot further identifies the Flat Heads as Cherokees and Catawbas (note 869).

353 The Cherokees of the Carolinas. Hodges, I, pp. 245-49.

354 The French name for the Chickasaws of northern Mississippi. "Glossary of Indian Tribal Names", *DCB,* III, p. xxxiii.

355 Casgrain, *Collection,* XI includes, on pp. 87-116, a "Journal de Niagara" recording events at the post, primarily Indian affairs, between June 6 and August 3, 1757. Although it is not signed, it was almost certainly kept by Pouchot and includes several of the incidents related here. A translation of this journal may be found in Appendix D of this volume.

dian officers captured & killed because their war party, of about 100 Indians, had abandoned them. M. Bellestre, junior, who was captured by the Cherakis, became a highly-esteemed chief among them & subsequently persuaded them to serve the French.[356] [97]

Let me recount at this point a singular example of presentiment. The mother of a Missisake war chief called Techicabavoui called on M. Pouchot on June 22 & said: "Father, I've come to tell you that my son has done well in battle." He asked her if a member of her son's party had returned. She said "no," but that ten days before she had felt a shuddering feeling in her breasts and that since that time she had felt nothing. In her view, this experience was unequivocal; she had always found it a sure sign. Sure enough, on July 5, Techicabavoui arrived back at Niagara with five scalps. One man in his party had been wounded.[357]

---

356 François-Louis Picoté de Belestre, second ensign in the Marine (1756) and son of François-Marie Picoté de Belestre, lieutenant (1751) in the Marine. *DCB,* IV, pp. 633-36; "Tableau ... des Officiers de la Marine", Brymner, 1886, p. clxxxiv.

Belestre, with a party of 40 Indians and 12 French, had left Fort Duquesne on May 17, 1757 for what proved to be an unsuccessful foray against Fort Cumberland. The Indians subsequently "separated themselves" to raid the more vulnerable frontier farms. The French, by this time numbering 10 men, were attacked while returning to Fort Duquesne. Two officers and three soldiers were killed. Belestre was captured and taken to Winchester, Virginia where he was questioned by English officials, including George Washington, on June 20. His statement is in "Papers Relating to the French Occupation in Western Pennsylvania, 1631-1764", *Pennsylvania Archives, Second Series, VI* (Harrisburg, 1876), pp. 562-63. French accounts are in Vaudreuil to Moras, July 12, 1757, *DRCHSNY,* X, p. 582 and "Journal de Niagara", Casgrain, *Collection,* XI, p. 99.

The French raiders had reached the headwaters of Turtle Creek, within twenty miles of Fort Duquesne, when they were surprised on June 5 by Lieutenant James Baker of the Virginia Regiment with 5 soldiers and 15 Cherokee warriors. One Cherokee, the Swallow, was killed by return fire from the French. See Baker to Washington, June 10, 1757 in Stanislaus M. Hamilton (ed.), *Letters to Washington* (5 vols.; Boston, 1898-1902), II, pp. 89-90 and Washington to Stanwix, June 15, 1757 in Worthington C. Ford (ed.), *The Writings of George Washington* (14 vols.; New York, 1889-93), I, pp. 447-49.

The two officers killed in the ambush were identified as second ensigns de La Saussaye and St. Ours. Both were on the Canadian establishment. Vaudreuil to Moras, July 12, 1757, *DRCHSNY,* X, p. 582. "Douville Lasaussaye" and "Chevl. St. Ours" were recorded as killed near the Ohio in 1757. "Liste des Officiers ... tués", Brymner, 1886, p. clxxv. Douville de La Saussaye was probably a son of Philippe Dagneau Douville de La Saussaye. *DCB,* IV, p. 192. Second Ensign St. Ours, judging by the use of the title "Chevalier" with his name, was probably a son of Captain Pierre-Roch de Saint-Ours Deschaillons. See Fauteux, *Les Chevaliers,* pp. 168-69. Neither appears in the "Tableau ... des Officiers de la Marine", Brymner, 1886, p. clxxxiv.

357 The return of this party is recorded in the "Journal de Niagara" where the leader's name is given as "Techicabaoui". Casgrain, *Collection,* XI, p. 101. Techicabavoui (fl.1757-1772) is probably the Tequakareigh, "a Chippaway chief" from the Toronto area, who came to

...(continued next page)...

On July 1, a great council was held at Niagara in which the Iroquois guaranteed to the Hurons, the Ouias, the Miamis & the Outaouais[358] with a beautiful wampum belt that they had taken up the hatchet on behalf of **[98]** their father & that they would never leave him again. These nations always distrust the Iroquois & do not like them. Each nation jealously upholds its superiority over the others and considers the Five Nations only as allies of the English. They have few links with the Iroquois and rarely frequent them. M. Pouchot gave them a joint gift of brandy, but in moderate quantities.

On the next day, the Ouias presented the same belt to the Iroquois covered with vermillion, which is an invitation to war. They recommended the Iroquois to obey the will of their father & not to listen to evil words. They added that they were very angry with themselves for listening to the Hurons who, just like the Iroquois, constantly made fine promises without ever doing anything. They invited them, **[99]** by means of the belt, to bring their father at Niagara some of the bad meat that he was so fond of. By this, they meant English prisoners.

They said to the Hurons: "You have a spirit in you; the Jesuits have given it to you. You pray and go to mass. We advise you to make good use of your spirit & to obey the will of our father[359] as we do, although we have no spirit."[360]

The Kicapous & the Mascoutens joined up with them to go to Mont-Réal.[361] The different nations left behind at Niagara their

---

meet with Sir William Johnson at Fort Niagara on July 30, 1759 and presented a speech on August 2. *SWJP*, XIII, pp. 117-19. He visited Johnson Hall on July 6, 1772 when his name was recorded as "Tes, capuech". *SWJP*, XII, p. 970.

358 The Hurons who visited Pouchot were an Iroquoian people residing near Detroit where they were also known as Wyandot. The Ouias were a subgroup of the Miami and had their villages along the upper Wabash River of Indiana. The related Miamis lived nearby in northern Indiana, southern Michigan and western Ohio. The Outaouais or Ottawa had villages near Detroit and on northern lakes Huron and Michigan. "Glossary of Indian Tribal Names", *DCB*, III, pp. xxxv-xlii and Tanner, pp. 40-41.

For details on these nations see Elisabeth Tooker, "Wyandot", in Trigger, pp. 398-406; Charles Callender, "Miami", *Ibid.*, pp. 681-89 (including information on the Ouia or Wea); Johanne E. Feest and Christian F. Feest, "Ottawa", *Ibid.*, pp. 772-86.

359 Note in original, p. 100 (a): *They thus signified the Governor of Canada or the King.*

360 Note in original, p. 100 (b): *By that they meant that they were not Christians.*

361 The Kickapoo and Mascouten were closely related peoples from the middle Wabash Valley of western Indiana. Tanner, pp. 40-41. See Charles Callender, Richard K. Pope and Susan M. Pope, "Kickapoo", in Trigger, pp. 656-67 and Ives Goddard, "Mascouten", *Ibid.*, pp. 668-72.

wives & children, 150 of them, **[100]** who remained until their return. However, hardly any trade goods & brandy arrived, which depressed the Indians. The resources of all the forts were exhausted by the great quantities of equipment that had to be issued to the warriors and by the presents that had to be given to the Indians who attended the council & often came back there in large numbers.

The regiments set out[362] around May 20 to go to Carillon where they worked on the preparations for the siege of Fort George. M. de Montcalm arrived there in June. The 4 French battalions, about 1600 men, the colonial troops, about 800, the Canadians, 900 strong, & 1900 Indians left Lachute in the early days of August to travel to Fort George.[363] M. le Chevalier de Levis, in command of a large **[101]** detachment of Canadians, Indians & *piquets* of troops marched along the right bank of the lake in order to be able to cover the disembarkation of the army, which was in boats.[364]

This officer arrived near Fort George without encountering obstacles other than the country itself, which was extremely difficult. M. de Montcalm had Indian canoes in the advance guard of his flotilla. On rounding a point, they observed an English detachment of about 150 men which was on patrol. The Indians, although in simple bark canoes, attacked the forward bateaux with hatchets & musket fire & plunged into the water to board them. The enemy was so taken aback by this boldness that 60 to 70 **[102]** of their men were either captured or drowned. The remainder, who were rather behind them, regained the beach of the fort as fast as possible without trying to help their comrades.[365]

---

362 From their winter quarters in different parts of New France.

363 Pouchot understates the size of the army assembled at Carillon for the attack on Fort William Henry. Bougainville's return of the whole force shows 2,561 men of all six *troupes de terre* battalions, 520 Marine troops, 188 artillerymen, 2,946 militia, 820 mission Indians, and 986 more from the western nations for a total of more than 8,000 men. Bougainville to Paulmy, Aug. 19, 1757, *DRCHSNY,* X, pp. 606-07. Non-effectives and troops left behind as garrisons somewhat reduced the number that departed the Lachute River, but a return taken at Fort William Henry on August 3 lists 7,626 officers and men, including 1,600 Indian warriors. *Ibid.,* p. 625.

364 The force led overland by Lévis numbered about 2,500 men. *DRCHSNY,* X, pp. 609-10. Lévis marched along a trail through the mountains on the west side of the lake.

365 On the morning of July 24, before Montcalm's army began its move up Lake George, French-led western Indians lying in ambush at Sabbath Day Point attacked 350 New Jersey

...(continued next page)...

The English had established a fortified camp right up against a height of land at the immediate rear of the fort. They had stationed 2000 men there & another 4 to 500 in the fort itself. From the evening of our arrival, we worked at setting up batteries to bombard the fort and the camp & we opened up a kind of communication trench which reached to about 150 *toises* from the fort. Our Indians & the Canadians spread out in the woods on the heights around the position in order to prevent an enemy retreat. Bombarded in their retrenchments & the little fort and seeing all communications intercepted, the enemy surrendered on August 10.[366] **[103]** Details of this siege have since been published.[367] We shall make mention of just one strange event to which their surrender gave rise.

The terms of the English surrender specified that their troops would not serve for 18 months against His Most Christian Majesty[368] & his allies and that they agreed to be sent back to New England.[369] The French were to escort them half way over the **[104]** Fort George

---

and New York provincials moving north along the lake in 22 boats. Led by Colonel John Parker of the New Jersey Regiment, they had planned to raid French outposts at Carillon. The Indians attacked in their canoes, and only Parker and about 100 of his panic-stricken soldiers escaped. Ian K. Steele, *Betrayals: Fort William Henry & the "Massacre"* (New York, 1990), pp. 87-88. Montcalm's report of the action and a partial list of English casualties are in *DRCHSNY,* X, pp. 591-93. Other French accounts are in *Ibid.,* p. 599 and Hamilton, *Adventure,* p. 142-43.

366 Note in original, p. 103 (a): *Although the fort only held out for six days of open trenches, Colonel Monro, in command there, nonetheless put up a courageous defense. General Webb, at the head of an army corps of 4000 men then in the vicinity, did not dare help him.*

George Monro or Monroe (d.1757), lieutenant colonel (1750) in the 35th Regiment of Foot, commanded Fort William Henry. See James A Holden and John A. Inglis, "The Lineage of Colonel George Monro", *Proceedings of the New York State Historical Association,* XIII, 1914, pp. 389-403. Monro survived the siege and the subsequent attack on the paroled garrison only to die less than three months later, on November 3. One source, quoted in Holden and Inglis, claims that he died of "apoplexy" on an Albany street.

Daniel Webb (c.1700-1773), was colonel (1755) of the 48th Regiment of Foot and brigadier in place of William Shirley. Webb was ordered home to England after his failure to support Monro's garrison at Fort William Henry. *DAB,* X, p. 573-74.

367 Contemporary accounts of the siege of Fort William Henry are in Mante, pp. 91-97 and Knox, I, pp. 67-70. Pouchot's coverage of the siege is the briefest in his narrative for a major event, probably, as stated, because of the wide publicity generated by the subsequent massacre.

Returns show 41 men of the garrison killed and 71 wounded during the siege with 13 French besiegers killed and another 40 wounded. *DRCHSNY,* X, p. 625.

368 King Louis XV. The original uses the abbreviation "S.M.T.C." for *Sa Majesté Très Chrétienne.*

portage.[370] As a consequence, they marched out of the fort in column, bearing their arms & baggage, accompanied by the escort detail. The Indians, attracted to them out of curiosity, although strongly warned by M. de Montcalm not to bother them, spread out in the woods of the gorge and followed them. As soon as their escort had left the English, some Indians tried to harry them, more to see how they would bear up than with any other end in view. They stole part of their baggage. Seeing the English troops hesitating over what to do & astounded by their cries, the Indians began to rob them. They were perhaps put up to it by their French interpreters who, annoyed at seeing the English withdrawing when they themselves had obtained no booty, as they had at the **[105]** Braddock incident, doubtless encouraged them to steal their baggage. In the end, they attacked them from all sides, fell upon their baggage & stripped them. Those who resisted were killed & others were led off as prisoners, 12 to 1500 in all. M. de Montcalm had almost all of them released, but they were entirely naked. The French officers and soldiers stripped off their own clothes so that they could cover themselves & they were sent away in greater security.

The situation of these troops was doubtless an extremely difficult one, as they were entitled to believe that the French would attack them if they fought with our Indians. It is certainly true that if they had shown greater firmness towards the first Indians who approached them with insults, they would surely have prevented a misfortune that they could not attribute to the French. Upon their arrival in **[106]** England, however, they were loud in their criticism of this infraction & would not adhere to the terms of the surrender. It is obvious that, without the concern shown by the French, not a single one of them would have returned home. Through their own experience, the English know that one cannot control that type of man. They behaved

---

369 Note in original, p. 104 (a): *Another essential term of this surrender, which was drawn up by M. de Bougainville, was a general exchange of prisoners captured on both sides in North America since the beginning of the war. The garrison of Fort George was included in this exchange.*

A detailed return of the garrison, taken on August 9, shows 2,331 men. These included regulars of the 35th and 60th Regiments of Foot, the Independent Companies and the Royal Artillery as well as American provincials from Massachusetts, New York, New Jersey, and New Hampshire. *DRCHSNY,* X, pp. 621-25.

The articles of capitulation of Fort William Henry may be found in *Ibid.,* pp. 617-18 and in Mante, pp. 93-94. Pouchot and his editor correctly represent the terms of the agreement.

370 The portage between Lake George and Fort Edward.

with the utmost bravery during the siege. Despite the cannon fire, they approached in their own way right up to beneath the fort and killed the sentries & anybody else who tried to communicate between the fort and the entrenched camp.[371]

Fort George was destroyed. All the ammunition and the armaments[372] were carried off to Carillon, where the army spent the **[107]** remainder of the campaign making constant raids along that frontier & working to perfect the state of the fort. The Indians set out to return to their own country. They carried with them a disease which killed many of them. Seeing freshly dug graves, a number of them exhumed the corpses to take their scalps. Unfortunately, some of them had died of smallpox. The germ immediately spread among the Indians. The

---

371 The Fort William Henry massacre has been treated in both history and fiction, and the attack on the paroled garrison provides Americans with one of their most familiar images of the French and Indian War. Montcalm, in his report to Paulmy of August 15, rather lamely admitted: "I cannot conceal from you that the capitulation has unfortunately suffered some infraction on the part of the Indians. But what would be an infraction in Europe cannot be so regarded in America." *DRCHSNY*, X, p. 598. The resulting outrage affected prisoner exchanges and colored English attitudes for the duration of the conflict. Fear of the Iroquois and of potential English vengeance for the Fort William Henry incident was very much on Pouchot's mind when he was forced to surrender Niagara in 1759 and Fort Lévis in 1760.

Gipson, like most other authors, is vague about the number killed, but estimates that about 400 prisoners were carried off by the Indians. He describes the campaign in VII, pp. 62-89. A recent and painstaking study of the affair is Steele, *Betrayals: Fort William Henry & the "Massacre"*. Steele carefully accounts for the captives and killed on pp. 133-44, places the event in the perspective of warfare of the time and analyzes subsequent historical study. He discusses Pouchot's account on p. 163. Several French versions of the campaign, the siege and the massacre are printed in *DRCHSNY*, X, pp. 597-651.

The rumor that French colonial officers initiated the massacre was also mentioned by Mante, pp. 95-96, who noted that "it has been said ... that a partizan who led the French savages gave the death halloo when the English marched out of the fort, to gratify the Indians in their lust for blood and plunder." Mante dismissed the rumors that this had been done on Montcalm's orders and described the French general's energetic efforts to intervene and save the prisoners. Bougainville, greatly distressed, confided in his journal: "Detestable position of which those who are not here can have no idea, and one which makes the victory painful to the conquerors." Hamilton, *Adventure*, p. 170.

372 Note in original, p. 107 (a): *There was a very abundant supply of both. We captured 36 thousand pounds of gunpowder, 23 cannon, 4 mortars, one howitzer, & seventeen* pierriers.

A return of captured stores, dated August 9, 1757, shows 23 cannon, three mortars (two burst), one howitzer, and seventeen swivel guns. The French carried 35,835 pounds of gunpowder to Carillon. *DRCHSNY*, X, p. 626.

373 The Potawatomi had villages at Detroit, St. Joseph's in southwestern Michigan and on the Chicago River of Illinois. "Glossary of Indian Tribal Names", *DCB*, III, p. xxix and Tanner, pp. 40-41. See also James A. Clifton, "Potawatomi", in Trigger, pp. 725-42.

Pouteotamis nation, one of the bravest & most faithful to the French cause, was almost wiped out by the epidemic.[373] Several chiefs greatly esteemed by the French were especially missed.

The two battalions of the Berri Regiment, engineers, artillery officers and recruits arrived at Québec in the month of July **[108]** with a dozen vessels loaded, so it was said, with provisions.[374]

The capture of Fort George filled the province of New York with consternation. It is certain that, had M. de Montcalm been able to keep the Indians, he could have devastated the entire region, including the capital, as it was entirely defenseless. However, in that sector, we were reduced to forming raiding parties of Canadians & our resident Indians.[375]

In August, M. de Vaudreuil sent some Abenakis to Niagara to become acquainted with the Loups of Theaogen, who are more or less of the same nation. They brought with them their missionary, a Jesuit.[376] In their council, the Abenakis presented a fine wampum belt

---

374 Berri (or Berry) was one of those regiments with three battalions. The second and third, each of nine 60-man companies, arrived at Québec in late July 1757. Chartrand, *The French Soldier,* p. 32; Hamilton, *Adventure,* pp. 138-39. A number of army artillery officers and a few enlisted men were also sent to supplement the colonial *canoniers-bombardiers.* Among them was Lieutenant Bonafous (note 512) who would later direct Pouchot's artillery in the defense of Fort Niagara.

375 Pouchot's mild criticism of Montcalm's inability to retain his Indian allies and follow up his victory by attacking Fort Edward echoes that of Governor Vaudreuil and other Canadian officers. See "Ministerial Minute on the neglect of the Marquis de Montcalm to attack Fort Edward" with supporting correspondence in *DRCHSNY,* X, pp. 659-66. Steele, p. 163, believes that Great Lakes warriors returning home by way of Niagara influenced Pouchot's opinion that their continued presence would have made Montcalm's army more successful.

376 These Abenakis were from the mission settlement of St. François-de-Sales or Odanak (near Pierreville, Québec) later destroyed in the autumn of 1759 by Robert Rogers. The mission's pastor was Father Pierre-Joseph-Antoine Roubaud (1724-after 1789). *DCB,* IV, pp. 685-87. He reported, on October 21, 1757, that he had left his post on July 12 to travel to Montréal and that "the purpose of my journey was simply to bring to Monsieur the Marquis de Vaudreuil a deputation of twenty Abnakis appointed to accompany Father Virot, who has gone to try to found a new mission among the Loups of *Oyo* [Ohio], or the beautiful river." R.G. Thwaites (ed.), *The Jesuit Relations and Allied Documents* (73 vols.; Cleveland, 1896-1901), LXX, p. 90.

Claude-Joseph (or Louis) Virot (1722-1759), was a Jesuit missionary assigned to the Abenaki. He served in the Ohio from 1758 until 1759 when he accompanied the relief expedition to Niagara and was killed in the French rout at La Belle Famille on July 24. Arthur Edward Jones, "Catalogue of Jesuit Missionaries to New France and Louisiana, 1611-1800", in Thwaites, LXXI, p. 178.

...(continued next page)...

to the Loups to convince them to listen to **[109]** the priest & accept him among them. The Loups replied that they were delighted that the Master of Life had granted them this opportunity to meet & to become associated with their kinsmen, whose words they listened to with pleasure. They said that they would defy the English, whatever advantage they were offered. Finally, they added that they would bear the belt to their nation, would invite everyone to listen to the missionary & would return in the spring to give their reply to the commandant. The Jesuit made a speech about the excellence of the Christian religion. One of them said that he had been baptized and that he was well aware that, in order to enjoy fulfillment in life, people should know about the birth of a child who had preached during his lifetime and that, at the age of 30, he had been put to death by having his hands **[110]** & feet pierced. It was he who taught about eternal life, which could not be obtained except through him. With regard to the Trinity, he distinguished God the Father as a great chieftain, comparable to a king, God the Son as a captain & God the Holy Spirit as the Church, or prayer. These three persons had created men as they were seen on earth, some red, some black, & some white. God had destined some of them for prayer, others for hunting and yet others for war but after that he allowed them to act according to their own free will without interfering in the world. These ideas are brought to the attention of readers to enable them to judge how the majority of Indians conceive of the great and sublime truths of our religion. The Indians in question seemed to have been indoctrinated by some English minister. **[111]**

Bougainville had learned of the proposed mission in February 1757 and attributed an ulterior motive to the Jesuits: "The pretext of these zealous missionaries was to get them [the Abenaki] away from commerce with the French and their brandy, and their object to profit from the lands which the Indians had cleared .... The Abnakis do not wish at all to consent to this change of abode." He described the pressure put on the Indians by the Jesuits but noted that the project had been abandoned, "at least for the present." In May, however, Captain de Lignery, commanding at Fort Duquesne, requested a missionary and some domiciliated Christian Abenakis to settle among the Delaware. He explained that the"Delawares regard the Abenakis as their brothers, they speak the same language." Hamilton, *Adventure,* pp. 83-84, 105.

The visit to Niagara was apparently successful, for Vaudreuil gave permission in January 1758 for Virot and a dozen converted Abenakis to establish their mission, despite the grumbling of the Delaware and Shawnee that "instead of trade goods and soldiers they are sent a black robe and some apostles." *Ibid.,* p. 196.

Pouchot and Bougainville associate the Abenaki with the Loups of Theaogen because they were related Algonkian peoples.

At that same period, there arrived at Niagara two hermits of some sort who had been captured by a party of Outaouais on the heights of Manenguelée.[377] The two men had been settled there for nearly 20 years with the consent of the nations of the Belle-Riviere. They told M. Pouchot that they followed the Roman rite in their prayers and that they had three monasteries containing 1500 monks in Pennsylvania.[378] Their founder was Frisham Cotre-Chiété. They added that the English had wanted to compel them to serve in the war and so the majority of them had been forced to disperse in the forest. They were three brothers whom the English had taken from their isolated community & kept [112] for a long period in prison in an attempt to

---

377 Pouchot probably refers to the high land between the Monongahela and Youghiogheny rivers. The prisoners had been taken on the Cheat River which flows from the south to join the Monongahela at the Pennsylvania line north of Morgantown, West Virginia.

378 Note in original, p. 112 (a): *They are the "Dumblers", a kind of lay brother. See what M. l'abbe Raynal says about them in the* Hist. pol. & phil., *vol. V, pp. 445* et seq.

The source cited is Abbé Guillaume-Thomas-François Raynal, *Histoire philosophique et politique des établissemens et du commerce des Européens dans les deux Indes* (6 vols.; Amsterdam, 1770).

Raynal and Pouchot's editor were referring to "Dunkers" or "Dunkards", names corrupted from the German *tunker* as in "dipper", "immerser" or "baptist". The name was popularly applied to members of the Church of the Brethren, a German religious sect, in reference to their practice of adult baptism. "Dunkard, Dunker", *Mennonite Encyclopedia* (Scottdale, PA, 1959), II, p. 109. The church had been founded in Schwarzenau, Germany in 1708 and was established in Pennsylvania by about 1718. Among their other tenets, the Dunkers rejected the use of violence. Rufus D. Bowman, *The Church of the Brethren and War, 1708-1941* (Elgin, IL, 1944), pp. 23-26.

The captives taken to Niagara were no longer technically members of the Church of the Brethren, however. In 1732 one Johann Conrad Beissel (1690-1768), who had been baptised into the church in 1728, split from the Dunkards and founded the Seventh-day Baptists. He established his group at Ephrata, Pennsylvania where the "cloister" (Pouchot's "monasteries"), including an impressive wooden community hall and other buildings, is preserved as a Pennsylvania State Historic Site. Beissel took the spiritual name Friedsam Gottrecht - Pouchot's "Frisham Cotre-Chieté". See Redmond Conyngham, "An Account of the Settlement of the Dunkers at Ephrata, in Lancaster County, Pennsylvania ... to Which is Added a Short History of That Religious Society by the Late Rev. Christian Endress of Lancaster", *Memoirs of the Historical Society of Pennsylvania, Vol. II* (Philadelphia, 1827), pp. 135-53; Bowman, p. 70; "Ephrata Cloister", *Mennonite Encyclopedia,* II, pp. 229-30. See also Walter C Klein, *Johann Conrad Beissel, Mystic & Martinet, 1690-1768* (Philadelphia, 1942).

The introduction to Conyngham's article describes the "long beards, capuchin habilments, and meagre visages" of Dunkers (actually Ephrata cloisterites) who had been seen occasionally in Philadelphia in the late eighteenth century. Conyngham, p. 135. "J.C.B", Pouchot's secretary in 1760, observed male and female "Dumplers" in New York wearing "long white robes and hoods." Stevens, *Travels in New France,* pp. 127-28. This apparel undoubtedly provided their "monkish" impression.

find out if they had any connections with the French & their Indians. Having discovered nothing, they had released them, with the exception of their third brother, who was still held prisoner at Williamsburg. The men seemed very genuine & truthful and were venerated by the Indians of those regions.[379]

The capture of Fort George instilled a sense of security in the other posts of Canada. In September, M. Pouchot had informed M. de Vaudreuil that the fort of Niagara & its buildings were completed and

[379] Pouchot's prisoners were yet one more step removed from the original Dunkers. During the 1730s four brothers who had come from Germany in 1725, Samuel, Israel, Emmanuel, and Gabriel Eckerlin (or Eckerle), gained influence in Beissel's Ephrata group. The resulting tensions caused Samuel, Israel and Gabriel to leave the cloister in 1745 to found a wilderness community on the New River near Blacksburg, Virginia. The Eckerlins briefly returned to Ephrata in 1750, but, finding that relations with Beissel had not improved, they moved west again. At first settling on the Monongahela near modern Morgantown, West Virginia, the danger presented by impending conflict in the West caused the local Delaware to move them to a more isolated spot on the Cheat River near Kingwood, West Virginia about 1752. There the three brothers were granted 5,000 acres by the Virginia government.

The outbreak of war and the defeat of Braddock in 1755 caused a change in frontier attitudes towards the Eckerlins. Suspicions arose that they spied for the French and Indians. Israel and Samuel were both detained for a time in 1756. One of the brothers, probably Samuel, was again arrested at Winchester in 1757. Sometime that summer, Colonel George Washington dispatched a force of 70 men to apprehend the other brothers and their followers. With the captured Eckerlin as a guide, the party arrived on the Cheat River only to find the settlement burned. It was later reported that the attack had been made by seven Ottawas led by a Frenchman. Emmert F. Bittinger, *Allegheny Passage: Churches and Families, West Marva District, Church of the Brethren, 1752-1990* (Camden, ME, 1990), pp. 19-22. Gabriel and Israel, who reportedly "with mad tranquility, remained seated at his writing table until the enemy captured him and tied his hands behind his back," were carried to Fort Duquesne and sold to the French. Klein, p. 163.

The Eckerlins arrived at Niagara after August 2, 1757 when the last entry was made in the "Journal de Niagara" in Casgrain, *Collection,* XI, pp. 87-116. Pouchot promptly sent them to Montréal where they arrived between September 1 and 10. Bougainville described them as "three hermits of some kind" while Montcalm's journal identifies them as Anabaptists. Hamilton, *Adventure,* pp. 180-81; Casgrain, *Collection,* VII, p. 301. The third prisoner was probably the Eckerlins' servant, John Schilling, who had been captured with them. Bittinger, pp. 21-22. Bougainville was not far from the truth when he speculated that the "hermits" were "probably some Anabaptists chased out of Alsace, where they had taken refuge." The Eckerlin family had, in fact, been expelled from Strasbourg in 1705. *Ibid.,* p. 19.

Gabriel, Israel and Schilling were held for awhile in the Jesuit College at Montréal and then moved to Québec. There they almost starved until permitted to solicit alms in the street. Some time later, probably in the spring of 1758, they went to France. The Eckerlins contracted an illness during the voyage and died in the hospital at La Rochelle. Schilling found his way to Strasbourg in 1761 where he found Samuel and told him of the fate of his two brothers. Klein, p. 163; Bittinger, p. 22.

its covered ways palisaded.[380] Since this fort was a very considerable one, because of its position and the large number of Indians who had dealings there & came from all parts to trade & form war parties, it was soon coveted by all the officers **[113]** of the colony. Moreover, they were extremely jealous that a Frenchman was a commander at a posting where they thought they might be able to do good business. M. de Vaudreuil was unable to resist their pressure &, against the advice of M. de Montcalm who understood perfectly well the importance of the fort, he had M. Pouchot relieved in October by M. de Vassan, one of the first & best qualified captains of the colony.[381] M. Pouchot, who had concentrated at Niagara on those functions which best suited the good of the service & who had never thought of deriving any kind of profit from his posting, left satisfied with his conduct there & taking the French *piquets* who had accompanied him.[382] The Indian nations were very unhappy to see him leave, as

---

380 Pouchot also ensured that the Minister of War was aware of his accomplishments. In a letter of November 3, 1757, he reported the completion of Fort Niagara to Antoine-René de Voyer d'Argenson, Marquis de Paulmy, (1722-1787; *Biographie universelle,* XLIV, pp. 148-49). Pouchot boasted of having done the work "with all the economy of a gentleman cultivating his farm," by which he had saved the King an estimated 50,000 *écus.* He also mentioned having sent a plan of Fort Niagara to the minister on October 18, 1756. Since that time Pouchot had rearranged some buildings and added two large barracks, a church, a powder magazine, and a storehouse for provisions and merchandise. *DRCHSNY,* X, pp. 667-68. The letter is printed in Appendix A of this volume.

The 1756 map was almost certainly that entitled "Plan de Niagara et des Fortifications faites en 1755 et 1756". The original is in the French Archives with a copy in the National Archives of Canada (NMC-26647). For further details of the work accomplished, see Dunnigan and Scott, pp. 24-27.

Captain François Le Mercier reported that the new works were sodded and that the place was "reported good." Fort Niagara had also been properly armed with 30 guns, a dozen of which were 12-pounders. Memoir, Oct. 30, 1757, *DRCHSNY,* X, p. 656.

381 Jean-François de Vassan or Jean-Baptiste Mutigny de Vassan (fl.1727-1763), a captain (1748) who, unlike most of his fellow officers of the Marine, had been born in France. He came to the colony in 1727. Following his relief from command of Niagara in 1759, Vassan fought at the Battle of Ste. Foy where he was wounded. Later implicated in the *affaire du Canada* for alleged irregularities at Beauséjour and Niagara, he was admonished in 1763 and later retired to Blois in France. Roy, "Les Commandants du Fort Niagara", pp. 195-99. Roy and Fauteux, *Les Chevaliers,* p. 154, both render his name as Jean-François. Jean-Baptiste Mutigny is the form used in the *DCB.*

382 Montcalm reported that there remained "only one solitary Commandant belonging to the Regulars on the frontier this winter," Captain d'Hébécourt at Carillon where the garrison was chiefly men of the *troupes de terre.* Montcalm to Paulmy, Nov. 4, 1757, *DRCHSNY,* X, p. 670. This would appear to indicate a return to the policy of the winter of 1755-1756 for appointing commandants. (see note 280).

they held him [114] in very high regard because of the excellent treatment they had received from him. They were not accustomed to having chiefs as impartial as he.

Two Iroquois warriors, who had arrived back from a raid, swam out to the ship to see M. Pouchot, although the ship had just embarked & was riding at anchor while waiting for a favorable wind. They expressed to him, in strong and moving terms, their sadness over his departure. They wanted to give him belts to persuade him to stay. When he told them that his general had ordered him to leave, they replied: "Our father no longer loves us then, since he is taking from us a chief whom we all love." He informed them that the general needed him at Mont-Réal and that he had enjoined all the [115] chiefs to treat them as well as M. Pouchot had, that he would always grasp their hand in friendship. We had a great deal of difficulty convincing them to go ashore. They informed him that never again would they return to the fort.

M. de Vassan, through a kind of misplaced economy, increased their discontent. When several raiding parties that had been out on campaign returned, they were greeted coldly & were upset by the thrifty sort of presents he gave them. Among them, it is a great failing in a chief to be miserly. One must appear to be generous, otherwise they will despise you. M de Vassan spoke harshly of M. Pouchot's prodigality, though it will subsequently be seen that the latter's behavior was amply justified by events.[383]

The trade at the post of Niagara [116], which was all conducted on the King's account, was quadrupled during that campaign, either because of the Indians whose curiosity attracted them there & who spoke highly to their tribes about the excellent treatment they received, or because those who came to form raiding parties also brought goods to trade. M. Pouchot was mainly concerned to ensure that the storekeepers were fair in their dealings with the Indians & that he could be satisfied during his stay there with the probity of the

---

383 A number of sources indicate that the Six Nations were not only disappointed at the departure of Pouchot, but unhappy with Vassan as commandant. By the spring of 1758, Montcalm claimed, Iroquois visits to Niagara had declined noticeably. Casgrain, *Collection,* VII, p. 350. He reported that even Vaudreuil admitted there was a problem and "has frequently told me that he wished him [Pouchot] still there." Montcalm to Paulmy, April 10, 1758, *DRCHSNY,* X, p. 693. In April 1758, Bougainville lamented: "Why as the result of an intrigue of the junior officers' mess have they relieved M. de Pouchot, whom the Indians adore, and replaced him with a man whose Spanish arrogance has no sympathy for their character and who makes them wait in his antechamber and then has them announced by his servant?" Hamilton, *Adventure,* p. 201.

employees in question. He provided M. de Vaudreuil with an exact account of the presents he had been obliged to give to the Indians on behalf of the service and of the situation of the stores so that his conduct of them could be rectified if it was not in conformity with the general's intentions. The latter, however, constantly approved of his management. M. Pouchot prohibited any dealings **[117]** between the French or the Canadians and the storekeepers, whether of purchases or resale, & forbade the acceptance of any foreign goods on the King's account. The stores received only what the intendant sent on the ships. This trade in foreign goods was the source of all the pilfering committed at the other forts. As a result, during M. Pouchot's stay at Niagara, no one made the kind of fortune so fashionable at the time, a fact which greatly displeased the Canadians under his command.

We said above that the ship was riding at anchor until the wind strengthened. It veered to the N.W. &, at 4 in the afternoon, we sailed out of the river. The wind freshened constantly and we were obliged to use the foresail with all the reefs taken in. From time to time we experienced violent **[118]** squalls with lightning.

Towards 11 in the evening, as one of these squalls approached, there appeared electric fires at the tops of the two masts & at the ends of all the yards. These fires were from 8 to 10 inches long & appeared 3 to 4 inches across and were very bright. The sailors call them "St. Elmo's fire." They greatly alarmed our crew. They first had the hatchways and the pump hoses closed, as well as any other places where they imagined that the fire might penetrate. M Pouchot, amused by all this, nonetheless reassured the passengers & when one squall succeeded another, he predicted there would be further fires, which happened twice more. They disappeared as soon as the cloud moved away from the vessel's area of attraction. By contrast, **[119]** they flashed suddenly as soon as the clouds approached. On each occasion the phenomenon lasted for about half an hour. At daybreak, the wind freshened so much that we were obliged to scud under bare poles before the wind. We reached Frontenac from Niagara in 21 hours. That is one of the shortest crossings that can be made. We travelled around 90 leagues in that time. M. Pouchot & the officers of his detachment took only two days to get to Mont-Réal, where they were very warmly welcomed by the two generals.[384]

---

384 The party arrived at Montréal November 3, 1757. Pouchot to Paulmy, *DRCHSNY*, X, p. 667. Pouchot must therefore have departed Niagara on October 30 or 31.

On November 29, M. Bellestre, a colonial captain, returned from a raid on the Mohawk or Des Agniers River. He had burned 12 or 13 houses in a German village near the fort defending it & had brought back with him about sixty **[120]** prisoners, men, women & children.[385] In this incident, a colonial lieutenant was gravely wounded.[386] **[121]**

• 1758 •

At the beginning of January, an English detachment of 150 men came to Carillon and killed 16 of our cattle.[387] In that month, one of

385 François-Marie Picoté de Belestre (1716-1793), lieutenant (1751) in the Marine. He was promoted captain in January 1759. Belestre would be the last French commandant of Detroit and was the father of Ensign François-Louis Picoté de Belestre (note 356). *DCB*, IV, pp. 633-36. A summary of his expedition and November 12, 1757 attack on German Flats, protected by Fort Herkimer, is in *DRCHSNY*, X, pp. 672-74. There is a journal in Casgrain, *Collection*, XI, pp. 127-42. These give the number of prisoners as 150.

The "lieutenant", identified as one Lormier, was wounded in the right side by a musket ball. *DRCHSNY*, X, p. 673. He was probably Joseph-Antoine-Guillaume Lorimier de Ramilly (b.1733), ensign (1757) in the Marine. "Tableau ... des Officiers de la Marine", Brymner, 1886, p. clxxxiv. See also "A General State of the Canadian Noblesse ... 1767", Douglas Brymner (ed.), *Report on Canadian Archives, 1888* (Ottawa, 1889), p. 44. Bougainville notes "Sieur Lorimier Jr." at La Présentation in April 1758. Hamilton, *Adventure*, p. 201. He was a son of Claude-Nicolas-Guillaume de Lorimier de La Rivière (1705-1770), captain (1749) in the Marine and, at the time, commandant of the military garrison of La Présentation. *DCB*, III, pp. 408-09.

386 Note in original, p. 121 (a): *Before finishing his account of this campaign, M. Pouchot ought to have spoken of the aborted attempts of the English against Louisbourg, which was the work of M. de Machault, Minister of Marine. Since several squadrons which appeared to have different destinations gathered at the port of Louisbourg, Admiral Holbourne could not attack the place. On September 25, his fleet was buffeted by a southerly squall which dismasted 11 of his ships & caused the* Tilbury *to founder on the rocks. The French squadron, under the command of M. Dubois de la Mothe, would have taken advantage of this disaster had it not been for the disease which was beginning to break out among his crews. It was more fatal to our navy than the loss of a battle. The ships only arrived back at Brest through a kind of miracle. They were so infected that convicts who volunteered to clean them out were offered their freedom.*

Jean-Baptiste de Machault d'Arnouville (1701-1794) served as Minister of the Marine from July 28, 1754 to February 1, 1757. *Biographie universelle*, XXV, pp. 624-26.

Francis Holbourne or Holburne (1704-1771), Vice Admiral of the Blue (1757). *DNB*, IX, pp. 1006-07. Only nine of his 20 ships weathered the storm of September 24 without serious damage. The ship of the line *Tilbury* (60 guns) was wrecked on Cape Breton Island. Many others had to be sent back to England for repair. Gipson, VII, p. 180.

This show of force by Emmanuel-Auguste de Cahideuc, Comte Dubois de La Motte (note 182) frustrated English hopes of an attack on Louisbourg in 1757, but typhus broke out among the sailors of his fleet, and he returned to Brest with 5,000 sick. *DCB*, III, p. 93. See Gipson, VII, pp. 103-17 and 186 for an overview of the French and English naval maneuvering off Louisbourg.

our cadets was killed & another wounded in a raid on Fort Cumberland in Virginia. Mouet, a colonial cadet,[388] whom M. Pouchot had sent out on a raid with some Indians, returned from Pennsylvania with 14 scalps or prisoners.

In February, M. de Langis, a colonial officer, came back to Carillon from a raid in the area of Fort Edward or Lydius, bringing with him 3 prisoners & 23 scalps.[389] The prisoners stated that 20 regiments had arrived from England & that great plans were afoot for the **[122]** forthcoming campaign (these troops were among those who had surrendered at Closter-Seven).[390] They added that their general, Loudon, had quarrelled with Shyrley & that they had both sent official complaints to England.[391]

---

387 The raid actually took place on December 24, 1757 and was led by Robert Rogers (note 394). John R. Cuneo, *Robert Rogers of the Rangers* (New York, 1959), pp. 66-67. He attached a note for the French commandant to the head of one of the oxen thanking him for the fresh meat and sending his respects to the Marquis de Montcalm. *DRCHSNY,* X, p. 837. Doreil rather stuffily regarded this as "an ill-timed and very low piece of braggadocio." *Ibid.,* p. 703.

388 François Mouet de Louvigny (1732-after 1788), ensign (1757) in the Marine. Fauteux, *Les Chevaliers,* pp. 236-37.

389 Jean-Baptiste Levrault de Langis Montegron (1723-1760), ensign (1753) in the Marine. *DCB,* III, pp. 399-400; Tableau ... des Officiers de la Marine, Brymner, 1886, p. clxxxiv. His raid is noted in several French sources, including *DRCHSNY,* X, pp. 691, 697, 703. Bougainville, however, wrote that they "learned nothing." Hamilton, *Adventure,* p. 195.

390 The prisoners and Pouchot counted all regular English battalions in America as of January 1758. This information was probably derived from the same source used by Bougainville, "An Almanac Printed in New York" dated January 1, 1758. Hamilton, *Adventure,* pp. 250-51. In 1757 seven regiments (2nd battalion/1st, 17th, 27th, 28th, 43rd, 46th, and 55th) had been sent from England for the attempt on Louisbourg, two new highland battalions (which after several earlier designations became the 77th and 78th) were recruited for American service and the 80th was authorized to be raised in America. Pargellis, *Lord Loudon,* p. 109n. Combined with those in America before 1755 (40th, 45th, 47th), those sent or raised in 1755 (44th, 48th) and 1756 (22nd, 35th, 42nd, and 4 battalions of the 60th), but deducting the 50th and 51st surrendered at Oswego, there were 22 regular battalions in the colonies. The almanac list omits the 28th and 80th.

None of these regiments had served at Kloster-Seven in Hanover (Germany). On September 8, 1757 the Duke of Cumberland had been forced to capitulate his army of Hanoverians, Hessians and a few Prussians to the French Marshal d'Estrées. This allowed the French to occupy Hanover, Hesse and Brunswick. The capitulation was repudiated by King George II, and English troops were soon thereafter introduced into the war in Germany in support of the Prussians. Gipson, VII, pp. 118-21, 128-29.

391 The hostility between Loudon and Shirley is discussed in Gipson, VI, pp. 162-211 and in Pargellis, *Lord Loudon.*

On March 1, a raiding party of 200 of our resident Indians & 40 Canadians set out from Mont-Réal. The Indians, upon their arrival at the fort,[392] requested provisions from the commandant, M. d'Hebecourt, and said that they wanted to rest a little before continuing their march. He ordered them to be given food together with a little brandy. The Indians returned to their encampment, where they began to drink. One of them, who did not want to participate in the raid, went into a trance.[393] After some time, he called the others to a council, told them them that he had gone into a trance and that the English had **[123]** sent out a raiding party destined for Carillon which must be in the vicinity. He then exhorted his comrades to leave the next day, which was what they did. The commandant, rather agreeably surprised by this prompt decision which would rid him of them, granted them what they requested. A few officers & soldiers of the garrison desired to participate in the raid. They set out in the direction of the lake. Three leagues further on, their scouts recognized numerous human footprints on the lake ice and returned to inform the main party of the fact. They immediately decided to go back into the woods through which the English had to pass. Our scouts, seeing the English company marching down a small hill, ran to warn their comrades who were close by. They arrived at the top of a small rise at the very moment when the English were at the **[124]** bottom of the hill. They attacked & in an instant had killed 146 of them. Only about fifteen out of a company of 200 escaped. Robert Roger, in command of them, left behind his uniform, his commission & his orders the better to flee.[394] Eleven officers or volunteers had joined this detachment, 4 of whom

---

392 Carillon.

393 Malartic calls him "an old sorcerer." *DRCHSNY,* X, p. 837.

394 Robert Rogers (1731-1795), captain commanding ranger companies. *DCB,* IV, pp. 679-83. The standard biography is Cuneo, *Robert Rogers of the Rangers.* Cuneo, p. 78, notes that Rogers threw off his uniform jacket which was picked up by the Indians. In a pocket was his original commission from General Shirley. Other French accounts of the action are in *DRCHSNY,* X, pp. 693, 697, 703, 838.

395 On March 20 Captain-Lieutenant Henry Pringle (d.1800) and Lieutenant Boyle Roche (both from the 27th Regiment of Foot) gave themselves up at Carillon. Burt Garfield Loescher, *The History of Rogers Rangers* (2 vols.; San Francisco, 1946 and San Mateo, 1969), I, p. 259. Brief biographies of both officers are in *DRCHSNY,* X, p. 714n. Roche (1743-1807) was the scion of a prominent Irish family and later active in politics as Sir Boyle Roche. *DNB,* XVII, pp. 66-68. Both were included in the 1759 exchange for Pouchot and other French officers.

belonged to the regiments recently arrived from England. Five of them went to Carillon to surrender, having got lost in the woods where they were starving.[395] In this skirmish, we had five Iroquois of the Falls killed and a Nepissin of the Lake[396] & three other Iroquois fatally wounded. This was another action in which the Indians displayed great vigor. They had to engage a detachment of elite volunteers, who were called "rangers." [397]

On April 1, the weather was fine. The sun had a circle almost **[125]** as large as the horizon. From time to time during the day, it was burning hot. About 9 in the evening a rainbow of boreal light formed from east to west. It was nearly 9 leagues in diameter, one of its ends touching the village of Lac des deux Montagnes, the other the village of Château-Gay. Nothing could have been more majestic. The sky was extremely clear in both directions and the rainbow was at least three times the size of normal ones. The light, which was very bright, intensified now and again, like a fire reanimated by the air. It was of a pale white color. The ends of the rainbow narrowed to a point, just like the wood of a bow. It dispersed in less than half an hour. The boreal light, which extended from south to north, shone more brightly in very large rays. It lasted for more than two hours.[398] **[126]**

---

396 "Iroquois of the Falls" were expatriate Mohawks from the fortified mission settlement of St. François Xavier or Caughnawaga (Kanawaké), founded in 1716 on the banks of the rapids opposite Lachine. Trigger, p. 471. "Nipissing of the Lake" were Algonquin and Nipissing people residing at the mission of Lac des Deux Montagnes (Oka, Québec) on the north shore of the lake at the mouth of the Ottawa River. The mission had been established in 1721. *Ibid.*, p. 780. See also R. Cole Harris (ed.), *Historical Atlas of Canada, Vol. I: From the Beginning to 1800* (Toronto, 1987), pl. 47.

397 Of the 181 rangers who fought the "Battle on Snowshoes" on March 13, 1758, 54 returned to Fort Edward . English accounts of the action are in Cuneo, pp. 74-79 and Loescher, I, pp. 245-60. Rogers' own version, in which he acknowledged 125 killed and missing, is in Robert Rogers, *Journals of Major Robert Rogers* (London, 1765; reprint, New York, 1961), pp. 58-74. Bougainville claimed that the Indians took 144 scalps and seven prisoners. He noted losses of only four Indians killed and 18 Iroquois and three French wounded. Hamilton, *Adventure*, pp. 198-99.

398 This rainbow stretched from the Nipissing village (note 396) at the mouth of the Ottawa River to Chateaugay, Québec at the mouth of the Chateaugay River on the south shore of Lac St. Louis.

Bougainville observed this natural phenomenon which he described as an "aurora borealis accompanied by a luminous band which crossed the sky from north-northeast to south-southwest. It ended on the south by a trail of fire which seemed to form a tail to it." Hamilton, *Adventure*, p. 200. Malartic also describes it. Malartic and Gaffarel, p. 166. Both officers were then in Montréal as was Pouchot, having wintered there following his relief from Niagara. The spectacle was an aurora borealis or the northern lights.

A party of Iroquois from La Présentation had decided to go on a raid to the Palatine village on the Mohawk River. Despite the fact that the majority of them soon gave up this foray, one of their chiefs wanted to pursue it and two other warriors joined him. They arrived together at nightfall beside the first house of the village, where there were 11 men on guard. They were gambling all unawares. Their arms were stacked along the wall inside the room, beside a window through which the Indians could see them. The chief, who had hidden right beside the window which was very low, suggested to one of the warriors that they should attack the men. They both refused, seeing that they were outnumbered. After further futile exhortations, the chief said to them: "When I set out, I committed my **[127]** body, and so, I'm going to attack. You can follow me if you wish." The man, who is about five feet, nine inches, young & robust, immediately jumped through the window, naked & painted black. He was holding his musket and uttering the customary shrieks. The English militiamen, who were very far from expecting this apparition, got up & fled in fright to a corner of the room. The Indian, yelling & moving around like a man possessed, saw the muskets, seized them and threw them out of the window together with his own, which he did not seem to care about. In the meantime, the two other Indians, seeing their comrade inside fighting the English with a knife in his hand (they always have one suspended round their neck), to prevent them from capturing him, came through the window shrieking. **[128]** The English, believing them to be more numerous than they were, humbly asked for quarter. Our three Indians tied them up & brought them back to Mont-Réal. This adventure would be unbelievable had not everyone seen the prisoners.[399]

In the early days of May, a detachment of 45 soldiers & one officer, which had set out in a bateau from Carillon to fetch some planks being

---

[399] Pouchot later (*Mémoires*, II, p. 177; note 798) identifies the leader of this party as Kouatageté. Ohquandageghte (fl.1757-1773) was an Onondaga and head chief of the village of Oswegatchie at the mission of La Présentation. *DCB*, IV, pp. 587-88.

Encouraged by two successful raids in February, 80 La Présentation Iroquois left for Schenectady about March 13. Doreil to Belle Isle, April 30, 1758, *DRCHSNY*, X, p. 703. The party struck between German Flats and Schenectady, and the former place was the setting for Kouatagete's exploit. Hamilton, *Adventure*, p. 197. The raiders returned to La Présentation early in May with three prisoners and 46 scalps. They brought their prisoners to Montréal, where Pouchot saw them, on the 15th. Their leader reported one Iroquois killed and an Onondaga wounded. Malartic Journal, *DRCHSNY*, X, p. 840; Malartic and Gaffarel, p. 170.

...(continued next page)...

sawn on the other side of the river, was fired on as soon as they set foot onshore by about forty Indians who killed or captured 17 men.[400] We learned at that period through a despatch from Niagara that the Indians called *Folles-Avoines* had killed 22 Frenchmen & looted the stores at the post of La Baye. It will soon be seen what restitution they made.[401]

Before beginning to recount **[129]** the operations of the campaign, we shall say something about the situation in Canada. All the raids which were being made in every sector had resulted in a huge consumption of provisions & other trade goods. The enemy captured 14 or 15 ships destined for the country, which had diminished supplies even further. In consequence, that winter wine cost 400 *livres* cash a cask and a *minot*[402] of corn one *louis* or six *livres* a bushel. The price of everything else rose in the same proportion.

Kouatageté's story was his version of an April 30 raid on German Flats by 80 Indians and four Frenchmen. Although most of the residents were warned and found shelter in Fort Herkimer, four families, two Indian traders named "Clock", and six teamsters were surprised in a house at 4:00 p.m. The "Waggoners" ran upstairs as the "Indians immediately rushed into the House, and killed and scalped all that were below." When the warriors attempted to ascend the stairs "they were knocked down by the Waggoners; they then fired up thro' the Loft, and soon were joined by more Indians, who fired many Shot quite thro' the House and proposed to set it on Fire." One teamster attempted to escape through a window and was killed while the other five "defended themselves with great Intrepidity, having killed one Indian" until a party of rangers forced the attackers to retreat. A total of 33 inhabitants were killed. The Indians later admitted six dead and nine wounded. Letter from Albany, May 13, *New York Mercury,* May 22, 1758. This account is also printed in E.B. O'Callaghan (ed.), *Documentary History of New York* (Hereafter cited as *DHNY*; 4 vols.; Albany, 1849-51), I, pp. 522-23.

The same incident was reported to Sir William Johnson by John Butler from Fort Herkimer on May 1. Butler heard the story from "Hendrick Clock" who had escaped when "about five a Clock in the afternoon a large party of Indians atackt the house on this Side the Fort." Clock (or Klock) believed himself to have been the only survivor and that the others were "all tacken" by a party of at least 90 to 100 Indians. *SWJP,* II, pp. 835-36.

400 Malartic records this attack on May 16. *DRCHSNY,* X, p. 840.

401 The Folles-Avoines ("wild rice people") or Menominee resided along the shores of Green Bay, Wisconsin. Fort La Baye was a small post situated where the Fox River enters Green Bay. Tanner, pp. 40-41. See Louise S. Spindler, "Menominee", in Trigger, pp. 708-24.

Bougainville recorded this incident between May 12 and 20 and noted that the Menominees had "besieged the fort at the bay for three days and destroyed a French family in Wisconsin." Hamilton, *Adventure,* p. 204. Malartic says they killed 11 Canadians, "missed the Commandant and pillaged a storehouse." *DRCHSNY,* X, p. 840. Pouchot describes their punishment in *Mémoires,* II, p. 9 (see note 523).

402 A measure of approximately three bushels.

The intendant was forced to obtain goods from private individuals in order to supply the posts. The Canadian officers in command there brought over from France on their own account the goods necessary for trade with the Indians, their equipment & the gifts destined for them. They exacted [130] a high price from the King. Their accounts were approved without difficulty because they were favored officers & they themselves certified the details of these transactions. It may well be imagined that they lost nothing on them. As a result they made large fortunes in two or three years. The French troops, who had none of these resources, suffered greatly from the increase in the price of everything. They received no supplies from the King's stores and were dependent upon their rations. As both officers and soldiers were obliged to purchase their supplies for the whole campaign out of their pay, they could enjoy in the field only what they were able to carry with them, whereas the Canadian officers had the resources of their domains. As a result, the campaign [131] was very costly for the French.

On May 19, eight ships arrived at Québec. Those of the chief supplier were due to travel in three convoys. The first vessels declared that the second convoy had been attacked by the English & that they believed many of them had been captured.

On June 5, prisoners taken by a Nepissin[403] war party said that the English army was beginning to assemble at Fort George. On the 7th, the Languedoc Regiment set out from headquarters for Carillon. On the 12th it was the Guienne Regiment, on the 15th Royal-Roussillon, the 17th La Sarre and the 20th Béarn. The La Reine & Berri Regiments had also left Québec to journey to the fort, as well as MM. de Montcalm and de Bourlamaque.

M. de Lévis had remained at Mont-Réal because, in accordance [132] with the campaign plan, he was due to set out with a detachment of 300 French troops & 900 Canadians, colonial soldiers & Indians, to enter New England by way of Corlac. The French *piquets* were from the La Sarre, Royal-Roussillon & Béarn regiments, each with 74 men, one captain & two lieutenants. M. Pouchot, as the senior captain, was in command.[404] On the 28th, a courier arrived from Québec with the news of the siege of Louisbourg & that five vessels of the line had anchored in the river[405] because they could not dock in the port. They were carrying supplies, ammunition & some recruits. All the officers

403 Note in original, p. 132 (a): *Resident Christian Indians.*

The Nipissings had taken nine prisoners who, Malartic reported, were shown to Vaudreuil at a council on June 4. *DRCHSNY*, X, p. 842.

had with them a large supply of trade goods. It was said that they [133] had sold some of them for a million.

On the 29th, an Indian war party returned to Carillon with 20 prisoners, including two officers. They stated that their army was 31 thousand strong and that its mission was to attack Carillon. Through intercepted letters, we learned that they were to feign an attack on the Ohio. The arrival of M. de Langis from Carillon partly confirmed this news.[406]

---

[404] In a letter to Belle-Isle of June 17, Lévis outlined the purpose and composition of his little army. It was to march from Lake Ontario down the Mohawk River towards Schenectady to induce the Iroquois to support the French, prevent the English from reestablishing a post at Oswego and create a diversion in support of Montcalm. The plan was Vaudreuil's, and the army was to number 3,000 men, 800 of whom were to be of the *troupes de terre* and Marine and the rest militia and Indians. *DRCHSNY,* X, p. 719.

Although the *troupes de terre* were to have provided a 64-man *piquet* from each battalion (with the two battalions of Berri forming one between them), La Reine had departed for Carillon before Lévis' detachment was organized. The actual *troupes de terre* component was thus six *piquets,* each of 67 soldiers, drawn from La Sarre, Royal Roussillon, Languedoc, Berri, Béarn, and Guienne for a total of 402 men. Each had a captain and two lieutenants except that of Guienne which had only one lieutenant. The *piquets* assembled at Montréal on June 22. Hamilton, *Adventure,* pp. 211, 214-15.

While Pouchot was the senior *piquet* captain, he was outranked by Etienne-Guillaume de Senezergues de La Rodde (1709-1759), lieutenant colonel (1747) of La Sarre, who volunteered to accompany the expedition. *DCB,* III, pp. 590-91. Pouchot probably had functional command of the regular detachment while Senezergues assisted Lévis as second-in-command.

Pouchot's experience at Niagara and Oswego and his demonstrated influence with the Six Nations are likely reasons for his assignment to the Corlac expedition. His lieutenants were Jacques de Pinsens (note 350) who had served with him at Niagara in 1756-57 and the Sieur de Totabel (d.1760), a lieutenant (1753) later killed at the Battle of Ste. Foy. Casgrain, *Collection,* VII, p. 365; Hamilton, *Adventure,* pp. 214-15; "Officiers du régiment de Béarn", p. 357.

The success of Vaudreuil's plan rested on two essential elements: secrecy and the consent of the Six Nations. To ensure the latter, two Iroquois from La Présentation, enjoined to exercise the utmost discretion, had been sent to negotiate with the Six Nations. One of them, identified as "Pierre", might have been Kouatageté (notes 399 and 798), fresh from his raid on German Flats. Although the two envoys kept their pledge of secrecy, and Vaudreuil only informed Montcalm of it when he needed the *piquets,* the La Présentation Iroquois reported that, in the Onondaga town, the "young men, the children who play ball games, even the English know that thou wishest to make a secret expedition in this direction." They maintained that the secret had been "badly guarded by others than ourselves." Hamilton, *Adventure,* p. 241.

[405] Note in original, p. 133 (a): *It was the squadron of M. de Chaffault.*

Captain Louis-Charles, Comte Du Chaffault had served as second captain of *Formidable* in Macnénara's 1755 squadron (note 181).

[406] Several items of intelligence, obtained in June, caused rapid changes in the French plan of campaign. By June 22 Montcalm had received word that the English would make another

...(continued next page)...

On the 30th, the *piquets* destined for Corlac were ordered to Carillon. They left the next day by bateau under the command of M. Pouchot. On the evening of the 4th, he arrived at St. Jean, 30 leagues by water from Montréal.[407] After giving some thought to the intercepted letter which M. de Vaudreuil had kindly sent him, he judged that Frontenac was threatened. He wrote to M. de Vaudreuil urging him **[134]** to have the boats that were there armed at once, as well as those that had been captured at Chouegen, and to order them to block the Chouegen river. Since the enemy could only have small bateaux, it would therefore be impossible for them to enter the lake. He could not get this plan implemented - and it would have undoubtedly saved the post - because Péan and his company had commandeered all the rigging for the vessels & had sold it for profit.[408] On that same

---

effort against Fort Duquesne. Doreil to Crémilles, July 28, 1758, *DRCHSNY,* X, p. 762. Information from prisoners taken by Langis on June 26 made it clear to Vaudreuil, as Montcalm had maintained, that Carillon was threatened. He therefore abandoned the Corlac expedition and ordered Lévis to rejoin Montcalm. See Vaudreuil's observations and his letter to the Duc de Noailles of Aug. 6, 1758, *Ibid.,* pp. 789, 809.

French estimates of the English force threatening Carillon were, like Pouchot's figure, greatly inflated. Vaudreuil thought it numbered 27,000, the size actually projected but not achieved. *Ibid.,* p. 809. On June 30, Bougainville, then just arrived at Carillon, believed that 20,000 to 25,000 British were poised at Lake George. Hamilton, *Adventure,* p. 221. The force assembled by Major General James Abercromby actually numbered 17,550, including provincial and Native American auxiliaries. See the order of battle in Appendix A of Donald E. Graves (ed.), *Fighting for Canada: Seven Battles, 1758-1945* (Toronto, 2000), pp. 367-69.

407 On June 30 Bougainville grumbled that the Schenectady expedition was the "Don Quixotry" of Vaudreuil who persisted in a "blind [belief] in the security" of Carillon. Hamilton, *Adventure,* p. 211. The news, received at Carillon on July 3, that "the American General" (Vaudreuil) had called off Lévis' feint, was most welcome to Montcalm's army. *Ibid.,* p. 223.St. Jean, Québec is on the Richelieu River. Pouchot's estimate of the water distance from Montréal is computed for descending the St. Lawrence from Montréal to Sorel and then moving up the Richelieu to St. Jean. This was presumably the route taken by Pouchot.

408 Captain John Bradstreet (note 451), who would lead the attack on Frontenac later in the summer, had submitted plans for an attempt on the place as early as 1756, and Loudon had approved the idea the following year. In March of 1758 he reported having raised 800 rangers for a strike at Frontenac. Bradstreet to Abercromby, March 24, 1758, Charles Henry Lincoln (ed.), "Manuscript Records of the French and Indian War in the Library of the Society", *Transactions and Collections of the American Antiquarian Society, Vol. XI* (Worcester, MA, 1909), p, 61. Bradstreet was building boats for this purpose in the spring of 1758, but the demands of the Lake Champlain campaign forced a postponement. Gipson, VII, pp. 236-38.

...(continued next page)...

evening, there arrived at St. Jean 240 Abenakis with an officer of the colonial troops. M. Pouchot invited them to accompany him the following day. They replied that they were waiting for the other Indians & for M. Rigaud, who was in command of the Canadians.[409] On the 6th, M. Pouchot **[135]** was about 3 leagues from Isle au Chapon[410] in Lake Champlain when he met a courier sent by M. de Montcalm to M. de Vaudreuil to tell him that the English army had disembarked at the portage. The occupants of a boat anchored off the island told M. Pouchot that they had heard heavy fire for nearly three hours. This news prompted him to stay there for no more than four hours, to allow his company to rest awhile. The troops had rowed 24 leagues in almost windless conditions. On the 7th, he set out again before daybreak. Near the Split Rock at St. Fréderic[411] he met a courier from M. de Montcalm who informed him the the enemy was at Lachute and that it was his duty to tell M. Pouchot **[136]** to make all haste. He asked if we had seen any of the enemy along the river and the commandant said none that he was aware of. M. Pouchot continued on his way & at 7:30 in the evening he arrived beneath

---

The intercepted letter passed on to Pouchot was probably part of the intelligence discussed above (note 406) and must have revealed the English boat-building activity. While it has not been possible to confirm Pouchot's warning to Vaudreuil, it seems plausible considering the captain's substantial experience and knowledge of the Lake Ontario frontier.

While Pouchot's contention that no action was taken to prepare the Lake Ontario squadron appears to be without foundation, the response was tardy and ineffective. Two of the larger vessels were operational that summer, and, on July 30, Intendant Bigot ordered the naval commander to arm a third. This had supposedly been accomplished by August 21, but the warships made no attempt to guard against an English raid from Oswego. Bigot to Lévis, Sept. 11, 1758, Casgrain, *Collection,* IX, p. 34. The small size of their crews, which Bougainville reported as only 10 to 12 men apiece, might have been a factor in this. Hamilton, *Adventure,* p. 275. The vessels were also much in demand for transporting stone from Frontenac to revet the earthworks of Fort Niagara. Severance, II, pp. 211, 239. See also note 514.

409 Note in original, p. 135 (a): *This account, although very detailed, becomes necessary to justify what the author will soon say about the late arrival of these troops.*

On the same day, Bougainville complained of the small number of Indians supporting the French army at Carillon. Hamilton, *Adventure,* pp. 221-22. Rigaud had originally been assigned to the Corlac expedition to command the Canadians and Indians. *DRCHSNY,* X, p. 719. Most of Rigaud's Canadians and Indians arrived at Carillon on July 12 and 13, too late for the battle (see note 446).

410 Schuyler Island, New York opposite Burlington, Vermont.

411 Split Rock, New York at the head of the narrows north of St. Frédéric.

Carillon.[412] Seeing some tents there, he thought the army had gone under canvas but he was informed that the encampment was on the heights 400 *toises* farther on & that the troops were busy retrenching.[413] This fact prompted him to climb up the hill immediately to inspect the works. He found a retrenchment well laid out for the shape of the terrain & a prodigious amount of work already done considering the time spent on it, which had not exceeded seven hours.

This retrenchment was an abatis of trees. The parapet was covered with interwoven branches **[137]**, which were disposed in such a way that their prickly ends were more difficult to cross than palisades & provided better cover from an assault. It concealed the retrenchments which the enemy could easily mistake for a simple abatis.[414] M. Pouchot was extremely satisfied with these arrangements. Upon meeting M. de Montcalm, who welcomed him as a man bringing him 300 élite troops (they had been especially chosen for the Corlac expedition), he was asked by the general how he liked the position. Pouchot replied: "General, since the enemy has not forced you off the

---

412 Bougainville was impressed by the diligence of the *piquets,* "advancing day and night despite contrary winds to join their comrades" who were about to be attacked. The *piquets,* he wrote, were "received by our little army with the same joy as were Caesar's legions by those Roman cohorts blockaded with Cicero by a multitude of Gauls." Hamilton, *Adventure,* p. 230.

413 The original text uses *retranchement* to describe the French defensive works at Carillon. The equivalent English word, "retrenchment", was a broad term describing "any work raised to cover a post, and fortify it against an enemy; such as fascines loaded with earth, gabions, barrels, &c. filled with earth, sand-bags, and generally all things that can cover the men, and stop the enemy: but it is more applicable to a ditch bordered with a parapet." Smith, *Dictionary,* p. 225.

Montcalm's advance lines were "made of tree-trunks laid one on the other, with overturned trees in front, the cut and pointed branches of which served the purpose of a chevaux de frise. "An Account of the Victory ... at Carillon", *Report on the Public Archives for 1929,* p. 104.

414 Pouchot's description is somewhat confusing. He refers to both the wall of horizontally laid logs and to a true "abatis" placed in the open ground before it. An abatis was made "by cutting down many entire trees, the branches of which are turned towards the enemy, and as much as possible entangled into one another. They are made either before redoubts, or other works, to render the attacks difficult." Smith, *Dictionary,* p. 1.

This obstruction was very effective. An English observer reported that "the ground in their front ... was covered with an Abbatis de Bois, laid so close and thick, that their works were really rendered impregnable." Knox, I, p. 191. John Bremner, a soldier of the 55th Regiment, observed that "the ground was clogged up by the Enemy with Logs and trees Intermixed with Brush which greatly Interupted the speedy and Regular march of our troops." Journal of John Bremner, New-York Historical Society, p. 4.

heights, they cannot reconnoitre your retrenchments." He seemed surprised & pleased with this observation.[415]

The enemy army, 22 to 25 thousand men strong, had been commanded since the departure of Lord **[138]** Loudon by Major General Abercromby, an elderly and very cautious officer.[416] In his army he had a young aristocrat, Lord Ho or Hau,[417] strong-willed and highly admired in the army in which he preeminently set the tone. In the month of April he had come with a detachment to reconnoitre our positions at Fort Carillon. He seemed entrusted with the direction of the entire plan of attack in this campaign. He had urged all the officers to wear the same uniforms as private soldiers for fear of what happened in the Braddock incident, in which we fired at the officers as a priority. He encouraged the whole army to cut off the brims **[139]** of their hats leaving only two finger-breadths of material. All the officers & soldiers were equipped with a kind of gaiter like those worn by the Indians & the Canadians, which were called *mituzzes.* Their haversacks were enveloped in a flap which they carried like the Indians & the Canadians. Each of them had 30 pounds of flour, one pound of gunpowder and four pounds of shot in addition to a full cartridge box. As a result, this army could last for a whole month without recourse to stores. Their canteens were full of rum. Both soldiers & officers mixed some flour with a little water. They kneaded it & cooked it into cakes on a stone or under embers. This arrangement was very good for a light expedition. The soldier has everything he needs & is not more

---

415 Montcalm and a number of officers, including engineer captains Nicolas Sarrebource Maladre de Pontleroy (1717-1802; *DCB,* V, pp. 738-40) and Jean-Nicolas Desandrouins (note 286), colonial artillery Lieutenant Louis-Thomas Jacau de Fiedmont (note 238) and Captain Louis-Philippe Le Dossu d'Hébécourt (note 275), commandant of Carillon, had selected the ground for these lines on July 1. Hamilton, *Adventure,* p. 222. Some work was begun on the 6th, but the troops did not begin to labor in earnest until the morning of the 7th when even "the officers, ax in hand, set the example" for the frantic construction. *Ibid.,* pp. 226-29.

Pouchot was probably accorded such a warm welcome because he and the *piquets* preceded Lévis and Senezergues. The two senior officers did not reach Carillon until 5:00 a.m. on the morning of July 8. Daine to Belle Isle, July 31, 1758, *DRCHSNY,* X, p. 815.

416 Major General James Abercromby (1706-1781) had succeeded Loudon late in 1757. *DCB,* IV, pp. 4-5.

417 Note in original, p. 139 (a): *The English accounts speak of Lord Howe & not of Lord Ho. M. Pouchot was very likely misled by the English pronunciation of this name.*

George Augustus Howe, third Viscount Howe (1724-1758) was colonel (1757) of the 55th Regiment of Foot and brigadier since December 1757. *DNB,* XXII, p. 875.

laden than normal. Both [140] officers & soldiers had only one shirt; doubtless they chose cotton ones of good quality. Lord Ho had set the example. He washed his dirty shirt himself, laid it out in the sun, and while waiting for it to dry, was perfectly satisfied with his waistcoat.[418]

This army, which expected to reach Mont-Réal without serious opposition, left Fort George on the 6th. About ten in the morning it appeared out in the lake. First we saw five barges then another 60, and this was the advanced guard of the army. In the end the lake appeared to be covered with boats, each of them holding 20 men or more. These troops were first observed by a detachment of our troops of 300 men, under the command of M. de Trépezec, a captain of the Béarn Regiment,[419] who was under orders to proceed as far as Mount

---

418 Howe's popularity and the radical modifications to uniforms and equipment he ordered in 1758 provide some of the most enduring anecdotes of the campaign. Details of uniforms and equipment are given in John Elting (ed.), *Military Uniforms in America: The Era of the American Revolution, 1755-1795* (San Rafael, CA, 1974), p. 4.

Pouchot's description of the practices inspired by Howe must have been based on English sources, either personal conversations or letters printed in the colonial newspapers. One frequently cited letter of June 12, 1758 stated, among many other details of the orders issued to Abercromby's army, that "You would laugh to see the droll figure we all cut; Regulars and Provincials are all ordered to cut the brims of their hats off." It also noted that, "No women to follow the camp, to wash our linen; Lord Howe ... has already showed an example, by going himself to the brook, and washing his own linen." *New Hampshire Gazette,* July 7, 1758. The same letter appeared in the *Boston Evening Post,* the *Boston News Letter* and probably other papers.

419 Le Sieur de Trépezec (d.1758), captain (1746) in the Béarn Regiment. He died of his wounds in the French camp at Carillon on July 6. "Officiers du régiment de Béarn", p. 357.

420 Mount Pélée or "Bald Mountain" is Rogers' Rock near Baldwin Landing, New York on the west side of Lake George.

421 The Camp of Contre-Coeur was located on the west bank of the outlet of Lake George north of Baldwin Landing.

422 Trout Brook, called by the French the Bernes or Bernetz River, joins the Lachute River (or Ticonderoga Creek) from the west between upper and lower La Chute Falls. Trépezec was probably trying to follow the valley of Trout Brook to avoid the enemy and was caught on the south or Lake George side of the stream. A manuscript "Sketch of the Country Round Tyconderoga" [1758] in the Germain Papers of the William L. Clements Library shows an "Indian Path" running from the lake near Tiroga Beach along the west side of Cooks Mountain into the valley created by Trout Brook, a logical route for Trépezec to avoid the English and find his way home. Another very similar sketch from the Thomas Gage Papers in the same library shows the movements of the forces of both sides and marks the Trépezec-Howe encounter as "D" just off the northeast corner of Bear Mountain and south of Trout Brook. Trépezec blamed his guide for losing the way. Journal of Operations, *DRCHSNY,* X, p. 722.

Pélée[420] and resist **[141]** with all his might any troops who might come across country, as we supposed they would. The enemy sailed up and down the lake until four in the afternoon looking for a landing place, which they managed by assaulting the camp of Contre-Coeur on the opposite side of the lake.[421] As a result of this maneuver, Trépezec's detachment found itself behind the enemy army. He was obliged to take to the woods in order to rejoin the army. He lost his bearings and after marching for a considerable time he ended up between the Bernes & Lachute rivers and positioned between the enemy army & its advanced guard, commanded by Lord Ho, which attacked him as the French were crossing dangerous rapids.[422] The French company was beaten. Only a score of men escaped **[142]** with a captain of the La Reine Regiment and Trépezec who was mortally wounded. The English took 110 soldiers & four officers prisoner. The remainder fell on the battlefield.[423] Lord Ho was killed in this skirmish. His death

---

423 Pouchot confuses Captain François-Joseph, Chevalier de Germain (note 274) of La Reine, who commanded a separate advance detachment, with the officer who escaped with Trépezec. Trépezec's companion was Ensign Jean-Baptiste Levrault de Langis Montegron (note 389) of the Marine who was also the errant guide. Journal of Occurrences, *DRCHSNY*, X, p. 845. The largest part of Trépezec's detachment, under Captain Bonneau, was cut off and fought until forced to surrender. See *Ibid.*, p. 722, 738, 747. An English account is on p. 735.

The English reported the capture of one captain, four lieutenants, two cadets, and 144 men. Abercromby to Montcalm, July 21, 1758, *DRCHSNY*, X, p. 774. The officers are listed as either killed or captured in the French casualty return. *DRCHSNY*, X, p. 800. They were: Louis de Preissac de Bonneau (1724-after 1789), captain (1754) in the Guienne Regiment. *DCB*, IV, p. 642; Pierre-François de La Rochelle (d.1760), lieutenant (1755) in the Guienne Regiment. "Les officiers du régiment de Guyenne", p. 193; the Sieur Bernard, lieutenant in the La Reine Regiment. "Officiers du régiment de la Reine", p. 385; Antoine de Jaubert (d.1760), lieutenant (1755) in the Béarn Regiment. "Officiers du régiment de Béarn", p. 354; and the Chevalier de Rezie (or Derésie), lieutenant (1757) in the Marine. "Tableau ... des Officiers de la Marine", Brymner, 1886, p. clxxxii.

The two cadets were from La Sarre. One was Jean, Chevalier de Prémillac (fl.1757-1760) noted in French reports as "appointed to a commission." He was an ensign as of November 1757. The other, Alexis Granet (fl.1758-1760), was said to be "designed for a commission" and received his promotion while in captivity. A few weeks after the battle at Carillon, Bougainville reported that Granet was known to be a prisoner or war and that "M. le Marquis de Montcalm has sent him money and will appoint him to a position in the regiment of La Sarre at the close of the campaign." Bougainville to Mme. la Marquise de Montcalm, July 25, 1758, *Report on the Public Archives for 1929*, p. 84. Granet was an ensign when wounded at Ste. Foy in April 1760. "Officiers du régiment de la Sarre", pp. 254, 258.

The defeat of Trépezec's detachment was the first significant English capture of French troops. Hamilton, *Adventure*, p. 193. The officers and cadets, with the exception of Lieutenant de Rezie, were all exchanged with Pouchot in 1759. List of French Prisoners, Dec. 14, 1759, Amherst Papers/8.

was an irreparable loss for the enemy since he alone knew the terrain they had yet to traverse before reaching Carillon.[424]

The regiments of La Reine, Guienne & Béarn, which were camped at the portage, remained in the presence of the enemy from the 6th to the 7th. As they were obliged to fight throughout this period, their sentries, made up of grenadiers and volunteers, engaged in skirmishes with the enemy & lost their camp with all the campaign supplies which were captured by the enemy when they were obliged to rejoin the rest of the army. This **[143]** misfortune would easily have been avoided if M. de Bourlamaque had allowed the camp to be struck while the regiments were containing the enemy. He refused to do so, saying that such a maneuver might discourage the troops.[425] This officer was so confident that he wanted to attack with these *piquets* and grenadiers an entire army corps of 18 thousand men, who were in front of him, although M. de Montcalm had ordered the camp to be struck.

These units withdrew on the 7th, when their advanced guard found themselves at close quarters with the enemy, who were pushing them back in very good order. The French army halted

---

424 English sources are unanimous in lamenting the death of Howe and its effect on the morale of Abercromby's army. "This noble and brave officer being universally beloved by both officers and soldiers of the army, his fall was not only most sincerely lamented but seemed to produce an almost general consternation and langour through the whole," says a note in Rogers, p. 82. Provincial Colonel Henry Babcock believed that "Never did an army gain more advantage in so little time, whilst the late Lord Howe was alive; but soon after that, we became a confused rabble." Letter from Babcock, July 10, 1758, Howard M. Chapin, *A List of Rhode Island Soldiers & Sailors in the Old French & Indian War, 1755-1762* (Providence, 1918), p. 13.

Despite the loss of Trépezec's detachment, the skirmish was helpful to the French defenders. "The disheartened English gave us twenty-four hours delay," wrote Bougainville, "and this precious time was the saving of us and of the colony." Hamilton, *Adventure,* p. 229.

425 On July 5 Bougainville noted that "Sieur de Bourlamaque has not thought it desirable to have his [camp] equipment sent away. He fears such action would have an appearance of timidity. However, all the troops, even the veterans who are preparing for battle, are getting rid of their equipment." Hamilton, *Adventure,* p. 226.

One wonders, from Pouchot's comments, if Bourlamaque's hesitancy to move his camp caused the loss of personal property or company stores for which the captain was responsible.

426 Bourlamaque withdrew his three battalions "in the best possible order" and rejoined Montcalm and his four battalions on the 6th. They crossed the Lachute River, destroyed the bridge near the falls and took up a position on the left (north) bank. Montcalm to Belle-Isle, July 20, 1758, *DRCHSNY,* X, p. 738. When Pouchot refers to the falls themselves, the form "La Chute" has been retained.

beneath La Chute.[426] M. de Bourlamaque wanted to stay there, while M. de Montcalm was rather undecided. All the officers were fearful of this position. We found ourselves in the bottom of a valley **[144]** surrounded by hillocks. The enemy who had beaten Trépezec could have arrived at any moment on those heights and cut off our retreat to Carillon. Finally, the Chevaliers de Bernés and de Montguy, of the Royal-Roussillon & Béarn Regiments respectively, both of them senior captains, impressed upon M. de Montcalm how dangerous the position was, as we could have been caught there without the possibility of putting up any serious resistance. He deferred to their view & the army withdrew to a position 400 *toises* from that place, at the far end of the height of land which from that point slopes down towards Carillon.[427]

On arriving there, the troops were ordered to construct the retrenchment, as has been described. The 8th was such a memorable day, in which 2992 men repelled more than 22000, that it deserves a detailed description. We shall first speak of the lie of the **[145]** ground, so that the reader can better understand the operation.

The ground which runs from La Chute to Carillon forms a high peninsula of which the top is fairly rounded & undulating and where there are at intervals a number of depressions & knolls. The road from La Chute to Carillon runs along this height. There are three to four torrents which flow down from the high ground into the Lachute River. The trail passes along these streams, which make the way subject to sharp bends for the best part of a quarter of a league. The slope of the ground towards the river in this area is quite sharp: on the left side, it is rather more gentle. At the lower part of the slope a few

---

427 Montcalm's army arrived at its new position on the heights at dusk on the 6th. *DRCHSNY,* X, p. 738. This camp is shown on several plans of the battle, including "The Attack of Ticonderoga" printed in Mante, p. 144 and "Attaques des Retranchemens Devant le fort Carillon en Amerique" drawn by Lieutenant Therbu and now in the collection of the National Archives of Canada (F950-1758). In the latter, the position is identified as the camp established on July 6 where the men remained under arms on the night of the 7th-8th. This plan is reproduced in Knox, I, p. 192.

Chevalier de Bernetz (fl.1755-1759) was commandant (1755), with the rank of lieutenant colonel, of the second battalion of the Royal Roussillon Regiment. "Officiers du régiment de Royal-Roussillon", *Le bulletin des recherches historiques,* LI, No. 12 (December 1945), p. 426. The Bernetz River had been named for this officer (note 1024). Jean-Joseph de Montgay (fl.1733-1760) was a captain (1743) in the Béarn Regiment. "Officiers du régiment de Béarn", p. 355. Neither officer is mentioned in official reports of the battle of July other than Montgay who was slightly wounded. *DRCHSNY,* X, p. 799.

hillocks are to be seen and then a flat area of about 200 *toises* which ends at the river.[428]

The left of our retrenchment went all the way down this **[146]** steep slope of the river. It also reached as far as the highest point of the terrain. Our right was on the heights which overlook the little plain 200 *toises* wide, where the trenches were barely begun.[429]

The colonial troops & the Canadians occupied this plain. The La Reine & Béarn regiments were on the heights of the right. The Guienne, Royal-Roussillon and Languedoc regiments, together with a battalion of the Berri Regiment were at the highest point and another battalion of Berri & the La Sarre Regiment occupied the slope on our left.[430] The retrenchment was about 980 feet in circumference.

---

428 Many of these terrain features are discernible in a plan of the march of the English army from John Almon's *Remembrancer* printed in Gipson, VII, p. 228. The last two sentences refer to the topography at the French lines before Carillon, and the directions given are as if approaching the lines from the west. The steep slope is above the mouth of the Lachute River while the "river" bordering the plain is Lake Champlain on the north side of the Ticonderoga peninsula. The narrow part of Lake Champlain south of Crown Point was called the St. Frédéric River by the French and Wood Creek by the English.

429 Here Pouchot gives directions from the defenders' perspective with the left of the lines ending above the Lachute River.

The frequently reproduced plan entitled "The Attack of Ticonderoga" from Mante, p. 144, provides a detailed view of the French lines at the time of the English attack. Another manuscript map in the National Archives of Canada, entitled "Plans des retranchements en avant du Fort de Carrillon attaqués le 8 juillet 1758" (D950-1758), also has details of the lines on the day of the battle. It has a particularly clear representation of the unfinished trenches on the "plain" at the French right.

430 Pouchot provides a straightforward description of the positions occupied by the French battalions during the battle of July 8. Only one battalion of Berri was in the main defense line, however. The other was left to garrison the fort, although its 50-man grenadier company served in the lines. Hamilton, *Adventure,* pp. 231-32. It was probably this company that Pouchot observed supporting La Sarre on the French left and identified as "another battalion" of Berri. French sources almost universally refer to the "1st" and "2nd" battalions of Berri, even though it was the second and third that had been sent to New France. The battalions were designated in the order of their seniority. It was thus the second that stood in the line and the third that held Fort Carillon. Summers and Chartrand, p. 39.

The plan of the battle in Mante, p. 144 and the French maps cited in notes 427 and 429 do not identify specific units, and some French accounts are imprecise about the exact locations of the battalions on the right. The version of Mante's "The Attack of Ticonderoga" reproduced in *DRCHSNY,* X, p. 726 identifies the individual battalions in the lines, but the unit symbols and labels appear to have been added to the plan in the nineteenth century. They are not on Mante's original. The *DRCHSNY* version errs in showing the first three French battalions, from right to left, as Guienne - Béarn - La Reine. Pouchot says the order

...(continued next page)...

On the night of the 7th, the enemy were busy opening up roads, retrenching the camp at the portage, the burned camp & that of La Chute. On the 8th, they moved forward to within 150 *toises* of our retrenchment, where they constructed a retrenched abatis and several other fortifications of the **[147]** same sort in order to facilitate their withdrawal.

M. de Montcalm was rather irresolute all that morning. He did not know whether to resist the enemy attack or to withdraw to St. Fréderic. It is certain that if we had been obliged to adopt the latter course, all his army would have been killed or captured, lacking a line of retreat. Moreover, he could not have got all his troops into the fort, nor could they have found any security there, since it was overlooked on all sides and surrounded by water. Around 8 o'clock, the general had even chosen two officers of the Béarn Regiment, La Parguiere & Tourville, the first to go and occupy Fort St. Fréderic and the other

---

was La Reine - Béarn - Guienne. His account is confirmed by the careful record of the French dispositions in the "Relation de Poulariès", *RAPQ,* 1931-32, pp. 81-82. Poulariès (fl.1755-1760) was then captain of grenadiers (1755) in the Royal Roussillon Regiment. "Officiers du régiment de Royal-Roussillon", p. 431.

Further confirmation that the order from the French right was La Reine - Béarn - Guienne comes from *aide-major* Malartic who noted that La Reine and Béarn (his regiment) were posted on a high spot on the right and were forced to construct epaulements to prevent being exposed to enemy fire. Late in the action, Lévis, the brigade commander, thought it necessary to support Guienne, which was being heavily attacked, by shifting La Reine to a position behind Guienne. Malartic and Gaffarel, pp. 183, 185.

Using Pouchot, Poulariès, Montcalm's report in *DRCHSNY,* X, p. 739, and Bougainville, the French dispositions (and unit strengths) from their right was: Marine troops and Canadians (400); Brigade of La Reine: La Reine (365), Béarn (410), Guienne (470); Brigade of Royal Roussillon: Royal Roussillon (480), 2/Berri (450); Brigade of La Sarre: Languedoc (426), La Sarre (460), grenadier company, 3/Berri (50). Doreil's account also agrees with Pouchot regarding the positions of La Reine and Guienne. Doreil to Moras, July 28, 1756, *DRCHSNY,* X, pp. 744-45. The right brigade was commanded by Lévis, the center by Montcalm and the left by Bourlamaque.

The strength given for each battalion includes the grenadier companies and the *piquets* recently returned to the army by Pouchot. They were not incorporated into their respective battalions, however, but placed to support the line, reflecting their élite composition. Two of the *piquets* were assigned to the reserve with the eight grenadier companies (including that of 3/Berri) and placed to support the left and center while the others went to the center with the Brigade of Royal Roussillon. Pouchot fought near the Royal Roussillon Regiment and was thus with the *piquets* in the center. Two companies of volunteers, Bernard's and Duprat's, guarded the area between the left of the retrenchment and the Lachute River. The whole force on the field, including a mere 15 Indian warriors, numbered 3,526. The garrison of the fort is not included in this total. Hamilton, *Adventure,* p. 231 and Doreil to Moras, *DRCHSNY,* X, pp. 744-45. Poulariès also places some of the *piquets* (he says five) in one of the salient angles near Royal Roussillon. *RAPQ,* 1931-32, p. 82.

to bake his supplies of bread.[431] When he was on the point of sending them off, he had a consultation with the other officers. He asked M. Pouchot if he thought that the enemy would attack in the course of the day. Pouchot replied: **[148]** "Sir, I believe they will. The enemy cannot know about the preparations you have made. They will imagine that they have only to exert pressure on the units which occupy the heights. They sense that if they dislodge you from them, they will be masters of everything. But your trench works are immune from assault, sir. Should they attempt it, you have high hopes of repelling them. If they don't do it today, they will no longer be able to in two or three days, because they will have to open up roads to bring in their artillery. In that case, your position would change & you would have time to make up your mind on the best course to take." This decided him to await the outcome of that day. The soldiers worked unremittingly at fortifying their retrenchment.

Around ten o'clock, there appeared **[149]** a platoon of troops on the mountain of the Serpent à Sonnette, who fired on our camp & the soldiers busy with the earthworks. They failed to inconvenience us. It was Johnson with some Indians who had arrived during that morning.[432] We had a flag brought into the retrenchment and the order was given that, if any serious alert occurred, it should be flown & a cannon fired, so that the troops would leave the works and return to take up their arms.

At one in the afternoon, the English attacked our main guard, made up of grenadiers & volunteers, who were stationed at the outer points of the cleared ground, which was not more than 40 or 50 *toises* wide and was situated beyond the retrenchment. They withdrew in very good order & held up the enemy for a long time. As soon as the troops, who were spread out doing different **[150]** jobs, heard all this sharp musket fire, they ran to take up arms without waiting for the

---

431 The Sieur de La Parquière (fl.1726-1760) was captain *en second* (1748). "Officiers du régiment de Béarn", p. 356. The Sieur de Tourville (fl.1747-1759) was also captain *en second* (1747). He was wounded at the Plains of Abraham on September 13, 1759. *Ibid.*, p. 357.

432 *Serpent à Sonnette,* literally, "Rattlesnake" Mountain, is Mount Defiance to the south of and commanding the Ticonderoga peninsula. Bougainville and other sources record the arrival of Sir William Johnson on July 8 with as many as 450 Indians of various nations. Hamilton, *Adventure,* p. 232.

433 Other sources vary slightly as to the time the attack commenced. Malartic says a little after noon. Malartic and Gaffarel, p. 183. Poulariès says between 12:30 and 1:00. *RAPQ,* 1931-32, p. 82.

signal, which was very opportune, as the front ranks of the enemy columns were beginning to converge on the cleared ground, which was strewn with felled trees.[433]

It would be impossible to praise highly enough the good will of the private soldiers who went to their positions of their own accord. Not one of them was dilatory. Four columns of the enemy emerged at around the same time. The one on the right[434] immediately attacked the La Sarre & Berri regiments and then the one in the center began to advance. The two on the enemy left marched very close to one another, one of them on the heights, the other along the side of the hill. The latter was relying on the fire of the higher column to be able to get close & force the abandonment of that part of the trench occupied by the Béarn & La Reine regiments, which the **[151]** higher column could see a little in reverse. They were closely massed in that place. It was the same on our left, where they could infiltrate under cover of a small screen of trees which came up to their shoulders.[435]

It can, however, be said that all parts of the retrenchment were attacked in succession with the utmost vigor. In between the columns, they had posted *piquets* whose fire greatly troubled the trenches. For four hours the attack was extremely heavy. The fire of our regiments was more sustained & brisk than might have been expected considering their numbers. If the abatis held up a little the advance of the enemy columns, they also contributed to saving them a lot of men, because they provided cover. There occurred in the heart of the fray

---

434 The English right.

435 The attacking English formations may be seen in the "Attaques des retranchemens" and "Plan des retranchements". A third French manuscript plan showing the lines and the attacking columns is "Plan du Fort de Carillon ... avec L'attaque des Retranchments ... Le 8e Juillet 1758" in the William L. Clements Library (4-L-9). Mante's "The Attack of Ticonderoga", p. 144, does not provide an accurate representation of the four-column assault.

The right column attacking La Sarre and Berri was composed of the 27th Regiment and the 1st and 4th battalions of the 60th. The 44th and 55th regiments formed the center. The 46th and the highlanders of the 42nd made up the left columns attacking Béarn and La Reine. Eyre to Napier, July 10, 1758, Pargellis, *Military Affairs,* pp. 418-22.

Secondary accounts of the battle for the retrenchment may be found in Gipson, VII, pp. 229-31, Stanley, pp. 180-81 and Sautai, *Montcalm,* pp. 68-83. Montcalm's detailed orders for the defense of the lines are printed in *Ibid.,* pp. 72-73.

More recent and very detailed studies are Ian McCulloch, "'Like Roaring Lions Breaking From Their Chains': The Battle of Ticonderoga, 8 July 1758" in Graves, *Fighting for Canada,* pp. 24-80 and René Chartrand, *Ticonderoga, 1758: Montcalm's Victory Against All Odds* (Oxford, 2000).

an event [152] which was almost unique and from which a lesson might be drawn on another occasion.

To amuse himself, M. de Bassignac, a captain of the Royal-Roussillon Regiment[436] attached his handkerchief, which was red, to the end of his musket & used it to signal to the enemy to advance. The head of the enemy column, which was in front of the Guienne Regiment, thought it was a flag of truce & that we wanted to surrender.[437] Very pleased to be rid of the position in which they found themselves, the enemy started to run towards the retrenchment, holding their muskets high across their chests with both hands & shouting "quarter." Our soldiers, who knew nothing about the handkerchief, thought that they were coming to surrender and climbing up onto the side of the retrenchment, [153] stood there with their arms to watch them arrive. This brought a brief stoppage in fire.

M. Pouchot, who was very close to this regiment and occupied an angle in the trench which the enemy were making great efforts to take, lacked powder & shot. He had asked for some from nearby positions. As he was greatly harassed by the enemy and feared that if his unit ran out of ammunition entirely his soldiers would be forced to abandon their position, he ran to M. de Fontbonne, the commander of the Guienne Regiment and asked him to hand over some of his.[438] He was thus there when this strange incident occurred. Surprised to see the

436 Jean d'Anglars de Bassignac (1725-after 1789), captain (1746) in the Royal Roussillon Regiment. [Aegidius Fauteux], "Officiers de Montcalm", *Revue d'Histoire de l'Amérique Française,* III, No. 3 (December 1949), pp. 371-73. He was wounded in the action of July 8. *DRCHSNY,* X, p. 751.

437 Here, and throughout his account of the red handkerchief incident, Pouchot clearly places the Guienne Regiment between Royal Roussillon and Béarn and thus third from the French right. The position of the *piquets* given by Poulariès fits nicely with Pouchot's account. It gave him a good view of the attack on Guienne as well as of Bassignac's actions in the position of the adjacent Royal Roussillon Regiment. *RAPQ,* 1931-32, pp. 81-82. As Bassignac was not captain of the Royal Roussillon *piquet* or grenadier company, he fought with the main body of his regiment. Hamilton, *Adventure,* p. 214.

438 Le Sieur de Fontbonne (fl.1723-1759), lieutenant colonel (1755) and commandant of the second battalion, Guienne Regiment. He was killed in the battle at Québec on September 13, 1759. "Les officiers de régiment de Guyenne", p. 192.

Montcalm's orders had stated both that "Ammunition will be served to each battalion" and that "It is forbidden to make any requisition by word of mouth." Sautai, *Montcalm,* p. 73. Pouchot's men must have been hard-pressed for him to leave his position to seek ammunition. Poulariès, however, provides an alternate or additional reason for Pouchot to have been near Fontbonne and his battalion. He reported that the *piquet de Béarn* was sent to support Guienne against the English attack. *RAPQ,* 1931-32, p. 82.

soldiers perched on top of the retrenchment, he noted at the same time the forward movement of the enemy. At the same moment, M. de Fontbonne was shouting to his soldiers: "Tell them to lay down their arms & their surrender will be accepted." M. Pouchot, who **[154]** judged from the enemy advance that they had quite different ideas in mind & that their sole purpose was to reach the retrenchment, shouted excitedly to the soldiers: "Fire, fire; Can't you see that these troops are going to overrun your position?" This idea suddenly brought the soldiers to their senses and, still standing, they fired a volley, shooting down two or three hundred of them. It is certain that this misunderstanding almost made us lose the entire advantage of the day. The English very rapidly retreated behind the abatis. The battle lasted another three and a half hours but we realized very well that their only aim was to hold out until nightfall. They have since reproached us for playing an unpardonable trick on them.[439]

---

439 If Captain de Bassignac was amusing himself, it was a cruel joke indeed. Throughout the war in North America the use of a red flag or handkerchief by the French signalled a desire to parley. Pouchot himself describes this on several occasions, and most sources mention the practice. Bougainville, for instance, carried a red flag when he took a message to the commandant of Fort William Henry in 1757. Hamilton, *Adventure,* p. 166. The corresponding signal for an English request to parley, as at Oswego in 1756 and Fort William Henry in 1757, was the display of a white flag. *Ibid.,* pp. 26, 169. It can only be supposed that these signals mimicked the national flags of the opponent.

French corroboration of the incident described by Pouchot comes from at least two other sources. Poulariès says that, about 5:30 p.m., some 200 enemy soldiers advanced without firing, presenting the butt-ends of their muskets and crying out that they wished to surrender. When they did not respond to French calls to put down their arms, Poulariès, then with the grenadier company of Royal Roussillon, ordered his men to fire. "A number were killed, and the rest disappeared," he reported. "Relation de Poulariès", *RAPQ,* 1931-32, p. 82.

Desandrouins' account is similar. He recorded an advance on the Guienne Regiment by a disorganized column of English soldiers calling *Bon quartier* and holding up their weapons. The French soldiers hesitated briefly, calling *Armes bas! Armes bas!* When the English did not lay down their muskets, the defenders reopened fire. Desandrouins dismissed the possibility that this was attempted treachery on the part of the English. Later, during a truce, English officers said that they had seen flags raised and lowered in the retrenchment and took this as a signal of surrender. When 20 or 30 of their men approached and offered quarter, the French fired on them unmercifully. Desandrouins speculated that the "signal" might have been the regimental colors of Guienne being raised and lowered to the cries of *Vive le Roy!* Gabriel, pp. 180-81.

Surprisingly, none of the English histories of the war mention this action or any ill feeling over its consequences. One account from outside the retrenchment, a "Letter from a Gentleman at Lake George, dated July 11, 1758" printed in the *New York Mercury,* July 24, 1758, supports the red handkerchief story: "About 3 o'Clock, just as the Regulars were retreating, our Regiment and those on the Left, threw in a very heavy Fire, intending to

...(continued next page)...

The enemy behaved with the greatest bravery in this attack, bearing the brunt of **[155]** terrible musket fire without flinching. Several of them were killed only 10 or 12 feet from the retrenchment. Our soldiers also behaved admirably, with great cheerfulness & *sangfroid.* As soon as they heard the enemy fire increasing, they shouted in unison: "Pay attention on the right, pay attention on the left."[440] MM. de Montcalm, de Levis, & de Bourlamaque, because of their care in sending up suitable reinforcements to the sectors that were being most vigorously attacked, deserve all kinds of praise, as do the individual officers for their skill in maneuvering the soldiers in their sector. The same justice is due to the colonial troops & to the Canadians occupying the retrenchment on the plain, which were beyond any possible reinforcement. However, by making constant small sorties, they deprived the English of any thoughts of advancing in that **[156]**

---

retire likewise very soon, and indeed some had already begun to retreat, which it is supposed the Enemy observed; they then hoisted English Colours and clubbed their Arms, shew'd themselves on their Breast-Work, and becken'd to us; on which the whole advanced briskly, but coming within 15 or 20 Yards from the Enemy, they struck their Colours, and threw in upon us a most terrible and heavy Fire, such as we had not yet experienced, which killed Multitudes, and obliged us to retire to recover ourselves from the Disorder we were thrown in."

Another Englishman, John Bremner of the 55th Regiment of Foot, wrote that "after 2 ... the Enemy struck their coulers & hoisted an English flag. Our Troops imagined the Enemy would give up the Fort. Marched close to the Intrinchments; the Enemy gave them a smart Volly & Hove a great number of grenade shells among them which distroyed a great Number of our Regulars." He considered this action a "diceet". Bremner, p. 5.

The most likely source of the English reproach mentioned by Pouchot is the conversation with enemy officers recorded by Desandrouins. This might have occurred on July 23 when an English flag of truce arrived at Isle au Mouton (Prison Island) at the north end of Lake George. Malartic and Gaffarel, p. 194. Pouchot also had ample opportunity, following his surrender of Niagara and Fort Lévis, to discuss the Carillon action with officers of the 42nd, 44th, 46th, 55th, and 4/60th, all of which participated in the battle.

440 Officers and men had apparently heeded Montcalm's admonition that the "soldiers cannot be too strictly warned that the great fault of regulars lies in hasty firing without aim .... The officers ... will see to it that the soldier fires slowly and they must urge him to take good aim." Sautai, *Montcalm,* p. 72.

441 Note in original, p. 157 (a): *The English estimated their losses at only 1800 men but their accounts are notoriously unreliable. Since their government has to be more sensitive to public opinion than in France, it is compelled to tell lies, either by inflating its victories or by diminishing its losses. Some writers of that nation have accused General Abercromby of failing in his duty by not sending forward his artillery, with which he could have destroyed the French retrenchments. That is false. Cannon have no, or very little effect on retrenchments of the sort. The recent battle at Savannah proves the matter conclusively.*

...(continued next page)...

sector, where they could have easily turned the flank of the retrenchment if they had known the lie of the land & the easy gains they could have made there. In this engagement, the English lost 4 to 5 thousand men and a large number of officers.[441] We lost 42 officers & five hundred soldiers killed or wounded. M de Bourlamaque **[157]** was seriously wounded in the shoulder.[442]

We spent the whole night under arms, fearful lest the English might wish to start the battle again the next day. This was not very likely, as they had lost at the head of their columns all their crack troops & officers. The Royal Scottish Regiment, which was attacking the angle on the right, lost 950 men out of 1400 & almost all its officers.[443] The next morning the enemy withdrew and went aboard their boats in great disorder. They abandoned 700 *quarts* of flour, after smashing some of the containers. We found in the mire of the road leading to the Falls more than 500 pairs of shoes complete with

---

Abercromby's return in Knox, I, p. 195 shows that his army suffered 554 officers and men killed, 1,357 wounded and 39 missing for a total of 1,950 casualties. These numbers agree closely with those reported by Abercromby to Pitt on July 12, 1758. *DRCHSNY*, X, p. 727. A list of officers killed and wounded is in *Ibid.*, pp. 728-32.

Pouchot's editor cites the Franco-American siege of English-held Savannah, Georgia in September-October, 1779 and the unsuccessful assault on the works of October 9. Boatner, *Encyclopedia*, pp. 982-88; Blanco, II, pp. 1499-1501.

442 Bougainville acknowledged 44 French officers and "nearly four hundred" men killed and wounded on July 8 and lists by name 37 French officers killed or wounded. Hamilton, *Adventure*, pp. 236-37. A combined French return for the actions of July 6 and 8 shows 16 officers killed, 24 wounded and 5 missing. Enlisted casualties were 527, although the number killed is not specified. *DRCHSNY*, X, pp. 750-51.

443 The 42nd Regiment of Foot was not granted the title "Royal Highland Regiment" until July 22, 1758 and so did not have a royal association at the time of the Battle of Carillon. C.C.P. Lawson, *A History of the Uniforms of the British Army*, Volume II (London, 1941), p. 72. News of this honor only reached the regiment through General Abercromby's orders at New York of November 24, 1758. The 42nd was also authorized an additional seven companies and was to be reorganized into two battalions of ten companies each. Orders, Dec. 3, 1758, Alexander Monypenny, "The Monypenny Orderly Book", *The Bulletin of the Fort Ticonderoga Museum*, XIII, No. 2 (June 1971), pp. 155-56.

The 42nd suffered terribly as a result of its determined attacks on the French lines. Although returns vary slightly, it appears that eight officers and 308 men were killed with another 17 officers and 316 men wounded for a total of 649 out of 1,100 engaged. Frederick B. Richards, *The Black Watch at Ticonderoga and Major Duncan Campbell of Inverawe* (Ticonderoga, NY, 1926), pp. 25, 52. This work includes extracts from several accounts of the action by officers of the regiment. A fascinating compilation of documents relating to the 42nd Regiment's participation in the battle is Nicholas Westbrook, " 'Like Roaring Lions Breaking From Their Chains': The Highland Regiment at Ticonderoga", *The Bulletin of the Fort Ticonderoga Museum*, XLI, No. 1 (1998), pp. 17-91.

buckles, a sure sign of the haste with which they departed. We picked up several soldiers in the woods who had got lost.[444] [158]

On the 11th, we began work on changes to the retrenchment, having had a good opportunity to discover the flaws.[445] On the 12th, M. Rigaud arrived with three thousand Canadians or colonial troops. He was followed on the 13th by 600 Indians. If these reinforcements had arrived on the day of the battle - which could have happened, since M. de Levis arrived in the morning &, as we have already noted, M. Pouchot had invited the Indians at St. Jean to follow him - then it is possible that these troops would have pursued the enemy as they withdrew. In that event few of them might have reached their vessels & a part of their army could have run the risk of perishing in the forest.[446]

---

444 The precipitous withdrawal of Abercromby's still-powerful army was a source of amazement to French and English officers alike. Captain William Hervey of the 44th Regiment of Foot recorded in his journal: "July 8th .... In the evening the army, to the surprize of most, retreated to the battoes. 9th. This morning the whole army moved off in great confusion, and Regulars, Provincials, Artillery and sick sailed mixed together, not stopping till they came to the end of the lake, the General among the first." Hervey, p. 50. The opinions of one outspoken officer, Colonel James Prevost (1725-1778), regarding the state of Abercromby's army following the debacle may be found in Ian McCulloch, " 'Believe Us, Sir, This Will Impress Few People': Spin-Doctoring—18th Century Style", *The Bulletin of the Fort Ticonderoga Museum*, XVI, No. 1 (1998), pp. 92-107.

On July 10 Pouchot personally observed the detritus of the retreating English army when Lévis took the grenadiers and *piquets* to reconnoiter the portage. They found flour and "shoes with their buckles." Malartic and Gaffarel, p. 191.

445 Malartic reported that this work commenced on the 12th. The original intent, he said, was to strengthen the wooden retrenchment with earth. *DRCHSNY,* X, p. 848. By the first of September, the wall had been almost completely rebuilt of horizontal logs in five levels with "five trees for a base, four above, three, two, and one." Hamilton, *Adventure,* pp. 267-68. An English traveler of 1767 found these defenses still intact and "built of large round logs, and are about eight feet thick at the bottom, narrowing to the breadth of one log at top. These logs are very large, and at the angles are morticed into one another; the lines are about man height." [Francis Grant], "Journal from New York to Canada, 1767", *Proceedings of the New York State Historical Association,* XXX, 1932, p. 320.

The lines were also extended from the right of the original position to the shore of Lake Champlain on the north side of the Ticonderoga peninsula. The results are shown in "Attaques des retranchemens" (Knox, I, p. 192). Pouchot again found himself assisting the engineers. By August 3 Malartic reported that the captain had "traced a redoubt to secure the right of the intrenchment". *DRCHSNY,* X, p. 850; Malartic and Gaffarel, p. 197.

Pouchot's fortification was at the water's edge and is shown as the battery or redoubt on the lakeshore labelled "M" in "Attaques des retranchemens". An undated *mémoire* entitled "Dispositions de Carillon", probably prepared late in 1758 or early in 1759, is critical of the field work called "Le Redan de Pouchot". Its most glaring deficiencies were that the position was too distant to be supported by the rest of the line and that it was unfortified and vulnerable on the lake side. "Les 'mémoires' du chevalier de La Pause", *Rapport de l'Archiviste de la Province de Québec pour 1932-1933* (Québec, 1933).

Through a deserter, we learned that on the day after the battle the enemy had formed a very large detachment & that their Indians, 500 **[159]** in number and all of them Iroquois, were withdrawing. Johnson had brought them, having promised that they would not fight, to show them how the English fought against the French. They were entitled to be displeased by the reception given them by the latter. The detachment mentioned above went off to attack Frontenac, having failed in its mission at Carillon.

On the 25th, under the command of St. Luc, a colonial captain, a raiding party of 300 Indians & 200 French or Canadians set out. It returned on August 1. It had attacked a convoy of 54 wagons, in which there were some provisions & a lot of equipment, which they pillaged. They brought back 60 prisoners & 110 scalps.[447] A few days before, a detachment of 500 men, under the command of **[160]** M. de Courte-Manche, had taken 40 scalps & brought five prisoners back to camp.[448]

---

446 The late arrival of this support and Montcalm's treatment of the Indians became a point of controversy between him and Governor Vaudreuil. See *DRCHSNY,* X, pp. 805-06, 810-11.

447 Luc de La Corne, commonly known as La Corne St. Luc (c.1711-1784), was a captain (1755) in the Marine. *DCB,* IV, 425-29.

Bougainville records St. Luc's departure date as the 24th with nearly 400 Indians. Hamilton, *Adventure,* p. 249. He returned on July 30 having attacked the convoy on the road from the Hudson to Lake George, between Fort Edward and Halfway Brook, on July 28. *Ibid.,* p. 253. Another source says that he captured 53 wagons with provisions and 230 oxen and took 80 scalps and 60 prisoners, "men, women and children" with the loss of one Iroquois killed and another wounded. *DRCHSNY,* X, p. 750.

From the English side, Captain Hervey reported an attack on 34 wagons guarded by 80 men. "Near 50" men and women were killed or taken, "oxen's horns cut off and their tongues cut out, a great quantity of men's necessaries destroyed." Hervey, pp. 50-51. Among the killed and captured were 30 women "that were going to the lake [George] as nurses." Pursuers, led by Major Robert Rogers (note 394) claimed one Indian killed and a Frenchman captured. *New Hampshire Gazette,* Aug. 11, 1758.

448 Jacques-François Legardeur de Croisille de Courtemanche (1710-1777) was a captain (1755) in the Marine. Fauteux, *Les Chevaliers,* p. 174. He had left Carillon on July 10 with a detachment of nearly 600 men, 400 of them Indians, to attack the fort at Halfway Brook (Glens Falls, New York). He fought an action there on July 20 and returned to Carillon "at full speed" with 15 to 20 scalps and 8 prisoners on July 21. Hamilton, *Adventure,* p. 245.

Hervey gives the English casualties as three captains, three subalterns and about 14 men killed. Hervey, p. 50. A very detailed account of the fight, which describes the stratagem used by the Indians, is in the *New Hampshire Gazette* of August 11, 1758. It lists casualties of three captains, two lieutenants, one ensign, and ten soldiers killed, two soldiers wounded and 13 missing, all from Nichols' Massachusetts Regiment.

For their part, the English worked at building a retrenched camp. A raiding party led by M. Marin, a colonial captain, came upon a corps of their troops, made up of 7 to 800 troops commanded by Roger. M. Marin captured a major of the English militia and a few others & took only two scalps. The losses of the English were estimated at 100 men by the French, who had four Indians killed & four wounded and six Canadians killed and six wounded, among them an officer & a cadet.[449]

One can judge from the report of M. de Longueil, who had been sent to the Five Nations, that they were little inclined to favor us. They were well aware of the **[161]** English advance towards Frontenac, which they did not reveal to him, telling him instead that the English wanted to rebuild the forts of the portage & river of Chouegen.[450]

---

449 This raid was led by Joseph Marin de La Malgue (1719-1774), lieutenant (1756) in the Marine. His promotion to captain came in January 1759, soon before he served with Pouchot at Niagara. *DCB,* IV, pp. 512-14. Marin had set out on August 4 with 50 Marine troops, 100 militia and 150 Indians. On August 8 he ambushed Rogers and a force of 700 rangers, Connecticut provincials and light infantry of the 80th Regiment of Foot near Fort Anne (Fort Ann, New York). After a spirited fight, Marin retreated in the face of superior numbers. Doreil to Belle Isle, Aug. 31, 1758, *DRCHSNY,* X, pp. 818-19 and Journal of Occurrences, *Ibid.,* p. 851. Bougainville reported 13 French dead and 10 wounded. They took five prisoners. Hamilton, *Adventure,* pp. 261-62.

A modern account of this fight is in Cuneo, pp. 89-90. A very defensive rebuttal, by provincial captains Stephen Maynard and Andrew Giddings, to charges that the Connecticut troops had not performed well was reprinted in the *New Hampshire Gazette* of September 29, 1758. The captains admitted English losses of five captured, 49 killed and 40 wounded. They also implied that any carelessness on the march had been the fault of the rangers and regulars of the 80th Regiment. On the morning of the ambush, they claimed, "the officers were so little apprehensive of an enemy near, that Major Rogers and a Lieutenant of the Light Infantry, upon a wager, fired their pieces several times at marks, and this was all the firing 'till the enemy fired upon us."

The major of the "English militia" was Israel Putnam (1718-1790) of the Connecticut provincials. He later gained fame as commander of the New England militia at the Battle of Bunker Hill and as a major general of the Continental Army. Boatner, *Encyclopedia,* pp. 902-04.

450 Paul-Joseph Le Moyne de Longueuil (1701-1778) was captain (1727) in the Marine and Governor of Trois-Rivières. *DCB,* IV, pp. 463-65.

Longueuil had been sent to the Six Nations early in July and returned from his mission before August 10. He reported that the Iroquois had not allowed him to proceed farther than Oswego. Hamilton, *Adventure,* pp. 223-24, 263; Casgrain, *Collection,* VII, p. 434. They assured Longueuil that they would send a deputation to Vaudreuil, that the English were only rebuilding the forts on the Oneida Carry and that, even if they tried to reestablish a fort at Oswego, they would not dare enter Lake Ontario in the face of the French vessels. The governor later claimed that these were assurances "on which ... I should not have relied had I any force." Vaudreuil to Massiac, Sept. 2, 1758, *DRCHSNY,* X, pp. 822-23.

On the 27th, the English, numbering about 3000 militiamen under the command of Bradstreet,[451] arrived via Corlack, sailed down the Chouegen river, crossed Lake Ontario & disembarked at Little Cataracoui.[452] The next day they mounted a battery behind an epaulement of the former retrenchments which had been allowed to remain. They soon opened a breach in the wall & the garrison, which was of 90 men including 30 voyageurs, surrendered on condition they could freely return to Mont-Réal. The English brought back with them some of the artillery[453] which we **[162]** had captured from them at Chouegen. Everything they were unable to take with them was smashed. They took away the bark *Marquise* & the brigantine. The remainder of our shipping was burned. It was said that they captured or set fire to nearly two million *livres* worth of goods. They also seized 2000 *quarts* of flour & 500 *quarts* of salt pork. After this expedition, they hurriedly withdrew. If we had been able to equip some vessels & if the advice of M. Pouchot had been followed, it is likely that this English expedition would have failed.[454]

---

451 John Bradstreet (1714-1774), then captain (1757) in the 60th Regiment of Foot with the local rank of lieutenant colonel in America. *DCB,* IV, pp. 83-87.

452 Cataraqui Bay at the mouth of Little Cataraqui Creek on the west side of modern Kingston, Ontario.

453 Note in original, p. 162 (a): *The artillery of this fort consisted at that time of 60 guns and 16 small mortars.*

On October 1, 1758 Captain Thomas Sowers reported the capture of 60 guns, ranging from four to 12-pounders, and 16 mortars in addition to the cannon aboard the French vessels. Most of the artillery was disabled by spiking the vents or knocking off the trunnions. Preston and LaMontagne, pp. 267-68. Many of these guns had been taken from Braddock in 1755 and at Oswego in 1756. Doreil to Belle-Isle, Aug. 31, 1758, *DRCHSNY,* X, p. 821.

454 The capture of Frontenac is discussed in Preston and LaMontagne, pp. 77-80 with related documentation on pp. 256-70. A detailed journal of Bradsteet's army has recently been published in Robert E. Mulligan, Jr. (ed.), "Colonel Charles Clinton's Journal of His Campaign in New York July to October, 1758 During the French War", *The Bulletin of the Fort Ticonderoga Museum,* XV, No. 4 (1992), pp. 292-315. A useful summary is Frederick A. Rahmer, *Dash to Frontenac: An Account of Lt. Col. John Bradstreet's Expedition to and Capture of Fort Frontenac* (Rome, NY, 1973). Bradstreet's force numbered 2,761 of whom only 181 were regulars and the rest provincials from New York, Massachusetts, New Jersey, and Rhode Island. *Ibid.,* pp. 21-22.

The French commandant was Pierre-Jacques Payen de Noyan et de Chavoy (1695-1771), elderly captain (1729) in the Marine. *DCB,* IV, pp. 611-12. The articles of capitulation are printed in *DRCHSNY,* X, pp. 825-26 and in Preston and LaMontagne, pp. 258-59. A list of names of the French soldiers and employees is in *Ibid.,* pp. 259-61. This shows five officers and 48 soldiers, all of the Marine, 24 King's employees and 27 voyageurs. There were also nine women and 13 children in the garrison.

...(continued next page)...

On the 30th, M. du Plessis[455] was detailed with 1800 men to take up a position at La Présentation & to cover this frontier. M. de Longueil, who had been sent on a mission to the Five Nations, could only go as far as Chouegen, as the Indians told him not to go [163] any farther because their own people were hunting & the English had 6 to 7 thousand men at the portage where they were rebuilding the forts. They informed him about the adventure at Carillon, which they had witnessed, adding that we were *Manitous* & that each of us must have had fifteen muskets. The enemy constructed a barge mounted with 18 guns at Fort George. From that time onwards they sometimes came prowling on the lake as far as the Isle au Mouton, which often meant we were on battle alert.[456]

On the 26th, we learned through a deserter that the camp of Fort George was made up of the fourth battalion of the Royal American Regiment, the 17th & 36th regiments, a Scottish battalion, two battalions of Boston militia and five hundred [164] men of independent companies, which could total in all 6 or 7 thousand men. We also

---

Nine vessels, comprising the entire Lake Ontario squadron, were taken. Sowers Report, Preston and LaMontagne, p. 268. Two were used to remove captured goods to Oswego and then burned. These were the schooner *La Marquise de Vaudreuil* of French construction and an English brig captured at Oswego in 1756. Rahmer, pp. 41-42. Regarding Pouchot's advice to Vaudreuil on the readiness of these vessels and their use in the defense of Frontenac, see note 408.

The capture of Frontenac, coming so soon after Montcalm's unexpected victory at Carillon and news of the surrender of Louisbourg, was a great blow to French morale, already beginning to erode due to the privations suffered by the colonists and their defenders. "Peace! Peace! My Lord, pardon me, I cannot too often repeat these words," Doreil wrote at the end of his account of the loss of Frontenac to Minister of War Belle-Isle on August 31. *DRCHSNY,* X, p. 822.

455 Captain François Lefebvre Duplessis Faber (note 280). French measures to block the St. Lawrence and reinforce Niagara following the loss of Frontenac are described in Vaudreuil to Massiac, Sept. 2, 1758, *DRCHSNY,* X, pp. 822-25.

456 The English had constructed the sloop *Halifax* at Lake George in August 1758. It was armed with 18 four and six-pounder cannon. Russell P. Bellico, *Sails and Steam in the Mountains: A Maritime History of Lake George and Lake Champlain.* (Fleischmanns, NY, 1992), pp. 75-76.

Isle au Mouton was Prison (or Prisoner) Island at the extreme northern end of Lake George. The English name originated when the French prisoners taken near Ticonderoga on July 6, 1758 were temporarily held there. Orders, July 7, 1758, "The Monypenny Orderly Book", *Bulletin of the Fort Ticonderoga Museum,* XII, No. 6 (October 1970), p. 440. One English incursion to the north end of the lake occurred shortly before October 16, 1758 when Adjutant Malartic reported the near approach of eight barges to Isle au Mouton. Malartic and Gaffarel, p. 205.

discovered that 5 thousand men had arrived at Albany from Louisbourg and that they had to report to the camp. The same deserter added that they were to attack Carillon & that they had mortars, 20 pieces of cannon of which 8 were 24-pounders, a large bark & two barges each of 48 oars.[457]

In the month of October, M. de Vaudreuil sent reinforcements of 1300 Canadians to Carillon. On the 26th we learned through another deserter that the English were striking camp in order to go into winter quarters, that they had evacuated Fort George and scuttled their barks.[458] So ended this campaign, which would not have been an unhappy one had it not been for the loss of Louisbourg & Isle Royale. **[165]**

The English fleet, after sailing out of Halifax, appeared off the coast of this island. At first, it made several attempts, all of them fruitless, to make a landing there. Fortunately for the enemy, they noticed that a cliff considered inaccessible had been left unguarded. They approached & arrived at the top without meeting any resistance. The outcome of this action and most of its details are known to everyone[459] **[166]** so we shall simply note a few anecdotes. Admiral

---

457 The 36th Regiment was not in America, so the deserter probably spoke of the 46th. The highland battalion was the 42nd. Plans were afoot for sending six of Amherst's battalions from Louisbourg for another attempt on Carillon, but concern about the strengthened French defenses prompted the decision of a council of war, held in October, to postpone a late season attack. Gipson, VII, pp. 234-36.

By the end of autumn the English had constructed, in addition to the *Halifax,* two row galleys and two "radeaus", barge-like vessels with high bulwarks designed to transport heavy artillery with the guns mounted and serviceable for defense. Bellico, pp. 76-77.

458 The English vessels were sunk in the fresh, cold water of Lake George to hide and preserve them for the next campaign. Several, including the *Halifax,* were raised in 1759. Amherst Journal, Knox, III, p. 37. The large radeau, *Land Tortoise,* was sunk too deeply, however, and remained to be discovered by divers in 1990. Bellico, pp. 77-85, 90-91.

459 Note in original, p. 166 (a): *A very detailed journal of this siege can be found in* Mémoires sur le Cap Breton. *The* Histoire philosophique & politique des établissements des Européens [Raynal], *Vol. VI, p. 241 et seq., should also be consulted. The state of the fortifications at Louisbourg, its remoteness from Canada & the weakness of our navy did not allow us to retain this stronghold any longer. It had to be evacuated, therefore, & its garrison transported to the mainland where it could have been usefully employed in defending itself against the English. If we had taken this decision, we might have avoided the loss of several vessels of the line.*

Pouchot's editor apparently cites [Thomas Pichon], *Lettres et mémoires pour servir à l'histoire naturelle, civile et politique du Cap Breton ... jusqu'à la reprise de cette isle par les Anglois en 1758* (La Haye, 1760).

...(continued next page)...

Boscawen[460] was patrolling in front of the fortress with his warships & all his transport vessels which were carrying all the army's provisions. An inexperienced officer[461] of the French navy asked for just two fire ships in order to destroy them. Although risking capture, he was **[167]** nonetheless almost certain of being able to succeed & thus forcing the English to raise the siege of Louisbourg as a result of their losses. M. Desgouttes,[462] in command of the French ships, did not favor this plan & refused to supply the means necessary for its accomplishment. It was doubtless not because he hoped to save the King's vessels. Several of them were burned in port by the English who even took one of them by surprise with their long boats. The crews had been sent ashore to defend the fortress. When it was on the point of surrendering, at a time when the English could not reach the port because of contrary winds, M. de Vauclein wished to persuade the French squadron to depart. Unable to achieve this, he set sail with his

---

Contemporary English treatments of the siege of Louisbourg, including the articles of capitulation, are in Mante, pp. 112-42 and Knox, I, pp. 205-59. The operation is covered in Gipson, VII, pp. 179-207. For more detailed treatments see John Stuart McLennan, *Louisbourg from its Foundation to its Fall, 1713-1758* (London, 1918) and Fairfax Downey, *Louisbourg: Key to a Continent* (Englewood Cliffs, NJ, 1965), pp. 149-83. A current and well-illustrated study of the campaign and siege is René Chartrand, *Louisbourg, 1758: Wolfe's First Siege* (Oxford, 2000). A comprehensive examination of the defenses of this North America fortress town is Bruce W. Fry, *"An Appearance of Strength": The Fortifications of Louisbourg* (2 vols.; Ottawa, 1984).

460 Edward Boscawen (1711-1761), Admiral of the Blue (1758). *DCB,* III, pp. 70-71. Boscawen's powerful squadron of 22 ships of the line and 15 frigates is enumerated in Mante, pp. 113-14.

461 Note in original, p. 167 (a): *It was M. Beaussier, a naval captain. Here he is quite improperly characterized as inexperienced, very likely because he had started in the port service, at that time separated in some sense from the navy.*

Captain Louis-Joseph Beaussier de Lisle (note 206) had served in the port service from 1732 and commanded *Defenseur* under Dubois de La Motte in 1755. He was captain of *Entreprenant* (74 guns) at Louisbourg in 1758. Beaussier wanted to burn the enemy transports anchored in Gabarus Bay. Hamilton, *Adventure,* p. 312.

462 Jean-Antoine Charry, Marquis Desgouttes (fl.1725-1764), *capitaine de vaisseau* (1746) and French naval commander at Louisbourg. Lacour-Gayet, p. 532. See also *DCB,* III, pp. 55, 72, 268.

French naval vessels at Louisbourg consisted of the ships of the line *Prudent* (74 guns), *Entreprenant* (74), *Célèbre* (64), *Capricieux* (64), and *Bienfaisant* (64), the frigates *Apollon* (50), *Aréthuse* (36) and *Fidèle* (36) as well as *Chèvre* (16), and *Biche* (16). All but *Aréthuse* were lost during the siege. McLennan, p. 263; Gipson, VII, pp. 204-05. *Apollon, Fidèle* and *Biche* were sunk to block the harbor entrance. Hamilton, *Adventure,* p. 312.

own vessel & returned to France.[463] All the others were set ablaze or captured by the enemy.[464] **[168]**

During this campaign, the English created yet another detachment, of 2000 men, to occupy the Belle-Riviere.[465] On 14 September, 800 Scots & militiamen, under the command of two majors,[466] approached at daybreak right up as far as the cleared ground created around Fort du Quesne, without being spotted. The militia major was hesitant to attack, but the Scotsman, Major Grant, reluctant to turn back without doing anything, had a small shed in an outlying area of the fort set on fire to provoke an engagement. The Canadians & a few Indians, who were lodged in huts around the fort, noted this unusual early morning fire and were curious enough to slide down into the brushwood in order to discover what was happening. They went one after the other. Since **[169]** the Indians & Canadians wear nothing

---

463 Jean Vauquelin (1728-1772), lieutenant commanding the frigate *Aréthuse* (36). He ran the blockade of Louisbourg on the night of July 14-15, 1758 and later commanded French vessels in the St. Lawrence in 1759-1760. *DCB,* IV, pp. 751-52.

464 The other French warships were doomed by the siege. *Entreprenant, Célèbre,* and *Capricieux* were destroyed by shells and fire on July 21. *Prudent* was burned and *Bienfaisant* was captured by a cutting-out party on the 25th. Gipson, VII, pp. 204-05. Louisbourg's governor asked for terms on July 26.

465 Earlier in this volume, Pouchot frequently uses "Ohio" in reference to the river. The alternate term, "Belle-Rivière", is more commonly employed from this point on and has been retained in this translation.

The new attempt against Fort Duquesne was led by John Forbes (1707-1759), colonel (1757) of the 17th Regiment of Foot. He had been appointed brigadier on December 28, 1757. *DCB,* III, pp. 218-19. Some of his papers have been published in Alfred Procter James (ed.), *Writings of General John Forbes Relating to his Service in North America* (Menasha, WI, 1938).

Forbes was assigned an army of over 6,000 men. Gipson, VII, p. 248. He gathered his forces and established a route during the summer of 1758. On September 3 his advance body crossed Laurel Hill and began to construct a post (later named Fort Ligonier) on Loyalhanna (or Loyal Hannon) Creek. Accounts of his campaign are in *Ibid.,* pp. 247-86, West, pp. 47-58 and Alfred Procter James, "Decision at the Forks," *The Western Pennsylvania Historical Magazine,* Vol. 41, Nos. 3 and 4 (Autumn 1958), pp. 39-56. Another is in Walter O'Meara, *Guns at the Forks* (Englewood Cliffs, NJ, 1965), pp. 183-210.

466 The major of regular troops was James Grant (1720-1806) commissioned in 1757 with that rank in the First Highland Battalion (Montgomery's Highlanders) or 62nd Regiment of Foot. The regiment was later renumbered the 77th. *DRCHSNY,* X, p. 903n; Boatner, *Encyclopedia,* pp. 442-44; Blanco, I, pp. 665-67; West, p. 48.

The "militia" major was Andrew Lewis (1720-1781) of the 1st Virginia Regiment of provincials. Boatner, *Encyclopedia,* pp. 618-19; Blanco, I, pp. 920-21. Lewis and Grant would later see service on opposing sides during the American Revolution.

more than a shirt in fine weather, they were very soon ready for action. Those who arrived first saw the troops and began to fire on them. The English beat the retreat, causing the alert to be sounded in the fort from which assistance was sent to the first men who had sallied forth from it. The enemy corps was so vigorously attacked that 250 scalps resulted and 100 prisoners were taken, among them six officers & the two majors. The remainder were pursued into the forest, where most of them perished.[467]

On 22 September, M. Aubry, a captain of the Illinois,[468] set out from Fort du Quesne with a detachment of about 600 men, Canadians & Loups Indians, in order to reconnoitre the English camped at **[170]** Loyal-Anon.[469] He found a small camp just beyond the retrenchments which was held by a corps of 2000 men. As soon as the vanguard of our detachment was discovered, the English sent out a captain & 50 men to reconnoitre it. They stumbled upon the entire detachment, which utterly destroyed them. In pursuing those who had been put to

---

467 Grant and Lewis left Loyalhanna on September 11 with over 800 men. Pouchot alludes to the piecemeal attack made on September 14 in the vicinity of Grant's Hill in the modern city of Pittsburgh. See Gipson, VII, pp. 268-70. Grant's report to Forbes of c. September 14 is in Sylvester K. Stevens, et. al. (eds.), *The Papers of Henry Bouquet* (5 vols.; Harrisburg, 1972-84), II, pp. 499-504. An account of the battle and subsequent captivity of one Virginia participant is Howard H. Peckham (ed.), "Thomas Gist's Indian Captivity, 1758-1759," *The Pennsylvania Magazine of History and Biography,* LXXX (July 1956), pp. 285-311.

An English return shows 273 killed or missing, including 22 officers. Stevens, *The Papers of Henry Bouquet,* II, pp. 508-09. The French at Fort Duquesne later listed seven of the officers as prisoners, one of whom had been carried to Detroit by the Hurons. Although Ensign Gist was not on that list, he too had been taken to Detroit. List of English Prisoners, [Sept. 22, 1758], *Ibid.,* p. 535. A French report claims five officers and 100 soldiers taken prisoner for eight French killed and another eight wounded. Daine to Belle-Isle, Nov. 3, 1758, *DRCHSNY,* X, pp. 884-85. Daine refers to prisoners in French custody and does not include those carried off by the Indians.

The five officers were sent to Canada and were among those exchanged for Pouchot's detachment in 1759. At that time, Majors Grant and Lewis, Captain Hugh McKenzie of the 77th Regiment, Ensign Edward Jenkins of the 60th and Ensign Henry Haller of the second battalion, Pennsylvania Regiment were returned to English lines. "List of Officers Returned from being Prisoners", Nov. 24, 1759, Amherst Papers/8.

468 Charles-Philippe Aubry (d.1770), captain (1750) of one of the Louisiana companies of the Marine. Knox, II, p. 185n; *DRCHSNY,* p. 901. The Illinois country was part of Louisiana, and troops and supplies from that region were very important to the French war effort along the Ohio. See Chartrand, *The French Soldier,* pp. 16-19 and René Chartrand, "The Troops of French Louisiana, 1699-1769", *Military Collector & Historian,* XXV, No. 2 (Summer 1973), pp. 58-65.

469 Loyahanna or Ligonier, Pennsylvania.

flight, the French came across the small camp, took it by surprise[470] & sent the garrison fleeing for the main retrenchment, which they only just reached in time. For two days, M. Aubry formed a kind of blockade around it. He had 200 cattle or horses killed. Almost all of our men returned on horseback. **[171]** They estimated enemy losses at 200 men while ours were one corporal and two soldiers.[471]

The enemy had another camp at Raiston[472] where General Forbes, acting as colonel of the Royal American Regiment, was commander in chief.[473] They came probing towards Fort du Quesne, which had to be abandoned because of lack of supplies. In the month of October,

---

470 Note in original, p. 171 (a): *A soldier entered a tent and found there an officer drinking his tea. He said to him: "What's this then? Your comrades are fighting & you are sitting here in comfort. You don't deserve to live." With one blow of an axe, he killed him.*

English sources do not mention this incident. The only officers killed were two Marylanders, Lieutenants Henry Prather of Joshua Beall's Company and Thomas Matthews of Richard Pearis' Company. Extract of a letter from Raystown in the *Pennsylvania Gazette* of October 26, 1758 in Stevens, *Bouquet Papers,* II, pp. 566-67. Documents do not record either the location or the circumstances of their deaths.

Several plans of Fort Ligonier indicate that the French attack came from the east and struck the English position near the encampment of the Maryland and Carolina troops. One of these (McDonald) shows the "Maryland Hospital" at the eastern extremity of the Pennsylvania and Maryland/Carolina camps. Pouchot's marginal comment may therefore document the death of one of the Maryland officers, either in a hospital tent or his own marquee. See James Burd's sketch of Loyalhanna, Theodosius McDonald's "A Plan of Fort Ligonier" and J.C. Pleydell's "Plan of the Retrench's Camp at Fort Ligonier". All are reproduced in Stotz, pp. 148-50.

471 Aubry's attack took place on October 12 and is discussed in Gipson, VII, pp 272-74. The English acknowledged losses of 12 killed, 18 wounded and 31 missing. Stevens, *Bouquet Papers,* II, p. 567. Henry Bouquet, disgusted by the affair, wrote to Forbes on October 15 that it "appears humiliating to me. A thousand men keep more than 1500 blockaded, carry off all their horses, and retire undisturbed with all their wounded and perhaps ours. This enterprise which should have cost the enemy dearly shows a great deal of contempt for us, and the behaviour of our troops in the woods justifies their idea only too well." *Ibid.,* p. 560.

Montcalm reported only two killed and seven wounded for the French. But he also noted the negative results of this victory for the defense of the Ohio - the Indians had "retired to hunt." The French general was mistakenly optimistic that "Fort Duquesne is safe for this autumn and winter." Montcalm to Belle-Isle, Nov. 15, 1758, *DRCHSNY,* X, pp. 900-01.

472 Fort Bedford, Bedford, Pennsylvania. Stotz, pp. 113-17.

473 Forbes was not colonel of the Royal American Regiment (60th) but rather of the 17th. His army included four companies of the 1st battalion of the Royal Americans, perhaps the source of Pouchot's confusion. Gipson, VII, p. 254.

M. de Lignery, in command there, sent back the Indians & French of Détroit, & some Illinois who had gone there to cover the position.[474]

On 27 November, M. de Corbiere, a colonial captain,[475] marched a detachment of 45 Indians, soldiers & Canadians in the direction of Loyal-Anon. He encountered a forward party, which he estimated at 7 to 800 men & attacked it. It withdrew in disorder to the camp. The French [172] took about a hundred scalps & seven prisoners. They pursued the enemy right up to their retrenchments, where they were fired on by the cannon.[476] These events must seem extraordinary, but readers will realize how they could happen if they consider that the English advanced only in the most fearful manner, and that when they were attacked, they could not estimate the numbers of their enemy because our men were always spread out and hidden behind trees. By contrast the English did not dare spread out in unknown territory & thus always remained in formation. As a result they were crushed by the fire of men who were excellent marksmen.

In the course of a second sortie, M. de Corbiere encountered the enemy three leagues away from Fort du Quesne. He went back to warn

---

474 The commandant was François-Marie Le Marchand de Lignery (note 328). Montcalm reported that the Illinois and Detroit detachments had left for home soon after Aubry's raid. Montcalm to Belle-Isle, Nov. 15, 1758, *DRCHSNY*, X, pp. 900-01.

475 Corbière (d.1760) was a lieutenant (1756) in the Marine. "Tableau ... des Officiers de la Marine", Brymner, 1886, p. clxxxiii. He was mortally wounded at the Battle of Ste. Foy on April 28, 1760. "List of French Officers Killed and Wounded at Québec", Knox, III, p. 147; "Liste des Officiers des Troupes ... de la Marine ... tués", Brymner, 1886, p. clxxxv.

476 Pouchot has confused two events. Forbes had taken possession of Fort Duquesne two days before the date given for Corbière's attack on Loyalhanna. The pursuit of the English up to their fortified camp can only be an episode from Aubry's October 12 action. A letter from Loyalhanna, dated October 14 and printed in the *Pennsylvania Gazette* of October 26, 1758, noted that "The Enemy attempted in the night to attack us a second time; but in Return for their most melodious Indian Musick, we gave them a Number of Shells from our Mortars, which soon made them retreat."

The Corbière raid took place a month later, on November 12. About three miles from Loyalhanna his 30 Canadians and 140 Indians encountered two provincial detachments totalling 1,000 men. One was led by George Washington. The provincials claimed one enemy dead and three prisoners, but Washington and his men were taken for French when they returned to their camp and were fired upon by fellow Virginians. *Pennsylvania Gazette,* November 30, 1758. See also Stanley, p. 189.

The Corbière raid had an unexpected and fatal consequence for the French at Fort Duquesne. Only the day before, General Forbes had reluctantly decided to end his campaign without another attempt on the French post. One of the prisoners taken from Corbière's party provided information about the weak state of de Lignery's garrison. Forbes accordingly reversed his decision and pressed on. West, p. 56.

M. de Lignery **[173]** who, finding himself short of both troops & supplies, immediately loaded onto his boats what remained of the munitions for the Illinois, together with the artillery, distributed the remaining trade goods to local Indians & himself withdrew with them to Sonnioto and the Riviere à la Roche.[477] M de Corbiere, after burning Fort Duquesne, took the land route to Fort Machault[478] with a few Canadians & soldiers.

The enemy arrived at Fort du Quesne where they found everything destroyed.[479] On noticing in the debris a few charred bones, which doubtless belonged to animals, they imagined that the French **[174]** had amused themselves by burning their prisoners, which is one of the most horrible calumnies that national hatred could ever have invented.[480]

One can see from the above details that a great many men, either Frenchmen or Indians, had been involved in this sector during the campaign, which meant a high level of consumption of supplies. However, goods & provisions continued in short supply. The commander often found himself obliged to buy back from soldiers & officers effects from the stores which had been sold or distributed.

---

477 The artillery was being sent towards Fort Massac (Metropolis, Illinois). The rivers are today the Scioto and the Great Miami in Ohio. Both flow south into the Ohio River. De Lignery himself withdrew up the Allegheny to Fort Machault.

478 The junction of French Creek with the Allegheny River (Franklin, Pennsylvania) had been occupied by the French in 1753. In 1755 they began construction of a stockade which was completed in 1757. Fort Machault was the third post on the route from Lake Erie to Fort Duquesne and served as a depot for the latter. Stotz, pp. 76-79.

479 Note in original, p. 174 (a): *It was 25 November that the English took possession of these ruins. General Forbes died of fatigue a little later.*

Forbes, in fact, died in Philadelphia on March 11, 1759. *DCB,* III, p. 218.

480 General Forbes' occupation of the ruins of Fort Duquesne on November 25, including the discovery of the remains of some of Grant's men, is described in Gipson, VII, pp. 283-84. Pouchot probably learned of the "calumnies" from letters to the *Pennsylvania Gazette* published in the issues of December 14 and 21. Two of these, both of which mention the torture of prisoners, remains at the site of Grant's defeat and the "unburied Bodies of those killed since, and strewed round this Fort," are printed in Stevens, *Bouquet Papers,* II, pp. 612-14. Letters from Bouquet himself are silent on the matter.

Ensign Thomas Gist witnessed the deaths of a number of prisoners, "massacred in the most barbarious manner that can be immajined," just outside the walls of Fort Duquesne. A few of the bodies were thrown in the river, he recalled, but the "rest was left in the field, about two or three hundred yards from the fort, and unburied." Peckham, "Thomas Gist's Indian Captivity", p. 294.

There arrived canoe loads of goods from different individuals who had received permission to sell to us. This they did at very high prices, because of the distances & the shortage of goods. Nevertheless, in spite of that, we were not better supplied. The loss of Frontenac increased all this disarray, **[175]** and that caused the loss of Fort du Quesne, which had cost so many millions. It is surely not an exaggeration to say that the total expenditures that year must have exceeded twenty million. Those who have verified the accounts are well qualified to judge if this is far from the truth.[481] The English crossed the Appalachian mountains again, leaving a force of 4 to 500 men at Fort du Quesne, where they began construction of the fort of Pittsburgh at the far end of the cleared ground around the original fort.[482]

They also posted 250 men to Loyal-Anon. They invited our Indians to remain in peace & to allow the English to fight with the French, who were dying of hunger & **[176]** had no more trade goods to give them. The Indians replied they would only listen when they were assured of their own safety.[483]

---

481 Note in original, p. 176 (a): *M. Pouchot is well below the true figure. The expenditures for that year were the highest of the war. They totalled twenty seven million nine hundred thousand francs.*

482 Forbes and his army departed the Forks of the Ohio on December 4, 1758, leaving Hugh Mercer and 200 Pennsylvania and Virginia provincials to erect a temporary fort. This was done in December 1758 and January 1759 on the edge of the river a short distance up the Monongahela from the ruins of Fort Duquesne. Although the site had been named Pittsburgh on November 27, the temporary stockade is commonly referred to as "Mercer's Fort". Construction of a permanent Fort Pitt began during 1759. Stotz, pp. 121-25.

Hugh Mercer (c.1725-1777) was then a lieutenant colonel of Pennsylvania provincials. In 1777, as a brigadier general of the Continental Army, he was mortally wounded at the Battle of Princeton. Blanco, pp. 1051-53.

483 Sentiments very close to these were expressed by Henry Bouquet in a conference with the Delaware at Pittsburgh on December 4, 1758. His speech was reported to de Lignery at Fort Machault by the Delaware chief Custaloga. The French officer, in turn, passed the information to Vaudreuil on January 4, 1759. Stevens, *Bouquet Papers,* II, pp. 621-24; De Lignery to Vaudreuil, [Jan. 4, 1758], *Ibid.,* pp. 624-26.

484 Shortages of provisions, especially during the winter, had become chronic as the war interrupted shipments from France and wore down the marginal ability of New France to produce sufficient foodstuffs. Complaints by *troupes de terre* officers about the expense of service in the colony grew louder at the end of the 1758 campaign. Montcalm wrote to Vaudreuil on October 8 lamenting that "it is impossible for our officers to live on their pay" and warning him of "the fermentation of their minds, the consequences of poverty and discouragement." *DRCHSNY,* X, pp. 886-87.

...(continued next page)...

In the final months of the campaigning season, provisions had already been in short supply, although more ships had arrived from France than in the previous years. The intendant offered to give 45 *livres* to captains & 20 *livres* to lieutenants in place of rations since he had no supplies of bread & salt pork to send them.[484] At the end of this campaign, M. de Levis took M. Pouchot with him in his bateau. Together with M. de Montcalm they visited those places which might constitute strong points. M. Pouchot suggested first the Riviere à la Borbue [sic], then they looked at Isle aux Noix and decided in its favor.[485]

As soon as the troops arrived back in **[177]** Canada at the end of November, they were billeted in the villages above & below Québec and right down to the far end of Mont-Réal Island. We stationed in the cities of Québec & Montréal just enough personnel to mount guard over them.[486]

During that winter, food was extremely scarce and the ration of bread was reduced to a pound & a half and that of salt pork to a quarter of a pound. When supplies of the latter ran out, the intendant suggested giving horse meat to the troops. We were obliged to accept this without grumbling.[487] Nonetheless, with greater economy they

---

The perspective of the governor and of the Canadians was quite different. Vaudreuil blamed Montcalm for the dissention and thought that the army officers wanted "the comfort of living in the Colony at small expense." Although he expressed his displeasure to Montcalm, the governor convinced the intendant to allow 30 *sous* per day to each captain and 20 to each subaltern. He believed this expedient was "much better than to have meat, bread and wine distributed among those officers at a low rate, since it would have cost at least three *livres* each a day." For the sake of fairness, he provided the same benefit to the officers of the Marine. Vaudreuil to Massiac, Nov. 4, 1758, *Ibid.*, pp. 885-86.

485 Although Lévis was in the vicinity of St. Frédéric during much of October, he does not mention taking Pouchot with him to reconnoiter positions. Lévis left St. Frédéric on November 4, arriving in Montréal on the 9th. Casgrain, *Collection*, I, pp. 156-58.

Pouchot's comments about the Rivière à la Barbue (Putnam's Creek at Crown Point, New York) and Isle aux Noix, found in Volume III of his *Mémoires*, nonetheless make it clear that he had carefully examined their defensive potential. Isle aux Noix, situated in the Richelieu below Lacolle, Québec, was chosen as the position from which to protect Montréal if Carillon had to be given up. It could easily have been examined by Lévis and Pouchot on their return to Montréal.

486 The Béarn Regiment was assigned to the district of Montréal. Bougainville to Crémilles, Nov. 8, 1758, *DRCHSNY*, X, p. 889.

487 The problems of food supply are discussed in detail in Stanley, pp. 191-200.

...(continued next page)...

could have provided sufficient salt pork since, when the thaw came, they were compelled to throw away quantities of it which had been spoiled. The quartermaster was thus charged with supplying horse meat. In consequence, he rounded up all the old nags of the territory to feed the troops. Hence, **[178]** as soon as we saw a worn out horse we called it a "Cadet." [488] To set a good example, all the generals served horse meat at their table.[489] M. de Montreuil, who also invited people to dinner occasionally, offered it to his guests. It cost 20 *sous* a pound. Even Péan, the Aide-major of Québec, had the impertinence always to have some placed on his table. What other term can be used to characterize the audacity of a man who had at his disposal all the resources of the country & who supplied provisions to the King for millions of *livres* and who yet affected this attitude towards officers & soldiers suffering hardship? On these tables they only served a quarter of a loaf. If **[179]** you wanted to eat more, you brought some in your pocket. Of course, these gentlemen ate very well indeed, but it was considered good form to act in this way.

The French officers tried energetically to satisfy their need from the inhabitants & with money they were able to obtain what they required. A bushel of grain weighing 45 pounds sold for thirty tournois *livres* & a cask of wine holding 220 bottles cost from 900 to 1200 *livres*. This was the situation in Canada during a winter when many Acadians fled there after the capture of Louisbourg. Reduced to

---

Horse meat was first issued during the particularly difficult winter of 1757-1758, and Pouchot's comments here might actually refer to that time rather than to the winter of 1758. In December 1757 the eight-day ration for each soldier was reduced to four pounds of bread, two of peas, three of beef, three of horse meat, and two of codfish. On December 9 Lévis, then at Montréal, found himself in the unenviable position of confronting his troops to convince them to accept horsemeat. The grenadiers of Béarn at first refused. Lévis then compared the colony to a besieged town and appealed to their sense of honor. The meat was accepted, and the grenadiers even reappeared at Lévis' quarters on Epiphany, 1758, to bring him a dish of horse "cooked after their fashion." Sautai, *Montcalm,* pp. 39-40. The incident is also described in Stanley, p. 194. Pouchot was then in the Montréal area as he was during the winter of 1758-1759.

488 Note in original, p. 179 (a): *The name of the quartermaster, who was the most notorious scoundrel yet to appear in America.*

Joseph-Michel Cadet (note 110).

489 Montcalm was among those who did so. On December 4, 1757 he wrote to Lévis that "Horse meat is eaten at my home in every way, except soup .... Small horse meat patties, Spanish style; horse meat stew; scalloped horse meat; horse meat steak on the spit with a good pepper and vinegar sauce; horses feet baked in bread crumbs; hashed horse tongue with onions; horse meat ragout; smoked horse tongue, better than that of moose; horse meat pie, like hare pie. This animal is much better than moose, caribou and beaver." Quoted in Stanley, p. 194.

four ounces of bread per day, they sought to satisfy their hunger in the sewers of Québec. Such was the fate of some of these settlers whose attachment to the mother country was only equalled by its indifference. It allowed them to die in poverty, while [180] detestable tax-farmers & squalid clerks grew rich at the expense of France through outrageous peculation.

The surrender of Closter-Seven & the capture of Louisbourg allowed for redeployment in New England that winter of 27 regiments of seasoned troops, including those who had arrived previously. We learned that the English were very busy with great plans for the forthcoming campaigning season. They worked at constructing new forts, those of George, Chouegen and Pittsburgh.[490]

The commander of the latter fort, in a council with the Indians of the Belle-Riviere, expressed his regrets at having quarreled with them & shed blood on their territory. He assured them that henceforth he wanted to live in peace with them and begged them to forget the past. He added that he was not angry at seeing them [181] attached to the French and that, although the French had withdrawn, he had no orders to attack the native peoples. According to him, the English had built a small post for trade only and that if they wanted to come and see him, he had prepared shelters where they would find everything they needed. This speech, which was full of pathos, greatly affected the Indians in question, the Loups & the Chaouanons. However, they replied that, before making up their minds, they would have to know what their other allies among the Indians felt about the subject. One can judge from this that the English had no intention of making great efforts in this matter.[491]

---

490 Pouchot again tallies all English regular battalions in North America. Based on Knox, a total of 25 were available for service in the campaign of 1759. Wolfe would have ten: 15th, 28th, 35th, 43rd, 47th, 48th, 58th, 2/60th, 3/60th, and 78th. Knox, I, p. 333. Three, less their grenadier companies which went with Wolfe, would remain at Louisbourg: 22nd, 40th and 45th. *Ibid.*, p. 333. Amherst would have eight: 2/1st, 17th, 27th, 1/42nd, 2/42nd, 55th, 77th, and 80th. *Ibid.*, pp. 466, 474n. Prideaux was assigned three: 44th, 46th and 4/60th. *Ibid.*, p. 480. The 1/60th Regiment was at Pittsburgh and other posts in Pennsylvania. This list does not include provincials, detached units or support troops. The same regular battalions are included in a "List of Regiments in North America, June 17, 1760" in *Collection de manuscrits*, IV, p. 266.

Construction of all three of the named forts did not begin until the summer of 1759. Fort George would be situated on the south end of Lake George, near the former site of Fort William Henry.

491 This is again a reference to Bouquet's conference with the Delaware at Pittsburgh on December 4, 1758 (note 483).

Johnson also held a great council with the Five Nations. He exhorted them to remove the Indians from our missions & he told them that it was absolutely imperative that they kill Onontio, the King of France, that spring & that they would arrive on Lake Ontario with 10 **[182]** thousand men in order to attack all our forts.[492]

As a conclusion to this account of the events of this campaign, I must mention the unfortunate episode of the ship *L'Aigle,* which set out from France in June loaded with arms and clothes for the French troops. This ship, of 50 guns, foundered at Mai-Catinat, at the entrance to the Saint Lawrence, after passing through the straits of Belle-Isle. The crew was fortunate enough to save many of the provisions & other goods. The authorities at Québec were informed about the accident & a ship was sent onto which was loaded everything that had been salvaged. This ship in turn was lost with everything that had been saved & this time only a part of the cargo of flour was taken from the wreck. This flour was loaded aboard the vessel which trades with the Mingan fort. This vessel too sank, at Kamouraska, & **[183]** the crew, which had numbered 280 men, was reduced to 80 as a result of these shipwrecks. Most of them arrived at Québec sick men.[493]

---

492 Pouchot may be combining two separate actions on the part of Sir William Johnson. On November 18, 1758 he presented a belt to two Iroquois, Red Head and an unidentified Seneca, at Fort Johnson. The belt accompanied a message "strongly pressing the Onondagas & Oneidas for the last time to withdraw their people from Oswegatchie & that if they would not comply & quit the French Interest ... they might repent it when too late." Journal of Indian Affairs, Nov. 18, 1758, *SWJP,* X, p. 57.

Johnson's great council with the Iroquois was held at Canajoharie, New York between April 4 and 21, 1759. In attendance were representatives of all Six Nations and some subsidiary peoples. The council is recorded in *DRCHSNY,* VII, pp. 378-94. At the conclusion of this assembly, Johnson gleefully reported that "It is with great pleasure I can Inform your Excellency, these Confederate Nations have, with every Mark of Unfeigned Zeal and Sincerity, declared their unanimous Resolutions of Joining in the present War against the French, [and] have Accepted & taken up the War Hatchet, which I threw down to them in Your Name." Johnson to Amherst, April 21, 1759, *SWJP,* III, p. 27.

493 *Aigle,* a small ship of the line, had been part of Macnémara's 1755 covering squadron at which time she was rated by Pouchot as carrying 54 guns.

Mécatina is near Cap du Gros Mécatina, Québec on the north shore of the Strait of Belle-Isle opposite Newfoundland. The Mingan fort was located farther west near Betchouane, Québec at Baie St. Laurent on the north shore of the St. Lawrence opposite Anticosti Island. Kamouraska, Québec is on the south shore of the St. Lawrence at the mouth of the Kamouraska River.

The wreck of *Aigle* occurred on August 8, 1758 on the shoals of Quincampoix, eight leagues from Mécatina, and only 30 quarters of flour were saved. The vessel *Légère* was sent

...(continued next page)...

In next year's campaign, the officer commanding the *Aigle* was entrusted with a fire ship to be used against the English fleet. He jumped overboard before reaching any of the English vessels & his crew all perished with the exception of five men.[494]

At the beginning of spring, it was necessary to stockpile flour & pork. The authorities requisitioned as much as they could from the habitants. Although the latter were highly paid from the King's treasury for these items, since the items in question could not be replaced from any other source, the settlers suffered a genuine shortage. And as the menfolk were fighting in the war during the campaign, they could not cultivate their lands as they would normally have done & thus did not harvest the same quantity of grain.

End of the First Volume [184]

---

to retrieve the crew. Hamilton, *Adventure,* p. 289-90. When Adjutant Malartic arrived at Québec on October 18, 1758 he was told that 224 men from *Aigle* were at Mécatina. By November 27 the equipment salvaged by the rescue vessel was only 30 leagues from Québec. Malartic and Gaffarel, pp. 211, 214. Unfortunately, the second vessel, presumably *Légère,* was wrecked at Bic, Québec. On November 27 Montcalm confided in his journal that "the equipment of the vessel *Aigle* seems destined for all the misfortunes possible." Casgrain, *Collection,* VII, p. 485.

Montcalm reported the wreck to Crémilles on October 21, noting that this was a grievous loss, and that the vessel had been carrying flour, pork, ammunition, muskets, brass guns, and mortars. *DRCHSNY,* X, p. 856. On the 27th he admitted to Belle-Isle that *Aigle,* which was also transporting clothing, was a total loss. *Ibid.,* p. 861.

494 Pouchot again confuses incidents and individuals. The fire ships released against the English fleet at Québec on June 28, 1759 were under the command of the Sieur Delouche who was accused of prematurely igniting his vessel. The enterprise was a failure. Knox, I, pp. 381-82, 382n. See also C.P. Stacey, *Quebec, 1759: The Siege and the Battle.* Revised edition, edited and with new material by Donald E. Graves (Toronto, 2002), p. 67.

Delouche was not, however, the unfortunate captain of *Aigle.* Montcalm identifies him as "M. de Laubara". On May 24, 1759, this naval officer, who had finally reached Québec, was sent with a detachment of sailors to man the French armed vessels on Lake Champlain. Casgrain, *Collection,* VII, p. 525.

"Laubara" was Joannis-Galand d'Olabaratz (d.1778), a former Louisbourg fisherman and privateer. He was commissioned in the French Navy as a fire-ship captain in 1748. His entry in the *DCB* says nothing of his connection with *Aigle* but indicates that he was recalled from retirement in 1758 to command vessels on Lake Champlain. *DCB,* IV, pp. 588-89. It is more likely that he was appointed to run supplies to New France. The *Aigle* fiasco made him a naval officer marooned in the colony without a ship and therefore available for duty on Lake Champlain.

Pouchot's erroneous association of Olabaratz with the failed fire-ship attempt (for which his navy rank would seem to have qualified him) might be attributable to the similar phonetics of "Delouche" and "de Laubara". Unfortunately, the military achievements of d'Olabaratz were no better than those of Delouche. On October 12, 1759 d'Olabaratz fled from English warships and ran his armed vessels aground on the shore of Cumberland Bay (note 535). General Amherst was aware that the commander of the French vessels, whom he identified as "Monsieur de la Bras", was "a Captain of Man of War." Amherst Journal, Knox, III, p. 52.

MÉMOIRES
SUR
LA DERNIERE GUERRE
DE
L'AMÉRIQUE SEPTENTRIONALE,
ENTRE
LA FRANCE ET L'ANGLETERRE.

Suivis d'Obſervations, dont pluſieurs ſont relatives au théatre actuel de la guerre, & de nouveaux détails ſur les mœurs & les uſages des Sauvages, avec des cartes topographiques.

Par M. POUCHOT, *Chevalier de l'Ordre Royal & Militaire de St. Louis, ancien Capitaine au Régiment de Béarn, Commandant des forts de Niagara & de Lévis, en Canada.*

TOME SECOND.

YVERDON.

M. DCC. LXXXI.

MEMOIRS
ON
THE LATE WAR
IN
NORTH AMERICA
BETWEEN
FRANCE AND ENGLAND

Followed by observations, of which several are pertinent to the current theater of the war, and fresh details on the customs and habits of the Indians, together with topographical maps.

By M. POUCHOT,
Knight of the Royal & Military Order of St. Louis, formerly Captain in the Béarn Regiment, Commandant of the Forts of Niagara & Lévis, in Canada.

VOLUME II

YVERDON
MDCCLXXXI

❁ *Memoirs On The Late War* ❁ *In North America, Between France & England*

• 1759 •

In the preceding volume we saw the fate of the goods destined for Frontenac. In order to replace these effects, recourse was had to merchants & private individuals. The little that had escaped the English in [1] the crossing from France to Canada was first sold to shopkeepers & other persons who had trading posts in the upper country to supply. They speculated in the goods among themselves at considerable profit. Finally, the last individuals to acquire them sold them to the intendant at the going price. It may well be imagined that the company & their friends were well informed about the purchases that would have to be made & monopolized the goods in order to be in a position to sell them to the King. If one adds to that the maneuvers to which people resorted at the trading posts, one will be able to estimate the kind of exorbitant price each item fetched & how many small fortunes were made among these dealers. Despite all these disadvantages, means were found to supply all the forts given the shortage of everything. [2]

Let us return to the military operations. In all the small war parties that took place, a few English prisoners were always taken and they revealed the preparation of a great campaign against Canada.[495] The commanders of the forts around Lake Ontario & the Belle-Riviere warned M. de Vaudreuil of the discontent of the Five Nations, which

495 Colonel James Montresor's plan for the 1759 New York campaign, dated December 29, 1758, is in *DRCHSNY*, X, pp. 907-12. Although many details would be altered before operations began, the main points of his outline would be followed. On March 15, prior to Pouchot's departure for Niagara, an Iroquois arrived at Montréal with information about the English plan of campaign. This included reference to the movement of three armies against Niagara, Lake Champlain and Québec. Malartic and Gaffarel, pp. 224-25.

496 "At last Captain Pouchot, of the Béarn battalion, is going to command at Niagara; he ought to have been sent off last fall; he was capable and agreeable to the Indians," wrote Montcalm to Crémilles on April 12, 1759. "Such was promised me," he further lamented of Vaudreuil, "but how resolve to dismiss a Canadian officer, however incapable and disagreeable to the Indians he might be?" *DRCHSNY*, X, p. 961. Vaudreuil had indeed agreed, after the loss of Frontenac, to send Pouchot and some *piquets* to relieve Vassan at Niagara. By the end of October, however, the governor had reversed his decision. See Memoir on Lake Ontario, [Sept. 12], 1758, *Ibid.*, p. 870 and Montcalm to Belle-Isle, Oct. 27, 1758, *Ibid.*, p. 860.

decided him to send M. Pouchot back to Niagara to command there.[496] M. Pouchot was very reluctant to accept this mission because he could foresee what must result from the shortage of resources in the country. M. de Montcalm tended to share his opinion.[497] However, he could not withhold the services of this officer from M. de Vaudreuil. They were to give M. Pouchot 300 French troops but, anticipating that they would be captured, they only gave him three *piquets* - 149 men in all. When [3] he took leave, M. Pouchot said: "It appears that we shall see one another next in England."[498] He left Mont-Réal on

497 Vaudreuil and Montcalm were deeply divided on the strategy for defending New France during 1759. Montcalm wished to withdraw from the West and hold the heart of the colony along the St. Lawrence River. Vaudreuil advocated further offensives in the Ohio Valley. Their dispute is summarized in Gipson, VII, pp. 381-84. A series of *mémoires* expressing the respective opinions of the two French leaders is in *DRCHSNY*, X, pp. 868-77.

Montcalm's position is outlined in his "Reflections" on the defense of the colony: "The war is entirely changed in this part of the world according to the manner the English are attacking us; nothing less is at stake than the utter and impending loss of the Colony or its salvation, that is to say, the postponement of its fall .... All private interests, every object of trade, ought to cease here; 'tis the trunk of the tree that's attacked; whatever concerns the branches is of the greatest indifference." *Ibid.*, p. 874. He nevertheless acceded to Vaudreuil's demands that Pouchot lead reinforcements to Niagara and, if possible, send them on to de Lignery on the Ohio.

Pouchot's contention that he shared Montcalm's views is not entirely true and very likely reflects an after-the-fact attempt to distance himself from Vaudreuil's position which encouraged him to send troops to the Ohio in June 1759. Niagara was therefore much less well-manned than Montcalm had hoped when the English appeared in July.

Montcalm's attitude towards Pouchot began to change at this time, probably because of the captain's support of Vaudreuil's plans or perhaps, ironically, because of Vaudreuil's favorable opinion of Pouchot. Negative comments about the captain began to creep into Montcalm's writings. "France never had more confidence in Marshal Saxe than the Marquis de Vaudreuil has in Pouchot," he wrote to Bourlamaque on March 23. Casgrain, *Collection*, V, p. 302. On June 1 Montcalm confided to his diary the sentiment that Pouchot had been "praised" in the cabinet (offices) of the governor and had thereby acquired the false belief that he had nothing to fear and could send troops to the Ohio. *Ibid.*, VII, p. 532. On July 16, after Niagara had been besieged, Montcalm wrote ironically to Lévis about "the Canadian arguments of Pouchot" that had resulted in stripping the post of many potential defenders. *Ibid.*, VI, p. 190.

498 This conversation is not recorded by Montcalm. The general was, however, under no delusions as to the fate of Pouchot's detachment if attacked in strength by the English. "If Niagara is besieged it will be taken. We must look forward to the siege, but it is not necessary to sacrifice too large a garrison," Memoir, March 21, 1759, Casgrain, *Collection*, IV, pp. 149-50.

499 Malartic confirms Pouchot's date of departure from Montréal. Repentigny had arrived there with 150 men on March 22 and moved up the river to Les Cedres on the 24th. Malartic and Gaffarel, p. 226.

...(continued next page)...

27 March with 157 Canadians under the command of M. de Repentegni, a colonial officer.[499] Having travelled four leagues over Lac St. François,[500] Pouchot noted that the ice on the lake was cracking & forming a channel in the middle. He immediately wrote to M. de Montcalm informing him of the fact, which meant that the river ice would imminently melt & that navigation would soon be possible. He only reached the top end of the lake with the greatest difficulty. As the ice gave way beneath the soldiers' feet, more than 30 Canadians fell in. Fortunately, they were able to climb back by clinging to the ice floes. Through some kind of miracle, no one was lost.[501] On 2 April, when above [4] the Long Saut,[502] he found bateaux from the fort of La Présentation which had come to meet him. He went on board and on the 4th he arrived at Pointe au Baril, three leagues above La Présentation.[503] He immediately went to the bay, where two barks were under construction, each of ten 12-pounder guns. They were very nearly ready to launch. He found one of them with its planking complete & the other with its ribs done. There and then, he had the number of laborers doubled to work on them day & night. He saw to the retrenching of the post, which a raiding party could very easily have burned, together with our vessels.[504]

Pierre-Jean-Baptiste-François-Xavier Legardeur de Repentigny (1719-1776) was a captain (1750) in the Marine. *DCB,* IV, pp. 448-49.

500 Lac St. François on the St. Lawrence above Montréal.

501 Pouchot sent a message about soft ice conditions to Montréal on March 28 and then set out from Les Cedres on the 30th. The warning caused Montcalm to order the *piquets* to postpone their departure. The *piquets* of Royal Roussillon and Guienne, with 200 Canadians, left Montréal on the 31st. Those of La Sarre and Béarn set out with another 200 Canadians on April 1. The last detachment of the Niagara reinforcement got underway the next day. On April 2 Montcalm received word from Pouchot that the ice of the St. Lawrence was breaking up. Malartic and Gaffarel, p. 226.

502 The long stretch of rapids above modern Cornwall, Ontario.

503 Pointe au Baril is now Weatherheads Point near Maitland, Ontario. The bay is identified in Volume III of Pouchot's *Mémoires* as Ance à la Construction. It is today known as Maitland Bay.

504 As early as September 2, 1758 Vaudreuil had reported to Massiac that his plans for reestablishing control of Lake Ontario included having "sloops built this winter, either to fight those of the English or to serve as transports on our lakes." *DRCHSNY,* X, p. 824. After some discussion, Pointe au Baril was selected for the shipyard site. Casgrain, *Collection,* VII, p. 483.

...(continued next page)...

On the 9th, we launched a bark which was named the *Iroquoise*. On the 10th, a courier arrived from Détroit. He announced that the Indians of the upper country intended to travel down to Mont-Réal. On the 11th, all of the [5] bateaux carrying the supplies & rigging arrived from Mont-Réal with the French detachments.[505]

On the 12th, the second vessel, named the *Outaouaise*, was launched. From this, it is easy to judge the diligence that had been deployed in their construction. They were rigged at once.[506] From the two

---

The military commanders consulted with naval constructors in the colony. Plans for the new vessels were drawn by René-Nicolas Levasseur (1705 or 1701-1784; *DCB*, IV, pp. 473-75), head of royal shipbuilding in New France. Levasseur interrupted work on the frigate *Québec* and dispatched a body of shipwrights under Louis-Pierre Poulin de Courval Cressé (1728-1764; *DCB*, III, pp. 539-40) to build two schooners pierced for ten 12-pounder cannon. Cressé had previously constructed a pair of armed vessels on Lake Ontario in 1755-1756. Levasseur to the Minister, Oct. 30, 1758, in Preston and LaMontagne, p. 270. Construction of the two new "barks" was well underway by the end of January 1759, and guns and ammunition for them were forwarded to Les Cedres on March 3. Malartic and Gaffarel, pp. 220, 222.

Pouchot reported that he had arrived at Pointe au Baril on April 5. Although he disliked the position, he proceeded to fortify it. *Ibid.*, p. 227. His primary concern was intelligence received at Montréal on March 5 that the English had plans for a winter raid from Fort Stanwix to burn the incomplete warships. *Ibid.*, p. 224-25. A crude sketch of the Pointe au Baril retrenchment is in Courville, *Mémoires sur le Canada depuis 1749 jusqu'à 1760.* The simple fortification was intended only to secure the shipyard against a *coup de main.* Casgrain, *Collection*, VII, pp. 509-10.

505 Rigging for *Iroquoise* and *Outaouaise* probably came from merchant vessels at Québec or Montréal. Montcalm Memoir, [Sept. 12], 1758, *DRCHSNY*, X, p. 871. The effort to reestablish control of Lake Ontario strained the resources of the colony. "The King's stores are stripped of all sorts of goods; the last detachment for La Présentation have [sic] emptied them," wrote Intendant Bigot to the Minister of Marine on April 15, 1759. *Ibid.*, p. 967.

A history of the construction and services of the two vessels is given in Ernest Green, "Corvettes of New France", *Ontario Historical Society Papers and Records*, XXXV (1943), pp. 39-38. This article should be used with care, however, especially regarding the vessels' British service following the campaign of 1760.

506 The vessels were of equal size and probably identical in most details. Each was fitted with two masts, but *Iroquoise* was rigged as a schooner while *Outaouaise* was completed as a brig. Casgrain, *Collection*, VII, p. 444. This decision was perhaps influenced by the rigging available. *Outaouaise* is depicted in a 1760 watercolor by Thomas Davies, entitled "A View of Fort La Galet", now in the National Gallery of Canada. See R.H. Hubbard (ed.), *Thomas Davies, c.1737-1812: An Exhibition Organized by the National Gallery of Canada* (Ottawa, 1972), p. 84 and Appendix D, Plate 12.

A conjectural profile of *Outaouaise*, based on the Davies watercolor, is in John F. Millar, *American Ships of the Colonial & Revolutionary Periods* (New York, 1978), pp. 212-13. While much of Millar's text is in error, he carefully scaled the vessel's length to 98 feet 6 inches overall and 84 feet between perpendiculars. Another source describes *Outaouaise* as "being a topsail [brig] of near 160 tons." When captured by the English on August 17, 1760, she was armed with one 18-pounder, seven 12-pounders, two 8-pounders, and four swivel guns. Knox, II, p. 541.

Frenchmen & two Indians belonging to a raiding party commanded by M. Villejoin, who was in the area of the Lac des Onoyottes,[507] M. Pouchot learned that there was no one at Chouegen & that the English were already massed at the portage,[508] where they were making all necessary preparations for the campaign. The Onoyottes had told him that Johnson was inviting all the Five Nations to come and hold council with the Onontagues, where their fire is lit.[509]

On the 25th, MM. Pouchot, Villars[510] & [6] Cervies,[511] all *piquet* captains, set out on the corvettes while their detachment went with M. de Bonna-foux, an artillery officer.[512] They arrived at Niagara on the 30th.

---

507 Michel Rousseau de Villejouin (fl.1750-1760), ensign (1754) in the Marine. McLennan, p. 344. See also *DCB,* IV, p. 688. This officer was operating in the vicinity of Oneida Lake.

508 The Oneida Carry had been refortified by the construction of Fort Stanwix, begun in August 1758. See Luzader, pp. 8-17.

509 Note in original, p. 6 (a): *That means their central meeting-place.*

Pouchot's map shows the principal Onondaga village. Tanner, p. 43, places it on the east side of Otisco Lake in Onondaga County, New York.

Johnson's council was not planned for the Onondaga village but rather for Canajoharie. It was held between April 4 and 21. See note 492.

510 Joseph Du Clercy de Villars (fl.1749-1760), captain (1749) in the La Sarre Regiment. There are several alternate spellings of his name. Villars had recently been granted the Cross of St. Louis. "Relation de Poulariès", *RAPQ,* 1931-32, p. 88. He took command of the grenadier company of his battalion on July 13, 1760. "Officiers du régiment de la Sarre", p. 260.

Montcalm had ordered that the two *troupes de terre* captains selected to accompany Pouchot to Niagara be "less senior" in date of commission. Montcalm to Lévis, Jan. 30, 1759, Casgrain, *Collection,* VI, p. 156.

511 The Sieur Serviès (fl.1755-1760), captain *en second* (1755) in the Royal Roussillon Regiment. "Officiers du régiment de Royal-Roussillon", p. 432. There are several alternate spellings of his name. He signed "Servies." Villars and Serviès to Amherst, Sept. 19, 1759, Amherst Papers/78.

512 Jean-Baptiste Bonafous (or Bonnafoux) de Caminel (1724-1802), lieutenant (1757 or 1758) in the Royal Artillery Corps. There are several variant spellings of his name. He entered the French artillery service as a cadet in 1746 and fought in the later campaigns of the War of the Austrian Succession. Fauteux, *Les Chevaliers,* pp. 188-89. See also Aegidius Fauteux, "Quelques Officiers de Montcalm", *Revue d'Histoire de l'Amérique Française,* V, No. 3 (December 1951), pp. 410-13 for further information on his military career.

Bonafous had come to New France in 1757 with the reinforcements noted earlier (note 374). He and his fellow officers of the Royal Artillery Corps were to supplement the colonial artillery officers. Bonafous was originally assigned to the company of colonial *canoniers-bombardiers* of Louis-Thomas Jacau de Fiedmont (note 238) and had charge of the artillery and fortifications of Québec and Fort St. Frédéric before going to Niagara. *Ibid.,* pp. 410-13. He and his fellow Marine and Army officers are named in a memoir on the artillery of New France, Jan. 11, 1759, *DRCHSNY,* X, pp. 935-37.

On May 2 & 3, M. Pouchot questioned prisoners taken on the Belle-Riviere, who gave him fairly exact details about the situation of the English in that sector. He learned that they had 4 to 500 men at Fort Pittsburgh, 150 at Loyal-Anon & that all regular troops had been recalled from them. These prisoners gave a fairly good description of the forts.[513]

On the 5th, M. de Vassan, in command at Niagara and relieved by M. Pouchot, left on the corvettes. On the 9th, M. Pouchot began work on repairing the fort, to which nothing had been done since his departure. He found the ramparts caved in, the protective sod [7] in a collapsed state & that the scarp and counterscarp of the ditches had become clogged with soil.[514] He also had blinds erected to act as cover in case of siege.[515] On the 11th, a raiding party arrived from Fort Bull bearing 6 scalps.[516] Joncaire,[517] who was with the Five Nations, sent M. Pouchot a message warning that the English were on the march

---

513 A winter garrison of 200 Pennsylvania and Virginia provincials had been left at the temporary fort at Pittsburgh. Stotz, p. 121.

514 The land defenses of Fort Niagara, as constructed in 1755-1757, were earthen walls faced with sod to impede erosion. Upon his return in 1759, Pouchot found two critical parts of these walls, the scarp (or escarp) and counterscarp in poor condition. The former term designates the outside of the rampart (i.e., the exterior face of the main walls), and the latter describes the outer wall of the ditch facing the scarp. Muller, *Treatise,* pp. 218, 215. The ditch (*fossé* in the original) was "a large deep trench made round each work, and the earth dug out of it, serves to raise the rampart and parapet." *Ibid.,* p. 217.

The deterioration of the scarp and counterscarp was probably due less to Vassan's neglect than Pouchot implies. The damage was likely the result of preliminary work on a project to revet the walls in stone. This effort had begun by May 3, 1758 when a lieutenant of the Royal Artillery Corps, Louis-Auguste-Victor d'Espinassi (fl.1757-1790; Fauteux, *Les Chevaliers,* p. 219), left Montréal for Niagara with 25 bateaux loaded with goods and provisions. Journal of Occurrences, *DRCHSNY,* X, p. 839. The stone was to be obtained from quarries near Fort Frontenac. D'Espinassi apparently began work, but his undertaking was aborted by the English capture of Frontenac. Vaudreuil to Massiac, Sept. 2, 1758, *Ibid.,* X, p. 823. See also Severance, II, pp. 211, 239. D'Espinassi had arrived in New France with Bonafous in 1757. Memoir on Artillery, Jan. 11, 1759, *DRCHSNY,* X, pp. 935-37.

515 Pouchot uses the word *blendage* [sic: *blindage*] which we have translated as "cover." His reference might have been more specific, however. In eighteenth century siegecraft "blinds" (*blindes*) were "wooden frames, composed of four pieces [of wood] .... Their use is to fix them either upright or in a vertical position against the sides of the trenches or saps, to sustain the earth .... BLINDS, are sometimes only canvas stretched to take away the sight of the enemy. Sometimes they are planks set up .... In short, they signify any thing that covers the labourers from the enemy." Smith, *Dictionary,* p. 30.

516 Pouchot presumably refers to one of the new outposts on the Oneida Carry because Fort Bull had been destroyed by de Léry's attack in 1756 (note 296).

& that the Indians had all come out in their favor.[518] We learned that during the month 14 French vessels had arrived at Québec for the quartermaster, as well as 4 frigates.

On the 14th, Pakens [a] Missisake[519] & 40 warriors arrived to form a war party. On the 17th, a Sauteur from Saguinan[520] announced that the commandant M. Bellestre was there & only waiting for M. Pouchot to summon him to come from Détroit.[521] The reinforcements would have arrived late, since they were more than 100 leagues away. Langlade, a colonial officer [8] resident in that region was due to go down to Mont-Réal with 1,000 Indians.[522] Among that number were the Folles-Avoines who, as has already been stated, had killed two Frenchmen. They brought with them the two men most responsible for the killing and handed them over to M. de Vaudreuil at a meeting in which they made restitution for the dead men. M. de Vaudreuil gave them back to the Indians to exact justice. They shot them both, a unique occurrence for the Indians since Europeans have been living in America.[523] Finally, all the nations of Upper Canada

517 Philippe-Thomas Chabert de Joncaire (1707-c.1766), captain (1751) in the Marine. *DCB,* III, pp. 101-02. See also Severance, II, pp. 441-47. Although Philippe is sometimes confused with his younger brother Daniel (note 331), Pouchot distinguishes between the pair by referring to Philippe as "Joncaire" and to Daniel as "Chabert."

518 Pouchot reported that the Seneca had told him that the English would attack Québec and Carillon but remain on the defensive at Fort Stanwix and on the Ohio River. This information was passed on to Montréal where it arrived, probably by the hand of Captain Vassan, on May 11. Malartic and Gaffarel, pp. 231-32.

519 This Mississauga chief has not been positively identified. He might be "Poton", described as "chief of Wapaskotiang" who met with Sir William Johnson at Johnson Hall on September 10, 1764. *SWJP,* XI, pp. 354-56. He visited Johnson again in 1772, *Ibid.,* XIII, p. 714. "Pakens" had been active in the French interest as early as June 8, 1757 when he led a war party from Niagara against the Pennsylvania frontier. Casgrain, *Collection,* XI, p. 94.

520 Chippewa from the lower Saginaw River area of Michigan. Tanner, p. 41.

521 François-Marie Picoté de Belestre (note 521) had been appointed commandant of Detroit in 1758. *DCB,* IV, p. 634.

522 Charles-Michel Mouet de Langlade (1729-1800/01) ensign (1755) in the Marine. Langlade had been appointed second in command of Michilimackinac in 1757. *DCB,* IV, p. 563. He arrived at Québec with 200 warriors on June 30. Malartic and Gaffarel, p. 245.

523 The attack on French residents of Green Bay is recorded in Volume I, p. 129 of Pouchot's *Mémoires.* The incident occurred in late 1757 or early 1758. Initial reports gave higher French casualties, Pouchot himself recording 22. See note 401.

...(continued next page)...

intended to contribute men to travel down along the St. Lawrence & Presqu' Isle.

On the 17th, there arrived a large deputation of Sonnontoins made up entirely of chiefs. They brought word from the Loups of Théaogen, who were our friends in 1757, & said **[9]** that they had set out to come and see their father Sategariouaen[524] but that a Sonnontoin had barred their way by means of a necklace (a ceremony to divert them from going to the French).

M. Pouchot reprimanded the chiefs for their reluctance to serve the French, after all the concern he had shown for them, which was very different from the treatment they had received from Johnson. He upbraided them for sending wampum collars to the Hurons & to the Outaouais in order to separate them from the French. These collars said: "My brothers[525] and my cousins,[526] we consider ourselves dead. The English & **[10]** the French cover our mats with blood. They are so large that we are crushed. Consider us dead men. But the smoke from our bones spreads over you and over all the nations of America & you in turn will die. We invite you to grasp each other by the hand in order to defend your mats & your lives. But do not rely on us, for we are dead."[527]

On the 18th, there arrived five Missisakes from Carillon bearing a scalp who said that they had seen a large number of troops & many bateaux at Fort George. A prisoner taken on April 21 said that the English were very weak in their forts on the Ohio, having lost **[11]** many men to disease caused by rotten food. At Loyal-Anon, there remained only 100 men fit

---

Pouchot places the surrender of the Folles Avoines in the wrong year. A party of that nation arrived at Montréal early in August 1758 with seven men who had been involved in the murders. Three were shot in the city square while the others asked to go to war to atone for their crime. Montcalm, then at Carillon, recorded the incident in his journal on August 13 and added that the submission of this distant nation reflected great honor on the French. Casgrain, *Collection,* VII, p. 432. The execution was also noted by Bougainville, although the modern English translation mistakenly says that the three were flogged. Hamilton, *Adventure,* p. 261.

524 Note in original, p. 10 (a): *This word means "the center of good dealings", the name given to M. Pouchot by the Five Nations.*

525 Note in original, p. 10 (b): *The Hurons.*

526 Note in original, p. 10 (c): *The Outaouais.*

527 Note in original, p. 11 (a): *They meant by this that they were inhibited by the forts and the armies of the English and the French.*

enough to serve.[528] On the 20th, there arrived a courier from the Illinois, bearing from France despatches for our generals & the intendant.

In that same month, M. de Langis burned 500 barges & the bark of the English on Lake George, capturing or killing 40 men who were guarding them.[529]

We were warned from France that an English fleet had set sail in the early days of February to besiege Québec with 10 thousand men under the command of General Wolfe.[530] Twenty-five thousand men were destined to penetrate into Canada by way of Lake George under General Amherst. A detachment was to be created for Lake Ontario.[531]

---

528 The prisoner was taken near Fort Ligonier on April 21. Lieutenant Colonel Thomas Lloyd reported that "a few days since a Party of Sick were surpriz'd on their Way to Bedford escorted by a Corporal and Eight [men], who brott off his whole Party & left the Sick to the Mercy of the Enemy[.] Eleven are killd or missing in this Affair." Lloyd also reported the deaths of 16 provincials in his garrison since April 1. Lloyd to Bouquet, April 21, 1759, Stevens, *Bouquet Papers*, III, p. 248. Other correspondence from Fort Ligonier during this period confirms that the garrison suffered greatly from sickness. See Lloyd to Bouquet, April 1, 1759 and Blane to Bouquet, March 20, 1759 in *Ibid.*, pp. 228-30 and 212-13.

529 Pouchot has again confused events. Ensign Jean-Baptiste Levrault de Langis Montegron (note 389) had indeed taken several English prisoners near Fort Edward sometime before May 19, 1759. Malartic and Gaffarel, pp. 234-35. He did not, however, destroy any vessels at Lake George. The English had intentionally sunk their sloop, radeaux and bateaux before withdrawing in the autumn of 1758 (note 458). Although a detachment sent by Montcalm from Carillon located the sloop *Halifax*, about 50 bateaux and some caches of supplies, the French were unable to retrieve the large armed vessel. Hamilton, *Adventure*, p. 292.

Pouchot was almost certainly confusing Langis' raid on Lake George with an earlier operation. Late in August 1758, following Bradstreet's capture of Frontenac, Langis had undertaken a reconnaissance of Oswego where he had seen "the wreck of our burned sloops ... as well as a great many barges." Journal of Occurrences, *DRCHSNY*, X, p. 854.

530 James Wolfe (1727-1759), colonel of the 67th Foot and major general commanding the expedition against Québec. *DCB*, III, pp. 666-74.

531 Pouchot credits a European source for this intelligence. Prime Minister William Pitt's instructions to Major General Jeffery Amherst for the 1759 campaign were drafted only on December 29, 1758. Gertrude Selwyn Kimball (ed.), *Correspondence of William Pitt* (2 vols.; New York, 1906), pp. 432-42. James Montresor's plan was submitted to Amherst in New York on the same day. It called for 10,000 men to operate on Lake Champlain and 9,500 for service on Lake Ontario. *DRCHSNY*, X, pp. 907-12.

It is likely that Pouchot was actually recalling information about plans for the 1759 campaign received from Europe in the summer of 1759 and sent to his attention in 1760. A copy of Montresor's plan had been obtained by the Minister of Marine and was forwarded to Vaudreuil in a letter of April 14, 1759. In preparing Pouchot for the 1760 campaign, Vaudreuil provided him with this intelligence. "I have lost no time in communicating to M. Pouchot the plan of attack by the upper part of the river drawn up by M. Montrezord, Chief Engineer of the Department of New-York, which you transmitted to me in your letter of the 14th April, 1759, in order that he may ascertain if the plan for this campaign be the same." Vaudreuil to Berryer, June 24, 1760, *DRCHSNY*, X, p. 1091.

We detached three thousand men [12] - French regulars, colonial troops & militiamen - under the orders of M. de Bourlamaque to cover Carillon. The remainder of our forces with MM. de Vaudreuil, de Montcalm, & de Lévis, proceeded to Québec to resist the English fleet. The forts on the lakes & the Belle-Riviere had no reinforcements other than those mentioned above.[532]

We shall begin the account of this campaign by the Carillon affair. In the early days of July, General Amherst[533] set out from Fort George with a corps of 12 to 14 thousand men & disembarked at the camp of Contre-Coeur. He approached the retrenched camp of Carillon which we abandoned. M. de Bourlamaque fell back on St. Fréderic, leaving a garrison in Fort Carillon in order to cover his retreat. The enemy [13] mounted several guns on the top of the retrenchments, from which they pounded the fort, which was soon evacuated after we had partly demolished it. Only a few men were captured in this retreat. M. de Bourlamaque also had Fort St. Fréderic blown up & withdrew to the Isle aux Noix in the St. Jean River, where he immediately built a retrenchment.[534]

General Amherst dug in at St. Fréderic, where he began construction of a fort & a retrenched camp. This frontier remained fairly quiet

---

532 Bourlamaque's force of 3,000 included the regular battalions of La Reine, 2/Berri and 3/Berri. Montcalm to Belle-Isle, May 8, 1759, *DRCHSNY,* X, p. 970. Bourlamaque's account of the 1759 campaign on Lake Champlain is in *Ibid.,* pp. 1054-57.

The reinforcements for the lakes and the Ohio were those men who went to Niagara with Pouchot.

533 Jeffery Amherst (1717-1797), appointed major general and commander of English troops in America late in 1757 in place of Loudon. *DCB,* IV, pp. 20-26.

534 Amherst's army numbered 11,376 men (6,537 regulars and 4,839 provincials) as of June 19. Knox, I, p. 474n. His troops moved down Lake George and landed on July 22 on the east side of the head of the Lachute River rather than at the Camp of Contrecoeur on the opposite bank. Bourlamaque withdrew his main force from Carillon on the 23rd leaving a rear guard of 400 men under Captain Hébécourt (note 275) who had commanded the post since its construction. He and his men slipped away on July 26 after igniting the powder magazine.

Fort St. Frédéric was also blown up by its garrison on the 31st and occupied by Amherst's army on August 4. For accounts of the campaign see Gipson, VII, pp. 361-65, Stanley, pp. 234-41 and Bellico, pp. 87-103. The operations before Ticonderoga are shown in a "Plan of Ticonderoga" published in the *Universal Magazine,* November 1759, p. 265. This has recently been reproduced in *The Bulletin of the Fort Ticonderoga Museum,* XV, No. 5 (1993), pp. 376-77.

By the St. Jean River, Pouchot means the Richelieu. A crude plan of the French lines on Isle aux Noix was printed in Courville and is reproduced in Stanley, p. 238 and Bellico, p. 105.

during the campaign after that event, apart from a few scalps that we took from the English. In the month of October, the English tried to advance in the region of the St. Jean River. A corps of 5 to 6 thousand men embarked & advanced to just below the Isles aux Quatre Vents, where they were [14] buffeted by such a violent wind that they lost several bateaux, which greatly disconcerted them. They returned to camp.[535]

On leaving Mont-Réal, M. Pouchot was under orders to abandon the outposts on the Belle-Riviere in the event that he was sure he was under attack. Having received no news that the enemy had come down the Chouegen River, he thought it feasible and without risk to send the troops and the officers with their provisions & the trade goods destined for the fort of Presqu' Isle, or Machault, where [15] M. de Lignery was in command. He hastened the departure of the convoy for this portage, having planned to destroy the enemy forts of Loyal-Anon & Pittsburgh on the basis of the intelligence he had received about the enemy's situation. After this expedition, there would have been nothing further to fear from this sector. All the troops & reinforcements heading there would have regrouped at Niagara in order to defend the fort.[536] The minds of the Indians were at that time in a great ferment, either for or against the French. Those of the Belle-Riviere invited those of Détroit to unite with them at Sandoské.[537] The latter refused to do so without the consent of M. de Vaudreuil.

Some Hurons & Iroquois who had been at Fort Pittsburgh, formerly Fort du Quesne, reported that they had met four Indians, Têtes Plates, a nation [16] hostile to ours, who told them to beware

---

535 Note in original, p. 15 (a): *The French had been forced to abandon the lake, after the shipwreck of several warships. General Amherst also set about opening up a communication route between Fort Carillon, or Ticonderago,* [sic] *and the provinces of New Hampshire and Massachusetts.*

Amherst began construction of a fort at Crown Point a short distance from the ruins of Fort St. Frédéric. On October 11 he at last moved north against the French. The next day, near the Isles aux Quatre Vents, today the Four Brothers Islands southwest of Burlington Bay and near the widest part of Lake Champlain, his boats took shelter from the French naval vessels in the mouth of the Au Sable River. There followed a period of bad weather, and the army returned to Crown Point by October 21. Gipson, VII, pp. 366-68; Bellico, pp. 101-02. Amherst's own version is in J. Clarence Webster (ed.), *The Journals of Jeffery Amherst* (Toronto, 1931), pp. 179-83.

The French warships were not wrecked accidentally. On October 12 the English vessels *Duke of Cumberland* and *Boscawen* drove three of the four French warships, *La Musquelongy, La Brochette* and *L'Esturgeon,* into Cumberland Bay (near modern Plattsburgh, New York) where they were scuttled by their crews. The English salvaged them and incorporated them into their own squadron. Only *La Vigilante* remained in French hands. Bellico, pp. 101-03.

of the English as the latter sought only to pit them one against the other in order to destroy them one after the other as soon as the French had been eliminated from the land. They added that some of them had joined in raiding parties against the English, with whom they had been at war.[538] Several nations had attended a council at Fort Pittsburgh. The commander (it was still Forbes)[539] had spoken with arrogance. "When I came to this territory," he said, "I thought that someone would contest my right to be here. [17] However, those who were here have fled like frogs at which a stick is thrown. As for you Indians, you allow yourselves to be deceived by the French for a piece of tobacco no longer than someone's finger. You take up the hatchet & when you lose a man you mourn him for years on end. We act quite differently. We learn of the loss of an army or a battle with as much pleasure as when we win a victory. We want to have for you the same friendship that we had for your ancestors the first time we set foot on your land. But if you fraternize with the Frenchmen, you will be dead & we shall strike in every quarter." The Indians replied:

---

536 The instructions given to Pouchot were carefully recorded by Desandrouins. Pouchot was to examine the St. Lawrence River for defensive positions, expedite the construction of the vessels at Pointe au Baril and then proceed to Niagara to take command. Once at his post, Pouchot was to ensure that supplies were forwarded to the Ohio and that his vessels guarded against an English descent of the St. Lawrence. He was to improve the defenses of Fort Niagara and mend relations with the Iroquois. From that point, Pouchot was given considerable latitude. If the English seemed poised to attack Niagara, Pouchot was to call in the colonial and Indian forces available in the upper lakes in order to defend his post. If not, then he could commit his men to offensive warfare along the Ohio. Gabriel, p. 256.

Pouchot's decision to send reinforcements to de Lignery on the Ohio was the only action of his North American service for which he was almost universally criticized by his fellow officers. His sensitivity about the matter is apparent here.

537 Sandusky Bay (Ohio) where the French maintained a small post on the south shore of Lake Erie at the mouth of the Sandusky River.

538 Note in original, p. 17 (a): *M. Pouchot doubtless is speaking here of the general uprising of the Indians of the Ohio and the upper country which took place in 1763 and 1764. This rebellion almost forced the English to evacuate the entire interior of North America.*

It is unlikely that Pouchot was confusing the hostility of the Têtes Plates or "Flat Heads" towards the English with the later "Pontiac Uprising" of 1763-1764. Some warriors might indeed have joined in raids against the English. The Têtes Plates encountered near Pittsburgh might also have simply been attempting to curry favor with the Hurons and Iroquois.

539 Forbes had departed Pittsburgh on December 4, 1758 and died at Philadelphia on March 11, 1759. The commandant at this time was Lieutenant Colonel Hugh Mercer (note 482). See Stevens, *Bouquet Papers,* III.

"Brother,[540] the Frenchman [18] is a hundred times braver than you. Your arrogance does not deserve a reply." They left him immediately.[541]

Certain Iroquois said to some Hurons that in the messages they had sent them[542] there were some things that were so thoroughly veiled that only the most subtle minds could understand them and that only the chiefs should deal with them. They added that they had resolved to impose peace on their lands & that in the first instance they desired to chase away the Frenchman, who was braver, and then the English. The Hurons replied: "Our brothers the Iroquois should be careful of what they do. Who [19] can keep the Frenchman in check? And even if you were strong enough to chase him away, when the nations come down from the north & say to you 'Come now, arise, here we are with the hatchet you sent us to kill the enemies of our father,' what reply will you give them? Take good care, brothers, not to do anything stupid." M. Pouchot reprimanded the Indian who had spoken before a senior Huron chief at Sandoské for the Iroquois participation in this council. The Iroquois denied having even thought of such an explanation. The Huron chief told him: "It is not surprising that these words were spoken at Sandoské, which was a fire lit without the consent of any nation. Only madmen and evil-doers remain where ill dealings are always [20] conducted." He added that his nation had promised not to listen to anything which emerged from such a place, as could be judged from the collars that had been sent to M. Bellestre. They had thus not deigned to reply.

On May 28, there came to Niagara four Tonniac[543] chiefs & 33 Goyogoins to hold council. M. Pouchot reproached them for sending collars in order to alienate other nations from our friendship. They replied with a collar: "Father, it is true that we have lost our minds. We are obliged to you for reminding us of the fact. We are well aware of

540 Note in original, p. 19 (a): *They always called the English thus, just as they called the French* father.

541 This Fort Pitt council has not been specifically identified. It might have been that held on May 12 between Colonel Mercer and a number of Delawares and "the Mingoes [Iroquois] from the Head of the Ohio." There is no indication, however, that the Mingoes walked out. Indian Intelligence, April and May 1759, Stevens, *Bouquet Papers*, III, pp. 279-80.

542 Note in original, p. 19 (b): *We have already mentioned above the wampum collars.*

543 It is uncertain what Indians Pouchot refers to here as the "Tonniac." They would seem to be Iroquois, and perhaps he means the Tuscarora or even Seneca.

the evil we have done and no one is more troubled than we ourselves. The Frenchman solicits us from one quarter, the Englishman from the other. Both of you give us [21] specious reasons. The Englishman tells us to beware of the French, who is cunning & tries to deceive us in a subtle manner. Each nation heaps presents upon us. For our part, what we know is that the master of life has given to us, the Indian residents, the island of America. We understand nothing of the claims of the English & the French. We are ignorant of their reasons for making war on one another. Our true intention is to remain neutral. You are both so strong that we consider ourselves crushed in spite of our own power. The Englishman constantly steals from us warriors lured by brandy & sumptuous presents. We have expressly forbidden them to join their cause & scold them roundly [22] when we learn that they have. You French rob us too, but we are indifferent to this. This is obvious from our relatives, who all make war in your cause." Their feelings could not have been expressed in a more naive manner. After begging with wampum strings[544] to have their implements repaired, they said: "Father, we know that the English have stolen Cataracoui from you, but that is not our fault. We had warned you that they would. If we learn that the Englishman is plotting anything against you, we shall inform you immediately, so that you are not taken by surprise. We beg you to cover your nipples with a leaf of Indian corn so that your milk will not be spilled. Then we shall be able to speak calmly of good matters."[545]

This Tonniac was genuinely [23] committed to the French. He was a highly intelligent man & greatly esteemed by his nation. But the general malaise had spread and they had been perverted. Fearful that some Iroquois favorable to the English had done something stupid in the fort, he prevented his fellows from drinking. It was the only time Indians refused drink, which was a clear sign of their mistrust.

---

544 The original word is *branches,* literally "branches." By this Pouchot means strings of wampum.

545 Pouchot's faith in the promise by the Iroquois to warn him of an English movement against his post, expressed here by Cayuga and "Tonniac" chiefs, would have disastrous consequences. Within the next five weeks an army would pass through the Iroquois country and appear without warning before Niagara. Lévis was one of many senior officers who believed that Pouchot had been deceived by the Iroquois and lulled into a false sense of security. Casgrain, *Collection,* I, p. 182. Pouchot was clearly overconfident in his belief that he had again brought the western Iroquois back into the French interest.

On June 1, the detachments & the ammunition for the Belle-Riviere left, under the command of M. de Montigny.[546] M. Pouchot handed him a letter for M. de Lignery, of which the following text is an extract. It will fill in some of the details of some subsequent events.

"M. de Montigny, sir, has now joined you. The difficulty of portaging has delayed him up to now. He has managed to overcome it and brings with him a sufficient quantity **[24]** of flour to enable you to wait for the reinforcements from the Illinois. I have had it put in sacks so that the difficulty of transporting it over the portages will not prevent it being carried in carts. I was only prepared to allow these provisions to travel with M. de Montigny, so that officers assigned to each bateau can see to their safety, & so that they will reach you in as good condition as when I sent them to you."

"I have also sent forty bales of assorted trade goods in excellent condition & ten cases of muskets. In addition, there are three hundred shovels, picks, & axes. They are indispensable tools for military operations & I imagine that your supplies of them are low, since you have just marched from Fort **[25]** du Quesne."[547]

"I flatter myself, sir, that you will want to maintain frequent contact with me. You will see that I shall do my very best to procure you every assistance within my power, whether for the good of the service **[26]** or in order to oblige you. I am, however, extremely short of supplies at the moment."

M. Pouchot also informed M. de Lignery that he had learned through a courier from the Illinois that 300 men under the command of MM. Aubry & the Chevalier Villiers[548] were due to arrive, together with 2 to 3 hundred thousand-weight of flour that had been left at the Miami portage. M. de Port-Neuf,[549] commandant of Presqu' Isle and

---

546 Jean-Baptiste Testard de Montigny (1724-1786), captain (1757) in the Marine. *DCB,* IV, pp. 733-34.

547 Note in original, p. 26 (a): *M. Pouchot added two 4 pounder fieldpieces which could be transported along the Belle-Riviere to Fort Pittsburgh. There were draught horses in that sector.*

548 François Coulon de Villiers (1712-1794), captain (1754) in the Marine. Villiers was an officer of the Louisiana establishment and spent much of his career in the Illinois country. He was one of a number of Illinois officers who served in the campaigns along the upper Ohio. Villiers was a member of a distinguished military family, and his brothers Joseph (note 165) and Louis (note 164) had both served in the Ohio Valley. *DCB,* IV, pp. 177-78.

549 Pierre Robinau de Portneuf (1708-1761), captain (1757) in the Marine and commandant of Fort de la Presqu' Isle since 1756. *DCB,* III, pp. 560-61.

responsible for controlling the portage, had been asked to fetch them immediately with his bateaux. M. Pouchot then turned to the operations of the campaign, according to the plan he had conceived on the basis of the English deployment on the Belle-Riviere.

"Upon their return, sir, your raiding parties should have informed you about the current deployment of the enemy from Raiston to Fort Pittsburgh. Following [27] the account of your latest prisoner,[550] they are, in my view, less well placed than we are in that sector & with little hope of receiving any substantial reinforcements."

"I beg you to interpret the views I have the honor to express to you here as the thoughts of a good friend. Even if we do not always think of everything, at least we may have a new idea."

"According to the above reports, sir, it seems that you may be in a position to undertake an offensive operation. I have requested M. de Montigny to assign to you, as soon as he arrives at Presqu' Isle, an officer & a number of very alert French troops, so that upon their arrival you may send out a party to scout the positions at Loyal-Anon & Pittsburgh & the roads around them and [28] take action on the basis of their reports."

"In his instructions, he should be ordered to observe the roads, those places suitable for encampment, and the passes suitable for ambushing convoys. He should also spy out the mountain tracks & the rivers they have to cross. All this will spare our detachments many unnecessary maneuvers.

"The officer in question must also study the extent of their forts, the kind of works used to defend them, those parts of the defenses not yet finished, the heights commanding them & those places where it might be possible to take up positions, either to fire on them or to seal them off. If, sir, you are so placed that you can march there with all your troops, as I hope will be the case, [29] you should carry with you the tools I have sent in order to be able, as of the first night there, to open up a trench, either of abatis of trees or a ditch with the earth thrown towards the fort. This entrenchment, as you know, must be sited as close as possible to the fort & in such a way as to cause the enemy the maximum harassment and to appear to cut off all his communications."

"If the enemy is in a poor position, he will be so taken aback that he will very likely seek to capitulate as soon as he is summoned to do so. Above all, it should be brought to his notice that if he allows the

---

550 Presumably the man taken near Fort Ligonier on April 21 (see note 528).

attack to begin, we shall not be able to save him from the hands of the Indians, as happened **[30]** after the surrender of Fort George.[551] The large number of Indians he will perceive in your ranks will doubtless reinforce this message. If you were fortunate enough to capture Loyal-Anon, you will certainly find that all the forts from Raiston to the Belle-Riviere will fall to you as well, as they will be abandoned to their own devices and unable to receive supplies of food."

"I believe, sir, given the present state of affairs, that that is the most appropriate course of action in the circumstances. If I had the honor to command in that sector, I should certainly do my level best to engage in some useful & striking operation. If these forts are such as we must suppose they are, they will doubtless fall to us. **[31]** If any body of troops proceeds to that sector, we shall be in a position to attack them in a part of the road of our own choosing. And if it was too large to be attacked, we would be well placed to destroy all of its baggage as it passed, which would soon reduce them to a desperate plight. Such, sir, are my general thoughts on the operations before you. Your campaign should be as lightly equipped & rapid as you can make it, whether to avoid the defection of the Indians or the danger of your supplies running out. You must also ensure that you can reach us in time if this sector is seriously threatened. The relative inactivity of the enemy seems to me to give us at least two months of security. You thus have **[32]** time at your disposal to do with as you judge most fitting."

An Iroquois arrived from Orange[552] and informed Joncaire,[553] who was with the Five Nations, that two war parties of Iroquois were being formed to strike at La Présentation & Niagara as revenge for the death of two Agniers killed by our Indians. M. Pouchot, who was absolutely convinced that the Iroquois had taken up arms against us,[554] had

---

551 In spite of his embarrassment over the actions of Montcalm's Indian allies at Fort William Henry in 1757, Pouchot was not averse to invoking the incident as a threat.

552 Albany, New York.

553 Note in original, p. 33 (a): *This colonial captain, a half-breed resident among the Iroquois, enjoyed a great deal of esteem among them. His brother Chabert and he had more than 60 dependents, children whom they or their fathers had procreated there.*

"Joncaire" was the elder brother, Philippe-Thomas Chabert de Joncaire (note 517).

554 If Pouchot had by this time awakened to the hostile attitude of the Iroquois, it was a recent revelation. Montcalm, at Québec, only recorded Pouchot's report of the fate of La Milletière and Joncaire on July 16. He sarcastically noted that the Niagara commandant "begins to doubt the good faith of the Five Nations." Casgrain, *Collection,* VII, pp. 574-75.

wanted to persuade him to return, together with his son-in-law & the Canadians they had with them. But their desire to trade [33] a cargo of goods which the son-in-law, La Miltiere, an officer from Languedoc,[555] had brought, prevented them from returning. La Miltiere & the French were captured by the Indian raiding parties, & Joncaire was obliged to escape to Niagara.[556]

Some Goyogoin chiefs informed M. Pouchot that Johnson had used very large collars of wampum to persuade their nation to follow him & had sent an invitation to our nations to imitate the Iroquois. They also warned us not to allow our troops to become separated from the main body, thus avoiding capture by the war parties that were to be sent. A very sizeable party was to be formed to pillage the fort of the

555 Honoré Dubois de La Milletière (or Miltière) (fl.1757-1761), lieutenant (1757) in the Languedoc Regiment. "Officiers du régiment de Languedoc", p. 285; Promotions in Languedoc, Jan. 28, 1757, Casgrain, *Collection,* III, p. 84.

La Milletière, like Pouchot, was one of those rare *troupes de terre* officers who demonstrated an aptitude for working with the Indians. This was certainly aided by his marriage, on September 19, 1757, to Gabrielle, daughter of Philippe Thomas Chabert de Joncaire. *DCB,* III, p. 102 and Severance, II, p. 443. Of La Milletière's prospects and talents, Montcalm noted, in April 1758, that he "is a young officer who has no property in France; is well liked by the Indians whom he often accompanies on an expedition, and has easily learned the Iroquois. The Marquis de Vaudreuil proposes to transfer him, some day, with his rank, to the Colonial troops." *DRCHSNY,* X, p. 699.

The lieutenant's connection to the Chabert family was apparently no guarantee against the jealousies that colored relations between *troupes de terre* and Marine officers. According to Bougainville, La Milletière, who had been adopted by the Indians, was sent by Montcalm with 16 Iroquois warriors "of his clan" in August 1758 "to follow the trail of M. de Joncaire." La Milletière was delayed for two days by Captain François-Pierre Rigaud de Vaudreuil (note 305), brother of the governor, because, Bougainville opined, "it is an attempt against the rights of the colony to send a French officer with the Indians." Hamilton, *Adventure,* pp. 264-65. It was perhaps at that time that La Milletière joined Joncaire at Niagara and in the Iroquois country. There is, however, some indication that the lieutenant was in Montréal during the winter of 1758-1759 and went to Niagara with Pouchot. Montcalm to Lévis, Jan. 30, 1759, Casgrain, *Collection,* VI, pp. 156-57.

556 La Milletière was delivered to British Brigadier John Prideaux (note 125) near Oswego Falls on June 26, 1759. He carried some useful papers but provided little intelligence about Fort Niagara. Prideaux to Amherst, June 26, 1759, Amherst Papers/46B.

The lieutenant remained a prisoner in New York when Pouchot and most of the Niagara officers were exchanged in 1759, probably because there were not enough captive English officers to be exchanged and not because of his relationship to Joncaire (who was also held) and the Iroquois. Although Lévis solicited La Milletière's exchange as late as June 1760, the formalities had not been completed before the fall of the French colony. The lieutenant remained in New York, returning to France with Pouchot and Bonafous in 1761. His exchange had been completed by that time. Lévis to Amherst, June 14, 1760, Casgrain, *Collection,* II, p. 344; Amherst to Lévis, June 21, 1760, *Ibid.,* p. 352; Bonneau to Choiseul, March 9, 1761, *Collection de manuscrits,* IV, pp. 307-08.

portage.[557] M. Pouchot sent 100 men to cover it and to scour the forest around it. However, several families of Sonnontoins assured him that they wanted to remain at Niagara because they came from that area. [34] M. Pouchot was not at all displeased by this, as he regarded them as a kind of safeguard against the enemy raiding parties who might well fear that we would take reprisals against their own people.

On the 17th, there arrived some Onontagues with scalps taken by a party from the Five Nations in the area of Loyal-Anon in an attack on a convoy of 16 wagons laden with provisions for the enemy. It had been escorted by 100 men, of whom 27 had been killed & 3 taken prisoner. The remainder had escaped into the forest. The wagons were burned & 84 horses captured. This party was under the command of M. St. Blin.[558] From this it may be gauged that many Iroquois still sided with us & that only fear of the English had decided the

557 Fort Little Niagara, the stockade situated above Niagara Falls at the head of the portage, had been constructed in 1751. The post was the concession of Lieutenant Daniel Chabert de Joncaire (note 331). Severance, I, pp. 374-76; Chabert, pp. 10-11.

558 The Sieur Duverger de Saint-Blin (1720-after 1774) was a lieutenant (1757) in the Marine. "Tableau ... des Officiers de la Marine", Brymner, 1886, p. clxxxiii. His name is often rendered as "Duverger Simblin." In 1763, in response to his implication in the *affaire du Canada,* he published *Mémoire pour le sieur Duverger de Saint-Blin* (Paris, 1763) in which he described himself as a "lieutenant of infantry" of the troops in Canada. A summary of his career is in Severance, II, pp.52-53 and in his own *Mémoire.* Saint-Blin was commandant of Fort Le Boeuf from 1756 until 1759. That post, the second in the chain of portage forts from Lake Erie to Fort Duquesne, had been established in 1753. Stotz, pp. 74-75.

Another biographical summary of Saint-Blin is in Roy, *Bigot et sa bande,* pp. 147-50. Roy identifies him as Paul-François Duverger de Saint-Blin and bases his account largely on the lieutenant's *Mémoire.* Roy gives his date of birth as 1696 which seems far too early for both the rank and the activity of this officer during the 1750s. A list of the "General State of the Canadian Noblesse", made in 1767, put Saint-Blin's age at 47. Brymner, 1888, p. 44.

Saint-Blin says his party included 40 Indians, and that he attacked 200 English. The lieutenant was wounded in the action. Duverger de Saint-Blin, p. 4. His attack on the supply train is also recorded in the anonymous "Mémoire du Canada", *Rapport de l'Archiviste de la Province de Québec pour 1924-1925* (Québec, 1925), p. 157. This source says he took four wagons of provisions and several prisoners.

The ambush took place on May 23 within three miles of Fort Ligonier where 100 Virginia provincials were escorting 12 wagons. Captain Thomas Bullet lost 40 men in the action, abandoned the wagons and fled to the fort. The English reported five wagons and the provisions destroyed and 50 horses carried off by the French. Clark to Hoops, May 25, 1759, Stevens, *Bouquet Papers,* III, pp. 331-32. Prisoners were also taken, and at least one of them, John Peter, Jr. (1732-c.1798), was among the English captives found in Fort Niagara following its surrender in July 1759. *New York Mercury,* Aug. 20, 1759. Peter had been carried to Venango and eventually taken to Niagara. Kenneth G. Robison, "John Peter, Jr. of Early Albemarle County, Virginia", unpublished manuscript. Copy in the Old Fort Niagara Association Library.

nation to come out in their favor. However, [35] these Indians carried out the general will of their nation with as much order as civilized nations & more secrecy than them.

In May & June, trade was very abundant because of the gathering of Indians who came from all parts of America to see their father Sategariouaen again. The trade at Niagara, which in a normal year, barely exceeded 150 bales, rose in those two months to seven to eight hundred. One may judge from this what it might have reached had the country been peaceful, for the Indians travelled there only with fear, in constant dread of the Five Nations & the English.

M. Pouchot sent a courier to warn M. de Corbiéres,[559] who was at Frontenac, M. de la Corne,[560] who was at La Présentation, & M. de Vaudreuil, about the La Miltiere affair [36] so that they would be on their guard against the Iroquois. However, the Sonnontoins from Sonnechio[561] sent wampum collars to M. Pouchot, by which they apologized for the capture of La Miltiere on their territory. Through the intermediary of M. Chabert, a colonial officer who was Joncaire's brother and highly esteemed by the Iroquois, M. Pouchot tried to persuade the Iroquois chiefs to come for a council to Niagara, in order to reduce the influence of Johnson. At this period he was informed that the Loups & the Chaouanons, who had seen the arrival of the detachment which was going to join M. de Lignery, had immediately asked the latter for some French troops to go and attack Fort Pittsburgh. He only detached M. Marin,[562] M. Rocheblave,[563] 3 Canadians & 280 Indians who went and harassed the forts which they found in a very poor state of readiness & which they would have captured if [37] they had had more French troops.

On the 27th, there arrived a troop of Missisakes whom M. Pouchot had sent out to gain intelligence about the English at Chouegen. These Indians had left on the bark *Outaouaise* which was assaulted by such a violent squall[564] that its mainmast & its bowsprit were broken.

---

559 Lieutenant Corbière (note 475) then commanded what remained of the French establishment at Fort Frontenac. Preston and LaMontagne, p. 468.

560 Captain Luc de La Corne (note 447).

561 Geneseo in the Genesee Valley of New York.

562 Lieutenant Joseph Marin de La Malgue (note 449).

563 Ensign *en second* Philippe-François Rastel de Rocheblave (note 340).

It had to put [38] into La Présentation for some time, which impeded patrolling on the Chouegen River & was in part responsible for the failure to detect the English advance. The Missisakes were with M. Blainville, a colonial cadet.[565] They only went 3 or 4 leagues up the river & found nothing. If they had gone up as far as the Falls, they would have seen the English, busy at the portage. The news that the enemy was not at Chouegen caused M. Pouchot to hope that there would be a further period without military operations. Above all, he believed that the English, before they came to Niagara, would establish a supply depot at Chouegen & would be compelled to retrench it there. They did not do this, however.[566]

On the 29th, a courier from Presqu' Isle announced that 100 Frenchmen & 150 Indians were due to arrive [39] from Détroit, plus 6 to 700 Indians with M. Lintot,[567] & 100 Indians with M. Bayeul[568] and then the convoy of M. Aubry from the Illinois with 6 to 700 from the detachments on the Mississippi. Some Indians from Michilimakinac[569] who arrived the same evening said that MM. la Verranderie[570] & Langlade had come down the Grande Riviere[571] with 1200 Indians - Christinaux, Scioux, Sakis, Folles-Avoines,

---

564 Note in original, p. 38 (a): *One remarkable thing was that the Indians, who had never experienced a storm on board a ship, were very deeply afraid and threw all their ornaments, arms & tobacco into the water in order to appease the Manitou of the lake. By chance, on the ship there was a Canadian passenger who was a genuine dwarf in stature. The Indians, who had not seen such a small man, took him for a Manitou & were only prevented with the greatest of difficulty from killing him & throwing him in the water like another Jonah.*

General Amherst learned of the damage to *Outaouaise* from a French deserter on July 17. The French soldier reported that "in coming to Oswego ran a Sloop on a Rock; were obliged to cut the masts to take her back to la Gallette to repair her." Webster, *Jeffery Amherst,* p. 140.

565 The identity of this young officer-candidate is uncertain. Since Pouchot uses the name "Blainville" instead of "Céloron", he probably does not refer to one of the sons of the leader of the 1749 expedition, Captain Pierre-Joseph Céloron de Blainville (note 154). In his list of officers of the garrison of Fort Lévis, he names them as "the two Celoron brothers." They were, additionally, both of higher rank. (note 786)

566 The advance elements of Prideaux's army reached Oswego Falls on June 24 and Oswego itself on June 27. Orders, June 23, 24 and 27, 1759, *SWJP,* III, pp. 52-53, 55-56.

567 Probably Daniel-Maurice Godefroy de Linctot (1739-1783), ensign (1759) in the Marine. Pouchot might also refer to one of his two brothers, Hyacinthe or Maurice-Régis, who were also Marine officers. *DCB,* IV, pp. 302-03.

568 François Odet de Piercot de Bailleul (1719-1782), lieutenant (1756) in the Marine. Fauteux, *Les Chevaliers,* pp. 186-87.

Sauteurs, & Renards.[572] If all these details are taken into account, the reader will judge that we ought to have been able to anticipate a happy outcome. It will soon be seen what all these announcements amounted to.

On July 6, the *Iroquoise* entered the river at four in the afternoon. M. Pouchot learned from this corvette that there were no English at Chouegen. If it had [40] patrolled during its voyage and moved in close to the south shore of the lake it would certainly have detected the enemy who were advancing inshore with their barges. It if had seen them, it was sufficiently well-gunned, with its 10 12-pounder cannon, to stop the army's advance or even to destroy it. The English would have been hard put to emerge unscathed, whether they continued their advance or turned back. This was a misfortune, since these ships were armed for no other purpose. The squall which one of them experienced was a contributory factor also. While one ship was journeying up and down the lake, the other was supposed to be on patrol.[573]

At six o'clock on the same day, a soldier hunting wild pigeons in the bush noted some Indians who captured two of his comrades. He

---

569 The French post at the Straits of Mackinac between Lake Huron and Lake Michigan. The site is in Mackinaw City, Michigan. The post at Michilimackinac had been established in 1715 or 1716. For a recent summary of its history see Donald P. Heldman and Roger T. Grange, Jr., *Excavations at Fort Michilimackinac: 1978-1979: The Rue de la Babillarde* (Mackinac Island, MI, 1981), pp. 15-54. The fort has been partially reconstructed to the period of the early 1770s and is operated as part of Mackinac State Historic Parks.

570 Louis-Joseph Gaultier de La Vérendrye (1717-1761), lieutenant (1757) in the Marine (*DCB,* III, pp. 241-44) and his elder brother Pierre (note 968).

571 The Ottawa River.

572 The diversity of the warriors led down the Ottawa River by La Vérendrye and Langlade (note 522) demonstrates the wide-ranging contacts of the French at Michilimackinac: Christinaux were Cree from the region south and west of James Bay; Scioux (Sioux) were the Nadouessioux or Dakota from the headwaters of the Mississippi and the Minnesota River; Sakis or Sauks came from the Mississippi River of northern Illinois; Folles-Avoines were the Menominee of Green Bay in Wisconsin; Sauteurs were Chippewa or Ojibwa from the margins of Lake Superior; and Renards were the Foxes or Mesquakies of the Wisconsin and Fox River corridor of Wisconsin. "Glossary of Indian Tribal Names", *DCB,* III, pp. xxxiv-xli; Tanner, pp. 40-41.

573 The master of *Iroquoise* was René-Hippolyte La Force (1728-1802). His was apparently a position appointed by the governor and not a military post. *DCB,* V, pp. 470-71. La Force's journal for the period July 6 through 14, 1759 was printed in Knox, III, pp. 233-41. A translation has been included as Appendix C of this edition of Pouchot's *Mémoires.* La Force's entry for July 6 gives no indication of concern about English movements.

immediately ran to warn M. Pouchot who straight away sent out a scouting party of about ten men [41], supported by 50 troops. These soldiers marched out in a light-hearted manner in the belief that it was only an Indian raiding party. Several of them were surrounded & were subjected to more than 200 rounds of musket fire. Five of them were captured & two wounded.[574] M. Pouchot had recommended this *piquet* not to become too deeply engaged, having judged that they were not equally matched. He had it return to the fort after ordering several cannon shots to be fired at the enemy. They responded with disciplined volleys from the edge of the cleared ground, which suggested to us that there were troops in battle order & that they were there in force. That night, M. Pouchot stationed guards to occupy the outer works of the fort.

It is necessary at this point to enter into some details about the state of the fort at the moment of the siege. M. Pouchot had just finished raising the height of the ramparts. The batteries [42] on the bastions were not yet completed. He formed them out of barrels filled with earth. Upon his arrival, he had ordered work to be started on protective blindage timbers, which were made of oak, 14 inches square & 15 feet long. He had them erected at the rear of the large house

574 The time of the attack on the pigeon hunters is given as 7:00 p.m. in an anonymous journal entitled "Siège de Niagara" in *RAPQ*, 1933-34, p. 92. The document might have been kept by Pouchot or his fort major. The La Force journal (Appendix C) says the surviving hunter returned to the fort at 7:30.

The *piquet* was led by Captain Serviès of the Royal Roussillon Regiment. Two of the captives from this series of skirmishes, identified in La Force's journal, were the Sieur Saumandre and Blois the Iroquois interpreter. Two others, the Sieurs St. Martin and Pramont were wounded but able to return to the fort. A slightly different list is given in the anonymous "Journal of the Siege of Niagara, Translated from the French", printed in the *New York Mercury*, August 20, 1759 and included as Appendix B of this edition of Pouchot's *Mémoires*. This says that "Messieurs Furnace and Aloque", interpreters to the Iroquois, two Canadians and two other "Gentlemen" were captured.

The captives and wounded appear to have been cadets of the Marine or garrison employees. François Premont, a commissary, was among those later exchanged with Pouchot in December. List of French Prisoners, Dec. 14, 1759, Amherst Papers/8. The "Furnace" identified in the 1759 English translation of the anonymous journal might actually be the "son of the elder Forge" named in La Force's journal. A cadet, identified as "Soumandre" but listed erroneously as having been captured on July 24, is certainly one of the men who fell into English hands on the 6th. *SWJP*, XIII, p. 115. The others have not been further identified.

575 As explained above (note 515), blinds were screens used to cover areas of fortifications or siege works from the view of an enemy. The "large house" was the stone building of Fort Niagara known today as the "French Castle." For a history of the building see Dunnigan, *Glorious Old Relic*.

beside the lake, which was the best protected place for a hospital.[575] He constructed along the walls of the powder magazine, in order to give protection to the walls and to be used as casemates, a large shed with beams shaped into a gable at the top. The arms and armorers were there. It will be noted that this type of fortification is excellent for field works in wooded country & that they could easily be used as barracks and magazines. A bomb falling on a sloping area does little harm, because this type of construction is very sturdy. **[43]**

The garrison was made up of 149 men, detached from the regiments of La Sarre, of Royal-Roussillon, of Guienne & of Béarn, under the general command of M. Pouchot, captain in the Béarn Regiment, of M. de Villars, a captain in the La Sarre regiment, of M. de Cervies, a captain in the Royal-Roussillon, of M. Morembert, a lieutenant in the Guienne regiment,[576] and M. Salvignac, a lieutenant of the Béarn regiment,[577] and La Miltiere, a lieutenant of the Languedoc regiment.[578] Add to this 183 men from colonial companies, under the command of M. de la Roche, a colonial captain[579] and of Cournoyer[580] & Larminac,[581] both lieutenants, of 133 militiamen and 21 artillerymen, under the command of M. Bonnafoux, a lieutenant in the Royal

576 Jean-Baptiste Roussel de Morambert (d.1760), lieutenant (1755) in the Guienne Regiment. He was mortally wounded at the Battle of Ste. Foy and died in the hospital at Québec on May 9, 1760. "Les officiers du régiment de Guyenne", p. 193.

577 The Sieur de Salvignac (d.1760), lieutenant (1753) in the Béarn Regiment. He was wounded at Ste. Foy and died in hospital at Québec May 3, 1760. "Officiers du régiment de Béarn", p. 357.

578 La Milletière was not present at the siege having earlier been captured in the Iroquois country. See notes 555-556.

579 Charles-René de La Roche Vernay (c.1728-1761), captain (1755) in the Marine. Fauteux, *Les Chevaliers,* p. 182. The list of French officers in the English translation of the articles of capitulation of Fort Niagara records his first name as "Oliver." *DRCHSNY,* X, p. 992. This is an error. The name in the French versions is actually signed "*Sieur* de la Roche Verney."

580 Jacques-Ange Hertel de Cournoyer (1727-1789), lieutenant (January 1, 1759) in the Marine. Fauteux, *Les Chevaliers,* p. 201. Cournoyer was probably the author of the anonymous "Journal of the Siege of Niagara, Translated from the French", printed in the *New York Mercury,* August 20, 1759. See appendix B of this edition of Pouchot's *Mémoires* for both the journal and further biographical information.

581 Jean-François de Larminat (fl.1757-1763), lieutenant (1757) in the Marine. He was a European-born officer who had arrived in New France only in 1757. Fauteux, *Les Chevaliers,* p. 193.

Artillery Corps. This number was increased by M. Pouchot to 100, drawn from picked troops and militiamen, in all 486 men & 39 employees, five of them women or [44] children. The women acted as nurses together with two ladies called Douville[582] & also sewed cartridge bags & made earth sacks.

On the 7th, seven barges appeared from the lake banks in order to reconnoitre the fort. We allowed them to group together and come in close. As soon as we saw they did not to wish to come any closer, we fired cannon at them, which made them put back into the lake again.[583] M. Pouchot immediately sent out a bateau to reconnoitre. It reported that it had sighted 15 to 20 barges, each containing 20 men, at the entrance to the small marsh.[584] M. Pouchot imagined at first

582 These ladies, unfortunately, are not further identified by Pouchot. Although Stanley, p. 217, says they were Gray Nuns, this was not the case. Captain Alexandre Dagneu Douville (note 714) was then commandant of Fort Rouillé at Toronto, and one of these women was his wife. It seems safe to assume that he had sent his family from Toronto to the relative security of Fort Niagara. Douville had wed Marie Coulon de Villiers, daughter of Nicolas-Antoine Coulon de Villiers (note 148), in 1730. Robinson, p. 140, says that Marie was Alexandre's first wife, and it is not clear if she was his wife in 1759.

On August 10, 1759 Captain Luc La Corne reported the arrival of 28 women and children of the former Niagara garrison at Isle aux Galops in the St. Lawrence River. Their leader was one "Mme Douville" who provided useful intelligence on the size and movements of the English force that had taken Niagara. Captain Douville, who had by this time evacuated Fort Rouillé and was with La Corne, was allowed to take his family and the other Niagara dependents to Montréal. La Corne to Lévis, Aug. 10, 1759, Casgrain, *Collection,* X, pp. 202-03. The report of the English officer who escorted the Niagara women and children from Oswego to French territory refers to "one of the women" as the wife of "Monr douVille governor of Tronto [sic]." Report, July 31-Aug. 7, 1759, Thomas Gage Papers, William L. Clements Library, Volume 3 (hereafter cited as Gage Paper/Volume number).

583 Both Captain La Force of *Iroquoise* and the anonymous French journalist of the siege also noted a sighting of English "barges" on the morning of the 7th. They attributed the boats' movements to a reconnaissance of the fort. See "Journal of the Siege of Niagara" (Appendix B) and the La Force Journal (Appendix C).

New York provincial Lieutenant Christopher Yates recorded an attempt by the English whaleboats to attack *Iroquoise,* but he placed it on the evening of July 6. "The whale boats went in order to catch the sloops [sic] but the sloops laid under the fort so that they could not catch them. The fort shot several cannons at the boats, shot one man, taking his leg right off." Austin A. Yates, *Schenectady County New York: Its History to the Close of the Nineteenth Century,* (n.p., 1902), p. 92. This is probably the incident described by the French observers on the morning of the 7th.

584 The "small marsh" was Four Mile Creek, New York, so called because it is located that distance east of the mouth of the Niagara River. The place was known to the French as *le petit marais.* The reconnaissance was made by a sergeant at 6:00 a.m. "Siège de Niagara", *RAPQ,* 1933-34, p. 92.

that it was the advanced guard of the English army. He despatched almost immediately a second boat to reconnoitre, under the command of the lieutenant of the bark.[585] The latter said that he had seen many barges and an encampment on the coast and that there appeared to be **[45]** many men & campfires on the land. A further scouting expedition, sent out two hours later, reported that it had sighted 16 barges & a single tent, but a lot of men walking along the coast.[586] The barges had all entered the small marsh & the army was camped in the woods.

M. Pouchot immediately sent a courier to M. Chabert, who was in command of the fort of the portage, to order him to withdraw along the Chenondac[587] if he was aware of any news of the enemy in the area of his fort, for fear that the position might be taken. The courier also carried a message ordering the withdrawal from Presqu' Isle of all the French & Indian detachments who might be there & also orders for M. de Lignery at Fort Machault to fall back on Niagara with all the French & Indian troops **[46]** under his command. He urged them to form a small advanced guard to observe if the little fort was abandoned & in that case to cross by the Chenondac and come and join him at Niagara, leaving behind only a detachment to cover their bateaux & their equipment.

At noon he ordered out the corvette *Iroquoise* with a month's supplies to patrol the area of the small marsh. The winds were from the S. & S.W. She went off to bombard the enemy camp.[588] During the day there appeared several scouts near a thicket 2 to 3 feet high, although M. Pouchot had had a part of it cut down as soon as he

---

585 The second scout departed at 9:00 a.m. "Siège de Niagara", *RAPQ,* 1933-34, p. 92.

La Force recorded on July 7 that "I twice sent my long boat on reconnaissance on the orders of M. Pouchot." On July 12 he identified his second-in-command as "the Sieur Corbin." This was probably Étienne Corbin (1706?-1768) who, with his father and brother, had worked in the royal shipbuilding program of the 1730s and 1740s. *DCB,* III, pp. 138-39.

586 The third expedition was dispatched at noon, led by the sergeant who had made the 6:00 a.m. reconnaissance. "Siège de Niagara", *RAPQ,* 1933-34, p. 92.

587 Chippawa Creek, also known as the Welland River, located just above Niagara Falls on the west bank of the Niagara River.

588 Pouchot was no sailor. La Force recorded winds from the northwest. The schooner captain's entry for July 7 describes a considerable exchange of fire with guns placed in the English camp.

The "Siège de Niagara" says that *Iroquoise* was ordered out at 2:00 p.m. *RAPQ,* 1933-34, p. 92.

arrived. Many Indians also appeared with the intention of sniping at us, but a few cannon shots forced them to withdraw.

In the evening, a Pouteoutamis & a Sauteur entered the fort. They had arrived from the portage fort. M. Pouchot **[47]** suggested to them that they should go out on a scouting expedition that night. He gave them a Huron from the fort as a companion. They went off along the lake banks as far as the woods right at the end of the cleared ground[589] and returned by way of its middle without having encountered anything. An hour before daybreak, the Pouteoutamis, who was extremely brave, went back by himself. He left by way of the lake banks towards the angle which it forms in front of the fort. He came across a canoe in which there were three men. He fired on the middle man, who was wounded. The other two returned his fire without hitting him and then fled. He went all the way around the cleared ground shouting taunts at the enemy Indians.[590] **[48]**

On the 8th, M. Pouchot sent the Indian, accompanied by two Frenchmen, with a letter to M. Chabert. When they were a league away, anxious about his situation, they sent back one of their number post haste to inform him that they had seen the tracks of about forty men in the forest. As these tracks led from the upper reaches of the river, M. Pouchot fearing that a number of the enemy had crossed it, which would have filled him with anxiety over the fate of those who had to come from that direction, sent out a scouting party to scour that part of the forest. But they found nothing after searching to the distance of one league.

At noon we signalled to the corvette to send in its long boat. The lieutenant who came said that the enemy had made an encampment on a small height of land, beyond the small marsh, in order to cover their bateaux. He estimated that there were 3 to 4 thousand men &

---

589 Note in original, p. 48 (a): *These cleared areas are called* déserts *in Canada.*

We have retained the word *désert* when it appears hereafter.

The three Indians left Fort Niagara at 8:00 p.m. and returned two hours later. "Siège de Niagara", *RAPQ*, 1933-34, p. 92.

590 Contemporary plans, particularly Pouchot's own "Niagara" of 1757, clearly show a prominent point on the lake bank about 800 feet east of the walls. This feature has long since been eroded into Lake Ontario. The "Siège de Niagara" places the exchange of fire between the lone Potawatomi and the canoe at 1:00 a.m. *RAPQ*, 1933-34, p. 92.

591 An abatis was an entanglement of tree branches placed before a fortification (note 414). Given the context here, it is likely that Pouchot referred to work parties cutting brush for the construction of fascines.

that they [49] were working hard at the edge of the cleared ground by the lake. There appeared to be 400 men busy constructing abatis.[591] M. Pouchot suspected that this might be the place where they were making a depot for their trenches. The guns of the corvette harassed them to such an extent that they were forced to abandon their camp in order to seek cover.[592] They fired 12-pounder cannon at the vessel. M. Pouchot ordered the corvette to take up station opposite the entrance to the small marsh to prevent the convoys from entering or bateaux from leaving to transport their artillery to the depot, which was a league & a quarter away from the marsh. This was sure to prolong their works. He ordered the captain of the ship, in the event that he was troubled by a squall, to sail back into the river & to lie in close to the sandbank beneath [50] the fort.[593] These precautions forced the enemy to undertake all operations on land & covered the place[594] which could easily have come under fire from the direction of the lake and the river.

In the afternoon several hundred Indians appeared in the thicket of the cleared ground and began to fire on the fort. They were dispersed with artillery using grapeshot. A number of them were killed. At sunset, M. Pouchot had a Frenchman & two Indians fetched from the other side of the river. The former was a brother of the quartermaster.[595] He had been sent to recruit Missisake Indians. He returned accompanied by one only. The others had gone back upon seeing the small fort that had been burned, imagining that the English had done it. The other Indian was an Iroquois sent by M. [51] Chabert[596] who in a letter announced his arrival for the following day. He had thrown

---

592 La Force recorded in his journal on July 8 that he "was not able to see what effect my fire had on them."

593 Pouchot refers to the low area between Fort Niagara and the river. The site is today occupied by a United States Coast Guard Station. This was the landing place for the post and was well-sheltered from both the lake and the English.

594 Pouchot uses the contemporary engineering term *place.* "PLACE, in fortification, signifies in general terms a fortified town, a fortress: hence we say it is a strong place." Smith, *Dictionary,* p. 205. For the sake of clarity, we have usually translated *place* as "fort."

595 The term translated as "quartermaster" is *garde magasin* in the original. The investigators of the *affaire du Canada* named two brothers, Curot l'ainé (the elder) and Curot le jeune as former *gardes magasins* of Fort Niagara. Roy, *Bigot et sa bande,* p. 12. From Roy's description, it seems that the elder was François Curot (b.1723) and the younger was Charles Curot (b.1733). *Ibid.,* pp. 197-99. The exchanged Niagara prisoners included a "Guard Magazin Curot." "List of the French Prisoners", Dec. 14, 1759, Amherst Papers/8.

into the Chenondac River everything he could, including 20 horses which belonged to him & some cattle that he had transported from Détroit at his own expense. He burned the portage fort since it was entirely untenable.[597] His brother **[52]** Joncaire had arrived the day before, guided by the Iroquois who had brought the letter.[598] M. Pouchot gave the latter a present.

Around 10 o'clock,[599] there appeared a white flag in the cleared ground.[600] M. Pouchot had it reconnoitered with caution. A captain of the Royal American regiment was brought to him blindfolded. He was led in by way of the thickest part of the underbrush. After his blindfold had been removed, he handed over in the commandant's room a letter from Brigadier Prideaux[601] in which the latter stated that

---

596 Note in original, p. 52 (a): *M. Chabert had the concession at the Niagara portage. The King was supposed to supply him at an agreed price with trade goods in order to pay the expenses involved in the portage. This officer was therefore well placed to make a lot of money, since he acquired at the fixed price goods which were more and more costly for the King. As a result, he made a substantial profit. However, it is certain that no officer was more zealous for the good of the service & that he was the most esteemed and essential in America in maintaining good relations with the Indians. In this he was much more trusted than Johnson among the Indian nations.*

See note 331 for references to Lieutenant Daniel-Marie Chabert de Joncaire de Clausonne and his own published *Mémoire.*

597 La Force and his crew observed smoke on the horizon of the Niagara escarpment on the morning of July 8. The captain correctly speculated that it came from Fort Little Niagara. La Force Journal (Appendix C).

598 Pouchot refers here to Joncaire's arrival at Fort Little Niagara following his close call in the Iroquois country. Joncaire and his brother entered Fort Niagara together on July 10. The "Journal of the Siege of Niagara" (Appendix B) agrees.

599 On the morning of July 9.

Pouchot's narrative does not clearly indicate the beginning of his entry for July 9, but it appears to be at this point. The "Journal of the Siege of Niagara" has no entry for July 9 and places the arrival of the flag of truce on the 8th. Both La Force and the "Siège de Niagara" put the event on the morning of July 9. *RAPQ,* 1933-34, p. 92. A summary of the siege in Lévis' journal says that Pouchot was "summoned" on July 9. Casgrain, Collection, I, p. 189.

From the English side, Lieutenant Yates says the call to surrender was made on "Monday the 8th." Monday was actually the 9th. Yates, p. 92.

600 Here is another of the many instances of the English using a white flag as a signal to parley (see note 439). This mimicked the flag flown over French colonial garrisons. For a concise account of French flags of the period see René Chartrand, "Les drapeaux en Nouvelle-France", *Conservation Canada,* I, No. 1 (1974), pp. 24-26 and Chartrand, *Canadian Military Heritage,* pp. 70-71.

601 Brigadier John Prideaux (note 125).

the king of England had given him command over Fort Niagara and that M. Pouchot must surrender it, otherwise he would compel him to do so with the superior force at his disposal. M. Pouchot replied that he did not understand English and had [53] absolutely no reply to give him. However, he understood the letter perfectly well.[602] The officer insisted upon the very large force that he had. M. Pouchot responded that the King had entrusted him with the fort, that he was well equipped to defend it & that he hoped M. Prideaux would never enter it. He added that at least before he did, M. Pouchot wanted to make their acquaintance and would surely earn their respect. He had the officer served breakfast & sent him back blindfolded to the place where he had been captured.[603]

After midday, La Force, the commander of the corvette, sent word to M. Pouchot that he could no longer see any barges on the beach nor any depots & only a few men around the banks. On receiving this information, M. Pouchot sent out a sergeant in a bateau who went up the other side of the river. He reported that he had seen [54] many men working at La Belle-Famille.[604] This led us to assume that they

---

602 The "Journal of the Siege of Niagara" (Appendix B) records only that Pouchot "seemed to be unwilling to receive the English General's letter."

603 For some reason Captain La Force, in his Journal (Appendix C), heard from his long boat crew that Prideaux's summons had been carried by an officer named "Blain" or "Blaine." This identification made its way into early secondary sources and is also in Courville's *Mémoires sur le Canada,* published in 1838 (p. 149). The only Royal American officer of that name, however, was Archibald Blane, (fl.1756-1759), lieutenant (1756) in the 60th Regiment of Foot. *Army List.* He was serving in Pennsylvania in July of 1759. Cochrane to Tulleken, July 12, 1759, Stevens, *Bouquet Papers,* III, pp. 405-06.

The bearer of Prideaux's summons was actually Walter Rutherford (fl.1755-1765), captain (1755) of the grenadier company, 4th battalion, 60th Regiment of Foot. *Army List.* Rutherford sold his commission on April 18, 1761. Rutherford to Johnson, May 12, 1761, *SWJP,* X, pp. 265-66. He then entered the Indian trade with John Duncan, another former officer, and unsuccessfully tried to obtain a monopoly on the Niagara Portage. Rutherford was still active in New York as late as 1765. Smith and Rutherford to Johnson, Feb. 5, 1765, *Ibid.,* XI, pp. 561-63.

Captain Rutherford left an account of his meeting with Pouchot. Breakfast, according to Rutherford, included drinking "an excellent Bottle of Claret and a glass of liquor with the Commandant." His complimentary appraisal of Pouchot's character was *que je crois etre bon soldat et Homme d'Esprit* ("I believe him to be a good soldier and a man of spirit"). Rutherford to Haldimand, July 14, 1759, Haldimand Papers, Add. MSS 21728.

604 Note in original, p. 55 (a): *This place is situated barely an eighth of a league away from the fort, on the right bank of the river above it. It was the most convenient place for cutting fascines, since there was no scrub among the large trees of the forest.*

...(continued next page)...

wished to open up the trenches that very night. When evening arrived there appeared some men in shirts at the edge of the *désert* who seemed to want to open up a trench. We fired three or four cannon shots at them & they withdrew, which indicated that that was not the place where they wanted to begin their trench.

The great calm which settled over the enemy during that day made us distrustful of their operations. In consequence, M. Pouchot stationed M. [55] Villars, a captain in the La Sarre regiment, in the demilune with 60 men;[605] in the retrenched place of arms of the covered way on the left, Lieutenant de Morembert & 30 men; in that of the right, Lieutenant Cornoyer & 30 men.[606] Captain de Cervies with 76 men occupied the salient from the covered way of the lake bastion up to the salient angle of the covered way of the demilune;[607] on the beach at the bottom of the banks of the lake bastion behind the palisade was Lieutenant de Larminac & 40 men; Captain de la Roche with 30 men was at the salient from the bastion of the covered way of the bastion

---

The location of "La Belle Famille", as it is today known locally, was on the east riverbank about one mile south of Fort Niagara. The exact site is believed to have been in the middle of the modern village of Youngstown, New York. The origin of the name and the reason the spot was so-called is uncertain. Contemporary maps do not record the name but seem to indicate that La Belle Famille was located near the edge of the forest. The maps indicate that La Belle Famille might have been partially cleared, but it was not a farm. The comment above, probably a marginal note by Pouchot, would seem to indicate the place was forested. For further discussion of the origin of the name and the exact location see Dunnigan, *Siege,* p.87.

605 The demi-lune or ravelin was the large triangular outerwork that still dominates the land defenses of Fort Niagara. A demi-lune was defined as "a work placed before the curtain [main wall] to cover it, and prevent the flanks from being discovered sideways; it is made of two faces meeting in an outward angle." Muller, *Treatise,* p. 217. The size of the detachment stationed there is indicative of its importance to the defense of the fort.

606 The covered (or "covert") way was the outermost defensive position of Fort Niagara. It was "a space five or six toises broad, going quite round the works of a fortification, and is adjoining to the counterscarp of the ditches, covered by a parapet 7 1/2 feet high, terminating in an easy slope towards the field." Muller, *Treatise,* p. 214. The "place of arms" of the covered way was that part of it "opposite to the re-entering angle of the counterscarp, projecting outwards in an angle." *Ibid.,* p. 228. The retrenched places of arms at Fort Niagara were smaller triangular fortifications. One was situated on each side of the demi-lune. Both were removed prior to the American Revolution. That of the right (river side) was reconstructed in the 1930s. The covered way from the salient of the demi-lune to the river has survived largely intact.

607 Serviès and his men were thus responsible for the northern half of the covered way. The size of his detachment is a further indication that Pouchot expected the English siege works to approach along the lake rather than the river side.

of the Five Nations;[608] on the flat space below this salient, behind the palisade which went down to the river, was M. Chabert with 60 men; on each bastion we stationed 25 men. [56] These different posts supplied all the sentries necessary. The 100 artillerymen were allocated to the batteries.[609] There remained only M. Bonnafoux, the artillery officer, & M. de Salvignac, a lieutenant of the Béarn Regiment, acting as major.[610] This was the order of battle every night during the siege. In the daytime, we tried to relieve the soldiers, in order to allow them to sleep, or we kept them busy at the various chores necessitated by siege operations.[611]

That night, M. Pouchot created a detachment of 30 volunteer scouts, among whom there were three or four Indians. They left via the right, the center & the left. They fired on a few enemy Indians who had crept up as far as a cemetery 50 *toises* away from the glacis.[612] A Huron from the detachment on [57] the right strayed from his party and was wounded by one of our own Indians when he tried to rejoin his detachment.

---

608 This appears to be a misprint or an omission that should probably read "Captain de la Roche with 30 men was at the salient from the salient angle of the covered way of the demi-lune to the covered way of the bastion of the Five Nations." This was the right flank position corresponding to that of Serviès. The smaller detachment reflects the lesser threat anticipated on the river side.

609 Pouchot notes earlier that his garrison included only 21 men of the Royal Artillery Corps and colonial *canoniers-bombardiers.* The other gunners were "picked men" of the regular units and militia.

610 Salvignac's position was that of "major" of the place and not the rank of major in the modern military sense. The appointment is described as: "*Town*-MAJOR, the third officer in order in a garrison, and next to the deputy-governor. He should understand fortification, and has a particular charge of the guards, rounds, patroles, and centinels." Smith, *Dictionary,* p. 157. Vauban commented that "If the fortress is small or mediocre the governor will need only two aides-de-camp." Vauban, p. 145. Pouchot had only Salvignac. He does not indicate who held the position of "deputy-governor" or second in command. It was very likely Captain Villars, the most senior of his *troupes de terre* officers.

611 A good idea of these duties may be obtained from Vauban's chapter on "Defense." Vauban, pp. 121-47.

612 This cemetery appears on most contemporary plans of Fort Niagara and exists today, known locally as the "1812 Cemetery", just southeast of the Old Fort Niagara parking area.

A "glacis" is "that part of a fortification beyond the covert-way, to which it serves as a parapet, and terminates towards the field in an easy slope at about 20 toises distance." Muller, *Treatise,* p. 223. The segment of Fort Niagara's glacis nearest the cemetery is still intact.

On the 10th, there was rain and mist at daybreak, which prevented us from discovering their order of battle until it was broad daylight. We then observed a parallel[613] more than 300 *toises* away which began around the middle of the front of the fortifications and extended to the left in the direction of the lake.[614] It began at a rather low-lying piece of land which was normally flooded but was dry at that moment because of the extensive drought, which facilitated opening up the trench that the English, without it, would have been forced to begin much farther along.

We bombarded the two ends of this parallel with four guns although it rained quite hard. The enemy appeared to be working **[58]** with great vigor. That night we fired our cannon at the left sector, because we estimated that they wanted to extend in that direction. At noon M. Chabert & his brother Joncaire arrived together with 70 other persons, many women and Indians, three Iroquois, among them Chief Kaendaé.[615] The Indians were quite calm.

On the morning of the 11th, we noted that the parallel had been extended a little towards the left. It was heavily bombarded. They could not extend it any farther. During the day they concentrated on

613 The "Siège de Niagara" indicates that the garrison was aware that the English had opened their trench at 8:00 p.m. on July 9. *RAPQ,* 1933-34, p. 93. Prideaux's orders for that day had ordered the trenching party to assemble to "brake ground" at 6:00 p.m. *SWJP,* III, p. 66.

"Parallels" were essential features of eighteenth century siegecraft. They were "a part of the trenches, which surround the whole front attacked, and serve to hold the soldiers, who are to protect and support the workmen: they are about 8 or 10 feet deep, and 16 or 18 feet wide." Smith, *Dictionary,* p. 233. Muller adds that "they were first invented or used by M. Vauban." *Treatise,* pp. 227-28.

Although Pouchot describes the English trenches as such, they were not in fact true parallels. What he observed on July 10 were termed "approaches" on English plans of the siege. These were "a kind of roads or passages sunk in the ground by the besiegers, whereby they approach the place under cover of the fire from the garrison." Muller, *Treatise,* p. 209. An approach might also be termed a "Sap ... a trench, or an approach made under cover of 10 or 12 feet broad, when the besiegers come near the place, and their fire grows so dangerous, as not to be approached uncovered." *Ibid.,* p. 230. Parallels were constructed at intervals to connect multiple saps or approaches, and it was expected that three parallels would be necessary in most sieges, placed at distances from the fortress of 300 *toises,* 160 *toises* and, finally, near the glacis. *Ibid.,* p. 228. Several plans illustrating parallels are in Vauban, for example, p. 64. The siege of Fort Niagara was simple enough to require only one sap or approach and only the most rudimentary parallels. The process of sapping at Niagara is detailed in Dunnigan, *Siege,* particularly pp. 48-50. A plan of the English approaches is on pp. 72-73.

614 Once again, as in his descriptions of the 1758 action at Ticonderoga, Pouchot usually refers to the defenders' right or left.

finishing it and we observed that they were mounting batteries. We harassed them as much as we could with the artillery.

In the afternoon, M. Pouchot wished to have the pickets removed from between the parallel & the glacis in order to construct some embrasures.[616] **[59]** He therefore detached a few men in order to support those who were to bring back the pickets. On their own initiative they pressed forward as far as the top part of the head of the enemy trench. They were followed by about sixty men who sallied forth from the covered way. They fired their muskets into the interior of the trench. The enemy there, who were relying on our small numbers, were mounting insufficient guard. They abandoned the head of this trench. A soldier hastened to inform M. Pouchot that there was no one there. Since he had a better knowledge of the enemy, M. Pouchot ordered the man to go and tell M. de la Roche, who was allowing himself to be drawn into a trap, to withdraw his party. In the meantime, all the soldiers & militiamen, despite their officers, jumped over the palisades of the covered way in order to follow the others. The garrison was on the point of **[60]** engaging the entire English army since at that moment their Indians, of whom there were at least 900,[617] & all their troops, came to do battle at the head of the trench. Thanks to the precautions taken by the officers, we were fortunate not to become too deeply involved. We contained the enemy by very heavy artillery fire which prevented them from charging our soldiers. However, the English could not help losing some men, since they

---

615 Kaendaé (fl.1757-1774) was a Seneca chief who would remain loyal to the French throughout most of the siege of Fort Niagara. His name is rendered "Cayendesse" in the "Journal of the Siege of Niagara" (Appendix B). He was probably the same Genesee chief whose name was recorded as "Kaanyes" and "Kaghnawais" when he signed peace treaties with the English on April 3 and August 6, 1764. *DRCHSNY,* VII, pp. 621-23, 652-53. He is also probably the "Kanaegi", identified as one of the "Headmen of Canadarago" who met with Sir William Johnson at Johnson Hall in October 1763 and again as late as January 1774. *SWJP,* X, p. 903; *Ibid.,* XII, p. 1053.

Kaendaé had committed himself to go to war against the English in November 1757, about the time Pouchot was relieved as commandant of Niagara. Bougainville described him, based on news from Niagara of November 21, as "Kouevidy, famous chief of the Five Nations." Hamilton, *Adventure,* p. 193. Montcalm recorded the name as "Kouindy." Casgrain, *Collection,* VII, p. 320.

616 "EMBRASURE, in fortification, is an opening, hole, or aperature in a parapet, through which the cannon are pointed to fire at the enemy." Smith, *Dictionary,* p. 81.

617 Pouchot's figure is quite accurate. English accounts indicate that Johnson was able to attract 945 Iroquois to support the Niagara expedition. Dunnigan, *Siege,* p. 35.

were forced to stand in the open. This caused them to remain under arms until nightfall.[618]

This episode occasioned another that was rather unusual. The Iroquois chief Kaendaé requested permission to go out to speak to the Indians of his nation. M. Pouchot thought it best not to refuse, more especially as he hoped by means of this Indian to cause at the very least the abandonment of **[61]** the army by a number of Sonnontoins. The Iroquois agreed to parley at the far end of the cleared ground. The result was that the Five Nations were to send two representatives to discover M. Pouchot's attitude towards them. They sent to ask for a safe conduct, on the word of honor of M. Joncaire, whom they regarded as one of their chiefs. They ushered him blindfolded into the room of the commander who recognized the nephew of the Tonniac who had left him five or six days before the arrival of the English. The representatives said that they did not know how they had become involved in the war and they were ashamed of it.[619] M. Pouchot asked them what cause for going to war he had given them and they stated they ought to remember that they had named him Sategariouaen ("the center of good dealings") & that he had **[62]** never deceived them. He expressed to them his surprise at seeing Iroquois in the English army, in particular several who had shown him a great deal of affection. He added that they could judge, from the manner in which he was fighting, that he did not intend to spare the enemy & that his heart bled to think that he might strike down men other than the white men with whom he was at war. He invited them to extricate themselves from the quarrel & assured them that he would no longer harbor any ill will towards them. He finished by warning them that all the nations from the upper country were due to arrive shortly to help him. If he then found himself faced with the prospect of shedding their blood, he promised them to use his authority to bring about peace. He presented them with a large wampum belt to take his words with them to their nation.

---

618 The "Journal of the Siege of Niagara" (Appendix B) provides additional details but indicates a much smaller scale action: "At Noon a Party of about 15 Men, Soldiers and Militia, went very nigh the Trenches of the Enemy, and perceived them sally out between 4 and 500, who came towards them at a quick Pace, but they were stop'd by our Cannon." The "Siège de Niagara" places the beginning of the incident at 10:00 a.m. *RAPQ*, 1933-34, p. 93.

619 The "Journal of the Siege of Niagara" says the two representatives arrived at 6:00 p.m. "Siège de Niagara" says 5:00. *RAPQ*, 1933-34, p. 93.

The Missisakes who were [63] present also wished to speak in their turn. They expressed to the Iroquois the pleasure they felt in hearing them speak of reconciliation. They added that their nation - and they were very numerous - would be delighted by this and they invited them never again to let go of their father's hand. For their part, they said, their decision was made, they wanted to die with their father and left to their own nation the duty of avenging their death.

The Pouteotamis said to them: "Uncles,[620] the master of life has brought us all together on this island (America). Who has more wisdom than our ancestors? Was it not they who first held out their hands to the French? Why would we be not bound to them? We do not know the English. We [64] are delighted that you intend to live in peace with our father. In this way, we shall continue to hold one another's hands." These speeches lasted until nine in the evening, when we led out the delegates blindfolded. They promised to come back with a reply the next day.

This meeting had caused a cease fire on both sides. The enemy took advantage of it as soon as it was dark to open up a trench about 40 *toises* in length, which they would perhaps not have done otherwise. This was a lesson for M. Pouchot.[621]

At daybreak on the 12th, we observed a very tall heap of earth 200 *toises* away which seemed to be prepared for a battery. We bombarded it very effectively with 11 cannon. They did not dare to push forward any sap [65] since they were vigorously bombarded as soon as they sought to begin it.

In the morning, Kaendaé once again asked permission to go out and hold council with the chiefs of his nation. M. Pouchot was not difficult about letting him go. He warned him, however, that he did not intend to stop any of his operations since the whites took advantage of these intervals to continue their work. He added that if his compatriots decided to come and speak to him, they should carry a small white flag & that, provided there were only a few of them, they would not be fired on and would be permitted to enter.[622]

620 Note in original, p. 64 (a): *A term which indicates respect and the degree of understanding which exists between these two nations.*

621 The resumption of English fire caused the first casualties in Fort Niagara. Four men were wounded on the night of the 11th-12th. "Siège de Niagara", *RAPQ*, 1933-34, p. 93.

622 The use of a small white flag by the Indian envoys is consistent with the parley procedures noted previously. For the Iroquois, the symbol might also have been related to the use of a large white belt to signify peaceful intent as noted in Pouchot's next paragraph.

At three in the afternoon, Kaendaé returned with an Onontague chieftain, called Collier Pendu,[623] & two Goyogoins. They presented to M. Pouchot a large **[66]** white belt[624] in response to his. They said to him: "We have listened to what you had to say. You spoke the truth. We have made up our minds. We shall leave the English army & as a proof we shall go and camp at La Belle-Famille." They thanked him for giving them such good advice & for harboring no ill will against them. They promised him henceforth to be peaceful. Kaendaé's council with the Iroquois had been held in the presence of Johnson, to whom the chief spoke haughtily, upbraiding him for launching his nation into evil doings. Johnson smiled and considered the reproach as a joke.

Through another belt, they asked that Kaendaé and the Iroquois women **[67]** and children who were in the fort be allowed to leave with Joncaire, whom they considered as one of their own, in case some cauldron[625] blew their heads off, especially Kaendaé's who was in charge of their dealings with the Indians of other nations, all of whose languages he spoke.

M. Pouchot replied that, since the women & Kaendaé were present, he would leave them free to respond & to take whatever course of action they desired. Kaendaé had assured M. de Chabert that he did not wish to leave us. He gave no reply. M. Pouchot went through the ceremony, in the presence of the deputy chiefs, of burying his body in advance in case misfortune befell him. This ceremony consists of placing a belt & a set of **[68]** equipment in front of one, as is done in a grave. This death cannot lead to vengeance since the man is content. Then the women & children presented M. Pouchot with

623 Although Pouchot identifies Collier Pendu or "Hanging Belt" as an Onondaga, he was actually a Seneca or Mingo living in the upper Ohio Valley. His name was Kaghswaghtaniunt (d. c.1762), and he was active as an English ally in both the 1759 and 1760 campaigns. *DCB,* III, pp. 319-20.

624 Note in original, p. 67 (a): *This is a sign of peace.*

About 4:00 that afternoon Pouchot sent this wampum belt to Captain La Force aboard *Iroquoise,* probably with the intention of forwarding it to Governor Vaudreuil. La Force described it as "a wampum collar that three of the Five Nations Indians currently with the English brought to M. Pouchot. This collar is of white porcelaine and was brought by an old chief to ask for peace with Onontio." La Force records the day's Indian council in some detail. La Force Journal (Appendix C).

625 Note in original, p. 68 (a): *This is the way they refer to bombs.*

The hollow, cast iron mortar bombs and howitzer shells used by the English bore some resemblance to cooking kettles, especially after they had burst.

wampum strings to guarantee that they wished to stay with us, their fathers, who always had pity on them.

The deputies also presented wampum strings on behalf of the nation of the Loups, or Moraiguns, who were in the council of the Iroquois, in order to encourage the Outaouais & other Indians to withdraw to the Fond du Lac[626] & to let the whites engage in warfare since they themselves intended to leave. These two statements seemed to M. Pouchot to be inspired by the English in order to turn away those nations which were on our side. M. Pouchot replied that he did not know the nations who **[69]** were sending these strings & that he would therefore hand them back. He said that, with regard to the Outaouais & other nations attached to us, they had no need of advice in order to know how they should behave towards their father, that at Niagara they were in their own home & that he found it very strange that people to whom they were not bound should seek to encourage them to abandon their house. The Outaouais replied to the deputies that they had come to die with their father & expressed to the Iroquois their delight at learning that they were going to leave the English. M. Pouchot did not wish to acknowledge the words of the Loups, which he felt were not their own.

The same deputies suggested that they should return that evening. M. Pouchot refused, assuring them that **[70]** he would be satisfied provided that they remained peaceful. He warned them that at night he could recognize no one and would fire in all directions, but that if a small number of them came by daylight he would receive them. He sent them away with a loaf each, as he knew that the English army was only eating flour cooked under embers in the form of biscuits.[627]

---

626 The extreme western end of Lake Ontario at Burlington Bay, Ontario.

627 The besieging army was indeed experiencing supply problems, a difficulty Brigadier Prideaux blamed in large part upon the same Indian allies Pouchot was trying to impress with his bravado: "As the Voracious Indians make a great Consumption of provisions I have sent De Normandy [to Oswego] for a supply which I beg you will forward as fast as possible as Circumstances may make them requisite." Prideaux to Haldimand, July 13, 1759, Haldimand Papers, Add. MSS 21728. The strength of Niagara's fortifications had been an unpleasant surprise for the besiegers, and it is likely that Prideaux, anticipating a siege of short duration, had provisioned accordingly. Dunnigan, *Siege,* p. 39.

The resupply of Prideaux's army was complicated by the French armed vessels. Quartermaster Allan MacLean reported that "their shipping makes them Masters of the Lake, we are forced to send all our Expresses by stealth." MacLean to Loudon, July 22, 1759, Loudon Papers, Huntington Library, San Marino, California. Despite thesedifficulties, De Normandy had managed to move three weeks' worth of provisions

...(continued next page)...

In order to explain all these negotiations, it is first necessary to note that the English were entrusting the Indians to provide covering fire for their workers at night.[628] Our fire from the covered way worried them a good deal. They had lost eight or nine of their men. M. Pouchot, who was well acquainted with the nature of these nations, was not unhappy to find the opportunity to rid himself of 900 men, whose activities he feared more than those of the English, because of **[71]** their numbers & the knowledge they had of the place. By retaining within the fort several of the chiefs, some women & a number of warriors from other nations, he ensured that, should any mishap occur, those same Indians would be responsible to their nations or to those they had offended. They were thus delighted to have the opportunity to remain neutral while waiting for the outcome. For their part, the English did not dare to stop the Indians participating in these meetings. They merely attempted to derive the maximum benefit from them.

Once the Indians had left, M. Pouchot immediately sent eight volunteers, under the orders of M. Cornoyer, who approached right up close to the battery.[629] There was little activity in the trench. Upon their return, we bombarded the battery **[72]** with heavy artillery fire and directed musket fire on the right and left from where their infantry would have to emerge.

---

from Oswego by July 21. MacLean to Haldimand, July 21, 1759, Haldimand Papers, Add. MSS 21728.

Prideaux's energetic transport officer was Daniel De Normandie (d.1760), recently commissioned ensign (18 May 1759) in the 44th Regiment of Foot. *Army List,* 1759. De Normandie drowned in Lake Ontario in June 1760 when, as he was trying to reach Niagara, "the wind blew hard & the whale boat opened from Stem to Stern." Webster, *Jeffery Amherst,* pp. 215-16.

628 Prudent siege practices called for covering parties to protect the workers at night. Their function was to form groups "of 10 to 12 men ... to range in the night between the Trenches and the Town, to watch and to discover the enemy, in case they come out, and give notice to the guard of the Trenches." John Muller, *The Attack and Defence of Fortified Places* (London, 1765), p. 49.

629 This incident identifies Lieutenant Jacques-Ange Hertel de Cournoyer (note 580) as the probable author of the anonymous "Journal of the Siege of Niagara" (Appendix B). "In the Night between the 12th and 13th," he wrote, "they fired many Bombs: I went with 30 Men to observe where the enemy wrought." The "Siège de Niagara" also indicates that the scouting party was larger than reported by Pouchot. It says 50 men left the fort at 10:00 p.m. Pouchot might have meant that only a few men accompanied Cournoyer right up to the battery. *RAPQ,* 1933-34, p. 94.

...(continued next page)...

M. Pouchot ordered the corvette to set sail and reconnoitre Chouegen & try to obtain news from M. de la Corne and from Mont-Réal. During the day the corvette opened fire on the enemy trench with considerable success. It left during the night.[630]

On the 13th, MM. Pouchot & Bonnafoux examined at daybreak the enemy works. They had only erected a six-mortar battery. It fired on us the whole day to little effect. During that day, we did not fire much with our own batteries as the enemy works were too well advanced to be destroyed. **[73]**

In the evening we observed a white flag & Indians on the other side of the river.[631] Kaendaé asked for permission to go and meet them, which was duly given. It was some of the Indians who had participated in the council. They asked to come to the fort during the night. M. Pouchot was not willing to give his consent. The fire from our batteries & that of our muskets were quite heavy but not as much as on previous nights since it was no longer necessary to intimidate the Indians who were covering the workers. The Indians announced to Kaendaé that they had all withdrawn to Belle-Famille & would remain neutral. They also told him of the news being spread in the English camp that they had beaten M. de la Corne at Chouegen.[632] **[74]**

On the morning of the 14th we noted a work extending 40 to 50 *toises* beyond the trench in the direction of the lake banks. Its far end

---

Cournoyer's action had only been made possible by the withdrawal of the Iroquois earlier that day. Although they had promised, on the 12th, to retire to La Belle Famille, plans of the siege give no indication of an Indian encampment there. Instead, they are shown on the left (south) flank of the main English camp where they remained until the evening of July 23 when they moved to the right of the army near the lake. See "Plan Of the Fort and Attack of Niagara ...." by Quartermaster Jones of the 46th Regiment, Great Britain, British Library, Add. MSS 57708/10.

630 La Force received his orders from Pouchot on the afternoon of July 12. He sailed from the vicinity of the fort about 5:00 in the evening. Around 7:00 he fired ten cannon shots at the English siege lines, then proceeded to Four Mile Creek and "fired 12 times at their camp." Having done little or no damage, *Iroquoise* sailed for Oswego. La Force Journal (Appendix C).

631 At 8:00 p.m. "Siège de Niagara", *RAPQ*, 1933-34, p. 94.

632 On July 5 Captain Luc de La Corne (note 447) attempted to surprise that part of Prideaux's army left at Oswego to guard communications and begin construction of a new fort. The English were commanded by Frederick Haldimand (1718-1791), lieutenant colonel (1756) in the 4th battalion, 60th Regiment of Foot. *DCB*, V, pp. 887-904.

...(continued next page)...

was 100 *toises* from the covered way. They immediately started work on a battery from which they began firing in the afternoon. Kaendaé & Chatacouen[633] asked for permission to go and speak to their people. M. Pouchot was not sure whether he should allow them to go, but the hope of learning some news decided him to grant them leave. They went to the Iroquois & English camp. They reported that they had seen about 1800 men, that one of their camps was at the small marsh & another closer to the trench and that they had observed 10 mortars & two batteries & 15 cannon, of which three were of very large caliber.[634] They also said that Johnson had persuaded the Indians to stay by [75] promising them the right to pillage the fort which was to be attacked in two or three days. It was finally noted that there was a shortage of food & that they were waiting for a convoy.[635]

From that day on, we no longer saw any Indians in the trench. The Iroquois asked to cross to the other side of the river for fear of the bombs. About a hundred had been fired during the day. M. Pouchot allowed them to cross the river with their womenfolk. He was very happy to be rid of them. They had taken from the Chenondac the cattle of M. de Chabert, saying that it was better that they rather than others should profit from them. They carried the meat to the English camp. The enemy were busy at their works. We directed heavy fire at

---

La Corne, with a formidable force of nearly 1,200 men, lost the element of surprise when his troops began to fire prematurely. See Gipson, VII, p. 357 and Stanley, pp. 219-20. Haldimand had about 1,300 officers and men of the 4th battalion, 60th Regiment of Foot and the New York Regiment. He lost two killed and ten wounded in the action. General Return, July 6, 1759, Haldimand Papers, Add. MSS 21687.

Pouchot provides further details of the French attack on Haldimand's camp at Oswego at the conclusion of his account of the siege of Niagara. See notes 709-713.

633 Although Chatacouen was probably a Seneca, he has not been further identified.

634 The heaviest guns of the English siege train were a pair of iron 18-pounders. The only other significant battering pieces were four light brass 12-pounders. Dunnigan, *Siege*, pp. 136-37.

The bombardment was beginning to have an effect on the garrison. Seven men were wounded on July 14 and two more during the following night. "Siège de Niagara", *RAPQ*, 1933-34, p. 94.

635 Further indication to Pouchot that the English supply problems continued. See note 627.

636 This man is described as a "deserter" in "Siège de Niagara", *RAPQ*, 1933-34, p. 94. Pouchot's map shows "Kanuagen" on the Genesee River downstream from the principal Seneca village at the site of Geneseo, New York. See also Tanner, p. 41.

the sector where, in our judgment, they wanted to emerge in order to extend the line in the direction of the lake. [76]

At daybreak (the 15th), they seemed to be working on another battery. Throughout the day they fired many shells with 10 mortars. A number of our men were wounded by shell fragments. In the evening a deserter arrived, a sort of Frenchman, who was with the Iroquois of Kunoagon.[636] He reported that the English army was made up of the regiments of Halket, Royal American, Loudon, Yorck & Gersey & of 900 Indians, Iroquois or Loups.[637] He also stated that there were three camps, one at the small marsh, one by the lake and the third inland.[638] The Indians were at Belle-Famille. He said that the English intended the next day to install their cannon into a battery which would consist of 15 guns. He added that they were very short of food and that the Indians were complaining of being starved. They were waiting for a convoy from [77] Oswego, where they had a very large camp. M de la Corne, who had tried to attack it, had been repelled.

On the 16th, it rained all day. Two barges appeared well out on the lake; the 12-pounder cannon could barely reach them. Their intention was to reconnoitre the fort. The enemy began musket fire from their trenches. They had installed saucissons[639] along the top of their trenches in order to provide cover to their musketeers.

---

637 This intelligence was not entirely accurate. Prideaux's army included the 44th (Abercromby's, formerly Halkett's) and 46th (Murray's) regiments and the grenadier and light infantry companies of the 4th battalion, 60th (Royal American) Regiment of Foot. Loudon was colonel-in-chief of the four battalions of the 60th. There were no New Jersey provincials with the army, but there was a provisional battalion of 720 men drawn from the New York Regiment. Dunnigan, *Siege,* p. 35; *Army List,* 1759. The English numbered 2,210 regular and provincial soldiers. Huck to Loudon, July 28, 1759 and MacLean to Loudon, July 22, 1759, both in the Loudon Papers. The figure for the Indian warriors is accurate, although there was not a significant number of Mohicans (termed "Loups Moraigan" by Pouchot). Dunnigan, *Siege,* p. 35.

638 The English had a post at Four Mile Creek with the main body encamped near the "tail" or beginning of the trench. The main, or "inland", camp was connected to the lake by a path. The camp "by the lake" must have been either a small post at the end of this path or the Indian's impression of the right (north) end of the main camp. See "Plan Of the Fort and Attack of Niagara ...." by Quartermaster Jones of the 46th Regiment, Great Britain, British Library, Add. MSS 57708/10.

639 Saucissons were fascines, "much longer than the common ones." They were used particularly to raise batteries and repair breaches in walls. Muller, *Treatise,* p. 231; Smith, *Dictionary,* p. 228.

On the 17th, because of the fog which is fairly rare in this part of the country, especially in the summer, & which lifts only very late, we were unable to see whether the enemy had completed any new works. They revealed their artillery with a cannon shot fired from Mont-Réal Point, on the other side of the river.[640] [78] It struck the chimney of the commandant's house & the ball rolled beside his bed upon which he had just lain down to rest. At that position, they had installed a battery of two large cannon & two howitzers. At the same time, they revealed two further batteries, one of five guns and the other of two large cannon & two howitzers. They all maintained heavy fire throughout the day. We fired back at them.[641] The battery on the other side of the river forced us to build epaulements & blinds,[642] because that part of the fort was only covered by a retrenchment, as has been described, and cannon shots fell on the reverse sides of the bastions & other defenses of the fort.[643] That night, we maintained a

---

640 "Siège de Niagara" records the first shot at 7:00 a.m. *RAPQ*, 1933-34, p. 94.

Montréal Point was opposite Fort Niagara at the mouth of the Niagara River. The extremity of this area, today known as Mississauga Point, is the site of Fort Mississauga, begun in 1814. Plans of the siege show the battery on high ground some distance upstream from the point.

641 The "Journal of the Siege of Niagara" (Appendix B) also notes that the English gunners "seemed most inclined to batter the House where the Commandant lodges." Pouchot and his officers resided in the large stone house or "French Castle" situated in the northwest corner of Fort Niagara. The chimney struck so dramatically was that at the northwest corner of the building. Dunnigan, *Glorious Old Relic*, pp. 28-32 and Dunnigan, *Siege*, pp. 61-64.

The exact armament of the English batteries is today uncertain and was, in 1759, constantly changing. Pouchot and other officers could, at best, estimate the number and types of weapons being fired at them. The "Journal of the Siege of Niagara", for example, recorded three guns and four mortars in the Montréal Point battery on July 17. The most dependable reference is "A Plan of Fort Niagara and its Environs with the Attack made thereupon in the Month of July 1759", Royal Collection, Windsor Castle, Cumberland Maps, XII 107 (MP 2129). This says that the battery was armed with four guns.

642 The original text says *des épaulements & des blindages.* Epaulements were breastworks used to cover an exposed point, particularly a flank, and were frequently used in fortification. Smith, *Dictionary*, p. 83. Blinds were structures used to shield troops or the interior of a fortification from observation. These were wooden frameworks that might support earth, such as in the side of a trench, or they might simply be covered with canvas to obscure the enemy's view. *Ibid.*, p. 30.

643 A plan of the post, entitled "Niagara", drawn soon after the conclusion of the siege, shows only a ditch and very light earthworks on the lake and river sides of Fort Niagara. Great Britain, Public Record Office, MPG 342. This is reproduced in Dunnigan, *Siege*, p. 65.

...(continued next page)...

very heavy musket fire from the outer defenses, & the enemy responded vigorously [79] until midnight, when they stopped. From time to time in the night they fired bombs & howitzers. M. de Morembert was slightly wounded.

On the morning of the 18th, the enemy had not advanced their earthworks as far as we could see. They seemed busy repairing the damage done to them by our batteries. A shell had set fire to one of their powder magazines. On that day, General Prideaux was killed in the trenches.[644] There was quite heavy fire from all directions & it increased in intensity in the evening, cannon, bombs & howitzers. This bothered us a good deal. A number of soldiers were wounded & several killed.[645] That night, believing that the enemy would break out on his left to form a forward zigzag, [80] or to open up a parallel, we maintained heavy fire, as did the enemy.

On the 19th, we discovered that the enemy had advanced their earthworks by about 30 *toises* along the lake banks by means of a double sap,[646] from which they opened up a zigzag boyau[647] almost

---

Despite the initial consternation over the English battery across the river, by the night of July 18-19 the writer of "Siège de Niagara" was able to report that the guns there had not had the effect that had been feared. Only two defenders were wounded on the 17th. *RAPQ,* 1933-34, p. 94.

644 Prideaux was not killed until late in the evening of July 20. Pouchot presumably inserted this information at a later time and incorrectly dated the event. According to the "Journal of the Siege of Niagara", the garrison did not learn until July 23 that the besiegers' "first or second in Command was killed by one of our Bullets."

The English general was killed accidentally. "It being almost dark, & the Genl. just passing one of the Batteries, as the Gunner was firing one of the Cohorns [mortars], the Shell struck him in the back of the Head, & put an immediate Period to his life." Ogilvie to Johnston, July 21, 1759 in Severance, II, pp. 296-97. Prideaux's second-in-command, Colonel John Johnston of the New York Regiment (note 125), had been killed earlier that evening. See Dunnigan, *Siege,* pp. 70-71.

645 When Pouchot uses the word *bombes* he refers specifically to the shells thrown by the English mortars. Howitzers could also fire shells.

The "Journal of the Siege of Niagara" (Appendix B) noted "one Soldier dismembered and 4 wounded by their Bombs" on July 18. The "Siège de Niagara" gives one killed and two wounded on that day. *RAPQ,* 1933-34, p. 94.

646 A "double sap" was a trench with a parapet on both sides instead of on just the side facing the besieged fortress. Smith, *Dictionary,* p. 228.

647 A "boyau", as defined in Smith, *Dictionary,* p. 34, was "a particular trench separated from the others ... [that] runs parallel with the works of the place, that it may [not] be enfiladed .... the boyau makes a communication between the trenches, and serves as a line of contravallation, not only to hinder the sallies of the besieged, but likewise to secure the miners."

...(continued next page)...

equal to the front between the two batteries. They worked away at this all day long, keeping up heavy fire with their cannon, mortars & howitzers. We responded vigorously with our artillery.

In the afternoon, the corvette appeared. It was beating to windward well out in the lake. At sunset, M. Pouchot sent out a bark canoe with seven men on board. It ran the risk of being sunk by enemy cannon fire; one of the balls destroyed the oar. As we assumed the enemy would advance even further, we kept up a heavy fire from the [81] covered way & the corresponding defenses.

On the 20th, at daybreak, we noted that the enemy had constructed the other branch of the zigzag which they had brought towards the left from our right beside the lake banks and very close to a ravine which extends 30 *toises* away from the left section of the covered way. They kept up a very lively musket fire until midnight. Our own small arms fire flagged a little towards dawn because our troops were exhausted and their weapons in poor condition. They worked all day at the trench & installed in it some musketeers who greatly harassed those who were servicing the battery of the lake bastion. A number of them were killed & wounded.[648]

That night, the canoe sent out to the [82] corvette returned. The vessel had brought dispatches from Mont-Réal & Québec. They were worried about us but unaware that we were being besieged. They sent news of the English operations at Québec.[649] Around ten o'clock, M.

The 30-*toise* double sap advanced from the second English battery and the boyau extending from the sap towards the lake bank are shown between key letters "H" and "F" on Demler's "Plan of Fort Niagara with its Environs, and the Attack made there-upon in the Month of July 1759", Great Britain, British Library, Add MSS 15535-H. Dunnigan, *Siege*, pp. 72-73.

648 The "Journal of the Siege of Niagara" records one killed and four wounded on the 20th.

649 Montcalm had only received news from Pouchot on July 16 that the English had appeared before Niagara. Pouchot's dispatch must have expressed his embarrassment at being surprised after having sent many of his troops to the Ohio. "Great regrets but too late," wrote Montcalm in his journal, "Behold the Niagara adventure!" Casgrain, *Collection*, VII, p. 576. Montcalm was then at Québec where the English fleet had arrived in June. Stacey, p. 66.

650 Pouchot, at this point, loses interest in the activities of the *Iroquoise*, and Captain La Force's own journal concludes on July 14. The "Journal of the Siege of Niagara" (Appendix B) notes that Pouchot sent two canoes aboard the schooner, one of which was to take news to Toronto and the other to continue on to Montréal with dispatches. La Force was then to cruise off Oswego. He sailed for Toronto on the 21st.

...(continued next page)...

Pouchot sent back the canoe carrying his dispatches for MM. Vaudreuil & de Montcalm.[650]

On the 21st, at daybreak, we observed that the enemy had dug the zigzag back from left to right and was aiming towards the salient of the demi-lune. They were unable to reach it because of our heavy fire all night, to which they responded with great vigor until one in the morning. This earthwork was around 70 *toises* in length. During this day, it appeared that they wanted to install at the end [83] of the boyau a battery facing the salient of the demi-lune. During that day, the enemy fire was lighter than the day before since they were busy working on their trenches & constructing their batteries. Nevertheless, their musket fire greatly troubled our batteries.

Around 7 in the evening, the enemy redoubled their rate of fire from this last parallel. It was very heavy until after midnight. We lost a number of men killed & wounded in the fort.[651] We replied with very heavy fire from our defenses & the covered way, where we had installed three cannon, which each fired 50 rounds loaded with grapeshot. A storm, which did not last long enough from our point of view & which would have flooded their trenches, put an end to this fusillade. **[84]**

On the 22nd, at daybreak, we thought that the enemy had extended a parallel along a ditch which was at the far end of the glacis,[652] but they continued to perfect the works & the two batteries. The one on the left, of 8 guns, was farther forward than the one on the right. Their fire from the trench on their right, directed at the lake bastion & our defenses on the left, was very heavy and caused damage there. They fired very few bombs.

Around 9 in the morning, they began to fire red hot balls from the battery situated on the other side of the river. The battery where their heavy cannon were placed did the same thing. Thanks to the precautions

---

The incomplete journal entitled "Siège de Niagara", covering the period July 6 through 19 and probably kept either by Pouchot or his fort major, Lieutenant Salvignac, was probably included among the documents sent aboard *Iroquoise*. It is preserved in the La Pause "Papers", *RAPQ*, 1933-34, pp. 92-94.

Pouchot's dispatches, containing news of events through July 20, reached Québec on the 31st. Malartic and Gaffarel, pp. 262-63.

651 Only one defender was killed by the besiegers' fire during the night of the 21st but another was killed "by our own cannon" according to "Journal of the Siege of Niagara" (Appendix B).

652 This ditch was a natural gulley in the lake bank noted by Pouchot on July 20 as situated some 30 *toises* from the glacis. The feature is still visible.

of M. Pouchot in setting up barrels full of water in front of all the buildings & stationing detachments of **[85]** carpenters with axes ready to rush to those places exposed to the flames, the fire failed to gain a hold, although it broke out in several places, even in the trade stores, which is not surprising as all those buildings were made of wood. The enemy were never able to perceive this.

They directed their fire at the battery on the lake bastion in order to prevent it from being served. Their fire was very heavy. M. Bonnafoux, the artillery officer, was slightly wounded & 10 men were killed or wounded.[653] The cannon & the howitzers put out of action three of the five guns which were on the same bastion. They destroyed the flanked angle of this bastion so that one could go right down to the berm.[654] The howitzer shells, which burrowed into the ground & exploded, tore up the newly laid sod & each shot **[86]** made craters 6 to 8 feet across. During the night, the enemy kept up a very heavy fire on our works from their parallel & fired ball & grapeshot from their batteries into the breach and at the bastion under attack.

It will be noted that once our batteries on the bastions, which were made of barrels filled with earth, were destroyed, it was necessary to make them out of sacks filled with earth which, placed in an overlapping pattern, formed merlons that were quite effective and easy to change according to the direction of the fire.[655] This method is very good for emergencies & useful if one has at one's disposal a large number of such sacks through diligence in the work that can disconcert enemy

---

653 Because of his duties with the garrison artillery, Bonafous (note 512) was exposed to fire throughout the siege. He reported head wounds from exploding bombs on July 17 and 19 and, on July 21, the lieutenant was struck in the second and third ribs by a long-range musket shot. The ball was spent, but Bonafous nonetheless received a "great bruise." This is probably the wound noted by Pouchot. Despite the pain of his injuries, the lieutenant remained at his post. Bonafous Memoir to Crémilles, 1761, quoted in Fauteux, "Quelques officiers de Montcalm", pp. 410-11.

654 Generally the "flanked angle" was the salient angle formed by the two faces of a bastion. Smith, *Dictionary*, p. 92. Niagara had only "half-bastions", however, and the English fire had probably concentrated on the angle formed by the juncture of the face and the flank of the Lake or Flag Bastion.

A "berm" is the narrow ledge at the exterior base of the parapet, four to eight feet in breadth, that helped prevent earth from falling into the ditch. It could also be used as a path by the defenders. *Ibid.*, p. 29.

655 A "Merlon" was the section of parapet between two embrasures. Smith, *Dictionary*, p. 175. Smith further noted that merlons were "better when made of earth, well rammed and beat close, than of stone, because these fly about [when struck by enemy shot], and wound those it should defend."

batteries. Unfortunately, our earth sacks were in short supply. Those that had been used were either torn, worn or [87] burned in service. Wadding for ramming the cannon was also lacking, as well as hay. The supply which M. Pouchot had set aside had run out. We used the mattresses from the beds, first the straw then the fabric.

During the night of the 22-23, the enemy advanced their trench as far as the salient of the covered way of the demi-lune. They subjected the breach to heavy fire from their artillery, both case shot[656] & ball. Their small arms were also active & they fired a great many bombs. We returned their fire from the fort but our weapons were in such poor condition that out of ten muskets scarcely one was serviceable & the next morning less than a hundred could be fired, despite the repairs we made on them each day. Seven blacksmiths or armorers were continually [88] employed making them serviceable. Domestic servants & the wounded had the job of washing them. As we have already noted, the women looked after the wounded and the sick or worked at sewing cartridge bags or earth sacks. During that day, M. Pouchot was obliged to leave only a small unit of soldiers in the branch of the covered way at the bastion under assault, since the Canadians refused to take up station there because of the enemy's heavy fire.[657] We tried to repair the breach & the palisades of the berm beneath, but with little success, despite the willingness of the ordinary soldier to work at it.

At 10 in the morning, there appeared a white flag on the track leading from La Belle-Famille to the portage. M. Pouchot signalled with another flag. It was four Indians sent by MM. Aubry & de Lignery. We [89] let them into the fort.[658] They handed over two

---

656 Pouchot uses the word *cartouche* by which he means case shot. It was "formed by putting a great quantity of small iron shot into a cylindrical tin box, called a canister, that just fits the bore of the gun." Smith, *Dictionary,* p. 141. When fired, the canister ruptured, spraying the balls in a pattern deadly at close range. The brass 12-pounders and six-pounders taken to Niagara by the English were all well-equipped with case shot. Dunnigan, *Siege,* pp. 136-37. Pouchot uses *gargouche* [sic: *gargousse*] when referring to a cartridge bag for gunpowder.

657 Pouchot's phrase is *dans la branche du chemin couvert du bastion attaqué.* He probably refers to that part of the covered way in front of the Lake Bastion. The difficulty of finding men to defend the most exposed part of the defenses had appeared as early as July 21 when the "Journal of the Siege of Niagara" (Appendix B) recorded orders that "The Company of Volunteers are always to pass the Night in the covered Way."

658 The "Journal of the Siege of Niagara" records their arrival at 8:00 in the morning. It was these Indians who brought the news that Brigadier Prideaux and Colonel John Johnston, first and second in command of the besieging army, had been killed in the trenches on the evening of July 20.

letters, one dated the 17th, & the other the 22nd July. In the first, addressed from Presqu' Isle, they acknowledged those of M. Pouchot of the 7th & the 10th. They stated that they had immediately left Fort Machault and that they believed themselves to be well placed to engage the enemy successfully, thus obliging them to raise the siege.[659]

Through these same letters, the gentlemen in question asked M. Pouchot for his advice on the best way to afford him assistance. The Indians told M. Pouchot that they had called at the camp of the enemy Indians with whom they held a council in the presence of Johnson. They also said that they had presented the Iroquois with five wampum belts on behalf of the nations who were to accompany M. de Lignery. Their message was that the Iroquois should withdraw, otherwise they would be attacked **[90]** just like the English. The latter assured them that they would not become involved in the dispute. We also learned from the same source that there were about 600 Frenchmen & 1000 Indians[660] and that when they had crossed the small rapids on leaving Lake Erie,[661] the river was so completely covered with bateaux & canoes that they seemed like a floating island.

M. Pouchot immediately answered these two letters, after deliberating in the presence of all **[91]** the officers of the garrison in order to benefit from their views. The reader will recall that M. Pouchot, in his letter of the 10th, had warned M. de Lignery that the enemy could well number 4 to 5 thousand men, not counting the Indians and that

---

659 Pouchot summarizes his letter to de Lignery of July 7 in his *Mémoires* (II, pp. 46-47), but he makes no mention of his communication of July 10.

Pouchot's call for relief had reached de Lignery at Fort Machault (Venango) on July 12. De Lignery immediately took his soldiers and as many Indians as would follow and set out for Niagara. Dunnigan, *Siege,* pp. 83-86.

660 Note in original, p. 91 (a): *Among this number there were 300 soldiers & militiamen whom M. Aubry had led from the Illinois and 600 Indians whom he had persuaded to follow him on his way there. M. Aubry, after a very difficult journey, had proceeded to Fort Machault, where he joined up with M. de Lignery. The latter assembled the Indians of the Ohio at Fort Presqu' Isle, from which they set out with M. Aubry.*

Colonial troops and militia from the colony of Louisiana, particularly the Illinois region, had supported their Canadian counterparts in the upper Ohio Valley for most of the war. Chartrand, *The French Soldier,* p. 19. Captain Barthélemy Macarty Mactigue, commandant of Fort de Chartres (Prairie du Rocher, Illinois) in the Illinois country, reported that he had sent the élite of his men. Vaudreuil to Berryer, June 24, 1760, *DRCHSNY,* X, p. 1091.

661 The rapids at the head of the Niagara River. These are visible today below the Peace Bridge connecting Buffalo, New York and Fort Erie, Ontario.

if he was not in a good position to subject them to a full-scale assault, he would have to proceed along the Chenondac in order to make his way to Niagara on the other side of the river. There he would be well placed to disperse the English stationed there - there were only 200 of them and they could not be very easily relieved. From there he would certainly have reached the fort because, after the defeat of that English corps, M. Pouchot would have sent bateaux to carry them back.[662]

M. Pouchot did not doubt that the English read his reply when the Indians returned [92] but he would have been satisfied if it had reached its destination. In this letter he begged M. de Lignery to recall that he had written to him previously. He informed him that the enemy were divided into three corps, one around the small marsh, which was guarding their bateaux, another in the middle of the forest near the depot of supplies for their trenches & the third close to La Belle-Famille. He stated that they currently numbered around 3 thousand together with 900 Indians and suggested that if M. de Lignery believed his column to be strong enough to attack one of these corps, that would be the best course of action he could take, since the enemy was extremely close to the fort and would not dare to withdraw troops from the trenches. He added that, were they to overwhelm one of these positions, that would presumably compel the enemy to raise the siege. Thus they should send [93] scouts on ahead of them and armed with their reports would be even better placed to decide what course of action to take.

Although the enemy saw this letter, they could nevertheless not foresee what decisions the leaders would take, & could take no precautions other than to be on their guard. M. Pouchot left M. de Lignery a free hand to act as his numbers allowed him. According to what M. de Portneuf, the commander at Presqu' Isle, had written to M. Pouchot, they might well total 2800 men, including 1800 Indians. M. Pouchot made four copies of this letter & handed one to each Indian, one of whom was an Onontague, the second a Loup from the Belle-Riviere & the third a Chaouanon. He did this to avoid jealousies

---

662 Pouchot refers to the English detachment manning the battery at Montréal Point. They were indeed badly isolated as their only means of communicating with the main body of the besiegers was by boat. Because of the high riverbanks, boats could only land on the east side of the river at a point nearly two miles upstream from Fort Niagara and the battery opposite it.

This paragraph is the only place in the *Mémoires* where Pouchot summarizes his letter to de Lignery of July 10.

between them & also so that, in the event that the English kept them, [94] at least one of them would get through. And that is what happened.

After taking refreshment, the Indians immediately left again, departing with the same ceremony over the flag. The English & the Indians who saw them leave did not trouble them. Consequently, M. Pouchot had no doubt at all that they held another council with the Iroquois in the presence of Johnson.[663]

Around two o'clock in the afternoon, the Onontague returned and said that he had lost his wampum - it is the same thing as a European who has lost a jewel. He returned, he said, to look for it & had entrusted another Indian with delivering his letter. As a result, M. Pouchot thought that the Indian was a spy rather than a friend and refused to trust him. Later events showed him to have been mistaken. Kaendaé, having got rather drunk, importuned [95] M. Pouchot all day long, siding at one moment with the English, at the next with the French. The Onontague remained very calm. He had great *sangfroid* and examined our works at the most dangerous spots without attempting to seek cover, despite the heavy fire of the enemy. He was perhaps the only Indian to have shown such marked bravery.

The enemy kept up a prodigious & extremely sustained fire with their artillery throughout the day. They destroyed the battery of the flag bastion. Along the entire length of its parapet, the defenses were no higher than two feet. It should be noted that as of the day before we had been obliged to create our embrasures out of bundles of furs for lack of other material & that we used blankets & shirts from the trade storehouse in order to ram the charges into the cannon. [96] We tried to install two cannon as a battery on the left side of the curtain with the aim of counteracting the enemy fire.[664]

We could no longer trust the Canadians to maintain fire at the enemy embrasures, which would have greatly incommoded them.

---

663 From his own sources and the visits by de Lignery's messengers, Sir William Johnson, who had assumed command of the army following the deaths of Prideaux and Johnston, was aware of the approaching relief force. On the afternoon of July 23 he dispatched the light infantry companies of the 44th, 46th and 4/60th regiments to take a blocking position at La Belle Famille. Dunnigan, *Siege,* pp. 82, 88-90.

664 "Curtain, is the part of the body of the place, which joins the flank of one bastion to that of the next." Muller, *Treatise,* p. 216. Pouchot's makeshift embrasures, directed towards the head of the English sap, appear on the plan "Niagara" drawn soon after the siege. Great Britain, Public Record Office, MPG 342; Dunnigan, *Siege,* p. 65.

The counter fire was too heavy for them. Wherever they were stationed, they sat down to get under cover & immediately fell asleep, in spite of everything the officers & sergeants did to encourage them to remain at their posts & maneuver. The remainder of the garrison, despite all possible good will, was no less harassed. Since the sixth of July, no one had gone to bed & it was necessary either to repair the earthworks, as already noted, or to busy oneself at other indispensable tasks. There were so few men left that we had neither [97] the time nor the leisure to sleep.

In the evening the enemy fire slackened a good deal, especially that of the cannon. They merely directed two guns firing ball & case shot at the breach to prevent us from repairing it. This slackening of fire made M. Pouchot suspect either that they wanted to raise the siege in order to go and meet the reinforcements or that they were preparing for a very sustained attack. We remained on our guard as much as possible. That night many men were wounded & several killed while attempting to repair the earthworks.

On the 24th, we heard a number of fusillades from the direction of La Belle-Famille. It was some Indians, scouts of M. de Lignery, who surprised an English sentry post guarding 22 bateaux which they had made into a ferry [98] to cross the river and communicate with the detachment at Mont-Réal Point. The Indians killed a dozen of the enemy and after cutting off their heads, they mounted them on some stakes. This event triggered a number of others. It encouraged the Indians to ask MM. Aubry and de Lignery to wait until they had spoken with the Iroquois in order to compel us to make peace with the English. M. de Lignery dissuaded them from this and wanted them to follow him, since he was on the very point of attacking. They refused to march and only around thirty of them, more determined than the others, followed M. Marin.[665]

M. Pouchot, hearing unexpected musket fire, immediately proceeded with M. Bonnafoux to the bastion of the Five Nations.[666] He observed some Englishmen who were fleeing rather hastily towards [99] their main sentry posts and troops who were marching from the position in the center towards the edge of the cleared ground in order

---

665 Captain Joseph Marin de La Malgue (note 449). His promotion had come in January 1759.

666 The bastion of Fort Niagara nearest the river. This contained the main gate and offered an unobstructed view up the river to La Belle Famille. Pouchot and Bonafous had grandstand seats for the battle.

to join them at the entrance of the La Belle-Famille terrain where we could see a little of the back of a retrenchment of abatis.[667] We aimed two cannon towards that point and fired two or three shots. At the same time M. Pouchot observed a few odd Indians with a white flag. He first assumed that they were Iroquois who wanted to do some act of bravado or to persuade someone to leave the fort. M. Pouchot had two cannon shots fired between the English & them in order to scatter them or, if they were on our side to alert them to the presence of the enemy & prevent them from advancing since, noting how few of them there were, he was afraid they might fall into the ambush. He warned M. Bonnafoux about this. **[100]** The result was nothing more than the unfurling of a large white flag. At the same time we saw a unit, all bunched together at the front proceeding in perfect safety along a track 7 to 8 feet wide. It seemed that, upon perceiving the enemy, to whom it was very close, it was trying to draw up in a tight battle formation, without ranks or files.[668] On their right appeared about thirty Indians who formed a front facing the left flank of the enemy. This battalion fired off one or two salvoes as it approached the enemy, who seemed to make a forward movement out of their abatis. Having

---

667 The abatis or breastwork of logs had been constructed the previous evening by the light infantry under the command of James De Lancey (1732-1800), captain (1759) in the 46th Regiment of Foot. *DRCHSNY*, VII, p. 402; *DAB*, III, pp. 213-14. The fleeing troops were probably survivors of the French attack on the English boat party, and the other marching troops were two detachments sent by Johnson to reinforce De Lancey. By the time the French appeared before the La Belle Famillie position, it was held by 464 men of the 44th, 46th, 4/60th, and New York regiments under the command of Eyre Massey (1719-1804), lieutenant colonel (1758) of the 46th. *DNB*, XIII, pp. 5-6; Blanco, II pp. 1030-31. The preliminaries of the battle are detailed in Dunnigan, *Siege*, pp. 90-92.

668 Pouchot's effort to warn the advancing French column was in vain. Information later gathered from French prisoners recounts the attempt: "When they [the relief force] were near the Fort, Monsr. Pouchot fired two Cross Bar Shot from the Fort towards them - Whether these Shot were intended to warn them, or were fired at our [English] Troops is uncertain: The French themselves were apprehensive that Monsr Pouchot might mistake them for Enemies, wherefore Monsr Du Coigne was sent to a Point of Land in Sight of the Fort to make a Signal with a Flag." The battle erupted before any further attempts could be made to communicate with Pouchot's garrison. Enclosure with a letter from Lieutenant Colonel Massey, Jan. 17, 1761, Great Britain, British Library, Add. MSS 57708/11. The site of Du Coigne's signal was probably Fox Point in modern Youngstown, New York.

Pouchot might have seen further white flags at this point. The head of the French column was marked by a pair of large white colors. Massey to Amherst, July 30, 1759, Amherst Papers/53. A French report says that the roughly 800 troops advancing on the English position moved down the road in a column 12-men wide. Macarty to Kerlerec, Aug. 30, 1759, France, Archives Nationales, C13A/41/103.

been greeted by a third salvo, they returned to it in great haste.[669] Thereupon the battalion advanced in order to enter the abatis but it was stopped by an enemy volley. They immediately **[101]** knelt in order to fire into the abatis. Throughout this time a great quantity of rain fell and made their firearms very wet. While one half of this battalion was firing, the other seemed to be withdrawing rather rapidly, the enemy having discharged two volleys at those who remained. Very few of them held their ground. About fifty of them seemed to be firing as they retreated kneeling on the ground to do so. At that point, the English ran out of their abatis almost in file, with fixed bayonets.[670] But from the lack of musket fire that we heard, we estimated that the entire battalion had withdrawn. It was to our eyes so small that through the rain we judged that it might have been M. Marin or some other officer who had come up to reconnoitre the enemy & had pushed forward as far as the abatis.[671] **[102]**

While this action was taking place, a sergeant in the covered way estimated from the silence of the trench that it was unmanned and asked M. Pouchot if he could make a sortie. Although M. Pouchot thought that the trench on the contrary was certain to be well guarded, he decide to please the soldiers by encouraging this challenge. He summoned 150 volunteers who were not otherwise occupied, except for their officers & sergeants, & ordered M. de Villars to lead them, enjoining him to leave the covered way only with great caution &

---

669 Pouchot might have observed an incident, reported by Colonel Massey, when, just before the battle, the light infantry companies were ordered to leave the shelter of their breastwork and form in the open with the rest of the English troops. Their captains objected to the order but complied. Then, according to Massey, "notwithstanding this as soon as I went to my post on the Right but ye Light Infantry went into their Breast Work again." Massey to Amherst, July 30, 1759, Amherst Papers/53.

670 The French soldiers were following orders to "squat down & to let the Heat of our Fire pass then to rise & push forward." The tactic proved ineffective because of the "Loss they sustained & by falling among our [English] Troops before they expected it. They were thrown into Confusion & obliged to retire." Enclosure with a letter from Lieutenant Colonel Massey, Jan. 17, 1761, Great Britain, British Library, Add. MSS 57708/11.

Massey succinctly described the performance of the battalion companies of the 46th under his immediate command: "I gave the word Fire and to do justice I never saw a Grand Division ... give so plump a Fire, [and] after firing Standing seven Rounds, I gave the word for the whole to advance by constant firing, which was done in great style and ye most of ye 46th fird sixteen rounds." Massey to Amherst, July 30, 1759, Amherst Papers/53.

671 A full account of the Battle of La Belle Famille, with contemporary plans of the action, is in Dunnigan, *Siege*, pp. 83-99.

when he gave the signal but to make a great deal of noise. He ordered him to station some men on the palisades, which could not fail to bring an enemy response, & to judge their situation. And indeed, when the English saw men bestraddling the **[103]** palisades, their entire trench immediately appeared filled with men naked to the waist & there were companies of grenadiers at the head of the trenches.[672] We fired several volleys at them with our cannon, which made them return to cover, and our sortie did not take place.

With the arrival of reinforcements, the Onontague who had come back recognized the troops of M. de Lignery and asked M. Pouchot for permission to go out and fight with them. M. Pouchot granted this request. He passed freely through the lines of the enemy, who doubtless paid him no heed. He joined up with our troops around noon. He subsequently returned around two o'clock & recounted the disaster that had befallen us, which we had difficulty in believing, as we imagined that the English had prompted him to tell us these things. He stated that our entire force had fled, that MM. Aubry **[104]**, de Lignery, de Montigny, de Repentigny were wounded and taken prisoner & that all the other officers & men had been killed.[673] We hoped that this man was not telling the truth.

---

672 These must have been the grenadier companies of the 44th and 4/60th regiments since that of the 46th was at La Belle Famille. *Dunnigan,* Siege, p. 92.

673 Note in original, p. 105 (a): *From M. Pouchot's narration and from the English accounts, it seems that our troops had fallen into an ambush prepared by Johnson.*

The disaster that befell Aubry and de Lignery was not an ambush in the usual sense. The French officers were well aware of the location of the opposing force although not of its strength. Lévis recorded in his journal that the "Ohio detachment .... wanted to march in platoon to pierce the circumvalation [siege lines]; it was invested and mauled." Casgrain, *Collection,* I, p. 191. French casualties were nevertheless enormous. Of some 800 men engaged, the most detailed French account records 344 "taken or killed." Macarty to Kerlerec, Aug. 30, 1759, Archives Nationales, C13A/41/103.

Sir William Johnson at first reported "all their Officers made Prisoners ... to Number of Seventeen." Johnson to Amherst, July 25, 1759, *SWJP,* III, p. 109. Later that day he revised his count to "all their Principal Officers & partizans to the Amount of 19." Johnson to Stanwix, July 25, 1759, *Ibid.,* p. 112. The list of prisoners in Sir William's journal includes 20 officers and cadets, a commissary and one surgeon. One of the cadets, however, is "Soumandre" who had been captured on July 6 (note 574), so the figure of 19 officers appears to be consistent. Johnson claimed that the French relief force had included 30 officers, three of whom escaped. The remainder (seven or eight) were killed. *Ibid.,* XIII, p. 115. Another source names 23 officers and sergeants who fell into Johnson's hands and describes the nature of their wounds. Four of these men, however, were either sergeants or non-military personnel. "Liste des officiers pris et blessés", Brymner, 1886, p. clxxiv. This

...(continued next page)...

When M. Pouchot saw this withdrawal, he ordered all the batteries that were still operational to redouble their fire in order to contain the enemy. They responded very vigorously, which caused us considerable losses. At four in the afternoon, the enemy recalled all their men to the trenches. An officer emerged in order to parley. We admitted him into the fort. He was bearing a letter from Johnson who had been in command of the **[105]** army since the death of Prideaux.[674]

In his letter Johnson told us to lend credence to what Major Hervey, the son of Lord Bristol, would say on his behalf. The major gave the names of all of the Canadian officers who were now prisoners. Although M. Pouchot had been briefed by the Indian, he feigned ignorance & refused to believe what he had just been told until these officers had been shown to a member of the garrison. He made this demand in order to have nothing with which to reproach himself. M. de Cervies, a captain in the Royal-Roussillon Regiment, went to the camp. He saw a wounded M. de Lignery & others in a bower near Colonel Johnson's tent. He had little chance to speak to them & returned to report to M. Pouchot.

This news, which had first been reported by the Indian & then confirmed by the officer, had diminished the courage of the **[106]** garrison to such a degree that M. Pouchot & the other officers had very great difficulty in keeping the soldiers & the militiamen at their posts which they were everywhere abandoning as if everything was all over. If the enemy had been able to see this disorder, they would certainly have taken advantage of it. The German recruits, who were

---

is also in the Haldimand Papers, Add. MSS 21687. A "Liste des Officiers ... tués", Brymner, 1886, p. clxxxv names six French officers killed. A detailed list of the captured French officers may be found in Dunnigan, *Siege*, pp. 130-31.

The English loss was 12 dead and 40 wounded. Huck to Loudon, Aug. 4, 1759, Loudon Papers.

674 The officer was William Hervey (1732-1815), captain (1756) in the 44th Regiment of Foot. A biography and his journals for several campaigns in North America are in *Journals of the Hon. William Hervey*. The captain was a son of John, Lord Hervey (d.1743), first Earl of Bristol. Pouchot refers to Hervey as a major because he had been appointed "brigade major" to Prideaux. Orders, May 5, 1759, Knox, I, p. 459. This was a staff appointment to the brigadier "to assist him in the management of his brigade." An experienced captain was usually selected for this post. Smith, *Dictionary*, p. 36.

Sir William Johnson, despite the misgivings of a number of regular army officers, had taken command of the army following the death of Prideaux on July 20. Dunnigan, *Siege*, pp. 71-75.

numerous within the colonial troops & who had come from France that year were the most mutinous.[675]

M. Pouchot assembled all the officers of the garrison to discuss the situation of the fort and to take the most appropriate decision. He allowed M. Bonnafoux, the officer best qualified to judge, to report on the fort's condition. We began with the covered way and it was agreed that, in view of the enemy's proximity, they would need no more than two days in order to capture it, **[107]** either through a sapping operation or by direct assault.[676]

We had only 110 men to guard the covered way, from the banks in front of the lake bastion as far as the salient of the demi-lune, & 25 men on the place of arms of the right who were guarding right up to the salient of the covered way of the bastion of the Five Nations. There were more than 8 to 10 foot intervals between the men along the front under attack. The weapons were in such poor condition that there were no more than 140 serviceable muskets. Most of them were without bayonets. Since the colonial soldiers & the Canadians were without them, we had adapted woodsmen's knives to fit onto the end of a stick as a substitute, which the men kept with them at their posts.

---

675 The last years of the war provide frequent references to Germans among the French troops, especially those of the Marine. See *DRCHSNY*, X, pp. 1074-75. There were Germans among the troops who attacked Oswego on July 5, and a number were later recruited by the English from among the Niagara prisoners. Haldimand to Amherst, July 7, 1759, Amherst Papers/9; Munster to Haldimand, Aug. 1, 1759, Haldimand Papers, Add. MSS 21728.

Pouchot's Germans are unlikely, however, to have been from the 400 recruits who had come from France with Bougainville in the spring of 1759. Hamilton, *Adventure*, p. 324. They did not arrive until May, and Pouchot earlier (*Mémoires*, II, p. 13) says that he received no more reinforcements after the men sent in March.

There is another possible source for the Germans of the Niagara garrison. In July 1757 Bougainville noted that 150 recruits from the "Foreign Volunteers" (*Volontaires-Etrangers*) had landed in Canada to help bring the companies of the Marine up to strength. Hamilton, *Adventure*, p. 144. The 2nd battalion of the Volontaires-Etrangers had arrived at Louisbourg in 1758 and was surrendered with that fortress. Chartrand, *The French Soldier*, p. 32.

676 Johnson was, in fact, contemplating an assault. By July 21 he hoped to soon have his third and most powerful battery in place only 140 yards from the covered way of Fort Niagara. He hoped this would force Pouchot to ask for terms but, "if not, I shall be oblidged to go to greater lengths, and attempt an Escalade, as I am determined to take the place if possible." Johnson's planning was influenced, in part, by a continuing shortage of ammunition and provisions in his army. Johnson to Haldimand, July 21, 1759, *SWJP*, III, pp. 107-08.

An "escalade" or "scalade" was "a furious attack upon a wall or rampart, contrary to form, and with no precaution; frequently carried on with ladders; to insult the wall by open force." Smith, *Dictionary*, p. 228.

We had burned 24 thousand pounds of powder, out of the 54 possessed by the fort.[677] There remained **[108]** very few balls of 4 & 6-pounder calibre. The supply of 12-pounder ball was exhausted. We thus could not hope to defend ourselves with any vigor. As we noted above, the ditches had no escarpment. The earth had caved in and the slopes were so gentle that one could run up & down them. To avoid this disadvantage, we had erected a palisade at the bottom of the ditch but since the enemy could go down it along its entire length, they would have been able to slaughter the whole garrison between the palisade & the covered way because, in hand to hand fighting, our men could not have been protected by the artillery on the flanks. Moreover, there then remained no more than sixty men inside the fort itself, not including the artillerymen. The palisades in front of the breach were all broken & it was very easy **[109]** to go down there from the breach, where two thirds of the face of the bastion were in the ditch. We had out of service or had lost 10 men from the La Sarre Regiment, 9 from the Béarn, 8 from the Royal-Roussillon, 13 from the Guienne, 43 colonial regulars, and 26 militiamen. In all 109 men killed or wounded together with 37 men who were sick.[678] Besides these losses & taking into account the enemy's numerical superiority, the fort could be very easily assaulted all along the river and from the lake banks.

---

677 An inventory of ordnance stores captured at Fort Niagara printed in the *Universal Magazine* of September 1759 showed that Pouchot's garrison still had 15,000 pounds of bulk gunpowder at the time of its surrender.

678 Note in original, p. 110 (a): *The garrison was only made up of 486 men, as has already been seen, and thus, there remained 340 persons able to bear arms. Why, then, have the English accounts stated that when the garrison marched out of the fort, the complement was 607 men?*

It has been impossible to reconcile French and English counts of the garrison. Throughout the siege, most of those in the besieging army believed Fort Niagara to be defended by 800 men. When the place surrendered, however, there was some surprise that "not above 610 Privates could be found." *New York Mercury*, Aug. 6, 1759. Johnson recorded in his journal that the prisoners totalled "607 men and 11 officers, besides a number of women, children, &c." *SWJP*, XIII, p. 114. His letter to Haldimand of July 26 included a virtually identical tally. *Ibid.*, III, p. 114.

The 1781 editor obtained his figure of 486 defenders by adding the detachments noted in Pouchot's earlier description of the garrison (*Mémoires*, II, p. 44). On pp. 56-57 of the same source, Pouchot's account of the disposition of his forces in the defenses totals only 476 men. To the number of soldiers may be added 39 civilians, including women, for a maximum total of 525 persons. It is not clear whether Pouchot's count includes the 70 individuals, including "many Indians and women," who entered the fort with Chabert on July 10. *Ibid.*, p. 59. It apparently does not, and therein lies the chief discrepancy.

...(continued next page)...

All these considerations made the officers of the garrison ask **[110]** M. Pouchot to consent to a capitulation. Up to that point he had said nothing. He begged these gentlemen to consider whether there might be any other course of action. They pointed out to him the exhaustion of the garrison, which had not slept for 19 days & had been continuously under arms or repairing the works. They also claimed that a delay of two or even eight days, even were that possible, could not save the place & could only lead to the pointless loss of even more brave men, more especially as no reinforcements could be hoped for from any quarter.

M. Pouchot, sensing the truth of these views, sent for the English officer & asked, as a condition of surrender, for the garrison to be able to march out with the honors of war and to have a safe conduct to Mont-Réal with its effects & those of the king in the shortest possible time, and all at the expense of H.B.M.[679] **[111]** There were comings & goings all night, as M. Pouchot insisted on sticking to his proposals. Colonel Johnson sent him a message that in all good faith these conditions were beyond his control. At daybreak, M. Pouchot wanted to send the officer away because he would have to be a prisoner and was prepared to gamble on the outcome. At that point, the entire garrison asked to surrender. The Germans, who were in a majority,

The greatest number that can be derived from French sources, however, is still only 595 individuals of all categories. Pouchot states that 109 men had been killed or wounded and 37 were sick by the afternoon of July 24. Subtracting these numbers from 486 gave the 1781 editor 340 persons "able to bear arms." Most of the casualties seem to have been wounded, however. French sources do not mention a death in the garrison until July 18, and the vague language regarding casualties allows a best guess of a maximum of 15 to 20 dead. This brings the total number in the fort at the time of its surrender, based on French accounts, to about 575.

English reports consistently say that 607 to 610 privates were taken in the fort. The discrepancy cannot be explained as other than an error by Pouchot unless additional small parties had slipped into the fort in the course of the siege.

679 The abbreviation for "His Britannic Majesty." The original text uses the initials "S.M.B." or *Sa Majesté Britannique.*

680 Pouchot implies that his intransigence was based on an unwillingness to become a prisoner of war, an event that would have made it necessary to await exchange. This was a serious matter for an eighteenth century officer because any delay in the exchange process would have the potential to retard his career advancement.

If such mutinous conduct was observed by Captain Hervey, it was not recorded in English accounts of the siege. Pouchot might not, in fact, have been quite as resolute as he claims. A letter from Oswego of July 28 claimed that the articles of capitulation had been drawn up by 8:00 or 10:00 on the evening of the 24th. *Pennsylvania Gazette,* Aug. 9, 1759.

mutinied, which unfortunately the English officer witnessed and was encouraged to show greater firmness.[680] It may be noted on this occasion that any commander who is in the position of having to surrender would be well advised to send away the hostages until an agreement is reached. M. Pouchot was thus forced to accept the following articles of surrender.[681] **[112]**

ART. 1 The garrison will march out[682] with its arms & baggage, drums beating, a fuse lit at both ends with one small piece of artillery, in order to go on board bateaux or other vessels or conveyances to be

---

681 Pouchot's version of the articles of capitulation includes some striking differences from other known copies, at least two of which must be considered "official." Several possible explanations exist for these discrepancies. Pouchot might have worked from an earlier draft in his possession. He might also have made additions from memory in an attempt to clarify what he believed to have been the intent of the agreement.

The articles of capitulation were drafted in French. Sir William Johnson sent a copy, in French, with the signatures of the garrison officers, to General Amherst on July 25. This is in Amherst Papers/39 and has been published in *SWJP,* X, pp. 121-25. Amherst in turn sent a copy, in French, to Prime Minister Pitt on August 5. The original is in the Public Record Office, C.O. 5/56. It was reprinted as an appendix to Knox, III, pp. 242-44. Both contain 11 articles with marginal comments. They are virtually identical and must be considered to represent the official version.

A contemporary English translation was printed in Knox's history. Knox, II, pp. 188-90. It has 12 numbered articles. The difference from the two versions above is easily explained, however. Knox simply assigned "XII" to the unnumbered item at the end of the articles stating that the document had been accepted and that the English would take possession of the fort. In all important details Knox's translation is true to the "official" copies.

Another English-language version was printed in *DRCHSNY,* X, pp. 990-92. Its context in that work, placed at the end of the account of the siege from the *Mémoires,* gives the impression that this version of the articles had also come from Pouchot. It had not, as there are only 11 and Pouchot lists 14. Rather, it appears that the copy printed in Knox was used with a few additional statements of Pouchot's tacked on. Pouchot's version was, of course, translated by Franklin Hough and printed in Pouchot, *Memoirs,* I, pp. 199-201, but Hough also added the marginal notes found in Knox. The Hough translation was reprinted in Severance, II, pp. 438-39.

According to a "Journal du Siège de Québec", the English also allowed Pouchot to send a copy of the articles of capitulation to France, "accompanied with a statement of his own conduct." "Translation of 'Journal du Siège de Québec'", Public Archives of Canada, *The Northcliffe Collection* (Ottawa, 1926), p. 255.

Pouchot's chief discrepancies from the *SWJP* version are noted below.

682 Note in original, p. 113 (a): *They would have specified by "the breach", which was very practical, had not the garrison prepared to leave by a gate opposite the breach.*

The breach had been made in the Lake or Flag Bastion situated on the Lake Ontario side of the fort. The gate through which the garrison prepared to leave was either the drawbridge of the Gate of the Five Nations or, more likely, that on the river side of the fort leading directly to the landing place below.

provided by the general of His Britannic Majesty, and to be transported to New York by the shortest route & in the shortest period of time.

2. The garrison will hand over its arms as soon as the journey begins & will retain its baggage.

3. The officers will retain their arms & equipment.

4. The French ladies & other women who are here will be sent back **[113]** as well as the chaplain.[683] The general of H.B.M. will provide them with the necessary means of transport & subsistence. Within the shortest possible time they will be taken to the nearest French post. Those who wish to follow their husband will be free to do so.[684]

5. The sick & wounded forced to remain in the fort will be able to leave with all their belongings. They will be taken with an escort, when they are able to bear the journey, to the destination of the remainder of the garrison. Meanwhile, they will be provided with a guard, so that they are not molested by the Indians, & they will receive medical treatment & subsistence at the expense of His Britannic Majesty.[685]

6. The commandant, all officers, the troops themselves & everyone in the service of the king, will march out of the place **[114]** without being subjected to any reprisal whatsoever and for whatever reason.

7. An inventory will be drawn up of the munitions & artillery supplies in the stores. The latter will be handed over in good faith, together with all other effects in the stores belonging to the king as soon as the surrender takes place.[686]

---

683 The name of the Fort Niagara's chaplain is recorded, in only one known source, as the Recollet Père Berey. "Memoire du Canada", *RAPQ,* 1924-25, p. 159. This was Claude-Charles-Félix de Berey Des Essarts (1720-1800), ordained in 1743 under the name "Brother Félix." Père Berey served as a military chaplain. It is uncertain when he came to Fort Niagara. At some time during the war he was wounded while administering the sacraments to a dying man, but it is not known if this occurred at the siege of Niagara. Abbé J.B.A. Alaire, *Dictionnaire biographique du clergé Canadien-Francais* (2 vols.; Montréal, 1910), I, p. 47.

Father Berey, one man and 26 women and children were released to a French party near Grande Isle à Chevreuil (Grenadier Island, New York) on August 4. Report of a Flag of Truce, July 31-August 7, 1759, Gage Papers/3. After his release, Berey resumed his duties with the troops at the ruins of Fort Frontenac. La Corne to Lévis, Aug. 10, 1759, Casgrain, *Collection,* X, p. 201.

684 The "official" versions add the note "Granted, except with regard to those women who are his Britannic Majesty's subjects." Knox, II, p. 189; *SWJP,* X, p. 122.

685 The last phrase relating to medical treatment and subsistence at English expense is not in the "official" versions.

8. The soldiers & militiamen will not be stripped nor will they be separated from their officers.

9. When the garrison marches from the fort, it will be forbidden to suborn the soldiers to make them desert.[687]

10. The garrison will be taken under escort to the place determined for its detention. The general will expressly order the escort to protect them from the Indians so that they are not molested when the garrison has **[115]** given up its arms in order to embark. The escort will be subject to the same orders throughout the journey.[688]

11. An exact list will be drawn up of the full names of soldiers in the different units, as well as of the militiamen & others in the king's service.

12. All employees, whatever their functions, will retain their equipment & will share the fate of the garrison.[689]

13. All Indians in the fort, whatever nation they belong to, will be free to withdraw without being molested.[690]

14. A report will be delivered to the general of His Britannic Majesty.[691]

---

686 A copy of this inventory was printed in the *Universal Magazine* of September 1759. The "official" versions of the capitulation include a marginal note that "the vessels and boats are included in this article." Knox, II, p. 189; *SWJP*, X, p. 122.

687 Pouchot's Article 9 is not in the "official" versions. It is possible that Johnson considered the point to be covered by Article 8 or that he simply rejected it.

The English, in fact, busily encouraged soldiers of the Niagara garrison, especially the Germans, to desert and enlist in the army of King George II. At least ten Germans joined the Royal Americans when the prisoners passed through Oswego, and Major Herman Munster was confident that the escorting officers would get more on the route to Albany. Munster to Haldimand, August 1, 1759, Haldimand Papers, Add. MSS 21728. General Amherst later wrote that more Germans could be obtained from the Niagara prisoners but that it was contrary to the cartel agreement signed in Europe. He nonetheless hinted broadly that "Protestant volunteers" might be accepted for English service. Amherst to Bradstreet, Sept. 25, 1759, Amherst Papers/58.

688 This is Article 9 in the "official" versions, and there are substantial differences in the language. See Knox, II, p. 189.

689 Pouchot's Articles 11 and 12 are combined as Article 10 in the "official" copies. Knox, II, pp. 189-90; *SWJP*, X, p. 123.

690 This is Article 11 in the "official" versions. A marginal note also specified, rather ominously, "Granted; but it will be advisable for them to depart as privately as possible." Knox, II, p. 190; *SWJP*, X, p. 124. It is likely that Kaendaé took leave of Pouchot under these circumstances.

691 This item is not in the "official" copies.

The exchanges of articles were signed respectively by the general and by all the officers of the garrison. Since M. Pouchot was the last to sign, the general suggested **[116]** he stipulate that his garrison should be taken to France. M. Pouchot did not agree & on the contrary fixed on a place where they would very likely be among the first to be exchanged.[692] This was duly done.

On the 25th, between 10 & 11 o'clock, the English sent 4 companies of grenadiers, 4 *piquets* & a regiment into the fort.[693] M. Pouchot had his garrison drawn up in battle order on the parade, their arms in hand & their haversacks between their legs. He asked the officers to stand with their own troops. We remained in that position for nearly 30 hours. M. Pouchot had warned them of the necessity of the maneuver in order to protect themselves against harassment by the Indians. He reminded them of the events at Fort George.[694] He advised them that, if an Indian approached to strike them or to steal any of their belongings, **[117]** they should give him a good kick in the belly or a punch in the stomach,[695] which was the safest way of keeping them in check. If this did not have the desired effect, it was better to die with one's arms than be tortured by them. Those orders were punctiliously obeyed.

The English had posted guards everywhere in order to prevent the Indians from entering the fort. They wanted to persuade the garrison to hand over their arms on the pretext that they would better be able to defend us. M. Pouchot firmly refused and assured them they would not be able to prevent the Indians from entering before our **[118]** departure. And indeed, an hour after the English had entered the fort, the Indians swarmed in on all sides & in less than half an hour there were more than 500 in the place. At first they were quite calm.

The French officers had taken the precaution of placing a part of their equipment in the powder magazine. Everything that had not

---

692 New York.

693 At 7:00 am on July 25, Lieutenant Colonel William Farquhar of the 44th Regiment of Foot led the grenadiers of the 44th, 46th and 4/60th to take possession of the gate of Fort Niagara. They were followed by the rest of the 44th Regiment under Major John Beckwith. Orders, July 25, 1759, *SWJP*, III, p. 81.

694 The attack on the paroled English garrison of Fort William Henry in 1757 by Indians allied with the French.

695 Note in original, p. 118 (a): *An Indian does not retaliate when struck in this way. His comrades do not side with him as they would if one used a musket, a sword or a bayonet.*

been locked away there was taken, either by the English officers or by soldiers designated for the purpose. M. Pouchot offered lunch to Colonel Johnson & some other officers. After the meal these officers took possession of all of the utensils & furniture.

The Indians were discreet enough not to take anything from the house or lodging of the officers until the latter had left. But immediately after their departure, they took everything, **[119]** even the locks & hinges of the doors. They broke everything they could not take away. They pillaged the king's effects from the stores, where there were still 500 to 600 bales of pelts.[696] We had used many of them for the battery merlons. They broke & wasted almost all the barrels of flour.

At first, the Indians sought to take the weapons from our soldiers & militiamen, who did not hesitate to resist as they were resolved, at the least provocation, to fight with them & with the English. It is not certain that they would not have expelled them from the fort, despite the enemy army. M. Pouchot would have seen to it that measures were **[120]** taken in the event they failed to respect the terms of the surrender. Upon seeing our resolve, the Indians came and consoled us rather than insulting us. They were almost all known to members of the garrison. Some chiefs said to M. Pouchot: "We have no evil purpose. Have no fear. It is to the English that we shall do harm."

Some English officers said that this was a fine opportunity to take revenge for Fort George. But one must do justice to the majority of them who in these first moments did everything possible to keep the Indians away. One of them was even wounded with a knife. In the outrageous things they said, the Indians did not spare them. Among others, the Onontague who has already been mentioned a number of times, while he was at Niagara said some very hard things to **[121]** Colonel Johnson who did not dare to get upset about it.[697]

---

696 Note in original, p. 120 (a): *They must have been worth a good deal of money to Johnson, who was the only officer known to these nations and who contrived to buy them back from the Indians with goods belonging to the King.*

697 Johnson indeed had a difficult time keeping order inside the fort on July 25. "All my attention is taken up with the Indians, that the Capitulation I have agreed to may be observed," he wrote to General Amherst later in the day. *SWJP,* III, 108-10. In subsequent correspondence with Amherst, the Marquis de Montcalm stated that, although the Indians had pillaged the prisoners, he was confident that the English officers had done their best to prevent it. He admitted, no doubt with thoughts of Fort William Henry, that such things were bound to happen with native allies. Montcalm to Amherst, Aug. 15, 1759, Casgrain, *Collection,* IV, pp. 255-56.

Certain English officers & soldiers took some hunting muskets from our officers & militiamen but by stealth rather than force. As of the 24th there was such heavy weather coming out of the northwest that we could not launch a boat, otherwise M. Pouchot would have been tempted to arrange the escape of a part of his garrison before surrendering the fort, which would not have been all that difficult.

On the afternoon of the 26th, the garrison marched out of the fort to assemble at the flat space with muskets shouldered, drums beating & with two pieces of heavy cannon at the head of the column.[698] As soon as the troops were in front of the bateaux upon which they were to embark, they laid down their muskets and set out immediately, although there was **[122]** still a very heavy swell on the lake.

---

698 Pouchot here implies that he exceeded the terms of the articles of capitulation which allowed the garrison to march out with only "one small piece of artillery."

The "flat space" where the garrison assembled is *le platon* in the original. It refers to the landing place below the river side of Fort Niagara, today the site of the Coast Guard Station.

699 Most of the French prisoners had been taken by the Iroquois in the pursuit that followed the action at La Belle Famille. The English soldiers were disgusted that their allies had remained neutral during the first part of the action, but, once the French column collapsed, they "fell on them like so many Butchers, with their Tomahawks and long Knives, hooping and shouting, as if Heaven and Earth were coming together, and kill'd abundance of the Enemy." Letter from Albany of August 6, 1759, *Pennsylvania Gazette,* Aug. 23, 1759. Lieutenant Colonel Massey later reported that the only part played by the Iroquois was to take prisoners and that his officers "did not see an Indian fire a Shot, nor appear, but to pull out the Prisoners, hid under Log's and Scalp'd poor Wretches after they had fallen by our hands." Massey to Pitt, July 11, 1760, Great Britain, Public Record Office, Chatham Papers.

The English officers did what they could to recover the prisoners. Captain Charles Lee of the 44th wrote that it "was with difficulty we restrain'd our Mohocks & Oneidas from sacrificing all their officers." Lee to Bunbury, Aug. 9, 1759, *Collections of the New York Historical Society for the Year 1871* (New York, 1872), pp. 20-22. Only Sir William Johnson could exercise the influence necessary to keep the promise made to Pouchot. This was not an easy task. On July 27 he confided to his journal that "I divided among the several nations, the prisoners and the scalps amounting to two hundred and forty-six, of whom ninety-six were prisoners. The officers I with difficulty released from them, by ransom, good words, &c." *SWJP*, XIII, p. 115.

Clearly, only the officers were recovered. This incident, and the attention it gained for Johnson, very likely influenced a historical painting begun about 1763 by Benjamin West. The work, entitled "General Johnson Saving a Wounded French Officer from the Tomahawk of a North American Indian", has usually been associated with Sir William's 1755 victory over Dieskau at Lake George. Helmut von Erffa and Allen Staley, *The Paintings of Benjamin West* (New Haven, 1986), pp. 210-11. The subject matter, as well as physical and symbolic details in the painting, however, more closely correspond to the events at Niagara.

We were not able to see the officers who were prisoners. Johnson had given his word that he would remove from the custody of the Indians those whom he had captured since, when they saw our men fleeing, the Indians pursued them & captured a great many.[699] On this occasion there occurred a tragic accident. Moncourt, a colonial cadet,[700] had taken a great liking to an Indian with whom he had become very friendly. The Indian, who was in the English army, saw that his friend was a prisoner and expressed to him how deeply he was affected by his situation. He said to him: "Brother, I despair at seeing you dead. Keep calm, though. I intend to prevent them from making you suffer." He killed him with a blow of his tomahawk, thinking to spare him the torments which are the normal fate of prisoners among them. **[123]**

The remainder of the troops who escaped from the battle withdrew to an island above the portage fort where Rocheblave had been left with 150 men to guard the canoes & the bateaux.[701] They retreated to Détroit, as did the garrisons of all the posts of Presqu' Isle & Fort Machault, under the orders of M. Belestre, who had not been able to be present at the action because of sickness.[702] Out of 400 men, more than 250 were killed, almost all of them colonial soldiers who were very brave and had served very well in these engagements. There were many Frenchmen from the Illinois killed or captured. All the prisoners were taken to New York like the Niagara garrison.[703]

---

700 Pierre Hertel de Montcour (d. 1759), son of Pierre Hertel de Montcour. *DCB*, II, p. 284 and P.G. Roy, "Les commandants du Fort Saint-Frédéric", *Le bulletin des recherches historiques*, LI, No. 9 (September 1945), pp. 318-19. Cadet Montcour had been serving in the Ohio and at Fort Niagara since at least the summer of 1757. Casgrain, *Collection*, IX, p. 96.

701 Grand Island, New York. The officer was Second Ensign Philippe-François Rastel de Rocheblave (note 340). Rocheblave retired "in good order" to Detroit. Malartic and Gaffarel, p. 298.

702 Captain (1759) François-Marie Picoté de Belestre, commandant of Detroit (note 385).

703 The 250 La Belle Famille casualties recorded here are less than the 344 reported in Macarty to Kerlerec, Aug. 30, 1759, Archives Nationales, C13A/41/103. Pouchot, however, gives only the number killed. Johnson counted 96 prisoners with the Indians. *SWJP*, XIII, p.115. Adding them to Pouchot's total gives a number very close (346) to Macarty's 344 "taken and killed."

Twenty-one French officers and cadets captured at La Belle Famille were sent to Oswego on August 4. Journal of the Niagara Campaign, *SWJP*, XIII, p. 120. A total of 31 men, including those officers, arrived at New York on the 24th. *New York Mercury*, Aug. 27, 1759. Captain de Lignery was not among them. He succumbed to his wounds at Niagara on August 29. Farquhar to Amherst, Aug. 29, 1759, Amherst Papers/21.

The latter experienced a comic adventure near Lac des Onoyottes.[704] Its escort was made up of 100 men of the Royal American [124] Regiment, 300 militiamen & a company of rangers, or scouts.[705] The soldiers of this company wanted to pretend that they had some Indians with them and to show how friendly they were. During the night they painted themselves & dressed up like Indians. They then entered the French camp carrying knives & war clubs and whooping like Indians on the attack. From their awkward manner, our soldiers immediately realized what was happening. They joined in with the rangers and began to dance, sing & yell after the manner of the Indians, which they did with the best grace in the world. The officers of the Royal American Regiment, with whom the French officers were then dining, noticed that the latter were smiling & joking about this act of bravado undertaken without their knowledge. They set about [125] the masquerade with their sticks and sent their soldiers back to bed ill-satisfied with their entertainment.

When the garrison had arrived at Fort Stenix the English, who had been recommended not to allow M. Pouchot to see the fort, obliged him to make a wide circuit with his garrison in order to ford the Mohawk River. As the water level was quite high, the escort wanted to return & pass by way of the fort, leaving us to cross the river by ourselves. On being warned of their intention, M. Pouchot entered the water fully dressed & was followed by the officers & his troops. The English, by now ashamed to retreat, did the same thing, uttering

---

704 Oneida Lake.

705 The Fort Niagara garrison was escorted to Oswego by Lieutenant Colonel Massey and 300 men of the 46th Regiment. *SWJP*, XIII, p. 114. Although Massey was to have continued to Albany, Sir William Johnson countermanded his original orders and directed that the New York Regiment take the prisoners from Oswego to Albany. Hervey to Haldimand, July 26, 1759, Haldimand Papers, Add. MSS 21728.

Receipt of the new order at Oswego caused Colonel Bartholomew Le Roux to vehemently protest the assignment of 200 men of his regiment on one more non-combat duty. Le Roux to Haldimand, July 29, 1759, *Ibid.* The New Yorkers were sent nonetheless, but a substantial part of the escort was also composed of regulars of the 60th Regiment. Munster to Haldimand, July 29, 1759, *Ibid.* Pouchot's statement of the escort appears to be accurate. If rangers were, in fact, involved, they were likely from Captain Joseph Waite's company.

706 This security-conscious officer was William Browning (fl.1740s-1767), major (1757) of the 46th Regiment of Foot. Army List; *DRCHSNY*, X, p. 730. He had served with the 46th since the 1740s and claimed to be the oldest major in Amherst's army. Browning to Amherst, Oct. 24, 1759, Amherst Papers/53.

many oaths about the precaution of the fort commander.[706] We shall not forget at this point the civility of M. Fech, a Swiss and a captain of the [126] Royal Americans who was in command of the escort.[707] He lent M. Pouchot about 25 *louis* in order to help the French officers subsist. It was an essential subsidy. At this time all of the officers were without money & if they had been obliged to exist on English rations, they would have eaten very poorly since they consisted solely of a pound of flour & a pound of bad pork. The English officers were fed by their sutlers[708] who also found them accommodations. They had the same rations.

The adventure of M. le Chevalier de la Corne has only been mentioned in passing.[709] Here is the story in greater detail. We noted that M. de la Corne was at Frontenac & La Présentation. He had with him in that sector in order to cover the entrance to the river, 4 to 500 Canadians [127] & a number of colonial soldiers. In the early days of July, he marched on Chouegen with all his troops together with the Indians of the mission accompanied by the Sulpician Abbé

---

707 Pouchot appears to identify Rodolp Faesch (fl.1755-1764), captain (1755) in the 60th Regiment of Foot, as commander of the escort. Faesch (or Fesch) was one of a number of Swiss commissioned in the Royal American Regiment. He was wounded during the 1760 campaign but remained with the regiment until 1764. See Sophia Fesch to Bouquet, Aug. 16, 1760, Stevens, *Bouquet Papers,* IV, p. 697 and *Army List,* 1759-1764. Captain Faesch had served in the French Army. He had also witnessed the events following the surrender of Fort William Henry in 1757 and had gone to Montréal as the official English witness for the redemption of the prisoners taken by the Indians. Steele, pp. 109, 134-35.

Unfortunately, Faesch does not appear to have been with Pouchot's escort. Command was given at Oswego to Captain George "Du Fez", another Swiss of the Royal Americans. With him went Lieutenant Ramsay and Ensign Monin (or Monins) of the regiment. Munster to Haldimand, July 29, 1759, Haldimand Papers, Add. MSS 21728. In his letter to Amherst of August 14, Pouchot mentions Captain "Dufes" as an officer "under whose conduct we have been" and praised his "attentions and courtesies." Amherst Papers/10 and Appendix A. Pouchot seems to have again fallen victim to phonetics and memory lapse. It is possible that he later retrieved Faesch's name from a copy of the *Army List.*

George De Fez (fl.1757-1760) was a captain (1757) in the 4th battalion, 60th Regiment. He had been wounded at Ticonderoga in 1758 and sold his commission in 1760. *Army List; DRCHSNY,* X, p. 731; Amherst to Bouquet, Sept. 24, 1760, Stevens, *Bouquet Papers,* V, p. 50. His subordinates were William Ramsay, lieutenant (1758) and John Monin (fl.1758-1762), ensign (1758). *Army List.*

708 The word in the original is *vivandiers.* The commandant of Oswego procured a *vivandier* to accompany De Fez who was to take Pouchot only as far as Fort Stanwix. The sutler was to continue to Albany. Munster to Haldimand, Aug. 1, 1759, Haldimand Papers, Add. MSS 21728.

709 La Corne had attacked the English at Oswego on July 5 (see note 632).

Piquet,[710] a famous missionary of the area who wanted in his zeal to be with his flock. They disembarked at the same place as M. de Montcalm at the time of the siege.[711] The English, in setting out to lay siege to Niagara, had left on the site of Fort Ontario about 5 to 600 men who had not yet had time to retrench. They had only erected a kind of enclosure around the camp with pork & flour barrels, of which this army corps had considerable supplies. As the detachment believed itself to be entirely safe, the majority of them were spread out in **[128]** the surrounding forest cutting wood for the retrenchments.[712]

M. de la Corne ordered forward a large body of his men to the site of Fort Ontario in order to reconnoitre the enemy. They shot the men working in the forest and reached the camp, which they found in great disorder. The guard & the elements remaining in the camp put up resistance to our scouts. If M. de la Corne had followed his advanced guard, the English would have been destroyed. The Abbé Piquet, who had heard the opening volleys, thought it his duty to address a brief exhortation & give absolution to his troops before they attacked. As a result, the opportunity was lost. The English ran to their arms & posted themselves behind their barrels. M. de la Corne moved up close to his detachment, which was spread out around the English, **[129]** but which could not advance any farther because of the enemy's great numbers. He tried to encourage them to renew the attack. Some of the Canadians, who were more desirous of retreating than of fighting, shouted that they were being completely cut off &, despite the officers, there was a frantic rush to regain the bateaux as fast as possible. The Abbé Piquet tried to stop them and was knocked over. He finally seized hold of one of them & cried: "At least save your chaplain!" There were few losses as the English had not pursued them.

---

710 François Picquet (1708-1781), Sulpician priest and founder of the mission at La Présentation at the site of Ogdensburg, New York. *DCB,* IV, pp. 636-38. An account of the founding of his mission is in Boyesen, "The French and Indian Settlement at Ogdensburg", pp. 4-10 and in Fortier, pp. 8-11.

711 In August 1756. The landing was made at Baldwin's Bay, one mile east of Fort Ontario.

712 The English detachment at Oswego under Lieutenant Colonel Frederick Haldimand numbered about 1,300. La Corne had a mixed force of Marine troops, militia and Indians totalling 1,191 men. Casgrain, *Collection,* XI, p. 215. Haldimand had been ordered by Prideaux to guard the line of communication with the Mohawk and begin construction of a replacement for Fort Ontario on the east side of the Oswego River. Orders, June 29, 1759, *SWJP,* III, pp. 58-59; Prideaux to Haldimand, June 30, 1759, Haldimand Papers, Add MSS 21728.

After the action they agreed that if the entire force had followed the first detachment, they would have very easily routed the English corps since they were surprised & disconcerted at the opening assault. Had this corps been beaten, Niagara would have been saved as their army would not have received the reinforcements of troops and the victuals that they sent.[713] **[130]**

M. Douville, a colonial captain in command of Fort Toronto, when he no longer heard the cannonades from Niagara, suspected that the fort was taken. He destroyed his own position, setting it on fire, & withdrew to Montréal so as not to be captured.[714] As has already been noted, this fort was only of use as a defense against the Indians & there were no more than 12 to 15 men garrisoned there. That was the outcome of the campaign in the upper country. Let us return to the operations in Québec.

The English fleet, 28 warships strong, some of them with three decks, accompanied by transport vessels carrying ten thousand troops under the orders of General Wolfe, appeared off the Isle aux Coudres[715] during the month of June. The enemy seized it & subsequently **[131]** occupied the Isle d'Orléans.[716] They found none of those imaginary difficulties upon which the Canadian seamen were so relying. When

---

713 For modern accounts of the battle see Gipson, VII, p. 357 and Stanley, pp. 219-20. A "Relation de la Campagne de M. le Chevalier de la Corne à Chouguen" is in Casgrain, *Collection,* XI, pp. 215-18. La Corne reported the loss of ten men killed, including an officer, and two officers and 20 soldiers wounded. Four of his men deserted. *Ibid.,* p. 218. All were Germans from the Marine troops. Haldimand to Amherst, July 7, 1759, Amherst Papers/9.

The French deserters told Lieutenant Colonel Haldimand that Picquet had "marched with them, 'till they made the attack. then gave them his benediction, and recommended it to them to give no quarter." Amherst Journal, Knox, III, p. 38.

714 Alexandre Dagneau Douville (1698-1774), captain (1756) in the Marine. *DCB,* IV, 191-92. Although an elderly officer who, according to the *DCB,* had retired on half pay in 1756, he is nonetheless identified by Robinson (p. 139) as the commandant of Fort Rouillé from 1757 until 1759 when he burned the post and retired with its tiny 15-man garrison. Two ladies named Douville are mentioned by Pouchot as being in Fort Niagara throughout the siege (note 582). They were returned to French territory under a flag of truce (see note 683 and Dunnigan, *Siege,* pp. 106-07). Captain Douville and his garrison had reached Isle aux Galops in the St. Lawrence River by August 10 whence Douville continued to Montréal with his family. La Corne to Lévis, Aug. 10, 1759, Casgrain, *Collections,* X, p. 202.

715 Isle aux Coudres, off Baie St. Paul, Québec, on the north shore of the St. Lawrence.

716 Note in original, p. 132 (a): *On the 29th & 30th of the month.*

Isle d'Orleans is in the St. Lawrence River just below Québec.

the fleet entered the basin of Québec, a few fireships were aimed at it, but in vain.[717]

MM. de Vaudreuil and de Montcalm positioned the Canadians & the regular troops, numbering 5 to 6 thousand men and comprising the La Reine [132] Regiment and those of Languedoc, La Sarre, Royal-Roussillon & Béarn, together with the colonial force of 1800 men in two battalions extending from the St. Charles River to the Montmorency Falls. They left a garrison in Québec. They erected redoubts along the river & retrenched. The majority of the enemy landed on the other side of the river with a great deal of artillery.[718]

By different maneuvers & also by means of their artillery which was heavy & supported by the guns of their ships which they brought in close to the coast, they tried to make us abandon the shore of the St. Charles River to enable them to pass it.

On July 31, they landed a large number of men near the falls at low tide in order to capture a redoubt which covered the channel & the

---

717 Note in original, p. 132 (b): *The English fleet, under the command of Admiral Saunders, was buffeted by a violent squall after the capture of the Isle d'Orléans. Several large vessels lost their anchor & a large number of transport vessels sank. Advantage was taken of this opportunity to release some fire ships during the night, but excessive haste, almost inevitable in that kind of operation, led to the failure of this one, which had been very well planned.*

Charles Saunders (c.1715-1775) was Vice Admiral of the Blue. *DCB*, IV, pp. 698-702. Saunders' formidable fleet included 49 warships. For a detailed analysis of the British naval force at Québec see Donald Graves' appendix, "The Royal Navy at Quebec," in Stacey, pp. 215-21. Among the heavy warships was *Alcide* (64), captured from the French in 1755 while escorting the first *troupes de terre* battalions to New France.

Numerous sources describe the summer-long struggle for Québec. Volumes I and II of Knox include his own narrative as well as many of the orders issued to Wolfe's army. The appendices printed in Volume III reproduce other documents relating to the siege. Two French journals of the siege of Québec are printed in *DRCHSNY*, X, pp. 993-1001 and 1016-46. Accounts appear as well in most of the French officers' journals and narratives cited in this edition of Pouchot. Many documents are printed in A.G. Doughty and G.W. Parmalee, *The Siege of Quebec and the Battle of the Plains of Abraham* (6 vols.; Québec, 1901). C.P. Stacey's *Quebec, 1759: The Siege and the Battle* is the best modern study. René Chartrand's *Quebec 1759* (Oxford, 1999) presents a detailed and heavily illustrated day-by-day account of the campaign with much supporting information on the military units involved.

The French fire ships were released on the night of June 28-29. Knox described them as "the grandest fire-works ... that can possibly be conceived," but they failed in their attempt to destroy Saunders' ships. Knox, I, p. 381. See also Stacey, p. 67.

718 The St. Charles River bordered the eastern side of the city of Québec. Montmorency Falls are just above the mouth of the Montmorency River where it joins the St. Lawrence below Québec. The Heights of Beauport between the two rivers were fortified by Montcalm's army. Wolfe's army encamped on the east side of the Montmorency.

center of the retrenchments. **[133]** All day long they kept up very heavy fire with 200 guns on the whole camp in order to cover the main body of the army which was creeping along the beach in order to capture the redoubt where there were 2 pieces of cannon in such poor condition that they were unuseable. Our troops gave such a good account of themselves along the whole sector that the English found no weak spot anywhere. They even had quite a lot of difficulty taking the troops, who numbered nearly 2,000 men, off the beach in boats because they were threatened by the rising tide.[719]

They abandoned this enterprise & positioned a corps of about three thousand men opposite Québec on the other side of **[134]** the river. A detachment of Canadian troops was formed, under the command of a colonial captain, M. Dumas, in an attempt to dislodge them. It had the same sort of success as that of M. de la Corne.[720]

The enemy placed a great many cannon & mortars in that position, which ruined & partly burned Québec during the month of August. Our army bivouacked every night, while almost every day the enemy tried some maneuver in an attempt to dislodge them & find some means of landing on that side of the river. As soon as they had taken the latter position, we were obliged to keep to the river above Québec. We built redoubts at the places we thought the enemy likely to land. They were defended only by units of 50 men. These places did not yet seem to be **[135]** in danger because of the situation of the enemy. These *piquets* remained in fixed positions for nearly three months, which are very bad tactics because in the long run there is an end to vigilance as all the troops become so fatigued.

MM. de Vaudreuil & de Montcalm, having learned in the month of August of the capture of Niagara, detached M. de Lévis with 5 to 600 men to march to La Présentation & establish a position capable of covering that frontier.[721] Once arrived at the site, he judged, in consultation with M. de La Pause, the *aide-major* of the Guienne

---

719 Note in original, p. 134 (a): *By their own admission, the English that day lost more than 500 men & a number of fine officers.*

The disastrous assault on the Beauport lines is described in Knox, I, pp. 449-56. He gives 443 officers and men killed, wounded or missing. See also Stacey, pp. 87-95, who cites Wolfe's account of casualties as 440 killed and wounded.

720 The raid, led by Captain Jean-Daniel Dumas (note 247) against English forces on Point Lévy on the night of July 12-13, 1759, fell into disorder and was a failure. See Stacey, pp. 78-79.

Regiment,[722] that the little Isle Oraquointon, above that of the Galots, was the most amenable to fortification in order to block the river.[723] It was La Pause who made up the mind of M. de Lévis. He himself wanted to lay out the fort, or redoubt, according to his own caprice, in spite of the views of M. des [136] Androins, the engineer, who was entrusted with building it. The latter was left to see to this task. M. de Lévis remained there until September, observing the movements of the enemy in that sector. During that time he kept his men busy speeding up the construction of the new fort. As soon as he considered it in reasonable condition, he returned with some of his men to rejoin the army at Québec.[724]

721 News of the fall of Niagara reached Montcalm at Québec on August 9, and Lévis was dispatched that night with 800 men to block an English descent of the St. Lawrence. Casgrain, *Collection,* VII, p. 587. The news apparently arrived in letters written by Pouchot from Oswego. "Relation de la campagne 1759", *RAPQ,* 1933-34, p. 115. Lévis took with him artillery Captain François Le Mercier (note 163), engineer Captain Jean-Nicolas Desandrouins (note 286) and Captain La Pause of the Guienne Regiment. The party reached Montréal on August 12 and left two days later for the rapids near La Présentation. Lévis Journal, Casgrain, *Collection,* I, p. 192; Gabriel, pp. 301-02.

722 Jean Guillaume Plantavit de Lapause de Margon, Chevalier de La Pause (1721-1804), captain (1751) in the Guienne Regiment. *DCB,* V, pp. 676-78. An extensive collection of his papers has been printed in *RAPQ,* 1931-32, 1932-33 and 1933-34.

723 Isle Oraquointon or Orakointon is today Chimney Island on the New York side of the St. Lawrence about three miles below Ogdensburg. "Les Galots" referred to a group of islands in the area, including "La Grande" Isle aux Galots (usually called Isle Picquet from 1759), today Galop Island, New York a mile below Chimney Island. The smaller Isle aux Galots is today Prison Island on the Ontario side of Galop Island. French sources sometimes employ the spelling "Galops." British maps normally use "Galops" as, for example, the "Plan of the attack upon Fort Levi" in Mante, p. 303.

Orakointon was selected as the site for the post ultimately named Fort Lévis, and the evolution of its name was part of the process of acquiring it for military use. Upon his arrival from Montréal, Desandrouins constructed temporary batteries on the upstream end of "la grande île des Galots" (Galop Island) to block the river while Lévis examined the ruins of Fort Frontenac. Gabriel, pp. 301-02; Casgrain, *Collection,* I, p. 197. By August 26 Lévis had returned to La Présentation where he visited the little island and decided to establish a fort on it. Later in the day he met in council with the clan mothers of the mission village of La Présentation (Ogdensburg, New York). In the course of the ceremony, Lévis was adopted by them and given the name *Orakouintone* or *Oracoüenton,* meaning *Le Soleil Suspendu* ("Hanging Sun") in their language. The little island was apparently then without a name, and the following day, August 27, the French took possession and christened it "Orakouintone." Considering that Desandrouins found Lévis' new title "pompous", the naming of the island that would be the site of Fort Lévis might have been somewhat tongue-in-cheek. Casgrain, *Collection,* I, pp. 198-99; Gabriel, p. 301.

In the meantime, the city was struggling, so to speak, with its artillery, which was very extensive, against that of the English. The latter were constantly patrolling with frigates & armed sloops in an attempt to penetrate at some point. They first tried to make a thrust between their camp & the city with some frigates. They sailed up the river, assisted by the tide & the wind, **[137]** despite the bombardment from the fort. Then, during the night, they sent bateaux carrying troops, escorted by heavy warships. M. de Vauclein, in command of two frigates blockading the river, engaged three warships in battle for seven hours. He was so badly mauled that his ships were lost.[725] The enemy, now masters of the river, burned some stores at Jacques Quartier, in which were stowed the personal effects of almost all the officers of the army.[726]

M. de Montcalm detached M. de Bougainville with all the grenadiers & volunteers from the army & hastily formed a cavalry unit of about 200 horse.[727] This corps, totalling a thousand élite troops, **[138]** proceeded to Pointe au Tremble, 5 leagues above

---

724 Lévis rejoined the main army on September 17 following the death of Montcalm and the withdrawal to Pointe aux Trembles. Malartic and Gaffarel, pp. 288-89. Desandrouins was left in command at Fort Lévis and continued his work there until relieved by Pouchot in the late winter of 1760. This freed the regular engineer to assist Lévis in besieging English-held Québec. *Ibid.*, pp. 301-02; Gabriel, p. 315.

725 Note in original, p. 138 (a): *It was in Canada, & not in France, that prams ought to have been built. They would have prevented the English from sailing up the river.*

A "pram" was a flat-bottomed boat or lighter that was especially popular for shipping cargo in the shallow waters of the Baltic and the Netherlands.

The naval officer was Captain Jean Vauquelin (note 463). Pouchot misplaces this event by a year and actually describes Vauquelin's action of May 16, 1760. Vauquelin, with the frigates *Atalante* and *Pomone* and some smaller vessels, was supporting Lévis' siege of Québec. When English and not French warships were first to reach the city that spring, the French withdrew. Vauquelin, commanding *Atalante,* fought a gallant delaying action against the English frigates *Lowestoft* and *Diana* and ran his ship ashore with colors flying before falling into their hands. See note 778. An account is in Stanley, pp. 249-50. Vauquelin's journal of events is in Casgrain, *Collection,* XI, pp. 263-71 with an extract in Knox, III, pp. 245-51.

726 The attack was made near the mouth of the Rivière Jacques-Cartier at Jacques-Cartier, Québec. The English raid was opposed by Bougainville and his mobile forces. Hamilton, *Adventure,* p. 320.

727 This "Corps of Cavalry" was raised during June 1759 and remained in service until the surrender of the colony in 1760. Chartrand, *The French Soldier,* p. 36. See also Eugène Lelièpvre and René Chartrand, "Corps of Cavalry, Canada, 1759-1760", *Military Collector & Historian,* XXVIII, No. 3 (Fall 1976), pp. 131-32.

Québec,[728] in order to prevent the enemy from landing in that sector, where they would have cut off all communications with the rest of Canada. The Guienne Regiment was sent to a position a quarter of a league above Québec along the river in order to be ready to support the redoubts already mentioned.

The enemy had more than 4000 men deployed in bateaux above Québec who were looking for some place to land between Pointe au Tremble & Québec. They constantly perceived detachments & cavalry formed up to oppose them. On September 13, at daybreak, these troops were sailing down the river again, despairing of finding any means of implementing their plans, when they went close to the redoubt **[139]** guarded by M. de Vercors[729] and noted a very steep cliff[730] which, they conjectured, would be undefended. One or two bateaux went in close and landed troops who scaled the cliff.[731] There was a Canadian sentry there, who fired his musket at them. Unfortunately, he was unable to withdraw to his post. In single file the English reached the top of the cliff. The position was so confident of its security that the majority of soldiers had gone off to cut hay or harvest wheat. Captain Vercors **[140]** was still in bed, where he was shot in the region of the ankle. All his men were scattered. The Guienne Regiment was informed rather late about this incident and then only by a number of men who had fled.[732]

The enemy, as one may well imagine, hastened to form up in the area of the place & even to transport 4 cannon there. The Guienne

728 Pointe aux Trembles, Québec, a short distance upriver from the city of Québec.

729 Captain Louis Du Pont Duchambon de Vergor (note 238).

730 Note in original, p. 140 (a): *This war furnished several examples of the sort. Almost all the English assaults were undertaken at places whose situation seemed to render them invulnerable to capture.*

731 Note in original, p. 140 (b): *Colonel Howe, in command of the light infantry & the Scottish Highlanders, climbed up the cliff with great dash & courage.*

William Howe (1729-1814), major (1756) in the 60th Regiment of Foot and commander of a light infantry battalion in Wolfe's army. He was a younger brother of George Augustus Howe (note 417), and later commanded British troops in America from 1775 to 1778. Boatner, *Encyclopedia,* pp. 522-26; Blanco, I, pp. 785-89.

732 Before dawn the Guienne Regiment was ordered to send two *piquets* to the Plains of Abraham. The whole regiment was ordered to march one-half hour later. Malartic and Gaffarel, p. 283.

The events of September 13, 1759 are detailed in Stacey, pp. 133-69. Knox's version is in II, pp. 94-06.

Regiment immediately stationed itself in such a way as to keep them under observation, having first sent a message to warn M. de Montcalm. It was then already close to 9 o'clock. The camp was left on alert & the army under arms. M. de Montcalm took with him the regiments & the colonial troops, together with about 3 to 400 Canadians. The remainder stayed behind to observe the enemy down by the river. M de Vaudreuil wanted them all to stay in Québec & to bring back **[141]** all the detachments. This view was doubtless the wiser one, but M. de Montcalm judged it more expedient to go & attack the enemy as they were landing. They had already completed the landing and had even organized themselves in formation. The general sent an order to M. de Bougainville to join up with him & set out with about 1500 men, among whom there were many Canadians who were dispersed among the regiments to swell their numbers. These men, who are only suited to petite-guerre[733] & who were in any case poorly armed since they only had simple hunting muskets without bayonets, were a hindrance to the operation.[734]

Some of the lieutenant colonels suggested to M. de Montcalm that it would be better to wait for the élite corps of Bougainville, all the

---

733 The French expression is *la petite guerre.* The form used here is the contemporary English equivalent. Like so many eighteenth century military terms, it is taken directly from the French and refers to a very specific style of warfare that had gained popularity in Europe well before Pouchot came to New France. According to Smith, "PETITE-GUERRE, is carried on by a light party, commanded by an expert partisan ... separated from the army, to secure the camp or a march; to reconnoitre the enemy or the country; to seize their posts, convoys and escortes; to plant ambuscades, and to put in practice every strategem for surprising or disturbing the enemy." This was known as "carrying on the *Petite-Guerre.*" Smith, *Dictionary,* p. 202. Although the formal concept had been developed in Europe, this was the natural form of warfare practiced in New France.

734 The practice of mixing Canadians into the regular regiments was a response to the lack of replacements for the Army battalions. Chartrand, *The French Soldier,* p. 32. As early as September 12, 1758 Montcalm had expressed the idea of mixing 15 militiamen into each company of the *troupes de terre.* He hoped that the Canadians would thus "imbibe the military spirit and perform the same service as our soldiers .... The advantage of this incorporation is, that each company of the Line and Marine will have with it and in it some excellent marksmen, some excellent canoemen and some excellent workmen; that, in emulation, the one of the other, the soldier and the Canadian will operate on each other as a spur to act well, and will mutually teach one another things they know; remedy the inconvenience of not having enough officers in the Colony to lead the Militia, watch them and derive advantage from them." General Reflections, *DRCHSNY,* X, p. 875.

The tactical relationship of the regulars sent from France and the Canadian militia is explored in Martin L. Nicolai, "A Different Kind of Courage: The French Military and the Canadian Irregular Soldier During the Seven Years' War", *Canadian Historical Review,* LXX, No. 1 (1989), pp. 53-75.

more so because the enemy had already completed their landing. He did not take this suggestion **[142]** well & marched at a very brisk pace towards the enemy, still in battle formation, across thickly sown wheat fields. Combined with the very rapid pace, this had the effect of leaving the soldiers winded. The officers were pessimistic about the outcome of such a hastily performed maneuver. Nonetheless, the army met up with the Guienne Regiment & formed up in battle order. Detachments of Canadians plunged into woods on the flanks of the enemy army & in a very short time killed a large number of them. General Wolfe was mortally wounded.

Our army, which was marching towards the enemy in uninterrupted haste, found them lined up behind tall fences which marked the boundaries of the fields with 2 cannon on their flanks. It was greeted with salvoes of case shot and musket fire, to which it responded while on the **[143]** march. But the enemy fire, which had torn gaps in the column, stopped it in its tracks & shook the nerve of the Canadians, who had little experience of being under fire without cover. They broke ranks & fled. The soldiers immediately retreated in confusion. M. de Montcalm, mounted on a horse, galloped up to stop them & rally them, when he was shot in the back. Many officers fell on the field of battle and others were captured, almost all of them wounded. The English pursued us vigorously right up to the walls of Québec.[735]

M. de Bougainville, who had marched immediately, attacked some enemy guard posts installed in houses behind their lines but, on hearing that the battle had been lost, he awaited orders instructing him what to do. M. de Vaudreuil believed that there was no better course of action than to **[144]** regroup his army &, by marching a little way up the St. Charles River, to go and take up a position at Pointe au Tremble. The entire camp was almost abandoned because they failed to inform anyone of this departure. The officers & soldiers lost their personal belongings & supplies there, which they could have taken with them.

They left in Québec 600 men in *piquets* as a garrison, under the command of M. de Ramsey, the major of the fort.[736] The next day, M.

---

735 Although he faults the haste with which Montcalm attacked Wolfe's army, Pouchot also lays much of the blame on the Canadians in the ranks.

A return of French casualties is in Casgrain, *Collection,* I, pp. 218-19 and in Knox, III, p. 143.

736 Captain Jean-Baptiste-Nicolas-Roch de Ramezay (note 232).

de Montcalm died of his wound like a true hero, that is to say, like a Christian hero, after writing to General Townshend,[737] who had just taken over command of the English, to recommend the French prisoners to his good offices. M. de Montcalm's intrepid nature, of which he had given so many tokens, did not desert him in his last moments. The love of glory was, in him, **[145]** second only to his devotion to the interests of the mother country. The purity of his intentions & his disinterestedness were always equal to his valor, on which he relied too much in that final operation.[738] His loss was keenly felt by his soldiers. His officers have given public expression to their regret & the esteem in which they held him.[739] **[146]**

General Wolfe died almost immediately. He was a most ardent warrior. He had asked **[147]** the admiral to make this one last attempt, as the English fleet wanted to go back, fearful of being caught by the bad weather which begins to prevail at that season. The general's last words were: "I die a happy man, because I can see the French fleeing."[740]

---

737 George Townshend (1723/24-1807), colonel and brigadier in Wolfe's army. *DCB,* V, pp. 822-25.

738 Modern historians have been less generous about Montcalm's qualities. See, for example, the analysis of his performance at Québec in W.J. Eccles, *France in America,* (Toronto, 1972), pp. 198-204, 208 and of his other attributes on pp. 188-89. Eccles' entry on Montcalm in the *Dictionary of Canadian Biography* mentions the differing historical attitudes about the general. Eccles himself flatly states that "He was a brave officer, of this there can be no doubt, but serious defects in his character made him unfit to command an army." Moreover, "when he was presented with an opportunity to destroy Wolfe's army, or at least avoid his own defeat, he threw it away and suffered one of the most disastrous defeats in history." *DCB,* III, p. 468.

739 Note in original, p. 146 (a): *It was as a result of their wishes and at the request of M. de Bougainville that the Académie des Inscriptions & Belles-Lettres created his epitaph in 1761. Some of those same officers have just supplied the subject for a print designed by the young Watteau & engraved by the Englishman, Mr. Chambers, in honor of M. de Montcalm. On it, the general is featured on his camp bed near his tent, supported by his adjutant, M. de Montreuil, his friend & the trustee of his last wishes, & by M. de Bougainville, his pupil & former aide de camp. It is at that moment that, sensing his imminent death, he begs his officers & friends to give him as a grave a nearby bomb crater. Some Indians are busy removing the bomb fragments from the crater. A group of officers & soldiers assembled around his bed are expressing their deep grief. In the middle ground, one can recognize the general officers, Sénézerque & Fontbonne, who commanded the two flanks of the army, were killed in the battle and brought by soldiers to the tent of the general headquarters, where one can already see several wounded officers. In the background, one can just distinguish a heap of dead & dying combatants, among whom one can make out young Wolfe, whom his officers are vainly attempting to revive. In the far distance, the unfortunate city of Québec is disappearing amid the flames that are being belched forth by the enemy fleet.*

...(continued next page)...

The whole French army calmly assembled at Pointe au Tremble where M. de Lévis had just arrived at the same time. He still found himself in command of around 5 thousand men who were in good heart since none of them blamed themselves for the setback. He decided to set out straight away and attack the English & sent a cavalry captain, M. de la Roche,[741] on ahead with his troop, carrying sacks of biscuits, to enter Québec. He was ordered to inform M. de Ramsey [148] about the imminent arrival of M. de Lévis & to recommend him to hold firm. This commander replied that it was too late, that he was discussing surrender terms, & indeed that he had given his word and that in any case he had no more supplies. M. de la Roche told him that he was bringing him some pending the arrival of M. de Lévis.

After winning the battle, the English were so astonished by the happy outcome that they were undecided about their appropriate

---

Bougainville wrote that, in 1761, "I got the *Académie des Belles-Lettres* of Paris to compose an epitaph for the Marquis de Montcalm, and I wrote to [English Prime] Minister [William] Pitt to arrange to get it cut on the tomb of the General at the Ursulines of Quebec. The answer of this minister was most amiable and did him as much honor as it did the General and the French troops." Hamilton, *Adventure*, p. 327. In December of the previous year, Bougainville had written to Montcalm's mother to request "his baptismal names, title, arms, the place and date of his birth" which he needed for his epitaph. Bougainville to Mme. la Marquise de St. Véran, Dec. 3, 1760, *Report on the Public Archives for 1929*, p. 90.

The "Young" Watteau, who composed a scene of Montcalm's death, was François-Louis-Joseph (1758-1823), son of Louis-Joseph ("Watteau de Lille", 1731-1789) and great-nephew of the renowned Antoine Watteau (1684-1721). His design must have been available by 1781 to receive notice in Pouchot's *Mémoires*. The footnote gives the impression that it had been very recently completed. A drawing corresponding to the editor's description, which apparently served as the basis for the engraving, survives in the National Gallery of Canada. *Master Drawings from the National Gallery of Canada* (Washington, 1988), pp. 191-93.

"Mr. Chambers", the English engraver, was probably Thomas Chambers (1724-1789). George C. Williamson (ed.), *Bryan's Dictionary of Painters and Engravers* (1816; Reprinted in 5 vol.; Port Washington, NY, 1964), I, p. 281. His print, if it was actually produced, is not well known. The William L. Clements Library in Ann Arbor, Michigan, and the National Gallery of Canada each hold examples of the more familiar "Mort du Marquis de Montcalm Gozon" engraved by Juste Chevillet (1729-1790). E. Bénézit, *Dictionnaire critique et documentaire des peintres, sculpteurs, dessinateurs et graveurs* (8 vols.; Paris, 1966), II, p. 481. Publication of Chevillet's print, which reverses the figures in Watteau's sketch, was not announced in Paris until July 3, 1783. *Master Drawings*, p. 191. Like Benjamin West's much better known "The Death of General Wolfe", the scene created by Watteau, Chambers and Chevillet bears little resemblance to the circumstances of Montcalm's death.

Etienne-Guillaume de Senezergues de La Rodde (note 404) was brigadier general and lieutenant colonel of the La Sarre Regiment. Mortally wounded in the battle of September 13, 1759, he died the following day. *DCB*, III, pp. 590-91.

Fontbonne (note 438) was lieutenant colonel of the Guienne Regiment.

740 Wolfe received two or possibly three wounds and died soon after. See Stacey, pp. 163-64.

course of action, whether they should withdraw or undertake the siege of Québec, which seemed to them likely to be a lengthy operation in view of the season. They were rather agreeably surprised to see that they were being approached with a proposal to surrender the fortress. The commander,[742] who was from Québec, could not refuse the earnest entreaties of the inhabitants, who were more concerned about saving their **[149]** possessions than their country. The English thus granted everything that was asked of them.[743]

M. de la Roche immediately left to report to M. de Lévis on his mission. He found him in the vicinity. The general hastened his march in order to arrive before or fight the English but, upon arriving in front of the city, he was very surprised to find it already occupied by the English army. He found himself compelled to go back to Pointe au Tremble with all his troops, who were deeply troubled by this unexpected event.

We may observe at his point that if M. de Montcalm had decided instead to come and station himself with his troops in front of Québec beneath the area of the citadel, he could have installed on the ramparts extensive artillery to protect it, at least two hundred guns. He could also have been **[150]** joined there by M. de Bougainville's detachment. The enemy would have been able neither to dislodge him nor lay siege to the place. In view of the season, they would not have been able to remain in that position for long & as soon as they attempted to re-embark they would have run the risk of suffering a major setback. The same comment applies to M. de Ramsey. His position, although a poor one, was protected against a *coup de main.* The enemy would have needed at least 3 to 4 days to install their batteries. When M. de Lévis arrived, he would have attacked them or, stationing himself in their vicinity, he would have impeded all their operations. Indeed, they would have had a great deal of trouble extricating themselves from the situation. Having failed to take Québec, England could not conceivably

---

741 Captain de La Roche-Beaucourt (fl.1756-1760), was an officer of the Montcalm Regiment of cavalry and had served as an aide to the general. He commanded the Corps of Cavalry in Canada. Lelièpvre and Chartrand, "Corps of Cavalry", p. 131.

742 Ramezay.

743 Bougainville, who tried to forestall the city's surrender, was disgusted that "This man surrendered without having tried a cannon shot, and the enemy not even having started their trenches." Hamilton, *Adventure,* p. 325. The articles of capitulation are in Vaudreuil to Berryer, Sept. 21, 1759, *DRCHSNY,* X, pp. 1011-13.

have undertaken new initiatives with any hope of success. The capture of [151] the city urged them on, despite themselves, to the efforts they made in 1760. They left a very large garrison in Québec, under the command of M. Murray.[744]

The French formed the bulk of their winter quarters at Pointe au Tremble and Jacques Quartier. These positions were fortified.[745] The regiments & the colonial troops were billeted in the villages & at Mont-Réal, where the generals and the intendant stayed. First of all, the exchange of prisoners was dealt with.[746] In November the officers

744 James Murray (1721/22-1794), lieutenant colonel (1751) of the 15th Regiment of Foot until October 1759 when he was named colonel commandant of the 2/60th Regiment of Foot. He had been one of Wolfe's brigadiers. *DCB,* IV, pp. 569-78.

745 The defenses at Jacques-Cartier were laid out on September 26. Construction began the next day on what Malartic described as a hornwork on a point. Malartic and Gaffarel, pp. 293-94. A "hornwork" was a simplified design with a single curtain connecting two half-bastions. Muller, *Treatise,* p. 223. It was useful for places where it was necessary to defend against attack from one direction only. Fort Niagara was also a hornwork.

The Jacques-Cartier fortifications are described in Knox, II, p. 333. When Amherst viewed them on October 2, 1760, he opined that "the Enemy had been at great pains to make a work of no consequence." Webster, *Jeffery Amherst,* p. 257.

746 Pouchot provides few details of his captivity following the surrender of Niagara. The prisoners' otherwise uneventful journey down the Mohawk River was marked by two incidents, the escape of Lieutenant Larminat (note 581) and several men, probably near Little Falls (see note 1502) and the drowning of eight prisoners and four New York soldiers of their escort at Schenectady. Pouchot and his garrison had reached Albany by August 12. Severance, II, p. 360 quoting an unidentified newspaper. They were at New York by the 20th. *New York Mercury,* Aug. 20, 1759. The newspaper described the prisoners as "robust, fat and spry-looking Men" whose appearance belied their recent participation in a desperate siege.

Pouchot was settled into his place of confinement, Jamaica on Long Island, by August 27. There, as senior officer of the prisoners, he corresponded with General Amherst regarding living conditions and exchange. Pouchot to Amherst, Aug. 27, 1759, Amherst Papers/10. Translations of his letters are printed in Appendix A of this volume.

The prisoners from Fort Niagara and La Belle Famille were widely scattered among a number of villages on Long Island. Pouchot and the other *troupes de terre* officers (Villars, Serviès, Bonafous, Salvignac, and Morambert) were sent to Jamaica, relatively close to New York City. The colonial officers of the garrison, except Larminat who was detained at Fort Edward after his recapture on August 22, went to Newtown (La Roche Vernay and Cournoyer) and Brookhaven (the Chabert de Joncaire brothers). Lieutenant La Milletière was confined with the Chaberts. The 15 officers and cadets taken in the battle of July 24 and subsequently sent to New York were kept in groups of different sizes at Southwold (Southold), Shelter Island, Southampton, and Westchester. The isolation of most of the Marine officers, far out on the eastern end of Long Island, was perhaps an effort to keep them from escape routes to Indian country. "List of the French Officers Prisoners of War", Amherst Papers/8.

...(continued next page)...

of the Niagara garrison left with those of the detachment of Trépézec, 15 of them in all, and accompanied by 250 militia, 50 French soldiers and the same number of colonial troops. The English refused to exchange M. Bonnafoux, the artillery officer, because they said they had no such corps themselves.[747] **[152]**

In the area of Sarratoga,[748] they met General Amherst with his army returning to the area. He sent his aide de camp, Abercromby, to M. Pouchot to hand over the letters from Canada.[749] The officers in

---

The 567 *troupes de terre,* Marine and militia personnel listed as prisoners, including 12 women and children and a number of employees, were more widely scattered. New York Lieutenant Governor James De Lancey, probably alarmed by the large number of enemy soldiers in his capital, reported to Amherst that he had sent 200 of them to New Jersey, 200 to Connecticut, 100 to Long Island, and quartered the remainder in the barracks of New York City. Webster, *Jeffery Amherst,* p. 162. The prisoners destined for Connecticut arrived at New Haven on August 23 where 100 were retained while another hundred were sent on to Fairfield. *Boston Gazette,* Aug. 27, 1759. Peter Pond, a Connecticut man who had served at Niagara as a sergeant in the New York Regiment, later encountered some of his former enemies. "On my Arival at Milford," he wrote, "I found Maney of the Prisners I had Bin so Industres in Captering ware Billeated in the town. I Past the winter among them." Mrs. Nathan Gillett Pond (ed.), "Experiences in Early Wars in America .... Original Journal of Peter Pond, Born in 1740", *The Journal of American History,* I, No. 1 (First Quarter 1907), p. 93.

Following common practice, the prisoners were offered the opportunity to work for local farmers or merchants. At first most of those sent to Connecticut, "chose to go to prison rather than be hired, the proposal of which they seemed to treat with disdain." All but two or three quickly changed their minds, however, and "hired themselves to Farmers and others ... rather than be confined in a Prison." *Boston Gazette,* Aug. 27, 1759.

747 An exchange of prisoners had been under negotiation since August. Montcalm to Amherst, Aug. 15, 1759, Casgrain, *Collection,* IV, pp. 255-56. Discussion centered on French troops taken at Carillon in 1758 and at Niagara rather than those from earlier in the war, and Amherst hinted that if arrangements could not be completed, the prisoners would have to be sent to Europe for exchange. Amherst to Montcalm, Sept. 10, 1759, *Ibid.,* pp. 256-58. Amherst had, in fact, already requested transports for that purpose. Saunders to Pitt, Sept. 5, 1759, Kimball, II, p. 162. He preferred, however, to keep the prisoners in America to encourage the return of "every Soul they [the French] have belonging to the King's Dominions." Amherst to Pitt, Jan. 9, 1760, *Ibid.,* p. 244.

Correspondence was interrupted by Montcalm's death, and it was not until October 29 that Governor Vaudreuil sent a new proposal to Amherst. The English general agreed to the terms on November 2. Malartic and Gaffarel, p. 299; Webster, *Jeffery Amherst,* p. 186.

The list of prisoners returned to Canada includes 14 officers, 3 commissaries, 1 "Guard Magazin", 1 surgeon, 3 volunteers, 53 enlisted men of the *troupes de terre,* 38 of the Marine, 100 militia, 6 workmen, 7 servants, and a woman and her daughter for a total of 228 individuals. "List of the French Prisoners Sent to Canada", Dec. 14, 1759, Amherst Papers/8. Included in this list was Lieutenant Larminat who joined the prisoners when they passed Fort Edward. Webster, *Jeffery Amherst,* p. 194.

...(continued next page)...

command of the English forts on their itinerary received them with very great civility. One cannot speak too highly of their welcoming attitude.[750]

At the falls of Carillon, the French officers were obliged to wait around for 7 or 8 days through the quibbling of the fort commander.[751] During this time Major Rogers arrived. He had formed a raiding party of about **[153]** 400 men, who had been to our mission of St. François on Lac St. Pierre. He found this Abenaki village entirely denuded of its warriors. He killed around thirty women & old people and brought back a number of young men prisoner. Since he lacked provisions, he

---

Four of the officers were from Trépezec's detachment and had been held since July 6, 1758 (note 423). The others, with the exception of a colonel and a major of the Detroit militia captured at La Belle Famille, were from the Niagara garrison. Listed among them is a Lieutenant Morandière, the circumstances of whose capture have not been ascertained. He was a Marine officer, and is specifically named as having been made prisoner by the capitulation of Niagara, but he appears in no other sources on the siege. "List of the French Officers Prisoner of War by the Capitulation of Niagara", Amherst Papers/8. This officer was Etienne Rocbert de La Morandière Ducoudray (1740-1781), recently promoted ensign *en pied* (January 1, 1759). Fauteux, *Les Chevaliers*, p. 221. An Ensign (*en second*) La Morandière was assigned to Fort Little Niagara as early as 1756. Casgrain, *Collection*, VII, p. 82. This was almost certainly the same man, who had a family connection with his posting. He was the brother-in-law of Lieutenant Daniel Chabert de Joncaire, commandant of Little Niagara. See *Severance*, II, p. 442. It seems most likely that La Morandière entered Fort Niagara with Chabert's party on July 10 or sometime soon after. If so, he escaped mention in Pouchot's *Mémoires* and did not sign the articles of capitulation. His identification by the English as a lieutenant would seem to be an error.

None of the Marine officers captured at La Belle Famille were exchanged, and most would remain prisoners until early 1761. Five of the Niagara officers were not exchanged: Bonafous, Joncaire, La Roche Vernay, La Milletière, and Cournoyer. Pouchot provides a plausible explanation for the detention of Bonafous - the English prisoners exchanged for Pouchot's party did not include an artillery officer. The exchange was an even one, and there were simply not enough English officers then prisoner in Canada to permit the release of all the French. "List of Officers Returned from being Prisoners in Canada", Nov. 15, 1759, Amherst Papers/8.

The 270 English prisoners included 16 officers, 137 regulars, 48 provincials, 12 sailors, 26 "Inhabitants" and 31 women. Most of the civilian men and women had been taken at German Flats in 1757. *Ibid.* Led by Major James Grant (note 466), the officers dined with Vaudreuil on November 7, left Montréal on the 11th and arrived safely at Crown Point on the 14th-15th. Malartic and Gaffarel, pp. 299-300; Webster, *Jeffery Amherst*, p. 191.

748 Note in original, p. 153 (a): *It was from that fort that the army which conquered the whole of Canada later set out & there too that the British empire in America only recently suffered a fatal setback, by a singular stroke of fate.*

The post of Saratoga was at modern Schuylerville, New York. The village was the site of the surrender of General John Burgoyne's army to the Americans on October 17, 1777. See Boatner, *Encyclopedia*, pp. 978-80.

divided his troops into several bands to return to Fort George. They all perished of exposure or hunger in the forest, except that of Rogers himself, who fortunately had a Loup Moraigan[752] as a guide. He only returned with 21 men, however, all of them haggard & emaciated.[753]

Since the ice was forming during the detachment's journey, the water was frozen solid around the middle of Lake Champlain & he[754] had an enormous amount of difficulty reaching land with his bateaux, which were frail and sliced by the sharp edges **[154]** of the ice. In the evening, the French soldiers were obliged to convert their bateaux into sleighs & journeyed in great peril on the recently formed ice. The ice often starred in cracks 4 to 8 feet long. After many difficulties, they

749 On November 29 General Amherst encountered the prisoners between Fort Edward and Saratoga. He was left with the impression that "they all dislike the march at this season & the major part of them, I believe would be glad to get off returning to Canada." Amherst Journal, Knox, III, p. 76.

James Abercromby (1732-1775) was a captain (1756) in the 42nd Regiment of Foot. He had previously served as aide-de-camp to his uncle, Major General James Abercromby, and took the same position with Amherst in May 1759. He was mortally wounded at the Battle of Bunker Hill. *DCB,* IV, pp. 3-4.

750 James Gabriel Montresor (1709-1776), colonel of engineers and commandant of the incomplete Fort George, received Pouchot's party on December 1. They dined with him that evening and shared breakfast the next morning before setting out on Lake George. G.D. Scull (ed.), "The Montresor Journals", *Collections of the New-York Historical Society for the Year 1881* (New York, 1882), pp. 3-4, 109. Montresor's hospitality was due, in part, to Amherst having ordered him to "receive the French Prisoners, shewing them civility." *Ibid.,* p. 108.

751 In contrast to Montresor's "welcoming attitude," Crown Point's commandant was decidedly uncooperative. The prisoners arrived at the Ticonderoga portage on December 4 where their escort turned them over to Sir Henry Seton (fl.1757-1766), captain (1757) of the light infantry company of the 17th Regiment of Foot. *Army List.* Finding Lake Champlain blocked by ice from Ticonderoga to Crown Point, Seton marched the prisoners overland by the newly constructed military road. Just short of his goal, however, he received a letter from the commandant stating that "it was very improper to march the French Prisonners to Crown Point" without Amherst's orders. Seton accordingly returned his party to the saw mills near Ticonderoga "and staid with them here to ye 13th" when he was allowed to proceed. Seton to Amherst, Dec. 17, 1759, Amherst Papers/78. Seton's instructions from Amherst, dated November 24, are in Amherst Papers/81.

The quibbling Crown Point commandant was William Haviland (1718-1784), lieutenant colonel (1752) of the 27th Regiment of Foot. *DCB,* IV, pp. 334-36. His attitude might have been influenced by his participation in the disastrous assault on Carillon in 1758. There is no evidence that Haviland awaited orders from Amherst before allowing Seton to proceed, and the general does not mention the incident in his journals. He reported to Prime Minister Pitt only that the prisoners were "by the Severity of the weather detained at Tionderoga [sic] from the 3rd to the 13th Decr." Amherst to Pitt, Jan. 9, 1760, Kimball, II, p. 244.

arrived at Mont-Réal on Christmas Eve. They were warmly welcomed because of the reinforcements they had brought. M. Pouchot was very cordially entertained by MM. de Vaudreuil & de Lévis.[755]

In the course of the winter, a number of small raiding parties were formed of Frenchmen, Canadians & Indians in order to harass the garrison of St. Fréderic & that of Québec. We also worked on preparations to win back the latter city in the immediate future. As soon as this was known by the English, they lived in perpetual anxiety which so exhausted their garrison [155] that they lost nearly 1500 men.

---

752 Pouchot uses the term "Loup Moraigan" on several occasions. By it he means a Mohican. Rogers' rangers had included Mohican or Stockbridge Indians since 1756. Cuneo, p. 42.

753 Angered by the capture of two regular officers sent through French lines under a spurious flag of truce, Amherst ordered Robert Rogers with 200 men to strike the Abenaki mission village of St. François-de-Sales or Odanak (near Pierreville, Québec). The village was on the St.-François River that flows into Lac St. Pierre. The party set out from Crown Point on September 13. After a difficult journey by way of Lake Champlain and an overland march, Rogers led the 141 men who had not dropped out into the sleeping village at 3:00 a.m. on October 6. Cut off from their boats, the rangers split into smaller parties and marched south through the wilderness towards the Connecticut River. Rogers reached Fort Number 4 (Charlestown, NH) on October 31 to obtain help for his starving men. The return march from St. François cost Rogers three officers and 46 enlisted men. Cuneo, p. 114.

The raid is one of the more controversial incidents of a long and bitter series of frontier wars between Indians and English colonists. Abenaki casualties are uncertain, but Rogers' first report to Amherst, brought to him at Crown Point by a wounded officer on November 7, claimed 200 killed and 20 prisoners. Amherst had learned of the attack on November 2 from a French officer who had arrived under a flag of truce. He said that Rogers had "burnt the settlement at St. Francis, killed some Indians, women & children." "I fancy he is mistaken about the women & children," the general wrote in his journal. Webster, *Jeffery Amherst,* pp. 188, 186. Although Amherst's orders to Rogers clearly stated that "it is my orders that no women or children are killed or hurt," they also included the reminder that "those villains have dastardly and promiscuously murdered the women and children of all ages" and enjoined him to "Take your revenge." Rogers, pp. 104-05. An account by Rogers' biographer is in Cuneo, pp. 100-16 with a different perspective provided by Stanley, pp. 239-40. The ranger captain's own account is in Rogers, pp. 105-15. He was careful to point out that the St. François Indians had, for years, "been killing people of all ages and sexes in a most barbarous manner." *Ibid.,* p. 111.

754 Pouchot.

755 Pouchot greatly understates the difficulty of his trip from Crown Point to Montréal. Captain Seton obtained permission to proceed from Ticonderoga to Crown Point on December 13. There Haviland furnished only 14 bateaux for the 228 prisoners and one for

...(continued next page)...

However, Canada was in an appalling situation because of the lack of food & goods of all kinds. In the winter, wine cost 2400 *livres* for a barrel of 240 bottles, brandy 1500 *livres* a *quart*,[756] salt 3 to 400 *livres* a box, wheat 30 to 48 *livres* a bushel weighing 45 pounds, mutton 3 *livres* a pound, horse meat 1 *livre* 4 *francs*, a beef carcass 4 to 500 *livres*, a calf 50 to 60 *livres*, a turkey 50 *livres*, a pair of shoes 30 *livres* &c. Everything rose to a quite arbitrary price. A cord of wood which normally cost six *livres* sold for from 80 to 100 *livres*. The intendant printed as much money as he could in order to counteract these prices but he never thought of taxing anything because all these increases were very much to his own advantage and **[156]** that of the society[757] round him. They were careful to monopolize both provisions & trade goods, which they then sold to the king and to private individuals. At least half of the habitants, who had been kept under arms throughout the campaign were living in want. Their wheat & cattle were taken to feed the troops. It is true that they received a good price for these commodities in paper money, which was the common currency, but it did not suffice to cover their basic needs. The discredit into which the currency fell made everything increase by 15 percent in 15 days.[758] **[157]**

This inflation continued right up to the surrender of Canada. In the summer, a barrel of wine increased in price to 10 thousand *livres*, &

Seton's escort. He reported that "even these were so much damaged by the Ice, that three of them sunk before I got eight miles from the Point, which oblidged me to give them [the French] the one I had & to return by Land. I left them out of sight of the Fort, and they were in a great hurry, and so very cold, that I believe they scarcely even looked back." Seton to Amherst, Dec. 17, 1759, Amherst Papers/78.

Seton's orders had called for him to escort the prisoners as far as "The Narrows", about 12 miles north of Crown Point and near the mouth of the Otter River. The convoy travelled down the west side of the lake, and Amherst had ensured their security after Seton's departure by ordering that no English parties be sent to operate in that area. Pouchot carried letters for Vaudreuil. Amherst to Seton, Nov. 24, 1759, Amherst Papers/81. In accordance with Amherst's instructions, Seton had Pouchot sign a receipt for the prisoners and 12 bateaux before he sent him on his way on December 14.

Pouchot arrived at St. Jean on December 20 with the officers and 168 soldiers or militia, a discrepancy in numbers that might indicate some desertion on the part of the militia. He proceeded to La Prairie where he found orders to go to Montréal, arriving there on the 24th. Pouchot commented to Adjutant Malartic both about his good treatment by the English and the severity of the weather on his trip from Crown Point. Malartic and Gaffarel, p. 303.

Neither Vaudreuil nor Lévis mention having entertained the returning captain.

756 Here Pouchot means the *quart* or *quartout* discussed in note 299.

757 Bigot's *grande societé*. See note 289.

everything else rose proportionately. One might ask how the troops, who only had their pay, managed. It was gambling that supplemented it. The highest stakes imaginable in France are nothing compared to the sums which were wagered. The intendant & the ladies of his society, as well as the Canadian officers, the majority of whom had made huge profits from their trading activities, lost large sums at the expense of the French officers.[759] Some of the latter took a great deal of money back to France. Some sold their ration of brandy at a very high price, & their furniture as well. Others, by wheeling and dealing, found a way to acquire small sums of money. Well-to-do citizens were delighted **[158]** to feed their defenders. Everyone was on very good terms with everybody else, drawn together by their common misfortunes.

M. Pouchot could cite one example & would have liked to be free to name the lady who was his benefactress.[760] Living alone, this lady said to him one day: "Sir, food is very expensive. It is very difficult to obtain supplies. Let us pool our resources. What you can contribute and what I have already will permit us both to live in greater ease." Since he arrived in the dead of winter, M. Pouchot had not been able to acquire supplies for himself. He thus thought himself fortunate to avoid the trouble of having to find them by contributing his money. In the two & a half months that he stayed in Mont-Réal, he very often urged her to take some money. She always replied, "At the end of the winter, we shall settle up." Upon his departure, he wanted to pay her his **[159]** share, which came to two thousand *livres*. Although pressing her very insistently, he was never able to persuade this generous woman to accept the sum. Several other officers enjoyed the same kind of generosity.

---

758 Note in original, p. 157 (a): *M. Murray profited from this cruel situation by selling supplies to the French, thereby amassing a great deal of money in a very short time. If the English believed that assistance of such a self-interested kind deserved a monument at Vauxhall, they were mistaken, or perhaps their vanity deluded them.*

The comments here are probably based on Murray's post-war attempts to stabilize the Canadian economy which ultimately proved costly to some of those who followed his advice. His actions included price controls and sales regulations. *DCB,* IV, pp. 572-73.

759 The uncontrolled gambling in the last years of the French colony is discussed in Stanley, pp. 210-11.

760 Pouchot's confidence remains unbreached. His "benefactress" has not been identified.

• 1760 •

At the beginning of March, MM. de Vaudreuil & de Lévis decided to send M. Pouchot out over the ice to take command of Fort Lévis on Isle Orakointon, near La Présentation, & to recall the engineer, M. des Androins, who had remained there since the month of September. They needed his skills for the siege of Québec, for which they were making every effort to speed up the preparations.[761] M. Pouchot was constantly aware of the difficulty of the task with which he had been entrusted, owing to the lack of resources he would require to **[160]** do a good job. However, his zeal for the good of the service enabled him to make light of the drawbacks. The generals promised to send him a corps of 12 to 1500 Canadians in the summer.[762]

He entirely lacked equipment.[763] Having lost a part of it at Niagara, he had been obliged to abandon the rest of it upon his return from New England. He thus had to reconstitute some gear and assemble a few provisions for the campaign, which cost him a great deal of money. As he had to travel across the ice, he asked for a blanket from the intendant, who had the barbarity to refuse. The quartermaster presented him with a 12-quart[764] cask of wine, which was a very substantial gift in the current situation. That was the only thing supplied by the King. **[161]**

M. Pouchot set out on March 17, together with the Abbé Piquet, the missionary from La Présentation, 5 men & 3 sleighs. Before speaking about the outcome of this journey, let me report on the result of the Québec expedition. On April 23, the ice began to break up.

---

761 Vaudreuil believed that Pouchot would be the best replacement for the engineer because he "combined within himself the qualities of a good Engineer, and had acquired experience in managing the Indian nations." Vaudreuil to Berryer, June 24, 1760, *DRCHSNY,* X, pp. 1089-90. The recall of Desandrouins is also discussed in Gabriel, p. 315. Fort Lévis was lightly manned that winter because Desandrouins had been forced to send 100 of his men to Montréal due to a shortage of supplies. Malartic and Gaffarel, p. 305.

762 No such commitment has been found in other sources, and it seems an unlikely promise considering the rapidly declining military resources of the colony. In May Governor Vaudreuil admitted that he had no forces to further augment Pouchot's position. Vaudreuil to Lévis, May 15, 1760, Casgrain, *Collection,* VIII, p. 184.

763 By this Pouchot means the personal field belongings of an officer.

764 The measure in the original is *pot* - "pot" or "jug."

The quartermaster or *munitionnaire* was Joseph-Michel Cadet, and this cask of wine is presumably what brought Pouchot to his table and earned him a mention in Querdisien Trémais' report to the Minister of Marine. See notes 110-111.

Each regiment & the whole army received the order to leave with its provisions & the artillery to undertake the siege of the place. Each troop carried its supplies in the bateaux designated for it. Their entire allowance for subsistence was 1/4 of a pound of pork and a pound & a half of bread per head. The bateaux followed the ice downstream & arrived at a place one league above Québec.[765] The enemy could not have expected our arrival, given the state of the river. Unfortunately, they were warned by three artillerymen who, having seen their bateau caught up in the **[162]** ice & smashed, climbed onto a large ice floe which carried them down as far as Québec, where they were arrested.[766] Apprised of the situation, the enemy sent out guards with the aim of making a retrenched camp near the fortress. Their guards were expelled from the houses they were occupying & the army advanced to spend the night 1/4 of a league from Québec. Rain was falling & there was a thaw. It is easy to guess the situation of our troops at that time, in fields covered with snow or muddy slush. On the morning of the 28th, the enemy moved up to occupy the camp they had laid out, their flanks supported by 12 cannon & some howitzers.[767] Our troops found them in battle order.

---

765 By April 17 Lévis reported available forces of 7,262 men, including non-combatants. Casgrain, *Collection,* I, p. 257. Lévis and his army arrived at Point aux Trembles, supported by Captain Jean Vauquelin's frigates *Atalante* and *Pomone* and smaller vessels, on April 24. Vaudreuil to Berryer, May 3, 1760, *DRCHSNY,* X, p. 1075. The army was disembarked a little farther down the river at St. Augustin in order to march overland towards the village of Ste. Foy. "Journal", *DRCHSNY,* X, p. 1080.

766 Both French and English sources mention this unlucky occurrence. A French "Journal" of the battle and siege notes: "A part of the ice that lined the river having broken loose on the morning of the 26th, carried away some bateaux loaded with artillery, a few of which were submerged; some of the gunners perishing and one of them floated on a cake of ice as far as Québec; the English Governor having learned from him our movements ... made his arrangements against being surprised." *DRCHSNY,* X, p. 1081.

John Knox reported that, about 2:00 a.m. on April 27, the watch aboard the sloop of war *Racehorse* heard "a distressful noise on the river." A boat was launched and recovered "a man whom they found almost famished on a float of ice." He proved to be "a Serjeant of the French artillery, who, with six other men, were put into a floating battery of one eighteen-pounder; that his bateau overset ... and his companions he supposes are drowned; that he swam and scrambled alternately, through numberless floats of ice, until he fortunately met with a large one, on which, though with great difficulty, he fixed himself." Drifting with the tide almost to Isle d'Orleans, he was moving upriver on the flood when rescued. The exhausted gunner was taken to Governor Murray "in a sailor's hammock" and provided useful intelligence. Knox, II, pp. 385-88.

767 The first detachment sent out by Murray was provided with 10 six-pounders. Knox, II, p. 388.

As they arrived, our troops formed from left to right, first the Guienne Regiment, then Béarn, Berri, La Sarre, Royal-Roussillon **[163]**, the Colonial,[768] Languedoc, the Canadians, La Reine, the cavalry & a few Indians. That was approximately our order at the start of battle. The English had ordered the occupation by the Scottish Frazer Regiment[769] of a house on their right which was within reach of both the Guienne & Béarn regiments, which led to an engagement on our left. These two battalions with their grenadiers dislodged them from the house at first, but it changed hands two or three times.

While the army was forming up, the English were firing case shot with their howitzers which killed a great many men. M. de Bourlamaque was wounded in the leg & his horse was killed. M de Lévis, who saw that his right was not moving up in regular formation, wanted to fall back on some hedges which were **[164]** behind him, in order to wait until they had all arrived & then to charge together. The Guienne & the Béarn regiments were advised of this, & since they felt themselves to be in a dangerous position were the enemy to return and occupy the house, they hesitated to execute the maneuver. The enemy interpreted the movement on our right ordered by M. de Lévis as a sign of retreat and moved forward in front of our artillery in order to pursue our troops. This placed Guienne & Béarn level with their right flank. The commanders of these two battalions, Manneville[770] & d'Alquier,[771] judged that that was the moment to charge them. It should be noted that these two battalions had already suffered heavy losses in the attack on the house, in which the Scottish regiment, which was very brave, was almost entirely wiped out, **[165]** as were the grenadiers of our two battalions. The latter attacked the English and the whole of our right

---

768 The Marine companies had been formed into a pair of battalions and fought at the Battle of Ste. Foy. One was led by Captain Vassan (note 381), Pouchot's predecessor as commandant of Niagara. He was wounded in the action. Journal, *DRCHSNY,* X, p. 1083.

769 The 78th Regiment of Foot (Fraser's Highlanders).

770 The Sieur de Manneville (fl.1734-1760), captain (1745) in the Guienne Regiment. "Les officiers du régiment de Guyenne", p. 193.

771 Jean d'Alquier (or Dalquier) de Servian (fl.1710-61), lieutenant colonel of the Béarn regiment. *DCB,* III, p. 12.

When he received Lévis' somewhat confused order to retreat, Lieutenant Colonel d'Alquier apparently realized the error. Rather than retreat, he shouted to his men (at some length): "It is not time now, my boys, to retire at twenty steps from the enemy. With bayonets upon your muskets, let us throw ourselves headlong amongst them; that is better!" Quoted in Stanley, p. 147. Another version of his words is in Malartic and Gaffarel, p. 317.

wing then marched on them once again all together. We immediately broke through their line. Those who had lagged behind quickly caught up & we captured all their artillery together with their equipment.[772]

We pursued them to beneath Québec, but not very vigorously. Our soldiers were exhausted by their discomforts & weakened by poor food. We have seen that they had left their quarters on the 20th & since that time they had been continuously without tents and exposed to the snow & rain. It is certain that, if they had been able to run, very few English would have returned to Québec & we would have remained masters of the place, which was only occupied by sick and crippled men.[773]

The enemy losses totalled **[166]** 1200 men killed, wounded and captured. On our side, we had 130 officers killed or wounded & 390 men. M. le Chevalier de Lévis commanded with intelligence & dash. Seeing that the enemy advance was making them lose a great advantage, he seized on that moment to order a charge, which was undertaken by our troops with the utmost vigor. There were a great many Indians in our army, but they did not want to participate. They found the task too hot & admired the fortitude of the regiments, who were cut down by the heavy artillery fire and nonetheless continued their advance.[774]

As of the next day, we opened up a trench in front of Québec, in the area of the citadel. A day-by-day account will not be given, as it is fairly well known.[775] After 8 or 10 days of work on the trench, **[167]** we

---

772 The French reported the capture of 20 guns and two howitzers. "Journal", *DRCHSNY,* X, p. 1084.

773 A return of April 24 listed 2,299 sick in the English garrison of Québec. Knox, III, p. 389n.

774 Losses were heavy on both sides. Murray reported 259 killed, 829 wounded, 13 prisoners, and 3 missing for a total of 1,104. Murray to Amherst, April 30, 1760, Knox, II, p. 397n.

Casualties among the French officers, although not quite as heavy as stated by Pouchot, were severe. Twenty-eight were killed and 68 wounded. Enlisted casualties were reported as 165 killed and 572 wounded for a total of 833. Casgrain, *Collection,* I, p. 269. Among those mortally wounded were lieutenants Morambert of Guienne and Salvignac of Béarn, Pouchot's comrades at the siege of Niagara and both recently exchanged with him. "List of the Officers Killed and Wounded", *DRCHSNY,* X, pp. 1084-86.

A contemporary French account of the battle is the "Journal of the Battle of Sillery and Siege of Québec", *Ibid.,* pp. 1077-86. Knox describes the action and its preliminaries in II, pp. 378-98. For modern accounts see Stanley, pp. 244-48, and Gipson, VII, pp. 438-40.

775 See "Journal", *DRCHSNY,* X, pp. 1087-89 and Knox, II, pp. 398-414.

bombarded the fortress with our artillery, of which we had very little. The guns were so poor that they exploded as soon as they became a bit hot. The artillery of the fortress, which was substantial & in very good condition, caused us heavy losses.[776] The English were on the point of surrendering. Their only hope lay in the arrival of a fleet. The French army was in the same position. People commonly said: "If our relief vessels arrive first, Québec will be captured & we shall be saved." Both sides were extremely impatient.

On the 12th, we learned of the arrival of English vessels in the river. On the 15th, some warships hove in sight, among them one of 74 guns & two frigates[777] which attacked two **[168]** of our own ships covering our supply dumps. After a long duel, they captured them, which forced us to raise the siege very hastily. We tried to spike our guns, which we could not take away with us for lack of limbers.[778] Our army fell back on Pointe au Tremble, leaving behind its hospitals in charge of religious establishments outside Québec. The English fleet brought

---

776 The French artillery was "composed of 6 iron 18 and 12-pounders, only one 24, and although the best in the Colony had been selected, the most of the guns were, on the second day unfit for service, and the remainder soon threatened to be in the same state." The French guns were also woefully short of ammunition. "Journal", *DRCHSNY*, X, p. 1087.

The English garrison was much better equipped, and they impeded French siege efforts. Not only was the construction of parallels and batteries very slow and difficult in the rocky ground, but the "enemy soon had 60 cannon unmasked on the attacked fronts. This artillery, served with the greatest activity, not only retarded the construction of the batteries, but also prevented the workmen transporting material; the balls plunging behind the heights, left no spot protected. The troops were even obliged several times to decamp." *Ibid.*, p. 1087.

French casualties during the siege were 73 killed and 133 wounded. Casgrain, *Collection*, I, p. 269.

777 Note in original, p. 168 (a): *It was the ships of Commodore Swanton, in advance of the squadron of Lord Colville, which had set out from Halifax on 22 April.*

Robert Swanton (d.1765), captain (1744) in the Royal Navy, commanded *Vanguard* and was accompanied by the frigates *Diana* and *Lowestoft*. *DCB*, III, pp. 605-06.

Alexander Colvill (or Colville), 7th Baron Colvill (1717-1770) was captain (1744) in the Royal Navy and had commanded the North American squadron since the departure of Vice Admiral Charles Saunders. *DCB*, III, pp. 131-33.

"We had hoped for a moment that they were French," wrote Adjutant Malartic of the warships. He was quickly disappointed. Malartic and Gaffarel, p. 323. John Knox expressed the "inconceivable satisfaction" of the garrison when *Lowestoft* was identified as English. The "Officers and soldiers mounted the parapets in the face of the enemy, and huzzaed, with their hats in the air, for almost an hour; the garrison, the enemy's camp, the bay, and circumjacent country for several miles, resounded with our shouts and the thunder of our artillery; for the Gunners were so elated, that they did nothing but fire and load for a considerable time." Knox, II, p. 415.

5 to 6 thousand troops to land who, with the seamen, totalled 8 to 10 thousand men.

Our army did nothing more than skirmish over small pieces of territory from its base at Pointe au Tremble until it was obliged to withdraw to the island of Mont-Réal. In the month of August, the English brought up their **[169]** frigates & armed launches. These were accompanied by troops marching along the river. They dislodged us from our positions one after the other. We abandoned them in order not to allow the troops to be surrounded. As the English marched through a village, they made the inhabitants swear the oath of allegiance. Doubtless they would have made swifter progress if they had not waited for the news of the other corps who were to attack on the other frontiers.[779]

The Lake Champlain frontier had not at first attracted the attention of the enemy. During the winter, 3 to 400 men had remained at Isle aux Noix. They worked at putting it in a better state of defense. It was from that position that, in the winter, they formed raiding parties which always brought back a number of prisoners. Langis made some further successful raids in the spring. This officer, **[170]** the best leader of raids among the colonial troops, was unfortunately drowned while trying to cross a river in a canoe with two of his men. The river was not completely frozen at the edges and a piece of ice broke off suddenly, fell on the canoe & swamped it.[780]

After the engagement at Québec, M. de Bougainville was sent with 1000 men to Isle aux Noix. Throughout the summer, he had no trouble. Only a few detachments appeared and they came to

---

778 Pouchot previously and erroneously describes this naval action as part of the 1759 siege of Québec. See note 725. It actually took place on May 16, 1760 when Captain Jean Vauquelin (note 463), with the frigates *Atalante* and *Pomone* and some smaller vessels, was attacked by the English ships. Both French vessels were lost after a hard-fought action. See Stanley, pp. 249-50 and Vauquelin's journal of events in Casgrain, *Collection,* XI, pp. 263-71 and Knox, III, pp. 245-51.

Having gained control of the river, the English vessels were able to trap the French bateaux at the Ance au Foulon: "'twas impossible to put our siege train on board; only our ammunition could be saved." The light artillery was also removed. "Journal", *DRCHSNY,* X, p. 1088.

779 For a summary of the 1760 campaign on the St. Lawrence see Stanley, pp. 250-53.

780 Ensign Jean-Baptiste Levrault de Langis Montegron (note 389). The accident occurred at Isle des Soeurs near Montréal, and Langis was buried at Longueuil on June 1. *DCB,* III, p. 400. The actual date of the accident is unknown, but Malartic recorded news of it on April 18. Malartic and Gaffarel, p. 313.

reconnoitre rather than to harass him. The English corps which was at St. Fréderic, made up of two regiments of militia, totalled 3 to 4 thousand men.[781] In the month of August, it set out by bateau & sailing-barge to proceed to the St. Jean river. We had erected **[171]** across the river bed a line of stakes, which was covered from the island. The English were obliged to set up batteries on wooden scaffolds on the land overlooking the island above the line of stakes because the terrain was flooded. After 2 or 3 days of bombardment from both sides, our garrison left the island & crossed the river to reach the forest. Often marching in water, it arrived at La Prairie. They had doubtless been ordered by MM. de Vaudreuil & de Lévis not to allow themselves to be captured & to come and reinforce Mont-Réal.[782]

The English having secured Isle aux Noix, proceeded to St. Jean & Chambly.[783] There were several skirmishes in the forest between St. Jean & the savannah of La Prairie. When the English had crossed this territory, the French troops entered the island of Mont-Réal. **[172]**

It is time to continue the account of the operations on the frontier of the upper reaches of the St. Lawrence, along which marched the great English army commanded by General Amherst. Since the events have not been recounted by anyone, we shall go into great detail in the narration.[784]

---

781 The English force on Lake Champlain was more formidable than simply a provincial army. Included were four companies of the 2nd battalion, 1st Regiment of Foot as well as the 17th and 27th regiments. Attached to it were five battalions of Massachusetts troops and one each from Rhode Island and New Hampshire. There were also five companies of rangers and one of Indians under Major Rogers as well as four independent companies. Knox, II, pp. 524-25; Hervey, p. 53. This army, numbering some 3,400 men, was commanded by William Haviland (note 751), by this time appointed brigadier. *DCB,* IV, p. 335.

782 For a narrative of the French defense of the Lake Champlain frontier see Stanley, pp. 253-54.

783 Chambly, Québec where a small stone fort was situated to guard the rapids of the Richelieu River. For an account of the history and development of this post see Cyrille Gélinas, *The Role of Fort Chambly in the Development of New France, 1665-1760* (Ottawa, 1983).

784 Note in original, p. 173 (a): *This section could have been shorter, but the pleasure of speaking of events in which one has participated oneself always induces the authors of historical memoirs to get carried away. One aspect of the detail given here, however, has the great advantage of allowing the reader to become better acquainted with the mind & character of the Indians than any travel literature could.*

The most comprehensive secondary account of the 1760 campaign on the upper St. Lawrence is Malcolm MacLeod, "Fight at the West Gate, 1760", *Ontario History,* LVIII, No. 3 (September 1966), pp. 172-94.

M. des Androins having left Fort Lévis, M. Pouchot remained there with 150 colonial soldiers or militiamen, six Canadian officers, M. Bertrand, the artillery officer,[785] the two Celoron brothers,[786] La Boularderie,[787] de Bleury[788] & the colonial cadet, de Poilly.[789] **[173]** There were also the captains of the two corvettes, La Force & La Broquerie,[790] and 180 crew members.[791]

The fort had nothing completed except the rampart faced with saucissons. The barracks, storehouses & officers' quarters, & other buildings pertaining to the fort were finished with wooden timbers

785 This officer, who would die during the siege of Fort Lévis, has defied identification. He was apparently a Canadian and not one of the regular gunners sent from France in 1757. It is possible, given the high proportion of very junior officers assigned to the Fort Lévis garrison, that Bertrand was not a lieutenant but rather one of the ten cadets of the *canoniers-bombardiers* and was therefore not named on the list of artillery officers in the colony. Memoir on Artillery, Jan. 11, 1759, *DRCHSNY,* X, p. 935-37.

786 The "two Celoron brothers" were probably Pierre-Joseph and Pierre-Hyacinthe, both sons of the late Captain Pierre-Joseph Céloron de Blainville (note 154) who led the expedition to claim the Ohio for France in 1749 and died in 1759. *DCB,* III, pp. 99-101.

The elder son, Pierre-Joseph de Céloron (1726-1801), was promoted captain in the Marine in 1760. His lieutenancy dated to 1755. Fauteux, *Les Chevaliers,* pp. 185-86; Theodore Calvin Pease and Ernestine Jenison (eds.), "Illinois on the Eve of the Seven Years' War, 1747-1755," *Collections of the Illinois State Historical Library, XXIX* (Springfield, 1940), p. xxix, n. 1.

The younger brother was Pierre-Hyacinthe de Céloron (b.1732), ensign (1756) and recently lieutenant (1760) in the Marine. Pease and Jenison, p. xxix, n.1. He is not listed among the ensigns in the "Tableau ... des Officiers de la Marine", Brymner, 1886, p. clxxxiv.

One of the Céloron brothers, recorded as "lieutenant in the Marine troops", served as Pouchot's fort major. Persis Yates Boyesen, "French Personnel at La Presentation." *The Quarterly,* XXXV, No. 1 (January 1990), p. 11. This was probably Pierre-Hyacinthe as his brother had been promoted captain on February 1.

787 This officer was probably one Le Poupet de La Boularderie (fl.1754-1760). Formerly of the Isle Royale establishment, he was a lieutenant (1758) in the Marine. McLennan, p. 343. He was probably one of the sons of Antoine Le Poupet de La Boularderie (1705-1771), a captain in the Marine who went to France after capture at Louisbourg in 1758 *DCB,* IV, pp. 468-69.

Lieutenant La Boularderie was serving at La Présentation as early as April 1759. Malartic and Gaffarel, p. 228.

788 He was probably one Sabrevois de Bleury (fl.1750-1760), ensign (1755) in the Marine on the Isle Royale establishment who was promoted lieutenant in Canada February 1, 1760. McLennan, p. 343; "Tableau ... des Officiers de la Marine", Brymner, 1886, p. clxxxiv. He might have been a son of Clément de Sabrevois de Bleury (1702-1781) who was very active in organizing transportation services on Lake Champlain. Clément had a son named Jean-Clément. *DCB,* IV, pp. 690-91.

De Bleury was active on the Lake Ontario frontier as early as September 1758. Malartic and Gaffarel, p. 206.

placed one on top of the other & covered by planks.[792] M. Pouchot, in order to render the position defensible, caused to be erected on the parapet, which was 18 feet wide, another of nine feet with timbers one on top of the other and filled with earth, which he had to have transported from off the island. He created embrasures within the parapet. Beneath the parapet he left a berm 4 feet wide on the outside, which featured a fraise. What remained of the first parapet on the inside was used as a banquette.[793] As a result, the parapet was 11 **[174]** feet high on the outside & 11 on the inside. The addition was indispensable to provide a certain amount of cover for the interior of

---

789 A cadet named de Poilly served at Louisbourg where he was made ensign *en second* by the governor in 1758 but later reverted to his former rank. McLennan, p. 346. Later in his narrative, Pouchot refers to de Poilly as a lieutenant. He too might have been a relative of a Louisbourg officer, engineer François-Claude-Victor Grillot de Poilly (1726-1761), who was captured there in 1758. *DCB*, III, pp. 267-68.

790 René-Hippolyte La Force, was master of the schooner *Iroquoise* (note 573).

Like La Force, Captain Joseph Boucher de La Broquerie (1708-1762) of the brig *Outaouaise* had also seen considerable service on Lake Ontario. In 1756 he commanded the schooner *Huron*. "State of Our Naval Forces on Lake Ontario" in Preston and LaMontagne, p. 254. He is perhaps best known for his "Carte du Lac Ontario ... 4 Octobre 1757." The chart includes insets of the four natural harbors on Lake Ontario (Oswego, Niagara, Frontenac, and Toronto) as well as renderings of the vessels of the opposing French and English squadrons. La Broquerie's chart served as the basis for a number of copies made early in the English regime. It is reproduced in Hulbert, I/3, cxxi 14. See also Green, "Corvettes of New France," p. 31.

Captain La Broquerie is described in one source as a "Canadian gentleman." Courville, p. 76. Connecticut provincial Colonel Nathan Whiting, who guarded him following his capture, considered La Broquerie "a discre[e]t modest man" and asked that his wife "Shew him Civility" should his place of confinement prove to be New Haven. Whiting to Whiting, Aug. 24, 1760 in Welles, p. 146.

791 Pouchot does not account for all the officers listed by the English following the surrender of Fort Lévis. The return of captives, dated August 25, 1760, includes two captains (Pouchot and Céloron), four lieutenants (Céloron, La Boularderie, de Bleury, and one unidentified) and two ensigns (possibly de Poilly and one unidentified). The composition of Pouchot's officer cadre – nearly all of the Marine, all quite junior and several apparently late of the Isle Royale establishment – attest to the shortage of officers with the French forces by 1760.

It would appear that the unidentified ensign was one Hertel de Rouville who signed a parole with the rank of ensign *en pied* at Fort Lévis on August 25, 1760. Amherst Papers/101. He is not identified in the "Tableau ... des Officiers de La Marine" in Brymner, 1886, p. clxxxiv. He was, in fact, the only garrison officer of the *troupes de terre* other than Pouchot. Jean-Baptiste-Melchior Hertel de Rouville (1748-1817), was a very young Canadian, only recently (June 1, 1760) commissioned ensign *en pied* in the Regiment de Languedoc. *DCB*, V, pp. 421-22.

...(continued next page)...

the fort, which was overlooked to a height of 24 feet by the terrain of isles à la Cuisse & de la Magdeleine.[794]

M. Pouchot also ordered construction of a gallery with pieces of oak 14 inches square by 10 feet long. It ran along the rampart & was used as a terre-plein, while the lower part acted as a casemate.[795] The batteries were positioned on this gallery or platform all around the island. He built an epaulement 4 feet thick out of the earth, most of which was dredged from the bottom of the river, since the island did not have 2 feet of earth as ground cover on slopes. An abatis of branches was placed over the front of the epaulement, which extended as far as possible [175] into the water in order to prevent longboats from landing. At the island's tip, this epaulement ended in a redoubt

---

As to the strength of the garrison, it was slightly larger than Pouchot says. Governor Vaudreuil reported that Pouchot's force, including the crews of the vessels, was only 316 men. Vaudreuil to Berryer, June 24, 1760, *DRCHSNY,* X, p. 1091. Lévis put it at 540 in the fort and "at the head of the Rapids", thereby including parties stationed outside Fort Lévis and farther down the river. Lévis to Belle-Isle, June 30, 1760, *Ibid.,* p. 1101. At the time of the surrender, there were 8 officers, 1 chaplain, 1 volunteer, and 237 soldiers and artificers of the Marine or militia in the fort. One officer and 11 men had been killed during the siege for a total of 259. In addition, the crew of *Iroquoise* was counted as 2 officers and 34 men. The previously captured *Outaouaise* had a complement of 2 officers, 34 sailors and 65 "Marines", presumably Marine regulars and militia. Three men had been killed when the English captured the ship. This gives a total of 399 men of all categories. There were 8 women and children in the fort. "Return of the State of the Garrison which Capitulated", Aug. 25, 1760, Amherst Papers/8.

792 The construction was *pieces sur pieces,* roofed with boards.

793 "*Banquette,* is a kind of step made on the rampart of a work near the parapet for the troops to stand upon in order to fire over the parapet." Muller, *Treatise,* p. 211.

794 The two islands lie just to the north of Orakointon (Chimney), and both were larger and more elevated. Isle à la Cuisse is the westernmost (upstream) of the two. It is today Spencer Island. Isle de la Magdelaine is today Drummond Island.

The islands are on the Canadian side of the river a short distance below Johnstown, Ontario, and both have been heavily altered by construction of the St. Lawrence Seaway and earlier canals. See Pouchot's own "Plan Des attaques du Fort Levis" printed as plate 2 in Volume I of his *Mémoires* and "A Plan of the attack upon Fort Levi" in Mante, p. 303.

795 "TERRE-plein, in fortification, the top platform, or horizontal surface of the rampart, whereon the cannon are placed, as well as the troops that defend the place." Smith, *Dictionary,* p. 243.

"*Casemate,* is a work made under the rampart, like a cellar or cave with loop-holes, to place guns in it." Muller, *Treatise,* p. 214. The casemates of Fort Lévis, like those in most North American forts, were intended to shelter people and supplies and were not pierced for cannon.

of wooden timbers pierced to allow fire from five cannon. On either side of the Island, two spaces in the shape of a jetty were left to allow the bateaux to tie up.[796]

All this work kept the little garrison busy. The troops were augmented by about a hundred militiamen throughout the campaign. As the majority of these militiamen had been pressed into service while bringing us food supplies, at least a score of them deserted and went back home again with the bateaux they were using to reach work sites off the island, since on the latter there was neither earth, nor stone nor wood. The ditches, which were 5 *toises* across, were only two **[176]** deep & flooded with water. Along a part of the epaulement, we were compelled to make banquettes out of oak chippings left over from squaring the timbers. The glacis was constructed with M. Pouchot's firewood. It helped to cover the front facing Isle de la Magdeleine, which was exposed. We collected all the pieces of old iron to be found among the debris of Fort Frontenac & also eight cannon without their trunnions. They were mounted on carriages without wheels in order to render them serviceable.[797]

---

796 The most detailed plan of Fort Lévis is that printed as the inset "Fort Levi" to "A Plan of the attack upon Fort Levi" in Mante, p. 303. This was based on an original drawing done soon after the battle. It has all the features described by Pouchot and includes a section clearly showing the additional parapet as well as the gallery serving as terre-plein and casemate. A copy of a manuscript plan of Fort Lévis, drawn in 1760 by Captain Jean-Nicolas Desandrouins, is in the National Archives of Canada (H12/1250).

Unfortunately, Chimney Island was badly damaged during construction of the St. Lawrence Seaway, most of the northern half having been cut away.

797 It was common practice to disable an artillery piece by knocking off the trunnions or lugs by which the gun was mounted to its carriage. This would have necessitated a jury-rigged carriage to secure the weapon when it was fired. The phrase used by Pouchot to describe the way in which he mounted these damaged pieces is *encadrés dans des affuts faits en crapauds.* The wheel-less bed for a mortar was known as a *crapaux,* and Pouchot's construction was probably similar. Graves, p. 26. Pouchot lacked the skilled workmen to construct proper wheels. He had requested a wheelwright, and efforts were made to find one for him in the vicinity of Québec. Vaudreuil to Lévis, May 2, 1760, Casgrain, *Collection,* VIII, p. 169.

The French had found salvageable artillery in the ruins of Fort Frontenac as early as September 17, 1758. Ensign de Bleury, later to serve with Pouchot at Fort Lévis, reported "six 12 pounders in good condition which the English, no doubt, overlooked ... the others either have their trunnions broken or have been spiked." "Journal of Occurrences", *DRCHSNY,* X, p. 854. These guns were recovered and taken to Orakointon by Lévis in August 1759. His men salvaged cannon balls and 12 guns intentionally damaged and abandoned by Bradstreet's army in 1758. Casgrain, *Collection,* I, p. 197.

...(continued next page)...

As soon as M. Pouchot arrived, all the Indians from La Présentation came to pay their compliments to him, together with Kouatageté, the famous warrior who captured the English guard post by jumping through the window.[798] He was receiving instruction with a view to being baptized. Although he served us rather well **[177]** subsequently, M. Pouchot was not able to persuade him to form raiding parties, on the religious grounds that killing is forbidden. He understood nothing of our distinctions in this matter.

On 30 March, there arrived an Onoyote chieftain named Tacoua Ouenda ("La Viande Tombe"),[799] who was a friend of the English. He asked to hold a council with Sategariouaen - M. Pouchot - & told him: "Father, I thank the master of life for granting me a fine day on which to arrive here in good health, so that I may have the pleasure of meeting my father & finding him also in good health. I have not been sent here by our chiefs, I have come here only to see you."

"I have always tried to engage in good dealings. I often used to go to Mont-Réal to see Onontio[800] & to speak with him **[178]** of good dealings. When I returned to my hut, I was always sweating & exhausted. People would laugh at me about it. Afterwards, I always

---

Escaped English prisoners reported, in June 1760, that several of the cannon mounted in Fort Lévis had broken trunnions. Haldimand to Amherst, June 29, 1760, Amherst Papers/9. Following the surrender of Fort Lévis, the English found twelve 12-pounders, two 8-pounders, thirteen 4-pounders, four 1-pounders (all presumably of iron), and four brass 6-pounders "besides several Guns with Trunions broke off." Webster, *Jeffery Amherst,* p. 239.

798 Although the warrior is not named, this incident is described in Volume I (See note 399). Kouatageté or Ohquandageghte (fl.1757-1773) was an Onondaga warrior and head chief of the village of Oswegatchie at the mission of La Présentation. *DCB,* IV, pp. 587-88. He is further identified by Boyesen, "The French and Indian Settlement at Ogdensburg", p. 8, as Pierre Otkuentaguette, baptized April 27, 1760. He was then 47 years of age which would place his birth about 1713. Kouatageté wed Cécile Seauentcaien in a Christian ceremony on May 10, 1760.

799 Teyohaqueande (fl.1756-1783) was an Onondaga warrior and sachem active in the English service for much of the war. He participated in the 1760 campaign. *DCB,* IV, pp. 734-35. Pouchot's rendering "Tacoua Ouenda" is phonetic and his identification of him as an Oneida is an error. His name translates as "the meat falls."

800 Note in original, p. 178 (a): *The Governor.*

"Onontio" was the title traditionally used by native peoples in referring to the Governor of New France. It was taken to mean "great mountain" and had been given by them to Charles Huault de Montmagny (c.1583-1653), first Governor and Lieutenant General of New France, 1636-1648. The name was used by the Indians as a respectful title for all subsequent governors. *DCB,* I, pp. 372-74.

rested tranquilly on my mat without leaving my village. Some days ago I set out to come hunting in this area. Our chiefs told me: 'Since you are going to be in the vicinity of Onontio, carry word on our behalf to our people at La Présentation & if you are not intercepted, go and see Onontio.[801] You are to tell him that we have been to consult Johnson, as we had informed those who brought us the views of the twenty nations, in order to learn from him the best way for us **[179]** to engage in good dealings.' I myself went to hold this council. I was entrusted with the mission in the absence of our chiefs. Johnson expressed to me his thanks that we wished to share with him this view of the nations & that he did not advise us to go to the home of our father at the Saut.[802] He added that those whose opinion it was that he should receive this message had spoken well, as well as the former chiefs who spoke of good dealings, but that he urged them to encourage the people at the Saut, & all the other nations who might wish to hold council, to come to the village of the Onontagues, the site of the original fire of the **[180]** nations,[803] from which they had taken logs to light fires elsewhere. This course of action was the best one."

He also said that "the Outaouais of Détroit had sent a message to say that they would shortly come to a council with the Sonnontoins & that if they made a special journey to the Saut they would find their huts empty. If they held council in different places, folk would not know about the various decisions. The Five Nations are determined to follow the advice of Johnson &, as soon as the ice thaws, to send delegates to the Saut to invite their brothers to go to the village of the Onontagues. There a **[181]** decision will be taken as to the best thing to do." One can see, from this speech, that Johnson was merely trying to dissuade the Five Nations & our allies from acting upon their feelings of friendship for us & from following the wishes of M. de Vaudreuil.

M. Pouchot replied in these terms: "I thank the master of life for having brought you here in good health & that I too am in good

---

801 Note in original, p. 179 (a): *This is M. Pouchot. In their confusion, they confer this title on all commanders.*

802 Note in original, p. 180 (a): *The Iroquois mission near Mont-Réal.*
St. Louis or Caughnawaga (note 396).

803 Note in original, p. 181 (a): *He wanted to indicate the seniority of that nation & its precedence over the others.*

The Onondaga village was the site of the central council fire of the Five (by this time actually Six) Nations of the Iroquois confederacy. Trigger, p. 492.

health, so that I may speak calmly with you of good dealings. I invite you to unblock your ears and listen to what I have to tell you. If you come on behalf of the chiefs, I am astonished that they have not given you wampum strings as a sure sign to me that you come on their behalf."

"It is thus to you alone I speak. The nations who [182] have passed this message[804] on to the Iroquois, had no intention of holding with them a council in which Johnson might play a significant part. You know that he is the enemy of your father Onontio. In addressing you, they thought that you would want to remain among the children of Onontio. Their words warned the Five Nations to beware of their brother, the Englishman, who brought dissension to the land, & that they invited the Five Nations to live in peace and not to quarrel either with their father Onontio or with them. This would surely happen to them if they did not become reconciled with their father whom they, as well as their warriors, had treacherously [183] killed while shaking his hand. This is the word from the nations of Détroit, & the French commandants in that region have had great difficulty restraining the nations, who wished to come and strike you. But your father still has a tender heart for children whose fear of the English has made them lose their senses. He prevented them from striking you. You must judge from the words which were addressed from the Sonnontoins, who were the first to bring dissension to the land. If you still consider yourselves to be the children of Onontio, you can accept no advice from your brother Johnson, who is his enemy. That is a very bad path to follow in order to bring peace to the land."

"I am well aware of the [184] savannah[805] into which Johnson & your dear brother, the Englishman, have thrown you. They treat you, nations friendly to them, worse than they do their dogs & their negroes. You do not have the right to sleep in their forts. They give you only a little poor quality liquor & when you are drunk, the Englishman casts you out of his house. I know that he has had some of you hanged & has smashed the skulls of others. You will not deny this truth: all the children of Onontio are free & tranquil in their own country. They have no business with such neighbors."[806]

---

804 Note in original, p. 183 (a): *This message had been sent by our Indians at the urging of M. de Vaudreuil, who had got them to agree to it.*

805 Note in original, p. 185 (a): *Muddy water, or flooded grassland.*

"For my own part - and you have named me 'the center of good dealings' - although I had noticed **[185]** that you were deceiving me, I nonetheless warned you what would happen if you let go of my hand. The warning and all the wampum belts you gave me to assure me of your affection did not stop you from striking me. Although you killed me,[807] you can see that I have come here to bring you back to your senses if I can. Before two moons are past, you could repent not having listened to a good friend whom your father Onontio has always sent before you to speak with you of good dealings. You say that you are men who depend solely upon the master of life. I am angry that you always choose the worst **[186]** path that will make you lose your freedom. If you wish to go and see your brothers at the Saut, follow your impulses like free people & your father Onontio will give you a good welcome. If you go there at the behest of your brother, the Englishman, in order to propose evil dealings to Christian Indians, you will gain nothing because they have made up their minds. You will see in Mont-Réal Indians of all the nations who feel the same way.

806 Pouchot raised a sore point, the insensitivity of many English officers to the Indians. His argument was strengthened by the general knowledge among his audience that, at the order of General Amherst, English troops had executed two Indians during the summer of 1759. The first of these, Elias Pegan, had been captured near Fort Herkimer on June 26, 1759 and was accused of being a French spy. Examination of Elias Pegan, June 27, 1759, Gage Papers/2. Amherst directed Lieutenant Montague Brown, commandant of the post, to "Order the Execution of Elias Pegan, by hanging him, or Shooting him to Death, whichever can be most conveniently accomplished." Amherst to Brown, July 2, 1759, Amherst Papers/80.

The second incident involved a Cayuga, accused of having murdered an English inhabitant of the Mohawk Valley. The Indian was apprehended and confined at Albany in May 1759. Amherst to Yates, May 26, 1759, Amherst Papers/79. On July 9 Amherst ordered the prisoner, whose name was Deceanachguneiga, to be hanged or shot. Three days later Lieutenant George Coventry reported from Albany that he had promptly carried out the sentence on the evening of July 10. Coventry recounted that "at the place of Execution he Confesd that he was the man that Murdered John McMickell, " and that "when tyed to the Stake he began his death song but had not time to finish it, he being immediately Shot." Coventry to Amherst, July 12, 1759, Amherst Papers/59.

Other officers were well aware of the potential damage these cases presented to English relations with the Iroquois during the critical 1759 campaign. Sir William Johnson and Brigadier John Prideaux, then deep in Iroquois country and committed to attack Niagara, were concerned that the execution of Deceanachguneiga might cost them their allies. They requested his release, but too late. Amherst to Prideaux, July 13, 1759, *Ibid.* General Amherst was already known for his hard attitude toward hostile Indians. Of the execution of Deceanachguneiga, he wrote: "I believe 'tis the best way of treating Indians." Webster, *Jeffery Amherst*, p. 135.

807 Note in original, p. 186 (a): *Captured at Niagara.*

The Frenchman sees all your machinations very clearly, but he pretends to ignore them. You may be duped by them. In the end, the French & the English will reach an agreement. But all the nations friendly to Onontio, who will be well aware of all the evil you have caused, will no longer leave you **[187]** alone, & we shall leave open all paths for them to strike you."

"Tell the chiefs of the Five Nations that I shall be very happy to see them here. I shall be able to give them a medicine which may perhaps open their eyes."[808]

The chief also added that when, in the same council, they asked Johnson to give them any news he might have, he replied that he knew of no news for the moment, and that they could all go hunting and that when the corn was as big as a man's hand[809] they had only to come and see him and he would then have news from the other side of the great lake.[810] At that point, he would tell them whether to prepare for war or to remain in peace. He also **[188]** warned them that, in the meanwhile, they had to hand over to him all the English blood[811] which they had among them, otherwise they would live to repent it. As a result of Johnson's advice, the Goyogoins passed a message to

---

808 Teyohaqueande left Oswegatchie about April 14 and proceeded to the English post of Fort Stanwix where he arrived on the 24th. There he recounted his version of Pouchot's remarks to John Butler, an Indian Department officer. He had been sent to the French post, Teyohaqueande said, by the Onondagas "to envite the Swegatchy Indians home." The expatriate Onondagas living at Oswegatchie refused to listen until "the french Officer" was present. Pouchot was agitated and "walked backward & forward all the time" during his speech. He remarked to Teyohaqueande that the Oswegatchies had been at the Onondaga village during the past winter and, no sooner had they left than "you run to Sr. Wm. Johnson with the news that passed there." Pouchot was just warming up. He continued: "look at me! (I am not Dead yet) in a great passion, and said you are the people that took Niagara from me, as the English could not have taken it, if you had not assisted. - therefore I am very angry with the Six Nations in my heart, and you shall see what it is to fight this summer." Pouchot's speech apparently had the desired effect. When the Indians met subsequently, without the French present, the Oswegatchies refused to leave and made it very clear that the Ottawa of Detroit were very angry with the Iroquois and would support the French. Butler to Johnson, April 24, 1760, *SWJP*, X, pp. 140-41.

Johnson was not at all perturbed. When he forwarded Butler's letter, he commented that "the French still continue their old custom of puffing, and threatening, wh[ich] I think at present ill becomes them." Johnson to Gage, April 27, 1760, *Ibid.*, p. 142.

809 Note in original, p. 188 (a): *At the end of May.*

810 From Europe.

811 Note in original, p. 189 (a): *The prisoners.*

their warriors to hide their hatchets until the middle of the summer & sent word to the other villages to do the same thing.

Since M. de Vaudreuil had recommended M. Pouchot to send on to him as much news of the enemy as he could, on April 1, he persuaded a chief from La Présentation, or Chouegatchi,[812] named Charles - one of the Indians who came to France in 1752 with the Abbé Piquet[813] - to go to Chouegen to trade, as if he were returning from a hunting expedition. M. Pouchot gave him a **[189]** number of pelts. This Indian was one of the shrewdest among them & spoke French fairly well. On the 19th, Charles returned. He reported that on his arrival at Oswego, he had been told to disembark at the old fort,[814] where an interpreter had been sent to ask him where he came from & the purpose of his visit. He replied that he was from Chouegatchi, that he was returning from the hunt & wanted to trade a few furs before returning to his village. The commandant[815] & several other officers came to see him. They told him that they would not let him enter the new fort,[816] because their interpreter was ill. He and his companions could, however, stay there unmolested & trade whatever goods they wanted to. The commandant told them that he deeply suspected them of coming on behalf of the French to reconnoitre their forts. They replied that they had **[190]** no aim other than to trade & that, if they received a favorable welcome, other Indians were preparing to come on their return from the hunt. They added that, if

812 The French rendering of the English "Oswegatchie", one of several names for the village at the site of the post of La Présentation or La Gallette.

813 Charles Sononguires (d.1767), an Onondaga war chief, was probably one of three La Présentation Indians taken to France by the Abbé Picquet in 1753, although there is some confusion with this identification. Boyesen, "The French and Indian Settlement at Ogdensburg", pp. 8-9.

On March 14, 1767 Sir William Johnson was informed that "the old Onondaga chief named *Sononguires*" had been "greatly battered" and that two of his ribs were broken, apparently at Canajoharie in the Mohawk Valley. Johnson sent rum and money to "cure his bones" and help him home but was informed on April 8 that Sononguires had died a few days previously. Journal of Indian Affairs, March 14 and April 8, 1767, *SWJP*, XII, pp. 288, 303.

814 The ruins of the 1727 Fort Oswego on the west bank of the Oswego River opposite Fort Ontario.

815 Lieutenant Colonel Frederick Haldimand (note 632).

816 The second Fort Ontario, on the east bank of the river, begun in 1759.

they had come on reconnaissance, they would have approached on the other side of the river &, after spying out what they wanted to see, they would have mounted an attack, as had happened the previous autumn.

Each side asked for news of the other. The English seemed fearful of being harassed by our two vessels[817] which were at Fort Lévis. They told Charles that a great fire[818] should be lit at Chouegen & that when all the army was together, they would prepare to march to Mont-Réal. They also said that they knew that the French had a **[191]** small fort on an island but that they would dispose of it en route as though it were a beaver colony & would cast it adrift. They had no intention of lingering in order to besiege it. Charles also saw at Chouegen some Goyogoin chiefs who told him: "The white men will fight once again this year. For our part, we intend to live in peace & to remain neutral." He did not notice at Chouegen any troop reinforcements or bateaux construction.[819]

On the 27th, a number of Missisakes arrived at thc fort. M. Pouchot wanted them to go out on a raiding party. They told him of the imminent arrival of other Indians from their nations. Kouatageté was baptized on that day. Charles, representing M. de Vaudreuil, was

---

817 *Iroquoise* and *Outaouaise.*

818 Note in original, p. 191 (a): *That is to say, a great council should be held.*

819 Sononguires had arrived at Oswego by April 15. The English were indeed suspicious of his intentions. Although the post interpreter, John Lottridge, was truly ill, there is no record of Haldimand himself visiting with the Oswegatchies. Instead, he sent a Cayuga sachem by the name of Taschlay to discover their intentions. Taschlay, who was probably the Goyogoin mentioned to Pouchot by Sononguires, refused to tell Haldimand what he had learned from the Oswegatchies until after their departure which must have been on April 15. Once they had gone, Taschlay reported that "those Indians had been sent by the Commandant of swegatchy, to view his Post, [and] that they had been given skins to be sold there." They boasted that the French would soon retake Oswego. Sononguires' "cover" as a trader had not been very convincing, however, especially once one of the Cayuga had pointed out to Haldimand that "the skins, were not of the Season, nor dressed in the Indian Manner, & that they were certainly taken out of some Magazine." Gage to Johnson, April 22, 1760, *SWJP,* III, pp. 223-24. Taschlay's report is in Lottridge to Haldimand, April 15, 1760, *Ibid.,* pp. 220-21.

The visit of Sononquires raised concern that the French were obtaining valuable intelligence. On May 7 Sir William Johnson wrote to Lottridge that it was "very wrong to suffer Caghnawageys or any other nation who are our Enemys to come to, and return from Oswego or any other of our Posts unmolested, much more so to trade there as I hear they have done .... it is giving the Enemy the greatest advantages Imaginable as by that means they may have constant Intelligence of every motion of ours .... If any should come there for the future ... it is your duty to acquaint the Commanding Officer." *SWJP,* X, pp. 143-44. This letter arrived at Oswego on May 14.

his godfather. He then came to the fort accompanied by all the chiefs & matrons of the council who had attended **[192]** the ceremony. M. Pouchot presented the new Christian with a fine blanket. They held a great council. Its aim was to send a large mission to the Five Nations to make up the minds of their uncles. It hoped to learn if the Iroquois intended to continue regarding them as relatives. They also wished to inform them that they had lit a fire at Chouegatchi, at the request of the entire nation who had asked permission of French generals. They thereby desired to have the opportunity to receive instruction in the Christian religion & to have a fire on their travels where they could light their pipes when they went to see their father, the Frenchman. They stated that they had been the first to come and live in that place & that, ever since they had become acquainted with religion, they did not wish to leave it. They added that, **[193]** as a sign of their resolution, they intended to sow their lands as usual & that, if people came and molested them, they would find that they were dealing with men.

M. Pouchot approved their intentions & suggested to them that, when anyone implied that they should return to the country of the Onontagues, they should understand that the English desired to catch them in the net in which were enmeshed the Five Nations, who now had good reason to repent because of the ill treatment they were suffering at the hands of the English.

As a result of the reports provided by a number of Indians & the women who had explored the whole fort at Chouegen, we learned that there might be 3 to 400 Englishmen there and that the earthworks were no further advanced.[820] There was a rumor that the Indians of the Belle-Riviere had destroyed **[194]** the fort of Pittsburgh but this turned out to be false.

On the 28th, two Indians set out for Chouegen & M. Pouchot equipped five Missisakes to go out on a raiding party. Kouatageté wanted to dissuade M. Pouchot from sending them, because they were envoys & also because if he provoked the English when his fort was not finished, that would prompt them to come and attack him all the sooner. M. Pouchot gave him to understand that, since these

---

820 This information was reasonably accurate. The garrison left in Fort Ontario for the winter of 1759-1760 numbered 613 officers and men of the 4th battalion, 60th Regiment of Foot and Royal Regiment of Artillery as of November 17, 1759. "Disposition of his Majesty's Forces ....", Gage Papers/4. By the end of May 1760, Lieutenant Colonel Frederick Haldimand's Oswego garrison had lost 135 men to scurvy. Webster, *Jeffery Amherst,* p. 207.

Indians were leaving spontaneously, that concerned only their own nation. He was unwilling to stop people of good will.

On the 30th, Kouatageté & 3 other chiefs came to inform M. Pouchot that three enemy Indians, Onontagues, had been in their huts to the south of the river for two days. Charles had warned the workers in that section to **[195]** beware of them, but they had said nothing. These Indians had found them so wary that they had not had the opportunity to strike. Kouatageté asked permission to go and speak to them. He persuaded them to come and spend a few days with an aunt they had in the mission. One of them, a son of Sononguires,[821] who was highly regarded by the English, had come to the fort the preceding fall to bring three workers there. He reported that the Cherakis had done a great deal of harm to the English in the Great Sabre region[822] & a number of other details we shall omit. **[196]**

On the 4th, two Missisakes arrived and announced that the chiefs of their nation desired to come and settle on the other side of the lake. They reported that the English had constructed a large vessel of 18 guns at Niagara the previous fall & that in the spring they intended to build an even larger one.[823]

---

821 The son of Sononguires has not been further identified.

822 Note in original, p. 196 (a): *The Indians refer thus to Virginia & Carolina where, as of 1759, the Cherokees had begun to make incursions. Mr. Littleton, the governor of the latter province, had not managed to stop them. Colonel Montgomery marched against them in 1760. After two fruitless expeditions & the loss of 7 to 800 men, he was obliged to return to base. The Cherakis took advantage of his withdrawal to occupy Fort Loudon & several other positions & to create further havoc. It was only in July 1761 that Colonel Grant forced them to sue for peace.*

William Henry Lyttelton (1724-1808) was governor of South Carolina. Appointed in 1755, he was captured by the French and did not reach the province until the following year. Lyttelton held the post until 1762. *DNB*, XII, p. 378.

Archibald Montgomerie or Montgomery (1726-1796) had been appointed lieutenant colonel commandant in 1757 to raise a regiment of Scots (Montgomery's Highlanders) that would become the 77th Regiment of Foot. *DNB*, XIII, pp. 749-50.

The 1760 campaign against the Cherokees is discussed in Mante, pp. 284-96 with a map of the region opposite p. 286. After a fruitless campaign, Montgomery and his troops departed Charleston, South Carolina on August 15, 1760 and returned to New York.

Fort Loudon was a stockaded post established among the Cherokee in 1756-1757. Roberts, pp. 742-43. The friendly relations with the English that had allowed construction of this fort later soured, and Fort Loudon was besieged by the Cherokee in 1760. The garrison was forced to capitulate on August 8. Allowed by an agreement with the Indians to withdraw to South Carolina, the troops were attacked on their march and all killed or made prisoner. Mante, pp. 294-96.

James Grant (note 466) had been exchanged for Pouchot in November 1759. He was promoted lieutenant colonel of the 77th Regiment in 1760. Boatner, *Encyclopedia*, pp. 442-44. For a brief summary of his 1761 campaign against the Cherokee see *Ibid.*, p. 220.

On 7 May, 2 Indians arrived from St. Régis,[824] having stopped en route at Chouegen, where they had remained for 7 days. The commandant had told them to have all their bateaux ready. **[197]** Onoroagon, an Onontague friend of the English,[825] told them that since Charles had entrusted him with sending him news, they should inform him on his behalf that the English army was beginning to assemble at Fort Stanwix, that the 18-gun vessel had arrived from Niagara and the other was to follow it immediately afterwards. They stated that Johnson would then hold a great council to bring the Indians together but that they were determined on this occasion to let the English fight by themselves.

On the 9th, all the chiefs from La Présentation came to the fort to find M. Pouchot. Among them was one named Saoten,[826] a great supporter of the English, whom he did not trust. He received them at

---

823 This intelligence was quite accurate. The construction of two heavily armed snows on Lake Ontario had been planned as part of the 1759 campaign. Amherst to Loring, May 22, 1759, Amherst Papers/64. Sir William Johnson ordered construction of the two 18-gun vessels at Niagara on July 29. *SWJP,* XIII, p. 117. Work on the first began almost immediately. Gage to Farquhar, September 12, 1759, Gage Papers/Gage Letterbook. She was launched about October 10. Farqhuar to Gage, Sept. 30, 1759, Gage Papers/3; Beckwith to Gage, Oct. 30, 1759, Gage Papers/4. The snow was christened *Mohawk,* but was also called *Mohawk's Revenge.* See entry, May 3, 1760, Journal of Nathaniel Fuller, William L. Clements Library.

A conjectural profile of *Mohawk* is in Millar, pp. 198-99. He scaled the ship's length to 95 feet, 2 inches overall with a deck of 77 feet. Most of his information on the vessel's history is in error, however. Millar's reconstruction is based on a pair of snows, undoubtedly *Mohawk* and *Onondaga,* shown in Thomas Davies' "West View of Oswego and Fort Ontario with General Amherst's Camps at Lake Ontario in the Year 1760". The watercolor is in the Library of Congress.

The second snow was begun at Niagara during the winter and launched on July 6, 1760. Fuller Journal. Ultimately known as *Onondaga,* she was pierced for 22 guns but armed with four 9-pounders and 14 six-pounders. *Mohawk* had 16 six-pounders for her 18 gun ports. Amherst Journal, Knox, III, p. 85.

824 The mission of St. Régis or Akwesasne at the mouth of the St. Régis River had been founded by Mohawks from St. Louis (Caughnawaga) sometime between 1747 and 1755. The modern reserve straddles the border near Hogansburg, New York. Trigger, pp. 471-73.

825 Onoroagon or Ononwarogo (d.1764) was a chief warrior of the Onondaga. Originally active in the French interest, particularly in convincing Onondagas to move to La Présentation, he shifted his allegiance to the English about 1755. He was busy in the spring of 1760 recruiting Iroquois warriors to serve with Amherst's army. *DCB,* III, pp. 494-95.

826 Saoten was probably Tsyonahady, an Onondaga listed among the Iroquois with Amherst's army in the summer of 1760. *SWJP,* X, pp. 182, 184.

the interpreter's hut & told them that he could not receive them in his own quarters because there were some among them whom he did not know and because they had painted their **[198]** faces in two colors and thus he could not know whether they were friends or enemies. He added that one of them had gone to report to the English on what was happening in his fort & had spoken ill of the Frenchman. They asked who this might be. He replied: "Saoten." The latter immediately responded that he had already heard it said that they distrusted him but that he had not wanted to leave for Chouegen without justifying himself. He wanted us to name the man who had accused him. M. Pouchot said that a little bird had told him.[827] At that point the Indian tried to justify himself & asked permission to return to Chouegen. He was afraid of being arrested.

Around that time Oratori,[828] another chief whom we did not trust, arrived from **[199]** Chouegen. He stated that Onoroagon had come to find him and tell him that Johnson was ready to set out to summon the Five Nations and that they were afraid that the Outaouais from Détroit might arrive and strike them in the Niagara area, which would prompt them to remain on their mats. On the 10th, M. Pouchot attended the marriage of Kouategeté for M. de Vaudreuil & gave him presents on behalf of the general.[829]

On the 14th, a Missisake arrived from Chouegen and said that the English army was expected there and that it was going to be stronger than it had ever been. Onoroagon had charged him with telling his brothers from La Présentation not to sow their seed because the English intended to destroy everything & those who did not wish to die should assemble at Toniata,[830] an island above **[200]** Fort Lévis. According to him, the English did not intend to linger at Fort Lévis. They were constructing rafts to transport artillery & would approach

---

827 Note in original, p. 199 (a): *An expression which refers to rumors that are circulating.*

828 Oratori has not been further identified. He might have been the "Orawaraghghte" who was providing intelligence to the English about French and Indian movements as early as 1757. See Butler to Johnson, May 24, 1757. *SWJP,* IX, pp. 772-73.

829 Pouchot signed as a witness along with Lieutenant Celoron (note 786) and M. de Lagarde (note 946), missionary assistant to the Abbé Picquet. The warrior was recorded in the mission register by his Christian name, Pierre Kouatageté. Boyesen, "The French and Indian Settlement at Ogdensburg", p. 8.

830 Grenadier Island, Leeds County, Ontario, about three miles below Alexandria Bay, New York.

from both sides in order to bombard the fort while with their barges they would come from all directions in order to capture it. On the 16th, the raiding party of 5 Missisakes arrived with 3 soldiers of the Royal American Regiment whom they had captured close to the small rapids of Chouegen while they were fishing, without anyone at the fort noticing. M. Pouchot sent the statement made by the prisoners to the general. They reported there were 5000 men at Chouegen.[831]

On the 18th, M. Pouchot held a great council with the chiefs & matrons of Chouegatchi in order to bring back there the families who had installed themselves at Toniata. He spoke to them thus: "Your father Onontio has [201] sent me here to look after you & to work at good dealings with you. But I have great difficulty in distinguishing the children of Onontio from those who do not love him. I have sent some of your finest men to Chouegen to obtain news. You have sent envoys to the Five Nations to find out if they rejected you. However, it pains me to see that you only think of going to Chouegen in order to seek brandy, which preoccupies you so much that it does not even occur to you that you are at war with the English. I know of five who are at Toniata under an English flag, doubtless raised to give them a greater sense of security. They are continually drunk, & when my party of Missisakes visited them, they attempted to encourage them to take [202] their prisoners back to Chouegen and told them that they would be given as much brandy as they wanted."

They thought seriously of sending out a group to extinguish all these fires & also those of La Présentation & make them all return, after the sowing season, to Isle Piquet where the mission had been established.[832] They decided that those who did not want to come would be left free to do as they wished, but that they would no longer be regarded as belonging to their village.

---

831 The returning war party was the one outfitted by Pouchot on April 28. They had taken their prisoners on May 12. Haldimand to Amherst, May 19, 1760, Amherst Papers/9.

Pouchot passed the information derived from his prisoners to Vaudreuil who received it at Montréal on May 22. Much of this intelligence, such as premature reports of a Spanish declaration of war against England, was inaccurate. Nor were there anywhere near 5,000 troops at Oswego by that time. More useful was the statement by one prisoner that neither provisions nor reinforcements had yet reached the post. Vaudreuil to Lévis, May 22, 1760, Casgrain, *Collection,* VIII, p. 199.

832 After the fall of Niagara in 1759, Picquet had relocated his mission from La Présentation to the big Isle aux Galots (Galop Island, New York), below Fort Lévis. La Corne to Lévis, Aug. 10, 1759, Casgrain, *Collection,* X, pp. 201-04; Boyesen, "The French and Indian Settlement at Ogdensburg", pp. 6-7.

A certain Indian called Sans-Souci, an Iroquois from the mission who had come from Chouegen, did not want to appear at this council. In the evening, he came to visit M. Pouchot who was spending some time inspecting the fort. The guard brought him. M. Pouchot scolded him for visiting Chouegen without giving him notice, also for speaking ill of the French & attempting to put their raiding parties [203] off the track. He denied everything & added that he could only be reproached for what he was accused of saying, that he was his own man and his heart alone knew on whose side he was. He stated that in casual conversation with the commandant of Chouegen, the latter had said to him: "Is it true that the commandant of Niagara is at Orakointon? He must wish to die, just as he did last year. This time, though, he will die with all the Indians who are with him. Within six days, the other vessel which is at Niagara is due to arrive & we shall set out straight away. We shall have an army of twelve thousand men & we shall go and lay siege to La Présentation. After surrounding it with our vessels & barges, we shall bombard his fort, laying waste all the [204] territory & islands in the vicinity. We shall hold firm and then we shall go to Mont-Réal." Sans-Souci also reported that more than 200 bateaux were stationed around the large vessel. He asked M. Pouchot why he had not installed his cannon on the ramparts. M. Pouchot answered that he did not lack guns, as he could see, but that he would put them into place when he was fighting with the English, because he did not want Sans-Souci to go and tell them how many he had & where they were positioned.[833]

---

833 Sans Souci has not been further identified. Pouchot implies that he was from the mission of La Présentation. He might have been one of a party of Oswegatchies reported by Indian Department Captain John Lottridge to have arrived at Oswego in two canoes before May 10. They provided intelligence of French activities at Fort Lévis and Québec. Lottridge reported that the "Swegatchies behaved extreamly well when they were here, and are much taken with their Usage at this Post, they also say they were all in our Interest, and never give ear to any thing the French may recommend to them." Lottridge to Johnson, May 10, 1760, *SWJP*, X, p. 146.

Parties of Oswegatchies continued to visit Oswego throughout the spring and summer. Sir William Johnson's letter to Lottridge of May 7 (note 819), which arrived at Oswego on May 14, caused the garrison to watch them more closely, however. Johnson's orders even led Lieutenant Colonel Haldimand "to prevent the Indians of Oswegatchy from carrying on any trade" at Oswego. Sometime between May 10 and 19 Haldimand nearly went too far with one party. He "had the Sachem of the Onondagas tell them that I even intended to have them arrested." When Lottridge advised Haldimand that such an act might have "unfortunate consequences," the commandant took a more diplomatic course. Haldimand to Johnson, May 19, 1760, *Ibid.*, p. 153.

On the 19th, M. Pouchot formed a raiding party of 14 Indians. The Indian envoys announced the return of their people who had settled on Toniata Island & that they had surrendered the English flag. One of them, who had arrived from Chouegen, said that it was the governor **[205]** from the Great Sabre[834] who was to command the army.

On the 27th, La Broquerie, who was to command the bark *Outaouaise,* arrived. On the 30th Oratori, who had come from Toniata, told M. Pouchot that Sans-Souci had returned to Chouegen and that he had been paid by the English to come to the fort and discover what was happening. He also announced that an Iroquois raiding party was due to arrive within eight days along the south of the river; they did not want to mass their canoes as Sans-Souci was to warn the English that they were outside the fort. They intended to strike at the lake of the Onoyotes. On the same day, an Indian who had arrived from Chouegen said that the commander wanted to encourage the Onontagues to form a raiding party to take **[206]** reprisals but that they had refused. They had stated that if the Indians of Chouegatchi were to strike, they would make up a party from the bear clan to strike at St. Régis. Sans-Souci had informed the English about the party of our Indians which was on the warpath.

On the 13th,[835] Kouatageté arrived from La Présentation, towing two bark canoes which he had captured from a party of eight Indians & bringing an Englishman who had come on a raid in the area of the fort. He was alone with some women in his canoe, having landed above the falls of the Chouegatchi River.[836] He encountered this raiding party & struck up a conversation with them with the following words: "The master of life has sent me a good dream for you & me. Since I have met you, I am very happy to tell you that you are all dead if you do not **[207]** withdraw as soon as possible. You believe you are far from us. You are in the French posts & you are without cover. Thus I advise you to withdraw as soon as possible."

---

834 Note in original, p. 206 (a): *General Amherst, the Governor of Virginia.*

Major General Jeffery Amherst (note 533) was appointed governor of Virginia in September 1759. This was merely a sinecure (worth £1,500 per year), and his primary duty remained that of commander-in-chief in America. *DCB,* IV, p. 25.

835 This date, *Le 13* in the original, is out of sequence and is probably a misprint for May 31.

836 Oswegatchie River.

The Indians replied: "We can see very well that we had a bad dream. Since we are without cover, we shall take your advice. But before that, tell us who you are, give us news & we shall tell you what we know." The other replied: "I am Kouatageté." They asked him if there were a large number of Outaouais. He replied that there were some, but that many of them had gone to Mont-Réal along the great river. He also said that the fort was finished & that the commander was only waiting for the English in order to **[208]** fight with them. They asked him if it was true that the French had not been able to recapture Québec. He replied that this was true.[837]

In their turn, they told him that the Englishman who was with them had made war on the great lake against Onontio[838] for three years, that he had captured 25 ships, that Onontio had none left and that as soon as another put into the lake, it too was captured. They added that the French had been to the country of the King of England, but that once having marched into it,[839] they had gone back home, telling the King of England to beware as they were going to kill him.

---

837 Kouatageté talked in far greater detail than Pouchot realized. Sir William Johnson reported the encounter (which he believed took place on June 2) to Amherst on June 10. *SWJP*, X, pp. 171-72. The general learned that "an Indian Party had come in who had met near Oswegatchi on the 2nd June one of the Head men of Oswegatchi who was formerly an Onondaga who said the French were returned from Quebec where they were repulsed with a great loss of men and Artillery, besides six ships they lost of which four were loaded with Provisions, the other two with ammunition." Their very accurate report contained even more detailed intelligence. Webster, *Jeffery Amherst*, p. 209.

The source of Johnson's information was a letter from Captain John Butler at Fort Stanwix. Butler had reported the return of one John McMicking, described by Johnson as "a young lad I sent out with a party of Indians from here [Fort Johnson] Some time ago." McMicking claimed to have met with "Tawangatha", the "Head man of the Swegatchy Indians, originally an Onondaga" within one-half mile of La Présentation. Johnson to Amherst, June 10, [1760], *SWJP*, X, pp. 171-72. McMicking was presumably the Englishman brought in by Kouatageté. If so, he must have been almost immediately released in order to return so promptly to Fort Stanwix. Kouatageté was clearly playing both sides, and Johnson noted that the warrior had "promised me verry well" when they had met "last Autumn while at Osswego." *Ibid.*, p. 172.

838 Note in original, p. 209 (a): *The King of France.*

839 Note in original, p. 209 (b): *The incursion into Ireland by Captain Thurot.*

The passage preceding this note is unclear. Although it seems, at first, to refer to the Englishman, possibly McMicking, brought in by Kouatageté on the 31st and that he was a sailor, it is more likely that the Indians were speaking of their English allies in general. Their news was of the naval war in the Atlantic ("the great lake") and probably of the defeat of the French fleet under Conflans at Quiberon Bay on November 20, 1759 (note 177).

...(continued next page)...

They had [209] not yet marched, though.[840] For their part, they were waiting for their vessels to sail up from Québec & after that the army which was assembling at St. Fréderic & the one at Chouegen would march without delay to Fort Orakointon. They added that the English were only waiting for their return in order to transfer their army to Fort Stenix and that Bradstreet had the job of bringing up the cannon from Orange.[841] They also told him that a member of their party had gone in the direction of St. Régis and that there was another party of Onontagues in the field which had run into the fire of one of our parties.[842] M. Pouchot judged that this might be Thibaut, a captain in the Rangers or woodsmen,[843] or a marine officer [210] who

---

The incursion into Ireland occurred soon after, led by François Thurot (1727-1760) a French privateer captain. Taillemite, p. 321. Sailing from Dunkirk in October 1759 with five frigates, one corvette and 1,500 soldiers, Thurot arrived at Carrick-Fergus and laid siege to the place. The town surrendered on February 20, but Thurot was unable to pursue his invasion and withdrew. His flagship became separated from the other vessels and was caught by English ships off the Isle of Man on February 28. Thurot was killed in the subsequent action. Gipson, VIII, pp. 24-27. A contemporary account is in Entick, IV, pp. 318-33.

Pouchot seems to have learned of these events through Kouatageté, for news of them had reached Lévis at Québec only a short time before. Swanton's squadron brought the first European news since autumn to the besieged English garrison of Québec. On May 9 the "London news-papers, fraught with the defeat of Conflans, Thurot, and many other interesting events, were sent to the French Generals, early in the evening, by a flag of truce." Knox, II, p. 416. Amherst's army would have had the same information and communicated it to the Indians.

840 Note in original, p. 210 (a): *This refers to the planned assault on England.*

841 Albany, New York.

842 Note in original, p. 210 (b): *It was made up of thirteen Abenakis.*

843 Pouchot refers not to a French officer but to an English ranger, Nicholas Tiebout (b.1739). Tiebout had enlisted in the New York Regiment on April 18, 1758 at the age of 19. *Collections of the New-York Historical Society for the Year 1891* (New York, 1892), p. 114. He was described as a lieutenant in the regiment in the autumn of that year when scouting from Oneida Station. Gates to Johnson, Oct. 4, 1758, *SWJP*, X, p. 23. Tiebout received a commission as second lieutenant in Captain Henry Wendell's ranger company on February 10, 1759. When Joseph Waite took command of the company in the spring of 1759, Tiebout remained as his lieutenant. Muster Roll of Waite's Company of Rangers, June 16, 1760, Amherst Papers/9; Burt Garfield Loescher, *The History of Rogers Rangers, Volume III: Officers and Non-Commissioned Officers* (Burlingame, CA, 1957), p. 66.

Tiebout was active in ranging work from Fort Stanwix and Oswego. Although Sir William Johnson called him a captain when he sent him to lead a party from Oswego to Oswegatchie on September 14, 1759, Tiebout did not rise above the rank of lieutenant. *SWJP*, XIII, pp. 142-43. His career ended at Fort Ontario on August 7, 1760 when, accused of "embezzling ammunition, discharging a man, and behaving beneath the character of an officer" he was found "not fit to be an officer" and was dismissed from the service. Hervey, p. 101.

wanted to reconnoitre the river by himself. They had left two Indians & an Englishman to guard their copper pot or stores. The chief of the party, called "L'Ecureuil Rouge", was the son of Collier-Pendu,[844] the others were Sonnontoins, Agniers, Onontagues & three Missisakes whom the English had arrested & sent to Johnson.

On the 4th [June], four Missisake chiefs came to the fort. They asked to hold a council in front of the orator from La Présentation. They presented four strings of wampum. Through the first, they said to M. Pouchot: "Since losing our father at Niagara, we have become like men who are numb. We do not know where to rest our head & we have lost our wits. We hear all sorts of arguments, without knowing **[211]** which ones to accept. Consequently, we who are speaking to you have remembered our father, who took pity on us. We no longer listened to anything else & we ignored the fact that there were fallen trees across the path[845] that led to our Father. We have come to see what he thinks."

By the 2d, they said: "Father, we are worthy of pity. We have no more munitions nor any means of cover since we have lost you. We hope that you will take pity on us. Our people have all died this winter. Hunger forced us to eat about ten people this winter (in the area of Lake Huron). We **[212]** are counting on you to take pity on us."

By the 3d, they said: "Father, we beg you to grant our wishes. We are worthy of pity. We ask to be able to approach you so that we may hear your wishes and so that you may tell us what should be done & what is to become of us. We desire to light our fire in this area."

By the 4th, they said: "Father, we are to speak to the Iroquois of Chouegatchi. We are very happy for you to hear what we have to tell them, so that you may be able to unite us in our views."

Through a wampum belt given to the orator from La Présentation,[846] they said: "Brother, we are worthy of pity. You see us here reduced **[213]** to very few folk to form a village, but if you listen to us, it could become very large. Since we lost our father, our destiny has been a

---

844 The son's name means "Red Squirrel." The biographical sketch of Collier-Pendu ("Hanging Belt"; note 623) in *DCB*, III, p. 320 mentions only three daughters and a stepson, Aroas or "Silver Heels" but admits that little is known of his family. Red Squirrel was presumably a Mingo.

845 Note in original, p. 212 (a): *The difficulties & encumbrances across their path.*

846 Identified later by Pouchot by the name "le Chevalier de la Grimace" (*Mémoires*, II, p. 250; note 885).

wretched one. We are like madmen, we do not know whom to believe. We hear all kinds of evil birds[847] which speak to us of one thing & another. We do not know to whom we should lend credence or what decisions we should make. When we are with you, we shall listen to our father together & hold each other's hand. We shall know what must be done & receive good counsel."

Through several other wampum strings, they continued: "We ask you to grant us a place where we may light **[214]** our fire, where we can hunt & fish to feed our families & together hear our father."

The orator took it upon himself to make their requests in a full council & to give them a reply the next day.

They told M. Pouchot that they had come on behalf of the little chief from the Fond du Lac,[848] who had sent them to find out what opinion he had of them. He would come and see him if the response was favorable. He told him that, the previous fall, he had been entrusted by Peminol[849] to carry to the nations of the upper country the belts which had been handed to him by M. de Vaudreuil but **[215]** that they had made no reply because they were like madmen & in disarray and because they were all dying from the large quantity of brandy which the English had sent them. He added that he himself had waited for them for more than a month at the upper reaches of the Machidache River[850] without seeing any of them appear and that he believed them all to be dead. Moreover, he would not hide from M. Pouchot the fact that there were a number of them whose minds had

---

847 Note in original, p. 214 (a): *Public rumors.*

848 Note in original, p. 215 (a): *The English would call him the king.*

The Fond du Lac was the western end of Lake Ontario near Burlington, Ontario. The "little chief" led a band of the Mississauga.

849 Note in original, p. 215 (b): *A chief who was closely allied to us.*

"Peminol" is a misprint. Pouchot later uses the spelling "Peminot" and identifies him as a Mississauga chief.

This important chief might have been the "Paapinass" or "Paupinnash" who met with Sir William Johnson at Johnson Hall on August 16, 1768 (*SWJP*, XII, p. 590), July 4, 1770 (*Ibid.*, p. 835) and again on July 6, 1772 (*Ibid.*, pp. 969-70). He is described variously as being from "Le Miscotaigna to the north of Ontario" and as a chief of "the village of Pemidashkoudoyan" or simply "Cheif of ye Messissagaes."

850 Note in original, 216 (a): *The most populous place of that nation.*

"Machidache" is one of the variant spellings for what is now the Nottawasaga River flowing into the bay of the same name at the southern end of Georgian Bay. The river was part of the portage route connecting Lake Huron with Lake Ontario at Toronto via Lake Simcoe and was in the heart of Mississauga territory. Tanner, p. 41.

been corrupted by the English and who concentrated on evil dealings. He admitted that the previous fall he had been to Niagara & that they had said to the commander: "Brother (I was obliged to call him thus) we have come here to find out what your opinion of us is. **[216]** You have taken the land of our father, where we could satisfy all our needs. We wish to know how you will treat us. We have no more munitions, nor clothes to cover our bodies."

The English commander threw them a wampum belt, which they were tempted not to pick up, and replied: "You should no longer look to any other quarter. You will obtain all your needs here. But, for the moment, we have nothing. In the spring, we shall receive supplies of everything." He sent them away after giving them a pound of gunpowder & some shot.[851] They had gone back to Niagara that spring to find out what effect his promises would have. The commander replied that he had not yet received any trade goods but that if they wanted brandy, he had plentiful **[217]** supplies. He ended up by giving them the equivalent of a small copper pot of the liquor, which they downed on the spot, since there was not enough to take away with them.[852]

They informed us that the ship under construction at Niagara was 13 fathoms long[853] & that the Sauteurs from Michilimakinac had been

---

851 Parties of Indians from the Niagara-Detroit region were continually coming to Fort Niagara during the autumn of 1759. The Mississauga chief probably described a meeting held a few days before September 20 when the commandant greeted "too Large Numbers of differ[n]t Nations of Indians, Missisages, Chipawas, Christinons, whose provisions, in our situation, I grudge very much, and give as sparing as possible, but at any rate they're severe upon us ... All make flaming professions, and probably might join us." Farquhar to Gage, Sept. 20, 1759, Gage Papers/3. Parties of Mississaugas also visited on October 6 and 10. The first group presented a large white wampum belt and the message that "Brother the belt I now shew, is You and my self Shakeing hands over one heart" which symbolized their friendship with the English. Falconer to Gage, Oct. 13, 1759, Gage Papers/4.

Several officers commanded Niagara during the autumn of 1759. The first was William Farquhar (d.1759), lieutenant colonel (1758) of the 44th Regiment of Foot. *Army List*; Knox, II, p. 186. Farquhar died at Niagara on October 11 and was temporarily succeeded by Thomas Falconer (fl.1755-1760), captain (1755) in the 44th. Falconer to Gage, Oct. 13, 1759, Gage Papers/4; *Army List*. He was soon replaced by John Beckwith (fl.1742-1775), major (1758) of the regiment. *Army List*; Beckwith to Amherst, Oct. 21, 1759, Amherst Papers/78. Beckwith arrived at Niagara on October 28. Beckwith to Gage, Oct. 30, 1759, Gage Papers/4.

Farquhar was probably the officer remembered by the Mississaugas, and he was a sick man by the end of September 1759. If he so rudely threw them a belt it must have been one of the last in the fort. Unprepared and poorly supplied for Indian diplomacy, Niagara had "Nothing here to give them When they come here, Nor Belt to return the Missasagues." Falconer to Gage, Oct. 13, 1759, Gage Papers/4.

to Niagara to see how they would be treated but that they had returned home very displeased, as the commandant had only allowed them to trade their goods.[854] They also said that all the nations around the lakes had formed an alliance[855] and that only the Missisakes had not yet joined it. They stated that a great Iroquois chieftain had come to bring word to the little chief & to urge him to attend council with **[218]** the Five Nations in order to find the means to restore peace to the land. The chief had refused, telling him that if some council was to be held, they should themselves come to find him and he would listen to them.

---

852 When the Mississaugas returned in the spring of 1760, Niagara had yet another commandant. He was William Eyre (d.1765), appointed to succeed Farquhar as lieutenant colonel of the 44th Regiment. Eyre arrived at Fort Niagara on November 27, 1759. Eyre to Amherst, Nov. 30, 1759, Amherst Papers/21. Brief biographies are in *DCB,* III, pp. 214-15 and *DRCHSNY,* X, p. 545n.

Local Indians again began visiting Fort Niagara as early as March 1760. Eyre found "the Messesages Our New Allies ... very well behaved, and the Reverse of the Others." Eyre to Gage, March 18, 1760, Gage Papers/5. By May at least one or two canoes of Indians were arriving each day to trade. Eyre to Gage, May 6, 1760, Gage Papers/6. The "Chief Warrior of the Missisagos" had come and gone by May 8. Food, ammunition and rum were being given as presents. The Niagara garrison was still short of everything, however. Eyre had told Henry Nellis, his Indian officer, "that the Smith cannot be spared to do any thing for the Indians, He had ordered me to give two pipe Tomohawks, One to the Chief Warrior of the Missisagos ... which I borow'd of Levi the Sutler." Nellis to Johnson, May 8, 1760, *SWJP,* X, p. 145.

853 The vessel then on the stocks at Niagara was the second snow, originally named *Apollo* and rechristened *Onondaga* on August 1, 1760. It was similar in size and design to *Mohawk,* and the dimensions reported to Pouchot are very close to the 77 feet scaled by Millar, p. 198-99. Work on the ship was proceeding slowly at the time of the Mississaugas' visit due to a shortage of skilled laborers and sickness among the ship carpenters. Eyre to Gage, May 8, 1760, Gage Papers/6.

854 These people were Chippewa from the post of Michilimackinac (Mackinaw City, Michigan). They were probably the "near thirty Indians from the Neighbourhood of the Detroit" who arrived at Niagara to trade on March 18. Eyre to Gage, March 18, 1760, Gage Papers/5. Lieutenant Colonel Eyre had a different impression of their attitude. He believed that "they all Seem to be well pleased with their New Acquaintance the English, And promise much that they Will always for the future look [to] and consider us as their best Friends." Eyre to Gage, May 8, 1760, Gage Papers/6.

855 Note in original, p. 218 (a): *Under Chief Pontiac.*

Pontiac or Obwandiyag (1712/25-1769) was an Ottawa chief resident in the Detroit area. *DCB,* III, pp. 525-31. The reference to the alliance is premature and possibly based on later accounts of the uprising against the English in 1763-1764. This involved most of the Great Lakes nations. Pontiac and his band of Ottawas played a leading role, although his "conspiracy" was probably not as closely controlled as has often been supposed. The standard work on the subject is Howard Peckham, *Pontiac and the Indian Uprising* (New York, 1947; 1970 edition).

After this had been thoroughly discussed, the Iroquois had decided to come and find him. They assured the little chief that they would inform him about the aims & results of the assembly. When the Iroquois party left for the Onoyotes Lake they were met by another party of Onontagues and they agreed to return.

On the 6th, the orator came, accompanied by the chiefs & the ladies of the council.[856] They told M. Pouchot that they had come to give their reply to the Missisakes in front of him. The orator was carrying strings of wampum[857] & a fine wampum belt **[219]** of about five thousand beads.

By the first string he said: "Brothers, I thank the master of life who has given us a fine day to see you in good health, as I am myself."

By the 2d string: "I am unblocking your ears, so that you may hear what we have to say."

By the 3d string: "This one is to empty your stomach of that evil bile which puts one in a black mood, so that you may listen to me calmly & with pleasure."

By the 4th: "I bury the death of all your warriors, since we live in times when all our mats are together and so that, once you have come out of mourning, you may listen with gaiety & contentment." **[220]**

By the belt, he said: "Brothers, we have listened carefully to your words. We are delighted that you have not lent your ears to evil birds & that you have listened to the old lady who reminded you to resort to your father. That is to show intelligence, as did our ancestors who are no longer on the earth & who thought only of engaging in good dealings. We have great pleasure in seeing what your feelings are, but we beg you to speak from the heart and not from the lips. In that case you may come, & promptly. We invite you to come very quickly and share one fire only with us & here you will listen to the words of our father, as do we who wish to die with him. We live on islands where you will find **[221]** enough to eat in the water & in the woods. Moreover, our father will provide us with the means to satisfy all your needs." That is the substance of the orator's speech, which was a long one. He was famous among them. He also handed over wampum strings on behalf of the missionary, in order to urge them to listen to the words of the master of life.

---

856 The clan mothers of the Oswegatchie village.

857 Note in original, p. 219 (a): *Wampum strings always constitute the preamble to a council.*

The Missisakes replied: "We thank you, bothers. To this point, we had never heard you speak in that fashion. We did not know you. We thank you very much for having taken pity on us & because you wish to share the same dish and the same *mikoine*[858] with us. I leave, delighted by the fact that you have listened to us; [222] I shall bear your words to our chief. If the master of life has spared him, you will see him here shortly."

The orator thanked him & said: "I am delighted that you have used the expression about having only the same dish & the same *mikoine.* That reminds me of the exact words of our ancestors. Onontio, through that expression, had urged the members of all the nations to consider one another as brothers & as his children."

M. Pouchot also handed over a belt for the small chief with an assurance of his good intentions in his regard & also sent a guarantee that he would see him with pleasure, since a father was always flattered to see his family gathered together, & that, provided he spoke from the heart, he would treat him well.

On the 9th, Sonnonguires arrived from [223] Chouegen. He announced that two regiments had arrived there.[859] An Onoyote reported that the Five Nations were to come to Chouegen without delay & that from that base they were to spread out all along the river of La Présentation[860] in order to cut off communications with Mont-Réal. He claimed that the French had promised to march within two months to ravage the country of the Five Nations and that since we had not gone there, they wanted to come themselves. They claimed that the Frenchman was no longer Onontio since, formerly, when they spoke of him all the nations were afraid of him. At the present time, there was no value in striking him. At one time, the battles had been fought on the nations' territories while the Frenchman had only risen once to strike. Now it was their turn. He then said that the English had more than [224] three thousand bateaux at Korlac, and

---

858 Note in original, p. 222 (a): *A spoon.*

859 Advance elements of Amherst's army had begun arriving at Oswego by the beginning of June. Substantial detachments of the 80th Regiment of Foot and a battalion of the New York Regiment had reached the post by June 14. Detached companies of the 44th Foot were also concentrating there preparatory to rejoining the seven companies of the regiment that had spent the winter at Niagara. Haldimand to Amherst, June 14, 1760 and "Return of the Detachment Gone to Niagara", June 14, 1760, both Amherst Papers/9; Wllyamos to Bouquet, June 11, 1760, Stevens, *Bouquet Papers,* IV, p. 593.

860 The Oswegatchie River.

that they were portaging the Lake of the Onoyotes with large bateaux at that very moment.

On the 12th, M. Pouchot held a council with the Indians of La Présentation. He upbraided them, saying that he was sad to see that the majority of them had allowed their hearts to be spoiled by the bad rum of the English. He said that their legs were numb, since they amused themselves by exchanging bad news at Chouegen, like old women, instead of bringing some fresh information on which he could rely.

The Indians decided among themselves to send Kouatageté, Oratori & 4 others as envoys to Chouegen, under the pretext of finding out from the Five Nations whether they rejected them or not or whether they were attempting to bring about a reconciliation between the Englishmen and the Frenchmen. Their real aim was to determine [225] what decision they could take. Although they were attached to the French, they nevertheless wanted to have a clearer idea of what their own fate would be in the event that the English invaded that region. Kouatageté, who was such a brave warrior, thought only of good dealings since he had become a Christian, claiming that his religion forbade the profession of arms.

M. Pouchot, who sensed that the English would either suborn or arrest these Indians, did what he could to dissuade them from going to Chouegen. He warned them that they did not know the white man, who would take them for spies. They could not imagine that they could be misunderstood to such an extent. M. Pouchot would have preferred to persuade them to undertake a raiding party in order to acquire some prisoners. It should be noted that, before [226] the capture of Chouegen, Kouatageté was a decorated chief, very much attached to the English, for whom he acted as a spy. When we were at Frontenac, the commandants of the fort used him when they wanted to buy or barter something with the English, like calicoes &c. This Indian was so vexed that they had allowed the fort to be taken from them that he abandoned them & came over completely to the side of the French.

On the 17th, Peminot, a Missisake chief who was a sworn ally, arrived with a large number of Iroquois & Népicings. He was bearing a belt on behalf of M. de Vaudreuil inviting the nations to go to Mont-Réal and oppose the English. All the Indians were undecided. They said that both they & we were imprisoned by the English as though on an island and that they did not know in which quarter they should strike in order to extricate [227] themselves. That expressed our own position fairly well. He promised to leave immediately in order to inform them but he said that he was afraid he might be unsuccessful because all the Indians had scattered in order not to be trapped. He added that some Pouteotamis had come to trade at Niagara and had said to the English: "We have come to see how you will treat us since you have ejected our father from this place. We are asking you for gunpowder & shot for hunting, and also for garments. But we have not come to form an alliance with you; we are still under the wing of our father. We are at war with you but necessity forces us to solicit our needs from you."[861]

On the 18th, M. Pouchot sent out [228] 100 men in the ships with a month's rations to go and cruise around Chouegen. Around this period, there appeared an enormous quantity of that kind of small moths which fly around at night & burn themselves on the candles. They are called "Manna." They fall just like snow. They are very tiresome, since they fall in the food. Above all at night, when they were attracted by the candlelight, it was difficult to write because these insects got in the way. Different types appeared every two weeks, some gray, some mottled, others yellow & white. They were followed by a kind of white gnat, which were bothersome because of their abundance but which did not sting. The rains killed them & the ground was so covered with them that there were up to two finger breadths [229] of them on all the ramparts. They were 3 & 4 inches thick on the boats where they rotted into slime. In the fort we were obliged to clear them away just as one clears away snow. These gnats are useful in one way, however; all those that fall in the river provide food, or manna, to the fish, which fattens them up at the season. As a result, the Indians' catch is very abundant, particularly of eels in the area of Toniata.[862]

---

861 The Potawatomi visit to Niagara might have occurred as early as October 10, 1759, when two young men of that nation, accompanied by six Chippewas and an Ottawa spoke with Captain Thomas Falconer. The Potawatomi had been sent "to see & hear if it was true, that ye English as they heard had received them Well, and made a peace with them ... that if wee Consented their Chiefs would come and speak with Us, and make peace as the Chipoas and Missisagues had done; that they did not speak for the Nations Afar off, but for their Castles near Detroit." The commandant had little in the way of presents to give them in return for their wampum string. Falconer to Gage, Oct. 13, 1759, Gage Papers/4.

862 Franklin Hough, on the advice of contemporary entomologist Asa Fitch, identified the moths as caddis flies and the white gnats as part of the genus *Chironomus*, an extensive group of aquatic midges. Pouchot, *Memoir*, I, pp. 267-68. Caddis flies (also known as May flies and part of the order *Trichoptera*) have aquatic larvae and are common in the Great Lakes and St. Lawrence.

The soil of this island is very thin and at that season the whole surface was covered by thousands of little toads. In the surrounding area we found morel mushrooms 5 to 6 inches thick & two, & sometimes up to 3 inches in diameter at the base. They were **[230]** delicious. At that time, M. de Vaudreuil sent about forty Abenakis from the lower reaches of the river. M Pouchot gave them the Isle des Galots to sow.[863]

On the 27th, an Indian from Chouegatchi brought in an Onontague from the Kenchiagé River.[864] He belonged to a raiding party sent out against us. It was made up of 3 Sonnontoins, 2 Goyogoins, 4 Moraigans[865] & 2 Negroes and of an Englishman & the son of Sonnonguires. He said that the latter would arrive the following day. Our Indian reported that the Moraigans had done their best to persuade the others to take his scalp but that the Onontagues were not willing to do so. They pointed out that it was not necessary, as in the wars between Indians, and that the white man was quite happy with prisoners in order to obtain information and that they might get some given **[231]** to them. As a result, they had let him go. The two Onontagues arrived, as they had relatives at La Présentation, & the remainder of the party was released. One can see how simple they were. They imagined that they would be told matters relative to their mission.

Our Loups, on their return from their raiding party, brought with them two English prisoners & one scalp. They were a captain of militia & his brother living on the Mohawk River. M. Pouchot had lodged with them when he was going down to New York and he had not been very well treated by them. He was barely able to recognize them. The Indians had dressed & painted them in their usual manner. They were nearly six feet tall. They were made to dance the customary dance of slaves, the *chichicoy*. It is an indispensable ceremony. **[232]** Even then, they were lucky to escape the normal beating between

---

863 Pouchot refers to the smaller of the Galots below Fort Lévis, today Prison Island, Ontario. Upon the arrival of Amherst's army the Indians abandoned both this and Isle Picquet (Galop Island). Webster, *Jeffery Amherst*, p. 234. Both had buildings and "plenty of corn, beans, squashes, and cabbages, with some hogs and fowls." Nathaniel Woodhull, "A Journal Kept by General Nathaniel Woodhull", *The Historical Magazine and Notes and Queries Concerning the Antiquities, History and Biography of America*, V, No. 9 (September 1861), p. 258.

864 The Black River draining into Black River Bay near Sackets Harbor, New York.

865 Mohicans. See T.J. Brasser, "Mahican", in Trigger, pp. 198-212.

setting foot on the island and arriving at M. Pouchot's quarters. He recognized them during the dance. He was not entirely displeased to see them thus mortified. However, they were lodged in the quarters of the garrison surgeon. M. Pouchot sent them food from his own table.

They stated that General Amherst was to command the army, which would comprise eleven thousand men. The general had a great deal of artillery & the army was marching every day towards Chouegen.[866]

M. Pouchot also listened to the son of Sonnonguires, who told him that all the Indians had sung their war chant against the French & that we should beware of them. He said that the Five Nations had carried wampum belts as far as the **[233]** Miamis to persuade all the nations to abandon the hatchet & that they all reached an agreement between themselves. M. Pouchot expressed doubt as to the truth of these details. The Indian replied: "That is typical of our father. He is always unwilling to believe what the Indians tell him." He had also said that there were only two regiments at Chouegen & that he did not know whether others were due to arrive there. This was contradicted by the statement of the Englishmen.

On the 30th Saoten arrived. He said that he had left the territory of the Onontagues eight days before. He had crossed the river near the Lake of the Onoyotes. For 20 days, he said, only the noise of men rowing had been heard along the river and eight bands[867] & eight chiefs had gone past. Wagonloads **[234]** of supplies were constantly on the move and he had seen many cannon, mortars & howitzers. He added that, according to what people said, there were very few men in the area of St. Fréderic and that, upon the arrival of the army, the Iroquois & the Moraigans were to gather at Chouegen. He reported that the English had constructed large bateaux to carry 40 men each, together with a big cannon. However, they were reluctant to transport a lot of artillery, although the Agniers had advised them to bring up

866 The prisoners were brought in by the party of 14 Indians sent out by Pouchot on May 19. Pouchot's report of this intelligence reached Montréal on July 1. Malartic and Gaffarel, p. 335. Vaudreuil sent the information to Lévis the next day. The governor put the size of the English army at 15,000 men, presumably the number provided by Pouchot. Vaudreuil to Lévis, July 2, 1760, Casgrain, *Collection,* VIII, p. 204. Pouchot might have later adjusted the estimate given here to the more accurate figure of 11,000 men.

867 Note in original, p. 234 (a): *Regiments.*
The chiefs were their commanders.

a lot because a number of guns would be lost in the river in the journey down to Mont-Réal.

He also told M. Pouchot that the Onontague chiefs to whom he had sent strings of wampum to remain at peace had urged him to reply in an ambiguous manner but that they should nonetheless be believed. Four great chiefs **[235]** had deliberated together, according to him, and agreed not to allow their warriors to follow the army; one of them had promised that he would do what he could to prevent them, although there were many in his tribe who were too fond of the English.

This same Indian also reported that the Five Nations were beginning to have second thoughts & feared that as soon as there were no more Frenchmen, the English intended to destroy them.[868] They were **[236]** currently surrounded by all their forts and could judge what their fate would be from the experience of four nations which had asked for gunpowder and had not even received twelve pounds worth. The chiefs, in his view, were hesitant over the proper decision to take & the young men refused to listen to them. They had also been warned by the Têtes-Plattes[869] that the English intended to destroy them. They had themselves attacked the **[237]** English, killing a large number of them and capturing several forts.[870] They had then returned to their huts, where they were waiting for news & a decision from the Five Nations, who had not answered them.

On July 1st, M. Pouchot sent the prisoners to Mont-Réal together with the intelligence he had acquired. Many other Indians were reporting along the same lines. They even specified the uniforms of

---

868 Note in original, p. 236 (a): *They had, quite rightly, always feared such a misfortune & had done their utmost to avoid it. Only Johnson had managed to dispel their anxiety &, during this war, make them forget their former political system. Before, they were thoroughly convinced that they could not relinquish it without very great danger to themselves. From the beginning they had perfectly well understood the necessity of putting France & England under the obligation of placating them, thus preventing one from prevailing over the other. In conformity with this principle, they had in 1709 destroyed the English army destined for the siege of Québec by poisoning the water of a river near which it was camped.*

In 1709 Colonel Francis Nicholson (1655-1727/28) had set out to invade New France along the Hudson-Lake Champlain route by way of Fort Anne and Wood Creek. Fatigue, lack of supplies and disease prevented any further advance than the head of Lake Champlain. *DCB*, II, pp. 496-99. Francis Parkman attributed the account of the poisoning of Nicholson's army to Pierre-François-Xavier de Charlevoix's *Histoire et description générale de la Nouvelle France* of 1744. Parkman, pp. 106-09.

869 Note in original p. 237 (a): *The Cherakis & Katabas.*

870 Note in original, p. 238 (a): *This fact was true and has already been discussed in a note.* See note 822.

each regiment. M. Pouchot, who had seen them, was well informed on the matter which enabled him to judge the truth of the reports.

On the 3rd, the son of Sonnonguires came to tell M. Pouchot that he was going back to his village & that he would henceforth live in peace. His father guaranteed his good intentions, [238] & to prove it he handed over to M. Pouchot several certificates from the Belle-Riviere. These had been given to him by a friend who was at the battle of Niagara, where he had taken them from some inhabitants of the Illinois, to whom they belonged. He promised that, sooner than make war against the French, he would go to the Têtes-Plattes, the former enemies of his nation, & that when the English army was ready to leave, they would come and warn us.

On the 6th, a detachment returned with its officer which M. Pouchot had sent to carry supplies to our vessels. It had been as far as the Ance au Corbeau[871] without coming across them, because they had gone to patrol the lake and reconnoitre what was happening at Chouegen.[872]

On the same day, the chiefs from La Présentation came to give M. Pouchot their reply to a very fine [239] wampum belt which had been sent them so that they would go on a mission to Chouegen to take some prisoners. They begged him to believe in their devotion to him, stating that they were very happy to have him as their guide & had never had a better father than he. However, by undertaking the mission, they would place a war club at the head of Kouatageté & his band. They urged M. Pouchot to be patient until they had received news of this chief and that he would have occasion to be pleased with them, more especially as they were encouraged & sustained by the nations of the lower country.

On the 13th, M. Pouchot sent a detachment to La Présentation, which had been abandoned by the Indians of the mission since the

871 Ance au Corbeau was the stretch of the northern channel of the St. Lawrence River from the vicinity of Wallace Island, Ontario to the east end of Howe Island (Isle Cochois).

872 *Iroquoise* and *Outaouaise* appeared off Oswego on July 6. One of the vessels was sighted again on July 11, and both came close to reconnoitre the English camps on the 12th. They were identified as a brig and a schooner. The two English vessels arrived from Niagara on the 14th and sailed the following day to seek the enemy. Webster, *Jeffery Amherst*, pp. 217-19. Their sortie was depicted in a watercolor of Oswego by Thomas Davies. The sails of the French vessels are visible on the Lake Ontario horizon. "West View of Oswego and Fort Ontario with General Amherst's Camps at Lake Ontario in the Year 1760," Library of Congress. The same incident was depicted on a powder horn owned by one William Starling and dated July 17, 1760. Charles E. Baker, "The William Starling Powder Horn and the Niagara Campaign of 1759," *New-York Historical Society Quarterly*, XXXII, No. 2 (April 1948).

winter, in order to bring back the planks & ironware for the use of the fort, also to dismantle it and destroy the buildings, **[240]** thus preventing the enemy from using it as a refuge.[873]

Towards noon, an Indian from Lac des Deux Montagnes who had arrived from Chouegen in three days reported that Kouatageté & his band had been arrested by the English & were being kept under close guard. The guards never took their eyes off them. He had asked several times to be set free, but always in vain. He then told us that the chiefs of the Five Nations, who were at Chouegen, had interceded on his behalf and that there had been several meetings on the matter in the quarters of the commandant.[874] The son of Sonnonguires, who had just arrived, had also made a similar request on a number of occasions, telling the English that he had come to Orakointon to strike his father but that, having been to see him, he had been well received & sent on his way without problems. The English could perfectly well do the same thing. **[241]**

This Indian reported that the great chief of the English was expected any day now at Chouegen, where 4 thousand men were camped. According to his description, they comprised the Royal-American Regiment, Gage's & two battalions of the Royal Scottish together with some Rangers. He added that they had set up a large camp at the Falls where the artillery was being portaged & that the English army was to number 15 thousand men under the great chief of all the English. Six days after his arrival, Johnson was due to join them with his Indians.[875] After that, the whole army would march. A soldier had assured him that the departure would take place within

873 Demolition of the small wooden fort was incomplete. Captain William Hervey reported the place merely "unroofed" when he saw it on August 17. Hervey, p. 58. Colonel Nathan Whiting observed that "The Enemy have some time deserted this place had taken the Roofs off the Houses but not Intirely destroyed them, & had hurt the Fort but Little though it Never was of much strength being stockaded only, And a blockhouse at each Angle of the Square." Whiting to Whiting, Aug. 24, 1760, Welles, p. 146.

Thomas Davies' watercolor, "A View of Fort La Galet", shows the fort intact and roofed, but this is certainly artistic license. Hubbard, Appendix D, pl. 12 and p. 84.

874 English sources do not indicate that Kouatageté and his party were harshly treated. The warrior had left for Oswego on June 12 or 13 and was greeted at the English post on the 20th. Lieutenant Colonel Haldimand reported that "There arrived yesterday evening 7 Indians who had left the environs of Oswegatchie (seven days earlier). They brought a *Calumet de Paix* [and] told me that they had hoped to find here the Chevalier Johnson with the Sachem of the Five Nations." Although they had preceded Johnson's arrival, the Indians seem to have hinted that they might cooperate in convincing the Oswegatchies to leave their village. Haldimand to Amherst, June 21, 1760, Amherst Papers/9.

about 10 days. The Indian had seen the artillery, amounting to 100 guns, which the English were assembling. He had noted 20 large caliber guns in brass, three of which were larger than the others. The English had constructed **[242]** 5 large bateaux, each with 13 oars on either side and a cannon in the stern.[876] He informed us that our ship had appeared in front of the fort & had fired several volleys at it. The English then sent out against our ship a large bateau, which the French had allowed to approach and had only fired on it when it was making back to harbor. They had then left to go to Niagara, where the English vessel had been waiting for some time for another to be completed, so that they could sail in convoy.[877]

---

875 Major General Jeffery Amherst joined his army at Oswego on July 9. Sir William Johnson arrived with most of the Iroquois on the 23rd. Webster, *Jeffery Amherst,* pp. 217, 221.

The 4th battalion, 60th (Royal American) Regiment of Foot, the 80th (Gage's) Regiment and the 1st and 2nd Battalions, 42nd (Royal Highland) Regiment formed part of Amherst's force at Oswego. When the army set out from Oswego in August, it numbered 5,586 regulars, 4,479 provincials, 190 sailors, and 706 Indians for a total force of 10,961. Hervey, pp. 56-57; Knox, II, p. 527.

Pouchot rendered the word "ranger" phonetically as *Renger.* These men were from Captain Joseph Waite's company.

876 The English train of artillery was considerable. Siege weapons included six 24-pounders, eight 12-pounders, two 10-inch mortars, and four 8-inch howitzers. The light artillery was listed as two 12-pounders, two 6-pounders, five 3-pounders, two royal howitzers (5 1/2-inch), four royal mortars (5 1/2-inch), and ten coehorns (4 2/5-inch). Hervey, p. 57.

Amherst had carefully considered his needs for an amphibious campaign against Montréal. Control of the water was a priority. Late in the winter he ordered the construction of some "half galleys", each to carry a heavy gun. Amherst to Loring, March 3, 1760, Amherst Papers/65. The design might have been a simplification of that drawn by Thomas Davies in 1759. Notes with his careful sketch, entitled "Plan, Elevation & Section of a Galliot ...." called for a "Prow with 22 Oars besides Main-Sail and Fore-Sail on Occasion, to be Mounted with one 18 or 12-Pounder [and] for a Chase two 6 or 4 Pounders 16 Wall Pieces or Swivels and 250 Men. This vessel in its Construction will partake of the Advantages of a Galley or Zebeque in its Agility and of a Sloop in its Sailing and defence with the Addition of a Prow for Boarding or Landing." Hubbard, Appendix D, pl. 8 and p. 82.

The galleys were intended to counter the French "boats" (the "Jacobites" discussed in note 878) and "for use where our vessels cannot go." Webster, *Jeffery Amherst,* p. 222. Six were projected, but only five had been completed when the army moved against the French. Knox, II, p. 539. Even then one was still incomplete, and Amherst had a howitzer (probably of 8-inch caliber) mounted on her in the St. Lawrence on August 15. The others carried 12-pounder guns. These were mounted in the bow, not the stern as reported by the Indian. Webster, *Jeffery Amherst,* p. 230. Each vessel had a crew of five officers and 25 artillerymen "exclusive of provincial rowers." Knox, II, p. 541. The full crew was thus probably about 55 to 60 men.

M. Pouchot sent the Indian off immediately to carry this news to the general. In the evening, M. Pouchot was informed of the arrival of the two French ships at Toniata. On the 14th, the long-boat of La Force arrived with letters describing his reconnaissance of Chouegen & [243] containing a plan of the enemy position which confirmed the statement of the Indian.

On the same day, at 2 in the afternoon, a very violent storm blew out of the N.W. accompanied by loud thunderclaps, one of which caused a most extraordinary phenomenon. It was a column of fire which, with the noise of a lightning strike fell into the river quite close to the other end of the island. The water spouted up in such a way that it caused a huge wave which, after washing over the entire far end of the island, receded. It carried away a landing pontoon, sank a Jacobite boat & flooded the others which were cast up onto the beach.[878]

877 When the French vessels appeared off Oswego on July 12, Amherst sent ten bateaux towards Niagara in the hope of luring the enemy closer to shore, but *Iroquoise* and *Outaouaise* sailed off in the direction of Niagara. The English snows *Mohawk* and *Apollo* left Niagara on July 13 and arrived off Oswego the next afternoon without encountering the French warships. On the 15th Amherst sent them in search of the enemy. Webster, *Jeffery Amherst*, pp. 218-19.

878 A "Jacobite boat" was an open gunboat armed with a single heavy cannon. These were employed extensively by the French on Lake George, Lake Champlain and at Québec and were named for their designer, artillery officer Louis-Thomas Jacau (or Jacob) de Fiedmont (note 238). On May 22, 1757 Montcalm recorded Jacau's proposal to arm boats with a six-pounder and two swivels each for use on lakes Champlain and George. Casgrain, *Collection*, VII, p. 207. His plan soon became more ambitious, and a 12-pounder mounted in a bateau was successfully tested on June 25. Malartic and Gaffarel, p. 117. In August "a bateau mounted with a 12-pounder fixed by Lieutenant Jacquot of the the Artillery" escorted French troops up Lake George to Fort William Henry. Bougainville to Paulmy, Aug. 19, 1757, *DRCHSNY*, X, p. 610.

Writing to Minister of Marine Berryer on June 24, 1760, Governor Vaudreuil noted that "There are some Jacobites" at Fort Lévis. *Ibid.*, p. 1091. Escaped English prisoners described them as "big bateaux", each with a cannon in the bow. Haldimand to Amherst, June 29, 1760, Amherst Papers/9. These were not as heavily armed as Jacau's originals, however. After the surrender the English discovered "five small row gallies, with three of our iron three-pounders, very good guns, in three of them; - the other two had French fours." The writer also noted that the French "did not make use of them" during the siege. Knox, II, pp. 554-55.

The damage caused by this sudden storm, which Franklin Hough suggested was a tornado (Pouchot, *Memoirs*, II, p. 11n.), might explain why the Jacobite boats played no role in the defense of Fort Lévis. When Pouchot belatedly sent Captain La Force to support *Outaouaise* on August 17, he described his vessels not as "Jacobites" but only as "four longboats armed with *pierriers*." It is also possible that Pouchot believed his gunboats to be outclassed by the English galleys.

The 16th marked the return of the detachment sent by M. Pouchot to take provisions to our ships, which weighed anchor immediately and returned to their patrolling, as they [244] had been ordered. La Force told M. Pouilly, the lieutenant of the detachment, that judging from the large number of barges he had seen in the river of Chouegen, he believed that the bulk of the army was there, & that, from the troops who had arrived in the interval of his two missions in front of Chouegen, he estimated that they would be ready to march in 8 days. M. Pouchot sent an Indian to Mont-Réal with this fresh report.[879]

On the 22nd, an Indian woman who had left La Présentation two days before said that, having arrived there at sunset, around 10 in the evening, five Indians accompanied by an Onontague & five Onoyotes, all of them naked and armed with muskets and a pistol as well as their tomahawks, had entered her hut & asked her a number of questions. They wished to know if we made sorties from our island, if [245] there were a lot of us in the fort, if any men arrived from Mont-Réal & if there were a large number of Indians. She replied that there were a great many men in the fort, that troops often arrived from Mont-Réal but that she did not know how many, since women did not trouble themselves with such matters. She also said that when the French went to work on the islands close to the fort, there were always a large number of them. They asked if they were far away from the fort & if it was possible to land on the island and undertake a raid. She replied that there was only one small landing place for the fort, and that it was very well guarded. They asked again if people often came from Mont-Réal and she replied that they did, but that they were always well escorted. They absolutely refused to say where [246] they had come from & how long they had been on the trail.

This woman asked the Indians for news of Kouatageté & his companions. At first they feigned ignorance of the fact that he was at Chouegen. So she said: "You left there some time ago, then." Then they replied: "Kouategeté is safe. No harm will be done to him & you will see him again soon. He will return with all the English chiefs & people will see him again when the army is on the march." She added that she & the other Indian women were expecting to go to

---

879 Distracted by the other two armies moving towards Montréal, Lévis and other French officers made little mention of reports from Pouchot during the summer of 1760. His July dispatch, however, might have been one of the pieces of intelligence on the "considerable force at Fort Chouaguen" that caused Lévis to believe the English were about to move. Lévis to Belle-Isle, Aug. 7, 1760, *DRCHSNY*, X, p. 1102.

Mont-Réal soon, since they were afraid. They assured her that that would be the wrong thing to do and that all they had to do was detach themselves from the French. They were thus urged to stay in the area of La Présentation & Toniata. In that case, they would come to no harm. [247] The Indians claimed that they were stronger & more numerous than the English & that they had come in force from all the nations with the express intention of preventing harm from coming to the Indians.[880] They left before daybreak, taking with them the woman's canoe. They had told her that they wished to remain hidden for another day, because if the French came again to demolish the fort,[881] they would perhaps have an opportunity to strike. She replied that it was our last journey. That proved that they had already been in hiding during the day but that they had not dared to attack our detachment. Every day, the garrison had at least 60 men working outside the fort, without the enemy managing to capture or [248] kill a single one of them, thanks to the precautions taken by M. Pouchot in urging his Indians to scatter over the surrounding territory and watch those who belonged to the enemy side. Since these men realized that they were discovered, they returned home as they did not wish to come to harm and were content to make war only at the expense of the French or the English.

On the 24th, there arrived a convoy of provisions from Mont-Réal. We were told that the English were above Richeliéce[882] & that it was feared that Amherst's army would join up with Murray's. They did not yet know whether the bulk of the army was in the area of St. Fréderic. However, all the news that M. Pouchot had carefully and diligently forwarded to them ought to have given them some fairly precise notions on the matter.

On the 25th, at ten in the evening, [249] the longboat of La Force arrived.[883] From the letters it brought, we learned that he had kept watch on the same camps at Chouegen & that, on the 22nd, he had

---

880 Note in original, p. 248 (a): *This was true.*

881 The fort at La Présentation. This further indicates that the place had not been completely dismantled by the French.

882 Murray's English army advancing on Montréal from Québec had reached the Richelieu Rapids of the St. Lawrence at Deschambault, Québec.

883 From the schooner *Iroquoise.* The small craft is described as a *canot* which can either be a ship's boat or a canoe.

encountered at Isle aux Galots an English vessel which had been joined by another. Our corvette then gave chase &, after losing sight of them, it had come to anchor at Toniata.[884]

On the 27th, seventy Indians, women, children & old people, left for Mont-Réal. Fear made them flee. On the 29th, at daybreak, the orator from La Présentation - called the "Chevalier de la Grimace" by the French because he had a very twisted mouth, but a very eloquent speaker for an Indian[885] - came to tell M. Pouchot that one of the Missisakes who lived among them had said that he had seen in the Cataracoui area ten bateaux with English troops on board. He had spoken to them. **[250]**

At 7 in the evening, there arrived eight canoes of Iroquois Indians who out of fear had abandoned their fishing at Toniata.[886] With them was the Missisake who has just been mentioned. He told M. Pouchot that, while fishing four days before in the Bay of Cataracoui,[887] he had

884 After his last contact with Pouchot, on July 16, La Force had led his vessels back to Oswego. The schooner and brig were sighted off the harbor on the morning of the 20th, and Amherst sent three boats to find the English vessels and order them to block the entrance of the St. Lawrence to prevent the French from returning to their base. Webster, *Jeffery Amherst,* p. 221.

The aggressiveness of La Force's "chase" of the English vessels is questionable considering that it concluded with *Iroquoise* and *Outaouaise* anchored far down the St. Lawrence and the English in control of Lake Ontario across which Amherst's army would soon travel. A message from English Commodore Joshua Loring, received at Oswego on the 24th, told a rather different story of the encounter that took place on the 23rd: "the 2 French sloops on seeing the Apollo bore down upon her, who lay to, but on sight of the Mohock they in confusion went off as fast as possible, and the wind dying away they by rowing got into the river St. Lawrence; otherwise had the wind continued 1/4 hour, our vessels would have been along side of them. (About 1 1/2 mile distance.)." Hervey, p. 55. Amherst was not pleased by Loring's lack of success nor by his excuses for letting La Force slip into the river. "So he will see them no more & a fine opportunity is lost," the general confided to his journal. Webster, *Jeffery Amherst,* p. 221.

The encounter took place in the vicinity of Isle aux Galots in Lake Ontario, today Galloo Island near Sackets Harbor, New York.

The English commodore, Joshua Loring (1716-1781), was a captain (1757) in the Royal Navy charged with naval construction and operations on the lakes of North America. *DCB,* IV, pp. 486-88.

885 This orator from La Présentation had been prominent in Pouchot's councils with the Mississauga on June 4-6. The "Chevalier de la Grimace" has not been further identified, although he might have been the Onondaga speaker Teyawarunte who brought Sir William Johnson news of the fatal injuries sustained by Sononguires in March of 1767. *SWJP,* XII, p. 288.

886 Note in original, p. 251 (a): *Famous for its eel-fishing.*

887 The mouth of the Cataraqui River at Kingston, Ontario.

seen the two English vessels which were anchored near Petit Cataracoui.[888] It had come into his mind to find out whether what was said was true, namely that the English did no harm to Indians. As a result, he had gone on board the big ship, which had three decks, ten guns on each side, a topmast & grapnels. He estimated that each ship had a crew of 100 men,[889] half **[251]** seamen and half militiamen. There were many officers. The Missisake added that when he was on the Isle des Cedres he had seen ten bateaux laden with troops go by.[890]

On the 30th, some more Indians arrived from Toniata and reported that they had heard the English passing in the night among the Thousand Islands a short distance above the Ance au Corbeau.

On August 1st, La Force sent his longboat to inform us that his ship the *Iroquoise* had run aground on a *poulier*[891] in the middle of the river above Pointe au Baril. M. Pouchot immediately sent out some bateaux to help refloat it. On the 5th, the ships went and anchored at La Présentation **[252]** & La Force came to the fort. His corvette was taking on 12 inches of water every hour & had 15 feet of the keel holed at the bow. We worked at repairing it as quickly as possible.

On the evening of the 8th, Kouategeté arrived from Chouegen after a journey of three days. He was accompanied by an Onoyote & an Agnier, representatives of the Five Nations, who had come to urge our Indians to remain neutral. Kouategeté told M. Pouchot that General Amherst had been at Oswego for 15 days, that he had seen

---

888 Cataraqui Bay at the mouth of Little Cataraqui Creek on the west side of Kingston, Ontario.

889 Note in original, p. 251 (b): *There were in fact 150.*

The Mississauga was correct. *Apollo* (renamed *Onondaga* on August 1) had a crew of 100 and *Mohawk* a complement of 90. Hervey, p. 56. These were sailors only, however, and did not include troops serving as marines. Webster, *Jeffery Amherst,* p. 223.

The Indian was probably taken aboard Loring's flagship, the snow *Apollo.* His estimate of her armament was close to the 18 guns actually carried, although the vessel was designed for 22. She had only one gun deck. The *grappins* or "grapnels" were probably her anchors.

890 Isle des Cedres is today Cedar Island just east of Kingston, Ontario.

The Mississauga had probably encountered a detachment of 134 men led by Samuel Wllymos, captain (1757) in the 4/60th Regiment of Foot. *Army List.* He had been sent from Oswego on July 21 to convey provisions to Loring's ships. Wllymos and Loring were unable to make contact, although, on the 26th, Loring had anchored "at the Island where Capt. Wllymos had been." The warships returned to Oswego on July 27. Webster, *Jeffery Amherst,* pp. 221-22.

891 Note in original, p. 252 (a): *A bank of large pebbles which forms in the river, like a rock.*

him & spoken to him several times.[892] He also said that their army was 10 to 15 thousand men strong and comprised 8 regiments, one of red tunics with blue facings, one red & yellow, one of Scottish, one of red with small black facings, Gage's regiment of light infantry, one blue & red & several wearing undress caps.[893] [253] He had counted 60 cannon. According to his report, four large-caliber guns had remained

---

892 Amherst had been at Oswego for some time longer than reported by Kouatageté. The general arrived on the evening of July 9 and had met the chief by the 17th when he wrote in his journal that "The Oswegatchis I dont permit to cross the river. Their head Warrior is supposed to have killed about 300 men, women & children. I believe the number is overstated, tho' I dont doubt he has killed many of the first & all of the women & children that have fallen into his hands." Webster, *Jeffery Amherst,* pp. 217, 219-20. There were still 15 Oswegatchies in the English camp on August 5, but on that day Amherst noted: "The Oswegatchi I permit to go home to day with two of our Indians to try to bring in all the Indians from La Gallette [La Présentation]. This the Indians desire." *Ibid.,* pp. 224-25. Hervey says they departed on the 6th. Hervey, p. 56.

893 Note in original, p. 254 (a): *Militiamen.*

The "undress caps" were probably cut-down cocked hats. These were worn by some of the provincials. Peter Pond later recalled that, on their arrival at Oswego, the men of his light infantry company (of a Connecticut regiment) were ordered to have their "Hats Cut Small that they mite be youneform." Pond, p. 93. Whether this alteration was authorized for entire provincial regiments or only their light infantry companies is unknown. Military cocked hats could be cut down into either jockey-style caps with a raised front piece or round hats with a narrow brim. It had been ordered for some troops on Lake Champlain in 1759 that "all Great Hats are cut so that the brims be 2 inches and a half wide." Hugh Hastings (ed.), *Orderly Book and Journal of Major John Hawkes* (New York, 1911), pp. 19-20.

Kouatageté might also have been describing the caps worn by the light infantry companies of the regular regiments. These were formed into a battalion under the general's brother, William Amherst (c.1732-1781), on August 6. He was then Deputy Adjutant General in North America with the rank of lieutenant colonel (1759). J. Clarence Webster (ed.), *The Journal of William Amherst in America, 1758-1760* (London, 1927), pp. 2-3, 60. Leather caps were also worn by all five companies of the 80th Regiment of Foot. Eric Manders, Brian Leigh Dunnigan and John Elting, "80th Regiment of Foot, 1757-1764", *Military Collector & Historian,* XXXVIX, #4 (Winter, 1987), pp. 172-73. Another article on the subject is George A. Bray III, "Leather Caps During the French & Indian War", *Military Collector & Historian,* XXXVIII, No. 2 (Summer 1966), p. 66.

Kouatageté was an astute observer. The regiment with blue facings was the 4th battalion, 60th Regiment of Foot, which had provided the winter garrison of Fort Ontario. It assembled in the encampment outside the fort on July 28. Webster, *Jeffery Amherst,* p. 222. That with yellow facings was the 44th Regiment, most of which had arrived from Niagara on July 31. *Ibid.,* p. 223. Another regiment with yellow facings, the 46th, arrived the day of Kouatageté's departure. Hervey, p. 56. The "Scottish" regiment was the 42nd, both battalions of which were in camp by August 4. Webster, *Jeffery Amherst,* p. 224. Kouatageté might have counted Amherst's third highland battalion, the 77th, with the 42nd. The unit with "small black facings" is unidentified. It is possible that Kouatageté misidentified the green facings of the 55th Regiment which arrived July 13. *Ibid.,* p. 218.

...(continued next page)...

at the Falls,[894] where they were constructing a road to bring them by land. He said that the portaging of the mortars was not finished and that he thought it would take about another ten days before they could march. He added that he had come upon the ships in the river & that they were working at fortifying Chouegen.

On the 10th, M. Pouchot went to Isle Piquet to attend the council of the representatives of the Five Nations.[895] They presented a very fine belt of wampum, a gift from Colonel Johnson, on which were designed the Englishman, the Five Nations & the three villages of our Iroquois missions, Chouegatchi, the lake and St Louis, with a man & a fine road leading from one to [254] the other, in order to invite our Indians to take it, to remain neutral, to allow the white men to fight since they would soon make peace, & to get out of their way or to come there unarmed. They assured our Indians that they would be well received, that Johnson & they had come on ahead of the army to see the white men fight. Johnson, they said, had never invited them to do anything other than watch in 1755, 1758 & 1759, as they could see from the battle involving M. Dieskau[896] and from Niagara where, despite the Five Nations, the French had fought without waiting for good dealings. Another great belt from these nations expressed the same thing & invited them to speak the truth with them, that is to say, to adhere to their viewpoint.

Next came strings of wampum on behalf of General Amherst urging them to pay attention [255] to what the belts said. By them, he promised that in five or six days he would be at Chouegatchi, that he was going to fight with the French & that only the master of life knew what would happen.

The response of our Indians was to urge the representatives to go to Mont-Réal and reach the end of the road they were following. They

---

Gage's light infantry was the brown-coated 80th Regiment. The blue and red uniform belonged to one of the provincial units, most likely Schuyler's New Jersey Regiment which had reached Oswego on August 5. Knox, II, p. 538; Hervey, p. 56.

By the time of Kouatageté's departure, Amherst's army included seven regiments of regulars and eight of provincials in addition to artillery and rangers. The force was 10,961 strong, Indians and sailors included, upon its departure from Oswego on August 10. Hervey, pp. 56-57. Pouchot's report that Amherst had 10,000 men, probably based on Kouatageté's intelligence, reached Montréal on August 10. Malartic and Gaffarel, p. 339.

894 Oswego Falls on the Oswego River at Fulton, New York.

895 The two Iroquois sent with Kouatageté by Amherst. See note 892.

896 The Battle of Lake George fought September 8, 1755.

said that they themselves no longer had a fire lit since their father and the Iroquois of the Saut had agreed that the words coming from the Five Nations would go directly to Mont-Réal without stopping with them.

The representatives, after a good deal of reflection on this reply, which they were not expecting, answered that these words had been given them at Chouegen by the Five Nations and that they had only been [256] sent here, without any orders to go on to Mont-Réal. Hence, they intended to return home.

M. Pouchot allowed their council to come to an end and then said to these Indians: "If you had gone to Mont-Réal, I would have nothing to say to you & I would have allowed your father to speak. But since you are going back home, I wish to tell you what is in my heart. I shall not give you any advice and in any case you would not listen to it. As a message from the man whom you have called "the center of good dealings", please assure your brothers, the Iroquois, that they have lost their minds, that with a little brandy, Johnson persuades them to follow him wherever he wishes, without their having any regard for the precipice to which he is leading them. He gets all those warriors who have not consulted their chiefs to join him. [257] You yourselves told me this in 1755. At that time he wanted to go to Mont-Réal to thrash the Frenchmen & give you his trade goods. A handful of Frenchmen kept him a prisoner in Fort George. In 1758, the same thing happened. You upbraided him because a small band of Frenchmen had chased away the English & you returned home full of shame. Is it not I who showed you at Niagara that you should not let go of your father's hand if you wanted to live in peace on your mats & no longer wanted them to be covered with blood? You listened to me then & you withdrew and allowed us to fight. Did Johnson respect good dealings when your chiefs and those who came from the Belle-Riviere wanted to work at restoring peace to the land? [258] He laughed at you because he was stronger. If the big canoes[897] of your father, the great Onontio, were not in poor condition so that he must find time to construct others, rest assured that his children, the French, would cover the whole of this land like the trees. The Englishman would soon be forced to go and hide in a corner of the country[898] which he stole from the Abenakis. The Frenchman had never sought any other

---

897 Warships of the French Navy.

898 Note in original, p. 259 (a): *Acadia.*

course of action than to take pity on his children & to supply their needs. He has not overturned your mats and your fires with armies to go and find the Englishman in his own country, for fear of killing you. You were never willing to prevent the Englishman from passing through your territory. And now you are surrounded by their forts which they [259] requested your permission to build in order to hunt beaver. Where will you now go in order to satisfy your needs? Look at the state of the Abenakis in their own country. They go to the water & into the woods in order to find something to eat & can no longer sow their Indian corn. They are the Englishmen's dogs; they are beaten with sticks and hanged. The same thing will happen to you when the Frenchman leaves you. When you remind the Englishman that he has promised to satisfy your needs, he will laugh at you. But when, instead, you had both the Frenchman & the Englishman as neighbors, they competed with one another to give you things. Wampum belts will be unnecessary to you to recall my opinions when you remember with the old folk the advantage you have lost." [260]

The representatives, although friends of the English, agreed that M. Pouchot was speaking the truth. They apologized because they no longer had the wits of their ancestors. The Indians from Chouegatchi heartily applauded the speech of M. Pouchot, who gave a present to the visitors & dismissed them.

On the 13th, five Indians brought M. Pouchot letters from M. de Vaudreuil. He announced that the English vessels were at the confluence of the three rivers at St. Fréderic & that the enemy were

---

899 The text is unclear at this point. Pouchot might mean that the English were massed at Crown Point, formerly Fort St. Frédéric. He might also refer to the town of Trois Rivières on the St. Lawrence River. The sentence was probably misprinted and intended to read "at the confluence of the three rivers and at St. Fréderic."

As late as August 7 Lévis believed the Lake Champlain army to be under Amherst's command. The English general, as Pouchot well knew, was at Oswego. The English force of 3,400 men advancing north from Crown Point on Lake Champlain was under the command of Brigadier William Haviland (note 751).

900 If *Iroquoise* had indeed been repaired, she saw no further service under French colors. Pouchot provides no explanation for this. On August 13 Indians from Oswegatchie reported to the English that "one of the vessels is damaged, her guns and men taken out." Webster, *William Amherst,* p. 63. General Amherst recorded that the French "were taking the Guns out to put on the Isle Royal [Orakointon]." Webster, *Jeffery Amherst,* p. 229. It seems likely that the arrival of the English army interrupted efforts to rearm *Iroquoise* or that, once the enemy had entered the river, she was no longer deemed useful. La Force and his crew of 36 sailors were in Fort Lévis during the siege. "A Return of the State of the Garrison", Aug. 25, 1760, Amherst Papers/8.

...(continued next page)...

preparing to march. They expected Amherst to come from that direction.[899]

By the 15th, the *Iroquoise* was repaired.[900] I must at this point recount an event worthy of narration. Seventeen militiamen had deserted some days prior to this, one of whom returned to the Isle aux Cedres, which was his home. His father, whose name was Bray, a fine old man, brought him back to do **[261]** his duty. He arrived on the 15th & left immediately. Unfortunately, the young man was killed.[901]

On the 16th, at seven in the evening, two Indians who had arrived from a hunting expedition, announced that the English army was camped at Pointe au Baril & its advanced guard was at La Présentation. It had marched along the shore past the *Outaouaise.* La Broquerie did

---

Iroquoise was scuttled at Orakointon alongside an unfinished hulk. "A Plan of the attack upon Fort Levi" in Mante, p. 303, shows the figures of two ships, one with masts and the other without, labeled "Two of the Enemy's Vessels" and distinct from another shown farther upstream and identified as "The French Vessel taken the 17th." The two ships are on the south side of the narrow, downstream end of Orakointon. This placed them out of the line of fire of the English batteries and ensured their preservation. An English officer later reported that "the two vessels of the enemy were behind their island, filled with Water, that we might not take them." Knox, II, p. 554.

On August 28 Amherst reported that his men had "Got the French Vessel up, a very good new Vessel; carrys 10 Guns." Webster, *Jeffery Amherst,* p. 240. *Iroquoise* was immediately repaired and renamed *Anson* after the famous English Admiral George Anson (1697-1762), First Lord of the Admiralty since 1757. *DNB,* I, pp. 503-08; Amherst to Deering, [Aug. 30, 1760], Amherst Papers/65. The vessel remained in service until mid-October 1761 when she again struck a ledge in the St. Lawrence, stove in her bottom and sank. Loring to Amherst, Nov. 13, 1761, *Ibid.*

The hulk lying alongside *Iroquoise* was a third French corvette, ordered built as early as September 24, 1758. Hamilton, *Adventure,* p. 279. Construction had begun at Pointe au Baril by April 12, 1759, and Montcalm expected her to be completed by the end of May. Casgrain, *Collection,* VII, pp. 507-08. For some reason, probably a shortage of men or equipment, she was never put into service. Intelligence reports during the summer and fall of 1759 reported the third vessel "almost Finish'd" and "ready all to her rigging." Examination of Elias Pegan, June 27, 1759, Gage Papers/2; Journal, Sept. 12, 1759, *SWJP,* XIII, p. 140. The hulk was probably taken to Orakointon when the Pointe au Baril shipyard was abandoned and burned in September 1759. *Ibid.,* p. 142. By the spring of 1760 the third warship, reported to be "very large," still had not been completed. Lottridge to Johnson, May 10, 1760, *Ibid.,* X, p. 146.

Raised by the English on September 22, the hulk was completed as a snow in order to utilize sails from the damaged *Mohawk.* The new vessel measured 56 feet on the keel with a beam of 22 feet and a depth of hold of 9 feet, 2 inches and thus was probably smaller than *Iroquoise* and *Outaouaise.* Christened *Johnson* for Sir William, she was later noted as "not finished when she fell into our hands." Sinclair to Amherst, Sept. 22, 1760 and Loring to Amherst, Jan. 12, 1761, Amherst Papers/65. *Johnson* remained in service on Lake Ontario until wrecked on the bar at Oswego on November 11, 1764. Scull, p. 320.

901 This unfortunate militiaman has not been further identified.

not write, however, although he fired 3 shots with his guns. M. Pouchot sent two Frenchmen & two Indians to canoe out to the ship to find out what the situation was. The captain sent him a message to say that the advanced guard of the enemy & their Indians had landed in great numbers at La Présentation, that he had them under observation & that the bulk of the army had camped at Pointe au Baril.[902]

On the 17th, at 3 in the morning, M. Pouchot despatched a courier to M. de Vaudreuil to inform him of this event.[903] Around **[262]** seven o'clock, in very calm weather, General Amherst launched an attack on the *Outaouaise* with six barges called *Carcassieres*,[904] each carrying thirty men & a 12-pounder gun. The ship was located in a position where the currents were not obvious. They surrounded the ship, which at first made them drop astern on the north shore. But a shore battery forced it to put out into the center again. After a cannonade lasting three hours, the ship was captured. M. Pouchot sent out a detachment of four longboats armed with *pierriers* under the command of La Force, the captain of the *Iroquoise*. But before they could reach the action, the ship had surrendered. M. Pouchot had hoped that the *Outaouaise* might at least reach the protection of the fort, which it would have done had it been able to position itself ahead of the currents.[905] **[263]**

On the 18th, the enemy set out from La Présentation in a brisk wind. At the place where the currents began, their entire army struggled in their bateaux for four hours. It was a splendid sight. M. Pouchot at first imagined that they intended to mount a major attack in order to land on the island. As a result, he had positioned 9 guns to bombard the top of the river & had sited others in the epaulement which could make the balls ricochet eleven times over the water. It is

---

902 The advancing army sighted *Outaouaise* on the evening of the 16th. Amherst ordered the main body to encamp at Pointe au Baril while he continued with the advance guard to Oswegatchie. On the appearance of the English, Amherst heard La Broquerie fire two signal guns. Later, "so soon as She saw the Fires of the Camp she fired three more." Webster, *Jeffery Amherst,* p. 231.

903 His dispatch reached Montréal on August 19. Malartic and Gaffarel, p. 341. This would be the last news from the besieged garrison, although Indians brought in a verbal report as late as the 30th. *Ibid.*, p. 344.

904 A "carcass" was an incendiary device usually fired from a mortar. Smith, *Dictionary*, pp. 50-51. Pouchot's description of these vessels as *Carcassieres* gives the term a meaning akin to "bomb vessel." Smith further notes that "There are other carcasses for the sea-service." Amherst's *carcassieres* had been previously described to Pouchot in reports from Indians (note 876).

certain that the enemy would have lost many men before effecting a bridgehead had they attempted to do so. They decided to slip along the northern shores with a wide interval between each bateaux in order to avoid the fort's artillery. Their flotilla was led by the *Outaouaise,* which they **[264]** had captured from us, at half a cannon's range as protection.

M. Pouchot could only harass their passage with 4 guns that could get them in their sights. We fired 150 shots at them without, as far as we could see, doing much damage because the brisk wind & the current made them sail very swiftly past the lines of sight.[906] Since M.

---

905 Amherst described the action in his journal: "At daybreak the Vessel began to fire and Col Williamson attacked her with the five Row Galleys. She was going up the River but the wind calmed and the Row Galleys behaved very well; fired 118 shot. The Vessel fired 72; had three men killed & twelve wounded and then struck. The Artillery had one Sergeant killed and a New York Provincial lost his leg. The Carriage of the Hautwitzer [the howitzer with which one galley was armed] gave way & hurt the Row Galley. The Vessel called the Outawas had one 18-Pounder, seven 12-Pounders & two 8-Pdrs with four Swivels. The five Row Galleys had four heavy brass twelves & one Hautwitzer. Col Williamson took the command of the Row Galleys to himself & executed it vastly well. On the Vessel sailing up the River I expected she would have been at our batteaus." Webster, *Jeffery Amherst,* pp. 231-32.

Royal Artillery Lieutenant Thomas Davies (c.1737-1812; *DCB,* V, pp. 226-28) depicted the action in his watercolor, "A View of Fort La Galet." Hubbard, p. 84. The artist commanded one of the galleys, Knox, II, p. 552.

The French prisoners were landed and confined at Oswegatchie until sent with Pouchot's garrison to New York. Webster, *Jeffery Amherst,* p. 235. Amherst gave their numbers as 27 sailors and 65 militiamen plus officers for a total of 100. *Ibid.,* p. 232. This varies only slightly from the 101 (La Broquerie, 1 master, 7 petty officers, 27 sailors, and 65 "Marines") in the return of troops surrendered. Amherst's figure for French casualties agrees with the official return. "A Return of the State of the Garrison", Aug. 25, 1760, Amherst Papers/8.

The attack was led by George Williamson (c.1704-1781), colonel (1759) in the Royal Regiment of Artillery. *DCB,* IV, pp. 772-73. *Outaouaise* was renamed in his honor and put into service with the English squadron. Hervey, p. 58.

Williamson was the least damaged of the English ships in the subsequent attack on Fort Lévis, and repairs were completed by August 30. Webster, *Jeffery Amherst,* p. 241. The brig remained in service on Lake Ontario only a short time, however. During a late-season attempt to convey provisions to Fort Niagara, she ran aground near that post on October 25. The cargo was salvaged. Walters to Amherst, Oct. 26, 1760, Amherst Papers/21; Walters to Bouquet, Nov. 8, 1760, Stevens, *Bouquet Papers,* V, pp. 102-03. *Williamson* proved to be so "full of shot holes and Leaks that her hull is mostly buried in the sand," and she had to be abandoned. Walters to Amherst, Nov., 1760, Amherst Papers/21.

906 Amherst's detailed account of the harrowing run past Fort Lévis is in Webster, *Jeffery Amherst,* pp. 232-34. While the general led his 1st brigade along the north shore, he sent the 2nd down the south shore under Lieutenant Colonel Haldimand. *Williamson,* commanded by Lieutenant Patrick Sinclair (note 910), went down the middle of the stream "to anchor within Random shot of the Fort .... He took the Fire from the Fort & returned their Shot."

...(continued next page)...

Pouchot knew many of the officers in the English army, several of them wished him good day as they passed by, while others shouted out to him to let them through, as they were his friends.[907] They saw him on the ramparts, but did not stop to greet him properly. The greater part of the army was camped at Pointe à l'Ivrogne. They also dispersed troops on Isle à la Cuisse, Isle de la Magdelaine & Isle des Galots.[908]

On the 19th, their regiment of artillery **[265]** left the Vieille Galette with all the land artillery & sailed past like the others to go and camp at Pointé à l'Ivrogne.[909] The ship maintained as heavy a covering fire as possible. We fired very little at the bateaux because we expected little success from that. We concentrated our fire instead on the ship.

The fort's guns were more effective than Pouchot realized. Amherst reported that "They cannonaded us briskly from the Fort & sunk one Row Galley, killed two men in another Row Galley. The Ball went through but did not hinder [it] from proceeding. One man in a whaleboat had his thigh shot off & seven were wounded. Some batteaus & several oars struck with the shot." Amherst was concerned that the garrison might "send out their Row Galleys or rather batteaus with small Guns" - Pouchot's "Jacobite boats" - but they did not appear.

907 Pouchot had met these officers following the surrender of Niagara and during his captivity in New York. Among the more senior were Sir William Johnson and Eyre Massey (note 667) with whom he had dined following the surrender of Niagara. Massey commanded the grenadier battalion in Amherst's army, one of the lead units in the dash down the river. Webster, *Jeffery Amherst,* p. 233. Pouchot had also met many of the junior officers of the 44th, 46th and 4/60th Regiments of Foot.

908 Pointe à l'Ivrogne is today Tuttle Point on the Ontario shore.

Isle à la Cuisse (today Spencer Island) is the westernmost (upstream) of the two islands to the north of Orakointon (Chimney Island). It was dubbed Amherst's Island by the English in honor of William Amherst, the general's brother and commander of the light infantry battalion (note 893). Orders, Aug. 19, 1760, Hervey, p. 112.

Isle de la Magdelaine (today Drummond Island) was renamed Dalyell's Island. Its namesake, James Dalyell (d.1763), was then a captain (1757) in the 80th Regiment of Foot. He would later serve as an aide to Amherst and was killed at Detroit while leading an independent command. *Army List*; *DRCHSNY,* VII, p. 547n. The island was named for Dalyell because he had led the detachment that took possession of it. Orders, Aug. 19, 1760, Hervey, p. 112.

Both islands are now on the Canadian side of the river a short distance below Johnstown, Ontario. They, along with Tuttle Point, have been heavily altered by construction of the St. Lawrence Seaway and earlier canals. All these topographic features are clearly shown in Pouchot's "Plan Des attacques du Fort Levis" printed as Plate 2 of Volume I of his *Mémoires* and in "A Plan of the attack upon Fort Levi" in Mante, p. 303. The English camps are depicted in the latter.

The English also placed detachments on both the Isles des Galots. The smaller is now Prison Island, Ontario. The larger, known by 1760 as Isle Picquet, is Galop Island, New York. Both of the Galots were found abandoned. The Iroquois burned the Abbé Picquet's chapel and the cluster of huts around it. Webster, *Jeffery Amherst,* p. 234.

Of 50 cannon shots that we fired at it, at least 48 crossed the body of the ship, which made them withdraw a little. Its captain, whose name was Smul, behaved with the utmost gallantry, pacing constantly up and down his deck in his shirt. Many of his crew were put out of action.[910]

The two other ships, one of 22 guns, 8 & 6-pounders, called the *Sonnontoin,* the other of 18 guns, 6-pounders, called the *Onoyote,* arrived in the evening and stationed themselves beside the first one.[911] **[266]**

---

909 Amherst's army included 13 officers and 124 men of the Royal Regiment of Artillery. Hervey, p. 56. By "land" artillery, Pouchot means the siege train of heavy guns. Amherst ordered the "Heavy Artillery" moved down to the islands after dark on the 19th. Webster, *Jeffery Amherst,* p. 234.

La Vieille Gallette, according to Pouchot's plan of the siege, was Dawson Point just below Johnstown, Ontario. "La Vieille" or "old" Gallette was the original stopping point above the rapids on the north shore. The place had come into use in the seventeenth century but, by the 1750s, had been replaced by the mission fort and village of La Présentation or Oswegatchie on the south shore. See Fortier, p. 4. Amherst says that his artillery was at Oswegatchie.

910 Pouchot seems, in this instance, to have over-estimated the accuracy of his shot. No casualties were reported by the English although, General Amherst wrote, "The Fort fired against the Vessel which the Vessel returned, but I directed he should rather avoid firing than keep it up." Webster, *Jeffery Amherst,* pp. 234-35.

Pouchot confused two Scots in his identification of the brave officer who had been appointed to command *Williamson* only the day before. *Ibid.,* p. 234. He was Patrick Sinclair (1736-1820), newly promoted lieutenant (July 27) in the 2nd battalion, 42nd Regiment of Foot. *DCB,* V, pp. 759-61. Sinclair and a number of other junior highland officers were selected to command armed vessels on Lake Champlain and Lake Ontario in 1759 and 1760, presumably because they had prior sailing experience.

In calling the ship's captain "Smul", Pouchot confused Sinclair with John Small who later escorted the Fort Lévis garrison to New York. Hervey, p. 60. Small (1730-1796) was also a lieutenant (1756) in the 2/42nd Regiment. He later served during the American Revolution as major commandant (1775) and lieutenant colonel (1780) of the 2nd battalion, 84th (Royal Highland Emigrant) Regiment of Foot. Richards, pp. 78-80; *Army List,* 1759-1796. For a biography see Kim R. Stacey, "Major General John Small, 1726-1796", *Journal of the Society for Army Historical Research,* LXXIV, No. 298 (Summer, 1996), pp. 102-04.

911 Loring, the tardy English commodore had "at last" caught up with the army at noon on the 19th. Webster, *Jeffery Amherst,* p. 235.

Pouchot, at this point, begins to confuse the names of the two English-built vessels. By *Sonnontoin* he means *Onondaga* (formerly *Apollo*) and by *Onoyote* he refers to *Mohawk.* Pouchot's ratings of their armament is based on counting their gun ports and not the actual number of cannon carried aboard each. According to Amherst, *Onondaga* was armed with 4 nine-pounders, and 14 six-pounders while *Mohawk* had 16 six-pounders. Amherst Journal, Knox, III, p. 85.

On the 20th, there was a great deal of movement in the enemy army. Many bateaux passed to & fro from their encampment to La Présentation. They also set up a camp for two regiments at Pointe de Ganataragoin[912] and the troops began to dig trenches in that sector as well as on the Isle à la Cuisse and the Isle de la Magdelaine. We fired several artillery volleys at them to harass the men digging. We were forced to economize very much on powder since we only had five thousand-weight when the enemy army arrived.[913]

On the 21st, everything appeared quite calm as the enemy were working hard at mounting their batteries. The ships even withdrew out of cannon range. We fired on the men digging but without much success because they were already under cover & the terrain **[267]** was higher than that of the island by more than 24 feet.

As of noon, we could already make out embrasures. When evening arrived, their ships undertook a substantial maneuver & we counted up to 36 barges each carrying at least twenty men, who went on board the three ships.[914] We judged from this that the attack would begin the next morning. As a result, we worked at constructing epaulements with timbers, in order to provide cover to those areas that we considered would be exposed to the enemy batteries. All the guns were loaded with ball and grapeshot and every man was ordered to spend the night at his post.

On the 22nd, at five in the morning, the three ships approached to within 200 *toises* of the fort & covered the entire space at the top of the river between the Isle à la Cuisse **[268]** & Pointe de Ganataragoin. We judged from this that we were going to be vigorously bombarded by the ships & the shore batteries. Between them, they formed a semi-circle around the fort. As a result, M. Pouchot ordered the artillery officer to move the guns of the batteries back so that they were in the cover of the merlons, thus preventing them from being dismounted. He also had the embrasures hidden with the ends of large pieces of

---

912 Chimney Point, New York. Two of the Connecticut regiments, Fitch's and Wooster's, had been ordered to establish this camp. Webster, *Jeffery Amherst,* p. 235.

913 The paucity of fire from the garrison was correctly interpreted by Amherst. "Mons Pouchot thinks it necessary I believe not to throw away Powder till our batteries are up," he wrote in his journal. Webster, *Jeffery Amherst,* p. 236. The following day Colonel Woodhull remarked that "the French still very careful of their powder and ball." Woodhull, p. 258.

914 Additional soldiers were being ferried to the three vessels. On the 23rd Amherst noted "having put as many Marks-men on board each ship, as the Commanding Offices chose to have." Amherst Journal, Knox, III, p. 89.

wood cut for the purpose. They could be cleared away by simply pushing them forward.[915]

As soon as the ships were in position, they began a very heavy & sustained fire with 25 guns. Almost at the same time, the enemy opened fire from the Ganataragoin battery with two 24-pounder & four 12-pounder guns, from the battery on the Isle à la Cuisse **[269]** with fourteen 12-pounder & 18-pounder guns, & from a third battery on the Isle à la Magdelaine with two 24-pounder & six 12-pounder guns. At the first volleys, M. Bertrand, the artillery officer, was killed instantaneously by a cannonball which struck him in the back while he was showing M. Pouchot the caliber of the balls.

A quarter of an hour later, they began to fire bombs from the Isle à la Magdelaine, where there were two 12-inch bomb mortars, six royal grenade mortars[916] & two howitzers. On the Isle à la Cuisse, there were six royal grenade mortars and at Pointe de Ganataragoin two 12-inch mortars, two royal grenade mortars & two howitzers. Altogether there were 75 guns firing at us.[917]

---

915 Pouchot has moved events forward by one day, perhaps because little of note occurred on the 22nd. English sources all say the 22nd was spent working on the batteries and shifting troops. The determined attack by the three warships took place on August 23.

Early that morning, Colonel Williamson requested permission to open fire from his unfinished batteries. Webster, *Jeffery Amherst,* p. 236. William Amherst reported that all three batteries commenced a bombardment. Hervey says they did so at 8:00 in the morning. Webster, *William Amherst,* p. 65; Hervey, p. 59. Soon after, the French "drew in all their guns." Webster, *Jeffery Amherst,* p. 236.

916 "Royals" were small mortars firing a shell 5 1/2 inches in diameter. Smith, *Dictionary,* p. 227. The English brought two royal howitzers and two royal mortars from Oswego. Hervey, p. 57.

917 Pouchot and his artillery officer could only estimate the number and caliber of the English artillery. They exaggerated somewhat, perhaps because of the rapid rate of fire from the enemy guns. The total number of siege and field pieces brought from Oswego was 45 of all calibers and types. Hervey, p. 57. Pouchot's count of guns in the batteries totals 50, so he presumably includes the guns aboard the ships and row galleys to reach his total of 75.

The battery on Isle de la Magdelaine, alternately referred to in Pouchot's *Mémoires* as "à la Magdelaine", (Dalyell's) and that on Isle à la Cuisse (Amherst's) were constructed for identical armaments of three 24-pounders, three 12-pounders, one 10-inch mortar, one 8-inch howitzer, and two coehorn mortars. Hervey, p. 59. One 12-pounder was found to have been left behind at Oswego, however, and Amherst wrote "we must put something else in the Room of her." It is not known which battery had its armament reduced nor what was substituted. Two 12-pounders were assigned to the battery on Pointe de Ganataragoin. Webster, *Jeffery Amherst,* p. 234. On August 23 the Ganataragoin battery was supplemented with two more 12-pounders taken from *Williamson* after the vessel had been badly damaged in that day's action. Two coehorn mortars and a "royal", probably a mortar, were also sent there. *Ibid.,* p. 238.

M. Pouchot received a very bad cut from a piece of wood 10 feet long & 14 inches square which was [270] flung in the air by the explosion of a 12-inch bomb and hit him in the back. The injury did not prevent him from doing his duty.

All these batteries fired with the utmost vigor and without interruption until noon, which sent debris & splinters flying around the fort. Each man remained under cover at his post; only the sentries kept the movements of the enemy under observation. Judging from our silence that we were perhaps in disarray, they brought their ships up to within the distance of a pistol shot of the fort. They were filled with troops, even the crows nests, & supported by the fire of all the shore batteries.[918]

Fortunately, they could only take up station one after the other around the fort, so that the first ship that arrived had in its sights the gate of the fort, which was also [271] raked by the battery on the Isle de la Magdelaine. M. Pouchot had had it concealed in advance by large pieces of blindages with just an opening at the side to let a man through.[919]

He judged that the enemy's plan was to mount a full-scale assault and indeed 3000 volunteer grenadiers & light infantry were already on board bateaux stationed behind the point of Isle à la Cuisse. Their

---

918 General Amherst had provided Commodore Loring with instructions the day before the attack: "My intentions were and are; that the two Snows should approach towards the Fort, putting themselves in a line with the Williamson. When the Batterys begin to fire .... I ... desire you to take the Vessels close on the Fort, to make use of your Marksmen to drive the Enemy from their Batterys, that I may push in Batteau Safely with Troops to Assault the Place." Amherst to Loring, Aug. 22, 1760, Amherst Papers/65. Amherst had ordered "as many marksmen on board each ship as the commanding officer chose to have to fire from the Tops." Webster, *Jeffery Amherst,* p. 237.

According to Captain Hervey, the decision to commit the ships was made at 11:00 a.m. on the 23rd. Hervey, p. 59.

919 *Mohawk,* commanded by Lieutenant David Phips, led the attack. Webster, *Jeffery Amherst,* p. 237. The vessels moved down the north side of Isle Orakointon between the French fort and the island batteries. This was the only way in which they could have fired on the gate. Three ships are shown along that side of the island in Pouchot's "Plan Des attaques du Fort Levis" in *Mémoires,* I, plate 2.

David Phips (d.1811) was commissioned lieutenant in the Royal Navy in 1750. David Syrett and R.L. Dinardo, *The Commissioned Sea Officers of the Royal Navy, 1660-1815* (Aldershot, 1994), pp. 358-59. Phips first appeared on Lake Ontario during the 1760 campaign. After the fall of Fort Lévis, he was ordered to take command of the wrecked *Onondaga.* Amherst to Phips, Aug. 30, 1760, Amherst Papers/65. His letter to Amherst of Oct. 19, 1760 is endorsed "Lieut Phipps – of the Navy." Amherst Papers/65.

mission was to emerge under cover of the fire from the three ships & the shore batteries.[920]

The maneuver of the ships immediately induced M. Pouchot to send out 150 men & 4 officers to deploy along the epaulement. He had each ship bombarded one after the other with five guns, the only ones within reach,[921] using ball [272] & grapeshot. He did not respond to the shore batteries.

Despite the superiority of the enemy fire, with our five guns & musketry, we forced the *Outaouaise* & subsequently the *Onoyote* to run aground half a league from the fort, near the Isles des Galots. One of them was put out of action permanently.[922] The *Sonnontoin,* of 22

---

920 The plan of assault was perilous and depended on the covering fire of the warships. Amherst ordered "the Grenadiers with a thousand fascines to be ready to assault the Fort, sending with them two Hautwitzers in the Row Galleys and 300 (not 3,000) men of the light Infantry & Gages [80th] to row in the fascines & secure the batteaus." Webster, *Jeffery Amherst,* p. 237. The grenadier battalion was commanded by Lieutenant Colonel Eyre Massey (note 667) and was composed of the grenadier companies of six of the regular regiments. The battalion mustered 24 officers and 568 men at the beginning of the campaign. For the attack Amherst ordered that "officers and serjeants only to have fire arms, the men swords, axes and tomahawks, and [dressed] in their wastecoats and forage caps." Hervey, pp. 56, 59. They were also provided with scaling ladders. Webster, *William Amherst,* p. 65.

General Amherst was confident that his plan was sound and that his men would "easily" carry the fort. Webster, *Jeffery Amherst,* p. 236.

921 This phrase (*les seules qui étoient à portée*) could also be translated as "within range", but, considering the short distance, it is more likely that Pouchot means that these five guns, probably sited in the two half-bastions on the north side of the fort, were the only guns that could bear on the English ships positioned off that side of the island.

The officers and men stationed outside the fort proper, behind the epaulement along the island's north shore, were to meet the anticipated landing.

922 *Mohawk* (*Onoyote* to Pouchot) was not at first supported by the other two ships and anchored, exchanging fire with Pouchot's guns for 45 minutes. Then a French shot "beat in a plank and risked her sinking." Lieutenant Phips ordered the anchor cable cut and drifted with the current below the island. *Onondaga* (*Sonnontoin*) and *Williamson* (*Outaouaise*) then moved in, but "an unlucky shot" cut the cable of the latter, and "Lt Sinclair was drove down the river and obliged to follow the Mohawk." Neither *Mohawk* nor *Williamson* figured in the rest of the siege being "very much shattered with Shot" and only one anchor remaining aboard *Mohawk.* Webster, *Jeffery Amherst,* pp. 237-38. Both were later repaired and returned to service.

Onondaga ultimately proved too battered to repair, although attempts were made to salvage her into the autumn of 1760. On October 19 Lieutenant Phips opined that she could not be raised, and on the 21st Loring admitted to Amherst that they had abandoned the attempt. Phips to Amherst, Oct. 19, 1760 and Loring to Amherst, Oct. 21, 1760, Amherst Papers/65. Lieutenant Sinclair was ordered to cut up the wreck. Sinclair to Amherst, Oct. 20, 1760, *Ibid.* This must have been done as *Onondaga* does not appear on a November 1762 "List of Vessells on the different Lakes in North America." *Ibid.*

guns, which attempted to approach too close to the fort, also ran aground. It was so badly mauled that it struck its colors. There were 350 men on board. The side of the ship facing the fort suffered particularly badly. Its gun decks were touching the water & the gun ports were just one gaping hole. The water which entered the ship made it lean over in the direction of the fort. M. Pouchot ordered the firing to cease in order to save powder. The second captain & the sailors came [273] to offer surrender. M. Pouchot kept them as hostages as he could not let them enter the fort since they would have been more numerous than the garrison.[923]

---

923 The performance of *Onondaga* and her commodore was much criticized and created further ill-feeling between Loring and General Amherst. The latter was already annoyed by the sluggish performance of the vessels on Lake Ontario and Loring's tardiness in finding his way down the St. Lawrence.

Amherst reported that, after the other two ships had dropped down the river, "The Onondaga very unwisely tryed to get farther off & the Stream drove her on nearer & aground instead of getting through as the others had done. In this situation she lay much exposed, but nothing could be done immediately to help her." About 5:00 in the evening a boat, believed to contain either Loring or Captain Thomas Thornton, was observed heading towards the French fort. It was fired on by the English batteries, but the gunners missed. They had better luck when another boat returned to the ship, Amherst supposed, to raise French colors since "Onondaga had struck hers." The general then sent a detachment of grenadiers to rehoist English colors and thereafter stationed boats and detachments of soldiers to prevent the French taking control of the disabled ship. Webster, *Jeffery Amherst,* pp. 237-38.

The action was an embarrassment to Amherst who was vague in his report to Prime Minister Pitt, saying only that "The going down of the Vessells to the Fort, was not Effected in the manner I could have wished." Amherst Journal, Knox, III, p. 89. Central to the issue was the lowering of the ship's flag. Captain Hervey reported that "the colours are struck and Mr. Thornton sent ashore." Hervey, p. 59. Amherst blamed Loring for "suffering the Colours to be Struck" and for "permitting the Boat to go on shore" but made no mention of Thornton's part in the action. Loring to Amherst, Aug. 26, 1760, Amherst Papers/65.

The army's reaction was almost universally unfavorable to Loring. It was at first believed that it was the commodore who had gone to the French fort. "Mr. Thornton ... looking at that distance so like Loring we thought at the batteries it was he," wrote one officer who also expressed the general opinion of the army: "I confess I am not conversant in sea-rules; but, if he and the crew could not stand it on board, they ought not to strike so near our batteries, but come over to us ... to wait an happier issue." Furthermore, the officer believed, the clumsy handling of the vessels "stopped the grenadiers from going on." Knox, II, p. 554.

Loring was stung when he learned that his conduct was "much blamed by the Army for permitting the Colours to be struck." On August 26 he sent a long report to the Admiralty with details of the action and a justification of his actions. He claimed to have been poorly supported by the army. After *Mohawk* and *Williamson* were disabled, Loring said, he sent a messenger to learn if the landing was to proceed. When told no, he requested boats to tow his ship off. They were not forthcoming. Thinking it "high Time then to save the Vessell," he attempted to sail away but was forced ashore by the current. Aground, unsupported and

...(continued next page)...

In between these actions, the enemy sought to emerge from the point of Isle à la Cuisse and mount an attack. Two guns which were aimed in that direction kept them in check & made them withdraw behind the point. It is safe to say that the state in which they saw their ships removed all their desire to advance.[924] The action lasted from five in the morning until half past seven in the evening without any

---

out of ammunition, he ordered his men to take shelter in the hold. At that point Loring blamed his crew for not continuing to resist. The men came to him "in a Body" and said that "they had fought bravely during the whole Action, and did not understand their being Sacrifised in this manner, and therefore insisted on striking the Colours." He refused, but the men lowered the flag anyway. When the crew attempted to send a boat ashore "to make the best Terms for themselves they could," Loring aimed a musket at them and threatened to "shoot the first man that layed his Hand on an Oar."

Within a few minutes, the commodore was persuaded to send an officer (Thornton) and two men to the fort in a bid for time, he said, in which to receive assistance. The French began calling for Loring to come ashore, which he refused, and soon after the two boatloads of grenadiers sent by Amherst approached. The French fire intensified, and Loring "had the misfortune to lose the Calf of my right Leg by a Cannon Ball." The grenadiers boarded *Onondaga* and raised English colors, but they were also driven below after sustaining casualties of about one-third of their number. Sailors and soldiers huddled in the hold until 8:00 p.m. when the wounded were sent to the English shore. About 2:00 a.m. the others were evacuated. Next morning, when it became clear that the crew had abandoned *Onondaga,* the French stopped firing shot at the wreck. Loring to the Admiralty, Aug. 26, 1760, in Knox, II, pp. 549-51n.

A letter dated August 25, containing very similar information and defending Loring's conduct, was printed in the *New York Mercury* of September 29. It was probably written by the commodore himself. Similar information is included in Loring's letter to Amherst of August 26. Amherst Papers/65.

Despite the controversy, this was a hard-fought action. Colonel Whiting noted that the ships "went within Pistol Shot, so that the Monsieurs had warm work yesterday." Whiting to Whiting, Aug. 24, 1760, Welles, p. 146. *Onondaga* had fired 892 shot, eight of her guns were dismounted, 15 men had been killed and 15 mortally wounded, and the vessel was nearly destroyed, "every Thing tore to pieces." *New York Mercury,* Sept. 29, 1760.

The officer sent to Fort Lévis and identified by Pouchot as the "second captain" was Thomas Thornton (fl. 1759-1760). He had commanded English vessels on Lake Ontario since the autumn of 1759. Gage to Beckwith, Nov. 6, 1759, Gage Papers/Oswego Letterbook. Thornton was actually captain of *Onondaga,* although Loring sailed aboard her in his capacity as squadron commodore. Amherst to Phips, Aug. 30, 1760, Amherst Papers/65. Thornton might also have been a Royal Navy officer or master as he is addressed as "Mr." in at least one case. Thornton to Loring, Dec. 24, 1759, Amherst Papers/65.

Captain Thornton claimed that Loring had ordered him to strike *Onondaga's* colors. Knox, II, p. 554. John Bremner of the 55th Regiment heard, however, that it was Thornton who lowered them. Bremner Journal, p. 10. Pouchot implies that Thornton's actions were independent of the Loring's orders. Amherst apparently accepted Loring's account since the commodore remained in his post. Thornton was not so fortunate. Following the surrender of Fort Lévis, he and three of his men were apprehended and "sent to the Provost Guard's." The officer was released on August 28 and sent home, dismissed from the service. Hervey, p. 60; Amherst to Phipps, Aug. 30, 1760, Amherst Papers/65.

interruption in firing. We had 40 men killed or wounded. It would be impossible to praise too highly the fortitude shown by officers, colonial troops & militiamen and above all by our artillerymen who were all [274] sailors. Three or four of the latter were priceless because of their skilled & nimble operation of their guns.[925] The enemy, like us, fired ball & grapeshot. M. Pouchot had had some old iron cut up by a blacksmith. We filled caliber bags[926] with this and added it to the ball, which had a most destructive effect on the ships subjected to our fire from the top of the ramparts. We filled their decks with holes.

One thing that amused the garrison at such a serious time was that the Indians had climbed up onto the trenches & batteries to view these ships in battle. They considered them as their own, because of the names they had been given & because they had an Indian painted on their [275] flags.[927] They uttered the most frightful shrieks when they saw the ships so badly mauled. The English had convinced them they would make us surrender with the ships alone. When the Indians saw

---

924 With all three vessels crippled, Amherst decided "not to pursue my Plan without the help of the Ships, as I must have lost a great many men and perhaps miscarried." Webster, *Jeffery Amherst,* p. 237.

925 The sailors were the 36 officers and men of the disabled schooner *Iroquoise.* Total casualties among Pouchot's garrison during the siege were 12 killed and 35 wounded, and most must have occurred during the close-range duel with the ships and their supporting batteries. "A Return of the State of the Garrison", Aug. 25, 1760, Amherst Papers/8.

926 Pouchot calls these *sacs de calibre.* It is not clear what he means, although, from context, it is probable that he refers to bags of a size fitted to the caliber of his cannon. A cartridge bag would have been described as a *sac à charge.* The bags held home-made grapeshot.

927 Amherst and Johnson had indeed used the two snows to further cement the support of the Six Nations. *Apollo* had been renamed *Onondaga* on August 1, and a great ceremony was held at Oswego to mark the event. "To please the Indians I desired them to christen the Snow and took all the Chiefs on board in the afternoon, as they had told Sr Wm Johnson they would like to have her called ONONDAGA. I had a large flag made with an Onondaga Indian painted on it. This was hoisted just as I christened the Snow by breaking a bottle at the head. Then Gages Regt fired a volley. The Fort fired a gun & the R[oyal] Highlanders fired a volley & the ONONDAGA answered it with 9 guns. All this pleased the Indians extremely & I had made them some speeches by Sr Wm Johnson. Gave them some Punch & they were greatly delighted with the whole, promised to be fast friends & said they were ready to go with me." Webster, *Jeffery Amherst,* p. 223. The choice of this name might also have been intended to influence the substantial number of Onondaga from the mission of La Présentation who had retained their loyalty to the French.

Pouchot implies that *Mohawk* flew a similar flag, and this does not seem unlikely, especially if Johnson wished to prevent jealousy on the part of the nation that had always been most firmly in the English interest.

them drifting broadside-on before running aground, their shrieks increased in volume & they shouted obscenities at the English. They said: "You did not want to kill our father[928] at Niagara; see that you capture him now. If you had listened to us, we would not find him here. A handful of Frenchmen are showing you up." However, this action reduced to rubble all the top parts of the parapets of half the fort and blasted away all the fascines or shattered all those facing Isle à la Cuisse & the ones defending the front of the two half bastions.

That night, M. Pouchot endeavored to repair with earth sacks the batteries of the bastion facing the islands [276] in order to make them serviceable. This bastion was on the point of collapsing. A man could have climbed up the ramp formed by the earth that had caved in.[929]

The enemy continued their bombardment throughout that night & at intervals fired from their batteries shots loaded with ball and grapeshot in order to inhibit our repair work.[930] We had 2 men killed & a number wounded.

On the 23rd, the enemy continued their bombardment & kept up a vigorous artillery fire all day. That night we suffered a similar bombardment & occasional volleys from their cannon, as we had the night before.[931]

On the 24th, they opened up with another battery to fire on the wooden redoubt at the far end of the island & catch with enfilading fire the retrenchment facing the islands.[932] Their batteries continued [277] to fire as heavily as on the previous days. Fire broke out in the ruins of the storehouse & in the commandant's apartment.

---

928 Pouchot.

929 The damage was to the northwest bastion, the one to the left of the gate when leaving the fort. When he visited Fort Lévis on August 26, Captain Hervey observed "a breach made in the bastion nearest to the batteries." Hervey, p. 60.

930 This was a common practice in siegecraft. Late on the 23rd, Amherst had ordered his batteries to "fire now & then a Shot during the night to hinder the Enemy from repairing their works." Webster, *Jeffery Amherst,* p. 238.

931 Pouchot is confused here. The action with the ships occurred on August 23, so he probably refers to events of the 24th.

932 The English unmasked no new guns on August 24. Pouchot describes the battery at Pointe Ganataragoin which had opened fire with the others on August 23. Shot from that position probably became noticeably more annoying on the 24th, however, after two 12-pounders from *Williamson,* two coehorns and "a Royal" had been added to it. Webster, *Jeffery Amherst,* p. 238.

Fortunately, we managed to extinguish it without the enemy perceiving the problems we were having. We had fired very little in order to conserve the small quantities of powder & shot that we had left.[933] The enemy batteries put out of action all the cannon of the bastion facing the island. The frames of the parapets were razed to within two feet of the terre-plein thus leaving the powder magazine, which was only constructed of large wooden beams, extremely exposed.

On the 25th, at daybreak, M. Pouchot opened fire with three guns on the batteries which were causing us the most trouble. These guns were the only ones which remained intact to cover the front under attack and one of them, the most important one, was even missing a third of its barrel, [278] having twice exploded. For lack of shot of the right caliber, we fired two or three small balls at a time. From the enemy's movements in their trenches, we could see that this manner of firing bothered them a good deal. But we were simply unable to destroy or even to damage their batteries to any great extent.[934]

Our firing put the English in a bad mood. In the afternoon they redoubled the bombardment from all their batteries & fired red-hot shot, *pots à feu* & carcasses. It was just too much for our wretched fort, which was nothing more than a shambles of lumber & fascines. The red-hot shot set fire to the long fascines of the interior revetment wall of the bastion, right down at the very bottom, where we were able to

933 "Monsieur Pisheaue who commanded Niagra Last year ... seems now determined to Rectify the error he then made of firing away the Ammunition too fast," Colonel Whiting wrote to his wife on August 24. Welles, p. 146.

934 Of the English accounts, only John Bremner mentions the "Brisk fire" of the French guns on the morning of the 25th. Bremner Journal, p. 23.

Pouchot's secretary, the unnamed gunner "J.C.B." [Joseph-Charles Bonin], attributed a different motive to this action: "Commander Pouchot, realizing further resistance would be futile, began at daybreak to fire salutes from the cannons as long as the powder lasted, doing this to use up the rest of his war munitions. This astonished the enemy, who had no other thought than that it was the birthday of the King of France, which was actually that day, and that the French were really not much concerned if they were thus celebrating. But they were mistaken; for the real purpose was to use up all the ammunition so that the enemy would not profit by it." Stevens, *Travels,* p. 119.

King Louis XV was born on February 15, 1710. Michel Antoine, *Louis XV* (Fayard, France, 1989), p. 14. The royal birthday, however, was celebrated on August 25, the feast day of Saint Louis, canonized crusader King Louis IX. Although J.C.B. was an artilleryman and claimed to have been placed in charge of the fort's powder magazine, no other source supports his statement that Pouchot was intentionally burning ammunition. Hervey found 30 barrels of gunpowder in the fort, and Amherst described the ammunition supply as "a great quantity of Powder & ball." Hervey, p. 60; Webster, *Jeffery Amherst,* p. 239.

extinguish it. That made us realize how badly damaged [279] the rampart was. *Pots à feu* twice more set the debris of the fort ablaze. Once again we managed to put out the fires with water from the shell craters.[935]

This prompted M. Pouchot, with the endorsement of all the officers of the garrison, to write to General Amherst to complain about this manner of making war, which was only used against rebels and not against a brave garrison that did not deserve such treatment. In reply, he sent his aide de camp with a kind of surrender document making us prisoners of war together with a threat that, if we did not accept within half an hour, he would continue as before.[936]

M. Pouchot received the officer & read M. Amherst's message to the officers of the fort [280] and the assembled garrison. The latter

---

935 *Pots à feu* or "fire-pots" were, like carcasses, incendiary devices fired from mortars. They were described as "small earthen pots, into which is put a charged grenade, and over that, powder enough to cover the grenade; the whole covered with a piece of parchment, and two pieces of quick-match across lighted: it breaks and fires the powder, as also the powder in the grenade, which has no fuze, that its operations may be quicker: it burns all that is near it." Smith, *Dictionary,* p. 89.

Although Pouchot specifically identifies the English incendiaries as *pots à feu* and carcasses, English accounts speak only of "firing from the batteries & throwing shells." Webster, *William Amherst,* p. 65.

936 Pouchot's letter of August 25 is preserved in Volume 101 of Amherst's papers and is printed in Appendix A of this volume. In it he mentions bombardment by red-hot shot.

Pouchot could not have been surprised by Amherst's response to his communication since English sources all agree with the general that "In the afternoon Mons Pouchot put a stop to our Preparations by Beating a Parley and sending me a Letter which I immediately answered & sent him terms of Capitulation by Capt Prescott for him to sign & send back to me, which he did." Webster, *Jeffery Amherst,* p. 239. Captain Hervey, who had carried Johnson's summons to Pouchot at Fort Niagara in 1759, noted that the time was 4:00 and that the French "hoisted the white flag on the bastion next to the batteries, and a red one near it." Hervey, p. 60. John Bremner says they raised "an English flag." Bremner Journal, p. 10. Even J.C.B. recorded that "the firing ceased and the flag was lowered" before Prescott arrived under a flag of truce. Stevens, *Travels,* p. 119.

Robert Prescott (1725-1816) was captain (1755) in the 15th Regiment of Foot and the general's aide-de-camp. Boatner, *Encyclopedia,* p. 887.

Amherst's reply included a demand for a decision within ten minutes, so Pouchot had substantially less time to consult with his officers as he states. Amherst to Pouchot, Aug. 25, 1760, Amherst Papers/101. Another English participant reported that "Ten minutes were allowed them to consider whether or not they would accept of these Proposals: They were thought best to accept of them." *New York Mercury,* Sept. 8, 1760.

The articles were terse and to the point. The garrison was to become prisoners of war; they were to give up the fort with all its armament and stores; deserters from the English army and Indians in the fort were not included under the articles. The signed copy was sent to Pitt and is printed in Knox, III, p. 257. Amherst's copy is in Volume 101 of his papers.

urged him very strongly to accept the demands, in view of the impossibility of preventing fire breaking out everywhere & of being able to avoid the flames because of the small extent of the fort & the obstacles created by the debris.

On the front under attack there remained only two serviceable cannon but no ammunition. The exterior batteries of the fort had been destroyed. Since they, as well as the epaulements of the retrenchment, were dominated from the islands, we had no protection from an assault.

On the morning of the 26th, when the enemy had entered the fort,[937] they were extremely surprised to see only a few soldiers scattered among the posts they were handing over & about sixty militiamen with a kerchief on their heads and wearing shirts **[281]** and their backsides bare in the Canadian style.[938] They asked M. Pouchot where his garrison was. He replied that they were looking at all of it. We had more than sixty men killed and wounded. All the officers had been more or less seriously wounded.[939]

The enemy admitted that, in sailing past to set up camp, they lost a *carcassiere* which sank & six of their bateaux were holed, among them that of General Amherst. We had kept a very special watch out for it. The general, in polite terms, reprimanded M. Pouchot for this, and M. Pouchot replied: "Sir, we wanted to render you the honors which are due to you."[940]

On the *Onoyote* which had run aground, the English had 128 men killed or wounded; on the *Agnier,* the captain wounded & about fifty men; on the *Outaouaise,* which they **[282]** had captured from us, they had lost 54 men & on the various occasions when it had sailed in front of the fort another 100. To this figure should be added the men lost

---

937 Lieutenant Colonel Massey with three companies of grenadiers immediately took possession of Fort Lévis which they entered at 8:00 p.m. on August 25. Hervey, and probably many other English officers, visited the fort the next day. Hervey, p. 60.

938 This might mean that they were wearing breechclouts.

939 French casualties were 12 killed and 35 wounded. One of the wounded was a lieutenant of the Marine. "A Return of the State of the Garrison", Aug. 25, 1760, Amherst Papers/ 8. Pouchot states earlier that he had himself been injured, although the commandant is not listed among the casualties. It is likely that the less seriously hurt were not included in the English return.

940 Amherst does not mention damage to his bateau on August 18. The *carcassiere* was raised on the 19th and repaired. Webster, *Jeffery Amherst,* p. 235.

on the batteries & in the trenches, which they never admitted to.[941]

Once the fort had been handed over, several colonels came to take M. Pouchot into the presence of General Amherst. They treated him in a very friendly manner. He had seen a number of them at Niagara & in New York. They were afraid that their Indians, who appeared very threatening & were angry at finding nothing in the fort after it had been pillaged by the soldiers, might seek to do him harm. He thanked them for their solicitude.[942]

When he reached the mainland, many Indians appeared. M. Pouchot went directly up to several chiefs whom he recognized. He said to them: [283] "You have killed your father, which is not what brave men do. So much the worse for you." They replied: "Do not be angry, father, you are going to the other side of the great lake.[943] We shall get rid of the English." The latter were surprised to see them so calm.

General Amherst had an hour-long conversation alone with M. Pouchot.[944] He endeavored to extract details about what remained to be done in the campaign. It may well be imagined that M. Pouchot did not portray his task as an easy one. He seemed, just like the rest

---

941 Pouchot here confuses the English snows even further. By *Onoyote,* which he earlier misidentifies as *Sonnontoin,* he means *Onondaga. Agnier,* previously called *Onoyote,* is actually *Mohawk.*

Amherst gave the total casualties for his army during the siege as only 21 killed and 23 wounded. If this includes the naval contingent, then most of the casualties must have occurred aboard *Onondaga.* Webster, *Jeffery Amherst,* p. 240.

942 Although General Amherst had intended to send the prisoners to Oswego on August 26, the garrison could not be evacuated from Fort Lévis until 2:00 on the afternoon of the next day. Amherst had prevented any Indians from entering the fort during the two preceding nights . Webster, *Jeffery Amherst,* pp. 239-40. An English participant noted that the Iroquois were so "exceedingly disappointed at not being permitted to butcher and scalp the late garrison of Fort Levis" that 20 whale boat-loads of them promptly deserted. Knox, II, p. 555.

Some pillaging was done by English personnel, and Amherst dealt with it severely. On August 27 a court martial found Christian Parkey, a sutler, and Abraham Willey and Jacob Miller, both spruce brewers, guilty of removing "eleven fire-locks and some goods out of Fort Levi." They were each awarded 300 lashes. Orders, Aug. 27, 1760, Hervey, p. 116.

By the morning of August 26 Fort Lévis had been renamed Fort William Augustus in honor of the Duke of Cumberland, a son of King George II and former commander-in-chief of the army. In celebration of the victory, the parole of the day was "King George." Orders, Aug. 26, 1760, Hervey, p. 115.

943 Across the Atlantic to France.

944 This took place over a meal. Amherst recorded that, on the 26th, "Mons Pouchot dined with me." Webster, *Jeffery Amherst,* p. 240.

of his army, to dread going down the rapids. From among the Canadians, they took 36 guides for their bateaux. The garrison & its officers were taken by way of Chouegen to New York.[945] M. Belle-Garde, a Sulpician [284] missionary from La Présentation, who had preferred to stay in the fort and look after the wounded, obtained permission to go to Mont-Réal with two or three women.[946] He was an admirable priest because of his enlightened religious zeal, which had brought him to Canada with the sole aim of converting the Indians. The English subsequently returned him to his mission post. The English army stayed nearly 15 days preparing for its onward journcy. Dcspitc their guides, however, some of whom may have chosen bad routes, they lost 80 bateaux at Côteau du Lac, together with their *carcassieres.*[947]

---

945 General Amherst says he selected 19 crewmen of the former *Outaouaise* to serve as guides for the passage down the river. Webster, *Jeffery Amherst,* p. 235. The return of prisoners notes that 22 were chosen. "A Return of the State of the Garrison", Aug. 25, 1760, Amherst Papers/8. Others might have been taken from the garrison.

Following the delay in their departure, the French garrison departed Fort Lévis for Oswegatchie on August 27 and set out from there for Oswego on the morning of the 28th. Amherst clearly remained concerned about his Indian allies. The prisoner escort of one subaltern and 30 men of the 46th Regiment and one subaltern and 50 provincials was supplemented for the first two days of the voyage by a captain and 50 soldiers of the 80th Regiment. Webster, *Jeffery Amherst,* p. 239-41. The prisoners were entrusted to Lieutenant John Small (note 910) of the 42nd Regiment who was to convey them to New York. Orders, Aug. 26, 1760, Hervey, p. 115.

Small reported his safe arrival at Oswego on September 2. He then proceeded with the prisoners and an escort of 50 provincials towards New York where he was to deliver his charges to Anthony Wheelock, Amherst's commissary of prisoners (note 1503). Small expressed some concern that the "few provincials I take with me to Albany will I hope be sufficient to prevent any Disturbance or Injury to the prisoners from any Stragling or Lurking Savages we may meet with in our Rout." Small to Amherst, Sept. 2, 1760, Amherst Papers/19. He had little concern that the French would attempt to escape. Most of the fort's officers had signed the articles of capitulation. Those who had not done so had given their paroles and promised that their men would not escape. Among them were La Broquerie, Guillemetteau and Mercier, all captured aboard *Outaouaise.* Captain La Force of *Iroquoise* and Ensign Hertel de Rouville gave paroles because they had not signed the surrender document. Cadet de Poilly also did so because he wished to visit his brother on the island of Grenada. Their paroles, all dated between August 17 and 25, are in Amherst Papers/101.

The prisoners arrived safely at New York about September 20-22. Stevens, *Travels,* p. 121; *New York Mercury,* Sept. 29, 1760.

946 Pierre-Paul François de Lagarde (1729-1784), a Sulpician priest. Much of the rest of his missionary career was spent at Lac des Deux Montagnes (Oka, Québec). *DCB,* IV, pp. 431-32.

A total of eight women and children were with the garrison at the time of its capitulation. "A Return of the State of the Garrison", Aug. 25, 1760, Amherst Papers/8.

The Chevalier de la Corne, who was observing the English with a militia corps on the heights of Les Cedres, withdrew in a staged retreat towards the island of Mont-Réal as soon as he was aware of their arrival.[948] The enemy [285] landed a quarter of a league above this place. They immediately sent representatives in order to draft the terms of surrender, which everyone knows about. All the troops & Canadian officers who wanted to leave the country[949] were taken to France on English vessels on condition that they no longer served in the war.

---

947 The army began moving down the St. Lawrence from Fort William Augustus on August 31, only six days after the surrender of Fort Lévis. Webster, *Jeffery Amherst,* pp. 241-42.

The accidents occurred on September 4 as the army descended the series of rapids between Coteau du Lac and Lac St. Louis above Montréal. Amherst reported 84 men drowned, 20 regimental bateaux, 17 artillery bateaux, 17 whaleboats, and one of his row galleys or *carcassieres* lost. *Ibid.,* p. 244. William Amherst noted: "The General accounts for our loss and that of the Provincials, as proceeding from opposite causes, ours [the regulars] from too much spirit, and theirs, from too great fear." Webster, *William Amherst,* p. 66.

948 La Corne (note 447) had been sent with 400 men to support Fort Lévis or defend the rapids of the St. Lawrence. Finding Fort Lévis taken, he withdrew. Casgrain, *Collection,* I, pp. 297-98. His report of the loss of Pouchot's fort reached Montréal on September 1. Malartic and Gaffarel, p. 345. La Corne took a position near Les Cedres, Québec. He withdrew early on September 4 as the English army encamped on Isle Perrot. Hervey, p. 62.

Amherst's route down the river to Montréal is shown on a plan of "The River Saint Lawrence from Lake Ontario to the Island of Montreal" printed in Mante, p. 301.

949 Note in original, p. 286 (a): *They were strongly urged to choose this option by the English, who wanted to get rid of as many of them as possible.*

The remnants of French forces were trapped at Montréal by the converging armies of Amherst, Murray and Haviland. Governor Vaudreuil asked for terms on September 7. Lévis protested, particularly as the English insisted that the French troops be prohibited from further service for the duration of the war. He asked to be allowed to retire to Isle St. Hélène with the *troupes de terre* battalions to fight on for the sake of honor. In this he was overruled by Vaudreuil who sensibly replied that "the interest of the Colony does not permit us to reject the conditions proposed by the English General, which are favorable to a country whose lot is confided to me" and ordered him to surrender. Lévis to Vaudreuil, Sept. 8, 1760, *DRCHSNY,* X, p. 1106.

The articles of capitulation for New France of September 8, 1760 are printed in a number of places including *Ibid.,* pp. 1107-20 and Knox, II, pp. 566-89.

At the time of capitulation, Lévis could muster only about 2,200 men of his eight *troupes de terre* battalions, including sick and wounded. Lévis to Belle-Isle, Nov. 27, 1760, *DRCHSNY,* X, p. 1123. Quartermaster Bernier's list added 650 men of the two Marine battalions. Letter from Bernier, Sept. 25, 1760, *Ibid.,* p. 1121. Their count excluded officers, and an English "List of the French Troops", dated at Montréal September 11 gives a total of 3,894 individuals of all categories, including women and children. Of these 2,384 were sergeants and private soldiers of the *troupes de terre* and Marine. Webster, *Jeffery Amherst,* pp. 249-50.

An embarkation return of the French troops at Québec is in *DRCHSNY,* X, p. 1127. It shows 230 officers, 2,405 soldiers and 507 sailors, servants, women, and children.

It may easily be imagined that in the course of this wretched campaign everything became extravagantly expensive.[950] The Intendant issued [286] promissory notes in order to cover all the extraordinary expenses occasioned by the shortage & cost of all goods. But he only converted them into bills of exchange for the most favored people & in exceptional circumstances in order not to astound the French government with these huge expenses. The habitants & other private individuals were left holding an enormous quantity of drafts & certificates which he refused to convert into bills of exchange. [287]

The English, once they were masters of Canada, sensed that they were in a strong position with regard to France and collected these promissory notes - at the cheapest possible price, as can well be imagined - & requested that they be redeemed. Their request was granted.[951] It is no exaggeration to say that the money that France was forced to pay, according to the agreement, came to 23 to 26 million *livres*. If the fear of having to pay was a consideration in the surrender of Canada, it was a mistake.

M. Pouchot & all the French officers, together with the French soldiers [288] & those of the colony, as part of the terms of surrender of Mont-Réal, were taken to France, while the Canadians were sent back home. The first of them left New York on January 1st & after a

---

950 Note in original, p. 286 (b): *The Minister of the Navy, M. Berryer, had decided to send some supplies but their price & the freight costs made him delay from one day to the next the departure of the transport vessels which, all because of him, could not enter the St. Lawrence in time & were set on fire in the Baie des Chaleurs. The Marquis de Vaudreuil had anticipated this lack of reinforcements. He ordered M. de Minville to go and rendezvous with them at the mouth of the river with a frigate. Fourteen English ships loaded with munitions for Quebec were captured but, once again, they had to be burned and no other advantage could be gained from their capture.*

The final attempt to reinforce New France using the frigate *Machault* and five armed merchant vessels is described in Stanley, pp. 259-61. Documents pertaining to the expedition and action at Baie des Chaleurs on the Restigouche River are printed in Knox, III, pp. 353-421. Archaeological investigation of the wreck of the *Machault* is discussed in Walter Zacharchuk and P.J.A. Waddell, *The Excavation of the Machault* (Ottawa, 1984).

This footnote seems to confuse events and participants. The captured English vessels were probably those taken on May 15 in the Gulf of St. Lawrence by François Chenard de La Giraudais, commander of the *Machault*. Stanley, p. 259.

951 Note in original, p. 288 (a): *By a secret decree, signed in Paris on 10 February 1763, the King promised to cover the bills of exchange & promissory notes issued to the Canadians for supplies delivered to the French troops, according to a settlement agreed upon, within a given period, according to distance and feasibility, &c.*

Payment on promissory notes issued by the colonial administration had been suspended before the surrender. Only a small portion of those notes remaining in the hands of Canadians still resident in the the former colony were ever redeemed. Eccles, p. 210.

very stormy crossing they arrived off Spithead, where they remained for 15 days, and then put into Havre de Grace on 8 March 1761.[952]

During the crossing, they saw three rather strange phenomena. The first was the way the ocean gave off sparks from the tops of the waves during an immense storm, just like lightning in very dark clouds. The night was a very dark one. The second was a rainbow, the two ends of which extended from port to starboard astern of the vessel & which followed its track like a tow-rope. The third consisted [289] of a beautiful lunar rainbow, very well formed but the colors of which were much less bright than those of a solar rainbow. There was a great deal of moon yellow in it. [290]

• FRAGMENT •
On the French Colony of Canada

Canada was first of all peopled by fishermen & private individuals who engaged in trade, that is to say commercial exchanges with Indian

---

952 The last two articles of the capitulation of New France addressed the situation of Pouchot and his Fort Lévis garrison as well as those captives from Fort Niagara and La Belle Famille who remained prisoners in New York. Unlike the troops surrendered at Montréal, they were not prohibited from serving for the duration of the war. Article LIV stated that "All the officers and soldiers of the troops in the service of France, who are prisoners in New England, and who were taken in Canada, shall be sent back as soon as possible to France, where their ransom or exchange shall be treated of, agreeable to the cartel; and, if any of these officers have affairs in Canada, they shall have leave to come there." Article LV provided that "As to the officers of the Militia, the Militiamen and the Canadians, who are prisoners in New England, they shall be sent back to their countries." *DRCHSNY,* X, p. 1120.

Execution of these terms began promptly. On October 3, after they had taken the oath of allegiance to King George II, "a Number of Canadians" boarded a sloop at New York to go to Albany and begin their homeward journey. *New York Mercury,* Oct. 6, 1760. The Canadians were sent to Montréal in the care of the ships' captains, La Force and La Broquerie. Wheelock to Gage, Nov. 5, 1761, Gage Papers/8. General Amherst saw 200 Canadians, including Captain La Broquerie of *Outaouaise,* arrive at Crown Point on October 22. He noted that they were to be "sent to their homes." Webster, *Jeffery Amherst,* p. 261.

It took longer to return the men of the *troupes de terre* and *troupes de la marine* to France. The prisoners had been settled in quarters around New York City and Long Island. The account of J.C.B. is very enlightening as to the manner in which they were held. Most were placed in taverns or private homes with little or no supervision. Stevens, *Travels,* pp. 122-33. Pouchot remained in the city and lived with the "English commissioner" - Anthony Wheelock, commissary of prisoners (note 1503). The French captain was on good terms with his host, and, through Pouchot's influence, J.C.B. obtained temporary employment as the commissioner's secretary. *Ibid.,* pp. 129-30.

...(continued next page)...

nations, by soldiers who had been discharged from the service and finally by persons sent there from France under *lettres de cachet*. A number of the latter remained there for three years before regaining their freedom. Others were there for life. Yet others, possibly the majority, had been taken there to settle by the landed aristocracy. **[291]**

The land had originally been conceded by the King to foreign missions, to the Sulpicians and the Jesuits, & to officers. In Canada, very little land, perhaps none at all, belongs to merchants or bourgeois.

On October 27 Wheelock published a notice that he had been ordered to assemble all French prisoners to send them to Europe. The Frenchmen were to settle their bills, receive their pay and otherwise close their personal affairs in New York. On November 24 it was rumored that transports that had recently brought 500 soldiers to New York were to convey the prisoners to France. *New York Mercury,* Oct. 27 and Nov. 24, 1760. In December a copy of the cartel or terms of exchange was received in New York. The French and English had ratified this agreement on February 6, 1759, and the receipt of the terms at New York apparently cleared the way for Pouchot's return to France. According to J.C.B., two transports were readied to sail on January 1. Stevens, *Travels,* p. 133. The "cartel of exchange" is printed in Knox, III, pp. 258-88.

On December 31 a total of 504 prisoners were embarked on the transports *James* and *Boscawen,* leaving behind 121 regular soldiers who had chosen to remain in America. The officers, including Pouchot, went aboard *James* the next day and sailed immediately. The two ships became separated and, after a stormy crossing, *James* took shelter in Portsmouth on February 15. J.C.B. says they sailed for France on the 26th and reached Havre de Grace (today Le Havre) two days later, landing the passengers on March 1. Stevens, *Travels,* pp. 133-35. The report by Captain Louis de Preissac de Bonneau (note 423), however, agrees with Pouchot that the vessel arrived on March 8. Bonneau, who had been sent by Lévis to New York in 1760 to arrange prisoner of war exchanges, reported the return of 196 sergeants and men of the *troupes de terre,* 107 of whom had been exchanged or ransomed. With them were Pouchot, Lieutenant Bonafous and Lieutenant La Milletière. Bonneau to Choiseul, March 9, 1761, *Collection de manuscrits,* IV, pp. 307-08. Bonafous and La Milletière had, by that time, been exchanged.

The remaining 304 prisoners of war were officers and men of the *troupes de la marine.* Among them were captives taken at Niagara, La Belle Famille and Fort Lévis who had not taken the oath of allegiance. The officers travelled with Pouchot aboard *James.* Captain Pierre-Joseph de Céloron, for example, left New York in January 1761 and arrived at Havre in March. Genty to Jumilhac, April 22, 1762, Ravaisson-Mollien, XVIII, pp. 297-98.

At Le Havre the soldiers of the Marine companies were discharged, and the regulars were sent to rejoin their regiments. Stevens, *Travels,* p. 135. Pouchot must have gone directly to Paris where, by May 6, he had solicited the "Testimonial" of former Governor Vaudreuil printed at the beginning of Volume I of his *Mémoires.*

953 The Carignan-Salières Regiment had been sent to New France in 1664 to defend the colony against the Iroquois. Once that service had been accomplished, 400 officers and men were granted seigneuries (for the officers) or land and remained to augment the colonial population. A further source of male immigrants was soldiers of the *compagnies franches de la marine* who were recruited in France and encouraged to settle in the colony at the conclusion of their service. Chartrand, *The French Soldier,* pp. 5-8, 10.

The factor which most contributed to the expansion of these communities was the disbanding of the Carignan Regiment; its private soldiers became settlers & its officers owners of land reserved for lay folk. They are the only sources of the current population of this huge country.[953] It seems strange, in view of the lack of aid & attention granted towards its development, that the colony, which was feeble for a long while and very often on the point of perishing from starvation as a result of the lack of assistance it received from France, should have nonetheless managed to reach a population of [292] about 30 thousand souls.[954] It may be inferred from this that both climate & soil are good & prolific. It is not extraordinary to find, between a grandfather & his grandchildren, about sixty persons.

The Canadians are well built, very robust & active, with an admirable capacity to endure hard work & fatigue, to which they are accustomed through long & arduous [293] journeys connected with their trading activities, in which a great deal of skill & patience are required. As a result of their travels, they are habitually rather idle because of the type of life that they lead during them. They are brave, fond of war and very patriotic. They have an extraordinary affection for their mother country. Their lack of knowledge of the world normally makes them liars & rather boastful as they are ignorant of almost everything.

There exists no country in the world where the women lead a happier life than Canada. The men have a great deal of consideration for them & spare them all the hard work that they can. It can be said that they deserve it, as they are well-mannered, attractive, spirited & inclined to intrigue. It is only through them that their husbands obtained the [294] posts which put them in easy circumstances & above the common run of people.[955] In the towns there reigns a tone of polite company which would hardly be suspected in such a remote

---

954 Note in original, p. 293 (a): *This is a gross error. From a census conducted around the middle of the century, the population of the colony totalled 88 thousand souls. The latest count, under Governor Carleton, reveals the population to be 153 thousand, of whom three thousand are English protestants who have settled in Canada since the peace. The latter control all the trade & are attempting to make themselves the sole masters of the country's administration.*

A capitation list of 1754 for New France gives a total population of only 55,000, but it does not appear to be comprehensive. *DRCHSNY*, X, pp. 271-75. A general census of the colony had been completed by April 1759. It counted 82,000 individuals in all of "Canada." Montcalm to Belle-Isle, April 12, 1759, *Ibid.*, p. 962.

955 Pouchot here refers to the trading concessions at the interior posts that were given to officers of the Marine. See his comments about Mme. Péan (notes 151-153).

country. The ladies dance & dress with great naturalness & without special instruction.

The Canadians are generally religious & have high moral standards. The voyageurs are unreliable when it comes to trade. The priests keep them on a tight rein because they are the temporal & spiritual masters in the country & managed to control even the general & the intendant; it was a great misfortune if one or the other of them failed to obtain their good opinion. The benefices are wealthy and portable. The bishop of the largest diocese in the world, that of Québec, had revenues of ten thousand *livres.* He was responsible only to the Pope. Since the death of M. de **[295]** Pombriant,[956] thc English have not filled the vacancy. The whole country is now under the spiritual direction of two vicars general.[957]

The Governor of Canada was also the Governor of Louisiana. Although endowed with ample authority with respect to the administration of the country & to negotiations with the Indians and foreigners, he was very much obstructed by the intendant, who was absolute master of the finances and charged with the administration of trade and justice. He was at the head of the sovereign council of the country.[958] **[296]**

---

956 Henri-Maire Dubreil de Pontbriand (1708-1760), sixth bishop of Québec. *DCB,* III, pp. 192-99.

957 Note in original p. 296 (a): *The famous bill of 1774 allowed Canadian Catholics to have a bishop, provided that he was not consecrated in France. All the protests & turmoil which the promulgation of this bill caused & continues to cause in England are well known. The bill justifies the reflections of the author of the* Observations sur le traité de paix conclu à Paris en 1763, *see pp. 80, 81, &c.*

The "famous bill of 1774" was the Québec Act. Its adoption on May 20, 1774 aggravated the tensions between the government and its North American subjects of the original English colonies. The act granted Canadians free practice of their religion, recognized French civil laws and land tenure and expanded the boundaries of Québec. Boatner, *Encyclopedia,* p. 909.

Pouchot's editor cites a work then only recently published in the same town as the *Mémoires:* Guillaume-Emmanuel-Joseph Guilhem de Clermont-Lodève, Baron de Sainte-Croix, *Observations sur le traité de paix conclu à Paris le 10 février 1763 entre la France, l'Espagne & l'Angleterre: relativement aux intérêts de ces puissances dans la guerre présente* (Yverdon, 1780).

958 The Governor General of New France was commander-in-chief of all French forces in North America and the nominal superior of the Governor of Louisiana. Chartrand, *Canadian Military Heritage,* pp. 107-08; Eccles, "The French Forces", *DCB,* III, p. xviii; *DCB,* IV, p. 664. At an Indian council held in Montréal in 1755, Vaudreuil identified himself to the Seneca as "Governor, Lieutenant-General for the King throughout the whole of New France, the territories and countries of Louisiana." *DRCHSNY,* X, p. 345.

The trade of Canada was conducted on the King's account by private individuals. The intendant had general responsibility for this area. The King had stores at Québec, Mont-Réal, St. Jean, Chambly & Carillon, & for the forts in the upper country at La Présentation, Niagara, Frontenac, the portage fort,[959] Presqu' Isle, the Riviere aux Boeufs & Fort Duquesne.

The Québec store was a warehouse to supply that of Mont-Réal. It also supplied the trade with our resident Indians, the Abenakis and others located down river. The Mont-Réal store supplied goods to all the posts listed above. Its direct trade with the Indians amounted to very little until the king appointed a chief quartermaster. These storehouses supplied **[297]** victuals & armaments either for trading purposes or for the service of the King. They also provided support to the artillery.

The King maintained in all these places storekeepers appointed by the intendant to whom they reported directly. The intendant had under him a naval supplier, based in Mont-Réal, for all relevant aspects of the upper country.

Armaments, victuals, & goods for trade or intended as presents to the Indians came from France on vessels loaded at the King's expense. All these commodities came from the Department of the Marine. There can be no doubt that a number of the officials were self-interested.

They sent out cargo, **[298]** which they acquired everywhere at the best possible price & for which the King was apparently paid at the going rate of goods in Canada. But the greatest problem was that they sent goods which were not at all appropriate for trading with the Indians, for example large mirrors mounted on morocco leather, silk material, squares of various other materials, handkerchiefs, stockings, in fact things that were rejects from the shops. The intendant, who had close connections with the navy, would not have dared to refuse all these goods & for form's sake he sent them out to separate stores where they rotted, or were stolen or used for other purposes. At the end of a given period, inventories of goods disposed of were drawn up. The money provided by the King found its way into the pockets of the suppliers **[299]** while the King had to cover any losses. Add to that the damage which becomes inevitable over such long journeys & also that

959 Fort Little Niagara above Niagara Falls. See note 157.

960 Wampum or beads.

which might be lost through theft. The suppliers thus enjoyed guaranteed profits & the King had to stand the losses, although in normal times the financial benefits of trade must have been very considerable, otherwise no private individual would have decided to engage in it, especially in such immensely remote regions.

The goods for trade with the Indians are hunting muskets, shot, bullets, powder, tinderboxes, gun flints, wad-extractors, knives, axes, copper pots, porcelain,[960] tumblers, men's shirts, & material trimmed with blue & red cloth for blankets and covers, **[300]** vermillion & verdigris, red, yellow, green & blue ribbons, English lace, needles, thread, bradawls, blue, white, and red ratteen for gaiters, woolen blankets in 3 & one-half, 3, two & one & one-half point needlework, Léon cloth, wooden-framed mirrors, flat hats with embroidered & false brims, plumed headdresses in red, yellow, blue & green, hooded greatcoats for men & children, crimped ratteen, both cheap & finely worked braids, brandy, tobacco, razors for shaving heads, porcelain-style glassware of dark wine color, combs, &c.

For these goods the Indians barter skins of deer, bear, beaver, otter, *pécans,*[961] **[301]** squirrel, mink, lynx, fox, muskrat, forest rat, wolf, caribou & moose. For bread, pork, salt, prunes and molasses, they also trade all kinds of meat & fish, bear grease, which is better than goose grease, & the down of waterfowl. All these different exchanges are made on the basis of the value of a beaver skin, which is normally judged equivalent to a bottle of brandy selling for 30 *sous.* A pound of beaver fur fetches 4 *livres* 10 *sous* & the skin weighs from 2 & one-half to 3 pounds. The price of our goods varies according to the remoteness of the market.

In the King's posts, the storekeepers had the sole responsibility for this trade & for accounting to the intendant for the profits. The commander of each fort had the right to see to it that **[302]** the Indians were not cheated & to take from the goods those items that he believed were necessary to offer to them as presents. The differing interests of these two persons often led to disputes. The governor almost always found himself in the wrong and would be recalled. To avoid these disadvantages, they usually came to an understanding and worked together over business matters.

The command of posts in the interior was given to favored officers. Rank counted for nothing. They took with them a storekeeper and

---

961 A variety of mink.

traded to their own advantage.[962] Since they were short of cash, they found merchants in Québec and Mont-Réal, who supplied them with all the necessary merchandise on credit. This was called "equipping" them. They **[303]** agreed over price & in return gave the furs to the merchants. Both parties made a handsome profit. The officers often had occasion to negotiate on behalf of the King with the Indian nations in the vicinity of their forts and gave some of their merchandise as presents. The cost was covered by the intendant, with the approval and on the orders of the governor. This led to many padded bills that guaranteed substantial profits to the commandants, especially in time of war.

The commandants, as well as private traders, had to acquire special passes from the government, which cost them 4 to 500 *livres*, to obtain permission to take their merchandise to the forts & to take responsibility for some goods on the King's account. This requirement was always a very considerable obstacle **[304]** to the development of trade and settlement in Canada, as they were obliged to obtain the passes each time that they wanted to journey into the interior of the country. The remotest forts of the north west were the most sought after because of the abundance and cheapness of furs & the high price obtained for the merchandise.

The third sort of trade was undertaken by traders or woodsmen who loaded up some canoes with goods, obtained their passes & went off to Indian nations beyond the range of our posts. They waited for the Indians to return to their villages after the hunt, or followed them into the bush, and came back after trading the goods in their canoes for considerable profits. Those, especially, who were in a position to buy their merchandise at first hand made a fortune very rapidly. But in order to achieve this, one had to make up one's mind to lead a life of hardship and drudgery. The goods obtained from the Indians through these types of trade might add up to 2,500,000 *livres* of business in France. **[305]**

To all the above details, M. Pouchot had added some observations about the utility which Canada would have been to France if its riches had been

---

962 Pouchot airs a personal grievance based on the unwillingness of Governor Vaudreuil to appoint a *troupes de terre* officer to command Niagara in 1755 (see note 280) and again in 1757 when he was relieved by Captain Vassan (note 381). The latter incident was attributed by Pouchot and other regular officers to an unwillingness by Marine officers to forego the lucrative trade concession at the post.

better known & if better use had been made of the great advantages offered by the nature of the soil and the country's geographical position. However, as the author had only sketched out this material, intending to return to it and deal with it in greater depth, & since his more recent **[306]** *remarks were not found among his papers, we thought it best to suppress the original text as too superficial and incomplete. In any case, he says nothing that the Abbé Raynal has not already noted and diligently discussed in his work,*[963] *where he is the first to be brave enough to protest against the unjust prejudices held by the public with regard to the French colonies on the North American continent. Attempts were made to justify these prejudices in a series of memoirs printed in the first volume of the* Ephémérides du citoyen.[964] *Just because the government had made some mistakes in the administration of its Canadian colony, was it really proper to conclude that the country was of no use and that we should congratulate ourselves on losing it? Nonetheless, that is what the arguments of our economist boil down to, he being a member of a kind of* **[308]** *political sect, people who confuse enthusiasm and reason & who, slaves of their own system, seek to subject everything to it, stinting neither paradox nor logomachy to that end.*

End of the Second Volume

963 Guillaume-Thomas-François Raynal (1711-1796). *Biographie universelle,* XXXV, pp. 261-66. His work was entitled *Histoire philosophique et politique des établissemens et du commerce des Européens dans les deux Indes,* published in 1770.

964 *Ephémérides du citoyen, ou Chronique de l'esprit national* was a journal published in Paris from 1765 to 1772 and again from 1774 to 1776.

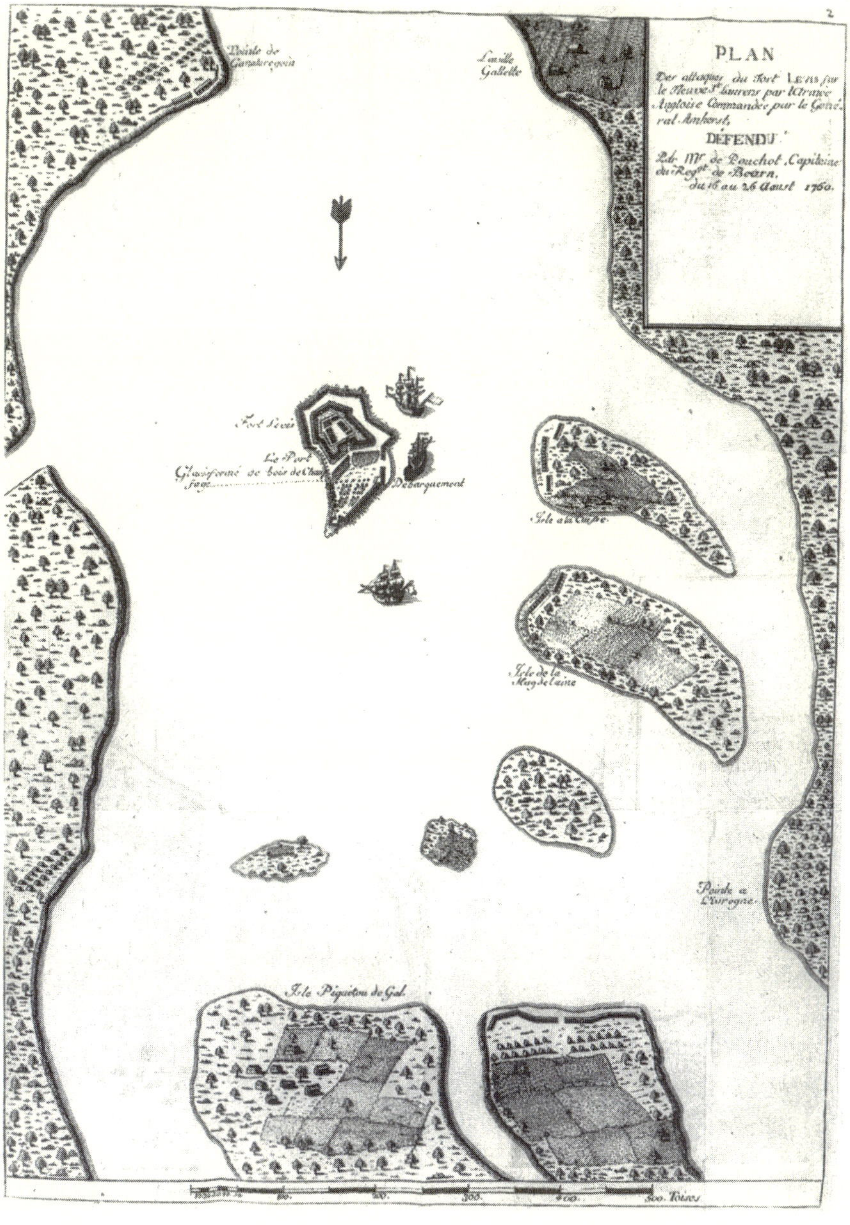

Pouchot's spirited defense of Fort Lévis was the subject of "Plate 2" of the *Mémoires*. It shows the positions of the English batteries and the three armed vessels that attacked the fort on August 23, 1760. *Courtesy, William L. Clements Library, University of Michigan.*

MÉMOIRES
SUR
LA DERNIERE GUERRE
DE
L'AMÉRIQUE SEPTENTRIONALE,
ENTRE
LA FRANCE ET L'ANGLETERRE.

Suivis d'Obſervations, dont pluſieurs ſont relatives au théatre actuel de la guerre, & de nouveaux détails ſur les mœurs & les uſages des Sauvages, avec des cartes topographiqués.

Par M. POUCHOT, Chevalier de l'Ordre Royal & Militaire de St. Louis, ancien Capitaine au Régiment de Béarn, Commandant des forts de Niagara & de Lévis, en Canada.

TOME TROISIEME.

YVERDON.

M.DCC. LXXXI.

MEMOIRS
ON
THE LATE WAR
IN
NORTH AMERICA
BETWEEN
FRANCE AND ENGLAND

Followed by observations, of which several are pertinent to the current theater of the war, and fresh details on the customs and habits of the Indians, together with topographical maps.

By M. POUCHOT,
Knight of the Royal & Military Order of St. Louis, formerly Captain in the Béarn Regiment, Commandant of the Forts of Niagara & Lévis, in Canada.

VOLUME III

YVERDON
MDCCLXXXI

❈ Note ❈

Since the map which M. Pouchot had drafted was too large, we have been obliged to reduce it in size in order to make it conform to the format in which his memoirs are printed. Although the map offered to the public is not so extensive as the original, it nonetheless incorporated the same information & has even benefited from additional care in the drafting. In any case, the topographical observations in some sense compensate for the unavoidable reduction in size.[965] [1]

❈ Preface ❈

In acquainting himself with the topography of North America, M. Pouchot omitted nothing. We found in his papers a general description of the continent - admittedly imperfect in several respects but finished & unusually accurate with regard to those regions which were the theater of operations in the late war. He had abstracted from that part of his work the observations that follow, which in effect serve as an explanation of his map. The particular care he deployed in composing them proves that the description of which we have spoken constituted the materials upon which he based his map. It is for that reason that we have refrained from publishing it.[966] [3]

However, we shall not hide the fact that, in this same description are to be found details which do not appear in the observations solely devoted to making known the principal communication routes from Canada to the English colonies for the use of military men who might one day serve in that region. Since the author based himself upon the reports of others, his account of the northern & eastern regions is less reliable. It has been possible to omit

---

965 Pouchot's detailed map of the seat of the war in North America was printed as Plate 3 of Volume I of his *Mémoires.* It is entitled "Carte des Frontieres Françoises et Angloises dans le Canada depuis Montreal jusques au Fort du Quesne." In the course of the topographical descriptions in Volume III, this work will be referred to simply as "Pouchot's map."
Pouchot's map was carefully copied in the nineteenth century and reproduced in *DRCHSNY,* X, p. 694. Although some of the key notations were translated into English, that version is largely true to the original.

966 The disposition of Pouchot's more detailed "general description" is unknown.

it without harming his reputation. Indeed, he himself admits that the extent and shape of Lakes Huron, Michigan & Superior are unknown & that we have barely the sketchiest notions about the regions situated to the north & the W.N.W. of those lakes, country inhabited by the Aniboels, [4] *Monsonis & the Cristinaux.*[967]

MM. de la Veranderie,[968] *both Canadian officers, penetrated into these regions about forty-two years ago. They travelled as far as 300 leagues or so to the West of Lake Superior & discovered a number of new lakes that flow into one another. They would have perhaps reached the Southern Sea had it not been for the obstacles that always stand in the way of such expeditions, which are nonetheless the only possible source of sure knowledge. The major difficulties are: 1st the lack of food & resources for every type of provisioning; 2nd the impossibility of communicating with the natives of the country, whose language becomes, beyond a certain point, incomprehensible even to the Indians who act as guides; 3rd the disinterestedness found so rarely in persons* [5] *who undertake such enterprises; the majority of them think only of going back once they have filled their canoes with the furs that are so abundant in those regions; 4th the unreliable or enigmatic reports of the Indians, who often seek to cheat travellers & get them lost in order to kill them & then pillage their possessions.*

It is doubtless to be hoped that such obstacles will disappear or diminish as North America is peopled & that communications between the different regions will then become more necessary. Such a revolution will be the work of freedom, which that vast continent will soon enjoy if the wishes of the whole of Europe are granted.[969] [6]

---

967 "Aniboels" were the Assiniboin, a Dakota-Siouian group allied to the Cree and, in the mid-eighteenth century, inhabiting the area to the north and west of Lake Superior. The Monsonis were a part of the Woods or Swampy Cree living between northwestern Lake Superior and James Bay. The "Cristinaux" were Cree whose territory then ranged from the southern coast of Hudson Bay as far west as Alberta. "Glossary of Indian Tribal Names", *DCB,* III, pp. xxxii, xxxvii, xxxiv.

968 Pierre Gaultier de La Vérendrye de Boumois (1714-1755), then a cadet in the Marine (*DCB,* III, pp. 244-45), and his younger brother, Louis-Joseph Gaultier de La Vérendrye (note 570).

969 Another reference to contemporary current affairs - the American Revolution.

❁ Topographical ❁ Observations

The different postings in which M. Pouchot, captain in the Béarn Regiment, served & the journeys he was obliged to make along the various communication routes from the French to the English possessions in America made him decide to provide the most precise details of them within his power.[970]

One will find here regions unknown even to the English in the country of the five Iroquois nations and the immense effort [7] required of them in order to reach the French territories.

M. Pouchot did not bother to detail, on the map he drafted, the region inhabited by the French in Canada, which can be found on other individual maps. He did not detail either the region inhabited by the English, which can be found on the maps of Mitchel[971] & above all on that of Evans,[972] which is the best.[973] [8]

---

970 Pouchot has recounted his "postings" and journeys in the preceding two volumes. His duties during the war permitted him to observe first-hand three of the five major routes discussed below. He served on Lake Champlain (route 1) as far south as Ticonderoga and northern Lake George in 1756 and 1758 and returned to Montréal by this route as an exchanged prisoner in 1759. He travelled the St. Lawrence River to or from Lake Ontario (route 2) on several occasions (1755, 1756, 1757, 1759, and 1760) and sailed down the east end of Lake Ontario to Oswego with Montcalm's army in 1756 and again as a prisoner following the surrender of Fort Lévis in 1760. Pouchot saw the country from Oswego to the "English possessions" (route 3) on two occasions as a prisoner of war in 1759 and 1760. Information on the fourth and fifth major routes, aside from his personal observation of Niagara Falls and the portage, was no doubt gathered from French and Indian informants during his time at Niagara.

971 John Mitchell (d.1768), English botanist, physician and cartographer who published "A Map of the British and French Dominions in North America" (London, 1755 with later editions in 1757 and 1782). A French edition was published in 1756. Mitchell also produced *The Contest in America Between Great Britain and France by an Impartial Hand* (London, 1757). *DNB,* XIII, p. 517.

972 Lewis Evans (c.1700-1756), Pennsylvania surveyor and geographer who produced several works including "A Map of Pennsylvania, New-Jersey, New-York, and the Three Delaware Counties" (1749) and "A General Map of the Middle British Colonies in America" (1756). *DAB,* III, pp. 206-07; Ronald Vere Tooley, *Tooley's Dictionary of Mapmakers* (Tring, England, 1979), p. 197. See also Evans' *Geographical, Historical, Political, Philosophical and Mechanical Essays. The First, Containing an Analysis of a General Map of the Middle British Colonies in America, and of the Country of the Confederate Indians* (Philadelphia, 1755).

M. Pouchot simply traced the course of the principal rivers which act as communication routes to the frontiers, together with the principal trading posts. [9]

Canada, although its surface area is very considerable, has very little communication with the English colonies, either because of the distances between the populated areas or because of the mountainous regions that separate them.

After very thorough research, M. Pouchot identified only five major communication routes, which he will describe individually:

1st. From the frontier of Canada, via Lake Champlain.

2nd. From the River St. Lawrence, from Mont-Réal to Chouegen.

3rd. From the Chouegen River to the English possessions.

4th. From Lake Ontario to the English frontiers, via the Casconchiagon River.[974]

---

973 Note in original, p. 8 (a): *The least effective map of Canada which originally appeared is without any doubt that of M. Delisle. After him, Pople, an English geographer, published a map in 20 sheets which included the whole of North America. M. Bellin listed all the errors on it. See his remarks in* Histoire de la Nouvelle France, *Vol. 5. Although he had at his disposal the admiralty charts, he himself is not free from error. He has been reproached for having relied too much on dubious observations and old maps & for his inability to profit from the discoveries of foreign navigators. The researches of M. Danville were much more precise. Although his maps of North America are not perfect, they nonetheless deserve a great deal of praise. He gave an account of his work in a letter addressed to M. Folkes & published in the* Mercure *of March 1751, p. 150. The merits of M. Green's map of America, published in 1753, cannot be omitted from the discussion. His compatriot, M. Mitchel, did little more than copy his predecessors in the eight sheets published in 1755. Before him, M. Evans had begun his excellent maps with that of Pennsylvania & New Jersey, which appeared in the year 1749.*

Guillaume Delisle (1675-1726) was appointed senior geographer to the King in 1718. *Biographie universelle,* X, pp. 333-35. See also Tooley, p. 395. Delisle's map was entitled "Carte du Canada ou de la Nouvelle France et des Decouvertes qui y ont été faites", published in Paris in 1703.

Henry Popple (d.1743) was a geographer, clerk to the Board of Trade and cashier to Queen Anne (1713). Popple's "Map of the British Empire in America" was published in 20 sheets with an index map in 1733. Tooley, p. 515.

Jacques-Nicolas Bellin (1703-1772) served as French royal hydrographer. *Biographie universelle,* III, pp. 582-83; Tooley, p. 49. His own "Carte de l'Amérique septentrionale depuis le 28 degré de Latitude jusqu' au 72 ... Avec une description geographique de cette partie de l'Amérique" was printed in Paris in 1755. It included his *Remarques....* on the map. The citation for Bellin's critique of Popple is to Pierre-François Xavier de Charlevoix, *Histoire et description générale de la Nouvelle France, avec le Journal historique d'un voyage fait par ordre du roi dans l'Amérique septentrionale* (6 vols.; Paris: 1744).

Jean-Baptiste Bourguignon d'Anville (1697-1784) was a royal geographer and cartographer who published his own map of North America in 1746. *Biographie universelle,* II, pp. 97-98; Tooley, p. 18. D'Anville's letter was directed to Martin Folkes (1690-1754), English antiquarian and scientist. Folkes was president of the Royal Society from 1741 to 1752 and had been elected a member of the French Academy in 1742. *DNB,* VII, pp. 361-62.

...(continued next page)...

5th. From Niagara to the Ohio, & from the Ohio to Pennsylvania & Virginia.

Before going into all **[10]** these details, we must first speak of the River St. Lawrence, from its mouth on the Gulf of the same name as far as Québec. The discussion will be brief, because that part of the river is more the concern of the navy than of land operations.

At the entrance of the River St. Lawrence, three leagues to the S. of Cap des Roziers,[975] is the Baye de Gaspé.[976] At its mouth, it is two leagues wide & is discerned from far off because of the white blocks cut into cliffs which are between its entrance & Cap des Roziers. It is possible to approach the shore without apprehension from the entrance to the N. where a little rock called "Fourillon" is located, which from far off resembles a sailing boat.[977] The tide carries ships away from the coast here.

To the S. where the cape extends farther into the **[11]** sea & is lower lying, there are rocks beneath the surface about half a league from the mouth of the bay which are dangerous when one is unaware of them.

"John Green" was the pseudonym of Braddock Mead (d.1757). In 1753 he produced a map of North and South America under the title *Remarks in Support of the New Chart of North and South America, in Six Sheets.* It was published in London with a second edition in 1768. Tooley, p. 262.

Mitchell's work was "A Map of the British and French Dominions in North America" (London, 1755).

Evans' map of 1749 was entitled "A Map of Pennsylvania, New-Jersey, New-York, and the Three Delaware Counties."

974 The Genesee River flowing north into Lake Ontario at Rochester, New York.

975 Cap des Rosiers near the tip of the Gaspé Peninsula. The use of capitalized abbreviations for cardinal compass points in the 1781 edition has been retained.

A useful source for identifying locations in the lower St. Lawrence River at this period is "A New Chart of the River St. Lawrence from the Island of Anticostia to the Falls of the Richelieu ... Taken by Order of Charles Saunders Esq ... London, Robert Sayer 16 February 1775", William L. Clements Library, Ann Arbor, Michigan. This has been used in conjunction with modern Canadian topographical maps and Harris, *Historical Atlas of Canada.* For locations above St. Barnabé see also the "Extract from a Sketch of the River St. Lawrence" by John Montresor printed in Knox, I, p. 360 and "River St. Lawrence from Montreal to the Island of St Barnaby ...." in Mante, p. 333.

976 The Baie de Gaspé at the tip of the Gaspé Peninsula. The Dartmouth River flows into the head of the bay.

977 Fourillon lies off Cap de Gaspé at the mouth of the bay of the same name. The 1775 Sayer chart notes the "White Rocks" off "Cap Fourillon."

Ships always anchor to the N. three quarters of a league into the bay, in 55 fathoms of water, at the range of a musket shot from land. The anchor does not hold very well because of the sloping sea bottom.

Four leagues farther in, at a place called "Penouille",[978] the anchorage is very good and the largest ships can anchor there in 12 fathoms from the sea bottom. There is a flat space there, which would be very suitable for building a good fortification.

The exit from this bay is rather difficult, since the ships have to wait for quite fresh winds off the land, without which departure is dangerous because of the currents which cause drifting **[12]** towards the rocks mentioned above.

Once one is out of the bay & heading in the direction of the St. Lawrence, it is possible to follow the line of the coast without fear at the distance of a carbine shot offshore in order to round the cape. Although it is said that there are reefs in those parts, none have been sighted, despite the fact that some ships have spent a whole day tacking there quite close to land.

Vessels which spend the winter in the bay have great difficulty continuing their voyage up the St. Lawrence to arrive before those that embark early in the season from Europe, because the N.E. winds which are prevalent in spring blow contrary to ships attempting to leave the bay & heap up the ice floes which arrive in the river.

Navigation of the St. Lawrence River, **[13]** although a wonderful experience, is nonetheless difficult when the winds are blowing from the N.E. (the most favorable for sailing up the river). The fog is very thick.

There are almost no anchorages on the south shore, which is the safer of the two. In order to anchor, one has to reach St. Barnabé or Bic.[979]

It is very probable that on the N. shore there are a number of good anchorages & fine harbors, but that shore is known very little and only random & very inaccurate soundings have been taken. One could cite, for example, the vortex that we feared as though it were Charybdis. The English anchored there; at the bar which we considered virtually impassable they found a channel 900 *toises* wide.[980] **[14]** Ships of 100 guns have negotiated the channel to the N. of the Isle d'Orléans as have the largest frigates & huge merchant vessels have sailed up as far as the rapids beneath Mont-Réal. All the above is enough to indicate

978 Penouille is on the north shore of the Baie de Gaspé.

979 Isle St. Barnabé lies off Rimouski, Québec at the mouth of the Rimouski River. Isle du Bic is at Bic, Québec at the mouth of the Bic River. Both locations are on the south shore of the St. Lawrence River.

the extent to which the banks of the river are poorly charted, since mistakes occur at places frequently navigated.

M. Pouchot noted that the Isle aux Coudres is a good defensive position for preventing ships from proceeding up river if batteries were placed on the Eboulements overlooking the narrowest part of the channel.[981]

If batteries of heavy cannon were also constructed at Cap Tourmente,[982] the site would be the only one capable of stopping the passage of ships. The current forces them to sail very close inshore. Nevertheless, because of that current, **[15]** they could not pause to bombard the site whether the tide was rising or falling. They would be heavily damaged as they passed. By its situation, this position can be rendered formidable since it is located on a steep rock very difficult of access all around.

The position of Québec is a very good one; it resembles that of Namur and is even better than it. But the fortifications were badly designed in relation to its position. The position has no natural disadvantages.[983]

---

980 The St. Lawrence proved to be a much less formidable obstacle than expected when the English fleet sailed up to Québec in 1759. Captain John Knox recorded his transport's passage of Pouchot's "Charybdis" on June 25: "At four P.M. we passed the Traverse, which is reputed a place of the greatest difficulty and danger, between the mouth of the St. Lawrence and Quebec: it lies between Cape Tourmente (a remarkably high, black-looking promontory) and the east end of [Isle d'] Orleans on the starboard side, and isle de Madame on the larboard .... the [French] Pilot declared we should be lost, for that no French ship ever presumed to pass there without a Pilot; 'aye, aye, my dear' (replied our son of Neptune) 'but d—— me I'll convince you, that an Englishman shall go where a Frenchman dare not shew his nose." Knox, I, pp. 373-74.

981 "Les Eboulemens" is shown as a place name opposite Isle aux Coudres on the maps of the St. Lawrence in Mante, p. 333 and Knox, I, p. 360. There is a modern town of that name. A January 1759 memoir on the defense of Québec recommended constructing batteries at "the Eboulement" ("landslip") at Pointe aux Bouleaux and near it at La Prairie on Isle aux Coudres. These would have been able to fire at any ships attempting to pass the Traverse. *DRCHSNY*, X, p. 928.

982 Cap Tourmente is on the north shore opposite the northeast end of Isle d'Orleans. Construction of batteries there to impede vessels approaching Québec had been considered since at least the fall of 1757. Hamilton, *Adventure*, p. 186.

983 Namur was a fortress town in modern Belgium.

Pouchot was only one of many critics of the defenses of Québec. See, for instance, Bougainville's 1757 opinion that "There are no other ways of defending it except by preventing the enemy from approaching it. The fortifications are so ridiculous and so bad that they can be taken as soon as they are besieged." Hamilton, *Adventure*, p. 186.

The French homesteads start at Cap Mouraska and continue right up to the city of Québec. Each village is three leagues distant from the next, measured from the middle of each village. No village in Canada is fortified as there is a space of two acres between each house.[984] [16]

That is all that will be said about the interior of the country. All that follows will deal with the lesser-known regions & the frontiers.

• Chapter I •
Concerning the Lake Champlain Frontier

There are two communication routes from the homesteads of Canada to Lake Champlain; the first is along the Sorel River,[985] which connects the lake to the St. Lawrence; the second runs across the St. Lawrence to La Prairie and then follows the land route from La Prairie to Chambly or St. Jean. The St. Lawrence is three leagues wide at Mont-Réal and La Prairie is located across the river opposite [17] the city. St. Jean is three leagues distant from La Prairie.[986]

Three quarters of a league from the village, the traveller crosses a little river with banks nearly 20 feet high[987] & after the crossing he enters an area of boggy meadows, called "Savannes", which are at that point a league in width. In this area one advances knee-deep in water & mud, but underneath the bottom is firm. The remainder of the way to St. Jean is through woods full of streams which render it almost impassable unless the track is kept in good repair.

This *savanne* could be drained quite cheaply if a ditch were dug to permit the water to drain away into the Sainte Therese & Sorel rivers. By the same operation the path through the woods would cease to be waterlogged.[988]

---

984 By "Cap Mouraska" Pouchot means Kamouraska, Québec on the south shore of the St. Lawrence. An excellent sense of the organization of the St. Lawrence settlements may be obtained from the highly detailed manuscript maps of the so-called "Murray Atlas", prepared by British engineers between 1761 and 1765. One copy is in the William L. Clements Library, Ann Arbor, Michigan.

985 Pouchot means the Richelieu River flowing north from Lake Champlain to join the St. Lawrence at Sorel, Québec. Earlier in his *Mémoires* (II, p. 171), he also calls it the St. Jean River.

986 The distances given by Pouchot are often faulty, and he usually overestimates. La Prairie de la Madeleine is about four and one-half miles above and opposite Montréal while St. Jean is about 15 miles from La Prairie.

987 Rivière Saint-Jacques which drains into the St. Lawrence below La Prairie.

There are thirty leagues from Mont-Réal **[18]** to St. Jean by the water route. One goes downriver along the St. Lawrence for 15 leagues before arriving at Sorel, where the mouth of the river of the same name is. It is as wide as the Saone in the area of Lyon[989] & deeper when the water is high in the spring. The current is fast-running. Boats coming from the St. Lawrence continue as far as the Chambly basin. The general direction of its course is N.N.W. & its meanders are 1/4 or 1/2 of a league in length. There is a considerable bend 4 leagues from Chambly, where it runs rather more to the N.E. The current is fastest at this great curve.

Above Chambly there are rapids for a distance of two leagues and, especially in the approaches to the village of Sainte Therese, the river is very flat and filled with boulders. At that place, one has to be **[19]** well guided if the boat is not to founder. Above Sainte Therese and as far as the area of St. Jean, the boats are emptied and then propelled by poles. The portage from St. Jean to Chambly is three leagues long.[990] After St. Jean and as far as the lake, the river is much wider, very smooth & deep enough to take quite large vessels.

The land around the river is very fertile and would be fertile above St. Jean too if it was cultivated. It is low-lying country, with many streams, which is flooded by the main watercourse. Part of it is waterlogged almost year round. Because of the streams, the river there is slow to freeze over & the ice is always dangerous.

---

988 The Sainte Thérèse is today Rivière L'Acadie which flows into the Richelieu below Chambly, Québec.

Bougainville placed the boggy meadows, which he recorded as "La Sahvane", three leagues from St. Jean. Pouchot had directed work parties on the road across the *savanne* in late August and September 1756 (note 324). Bougainville, in referring to further work on this strategic highway in October, described it as "abominable" and noted that "Until they give it [the water] a place to drain off, the road will always be impracticable. "Hamilton, *Adventure,* p. 65-66. The problem had been chronic. When Peter Kalm used the track in 1749 he found it "unrivaled in wretchedness, wet and winding so that my horse sank in the mire up to his belly in most places." Peter Kalm, *Peter Kalm's Travels in North America; The English Version of 1770.* Edited by Adolph B. Benson (New York, 1987), p. 555.

989 Lyon is at the confluence of the Saône and Rhône in southeastern France about 60 miles from Pouchot's birthplace of Grenoble.

990 The portage began at Chambly and continued through Sainte Thérèse, today only a site on the river between Chambly and St. Jean, Québec and near Isle Ste. Thérèse.

General Amherst described the portage in 1760: "Went 2 leagues to St Therese, the Carrying place from Chambly. The boats are hauled up empty about a mile and when there is much water in the river they go down loaded. From St. Therese I went 2 leagues further to St Johns in whaleboats." Webster, *Jeffery Amherst,* p. 259.

Five leagues above St. Jean is Isle aux Noix, which the French retrenched in 1759. The river around this [20] island is the same width as the range of a musket. The land & forest all around are waterlogged to the depth of two feet when the river level is low. We blocked off the river with a stockade defended by the retrenchments of the island. It is the only post capable of covering the colony once the enemy is master of St. Fréderic. There is no way round the position and the enemy cannot bring up his artillery by land. During his journey from Carillon in 1758, M. Pouchot had pointed out the position to M. le Chevalier de Lévis. It can hold two to three thousand men in an emergency.[991]

A league above this, there are a number of small islands covered with reeds but barges can pass with great ease. La Prairie à Boyleau & Pointe au Moulin Foucaut are the only dry spots [21] for comfortable campsites.[992]

---

991 Isle aux Noix, identified as "Isle au Noys" on Pouchot's map, is just below Lacolle, Québec. Pouchot's claim to have pointed out the position to Lévis is discussed in note 485. For a discussion of the post, consult André Charbonneau, *The Fortifications of Île aux Noix* (Ottawa, 1994).

Isle aux Noix was the site of a series of eighteenth and nineteenth century fortifications and is today dominated by Fort Lennox, constructed after the War of 1812. It is maintained as a historic site by Parks Canada.

992 A useful map for determining locations of the places named by Pouchot along Lake Champlain is William Brasier's "A Survey of Lake Champlain, including Lake George, Crown Point and St. John", drawn in 1762 and published in 1776. An original is in the Newberry Library in Chicago, but the map has been reproduced in Kenneth Nebenzahl and Don Higginbotham, *Atlas of the American Revolution* (New York, 1974), pp. 62-63. It will be cited as Brasier, 1762. Another is "A Topographical Map of the Hudson's River ... from Sandy Hook, New York and Bay to Fort Edward, also the Communication to Canada by Lake George and Lake Champlain .... by Claude Joseph Sauthier" (London, Oct. 1, 1776). It will be referred to as Sauthier, 1776. Current names of locations on Lake Champlain have been derived from modern navigational charts, such as *The Northwest Passage Cruise 'n Chart* (Albany, 1974), and from the *New York State Atlas & Gazetteer* (Third Edition; Freeport, ME, 1991).

The small islands are Hospital and Ash, both so-called on Brasier, 1762. They lie between Isle aux Noix and the Québec-New York boundary. The former got its name because the English established a hospital there during the 1760 siege of Isle aux Noix. Webster, *Jeffery Amherst*, p. 260.

La Prairie à Boyleau was four leagues above St. Jean. Bougainville, who called the spot "Boileau's Camp", identified it as a place where there had been "an establishment belonging to this Boileau" who, along with most French settlers on Lake Champlain, had abandoned it at the outbreak of the war. Hamilton, *Adventure,* p. 125. The place would have been in the vicinity of Isle aux Noix.

...(continued next page)...

When entering Lake Champlain by boat, one sails by way of the right bank as far as Pointe au Fer, from there to Pointe Skenoncton, and then one crosses to the Valcourt islands. It is possible to go by way of the west bank but there is such a large bend that more than two leagues are lost.[993]

Just above the Valcourt islands, on that side, is the Riviere au Sable.[994] It is possible to land at any point of the first Valcourt island and there are several good, well-sheltered harbors on the second island facing the main river.[995] More than 200 bateaux could be docked in these various harbors. The remaining shores of this island consist of steep rocks.

Four leagues farther on is a sort of rocky cape and a little bay that could shelter two or [22] three bateaux in the event of bad weather.[996]

Opposite is Isle au Chapon.[997] There is a little hook of land facing the cape mentioned above where bateaux can easily land & find shelter. The rest of the shore of this island consists of boulders. However, one can put into land when sheltered from the wind.

The cape which I have just mentioned is the farthest point reached by the mountains.

From Isle au Chapon, the boat proceeds to the headland of the Isles des Quatre Vents &, if the weather is fine, sails by way of these islands.[998] In 1759, General Amherst, who wanted to attack Isle aux

---

Pointe au Moulin Foucaut is "Windmill Pt." on Brasier, 1762, today Windmill Point, Vermont opposite Rouses Point, New York. Bougainville spelled the miller's name as Foucault. Hamilton, *Adventure,* p. 125. The mill was a stone structure with nearby habitations that had been abandoned during King George's War. The settlers had not returned by 1749, and their houses had been burned by the Indians, but the mill was intact. Kalm, p. 395.

993 Pointe au Fer is three miles south of Rouses Point, New York. Pointe Skenoncton, labeled "Pt. Aquinonton" on Brasier, 1762, is Cumberland Head which forms the headland of Cumberland Bay at Plattsburgh, New York.

Rather than follow the shore of Cumberland Bay, boats would cross from Cumberland Head to Crab Island (Isle St. Michael on Brasier, 1762) and Valcour Island.

994 Au Sable River.

995 The first is Crab Island in Cumberland Bay, and the second is Valcour Island.

996 Franklin Hough in Pouchot, *Memoirs,* II, p. 65n., identified this bay as being near Port Kent, New York. The cape is probably Trembleau Point where Trembleau Mountain meets the lake.

997 Schuyler Island, New York.

Noix, came with a detachment of 5 to 6 thousand men and camped on this headland. He was trapped for five or six days there by a N.W. wind [23] and lost a dozen bateaux. The winter season, which had just begun, compelled him to go back.[999]

It is possible to camp beside a river at the far end of the bay, near Le Rocher Fendu, to the south. The source of this river is in the area of Lac St. Sacrement. Enemy raiding parties sometimes chose this route to reach Lake Champlain.[1000]

After Le Rocher Fendu, the lake resembles a river & the contour of the mountains forms a handsome basin as far as St. Fréderic. The W. bank is very mountainous; reference to the map will provide a better idea of the configuration of the land than descriptions. On the E. bank, there are several bays good for campsites.[1001]

One rarely follows the E. shore of the lake, whether going up or going [24] down it. It is even important when sailing down it to keep to the left, or W. shore, otherwise one could run the risk of getting lost in the Baye de Missiskouit.[1002] The banks of Lake Champlain are large uninhabited plains. The earth is very fertile and the trees are splendid & suitable for the construction of ships.

Before the war, the area around St. Fréderic was inhabited. This fort, built on a peninsula, as can be seen from the map, was a redoubt of masonry to which a stone enceinte[1003] without an earth embank-

---

998 The headland is "Pt. Sabousin" on Brasier, 1762, today Willsboro Point. "Isle au 4 Vents" on Pouchot's map - literally "Four Winds Islands" on Brasier, 1762 - are today known as "The Four Brothers."

999 See note 535.

1000 Pouchot refers to the Bouquet River which enters Lake Champlain at Willsboro, New York. "Le Rocher Fendu", Split Rock, is at the tip of Split Rock Point, five miles farther south.

The Bouquet River has its source far from Lake George and was therefore not a practical route for enemy raiding parties, except possibly the Iroquois. The origin of Pouchot's statement is almost certainly an October 1756 incident when a French party discovered four English boats abandoned on Lake Champlain north of Fort St. Frédéric. Information provided by the Indians originally led the French to speculate that these had come down the Bouquet River. The boats were, in fact, from a party led by Robert Rogers. They had been portaged into Lake Champlain over the mountains east of Lake George and had slipped past both Carillon and St. Frédéric. Hamilton, *Adventure,* pp. 46-47, 50-51.

1001 Kingsland, Porter, Field, and Button bays, all on the Vermont shore.

1002 Missisquoi Bay straddles the Vermont-Québec boundary at the northeastern end of Lake Champlain.

ment had been added. The wall of the enclosure was two feet thick at the most. M. de Bourlamaque had the fort blown up in 1759 in the course of his retreat from Carillon to Isle aux Noix.[1004]

The English have constructed a [25] substantial fort there on a site where there was originally a windmill. It is a pentagon of 80 to 100 *toises* on each exterior side, entirely built of wood. The timbers of the external revetment are three feet square in section, lashed together with ropes & the space between is filled with beaten earth & there is a good ditch.[1005]

On small rocks in the area they have built redoubts, or blockhouses, according to the system of the Maréchal de Saxe.[1006] These rocks form

---

1003 "ENCEINTE ... is the interior wall or rampart which surrounds a place, sometimes composed of bastions or curtains, either faced or lined with brick or stone, or made only of earth." Smith, *Dictionary,* p. 81.

1004 Fort St. Frédéric stood near the tip of the Crown Point peninsula on the New York side of the lake. The stone redoubt and fort were constructed by Gaspard-Joseph Chaussegros de Léry (1682-1756; *DCB,* III, pp. 116-19) in 1734-37. Its ruins are preserved today in Crown Point State Historic Site. The fort was blown up by the French on July 31, 1759 and occupied by Amherst's army on August 4. See note 534.

1005 Although Pouchot personally observed the new works at Crown Point in December 1759, he was mistaken about the location of the English fort. It was placed on an elevation a short distance southwest of the ruined French post and not on the site of a defensible stone windmill constructed during the French occupation and also destroyed in 1759. The English built one of three advanced redoubts on the windmill site east of the old French fort on the edge of Lake Champlain. The position is today marked by the Champlain monument lighthouse. The massive ruins of Amherst's fort at Crown Point are preserved in Crown Point State Historic Site.

On August 5, 1759, General Amherst ordered the fort laid out by Lieutenant Colonel William Eyre (note 852). Strategically, Amherst noted, "this Post secures entirely all His Majesty's Dominions that are behind it from the Inroads of the Enemy and the scalping party's [sic] that have infested the whole Country, and it will give great peace and quiet to the King's Subjects who will now settle in their habitations from this to New York." Amherst Journal, Knox, III, p. 48. From a tactical perspective, he was equally pleased with the site and considered it "the first situation I think I have seen in America that is no where commanded." Webster, *Jeffery Amherst,* p. 154. Although Amherst had hoped for a design that could be completed expeditiously, Eyre's creation was immense. In September, with winter looming, the general feared that his engineer had "overdone it." *Ibid.,* p. 169. Work would continue at Crown Point until 1762, but the place would never be completed as planned.

A number of plans exist for this post. There is a printed "Plan of the New Fort & Redoubts at Crown Point" in Rocque. A design of the entire defensive system is William Brasier's "Plan of the New Fort at Crown Point on Lake Champlain, and Environs; with its several Redoubts began in August 1759." Versions of this are in the collections of the British Library (Crown Maps, cxxi 121.51), the Library of Congress and Crown Point State Historic Site.

a kind of enceinte around the fort. The highest of them might be 30 feet tall. The outer slopes facing the open country make up a glacis but with very little earth.

The space between the bay and the river is enclosed by blockhouses 100 *toises* apart with a wooden retrenchment between every two of them.[1007] This fort is far more advantageously situated for [26] the English than for the French colony because of the difficulty of landing troops to attack it.

Two leagues above St. Fréderic & on the same side is the Riviere à la Barbue.[1008] It has a sandy bottom. Its depth is 4 feet at low water and at least 7 in spring. The river bed is covered with reeds & encumbered with very thick water willows. Only one very narrow channel had been identified. The width of the river exceeds the range of a musket. On the St. Fréderic side the bank is more than 40 feet high. At the surface, the earth is level and the trees are free of undergrowth & well spaced.

By travelling for a league up this river on the same side, one comes across a very high & very steep mountain.[1009] The river flows down through these mountains, [27] which an army would have great difficulty in crossing. The river is 7 to 8 leagues long.

On the opposite bank is a tongue of land which marks the end of the mountain range. It would not be possible for a body of men of any size to camp there.[1010]

1006 Pouchot's original text uses the phonetic term *blachouses.* The structures described here, however, were open log, earth and stone redoubts placed to cover the land approaches to the new fort. They were ordered begun on August 24 and were named, from west to east, "Gage's Redoubt", the "Light Infantry Redoubt" and the "Grenadier Redoubt" for the units that constructed them. Amherst Journal, Knox, III, p. 55. They are shown on the plan printed in Rocque.

Maurice, Comte de Saxe (1696-1750) was a German-born professional soldier in the French service and the most successful of the generals of Louis XV. *Biographie universelle,* XXXVIII, pp. 156-61. Saxe's *Les Rêveries ou mémoires sur l'art de la guerre de Maurice, Comte de Saxe,* written in 1732, was published posthumously in the Hague in 1756. Saxe preferred the retrenchment of strategic places in open country to the use of towns as fortresses. Duffy, p. 154.

1007 A line of wooden blockhouses was sited across the base of the Crown Point Peninsula between Bulwagga Bay on the west and Lake Champlain ("the river") on the east.

1008 Putnam Creek at the village of Crown Point, New York.

1009 Putnam Creek rises in the hills northwest of Ticonderoga. Its south branch has its source near Buck Mountain in the Town of Ticonderoga, and that is probably the height described.

The Carillon River at this point is no wider than the range of a musket.[1011] It is the narrowest point of the river between Carillon and St. Fréderic. This post cannot be bypassed & covers St. Fréderic. Behind it there is a land communication route linking it with the fort. There is a bight, or bay, which starts at St. Fréderic and ends there. Bateaux can travel round the bay between the two without being seen.[1012]

Batteries placed in the angle of the river mouth described above can fire the length of the Carillon River as far as the peninsula.[1013] Above it is the best position that exists [28] along the whole course of the river for preventing entry into Lake Champlain. Quite large vessels can sail up as far as Carillon & bateaux right up beneath the falls.[1014]

The English have constructed a roadway from the falls to St. Fréderic. Cannon could be transported along it. They have also built a road from Carillon to the fort called "No. 4" in Connecticut. It is about 35 leagues in length and they travel along it in 4 days. The militiamen who returned to New England from their camp beneath

---

1010 A series of elevations - Miller Mountain, Breeds Hill and Sugar Hill - run generally north and south along the west shore of Lake Champlain terminating at the south side of the mouth of Putnam Creek.

1011 The narrow southern end of Lake Champlain as far north as Crown Point was often referred to as the Carillon River or, by the English, Wood Creek and even South River. Bougainville calls the part south of Ticonderoga the Chicot River.

When Pouchot speaks of the length of a musket shot, he presumably does not mean accurate range. The narrowest part of the "Carillon River" between Putnam Creek and West Bridport, Vermont is still about 400 yards.

1012 Pouchot was particularly interested in this place, having reconnoitered it in 1758 as a possible fall-back position in the event of a retreat from Ticonderoga (note 485). The contour of the west bank of Lake Champlain, covered by the narrows at Putnam Creek, offered shelter to boats moving between there and Fort St. Frédéric at Crown Point.

When Bourlamaque retired northward in the face of Amherst's advance, late in July 1759, he believed his army was too small to successfully hold the extensive position at Rivière à la Barbue and used it only to cover his further retreat from Fort St. Frédéric to Isle aux Noix. Gabriel, pp. 284-85.

1013 That is, from the mouth of Rivière à la Barbue (Putnam Creek) as far as "Presquile" on Brasier, 1762 - modern Fivemile Point on the Vermont shore.

1014 La Chute Falls on the Lachute River near Ticonderoga.

1015 Fort Number Four, located at Charlestown, New Hampshire, was established in 1743. Robert B. Roberts, *Encyclopedia of Historic Forts: The Military, Pioneer, and Trading Posts of the United States* (New York, 1988), pp. 500-02.

...(continued next page)...

St. Fréderic in 1759 took this road & during the summer they drove cattle from those provinces along the road for the use of the army.[1015]

The fort of Carillon[1016] [29] is a 45-*toise* square measured along its external sides and is constructed of 14 to 15-inch square timbers piled one on top of the other. The parapets are 12 feet thick and banked up with earth & pebbles excavated from the mountain. A demi-lune on the facade looking towards the mountain ridge, a ditch 5 to 6 *toises* wide with its covered way & a glacis at the back of the fort at the far end of the rising ground are its external fortifications. There is also a redoubt which commands the water.[1017]

Both roads are shown on Brasier, 1762 and Sauthier, 1776. That to Number Four is also represented by a dotted line terminating in a fort symbol on the Connecticut River on Pouchot's map. Both roads were begun soon after Amherst's army occupied Crown Point on August 4, 1759. On the 5th the general ordered 100 of his rangers to cut a road to Ticonderoga "that cattle may be drove from thence." Construction of the route to Fort Number Four began three days later when Amherst sent out another party of 200 rangers "to cut a Road to open a communication from New England & New Hampshire to Crown Point." The track led southeast from Crown Point to meet Otter Creek in Vermont at a point due east of Ticonderoga. It thereafter followed the stream. Webster, *Jeffery Amherst,* pp. 152, 153.

Amherst had suggested sending his New England provincials home by way of Number Four as early as September 1759. Their officers agreed, and most units marched in the second week of November after turning in the "Tents, Arms, & Cartridge Boxes" received from the King's stores. "As to their bayonets," observed the general, "they have lost most of them." *Ibid.,* pp. 172, 185, 190-91.

1016 Note in original, p. 29 (a): *In English "Ticonderoga" and in the language of the Indians "Teaonteaogen."*

The fort of Carillon was begun by the French in the autumn of 1755 to control the point where the waters of Lake George join Lake Champlain. Lotbiniere to Argenson, Oct. 24, 1755, *DRCHSNY,* X, pp. 367-68. A 1758 appraisal of the fortifications by Captain Nicolas Sarrebource de Pontleroy is in *Ibid.,* p. 720. Carillon was abandoned to the English on July 26, 1759 after having been set afire by its retreating garrison. Amherst ordered it reconstructed on the French plan to "save great expense & and [sic] give no room for the Engineers to exercise their genius which will be much better employed at Crown Point." Webster, *Jeffery Amherst,* pp. 145-47, 148.

Pouchot's description gives the length of each exterior side of the square. Pontleroy says the square was 54 *toises* on a side, possibly indicating a misprint here. Modern surveys make Fort St. Frédéric at Crown Point 45 *toises* square, so it is also possible that Pouchot confused the two posts. A contemporary printed "Plan of the Fort at Tienderoga at the Head of Lake Champlain, 1759" in Rocque.

A concise account of the post is S.H.P. Pell, *Fort Ticonderoga: A Short History* (Ticonderoga, 1966). The restored and reconstructed features of the fort are preserved today as part of the Fort Ticonderoga Museum in Ticonderoga, New York.

1017 A number of redoubts had been constructed to the north and east of Fort Carillon to improve its command of the water or to strengthen its outer defenses. Pouchot probably

...(continued next page)...

This fort is built on a rock which is very steep on almost every side. The side most likely to be assaulted is dominated by the hill, on which the retrenchment is constructed, set back 400 *toises.*[1018] Around the fort, in the area just described, there is no earth to open up trenches because it has all been used to form the glacis. **[30]**

By occupying the top part of the retrenchment & opening up trenches down by the river, it would be possible to mount a successful attack if one had earthworks for cover; & from Pointe au Diamant, bombardment of the fort with artillery is very feasible. This position defends the bay & the La Chute River as well as Carillon does, but it cannot block the land route to St. Fréderic.[1019]

Opposite Carillon is the mountain named Serpent à Sonnette, whence Carillon can also be bombarded with artillery.[1020] At the entrance to the bay, there is a rather difficult track onto the Serpent à Sonnette, which links up with Lake George or St. Sacrement.

The English have erected a fine sawmill at La Chute & also **[31]** built a blockhouse capable of holding 100 men & 4 pieces of cannon.

---

refers to the structure identified as "F. The Redoubt"" on the printed plan in Rocque and as "G. Lower Battery" on "Plan of the Fort at Tienderoga and Environs at the head of Lake Champlain, November 1759" by Diedrich Brehm and William Brasier. Hulbert, I/3, p. 46 (Crown Maps, cxxi/106).

1018 The field defenses constructed and defended by Montcalm's army in 1758 and improved following the battle of July 8. See note 413.

1019 Pointe au Diamant is now Mount Independence on the Vermont shore of Lake Champlain. The position was fortified by Continental forces during the American Revolution to more effectively block the lake. Boatner, *Encyclopedia,* pp. 1103.

1020 Literally "Rattlesnake" Mountain, known today as Mount Defiance and situated across the water to the southwest of Ticonderoga. In 1777 the British placed artillery near the summit and forced the evacuation of Ticonderoga. Boatner, *Encyclopedia,* pp. 1103-07.

1021 Pouchot had ample time to observe the Ticonderoga portage when he and his detachment of exchanged prisoners were detained at the sawmill in December 1759 (note 751). La Chute Falls are on the Lachute River which drains Lake George into Lake Champlain. The river is also known today as Ticonderoga Creek. The road ran along the south and east side of the river to Lake George. The area as it was in the summer of 1759 is shown on a "Plan of Ticonderoga" printed in *The Universal Magazine* for September 1759. This has most recently been reproduced in *The Bulletin of the Fort Ticonderoga Museum,* XV, No. 5 (1993), p. 376.

1022 Note in original, p. 32 (a): *This is, I believe, "Mount Independence."*

Pouchot's editor was mistaken here. The features described lie on the north or Ticonderoga side of the Lachute between the retrenchment and the bridge across the river.

...(continued next page)...

They have also shortened the portage road by nearly a quarter of a league.

The road is good & fairly straight; it runs along a mountainside and the slope is very gentle. It is bordered on one side by the Serpent à Sonnette & on the other by the river La Chute.[1021]

Before arriving at La Chute from Carillon, one comes across a ravine that runs along almost the entire length of the high ground of the isthmus. It is both very deep & very steep on the Carillon side. On the left there is a hill which covers the track leading to La Chute & on the right the screen of trees [32] overlooks a river & a bay of the Carillon River.[1022]

This is the best position to occupy with an army corps, which would cover Carillon and the entire course of the river & would not be seen from the other side, which is the case with the present system of retrenchments. The English have cut down almost all the trees in that sector, on the portage road, at the foot of the Serpent à Sonnette & on the Cap au Diamant.

When one goes to Lake George[1023] along the right bank of the river above La Chute, one comes across the Riviere de Bernes, which is narrow but deep and so difficult to cross that it could be blocked.[1024] By proceeding along it, one reaches the trail of the Arbres Matachés.[1025]

The road, ravine and hill between the road and the river are shown in a "Map of Ticonderoga, Mount Independence and the adjacent country" published in Philadelphia in 1778. It is reproduced in Nebenzahl and Higginbotham, pp. 96-97. Also shown is another sizeable hill commanding the river crossing. This was named Mount Hope and fortified by the Americans in 1776. Boatner, *Encyclopedia,* p. 1103. It is possible that Pouchot's description is skewed, and Mount Hope is the height referred to.

1023 Lake George was known to the French as Lac St. Sacrement.

1024 One would encounter Trout Brook, referred to in French sources as the Rivière de Bernes or Bernetz, on the right bank when proceeding upstream towards Lake George. The creek was named in 1756 for Le Chevalier de Bernetz, commandant of the Royal Roussillon battalion (note 427). See La Pause's "Mémoire et Observations", *RAPQ,* 1931-32, p. 40.

1025 A trail that ran from Carillon to Ganaouské (Northwest) Bay and then on to the south end of Lake George. It was situated behind the mountains on the west side of the lake. The route was used successfully by Lévis' corps of Montcalm's army in 1757 and is shown beginning at the Bernetz River on a map of Lake George and southern Lake Champlain in Mante, p. 33. Hamilton's translation of Bougainville gives the meaning of *Arbres Matachés* as "war painted trees", so the trail would seem to have taken its name from the way in which it was marked. Hamilton, *Adventure,* p. 15n.

This Arbres Matachés Trail is probably that which Bougainville, in 1758, called "the Mohawk road." *Ibid.,* p. 266.

Lake George is barely more than a league & a half in width and about ten leagues long. It is [33] surrounded by very steep mountains, especially those on the left. In the journey from Carillon to Fort George,[1026] they are almost impassable, even for travellers on foot. It is possible to cross those on the right, although they are very difficult. The detachment under the command of M. le Chevalier de Lévis crossed them in 1757 to go and invest Fort George.

At the entrance to the lake, we had an observation post called the "Camp de Contre-Coeur." It was not well-sited, as it could be outflanked by way of both the Arbres Matachés & the lake. It is the place where the English landed in 1758, at which time it was not occupied.[1027]

The position would have been better if it had been a little farther forward, at the foot of Pelée Mountain. A post on that mountain would have been very useful. It [34] would not have been possible to outflank it on the landward side although there was a risk of being outflanked by way of the lake unless our side had superior vessels on it.[1028]

The northern cape of Ganaouské Bay would be an excellent position for preventing passage along the lake. The camp is very secure & cannot be outflanked. The lake is very narrow at this point. By placing artillery on the two little islands close by, one could cover the whole lake.[1029]

1026 Identified as "F George" on Pouchot's map. The name was used by the French to describe several fortifications, including Fort William Henry (1755-1757), constructed by the English at the south end of Lake George on the site of Lake George Village, New York.

1027 The Camp of Contrecoeur was established at Montcalm's order in July 1756 to guard the left (west) bank of the foot of Lake George. It was named for Captain Claude-Pierre Pécaudy de Contrecoeur (note 162) who was to command there with 300 men. A camp of 500, under Captain Louis La Corne (1703-1761; *DCB,* III, pp. 331-32), was to protect the right bank. Casgrain, *Collection,* VII, p. 73. Although Contrecoeur's camp was withdrawn in October, the place continued to be known by that name and was the site of the English landing in 1758.

The camp was at Howe's Landing, New York on the west side of the lake near the head of the Lachute River and below Cooks Mountain.

1028 Pelée Mountain was Rogers' Rock or Rogers' Slide on Trumbull Mountain near Baldwin (Howe's Landing), New York. The map in Mante, p. 33, shows that the Arbres Matachés Trail passed much closer to this position but was on the opposite side of Cooks Mountain from the Camp de Contrecoeur.

1029 Identified as "Cankusker Bay" or "North Arm" on Brasier, 1762, it is today known as Northwest Bay. The cape or peninsula is formed by the Tongue Mountain Range and terminates in Montcalm Point. Turtle and Big Burnt are the largest of the islands dotting the narrows between Montcalm Point and the east shore of Lake George.

The position of Fort George, which we captured & destroyed in 1757, is on a kind of saddle. The English had retrenched the peaks in order to make a retrenched camp for fear of being outflanked by way of the bay, which has a defensive position called Le Rocher à la Reine.[1030]

In 1759, the English began construction of a square fort measuring about 80 [35] *toises* on each exterior side. The lower part of the rampart is more than 18 feet thick and made of masonry. The parapet is of timbers placed one on top of the other, all well cut to size and banked up with earth twelve feet thick. In December 1759 there was already a bastion completed, entirely casemated like a redoubt.[1031] Doubtless the others are planned to be in the same style. Beneath the fort and to protect the boat landing, there is another, and much smaller, square fort which the English have built since the demolition of the previous one. It is constructed of horizontal timbers. On the upper part there is a fraise, which extends outwards towards its base, & a piece of wood

---

1030 Fort William Henry and the retrenched camp were each placed on high ground separated by a low, swampy area. The camp was located east of Fort William Henry on the spot where Johnson had fought his battle against Dieskau in 1755. It is shown on most plans of the battle, including that in Rocque, "A Plan of Fort William Henry and the English Camps & Retrenchments with the French Camps and the Attack thereupon." There is also a plan in *DRCHSNY,* X, p. 600 and several in Steele, *Betrayals.*

"The bay" is South Bay or Wood Creek, the very narrow part of Lake Champlain leading south from Ticonderoga to modern Whitehall, New York. Rocher à la Reine or "the Queen's Rock" was a distinctive landmark shown on the map in Mante, p. 33 as "Two Rocks" and on Brasier, 1762 as "Narrows between two perpendicular Rocks." The spot is "The Narrows of Dresden" just south of Dresden, New York and Cold Spring, Vermont. Dieskau's route from South Bay to the portage road between Fort Edward and Lake George is shown on the map in Mante, p. 33.

1031 Pouchot observed the new Fort George when he stayed there the night of December 1, 1759. The fort had been laid out by Colonel James Montresor (note 750) on June 22 on the site of Johnson's 1755 victory and the retrenched camp of 1757. Scull, p. 78. Work continued during 1759-1760, but the imminent English victory caused the original plans to be greatly reduced. On September 8, 1759, Amherst ordered Montresor "to shut up the Citadel Bastion by a retrenchment at the Gorge, that it will then be a small Fort of itself which with the one by the Lake side will be sufficient, as that Post is no longer a Frontier, and the Fort may be finished hereafter." Webster, *Jeffery Amherst,* p. 166. The orders were received two days later, and Montresor proceeded to complete the truncated fort for a garrison of 150 men. Scull, p. 96.

Fort George would never be completed as planned. Late in October 1760 General Amherst inspected the works while on his way from Montréal to New York. "The bastion enclosed at Fort George is very neat," he noted, "mounts 15 Guns, is very small and a bad defence, but 'twas the shortest, cheapest & best method of finishing what was begun of the Fort, as completing the whole according to the first plan is not necessary & would have cost a large sum." Webster, *Jeffery Amherst,* p. 262.

...(continued next page)...

providing views over the surrounding area & which covers the top of the parapet as a substitute for battlements. The position of the previous fort, now demolished, is indicated on the map.[1032]

The portage road is very [36] good for all types of vehicles, although the landscape is rather hilly, which rendered it favorable to the ambushes set by the raiding parties which we used to send & who got there by way of the bay.[1033]

Towards the middle of the portage, next to the shed, there is a small fort of upright stakes which acts as a storehouse & protects the convoys; it can hold 100 men.[1034] A league & a half from this fort, the road follows the Hudson, or Orange, River. On the top of the hills a league and a half beyond that, one comes across the fort that we call "Lydius" & the English "Edward."[1035]

The fort was named for King George II on July 28, 1759. Knox, I, p. 510. Fort George was abandoned by British troops on August 16, 1768. List of Officers [Commanding]..., Dec. 25, 1767, to Dec. 24, 1768, Gage Papers/139. American forces reoccupied the position during the War for Independence. A "Plan of Part of Fort George with the Barracks &c. Erected in the Year 1759" printed in Rocque shows the trace of the walls and the single bastion completed by the time of Pouchot's visit. The ruins of the "Citadel Bastion" are preserved in Lake George Battleground State Campground.

1032 The previous fort was William Henry, destroyed by the French in 1757.

On June 28, 1759, Lieutenant Colonel William Eyre (note 852) laid out a "Post of Logs for 100 men" between the new fort and Lake George. Scull, p. 79. This and a smaller redoubt for 20 men were intended to provide security once Amherst's army moved on to Ticonderoga. Webster, *Jeffery Amherst,* p. 130.

The Rocque plan of Fort George shows, in the position described by Pouchot, a "Stockaded Fort erected to serve during the time the other was building." An engraved view of the structure, published in the *Universal Magazine* of September 1759, *The Bulletin of the Fort Ticonderoga Museum,* XV, No. 5 (1993), p. 376 and Bellico, p. 86, also describes it as "A Stockaded Fort." The engraving itself, however, clearly shows six courses of horizontal timbers topped by a vertical picketing or fraising tied together at the top with a ribband, as described by Pouchot.

1033 The portage road connected Lake George with Fort Edward on the Hudson River. "The bay" is, once again, South Bay on Lake Champlain.

1034 There was a post at Halfway Brook, today Halfway Creek on the outskirts of Glens Falls, New York, during the 1758 campaign. The fort described by Pouchot was a new one, however. It was ordered to be laid out by Lieutenant Colonel Eyre on June 13, 1759, and was completed on the 20th. Webster, *Jeffery Amherst,* pp. 118-19, 123. The post is labelled "Fort Amherst" on Sauthier, 1776. It was described as "a rectangular stockade" and mounted three four-pounder cannon. Gary S. Zaboly (ed.), "A Royal Artillery Officer with Amherst: The Journal of Captain-Lieutenant Henry Skinner, 1 May-28 July 1759", *The Bulletin of the Fort Ticonderoga Museum,* XV, No. 5 (1993), p. 367. The place was also known as the "Seven-Mile Post." Knox, I, p. 470.

...(continued next page)...

This fort is a square of 40 to 42 *toises* on each exterior side. One side is on the river bank. The ditch is about 5 *toises* across and quite shallow. The rampart of the fort is partly [37] of earth faced with saucissons, partly of earth faced with timbers piece on piece. The parapet is of cases of wood banked up with earth, with a fraise around the cannon on the flanks & on the points of the bastions. The embrasures were closed off with chevaux de frise.[1036] A little stream flows the length of these fortifications at the lower part of the fort, which is within cannon shot from every direction. Blockhouses have been built to provide cover for it, thus forming a retrenched camp. On the island facing the fort there is a poor system of trenches which is also dominated from all sides. There are a number of barracks.[1037]

The blockhouses mentioned above are wooden redoubts consisting of two parts in the shape of a square (the best examples are those at St.

Pouchot fails to mention two other small forts that stood along the portage road in 1759. Moving south from Lake George one would first have encountered the "Three-Mile Post" or Fort Williams, begun as a temporary retrenchment of felled trees on June 19, 1759, but soon completed as "a stockaded fort." Zaboly, p. 368; Knox, I, p. 473. On the Hudson River side of the Halfway Brook post, about three to four miles from Fort Edward, was the "Four-Mile Post", "a small rectangular stockade with two bastions and a ditch" begun May 29, 1759. Knox, I, p. 465; Zaboly, pp. 366-67.

1035 Pouchot's map identifies the fortification as "F Williams or Lydius." The former name would seem to indicate some confusion with the stockade three miles south of Lake George. That was named for Colonel Ephraim Williams (note 269), killed in the opening action of the Battle of Lake George. See Zaboly, p. 368.

1036 Chevaux de frise are defined as "large joists or beams stuck full of wooden pins armed with iron, to stop breaches, or to secure the passages of a camp, against the enemy's cavalry." Muller, *Treatise*, p. 214.

1037 Fort Edward was an irregular square post on the left bank of the Hudson River at the foot of the portage to Lake George. Construction began on the site of the trading post of John Henry Lydius in 1755 during Johnson's expedition to Lake George. The name of the original owner seems to have stuck in the minds of French observers. The new fortification was initially named for Phineas Lyman but soon renamed for Edward, the Duke of York. The lower or downstream end of the fort was protected by Bond Creek. For a history of the post see William H. Hill, *Old Fort Edward Before 1800* (Fort Edward, 1929). Fort Edward was abandoned on June 15, 1766. List of Officers [Commanding]..., Dec. 25, 1765, to Dec. 24, 1766, Gage Papers/139.

Rocque includes two renderings of Fort Edward and its surroundings: "A Plan of Fort Edward" and "A Plan of the Royal Block House with the Environs of Fort Edward." These also show Rogers' Island situated in the Hudson facing Fort Edward, the site of hospitals and barracks for Rogers Rangers and other troops. Pouchot passed through Fort Edward in the last days of November 1759. An excellent plan of the fort and vicinity, including its outlying blockhouses, is "A Plan of Fort Edward and its environs on Hudsons River" in Mante, p. 34.

Fréderic). They are surrounded [38] by a ditch 15 feet wide. The earth thrown up from digging the ditch has been shaped into a glacis. A sloping stockade has been built on the top of it. The ground level, which is loopholed, acts as a guardhouse. On the upper part, the angles of the square are situated in the middle of the face of the lower square, which gives the whole fort an octagonal shape. There are usually cannon embrasures on the upper part, in addition to the loopholes.[1038]

The country around Fort Edward, although mountainous, would be suitable for cultivation. It is a rather agreeable landscape, situated on uplands.

A league above the fort, the river is not navigable because of its swift flow; this is the point where it emerges from the mountains. It is the range of a musket in width & is also deep.[1039]

A quarter of a league below [39] Fort Edward, one crosses the river on a wooden bridge & the road continues for a league over low, marshy ground. Over the two remaining leagues from there to Fort Müller, one has to skirt the bottom of a hill which has been cut into to make the road. The ground shifts because of the river and it has been necessary to lay logs across the road in order to consolidate it. This is a substantial undertaking.

Fort Müller is a small square capable of holding 200 men and constructed partly of earth, partly of horizontal timbers. It seems to have been recently built and was not yet finished when M. Pouchot passed that way. It is located in low-lying, marshy, very muddy country. At a musket shot's range from the fort, one rejoins the river and it is at that point that the road described above begins.[1040]

1038 A ground plan and elevation of one of these blockhouses is in the Skene Papers of the Fort Ticonderoga Museum. This "Plan of Fort Edward Blockhouse" is reproduced in *The Bulletin of the Fort Ticonderoga Museum,* VII, No. 4 (July 1946), p. 8.

1039 The Hudson River became unnavigable at Hudson Falls, New York.

1040 "F Müll" on Pouchot's map, located on the west bank of the Hudson opposite the modern village of Fort Miller, New York. A post might have been situated there as early as 1755. Roberts, p. 566. Hill's *Old Fort Edward,* pp. 195-202 says the fortification was constructed in 1758 and named for Stephen Miller (1727-1814), then a major of Massachusetts provincials. A "Plan of Fort Miller and the Environs on the West-Side of Hudson's River" and "Plan of Fort Miller with the Barracks" are in Hulbert, I/3, pp. 19-20 (Crown Maps, cxxi 71 & 72). These show a square fort as part of a larger retrenchment built in the manner described by Pouchot.

1041 Today Schuylerville, New York. It is shown as "Sarasto or Fort George" on Pouchot's map.

It is twelve miles from Fort [40] Müller to Saratoga, or Sarasto.[1041] A league & a half before arriving at this latter fort, the river passes between two fairly high hills,[1042] which makes for a good natural outpost. The mountains to the right & left of this valley are very high & the river waters are splendid.

Saratoga is located at the far end of a prairie in a bend of the river. It is built of earth faced with saucissons & can hold a garrison of 300 men.[1043] At the range of a cannon shot from the fort, there is an elevation which dominates it; it is made of sod, rubbish & rocks.

From Saratoga, one continues to follow the river over a kind of prairie [41] or pasturage. Between there and Stil-Warter one comes across two sets of rapids. At the first there are sawmills and they used to build bateaux there. One can get very close to the rapids by water & beneath them one can immediately get into the boat again. The same comment applies to the second rapids.

Stil-Water is a little smaller than Saratoga & constructed in the same way. It is nothing more than a large star-shaped redoubt made of earth and faced with saucissons. It has a ditch 18 feet wide with a fraise. The fort is located in a prairie & dominated at the range of a musket by an elevation which runs all the way around it. Three to 4 thousand men could be stationed behind the elevation, which makes it a very poor position. This is the depot for food and other goods sent

---

1042 Note in original, p. 41 (a): *It was along this gorge that the luckless Bourgoyne expected to make his retreat. But when the Americans appeared at the head of it, he was obliged, on 16 October 1777, to sign the surrender which everyone knows about.*

British Major General John Burgoyne (1722-1792). *DNB,* III, pp. 340-42 and Boatner, *Encyclopedia,* pp. 130-43. The main actions of what became known as the Battle of Saratoga were fought north of Bemis Heights, the "fairly high hills" described by Pouchot. Burgoyne's army attempted to retreat northward, but was forced to lay down its arms at Schuylerville or Saratoga on October 17, 1777. *Ibid.,* pp. 978-80.

Since the hills at Bemis Heights are well below Schuylerville, Pouchot was describing them as if ascending the Hudson River from the south, the direction of his trip in November 1759.

1043 The fort at Saratoga was located on the north bank of the junction of Fishkill Creek (today Fish Creek) and the Hudson. The old blockhouse on the site was pulled down in August 1757, and a new fort was constructed under the direction of Colonel James Montresor. Scull, pp. 41-44.

The post at Saratoga is referred to in Sauthier, 1776 and other sources as Fort Hardy. Rocque includes "A Plan of the Fort at Saratoga" based on a 1757 plan of the same title in Hulbert, I/3, p. 19 (Crown Maps, cxxi 97). Both indicate a construction similar to that described by Pouchot. There is a copy of a sketch entitled "Saratoga Block House" in Scull, p. 40, but this appears to be the post replaced in 1757.

from Albany to be transported to Fort George.[1044] The journey from [42] Albany is by water, in flat bateaux which look like ferry-boats but are propelled by sail.

The river is tidal to this point. Transport from this fort is accomplished by land because of the two rapids & a number of reefs situated in the river. The latter is wide, swift-flowing & deep.

The entrepôt of vehicles carrying goods from Stil-Water to Fort George is a position called Halfmund, or half moon. In 1759, there were 400 carts chartered for the use of the army and paid for at the rate of 12 *livres* per day, plus subsistence for the men & horses, whether they were employed or not.

From Fort Edward to Stil-Water, the valley is narrow and the mountains are very high, especially on the Connecticut side.[1045] The slopes are steep & yet [43] could be cultivated. From Fort Edward, there is a road that crosses into Connecticut & to Boston.

At Stil-Water, the mountains start to diminish in size & cultivation begins. Halfmund is a poor redoubt, at the far end of a stretch of prairie, at the confluence of the Mohack River, which forms a half circle that has given it its name.[1046]

At the far end of the prairie, which might be a quarter of a league in length, one climbs up a fairly steep hill and two miles from there one

---

1044 Noted as "Stil Water" on Pouchot's map, this post was at Stillwater, New York, across the Hudson from the mouth of the Hoosic River. It is called Fort Winslow on Sauthier, 1776.

An old fort at Stillwater was also rebuilt by James Montresor in August 1757. Scull, pp. 30-35, 45-47. A section through the new works is reproduced on p. 45, and the wall construction conforms to Pouchot's description. Another plan in Scull, p. 17 and a "Plan of Still-Water with its Blockhouse & ca." in Hulbert, I/3, p. 16 (Crown Maps, cxxi 104) differ from Pouchot's remarks and represent the earlier fortification.

1045 The east side of the Hudson, actually part of the Province of New York.

1046 "F Half Mond" on Pouchot's map is shown on the north bank of the Mohawk at its junction with the Hudson in Waterford, New York. The post at Halfmoon was laid out by James Montresor in October 1757. Scull, pp. 34-35. Hulbert, I/2, pp. 43 and 44 has a "Plan of the Post at the Half Moon" (Crown Maps, cxxi 43), depicting a stockaded fort, and a "Plan of the Buildings at Half-Moon" (Crown Maps, cxxi 44) drawn before construction of the stockade. Both Pouchot and the Crown Maps place the fort at the junction of the two rivers at what is today Halfmoon Point across the north branch of the Mohawk River from Peebles Island.

Local histories, however, place the Halfmoon Fort farther up the river, one mile north of modern Broad Street in Waterford. Sydney Ernest Hammersley, *The History of Waterford, New York* (Waterford, 1957), p. 49. This would seem to be the site of former posts, however, dating to as early as the 1680s, and not the structure built by Montresor and shown in the Crown Maps. Pouchot's description also argues for the Halfmoon Point site.

crosses the Riviere des Agniers, or Mohack River, on a ferry, half a league from the falls of that river. On the bank opposite the ferry was a redoubt begun in horizontal timbers to protect the crossing against our raiding parties.[1047] The falls of the river are very fine. They are [44] almost exactly 75 feet high.[1048]

For nearly two leagues, the road winds across hillsides before reaching the bottom of the Hudson Valley. This area is very good for ambushes.

On arriving at the bottom, one continuously follows the River Hudson at the foot of some low, but rather steep, hills. They are higher on the other side of the river.

Almost at the edge of the outskirts of Orange, one crosses a bridge over a stream on which there is a great amount of mill machinery.[1049] Albany, or Orange, is built on the slope of a hill which ends at the Hudson River, formerly called the Albany or Orange River. It is shaped like a triangle, the base [45] of which is a fine quay built along the riverside. With its jetties, this makes a fine port. Barks, snows & schooners sail up the river from the sea to Albany, which is a thriving commercial center.

At the apex of the triangle is a citadel faced with stone. Its outside measurements make up a square of 40 *toises* on a side, with a simple ditch about 20 feet wide without a glacis. It is overlooked and it is possible to approach to within pistol range on the south side by means of hollows in the ground.[1050]

1047 The crossing of the Mohawk, known as "Loudon's Ferry", was above Cohoes Falls and just above the point where the New York State Barge Canal today cuts east to avoid Cohoes Falls. The redoubt was on the Albany side of the ferry near Crescent Station, New York.

1048 Note in original, p. 45 (a): *And not 50 feet, as M. de Buffon believed, see the* Histoire naturelle, *vol. II, p. 89.*

The Mohawk River rises in the southwestern Adirondacks and joins the Hudson at Waterford, New York. Cohoes Falls, a short distance above the confluence, interrupted navigation and made it necessary to portage goods from the Hudson at Albany to navigable water on the Mohawk at Schenectady. Cohoes Falls are 70 feet high.

Georges-Louis Leclerc, Comte de Buffon (1707-1788) was a French biologist. Stanley J. Kunitz and Vineta Colby, *European Authors, 1000 to 1900* (New York, 1967), pp. 130-31. The reference is to his *Histoire naturelle générale et particulière.* (15 vols.; Paris, 1749-1767).

1049 Identified as Patroon's Creek in Albany, New York by Hough in Pouchot, *Memoirs,* II, p. 79n. The watercourse and mills were located at the site of modern Tivoli Street in Albany.

1050 Fort Frederick was constructed in 1702, and its demolition began in 1784. Roberts, pp. 552-53; Joel Munsell, *Collections on the History of Albany* (Albany, 1867), II, p. 239. By

...(continued next page)...

Beside the city, there is a very fine hospital built of wood by General Loudon which is capable of holding about 1500 patients. The streets of Albany are impressive and wide, well conceived & aligned but unpaved and thus muddy. The houses are soundly constructed in the Flemish style. The city might well [46] have a population of five to six thousand souls, almost all of them Dutch or Flemish in origin.[1051]

On the north side of the city, there is a deep ravine which starts in the area of the citadel. Fortifications - a good ditch & a stockade - have been built on the edge of it. The remainder of the precinct is enclosed by large upright stakes, more than a foot in diameter & 15 to 16 feet long. In spite of all that, the fort is still vulnerable to attack. On the other side of the river, there is a fairly good road which leads from Orange to Connecticut and Boston.

The Hudson River is a good quarter of a league in width and remains so to beyond the Issenglas Mountains.[1052] Its current is gentle & the river is quite deep with hardly any reefs. There is no safer form of [47] navigation and so the vessels which come up to Albany from New York normally have no more than a three-man crew. There is almost everywhere a good bottom for anchoring on both the E. and the W. banks. They choose the side according to the winds.

The banks on each side of the river are high & form a chain of hills that are covered with poor soil. The houses are separated from one another by about three quarters of a league. Except for a few houses of wealthy individuals, the region has a poor & deserted air, rather like the poor mountainous areas of Europe.

Along the course of the river, there are several mouths of tributaries which do not appear to be navigable, & also a number of poor villages. It is said that the interior of the country along the banks of these smaller rivers is more [48] populous, especially along the River

---

October 1785 Fort Frederick was described as "now destroyed." Antonio Pace (ed.), *Luigi Castiglioni's Viaggio* (Syracuse, 1983), p. 89.

Rocque includes "A Plan of the City of Albany" and a "Plan of Fort Frederick at Albany." The latter includes a section and elevation.

1051 Pouchot refers to the Dutch style of architecture prevalent in Albany during the eighteenth century. In 1755 there were reportedly 378 houses in the city with a permanent population of 2,972 "Regular Inhabitants in Families." Thomas Pownall, *A Topographical Description of the Dominions of the United States of America* (Pittsburgh, 1949), p. 38. Pouchot's 1759 and 1760 impressions of the town were surely influenced by an inflated wartime military and commercial population.

1052 The Highlands of the Hudson.

Sopus.[1053] However, the country is not very promising, as it is very hilly & covered with boulders or rocks.

Six leagues below Albany, there are two islands which create so many reefs right across the river that laden vessels can only pass at high tide. There are two channels, the first going straight towards the left-hand island & veering sharply around it; this is the worse of the two. The other runs between the right-hand island & the western bank on which a village is located. The ship is sailed directly towards the village and then changes course towards the island, the shore of which it then follows.[1054]

Although there is a great deal of traffic at this point, vessels often founder on the mud, but not dangerously. It is called Devil's Reef.[1055] [49] It is the only noteworthy difficulty encountered in this part of the voyage. A phenomenon to be seen here is the prodigious quantity of sturgeon constantly jumping out of the water in the summer.

Twelve leagues from this channel one can see on the right a chain of high mountains called the Kaatskills, which extend well into the upper part of Pennsylvania. They are the highest in the whole region & are comparable to our Alps, except that they do not retain snow on their summits in summer. They are very steep & are nothing more than rocks covered with forest.[1056]

About fifteen leagues farther on are mountains called the Isinglas which, although not high, descend almost sheer into the river. They are in general chains of rocks covered with rather poor woodlands, which [50] are commercially exploited in New York.

1053 The Esopus Kill or Esopus Creek enters the Hudson at Saugerties, New York.

1054 Pouchot's directions here are for descending the Hudson River. The village is Coeymans, New York on the west bank, although it is only about 15 miles below Albany, a distance incorrectly estimated by Pouchot. The islands lie above the village. Upper Schodack, that farthest upstream, is the left-hand island. Lower Schodack is that on the right. The dangerous channel between the two has been filled in.

1055 Pouchot probably refers to a feature still noted on charts as "Hell Gate" on the east side of Lower Schodack Island and at the base of the former channel between the upper and lower islands. On May 11, 1759, Colonel James Montresor recorded an incident with Hell Gate as he was traveling north towards Albany. Having passed Kinderhook they "Ran the sloop aground at 12 o'clock at Hell Gate between Queeman's [Coeymans] Islands, and saw several Sloops aground in the western Channel. We were not aground above an hour, but got off as the Tide was rising." Scull, p. 70.

1056 Pouchot describes the vicinity of Kingston, New York. The Catskill Mountains are located west of the Hudson. He probably did not intend to imply that they were as tall as the Alps but rather that they were the most significant mountains in this part of the country.

Over a distance of 4 leagues the river winds between these mountains. It is possible, however, to anchor and find shelter in a number of inlets. If they were not reached in time, there would be danger in bad weather. In this area the river continues to be more or less of the same width. The current is quite strong and so vessels only sail with the tides, which rise & fall fairly rapidly. They anchor when the tide is unfavorable, as they do throughout the voyage, unless a good wind counteracts the current.

A natural separation of the country occurs there, which could be called the upper & lower countries. Excellent forts could be built there to cut communication with the **[51]** low country & the sea. Above all, at the entrance to the gorge there is a little island which blocks the river & is not overlooked from the surrounding country.[1057]

On leaving the mountains, the river forms on the left a small bay, which might be taken for the main channel of the river on the return journey to judge from the way it flows through the rocks.[1058] On the right as one leaves there are very high bluffs for two or three leagues.

As one leaves this pass, the countryside seems to be just as agreeable, consisting of a fine plain with a pleasant, well-cultivated landscape & solidly built houses. From this point right down to New York, the river continues to be a league in width for most of the time.

The mountain chain described above runs E. & W. through the English **[52]** provinces, at approximately the same distance from the sea. The mountains prevent the other rivers of the region from providing communication between the sea and the interior of the continent, as will be seen later.

Only the River Hudson permits profitable navigation to the interior. At the point where the tide ends on this river, the mountains

---

1057 Pollepel Island, known to many as "Bannerman's Island", is situated in the river opposite Cornwall-on-the Hudson, New York at the upstream side of the Highlands. Pouchot was interested in a defensive position blocking the river against an attack from the north. During the American Revolution, Continental forces, concerned about enemies to the south, fortified the bend in the river farther south at West Point and Constitution Island. See Boatner, *Encyclopedia*, pp. 1187-88.

1058 Peekskill Bay at Peekskill, New York. Peekskill Hollow Creek falls into the northeast corner of the bay.

1059 Note in original, p. 53 (a): *An important remark for estimating the elevation & gradient of land on the North American continent.*

The Delaware rises in the Catskills and flows south into Delaware Bay. The Susquehanna has its source in Otsego Lake on the western edge of the Catskills. It flows south into Chesapeake Bay.

above provide the sources of the Delaware & the Susquehanna rivers.[1059]

The course of this river forms the finest approach to that part of the American continent called "Canada" which could communicate with Europe by way of the river all year round, something that is not possible along the River [53] St. Lawrence. By sailing along the Riviere des Agniers, you arrive without any great difficulty at the interior & the lakes.

The province of New York extends for twenty miles on either side of the Hudson River, the entire length of the Riviere des Agniers, or Mohacks River, & also includes Longue-Isle, or Long Island. New York, or La Menade,[1060] is a very fine city on a sort of island formed by a little arm of the Hudson River which stretches out into the sound separating Long Island from the mainland. The streets of the city are very wide, all paved and some of them have avenues of trees. The houses are built in the Dutch style, many of them in wood, & are of handsome design. The city is agreeable & bustling with commerce. Everyone seems to be [54] well off. The population would be 15 to 18 thousand souls.[1061]

The main docks of the city, facing Long Island, are always full of merchant ships constantly coming & going. There are normally 250 to 300 in the port. Vessels of more than 30 guns cannot anchor there. There is a rather deeper bottom on the Hudson side. The docks on that side are much less used, however, because there is no shelter from the S. winds. The reefs that are to be found from Sandy Houc[1062] onwards doubtless prevent vessels of greater draught from sailing up the river.

At New York, there are coastal pilots who are engaged to conduct ships from the narrows of Sandy Houc[1063] up to New York. [55] They

---

1060 A popular French name for Manhattan.

1061 Pouchot had an opportunity to observe this bustling seaport firsthand following his arrival in New York as a prisoner of war in August 1759 and again during the autumn of 1760. Thomas Pownall gave the 1755 population as about 15,000. Pownall, p. 44. A 1756 census of the "City and Co. of New York" set the population at 10,768 white and 2,272 black inhabitants. *DHNY,* I, p. 696.

Pouchot's secretary, J.C.B., had a similar impression of the city in September 1760. He described the area near Broadway: "This city has only one fine street, in the center of which they have market and butcher's stalls. Most of the houses on this street are of brick. This is the street I have called the 'grand plaza'. On the other streets the houses are of wood." Stevens, *Travels,* p. 122.

1062 Sandy Hook.

are very well paid. Along the quays of the city, little spurs have been constructed to act as docks for ships that anchor right inshore. They are unloaded using planks or gangways.

The city is not fortified. It only has a citadel located on the headland overlooking the two main channels.[1064] It is a square of about sixty *toises* measured on each exterior side, faced with solid masonry. It has neither ditch nor covered way but is well equipped with cannon.

Along the front on the headland, they have constructed on **[56]** outcrops of rock a wall 12 feet thick, forming a retrenchment & a sort of fausse braye to the citadel, in which there are 90 cannon, from 12 to 24-pounders, deployed as a battery. The gun platforms are all large flagstones. The guns are all mounted on naval gun carriages and cover the entrance to the sound & a little island used as a quarantine station.[1065]

Ships could sail up the river by keeping rather closer to the W. shore & troops could disembark above the city, which is defended on the landward side only by upright stakes, as at Orange. That site could be very well fortified since it only has one front facing land, which is a considerable advantage. Its natural position permits crossfire **[57]** over low-lying ground beneath it & its sides command the river and are 30 to 40 feet higher than it.

The warships proceed no farther than Sandy-Houc,[1066] which is situated ten to twelve leagues below New York. Anchorage there is

---

1063 Note in original, p. 55 (a): *The spelling of this name here conforms to the pronunciation. The French equivalent is* un crochet de sable *and the name refers to the shape of this headland.*

1064 Note in original, p. 56 (a): *One must always remember that the author is speaking of the state of these regions at the time of the late war.*

Pouchot means that New York had no walls like the important urban centers of Europe and the chief towns of New France, Montréal and Québec. New York's citadel was Fort George, site of a fortification from 1626 to 1790. Roberts, pp. 599-600. Fort George and the Battery at the tip of Manhattan Island provided the main defense of the harbor in 1759-1760.

1065 This was, of course, the Battery of Manhattan Island. Rocque includes "A Plan of the City of New-York" based on a 1755 survey and showing the defenses of lower Manhattan. John Montresor's "A Plan of the City of New York and its Environs" (1776) depicts the city 17 years after Pouchot saw it but provides good detail of Fort George and the battery along the shore in front of it. Nebenzahl and Higginbotham, pp. 84-85.

Pouchot compared the battery wall to a "fausse braye", a "low rampart going quite round the body of the place" and an element of fortification largely abandoned by the middle of the eighteenth century according to Muller, *Treatise,* pp. 219-20. The "entrance to the sound" is the East River, and the "little island" with a quarantine station is Governor's.

very good, sheltered from the south winds by several small mountains which form a cape.[1067] There is a large reef at that point which extends quite far into the sea. It originates at Long Island & forces ships entering to secure an anchorage, as well as those leaving to proceed with great caution.

When coming in, a ship steers directly towards the small mountains, after which it arrives at **[58]** the sandy cape which it follows at a distance offshore of a musket shot, all the time taking soundings. As soon as the cape is rounded, it is in deep water again.

When sailing up the river to reach Isle des Etats,[1068] a ship has to make several detours. It must hug the shore of Gersey most of the time, taking the N.E. coast of the island & from there keep to the middle of the river as far as the top part of the sound[1069] or the citadel, then enter the sound, which is the port. On travelling from Rhode Island to New York by way of the sound, ships encounter a narrow strait called "Heltgat", which can only be crossed at high tide because of the currents & whirlpools created there that force ships onto the rocks. This place is three leagues above New York.[1070]

Above Isle des Etats, there is **[59]** a little island on which stands a country house. A supply base could be installed there.[1071]

The country to the E. of York, in the direction of Connecticut, is very hilly and everywhere cultivated. I shall not enter into any detail about this land, except to say that all of it is in general divided into

---

1066 Note in original, p. 58 (a): *Admiral Howe showed in 1778 that they could anchor much farther up river.*

This refers to the 1776 activities of Admiral Richard Howe (1726-1799) in support of his brother William's (note 731) attack on New York. See Boatner, *Encyclopedia,* pp. 797-801. The Howes were brothers of George Augustus Howe (note 417) killed near Ticonderoga in 1758. For Richard Howe see Blanco, I, pp. 777-84.

1067 The "Highlands of Neversink" on Sauthier, 1776. The heights were near modern Navesink, New Jersey. See also Pownall, p. 41.

1068 Staten Island.

1069 Upper Bay or New York harbor.

1070 Here "the sound" refers to Long Island Sound, connected to the East River by Hell Gate. Most of the contemporary features of the New York area are shown in "A Plan of New York Island, with part of Long Island, Staten Island & East New Jersey" (1776) in Nebenzahl and Higginbotham, pp. 84-85.

1071 Bedlows Island, today Liberty Island, site of later fortifications such as Fort Wood which now provides the base for the Statue of Liberty.

shires or counties, that the farms are widely scattered - each one is 300 *arpents*[1072] - and that they all abut one another. The towns, or administrative centers, of each county are an accumulation of houses but of no great size.

When the government of each province raises militia, they are only under the colors for six months. Officers are only commissioned for the same period, which means that they are not very highly regarded. **[60]** Often a man may be an officer for one campaign, go on the next one as a private soldier and then become an officer again, &c.

The male population of the country is divided into companies of 100 men. When making up battalions, they form them out of a certain number of men from these companies. Every inhabitant can pay someone to substitute for him in the militia by covering his wages for the six months of the campaign from May to the 1st of November. Some inhabitants have given as much as 80 *piastres* to their substitutes, several of whom assured M. Pouchot that they had earned as much as 12 to 1500 *livres* by this means. The militia are almost all made up of men prepared to be hired in this manner.[1073]

The population of the region can be estimated from the following data. During the war against Canada, they recruited 12 men from **[61]** each company. New England and Connecticut supplied 7000 men, New York 2900, New Jersey 3000 and there was a proportional number from all the other provinces.[1074]

Long Island varies from 2, 3, 4, & five leagues in width and is thirty leagues long. Half of the island, especially on the New York side, is flat, very agreeable, well-populated &, although the soil is sandy gravel, it is nonetheless quite fertile. The other part is hillier & less fertile. A great deal of corn is sown, but not much wheat. They raise a large number of cattle and there is in consequence a considerable trade in salt meat with the islands of America.[1075] There are as many

---

1072 An archaic French unit of land measure equivalent to about one acre.

1073 Pouchot describes both the common militia system of the English colonies and the manner in which temporary provincial regiments were raised annually for service during the war. In the earlier volumes of his narrative, Pouchot's references to "militia" on campaign are to provincial regiments and not to the common militia available for home defense.

1074 For a discussion of the role of provincial forces in the war against the French see Douglas Edward Leach, *Roots of Conflict: British Armed Forces and Colonial Americans, 1677-1763* (Chapel Hill, NC, 1986), pp. 107-33.

1075 The West Indies.

inhabitants on this island **[62]** alone as there are in the whole of Canada.[1076]

Not much wheat is sown in the provinces of New England & Connecticut but enormous numbers of cattle are raised, which account for their large trade with the islands.

New Gersey is their granary. This province is almost entirely made up of flat lands crisscrossed by small tidal rivers. These conditions favor the import & export of goods. Large numbers of cattle are raised there also. There are iron & copper mines and the guns for **[63]** their merchant vessels are manufactured there. There is no large river in the province but several good wharves, mainly made of wood, where, so they say, the largest vessels can anchor.

As to the Connecticut River, although it is quite deep & has a current that favors navigation, it is of little use to them since it is barely tidal beyond its mouth. In any case, there are 4 or 5 rapids, or falls, several of which make portaging necessary.[1077]

The Delaware River is used for the internal navigation of Pennsylvania. However, it is no less difficult than the Connecticut River & has no tributary leading to the Canadian frontier. The same applies to the Schuilkill river, which is shallow and also has several rapids.[1078] **[64]**

In the province of Pennsylvania, large ships sail up as far as the top of Delaware Bay, five leagues below Philadelphia.

The Loups nation used to go down to the area of the sources of these two rivers in order to attack the English homesteads of the province & did a great deal of damage since the homesteads were very remote from one another, especially on the frontiers of the inhabited regions. **[65]**

1076 Note in original, p. 63 (a): *That is not correct. In a manuscript note, M. Pouchot claims that there are more than 30000 souls on Long Island. However, in Canada at the end of the late war, the population exceeded 90000.*

At the time of the war, the population of Canada was in the range of 82,000 to 88,000 persons (note 954). The 1756 census for Long Island (King's, Queen's and Suffolk counties) counted 19,724 whites and 4,059 blacks. *DHNY*, I, p. 656.

1077 The Connecticut River flows south from the Vermont-Québec border through western Massachusetts and Connecticut to Long Island Sound.

1078 The Delaware flows south from the Catskills past Philadelphia to Delaware Bay. It is joined below that city by the Schuylkill River.

• Chapter II •
The River St. Lawrence, from Mont-Réal to Chouegen

Although the River St. Lawrence is very well known, no one has ever spoken other than superficially about navigation along it from Mont-Réal, where its rapids begin, to Lake Ontario. We shall provide particulars about the voyage & the difficulties encountered in the course of it.[1079]

The first point to be noted is that the river is only navigable at the earliest around April 15, the time when the ice separates or melts. It is normally the March full moon which determines the exact season **[66]** and whether or not it will be early.

The rivers of Canada begin to freeze over around the 1st of December, and sometimes earlier when the winds are from the N.W., but in general the ice is only set hard around Twelfth Night.[1080]

During the great freeze-up it is possible to transport over the river ice all kinds of things, even artillery, from Québec and Mont-Réal as far as Les Cedres,[1081] except down rapids which never freeze over. At those places, though, good overland roads have been built.

Loads of goods sent out during the winter can save fifteen days in navigating through the upper country because Lac St. François[1082] thaws before the **[67]** river & as soon as the lake is unblocked, which sometimes occurs within 48 hours, the way through to Frontenac is also clear because of the difference in the climate.

From Québec to Mont-Réal, there is no real difficulty in navigation except for the Richelieu Rapids, where the tides cease to be appreciable. It is possible to sail back over the rapids in a stiff breeze.[1083]

1079 Sources for determining locations in the upper St. Lawrence River include "The River Saint Lawrence from Lake Ontario to the Island of Montreal" in Mante, p. 301; the "Murray Atlas" in the William L. Clements Library; *Maps, Reports, Estimates, &c. Relative to Improvements of the Navigation of the River St. Lawrence and a Proposed Canal Connecting the River St. Lawrence and Lake Champlain* (Toronto, 1856); current Canadian topographical maps and Canadian and United States navigational charts.

1080 The twelfth night after Christmas and the eve of Epiphany, January 6.

1081 Les Cèdres, Québec, above Montréal.

1082 Note in original, p. 67 (a): *This lake is, in fact, nothing more than a widening of the river.* Lac St. François is a widening of the St. Lawrence above Montréal.

1083 These rapids were at a bend in the river near Deschambault, Québec.

The winds in Canada blowing more often from the S.W. than the N.E., voyages upriver take much longer, wherever one is in the country, than the return journey. Ships can anchor at many points of the river because there are a great number of islands where they can shelter from the winds.

Frigates can sail up as far as Sorel & large **[68]** merchant vessels can reach the foot of the Ste. Marie Rapids a quarter of a league below Mont-Réal.[1084] They can anchor between the Isle Ste. Héleine[1085] and the north shore.

Ships sometimes take fifteen days & even a month waiting for a N.E. wind which is sufficiently strong to carry them past the rapids. The usual journey between Québec and Mont-Réal is by schooner.

Because of its position, Mont-Réal could be well-fortified: there is a stream and shallows between the mountain & the city. However, it is overlooked within cannon range by a screen of trees at the foot of the mountain. But since the site is in the center of the colony, it needs no fortification other than the island.

The place is girded by a wall **[69]** without an embankment, 3 to 4 feet thick at the bottom tapering off to 18 inches at the top. Its design serves no useful function; its only advantage is to afford protection against a surprise attack.[1086]

The island of Montréal could be well-defended, as one cannot land at all points and because of the rapids & currents all around it.

Its position is an admirable one because of the beauty of its landscape & that of the surrounding area, which consists of very large plains. It is an essential position, Mont-Réal being a necessary depot by virtue of the change from the St. Lawrence to the Ottawa River (or Grand River).

1084 Sorel is at the mouth of the Richelieu River. The rapids are shown as "Rapids of St Helene" on the plan of the River St. Lawrence in Mante, p. 333.

1085 Isle Ste. Hélène just below the eighteenth century town of Montréal.

1086 The fortifications are shown in a "Plan of the Town and Fortifications of Montreal or Ville Marie in Canada" in Rocque who copied a common contemporary survey of the city. Another is in the "Murray Atlas." These defenses were poorly regarded. Even Lévis, in arguing for further resistance against Amherst's army in 1760, admitted that they were "very bad and incapable of sustaining a siege" although the city was "safe against all surprise and cannot be taken without cannon." Lévis to Vaudreuil, Sept. 8, 1760, *DRCHSNY*, X, p. 1106.

1087 Sault St. Louis or the Lachine Rapids at the foot of Lac St. Louis. These are today bypassed by the Lachine Canal.

The second rapids encountered are the Sault St. Louis, two leagues above Mont-Réal.[1087] They [70] extend over a league. Travellers consider them to be the worst on the river before La Présentation. On the north shore, the empty bateaux are lifted and manhandled into a little channel beside a mill, called Lachine because it belongs to the Sulpicians.[1088] They are towed for a league to a place higher up where a general depot had been constructed, with stores for the goods that have to be brought by land from Mont-Réal to the village of Lachine.

The portage road is very bad because of the mud, especially in spring when the traffic is heaviest. The road could have been made very good by digging ditches for drainage. This deficiency has caused a great deal of expense in damage to [71] vehicles, delays & clogging of the route.

If the population was greater, they could build a canal running from Lachine to below Mont-Réal along the course of that little stream between the screen of trees & the city. It would go to below the Ste. Marie rapids and avoid this portage that extends for three leagues.

The bateaux used for navigation from the head of this river can carry six thousand-weight. They are especially built to stand up to the necessary strain of going up the rapids. The ones lately built by the English are larger & lighter, but they cannot sustain this sort of navigation after their initial journeys. They always fill with water because of the buffeting they receive. The [72] French bateaux are far more serviceable. The English do not fit a sail to their bateaux, which is in fact absolutely essential in many circumstances. However, they are equipped with good oars made of ash. Those of the French are of pine, very poor & badly made. We used to go through a very large number of them.[1089]

1088 Lachine, Québec on the island of Montréal. The name, which translates as "China", originated in the seventeenth century when explorers still hoped that the St. Lawrence led across North America to Asia.

1089 Bateaux were flat-bottomed boats, pointed at each end. Contemporary documents confirm that there was a significant difference between English and French designs and that the latter was considered more useful on the St. Lawrence and the Great Lakes because of its durability. Prior to his 1760 campaign, General Amherst commented that "I think our boats are not Strong enough for the Navigation of the Lakes and that we shall be obliged to build Some, near the same model of the french." Amherst to Loring, Feb. 18, 1760, Amherst Papers/65. The preference for the stronger French type continued well after the war. When boats were needed for the Niagara garrison in 1767, French bateaux were deemed "properest" because "the Schenectady Battoes are so very Slight built, that a few trips up the Rapids knocks them to pieces." Brown to Gage, April 25, 1767, Gage Papers/64; Brown to Gage, Aug. 18, 1767, Gage Papers/68.

...(continued next page)...

Bateaux departing from Lachine follow the north coast as far as a league from the church of Pointe Claire.[1090] Poles are still used because of the currents which are strong, especially around the headlands.

If you want to travel via Chateau-Gay,[1091] you cross the river at that point. If you want to travel via the cape of Isle Perraut,[1092] you go on as far as the church of Pointe Claire. From Isle Perraut, you cross to [73] Les Cascades. The first waterfall you encounter is small but extends over the entire width of the river.[1093]

The Swedish traveler Peter Kalm had an opportunity to compare English and French bateaux in 1749. The former, as constructed at Albany, were 18 to 24 feet long and made of boards of white pine. Kalm implies that the French bateaux he saw on Lake Champlain were larger, and, more significantly, he noted that "the bottom is made of red, but more commonly of white, oak which shows better resistance when it runs against a stone than other wood." To save weight, the sides of French bateaux were made of white fir. Kalm, pp. 333, 381.

Further clarification of the difference between the two types is provided by an English memo on the dimensions and costs of French bateaux when constructed at Detroit. The specifications are compared to those for "4 handed" (four-man) and "3 handed" English models. Contrary to Pouchot's comments, the French boat (at 36 feet in length x 6 feet, 4 inches in breadth x 2 feet, 5 inches in depth) is larger than both the four-man (32 feet x 5 feet, 6 inches x 2 feet, 5 inches) and three-man (30 feet x 5 feet x 2 feet, 5 inches) English types. Even more significant, the French craft were three times as expensive at £36-13-4 (New York currency) each compared to £13-6-8 and £10 respectively. The difference must be due to the robustness of the French construction. The document is undated but is part of a collection relating to the period 1772-1783. Frederick MacKenzie Papers, William L. Clements Library, Volume B, p. 171.

Estimates prepared in Montréal in 1771 claimed that the cost difference in constructing French and English bateaux was not great but that "the French *Boat* is much the Largest and Strongest." Furthermore, the English version was built "Clinch Work", presumably nailed with the ends clinched, and "Cannot answer in the Rapids." Maxwell to Gage, March 16, 1771, Gage Papers/100.

The construction of the English bateaux used on the Mohawk River is discussed in Robert E Hager, *Mohawk River Boats and Navigation Before 1820* (Syracuse, 1987), pp. 20-45. Contrary to Pouchot, Hager maintains (pp. 42-44) that masts and sails were commonly use on the English boats. For further information, including details of original French and English bateaux recovered at Québec and Lake George respectively, see Kevin J. Crisman, "Struggle for a Continent: Naval Battles of the French and Indian Wars" in George F. Bass (ed.), *Ships and Shipwrecks of the Americas: A History Based on Underwater Archaeology* (London, 1988), pp. 130-37.

1090 At Pointe Claire, Québec on the island of Montréal and the north side of Lac St. Louis.

1091 Chateaugay, Québec on the south side of Lac St. Louis at the mouth of the Chateaugay River.

1092 Isle Perrot. The cape is Pointe du Moulin.

Quite near land on the W. side, where the number (1) appears on the map, there is a channel in the rock of which the waterfall is composed and bateaux going down river pass that way. Men on the rock, up to their waist in the water, guide the bateau by pulling it towards them. From there it is towed & poled to a point about a musket shot higher up, where there is another set of rapids, though less difficult than the first.[1094]

On the way down river, it is possible to jump[1095] the rapids if you know about the two channels on the eastern shore beyond the island. Normally the rapids are jumped by way of the channel along which the bateaux are hauled.

The 3rd set of rapids is called "Le Trou".[1096] The bateau is half unloaded & [74] the goods are carried to a place 150 feet above this rocky cape. You pass by bringing the bateau right up against the earth bank, indicated by the figure (3). The bateau has to be controlled by a rope held by several men. Other men go into the water up to their shoulders in order to move it forward & haul it round the headland.

The river, which at this point is encumbered by large rocks beneath the water, hurtles boiling into what look like abysses. One of them especially forms a huge pit. Beside it is a thin stream of water under such pressure that the bateau passes along it as though ferried on the water's back. Failure to take that stream could mean falling into the abysses. Very few of those who have fallen in have escaped, and, as a result, this place has been named *Le Trou.* The different rapids are called "Les Cascades." [75]

A league below the village of Les Cedres is a headland where the river seethes mightily.[1097] The bateaux have to be brought up very close

---

1093 Les Cascades were the first rapids above Lac St. Louis and the beginning of the route to Lake Ontario. The name often refers to the entire series of rapids from Isle des Cascades to Pointe Coulonge. The rapids and islands are in the vicinity of Pointe des Cascades, Québec. These and the other rapids to above Couteau du Lac are today bypassed by the Beauharnois Canal of the St. Lawrence Seaway.

1094 The numbers in parentheses in the text are keyed to Pouchot's map. The channel and rapids lie along the shore of Pointe des Cascades.

1095 "Shoot" the rapids.

1096 Le Trou, or "the hole", was the rapids between Pointe des Cascades (the rocky cape) and Isle des Cascades. It is shown as "R. le Trou" in the "Murray Atlas." Modern maps show these rapids as Les Faucilles.

1097 Probably Pointe à Delisle on the north side of the river.

to the shore. A channel was dug to avoid the fierce currents but since it was not finished there is often no water in it & it is more trouble than use. This passage, marked (4) on the map & called "Le Buisson",[1098] is one of the most tiring for the boatmen because of the lack of water. From that point the bateaux are moved by pole as far as just below Les Cedres, where the goods are unloaded and carried half a league overland to the top part of the village. From there the bateaux are towed. Men get into the water to control them, especially around the mill belonging to M. de Longeuil.[1099] There are some very bad reefs at that point because the river **[76]** is very shallow & flows over boulders or rocks, making progress downriver dangerous & difficult.[1100]

Above the mill, there is a second reef less fearsome than the first. If instead of this mill they had constructed a small canal on the near side of the islet on which it is situated, they would have spared travellers a great deal of hardship.

The site of the church of Les Cedres would be a good place for a fortified position at the head of the rapids as at that point the terrain forms a natural fortification. There is a great deal of earth that would be easy to move.[1101] **[77]**

An encampment sited at this place would defend the entrance to the colony very well. Since the enemy would be absolutely unable to

---

1098 Le Buisson, "the thicket", was the rapids spanning the river from Pointe à Delisle to Pointe du Buisson. Modern topographical maps show it as Rapides des Cascades and Rapides du Rocher Fendu. It is shown as "River du Buisson" on the "Murray Atlas."

1099 Paul-Joseph Le Moyne de Longueuil, Chevalier de Longueuil (note 450). "Pt. au Moulin" is shown on *Maps, Reports, Estimates,* 1856, and the place name still exists although separated from the north mainland by the canal behind the Barrage [Dam] Les Cedres.

1100 The rapids at Les Cèdres, Québec.

1101 Note in original, p. 77 (a): *The English have since built a fort at Les Cedres, where Major Sherburne was unable to resist their assault for any length of time in 1776. This occurred after the siege of Québec by the Americans, who had previously taken possession of this fort, had been raised.*

In two separate actions on May 19-20, 1776, a British and Indian force from Oswegatchie attacked the small post held by American troops at Les Cedres (the Cedars). The fort, under Major Isaac Butterfield, surrendered on the 19th. A relief force from Montréal, led by Major Henry Sherburne, was defeated on the 20th. Boatner, *Encyclopedia,* pp. 191-92; Blanco, I, pp. 281-82.

There was a poor fort at Les Cedres in Pouchot's time. Bougainville saw it in 1756 and described it as "a little stockaded fort" that was "almost abandoned." Hamilton, *Adventure,* p. 15.

get downriver of this position, they would have to cut their way through the forest for at least four leagues in the vicinity of Vaudreuil.[1102] It is inconceivable that they would dare to do this while leaving this post intact behind them.

After the place of re-embarkation, the bateaux are propelled by pole up to the portage on the bank of the lake indicated as (5) on the map. It is a headland where the waters are so agitated & seething that it is necessary to unload the bateaux. This portage is sixty feet long and the men have to get into the water in order to move the bateau forward & maneuver it round the headland.[1103]

After that the oars are used to reach another headland called "Pointe du Diable",[1104] where the vessel is again towed. If **[78]** by mischance, the bateau yaws[1105] at this headland, the current carries it into a stretch of extremely turbulent water where travellers inevitably perish. This happened to a number of men who tried to get past the point using the pole.[1106]

The island, indicated as (6) on the map[1107] and situated above this dangerous place, is extremely useful for defending the rapids both to right and left, upriver and downriver. It is approachable from both directions & another very good place for defending the colony. The enemy could not use the river and would find it equally impossible to

---

1102 The seigneury on the mainland opposite the west end of Isle Perrot. It is now Vaudreuil, Québec.

1103 The portage is shown near "Rapide de la Grande Bature" in the "Murray Atlas." It was probably modern Pointe à Valier on the north shore of the river.

1104 Shown as "Pt. au Diable" (aptly, "Devil's Point") just below Coteau du Lac in the "Murray Atlas", the place retains that name today.

1105 Note in original, p. 79 (a): *This refers to the movement of the bow when the current makes it veer towards open water. In those circumstances, control becomes impossible and one has to allow the vessel to go with the current.*

1106 Pouchot does not specifically identify this as the Rapides du Coteau at Coteau du Lac, Québec, the last major barrier on the river before Lac St. François. Bougainville described Coteau du Lac in 1756 as "a portage of two or three arpents [380-570 feet] past Pointe au Diable. The waterfall at Coteau du Lac is one of the hardest to pass. It is three leagues from Pointe à Coulonge." Hamilton, *Adventure,* p. 15.

1107 Probably Isle Arthur, located in the river between Coteau du Lac and modern Isle Salaberry. This might be the site of the "painted trees" or arbres matachés described by Bougainville in 1756 as a landmark at the entrance to Lac St. François. Hamilton, *Adventure,* p. 15.

carry the boats through the woods as far as [79] Les Cascades. The island is well wooded & quite extensive.

The bateaux are propelled by poles along the shore of the lake[1108] & in a few places the oars are used. The current is very strong & the shoreline clogged with trees that have fallen into the water. The island marked (7)[1109] is remarkable in that when going along the shore of the lake, you have to seek out a current that carries vessels directly towards the island and thence into the channel that leads to Les Cedres, otherwise you would get caught up in very turbulent waters which would mean certain death.

When travelling down to Mont-Real, General Amherst's army lacked sufficient guides and at this point lost 80 ordinary bateaux & 4 of the sort known as *carcassieres,* each of which had mounted on it a 12-pounder cannon of brass. Even if there were only 4 men in charge of each bateau [80], that would make at least 336 men who died.[1110]

Lac St. François is seven leagues long & three or four wide. At the entrance to the lake, on the north side, is Ance au Bateau,[1111] which is always the route taken. From there, boats can be rowed, or the sail can be used. Two leagues farther on is Pointe au Banc, which is the normal place to camp.[1112] The soil is very good & the place would lend itself well to homesteading.

If a stop is not made there, one has to cross the lake to find a camp site, because all the bays extend well inland & the country is very marshy. Pointe Mouillée,[1113] indicated as (8), is a flat headland jutting out into the lake. The ground is very waterlogged. A brief stop is sometimes made there.

Farther on is Pointe [81] à la Morandiere,[1114] marked (9) on the map. It is a narrow strip of land where camping is possible but only if

---

1108 Lac St. François.

1109 Probably modern Isle Lalonde at Coteau Landing, Québec near the beginning of the old Soulanges Canal that formerly bypassed all the rapids as far as Lac St. Louis.

1110 Amherst's army lost 84 men, 54 boats, and one row galley (note 947).

1111 Ance au Bateau is shown as the north channel between Macintyre (Lalonde) and Giroux islands in *Maps, Reports, Estimates,* 1856.

1112 Identified as "Pointe au Baudet" on Pouchot's map, "Point Bodet" in Mante and "McGee's Pt." in *Maps, Reports, Estimates,* 1856, the place is today Pointe Beaudette at the mouth of Rivière Beaudette. Amherst's army camped there on its way to Montréal in 1760.

1113 Still known today as Pointe Mouillée.

there are very few men, as the site is extremely small. The trees on this N. shore are Cedars & Pines which have their roots in the water. The interior here is thick with dead and fallen trees.

From Pointe à la Morandiere, navigation is constantly encumbered by rushes. Caution is necessary to keep to the N. without getting too close to land in order to identify the correct channel of the river. After the rushes, the boat travels among a number of beautiful islands, which are called "Les Chenaux."[1115] After these islands, you cross to the S. shore if you want to visit the mission of St. Régis, recently established by the Jesuits though with very few missionaries.[1116] The lands in the surrounding area would be **[82]** admirable for cultivation. It is very fine hunting country.

Opposite St. Régis, which is the side always taken, the land is very high & a little farther along there is a very rugged double headland called "Pointe Maligne"[1117] and marked (10) on the map. Boats have to be hauled along this stretch. From there one reaches Milles Roches, indicated as (11). It is the waterfall of Long Saut by way of the side channel & the N. route. The river, which makes a large bend at this point, is encumbered with huge rocks. A channel has been made in order to avoid a long detour.[1118]

At the entrance to Mille Roches is the lower part of the island to the N. of the S. route going down Long Saut.[1119] It is possible to land

---

1114 Pointe à la Morandiere is shown as "Pointe a la Maudie" on Pouchot's map. It was probably Faulkner's Point near South Lancaster, Ontario. Modern topographical maps show rushes becoming common at this point.

1115 The channels among the islands at the head of Lac St. François.

1116 The mission of St. Régis or Akwesasne was established at the mouth of the St. Régis River between 1747 and 1755. The modern reserve straddles the New York-Québec boundary near Hogansburg, New York. Trigger, pp. 471-73. Like the other missions of the St. Lawrence, St. Régis was fortified with a stockade constructed in 1755. Hamilton, *Adventure,* p. 15.

1117 Pointe Maligne was located just above Cornwall. Pouchot, *Memoirs,* II, p. 100n. It was near the site of the Moses-Saunders Power Dam that today forms the artificial Lake St. Lawrence. Construction of this structure and the Long Sault Dam in the 1950s flooded the Long Sault rapids and completely altered this part of the river.

1118 Long Sault was the stretch of river and rapids above Cornwall, Ontario. The rapids were on both sides of Barnhart Island, New York. Both channels have been much changed by the formation of Lake St. Lawrence. Milles Roches was near Long Sault, Ontario between Sheek Island and the north shore. Pouchot, *Memoirs,* II, p. 100n.

1119 Barnhart Island, New York.

on the lower part of this island. If it were manned with fusiliers, they could prevent [83] bateaux from effecting a passage.

From Mille Roches, one proceeds to Le Moulinet,[1120] indicated as (12). It is necessary to sail round two little islands (the waters are very calm) and after that one enters the arm of a river which is very rough. Apart from the use of poles, some men have to get into the water up to their shoulders in order to guide the boat along. A channel has been made to allow passage. From there, you reach a little island on the right,[1121] thus arriving at the foot of Long Saut.

Long Saut is a good quarter of a league in length to the N. & three leagues long as one goes down towards the S. The water boils like the sea in a storm. Although the current is very strong at the N. point of passage it is nonetheless possible to haul the bateaux up it with 4 or six men on the rope & two men [84] to guide it. Fortunately, the currents always carry vessels towards the land. There are a number of rocks in the water which render passage difficult. They could be removed to make a road along the shore, which would greatly diminish the problems of hauling the vessels. Travellers normally camp on the top of Long Saut. The country is full of splendid free-standing woods & would be very suitable for cultivation.

The river above Long Saut has a very rapid current, especially where the land slopes, which it does frequently. At these points, vigorous use of the pole is always necessary. No (13) is Pointe Ste. Marie, one of the most noteworthy headlands.[1122]

No. (14) is Isle au Chat,[1123] also to be noted because it is beneath that island that the crossing to the S. is made in order to go down Long Saut. [85]

On the island lower down, there is a headland which would cover with artillery both the N. & S. passages and where troops could camp.[1124]

1120 The vicinity of Le Moulinet is marked by Moulinette Island in Lake St. Lawrence near Long Sault, Ontario.

1121 These islands were in the vicinity of Long Sault, Ontario.

1122 Probably Farran Point at Ingleside, Ontario.

1123 Wilson Hill Island, New York.

1124 Franklin Hough identified the island as Chrysler's, now submerged. Pouchot, *Memoirs,* II, p. 102n.

No. (15) is Pointe au Cardinal,[1125] equally worthy of note because of its very fast current. Apart from the current, there are trees which have fallen from the bank and make navigation extremely awkward.

No. (16) is Rapide Plat,[1126] the swiftest current along these ridges. It is not dangerous in either direction, however. There is a large eddy which brings your boat right up to its foot and from there you use the pole to go up it so as not to yaw.

No. (17) is Pointe aux Iroquois,[1127] which is not too arduous. It is a place to be noted because the traveller [86] going in either direction almost always stops there.

Les Galots are two very steep ridges.[1128] In this stretch the river seethes as it rushes down the slope. The boat is steered close to the bank along the first ridge & then comes across a kind of stone jetty, from which point the vessel is hauled. One must take care to keep the bow of the bateau towards the land, otherwise there is a risk of being carried away by the current. The second of the galots is not as long as the first one. Within a musket shot of that point and above it there is a bay called "aux Perches",[1129] because it is there that one stops using the poles. Since there are no more rapids, only oars or the sail are needed. While descending Les Galots, one follows the currents to the open lake.

No. (18), opposite this bay, is Isle aux Galots, which might be 700 *toises* in circumference. The only place [87] to land is a stretch of bank 150 *toises* long at the top part of the island because the currents meet below it. The island covers the N. passage. We retrenched it in 1759.[1130]

No. (19), beside it, is the island called "Piquet"[1131] because the missionary of that name fled there with the Indians settled at La

---

1125 Pouchot's description places Pointe au Cardinal near Morrisburg, Ontario. He seems to have been in error. Bougainville notes Pointe au Cardinal as being above Pointe aux Iroquois which would place it, logically, at modern Cardinal, Ontario. Hamilton, *Adventure*, p. 16.

1126 Rapide Plat is marked by Rapide Plat Point about three miles above Morrisburg.

1127 Iroquois, Ontario, site of the Iroquois Lock, uppermost of the St. Lawrence Seaway locks along the river.

1128 The rapids on either side of Galop Island, New York before the area was heavily altered by construction of the Iroquois Dam and seaway.

1129 Pouchot describes the bend in the river above the rapids and between Prison Island and the Ontario shore. Bougainville also noted that this was the point beyond which it was no longer necessary to pole bateaux. Hamilton, *Adventure*, p. 16.

Présentation. If a camp with artillery was installed on the island, & there were troops occupying Isle aux Galots, it would not be possible to go down the river.

This post would be the best for stopping the enemy, provided there were enough men to guard the two islands. Isle Piquet is a league in circumference. It is only possible to land there at a few places that are easy to defend. One can approach from the top or the bottom end. The island is heavily wooded.

It is possible to go up & down [88] the S. side of the river without difficulty. This passage has only been known since 1759. In 1760, a detachment of the English army camped there when engaged in the siege of Fort Lévis.[1132] Between Isle Piquet & those we shall shortly discuss there are a number of little islands of no great consequence.[1133]

No. (20), Isle à la Cuisse,[1134] is a quarter of a league in circumference, quite high in the middle & capable of holding a camp of 1200 men. It could be well retrenched. Within half the range of a musket, it covers the N. side & with Fort Lévis would provide an excellent defense of the entire river passage. It was from this point that the enemy launched their main assault against the fort. They had placed

1130 The islands at this point of the river are referred to collectively on Pouchot's map and in other sources as "Les Galots." Some French accounts use the form "Galops." The smaller island, shown as "Galops I" in "Plan of the attack upon Fort Levi" in Mante, p. 303, is today Prison Island, Ontario. Following the capture of Fort Niagara in 1759, the engineer Desandrouins built "several batteries" at the point of "la grande île des Galots" to block the route to Montréal. Gabriel, pp. 301-02. This was the larger island which Pouchot identifies as "Picquet."

1131 Today Galop Island, New York, split in two and most of its other outlines altered by construction of the St. Lawrence Seaway. A small English post is shown on the island in "Plan of the attack upon Fort Levi", Mante, p. 303.

1132 A detached camp is shown on the south mainland in "Plan of the attack upon Fort Levi" in Mante, p. 303 at the point opposite the southwest end of Galop Island.

1133 Several of these small islands, all shown on the plan in Mante, survive on the Canadian side of the boundary as Grass, Pier and Lame Squaw. That closest to the north mainland, called Duck Island in 1856, has been incorporated into Drummond Island along a canal known as the "North Channel." On the United States side, the Tibbets Island of 1856 has been incorporated into the southwest end of Galop Island.

1134 Isle à la Cuisse is today Spencer Island, Ontario. Its southern tip, site of the English battery, has been cut by the North Channel. See note 908.

1135 Today Drummond Island, Ontario. See note 908.

14 guns & six mortars in a battery. The island is higher by at least 24 feet than Isle Orakointon **[89]** on which Fort Lévis is situated.

No. (21), Isle Magdelaine,[1135] is a little larger than Isle à la Cuisse. It too overlooks Fort Lévis & covers the whole island in enfilade. The enemy placed there 8 guns, 8 mortars and two howitzers.

No. (23), Pointe à la Corne,[1136] could be well retrenched to cover this front if the Isle à la Cuisse & that of Fort Lévis were also occupied.

No. (22), is Pointe à l'Yvrogne.[1137] The main camp of the English & the headquarters of General Amherst were on this headland.

Orakointon is a little low island, barely above water level.[1138] Fort Lévis occupies two thirds of it. The fort is a redoubt with a circumference of 108 *toises.* The front on which the gateway is situated **[90]** consists of the head of a hornwork 42 *toises* on its external side. The gateway is perfectly enfiladed by the Isle à la Magdelaine. The two long sides are of unequal length, the northern side being the longer. They end at a small flank about 5 *toises* long. The back part is made up of three faces, like the three external sides of a hexagon.[1139]

The rampart is 27 feet thick at the bottom narrowing down to 18 feet at the top and faced with saucissons. The exterior height of the rampart is 7 feet & it is eleven feet on the interior side.[1140]

1136 Pointe à la Corne was presumably named for Captain Luc La Corne (note 447) who had guarded the upper reaches of the river in 1759. It is today Dawson Point in Johnstown, Ontario.

1137 Tuttle Point, Ontario, much changed by construction of the North Channel, was the site of Amherst's main camp in 1760. The place, "Drunkard's Point" in English, is alternately spelled *à l'Ivrogne* by Pouchot.

1138 Chimney Island, site of Fort Lévis, has also been heavily altered by construction of the seaway. Its north half has been cut away.

1139 Pouchot's description is very precise. See the plan and section entitled "Fort Levi", inset on "A Plan of the attack upon Fort Levi" in Mante, p. 303. Pouchot's own plan of the fort and islands was printed in Volume I of his *Mémoires* and is reproduced in this edition.

It is apparent, throughout Pouchot's description, that he is criticizing both the location and the design of Fort Lévis. In so doing, he questions the skill and judgement of Captain Jean Guillaume Plantavit de Lapause (note 722). Pouchot claims that Lapause convinced Lévis to place the fort on Isle Orakointon and "himself wanted to lay out the fort, or redoubt, according to his own caprice, in spite of the views of M. des Androins, the engineer, who was entrusted with building it." Pouchot, *Mémoires,* II, pp. 136-37.

1140 This might be a misprint. While the terre-plein on the interior appears to be about 11 feet above the ground, the exterior side of the rampart, exclusive of the parapet, is only slightly lower, perhaps 9 feet. See "Fort Levi" in Mante, p. 303.

Above it wooden cases have been added to form a parapet nine feet wide at the bottom and seven at the top. The height on the interior side is 6 feet. There is a fraise between the parapet & the rampart. The ditch is five *toises* wide & two feet deep with a foot **[91]** of water in it. On the hornwork side, the ditch is lined with a sloping palisade held in place with wooden pegs that are not especially secure because they are not buried very deep in the ground.[1141]

Around the N. part of the island an epaulement has been built measuring 9 feet at the bottom and tapering off to 5 or 6 feet at the top. On the N.E. headland we erected a redoubt of 18-inch timbers pierced for five cannon.

The S. side, where the dock is, was closed in by a palisade right up to the foot of the glacis which consisted of a woodpile for the use of the fort. **[92]**

The S.W. headland was an epaulement rather like the parapet of a covered way. All around the island we placed an abatis of the tops of trees which extended for 15 feet out into the water. We left a space 40 *toises* wide for landing on the north as well as all that part from the fort to the lower part of the island.

The fort is also overlooked by a point of land to the south, called "Ganataragoin", 450 *toises* distant from the island. The enemy placed 4 guns, 4 mortars & two howitzers there, which permitted them to

---

1141 Note in original, p. 92 (a): *All these fortifications could not have lasted very long. Indeed, we believe that since the late war the English have abandoned this fort, which was of no further use to them.*

Following the surrender of Fort Lévis, the British renamed the post Fort William Augustus. The main garrison was stationed at the island fort, although a detachment was also maintained at Oswegatchie on the mainland. Amherst to Gage, Nov. 17, 1763, Clarence Edwin Carter (ed), *The Correspondence of General Thomas Gage With the Secretaries of State, and With the War Office and Treasury* (2 vols.; New Haven, 1931), II, p. 210. In December 1765 General Thomas Gage proposed to abandon Fort William Augustus in the spring. "Report of the Forts in North America ...." Dec. 18, 1765, *Ibid.*, p. 322. The post was officially closed on May 20, 1766, and its function as a supply forwarding point for the Great Lakes forts was transferred back to Oswegatchie. List of Officers [Commanding]...,Dec. 25, 1765, to December 24, 1766, Gage Papers/139. When Francis Grant stopped at the island on June 15, 1767, he observed that the "fortifications are strong and regular, but now have fallen to decay. A Corporal and two men from Oswegatchie guard the stores." Francis Grant, p. 306. Although the buildings were maintained for awhile, and Fort William Augustus was included on returns of barracks stores as late as 1773, the position was not refortified during the American Revolution. "Return of Barrack Bedding and Furniture at Oswegatchie and Fort William Augustus 25th March 1773", National Archives of Canada, MG 23, G 111,7, M-479.

Once again, Pouchot's description of defensive features of the fort corresponds very closely to the English plan and sections.

cover the island with enfilading fire from the S.W. to the N.E. At the same place, opposite the Isle Orakointon, there is a little river of the same name as the point of land mentioned above. It is quite wide & deep for more than a league & a half. If a redoubt [93] and a camp were located there, they would provide excellent defense for the S. part of the river.[1142]

If the islands described above were to be occupied together with Pointe à la Corne, they would be the only sustainable positions at the head of the rapids. At Pointe de Ganataragoin, the current is strong & runs along that coast.

In front of Isle Orakointon, the river has a good current & at the lower part of that island forms a large eddy towards the S., which makes for a good anchorage. Vessels can spend the winter there without problems, but a stiff N.E. breeze is necessary if they are to sail against the current, which begins at Pointe de Ganataragoin.[1143]

Vessels could certainly reach as far as the front of Isle Piquet but the anchorages are worthless & the [94] currents on both the right and the left sides are very strong.

La Présentation or Chouégatchi, is a settlement of Iroquois established by a Sulpician father, the Abbé Piquet. We built a square fort there, the buildings of which formed the bastions. The curtains were of large upright stakes 15 to 16 feet high. The missionaries, the commandant, his little garrison & the storekeeper for the trade of the missionaries appointed by the King all occupied the four main parts of this facility. In 1759 the mission, which involved a good many people, withdrew to Isle Piquet & the fort was demolished so that the enemy could not use it for military purposes. It was a good place for a mission because the land is excellent for cultivation.[1144]

1142 The south headland is Chimney Point, New York. Its battery is discussed in note 917. The Ganataragoin River is today Tibbits Creek.

1143 The French warships probably wintered there in 1759-1760 after the abandonment of the Pointe au Baril shipyard. The damaged *Iroquoise* and the unfinished hulk (later christened *Johnson* by the English) were intentionally sunk at this spot before the siege and are so indicated on "A Plan of the attack upon Fort Levi" in Mante, p. 303. The English discovered the vessels "behind their island, filled with water." Knox, II, p. 554.

1144 Several plans and elevations of the fort at La Présentation, the originals of which are in the French Archives, are illustrated in Boyesen, "The French and Indian Settlement at Ogdensburg" and "French Personnel at La Presentation." These show the four houses, laid out as covered bastions.

...(continued next page)...

One can go some distance inland by way of the Chouegatchi River. [95] The interior of this region is very little known to the Canadians. The Indians only use it for hunting.

There is a reef in the river almost at the end of the Indian village where the Abbé Piquet had set up saw mills. Vessels can anchor in front of the village but not in safety, because of the winds & because trees flow down the river when it is in flood.

For about twenty leagues, the river lends itself well to navigation but the rest of it can only be managed in canoes & by portaging. It extends well into the highlands. Our raiding parties sometimes used to take this route when they were attacking the English frontiers.[1145]

Behind the fort of La Présentation, there is a knoll on which [96] a town or a village could be built. The position is a very advantageous one.[1146]

The St. Lawrence River is beautiful &, as far as two leagues above Pointe au Baril, its banks in this area are very fine for cultivation or for hunting & fishing. Fish are very abundant.

Along this stretch, the river is not more than a quarter of a league wide & it cuts a very straight course for eleven leagues, from the islands above Les Galots up to Toniata.[1147] It is not encumbered with a lot of islands & the water is very deep.

Three leagues above La Présentation, on the north bank, there is a headland called "Pointe au Baril."[1148] It provides a good view over the river & would protect vessels stationed there to defend it. A camp can be [97] located there to advantage because a league & a half farther on

---

As to the disposition of the original French fort, English sources are unanimous in saying that it was only slightly damaged by the removal of the roofs of the buildings (note 873). A 1767 description suggests that the old structures were preserved and later reoccupied: "Oswegatchie, called by the French La Gallette, is a small Fort on the South side of the river, consisting of four blockhouses joined together by rows of piquetts, mounting four pieces of artillery. It was built as a trading house, and since the abandoning of Fort William Augustus, six miles lower, the store for the provisions for the upper posts are kept here." Francis Grant, p. 305.

1145 The river is today the Oswegatchie. Lieutenant Joseph-Gaspard Chaussegros de Léry used this route for his successful 1756 attack on Fort Bull at the Oneida Carry (note 296).

1146 The village of Ogdensburg, New York was later built on both sides of the mouth of the Oswegatchie River.

1147 Grenadier Island, Ontario.

1148 Weatherheads Point at Maitland, Ontario.

the banks are rocky inlets where the enemy could not land in force. These inlets continue as far as the Ance au Corbeau.[1149]

Close to the headland marked (24) on the map is a bay called "Ance à la Construction" since 1759, when ships were built there. It is very good for ship construction. The water is deep right up to the shore, the forest is within easy reach & it is possible to dig a system of trenches to cover the shipyards.[1150]

A league & a half above Pointe au Baril is a little island marked (25) which is about 500 *toises* in circumference. It is a rock on which a fort could be built. It has a good view over the river down to Toniata & could cover it well with its artillery. There is an excellent **[98]** anchorage on the lower part of the island. We sent our ships to take up station there and keep the river under observation.[1151]

From the top part of this island as far as the Bay of Niaouré,[1152] the south shore of the river is low, with many tributaries flowing into the main river and full of marshy bays & very dense forest.

Five leagues from Pointe au Baril is the island of Toniata. The main channel of the river runs between the island & the south shore. The northern part of the river is clogged with rushes. In the summer it is famous for its eel fishing.

The island of Toniata[1153] is three leagues long & three quarters of a league wide. Its soil **[99]** is good for cultivation as is that of another island situated between the north shore of the river & Toniata. That

---

1149 These inlets begin just above Brockville, Ontario.

1150 Ance à la Construction is now Maitland Bay on the downstream side of Weatherheads Point. *Iroquoise, Outaouaise* and the unfinished vessel (later *Johnson*) were constructed there in 1759. A crude plan of the shipyard fortifications, constructed under Pouchot's direction, is in Courville.

1151 McNair Island, largest of the Three Sister Islands between Morristown, New York and Brockville, Ontario. Pouchot means that guns placed on the island could cover the width of the river at this point, not that they could reach Toniata. The latter island is actually "up" the river from this spot.

1152 On Lake Ontario (note 302).

1153 Note in original, p. 99 (a): *M. de Frontenac had ceded this island to an Iroquois. The latter sold it soon afterwards for four jugs of brandy to a Canadian who would doubtless have sold it in turn for a beaver skin.*

Toniata is today Grenadier Island, Ontario. Louis de Buade de Frontenac et de Palluau (1622-1698) was Governor General of New France from 1672 to 1682 and again from 1689 to 1698. *DCB,* I, pp. 133-42.

...(continued next page)...

island is a league in length & a quarter of a league wide.[1154]

At their top end is a small channel with very little water and full of rushes. It is called "Le Petit Détroit." Bateaux take this route when travelling up the river in order to avoid the currents.

Care must be taken not to go astray among side channels in the rushes, otherwise one would never get out & would end up getting stuck in the mud. At the Petit Détroit, the ceremony to baptize those who have never travelled up the river takes place.[1155]

A league & a half from there and a little farther on begin the Thousand Islands which extend over at least three leagues. They consist of an infinite number of little rocks **[100]** covered with trees which in several places leave a wide enough channel to pass. In other places, vessels pass between two islands almost touching each shore. They are very safe. There is always deep water all the way round each island & very little current.

After three leagues, the islands begin to get larger. One must take good care not to get lost. The bateau follows the channel nearest to the N. shore where note must be taken that several of the channels end up in marshes near the bank.

It is necessary to veer almost due N. in order to enter Ance au Corbeau, which is large & beautiful. A passage is effected between its S. headland, which is very narrow, & a small island where there is quite a strong current. From there one sails along the coast of Isle au Citron,

---

The story of the sale of the island comes from Pierre-François Xavier de Charlevoix, *Histoire et description générale de la Nouvelle France, avec le Journal historique d'un voyage fait par ordre du roi dans l'Amérique septentrionale* (6 vols.; Paris: 1744). An English edition was published under the title *Journal of a Voyage to North America. Undertaken by Order of the French King. Containing the Geographical Description and Natural History of that Country, particularly Canada. Together with An Account of the Customs, Characters, Religion, Manners and Traditions of the original Inhabitants* (2 vols.; London, 1761). This will be cited hereafter as Charlevoix, 1761. The Toniata island story is found in I, pp. 296-97 of the English edition.

Those footnotes that cite incidents related in Charlevoix appear to have been the work of Pouchot's 1781 editor.

1154 Tar Island, Ontario.

1155 Le Petit Détroit is today Tar Island Narrows between the island and the north mainland. Bougainville commented, when he passed in 1756, that a feature of Le Petit Détroit was Pointe au Baptême, "so called because of a baptism carried out there as they do on the Grand Banks." The custom was similar to that of sailors crossing the equator for the first time. He added, "notice in passing all the naval customs established in this colony." Hamilton, *Adventure,* pp. 17-18.

which is a good league in length. The island is beautiful & well wooded.[1156] [101]

There one sails across the bay for two leagues in order to reach Isle Cochois. This island is three leagues long & a half a league wide. It abounds with game & fish.[1157]

The outlook from the lower part of this island with the neighboring islands & the N. shore presents a most delightful spectacle because of the beauty of the channel. This area seems to be very suitable for cultivation & good for hunting and fishing.

Fort Frontenac is three leagues away from that point. Just before arriving at the Pointe de Mont-Réal,[1158] which is the southern headland of the Baye de Cataracoui,[1159] one comes across a fairly good, deep bay.[1160]

The Pointe de Mont-Réal would be an advantageous site for a camp. It can only be approached along a front which would require the enemy to make a long detour in order to reach it. The shore slopes down towards the headland. [102]

1156 Ance au Corbeau is marked on Pouchot's map. It is the broad stretch of the northern channel of the river from the vicinity of Wallace Island, Ontario to the east end of Howe Island (Isle Cochois).

Pouchot's directions for entering the Ance au Corbeau from the east (or northeast) are confusing, and his "north" direction might be an error. When moving upstream along the Canadian mainland of the St. Lawrence, one would turn almost due south at the northwest corner of Georgina Island, Ontario, near the modern Thousand Islands Bridge, and pass between Georgina and Brau Island. That would lead to the channel along the north shore of Hill Island, Ontario and Wellesley Island, New York. The two could easily be mistaken for a single land mass to form Pouchot's Isle au Citron.

1157 Howe Island, Ontario.

From this point and around the margin of Lake Ontario, several contemporary charts are helpful. One is "Carte Du Lac Ontario ... 4 octobre 1757" by the naval officer La Broquerie (note 790). It is in the British Library, Crown Maps, cxxi 14. Another is "Lake Ontario", a chart from the earliest days of the English occupation of Lake Ontario, probably c.1760, and based on La Broquerie's work. One of several versions is in the William L. Clements Library. There is also a simple map, entitled "Lake Ontario to the Mouth of the River St. Lawrence" printed in Mante, p. 30. Modern maps include the *New York State Atlas & Gazetteer,* current Canadian and United States navigational charts, and modern Canadian topographical maps.

1158 Point Frederick opposite Kingston, Ontario.

1159 The mouth of the Cataraqui River at Kingston.

1160 Pouchot probably identifies Deadman and Navy bays as a single feature.

Cataracoui, or Frontenac,[1161] was a square stone fort, without an embankment, with walls 19 inches thick, the exterior sides of the square measuring 42 *toises,* but with very short flanks and with wooden scaffolding as a terreplein. The fort is overlooked on the landward side at half the range of a musket & the surrounding terrain consists of natural screens running into one another. Since these all overlook the fort, they prevent it from being turned into a good position except at very great expense.[1162]

The anchorage, which is right up against the fort, is excellent for **[103]** ships, which can winter there. Near the entrance to the N. side of the bay, there is also an inlet suitable for shipbuilding. The interior of the bay is a kind of marsh inhabited by a great abundance of waterfowl. The land of the surrounding area does not possess topsoil to any depth but it is good for cultivation. The soil of the interior is very good.

The bay has the disadvantage of not being located on the lake & thus of not being able to observe what is passing along it. The shore outside the bay is very rocky and very difficult for effecting a landing. If it is impossible to enter the main bay, one has to use the Baye du Petit Cataracoui.[1163]

Petit Cataracoui has the same entrance as the main bay & is only a quarter of a league in extent. The bottom of the water is full of rushes. This first bay is of some consequence, because the enemy can come and land without being seen **[104]** from Frontenac, which they can easily reach as it is barely a league away. This is what Bradstreet did in 1758 when he came with 4000 men to attack the fort, which only had a garrison of 50 men plus thirty voyageurs who happened to be there also.

A quarter of a league from Petit Cataracoui, there is a wide but shallow bay called "Ance au Sable." It is the place where the builders of Frontenac came to get sand for the fort.[1164]

1161 Note in original, p. 103 (a): *Cataracoui is the name of Frontenac Bay and that of the fort built in 1672 on the orders of the Comte de Frontenac, subsequently abandoned and then re-established in 1695 in accordance with the plans of that Governor of New France.*

Rocque includes a "Plan of Fort Frontenac" showing it at the time of the British attack in 1758. See note 240.

1162 Note in original, p. 103 (b): *This fort had only been established in order to keep the Iroquois in check.*

Pouchot's evaluation of the natural position of Fort Frontenac might be an apology for his inability to construct effective retrenchments there in 1755.

1163 Cataraqui Bay at the mouth of Little Cataraqui Creek on the west side of Kingston, Ontario.

A league & a half farther on is another bay formed by the mouth of a river. The shores of the bay are high & covered with boulders. Bateaux could not safely stay there.[1165]

Two leagues away, along the north shore of Frontenac, there are three little islands called "Tonégignon", which have been abandoned by the Indians. **[105]** It is difficult to pass between these islands & Isle Tonti because of a considerable sandbar blocking almost the entire distance between them. One has to steer between the two little islands to the N. & S. in order to reach Isle Tonti, which is three leagues long & a league & a half wide in several places.[1166]

Vessels sail along the N. coast of this island almost to its tip, then stand off the island while proceeding directly to Petit Cataracoui. There is a tiny tree-covered island of rocks which should not be approached too close because there are reefs, especially around the upper part.[1167]

Bateaux traverse along the coast of the Baye de Quinté, the mouth of which is a league wide. You leave the bay on the right if you do not want to use the portage, which is fifteen leagues inside the **[106]** bay. By using it, you avoid the passage round the large peninsula, which is not too pleasant. The portage is close to a league long, over sand all the way.[1168]

You follow the coast of the peninsula for two & a half leagues, after which you cross the bay, which is three leagues across the mouth & which extends five leagues inland. It is not known whether it would be suitable to anchor there. The N. headland is a rock. The whole peninsula is covered with a splendid forest.[1169]

1164 Sand Bay on the west side of Caruthers Point which separates it from Cataraqui Bay.

1165 Noted as "B. de Tonneguignon" on the c.1760 "Lake Ontario." It is today Collins Bay.

1166 "Les Isles de Tonnégüignon" appear on the c.1760 chart. They are now Brothers Islands between Amherst Island (Isle Tonti) and Collins Bay. The sandbank is marked on modern charts as Amherst Bar.

1167 Salmon Island.

1168 The Bay of Quinte retains its French name. The portage back to Lake Ontario across the narrow neck at its western end is shown on Pouchot's map and on "Lake Ontario" in Mante, p. 30. The crossing was made near Carrying Place, Ontario.

1169 Pouchot here describes the passage on the lake side. Following the south shore of Cressy Point (the northern headland), a boat would cross Prince Edward Bay to Long Point (the southern headland).

A quarter of a league from the S. headland is a kind of strait. You steer close to Isle d'Ecoui, behind which there is a good anchorage. On the side of the open lake, there are two sandbars at the surface of the water called "Les Goëians."[1170]

The whole of the north coast of Lake Ontario is dotted with headlands every quarter or half a league, which form **[107]** reefs quite far out into the lake. It is difficult to sail around them when there is a little wind. The rocks are flat.

Two leagues from the Ecouis, there is an inlet with a mouth two leagues wide.[1171] It extends inland for nearly a league. The northern part of it is sandy, but the water is not sufficiently deep to act as an anchorage. The rest of the shore consists of flat rocks or pebbles.

At the S.W. extremity of the inlet is Pointe aux Gravois,[1172] which is where one anchors. For two leagues, the coast runs from N.E. to S.W. Vessels continue along the shore of Pointe aux Gravois, which is flat rock.

On returning from this headland towards the S.W. & in the W. sector at the first inlet, there is a sandy bottom where it is possible to anchor. The second inlet has a pebble bottom.[1173] **[108]**

From there, you pass by Pointe du Détour, which is the headland that extends farthest out into the lake.[1174] Its bottom is of flat rock and it is extremely difficult to get round it when the breeze is a fairly stiff one. The rollers are very dangerous because of the shallows.

After this headland, one comes across a number of large inlets extending inland for half a league. There is one, before you arrive at

---

1170 The southern headland (Long Point) is tipped by Point Traverse on the north and Prince Edward Point on the south. Isle d'Ecoui is Timber Island off Point Traverse. It is one of the "False Ducks", the other being False Ducks Island. The pair are shown as "Les Isles des Coüiu" on the c.1760 chart.

Les Goëians were Gull Bar and Traverse Shoal off Prince Edward Point.

1171 The large inlet is much farther than two leagues from the Ecouis, more like eight (24 miles). It is known today as Soup Harbor situated between Point Petre (shown as Pointe à la Barque on the 1757 and c.1760 charts) and Salmon Point.

1172 Pointe aux Gravois is so called on the 1757 and c.1760 charts. It is today Salmon Point.

1173 The second inlet is Athol Bay between Salmon Point and West Point.

1174 Although the 1757 and c.1760 charts would seem to indicate that this is modern Huyck's Point, Pouchot's description can only refer to West Point on the northwest side of Athol Bay. Seen on a modern chart, it does not extend the farthest into the lake, but might well have appeared so to Pouchot.

Ance des Dunes, which is half a league deep with a sandy bottom.[1175] But the W. shore is rock, just like all other headlands where the bays have a pebble bottom.

The Ance des Dunes[1176] is three leagues from headland to headland. The wind has formed sand dunes, just as at Dunkerk, which separate the lake from a marsh that extends three leagues inland.[1177] It abounds in waterfowl.

The shore of the lake as far as Pointe de Quinté[1178] is unrelievedly rocky. **[109]** On the inlets at the headlands there is a sandy bottom where it might be possible to anchor. There are also good anchorages around Isle de Quinté. The island might be three leagues in diameter.[1179]

From Pointe de Quinté, you enter a bay which extends five leagues across to the peninsula. The shore in the inland part of the bay is all sandy.[1180]

About two leagues from the peninsula, you come across the portage of the interior of the Baye de Quinté. It is necessary to sail well offshore of the peninsula, because inshore there are rushes. From that point, you have to portage over sand for 300 paces in order to arrive back at the lake.[1181]

The peninsula of Quinté was an island which was joined up to the mainland by sand & gravel that the S.W. winds blew into the bay of the S.W. **[110]** sector.[1182] The land all around is excellent. The

1175 East Lake, separated by a broad bar from Athol Bay.

1176 Wellington Bay.

1177 Dunquerque or Dunkirk is on the Channel coast of France. The marsh is West Lake, separated from Wellington Bay by a bar. The coast is noted on modern charts as "The Sandbanks" and is today an Ontario Provincial Park of that name.

1178 Huyck's Point.

1179 Nicholson Island. It is much smaller than Pouchot's estimate, more like two miles in circumference. The 1757 and c.1760 charts show two Isles de Quinté. The second is Scotch Bonnet Island.

1180 This is shown as "Bay de Quinté" on the c.1760 chart. It is that part of Lake Ontario between Huyck's Point and Presqu' Isle. The inshore part is Wellers Bay.

1181 The portage was east of (before) Presqu' Isle at the narrow neck of land between the modern Bay of Quinte and Young Cove on Wellers Bay on the Lake Ontario side. The distance is more like a mile and one-half. It is today near Carrying Place, Ontario.

1182 This is the Presqu' Isle de Quinté of the 1757 and c.1760 charts. It is today Presqu' Isle, Ontario forming the bay of the same name.

lowlands right up to the mountains, which are not very high, constitute a fine plain watered by two rivers that are marked on the map.[1183] It would be a delightful place to live. There are considerable quantities of game & fish and as a result the place is constantly frequented by the Missisake Indians.

From the peninsula as far as the Ganaraské River, the land around the shore would be the most suitable for cultivation of any to be found between there and Frontenac. The Ganaraské & Saumon rivers are only noteworthy because they abound in fish.[1184]

The Petits Ecors are almost sheer bluffs 40 to 50 feet high.[1185] They form little capes & bays; inland there are **[111]** river mouths and marshes. One can only put to shore in the interior of the bays.

After getting round the Petits Ecors, you encounter a large bay which is two leagues across. The river in the interior of the bay seems quite large.[1186] Its mouth is filled with rushes right up to its entry into the lake. This is very rare because, throughout that area, they are almost all blocked up with shingle which normally leaves only a narrow channel linking up with the lake. At that point, one can catch prodigious quantities of fish which at certain seasons swim up the rivers from the lake.

At the point where the Grands Ecors begin, there is a very considerable river mouth.[1187] The Ecors are indented, almost sheer cliffs from 80 to 100 feet **[112]** high. They extend over five leagues. At the end of this stretch of land, there is a sandy cape covered with trees which forms a peninsula. Behind there is a large bay partly filled with reeds.[1188] Vessels can anchor & spend the winter there.

1183 The Ganaraské and Saumon rivers noted below.

1184 The Ganaraska River enters Lake Ontario at Port Hope, Ontario. The Saumon is today Oshawa Creek at Oshawa, Ontario.

1185 Both Pouchot's description and his map would seem to place the Petits Ecors west of the Saumon River between Oshawa and Pickering, Ontario. The shoreline in that area is not marked by "sheer bluffs", however. The Petit Ecors were more likely the bluffs that begin just west of the Ganaraska River and end well before Oshawa Creek (Saumon River) near Newcastle, Ontario.

1186 This is the considerable bay off Fairport Beach, Ontario. While the river in the interior of the bay is not very large, its mouth at Frenchman's Bay, would have made it seem so. It is also possible that Pouchot refers to Duffins Creek at Squires Beach, just east of Frenchman's Bay.

1187 The Rouge River at Rosebank, Ontario marks the beginning of the Scarborough Bluffs.

At the cape of the peninsula, there is a good anchorage & at the far end of the bay a river suitable for the construction of a mill. In the surroundings there are very fine pine woods. You have to portage when travelling by canoe from the far end of the bay to the Ecors.[1189]

The fort of Toronto is at the far end of the bay, on the hill which is quite high & covered with flat rock. Vessels cannot approach within cannon range of it. This fort, or outpost, was a square of 30 *toises* on each external side. The flanks were 15 feet long. The curtains constituted the **[113]** buildings of the fort. It was made of logs and rather well constructed but was only useful for trading purposes.[1190]

A league to the W. of the fort is the mouth of the Toronto River, which is fairly wide. This river provides communication with Lake Huron if a 15 league portage is made. It is used by all the Indians who come down from the north.[1191]

The other rivers which join up to drain into the Fond du Lac appear to be quite large. They are above all advantageous for hunting & fishing.[1192]

The Fond du Lac is made up of a bank of shingle two leagues long. It separates the main lake from a little one which is almost entirely filled with reeds. At the far end of it, there is a river that ends in falls.[1193] The place is of very great interest because of the large quantities **[114]**

1188 The peninsula that sheltered the bay that would become Toronto Harbor has been much altered. It is today broken up into the Toronto Harbor Islands and the Outer Harbor East Headland.

1189 This portage was across the neck of the Toronto peninsula.

1190 Fort Rouillé, constructed in 1750-1751 and destroyed by its garrison following the surrender of Niagara in July 1759. See note 348.

1191 The Humber River was the beginning of the portage from Toronto to Georgian Bay via Lake Simcoe. See Robinson, pp. 195-208 for a discussion of the portage.

1192 The larger of these rivers, from east to west, are Mimico Creek, Etobicoke Creek, Credit River, Sixteen Mile Creek, and Bronte Creek. Pouchot's map identifies (also from east to west) R. au credit", "R. au atoka" and "R. au Gravois." Several of these are shown on the c.1760 chart of Lake Ontario, but only "R. au Credit" is named. The 1757 chart shows Sixteen Mile Creek as "R. des deau follis" and Bronte Creek as "R. au gravois."

1193 A substantial bar separates Burlington Bay from Lake Ontario. Pouchot's map identifies the river as "R. du Fond du Lac" and indicates a portage to the Thames. It is Spencer Creek which falls from the Niagara Escarpment to drain into Burlington Bay (Hamilton Harbor). The falls are probably Websters Falls near Glenwood Heights, Ontario.

of waterfowl which frequent it, for example, ducks, teal, bustards, geese & swans. You can shoot them very easily as they fly past by standing on the rocks of the falls.

The river extends well into the hinterland & provides communication with two rivers by means of portage. One of them flows into Lake Erie[1194] and the other, about sixty leagues long, into Lac Ste. Claire above Détroit. The land here is very beautiful & excellent for hunting. The river, for which M. Pouchot was never able to obtain a name, has no rapids & provides easy navigation along its whole course.[1195] The Indians, or Canadians, sent from Niagara to Detroit in winter opt for this route & usually take ten days to arrive; they reckon on a journey of a hundred leagues from Niagara to Détroit via this route. **[115]**

Between Fond du Lac and Niagara, a distance of 15 leagues, there are several rivers. These rivers almost all flow down from the curtain of high ground called "Les Cotes",[1196] which end at the river of the Fond du Lac. The space between Les Cotes & the lake is a very fine plain covered with trees. There are pines in the area of Le Grand Marais[1197] & the Marais aux Trois Sorties[1198] which are convenient for use at Fort Niagara.

This type of tree is rare in the region, where the trees are normally several species of oak, walnut, chestnut, *bois jaune,*[1199] which is excellent for panelling in construction. One also finds a type of black walnut

---

1194 The Grand River flowing south into Lake Erie.

1195 Until renamed the Thames in the 1790s, the river flowing southwest across Ontario to Lake St. Clair was known as Rivière à la Tranche.

1196 The Niagara Escarpment.

1197 Pouchot's map shows Le Grand Marais as the westernmost of the major ponds on the north shore of the Niagara Peninsula. The 1757 and c.1760 charts both place it closer to Niagara. While Pouchot might have identified the pond at Jordan Harbor formed by Twenty Mile Creek, that was not a convenient distance from which to move pine to Niagara. Le Grand Marais was more likely Martindale Pond at Port Dalhousie, Ontario, about 12 miles west of the Niagara River.

1198 Pouchot's map shows only a "Marais des 2 Sorties" and places it east of Le Grand Marais. He probably confused the name with that of Rivière des Deux Sorties to the east of the Niagara River. The 1757 and c.1760 charts show a "Marais a la Biche" west of Le Grand Marais. All three sources probably refer to the same pond which is formed by the junction of Fifteen Mile and Sixteen Mile creeks at that distance west of Niagara. The pond has multiple exits.

Pouchot's map also shows a "Marais Normand" even closer to the Niagara River. This is given as "Marais Marot" on the 1757 chart and is not identified on the chart of c.1760.

which makes very fine furniture, beech, plane-trees and maple. The two latter trees make very good sugar, which is less acidic than cane sugar. [116]

In the N. part of the Toronto region, the more common trees are pine & cedar because of the proximity of the mountains. The latter are not as high as the Vauges,[1200] but are covered with fine woods & excellent soil. They are not so cold as those in the Carillon area.

Before 1754, travellers hardly ever took the N. shore of the lake where there are, however, a greater number of harbors than on the S. shore for a considerable number of bateaux. The journey to Niagara is rather longer ever since the northern route has been preferred, even though Chouegen no longer existed as a threat.

We shall defer the description of Niagara until the chapter on the Belle-Riviere in order to follow the south shore of the lake.

The shore from Niagara as far as [117] the Grand Riviere aux Boeufs[1201] runs E. & W. for nearly 24 leagues. This shoreline is elevated, its cliffs being generally 30 to 40 feet high. The rivers found there do not extend very far into the hinterland.

The Petit Marais a league & a half from Niagara is a little bay that can accommodate 2 to 300 bateaux. The English landed there in 1759.[1202] The rivers called Aux Ecluses & Des Deux Sorties which are 5 to 6 leagues from this spot are only noteworthy because pines are to be found there.[1203]

1199 Franklin Hough identified this as a whitewood or tulip tree. Pouchot, *Memoirs,* II, p. 121n.

1200 The Vosges Mountains in eastern France. Pouchot greatly overestimates the height of the "mountains" north of Toronto.

1201 Oak Orchard Creek in Orleans County, New York.

1202 Four Mile Creek, so-called because of its distance east of the Niagara River.

1203 Rivière aux Ecluses and Rivière Des Deux Sorties enter Lake Ontario at Wilson, New York. The former is Twelvemile Creek and the latter, which periodically has two mouths separated by Sunset Island, is known as East Branch. The English quickly learned of the pines growing at that point, and they were used for masts for the snow *Onondaga,* completed in the summer of 1760, and probably for the *Mohawk* built the previous fall. Ship carpenter Nathaniel Fuller noted in his journal for June 9, 1760 that he and a work party "Went Twelve Mile Down the Lake" after "masts and Yards for the Vessel a Building." Fuller Journal.

Pouchot fails to mention another large creek shown on the 1757 La Broquerie chart and the c.1760 chart. This is noted on both as "R. a la Fourche" and is today Eighteenmile Creek at Olcott, New York.

Above the Riviere aux Boeufs, on the ground beyond the shoreline, there is a little mountain which appears round and which is called "La Butte à Gagnon."[1204] It is a landmark that indicates to vessels on the lake that they are 15 leagues away from Niagara. Upon arriving opposite [118] it, vessels put out into the lake as far as they can so as not to overshoot the mouth of the Niagara River, which is only perceived after one has passed it. Vessels would find themselves in very great difficulties if they were unable to enter the river mouth. Since the winds from the N.E. are normally very brisk, no shelter at all would be found between Niagara and Fond du Lac, which would thus force them to keep to the N. of the lake. If they navigated in that area, the winds from the W. & above all the N.W., which are very severe, would cast them up on the S. shore.

Since the shoreline forms a uniform curtain everywhere, there is no other landmark for establishing one's position than this hillock.

The crossing by ship from Frontenac to Niagara normally lasts 4, 6 or 8 days unless one encounters a wind from the [119] N.E., which normally only blows during the phases of the moon. To cross from Niagara to Frontenac, it is barely necessary to spend a night afloat. The winds are almost always from the S.W. & brisk.

The mouth of the Riviere aux Boeufs is good for putting ashore; but from the direction of Niagara, one has to sail out into the lake in order to pass the cape, because of a long & dangerous reef which lies to the west.[1205] From Niagara up to this river, there is very little, indeed no shelter for any considerable number of bateaux. After the river, the shoreline of the lake is not so high. It turns towards the south east & forms shallow bays of a sort which are about a league across.[1206]

1204 Golden Hill at Thirtymile Point at the eastern edge of Niagara County.

1205 The cape is Thirtymile Point. To the west of Oak Orchard Creek (Rivière aux Boeufs) and about five miles east of Thirtymile Point, shallows extend some distance into Lake Ontario with depths of only 13 feet as far as two miles from shore. A mile off the beach the water is as shallow as six feet. This is probably the reef. It does not appear on either the 1757 or c.1760 charts.

1206 The last eastbound landmark was Bogus Point in Monroe County, New York. This is noted on the 1757 and c.1760 charts as "Pt. Autourtre." From Bogus Point the shore turns southeast past Braddock Bay and Cranberry, Long, Buck, and Round ponds.

Braddock Bay would seem to take its name from the unfortunate General Edward Braddock (note 246). The earliest English charts of the area, however, identify the place as "Brideauxs" or "Brideauts" Bay, indicating some early confusion between Braddock and the deceased commander of the 1759 Niagara expedition, Brigadier John Prideaux

...(continued next page)...

Just before arriving at Fort des Sables, one comes across the mouth of the Casconchiagon River. **[120]** Its entrance forms a rather extensive and deep bay. At the entrance there is a reef which is difficult to negotiate.[1207]

This is the river of the area that extends farthest inland. It has three falls along the escarpment which are almost as beautiful as those of Niagara.[1208]

It is necessary to go right to the far end of the Baye des Sables[1209] in order to begin navigating the Casconchiagon River. There is a portage 3 leagues long which is the most convenient route. Details of this navigation will be given in a separate chapter so as not to interrupt the description of the shores of the lake.

Fort des Sables is nothing more than very high sand dunes which run round the bay of that name. The latter extends inland for three leagues and has good deep water. From the bay, the land up to the foot **[121]** of the escarpment is very low and marshy and is thickly wooded.

The Baye des Goyogoins is fairly beautiful & extensive.[1210] Les Boucauts is a bay filled with little islands and rather large sand dunes

---

(note 125). This might have occurred because German and not English was the native language of Ensign George Demler who drew the maps. The names appear on Demler's "A Sketch of the South Shore of Lake Ontario" in the William L. Clements Library and "A Sketch of Lake Ontario Between Oswego and Niagara" in the British Library, Crown Maps, cxxi 15.1.

1207 "Fort des Sables" was the name for the bar across the mouth of Irondequoit Bay east of Rochester, New York. The place was so-called as early as 1702 and appears on a French map of Lake Ontario from 1728. Severance, I, pp. 200, 236. The name might have been applied in reference to a temporary fortification constructed there in 1687 during Governor Jacques-René de Brisay de Denonville's military expedition against the Seneca. That post was maintained for only a few days. The spot was later a location for occasional commercial activity. English reports of 1716 described "a Trading house with 5 French Traders & a Smith" at Irondequoit Bay. Charles Howard McIlwain (ed.), *An Abridgement of the Indian Affairs ... Transacted in the Colony of New York from the Year 1678 to the Year 1751 by Peter Wraxall* (New York, 1968), p. 113. Charlevoix, who passed Irondequoit Bay in 1721, mentions neither the bay nor an establishment there.

The Casconchiagon is the Genesee River. The coast between its mouth and Irondequoit Bay forms a wide and deep bay.

1208 Pouchot notes the three falls of the Genesee River at Rochester, New York.

1209 Irondequoit Bay, four miles east of the mouth of the Genesee River. The portage circumvented the falls.

1210 Sodus Bay in Wayne County, New York.

covered with trees. The cliffs go sheer down into the water.[1211] If soundings were taken there, there is reason to believe that excellent anchorages for vessels could be found between the islands. The surrounding land is high and sandy & the escarpment is close by.

From this bay as far as Chouegen, which will be discussed in the next chapter, the lakeshore is pebble beach mingled with rocks.

The land beyond Chouegen, as one proceeds down the lake, is higher & the lakeshore is [122] mainly rock as far as Pointe au Cabaret.[1212] This is a fairly long cape which falls sheer down into the water from a height of 30 to 40 feet. It is the most prominent cape of all.

Half a league to the E. of Chouegen there is a small sandy-bottomed bay where M. de Montcalm landed & camped prior to the siege of Chouegen in 1756.[1213] The English have since made a *désert*, or cleared ground, there and erected redoubts which overlook the bay.

In the course of this navigation, one can enter by bateau the Riviere à la Planche,[1214] in the Indian language *Tensaré Negoui,* & the Riviere à la Grosse Ecorce,[1215] or *Cassontachégona.* These rivers do not extend very far inland.

The Riviere à la Famine, *Keyouanouagué* in [123] the Indian language, does extend a long way into the interior & almost reaches

---

1211 Little Sodus Bay in Cayuga County, New York. To the east, the coast is still known as "The Bluffs." The 1757 and c.1760 charts show "Les Boucauts" or "The Casks" as a stretch of shoreline and not a bay, so the cliffs were probably the primary landmark.

1212 Nine Mile Point in Oswego County, New York.

1213 Baldwin's Bay at the mouth of Wine Creek on the eastern edge of Oswego.

1214 Riviere à la Planche was the second major creek east of Oswego. It appears on the 1757 chart (where it is transposed with Rivière à La Famine) and on that of c.1760. It is also shown on "A Map of the Province of New York ... by Claude Joseph Sauthier", 1776, in Nebenzahl and Higginbotham, pp. 158-59. The river is usually shown entering Lake Ontario through a substantial swamp and was probably modern Grindstone Creek in the Town of Richland, Oswego County.

The name seems to have changed by Pouchot's time. In 1721 Charlevoix knew it as La Petite Famine, perhaps explaining the confusion of names on the 1757 chart. Charlevoix, 1761, I, p. 313.

1215 Riviere à la Grosse Ecorce is west of Riviere à la Planche on the 1757 chart and is the first large creek east of Oswego. The 1776 Sauthier map identifies it as "Cartargaren or R au Sable." The chart of Lake Ontario in Mante, p. 30 shows it as "Salmon River" or "R. au Sable" between Oswego and Riviere à la Planche. This was the Salmon River in the Town of Mexico, distinct from the nearby Salmon River that joins Lake Ontario near Port Ontario, New York.

the portage on the heights.[1216] This was the route usually taken by our raiding parties to go along the frontier, thence along the lake and via the Riviere des Onoyotes, in order not to be observed.[1217]

From Pointe au Cabaret as far as the Riviere à M. le Comte,[1218] the shore forms a great arc of sand, with tree-covered dunes. The land behind is made up of marshy plains right up to the banks of meandering rivers.

Between the Riviere au Sable & that of la Famine, there is a small river called in the Indian language *Canogatiron.*[1219] The Riviere au Sable,[1220] in Iroquois [124] *Etcataragarenré,* is noteworthy because at the top part of its S. tributary, called *Tecanonouaronesi,* is the place from which, according to Iroquois tradition, they all came or rather, in their own conception, where they were born.[1221]

Between the Riviere aux Sables & that of M. le Comte is the little river called *Outenessouéta.*[1222] The Riviere à M. le Comte provides good shelter for bateaux because of a sandy hook of land formed by the river mouth.[1223]

It is possible to canoe along all these rivers & the surrounding country is very good for hunting. Niaouré, or *Neyaouinré,* Bay extends inland for five leagues. A number of substantial rivers flow into it.

1216 Note in original, p. 123 (a): *So called since M. de la Barre, the governor of Canada, almost lost his entire army through hunger on the banks of this river while going to make war against the Iroquois in 1684.*

Riviere à la Famine is the Salmon River that joins Lake Ontario at Port Ontario, New York. It is the largest of the streams in the southeastern corner of Lake Ontario.

Joseph-Antoine Le Febvre de La Barre (1622-1688) was Governor General of New France from 1682 to 1685. When he led an army of over 1,300 French and Indians against the Iroquois during the summer of 1684, sickness and lack of supplies turned the campaign into a fiasco. La Barre was unable to attack the Iroquois. *DCB*, I, pp. 442-46. See the summary of his expedition in William D. Le Sueur, *Count Frontenac* (Toronto, 1964), pp. 183-86.

Pouchot's editor again drew on Charlevoix for the information in his footnote. See Charlevoix, 1761, I, p. 314. He called the river "La Grande Famine" to distinguish it from "La Petite Famine" (Rivière à la Planche) closer to Oswego.

1217 The upper reaches of the Salmon River come within about 15 miles of the Oneida Carry between the Mohawk River and Oneida Lake. Pouchot describes a route that would have allowed parties to move south to Oneida Lake and then west to the Oneida River. This could have been used for the purpose of interrupting English traffic with Oswego.

1218 This is identified as "R. à Mr. l'Comte La Motte" on the c.1760 chart of Lake Ontario. It is now Stony Creek in the Town of Henderson, Jefferson County, New York. The river probably took its name from Dominique La Motte de Lucière (1636-1700). *DCB,* I, pp. 415-16. La Motte was of noble birth and had been a companion of La Salle during his 1678-1679 expedition to Niagara. La Motte was sent from Frontenac to Niagara by land through the Seneca country and would have passed Stony Creek. Severance, I, pp. 43-45.

There are excellent anchorages for large vessels. The best of them is to be found between [125] the little islands & that almost circular stretch of land where M. de Montcalm came and camped with his army before proceeding to Chouegen.[1224]

This seems to be the best place to build a fort at the E. end of the lake. The position is connected to the mainland only by a causeway of shingles. The lake is sufficiently shallow to allow access to the shore only by vessels no larger than bateaux. It would be very easy to fortify & would protect ships at anchor. The land around the bay is excellent for cultivation and fish & game are abundant.

There are two large rivers along which one can easily reach the English roads to Chouegen & keep watch on them much better than if one were based at Frontenac.[1225]

There is a good external anchorage at the Isles aux Galots, with [126] easy access to the post & favorably sited for navigating the lake. From that point, one is well placed to sail along the Chouegen River when circumstances require it.[1226]

1219 Rivière Canagatiron was possibly Little Sandy Creek draining into Lake Ontario through North Pond in the Town of Sandy Creek, Oswego County. The 1728 de Léry map of Lake Ontario (Severance, I, p. 236) shows "R. Catagarène" as the northernmost creek in this part of the lake - probably Little Stony Creek. The 1757 chart agrees and calls it "Cataragaren." Pouchot places the river farther south. The 1776 Sauthier map shows only an unnamed small river between Au Sable and La Famine.

1220 Rivière au Sable is shown on the c.1760 chart. It is today Sandy Creek in the Town of Ellisburg, Jefferson County. English plans, such as that in Mante, p. 30 and Sauthier, 1776, call it the Assumption River.

1221 Pouchot's information might have referred to the southern branch of the upper reaches of Sandy Creek in the Town of Rodman, Jefferson County.

1222 Little Stony Creek in the northern part of the Town of Ellisburg, Jefferson County. Charlevoix (1761, I, p. 313) calls this the Assumption River. Later maps, such as Sauthier, 1776, apply the name Assumption (or Ganentouta) to what must be Rivière au Sable (Sandy Creek).

1223 Probably Sawyer Point.

1224 Association Island at the tip of Stony Point. See note 302.

1225 The Black River is the only stream of any size emptying into Henderson Bay that would have offered a route to English territory and the Oneida Carry. The second river referred to is probably Rivière à la Famine (Salmon River).

1226 The Isles aux Galots are Stony Island and Galloo Island situated to the west of Stony Point. The "post" is that suggested by Pouchot for Association Island. Charlevoix (1761, I, p. 313) says that one could see Chouegen or Oswego from the Isle aux Galots.

Ships which intend to enter the river from the S. shore of the lake pass between the Grande Terre & Longue-Isle, which is called the "Chenal de la Galette."[1227] They have to sail along the same channel to reach Frontenac or between Isle à la Forêt & Isle Tonti.[1228]

Bateaux leaving Frontenac to go to Chouegen sail between Isle à la Forêt & Longue-Isle. Vessels must stand off the latter island because there is always a heavy swell there. When the wind blows, there is no shelter. From there one crosses to the Isle au Chevreuil & the cape of Niaouré Bay. There is a good **[127]** bay on the lower part of the Isle au Chevreuil.[1229]

The capes that extend into the lake from Longue-Isle are made up of flat rocks or pebbles. All these islands would be excellent for cultivation.

There is a hook of land on the lower part of the Isle aux Galots, close to shore, where vessels can shelter in severe weather. There is a reef that abuts the E. cape which ships must stand off before sailing back to the island. The anchorage is very good for ships.[1230] **[128]**

• Chapter III •

On communication between the Chouegen River and the English possessions[1231]

Chouegen, lately rebuilt,[1232] is constructed on the site of the former Fort Ontario. The English have given it the same name. It is in the shape of a pentagon, each exterior side measuring 80 **[129]** *toises.* It is partly made of earth faced with saucissons overlooking the lake. The

---

1227 Pouchot at this point gives directions for entering the St. Lawrence, not the Oswego River. The entrance to the St. Lawrence is marked by the mainland ("Grosse Terre") at modern Cape Vincent, New York on the right and Wolfe Island, Ontario on the left. The latter is "Longue-Isle" on Pouchot's map, although other contemporary charts refer to it as "La Grande Isle." The c.1760 chart labels the body of water between Wolfe Island and Cape Vincent as "Chenail de la galette ou du Sud."

1228 To reach Frontenac it was necessary to either enter the St. Lawrence and round the eastern end of Wolfe Island or skirt the western end of Wolfe and pass between the much smaller Isle à la Forêt (Simcoe Island) and Isle Tonti (Amherst Island).

1229 Isle au Chevreuil is Grenadier Island, New York. By the cape of Niaouré Bay, Pouchot probably means the southern tip of Point Peninsula. From there one would cross Henderson Bay to Stony Point and the site of Montcalm's camp on Association Island.

There is a very distinct though seemingly exposed bay at the southwestern end of Grenadier Island. Another more sheltered bay is on the east end of the island.

remainder is encased in timbers about 3 feet square. The parapets might be 12 feet thick. The terre-plein is a floor made of large beams about fifteen inches square. The space underneath these floors is made up of quarters or casemates. The ditch is at least five *toises* wide and there is a glacis. No external earthworks have been observed. In 1760, the fort was almost finished.[1233]

1230 The hook of land is probably the point and bay at the southwest end of Stony Island off Stony Point. The c.1760 chart shows prominent reefs at each end of the island.

1231 Identifications for this segment of Pouchot's narrative are based on "A Map of the Province of New York... by Claude Joseph Sauthier", 1776, in Nebenzahl and Higginbotham, pp. 158-59; on the map entitled "Communication between Albany and Oswego" in Mante, p. 61; on the modern maps in the *New York State Atlas & Gazetteer;* and on the modern charts in *The Grand Canal Cruise 'n Chart* (Albany, 1974).

Another French description of the route from Oswego to Albany, probably prepared in 1757, is in *DRCHSNY,* X, pp. 674-80.

1232 Note in original, p. 129 (a): *Originally the fort consisted of nothing more than a trading post which the Iroquois had allowed the English to build in 1713. It was transformed into a fort in 1727 thanks to the cunning of the English who since that time have not ceased to increase its size. It had been erected on French territory. M. le marquis Beauharnois, the Governor of Canada, had protested this manifest usurpation.*

Charles de Beauharnois de La Boische, Marquis de Beauharnois (1671-1749) was Governor General of New France from 1726 to 1747. *DCB,* III, pp. 41-51.

A regular trading post did not exist at the site until the 1720s. See Mohar, "Oswego's First Permanent Settlement." The date 1713 might indicate confusion with English claims of influence over the Iroquois country stemming from the Treaty of Utrecht signed that year.

The date for establishment of a military post is correct. In a letter of July 20, 1727 to Governor Thomas Burnet of New York, Governor Beauharnois protested the English construction of "a Redoubt with machicoulis and full of loopholes and other works belonging to a fortification ... in which you have placed a Garrison of regular forces." *DRCHSNY,* IX, pp. 969-70. The French governor also sent a summons to the English commandant at Oswego on July 14. *Ibid.,* pp. 973-74.

1233 The new Fort Ontario had only just been begun when Pouchot passed through Oswego as a prisoner of war in August 1759. He observed the nearly completed fortifications in September 1760 on his way from Fort Lévis to New York. The use of saucissons on the lake side was a temporary expedient soon to be replaced by a continuation of the earth-filled logwork wall.

The 1759 Fort Ontario was on the east bank of the Oswego River, site of the first fort of that name destroyed by Montcalm in 1756. The place was demolished and reconstructed on several more occasions, each time reusing the 1759 trace. See Bertsch, pp. 122-27. A contemporary printed plan of the new fort is "A Plan of Fort Ontario" in Rocque. A spectacular perspective view by Francis Pfister and a section through the post by Thomas Sowers, both giving an excellent idea of logwork fort construction, are preserved in the British Library. See "Perspective View of Fort Ontario", 1761, Crown Maps, cxxi 94 and "Section and Elevation of Fort Ontario", Crown Maps, cxxi 95.

...(continued next page)...

Around the fort, at the extreme range of a musket, the English have constructed four very well finished blockhouses. One of them overlooks the slope already mentioned, the other the river. The fort could be bypassed at cannon range. On the side of the upper part of the river, there is a kind of escarpment **[130]** overlooking the fort where it would be easy to open up trenchworks. The ground slopes down towards the fort.

The entrance to the Chouegen river is narrow because of a rock under the water in the middle of the river mouth. A little farther upriver, at a point of two banks of shingle, the passage is narrow & quite difficult. Nonetheless the English have sailed ships of 22 guns up thc rivcr.[1234]

At the far end of the passage, there are two hooks of land which can be used as docks of a sort. They anchor all their bateaux there, sheltered from the river when in flood. Out of the one beneath the fort they have made a jetty in wood & stone, in order to contain and close off better the water.

The rapids begin half a league from the fort, at the first bend of the river. These rapids are fordable. Bateaux can be brought up them by **[131]** pole, if they are first emptied, with four men in the large bateaux & two in the small ones.

These large bateaux, loaded, can carry up to 20 men, while the small ones can take 7 to 10 men when empty. Besides these bateaux, the English have long-boats for hunting whales, which they call "Woel-Bot;" [1235] they are very light on the oar but are of no use for navigating

---

The final version of Fort Ontario, constructed in two major phases between 1839 and 1870, is preserved today in Fort Ontario State Historic Site in Oswego, New York. The summary of the early history of Oswego given by Pouchot's editor is essentially correct. The original fort at Oswego, constructed in 1727 and also destroyed in 1756, was on the west side of the river. See Bertsch, pp. 109-20 and Roberts, pp. 570-72.

1234 A reference to the large armed vessels based at Oswego during the campaign of 1760 and specifically to the snow *Onondaga* which Pouchot consistently rates at 22 guns. He also observed the large English snows constructed at Oswego in 1756.

1235 Whale boat.

The Oswego River was very hard on whale boats and bateaux, especially below Oswego Falls. William Amherst noted, in 1760, that "navigation from the Oswego Falls here [to Oswego] is very bad for the boats, destroying numbers, the greatest part of the way being rifts & the current running strong." Webster, *William Amherst,* pp. 58-59. His brother, the general, believed that, because of the damage always incurred, "it appears to me to be a very extravagent measure to build the batteaus & whale boats at Albany. It would be much cheaper & better to build them at the Lake [Ontario]." Webster, *Jeffery Amherst,* p. 217.

these rivers, especially when the water is low. You often have to get into the water and haul them, and they are not made for that.

The river bottom is full of small rocks, which have to be avoided. It is rather like the rapids of Chambly, but the river is not so wide there. On both sides, the banks are very high. **[132]**

There is a footpath which runs along the left bank of the river in the direction of the old fort of Chouegen for a distance of three leagues. The forest is very dense. The terrain is covered with humps & hollows which would favor ambushes.

At the end of these three leagues, the river is navigable, but every league or so there are bends with reefs that are very difficult to pass. The bateaux have to proceed in line ahead & be poled vigorously. If the bateaux yaws, the men get into the water.

Above the point where one re-embarks, the river becomes wider & the water smooth. The countryside is flat & covered with fine forests. There is quite a strong current in the crook of each bend. Each twist and turn runs for about a mile along the same compass bearing. The river flows generally in an E.N.E. direction.[1236] There are several islands in the river's main channel. **[133]** The one where M. de Villiers attacked Bradstreet is five leagues above Chouegen.[1237]

Above this place the islands become more frequent & are to be found at every mile stretch. At the foot of the falls,[1238] there are many islands. One must steer along the N. bank, otherwise the bateaux will run aground.

You must get out of the bateau at a distance of a good musket shot from the portage & bring it up by pole in the current as far as the falls

---

1236 An error or misprint. The Oswego River flows north-northwest.

1237 The action took place on the river near Battle Island, about seven miles above Oswego in the town of Minetto, Oswego County. On July 3, 1756 Captain Louis Coulon de Villiers (note 164) attacked a large convoy of empty boats returning from Oswego under the command of Captain John Bradstreet (note 451). Montcalm reported that the French and Indians encountered 300 to 400 bateaux and took 26 scalps and 50 prisoners. Montcalm to Argenson, July 24, 1756, *DRCHSNY,* X, p. 434.

English accounts speak of 300 bateaumen attacked by 700 French. The English reported losses of 40 killed and missing and 24 wounded. *New York Mercury,* July 19, 1756. Three versions of the battle and subsequent pursuit of the French were published in the *New York Mercury* of July 19, July 26 and August 2. These have been reprinted in *DHNY,* I, pp. 482-87. Another is in Shirley to Fox, July 26, 1756, in Lincoln, *Correspondence of William Shirley,* pp. 488-90. For a detailed secondary account see Parkman, pp. 277-78.

1238 Oswego Falls at Fulton, New York.

where there is a good corduroy road for manhandling the bateaux. A hundred paces above the falls, the water is good.

At this portage, the English have constructed a star-shaped fort of stakes 15 feet high & a foot in diameter. The fort is overlooked from the N.E. at half musket range. It can hold 100 to 150 men. **[134]** They have built a number of sheds for storing goods.[1239]

The river above the portage is fine & wide, like that of Sorel, and without any appreciable current. The bends extend for about a quarter & one half a league. Three reefs are to be found up to the fork of the Sonnontoins & Chouegen rivers.[1240] The one which is half a league from this confluence is the most extensive.

It should be noted that along its entire length this river is not very deep. The bottom is full of flat stones covered with very slippery mud, which compels the use of poles & oars with iron tips when navigating this stretch.

The River of the Five Nations, or of the Sonnontoins, is a fine one & a little wider than the Chouegen River. **[135]** The water is deep throughout its course and navigation right up to its end is safe. This river links up with a number of lakes & with the different Iroquois nations, as can be seen from the map. The land around it is excellent and covered with fine forests.

At the confluence of the two rivers there is a fort with four bastions measuring 40 *toises* on each exterior side and built of logs. There are four large sheds in the fort, which is situated on the east side of the river. The surrounding country is quite flat.[1241]

Three leagues above the confluence, there are two reefs that are not too extensive. Three quarters of a league before the Lac des Onoyotes,[1242]

1239 A post for 100 men was ordered constructed at Oswego Falls in 1759 to support Prideaux's expedition against Niagara. It was to have been a retrenchment enclosed by a ditch with a blockhouse in the center. Scull, p. 72. The "Sketch of the Stockade Fort, at Oswego Falls" printed in Rocque shows an eight pointed star with what might be a central blockhouse. Four storehouses were arranged around the central building. The plan also shows the carrying place around the falls. In 1760 General Amherst considered the post "but a bad one" and reported that the boats were moved over rollers for 60 yards without being unloaded. Webster, *Jeffery Amherst,* p. 217. The fort and storehouses at Oswego Falls were evacuated by the tiny garrison on April 23, 1766. List of Officers [Commanding]..., Dec. 25, 1765, to Dec. 24, 1766, Gage Papers/139.

1240 The Oneida River, which drains Oneida Lake, joins the Seneca River at Three Rivers, New York to form the Oswego. The place is identified as "Three Rivers" on the map in Mante, p. 61 which labels the modern Oneida River the "Onondaga." The Seneca River drains several of the Finger Lakes.

there is a barrier of flat rock which only allows passage along the middle of the river. One has to stand up to one's waist in the water in order to get by it. **[136]** The English have placed large trees in the river in such a way as to guide the water along the channel. This is the worst reef on the river.

At the entrance to the lake, there is another reef, but this one can easily be negotiated provided a little care is taken. There is a fort at the entrance to the lake which acts as a warehouse. It is an earth entrenchment faced with saucissons, with a poor fraise and a ditch a dozen feet wide.[1243]

The English have constructed two large flat bateaux for transport on the lake. The New Gersay militia, on their return from Canada in 1760, crowded onto one of these bateau in order to be among the first to be ferried across and were caught in a squall on the lake, which is dangerous because the water is not very deep. The vessel **[137]** capsized & more than 200 men perished.[1244]

1241 This identification is clearly an error. There was no fort at Three Rivers. Captain John Bradstreet had recommended that one be built there in 1756 to protect convoys bound for Oswego, but it was not constructed. Gipson, VI, p. 198. Colonel James Montresor's 1759 orders substituted posts at the west end of Oneida Lake and Oswego Falls. Scull, p. 72. When General Amherst descended the Oneida River in July 1760, he encamped at Three Rivers and observed only "an Indian with his Squaw, two children, a dog, a small canoe, Gun & fishing rod." Webster, *Jeffery Amherst,* p. 217.

Pouchot's detailed description indicates confusion, perhaps during later review of his notes, with the forts at Oswego Falls or the west end of Oneida Lake, both of which enclosed four storehouses.

1242 Oneida Lake.

1243 A fort for 150 men was ordered built at "the West end of Oneyda Lake" in 1759. Scull, p. 72. Although it too was intended to have a central blockhouse, Rocque's "Sketch of Fort Brewerton at the West End of Oneida Lake" shows four "L"-shaped storehouses within an eight-pointed enclosure. The space in the center might have been intended for the blockhouse. The William L. Clements Library has two slightly different manuscript plans of Fort Brewerton (Small 1758 [sic]; Small 1760). In 1760, General Amherst believed that the "post on the West End is badly chose but not worth while now to alter the situation." Webster, *Jeffery Amherst,* p. 217. Fort Brewerton was evacuated and abandoned on April 25, 1766. List of Officers [Commanding]..., Dec. 25, 1765, to Dec. 24, 1766, Gage Papers/139.

1244 The New Jersey Regiment, composed of 26 officers and 496 men, formed part of Amherst's army on Lake Ontario in 1760. Hervey, p. 56. It was commanded by Colonel Peter Schuyler (1710-1762). *DCB,* III, pp. 587-89. Despite the reported magnitude of this tragedy, however, no account of it has been found in official records or colonial newspapers.

...(continued next page)...

Lac des Onoyotes is eight leagues long, two & a half leagues at its widest point & in general from a league to a league & a half wide. The two shores appear to have no banks. The surrounding country is low-lying & bordered by rushes.

The Indians only travel in elm-wood canoes on this lake,[1245] which freezes over every winter. The ice begins to break up in March or at full moon in that period. The ice does not go away, which tends to delay the ability to sail on the lake. Three or four leagues to the right and overlooking the lake, there are some very tall, though fairly rounded, mountains. They are the Goyogoin Mountains.[1246]

There is a sandbar at the entrance to the Woods-Orick River. One can never pass it without the bateau touching bottom. To pass, it is [138] necessary to steer directly towards the fort & the vessel returns to the river, keeping to the right rather than the left bank.[1247]

The impatience of the provincial troops to return home at the conclusion of the 1760 campaign is indicated in a series of entries in Amherst's journal. The New Jersey Regiment had participated in the capture of Montréal where, on September 10, it was ordered back to Fort William Augustus with the New York troops to complete repairs to the former Fort Lévis. Webster, *Jeffery Amherst,* p. 247. On September 18, the 80th Regiment of Foot was sent from Montréal to relieve the provincials at Fort William Augustus so they could be sent home. *Ibid.,* p. 253.

On November 10, soon after General Amherst had reached Albany, he complained that "The Provincials from Fort Stanwix & Oswego [mostly Connecticut units] coming in daily. They did not wait for Col Schuylers passing first as I ordered." *Ibid.,* p. 263. On the 15th Schuyler, "who brought up the Rear of the Provincials" with his New Jersey Regiment, arrived at Albany giving Amherst "a very bad account of the irregularities of the New York troops." Webster, *Jeffery Amherst,* p. 263.

Two days later Amherst talked with Colonel Isaac Corsa (fl.1755-1760) who commanded the second battalion of the New York Regiment. Knox II, p. 528. Corsa reported 89 of his men "drowned, killed and died," presumably during the whole campaign. Amherst sourly noted "I reckon above half the Provincials who are lost are by mere negligence." *Ibid.,* p. 263. For further details of Corsa's military service see the New York muster rolls in *Collections of the New-York Historical Society for the Year 1891.*

Amherst does not mention an accident on Oneida Lake, and a number of provincials were among the troops killed at Fort Lévis or drowned during the descent of the rapids above Montréal. The context of the general's remarks indicate, however, that it might have been New York rather than New Jersey troops who encountered difficulties on their return from Lake Ontario. It is also apparent that, whatever the nature of the incident on Oneida Lake, the casualties were not nearly so severe as recounted by Pouchot.

1245 The use of elm bark canoes was common among the Iroquois. For a detailed discussion of these watercraft see Edwin Tappan Adney and Howard I. Chapelle, *The Bark Canoes and Skin Boats of North America* (Washington, 1983), pp. 212-19.

1246 Hough in Pouchot, *Memoirs,* II, p. 136n identifies these "mountains" as the highlands of Madison County to the south of Oneida Lake.

On the land to the west the English have built a large redoubt, entirely faced with logs. Above it they have constructed a machicolation. It is a very large storehouse for everything that passes along the river.[1248]

Vessels come and load up at the first bend of the river, where large warehouses have been built. The fort is situated on a peninsula formed by a meander of the river.

At the end of the lake to the W. there is a river on which there are Onoyote villages.[1249] On the river, at the end where the entrance to the lake is, there are Onontague villages. The one called *Cassouneta* was once laid to waste by M. de **[139]** Vaudreuil.[1250] It was at that time on the bank of the stream. This is the village from which they derive their name of Onontagues.

1247 Wood Creek, flowing into the eastern end of Oneida Lake, was the water connection to the Oneida Carry. The post on the east end of the lake was constructed on high ground in a bend on the south side of the creek. The site is in Verona Beach, New York. The sandbar is today kept from the outlet of the Erie Canal by a long breakwater. In 1759-1760 boats avoided the sandbar by steering for the fort and then moving up the east side of the creek mouth.

The post, labelled the "Royal Blockhouse", is incorrectly shown at the mouth of Oneida Creek, two miles south of Wood Creek, on the map in Mante, p. 61.

1248 The post at the east end of Oneida Lake was another of the communication forts ordered built in 1759. It was designed for 50 men and seems to have been the only one of the three between the Oneida Carry and Oswego that was actually constructed according to the specifications of Colonel James Montresor. He called for "retrenchments with a Ditch & a Block house in the Center with Flankers at each opposite angle on which swivel Guns are to be mounted." Scull, p. 72. The blockhouse, a plan and elevation of which are shown in Rocque's "Sketch of the Blockhouse at the East of Oneda [sic] Lake", has the typical machicolated or overhanging second story. Two small "flankers" on the ground floor display ports for swivel guns.

A traveller who visited the post in July 1765 described the building as "a kind of wooden castle; proof against any Indian attacks." He reported it had been abandoned by the army and was then occupied by a settler. [Ralph Izard], *An Account of a Journey to Niagara, Montreal and Quebec in 1765; or "'Tis Eighty Years Since"* (New York, 1846), p. 7. This must have been a temporary situation, for the blockhouse was not officially given up until April 23, 1766, when its garrison was withdrawn. List of Officers [Commanding]..., Dec. 25, 1765, to Dec. 24, 1766, Gage Papers/139. By May of 1767 the building had been burned by the Iroquois. Francis Grant, p. 189.

1249 Pouchot means to the west (actually south) of the mouth of Wood Creek. The Oneida village is shown on Oneida Creek on Pouchot's map and on that in Mante, p. 61. See also Tanner, p. 41. This was in the vicinity of Oneida Castle, New York. General Amherst visited the "Oneidas Castle" in July 1760. He found it in "a pretty situation on a Hill about four miles from the block house." The village then had about 60 warriors. Webster, *Jeffery Amherst,* pp. 216-17.

The Woods-Orick River is not thirty *toises* wide at its mouth. The water is very black & unsuitable for drinking. This river is very winding & quite deep for four or five leagues & has hardly any current. Its bottom is of sand and mud.[1251]

Its meanders do not extend over a distance of more than a musket shot & the farther one travels up it the shorter they are. The river is so narrow that a fallen tree would span it & one could cross over on it. In 1756 the English constructed an abatis a mile long there **[140]** in order to provide cover against the French, who had captured Chouegen. They had a great deal of difficulty in clearing it, although there is still only sufficient space to allow the passage of one bateau at a time and all vessels often touch the sides & scrape against the muddy bottom for lack of water.[1252]

1250 Note in original, p. 140 (a): *The first of the family to be Governor of Canada & whose numerous offspring have continued to render signal service to the State.*

Pouchot's map places the Onondaga villages on Onondaga Lake at the site of Syracuse, New York and between the lake and Chittenango Creek. That stream empties into the south side of Oneida Lake near its western end. The 1776 Sauthier map places "Onondaga the Meeting Town of the Six Nations" near the south end of Onondaga Lake. It also shows "Onnosarage Castle" on the Tuscarora Kill (today Chittenango Creek) and "Canesraoa" farther east on Caneseraga Creek. Nebenzahl and Higginbotham, pp. 158-59. See also Tanner, p. 41.

Philippe de Rigaud de Vaudreuil (c.1643-1725) was later Governor General of New France from 1703 to 1725 and the father of the Governor Vaudreuil known to Pouchot. *DCB*, II, pp. 565-74. The details of his action against the Onondaga village had become somewhat garbled by the time Pouchot heard them. Vaudreuil, then commander of the troops in Canada, participated in Governor Frontenac's 1696 campaign against the Iroquois. On August 4 the French army found the principal Onondaga town burned and abandoned. The troops proceeded to cut down the standing corn. Vaudreuil was detached to destroy the crops and village of the Oneida. Despite an offer of peace from the villagers, he carried out his orders and the army departed for Montréal on August 10. Le Sueur, pp. 350-53.

The Onondaga village of 1696 was on Onondaga Lake, while that of the Oneida was on Oneida Creek. Tanner, p. 34. In the 1750s a Tuscarora village called Canaseraga was located on Canaseraga Creek flowing into the south side of Oneida Lake between Chittenango Creek and Oneida Creek. *Ibid.*, p. 41. Pouchot might have confused this with the 1696 village.

1251 Much of the lower part of Wood Creek has been incorporated into the Erie Canal.

1252 In 1756, following the French capture of Oswego, General Daniel Webb (note 366) ordered the Oneida Carry abandoned and Wood Creek blocked with fallen trees. Gipson, VI, pp. 194-95. As late as the summer of 1760 General Amherst observed the progress of the 55th Regiment of Foot greatly slowed because Wood Creek was still choked with logs. Webster, *Jeffery Amherst,* p. 216.

Pouchot's map labels Wood Creek "R. au Chicots" - "River of Stumps."

This river is the most advantageous place for cutting off English communications with the lakes by means of abatis which could be made of such a kind as to block up the river bed. A land route would be very long & very difficult to construct, because this country is studded with areas of marshy low ground.

Three leagues from the source of the Woods-Orick, there is a little fort of upright stakes covering the sluices built to retain the water & to assist the navigation of laden bateaux. When **[141]** the water is not released, it is necessary for men to get into the water and haul their bateau over the shingle. On occasions there are no more than six inches of water.[1253]

These sluices are not sufficiently well guarded and as a result they can be destroyed at the same time as the abatis are made. From the highest point of the area, you can see that the countryside is full of marshy hollows. The forest is dense and full of pines.

A number of rivers running in different directions have their source in this high ground. A quarter of a league from there is the beginning of the Riviere des Agniers, or Mohack River, which is larger & much deeper than the last one referred to. Near its source, the water comes

---

1253 The navigation of Wood Creek was enhanced by dams and sluices to raise the water level. When Amherst crossed the portage, in July 1760, he saw a sluice at Fort Bull with a new one under construction a short distance below it. Another dam was still farther downstream where Canada Creek joined Wood Creek. Webster, *Jeffery Amherst,* p. 216.

Pouchot describes only one post on Wood Creek. Amherst, however, refers both to "Fort Bull" and to a stockade at the junction of Canada Creek and Wood Creek. *Ibid.,* p. 216. Following the destruction of the original Fort Bull by de Léry's raiders in March 1756, the English built a replacement post of the same name. This was burned in August 1756 and not rebuilt. Amherst's "Fort Bull" of 1760 was simply a place name. Pouchot's subsequent observation that the system of sluices was inadequately guarded was due to the lack of fortifications at the original Fort Bull site in 1759-1760.

The fort described here by Pouchot was the "stockaded Post" mentioned by Amherst as protecting the dam at Canada Creek. It was constructed during the summer of 1759 at the junction of Canada Creek and Wood Creek west of Rome, New York. Hagerty, pp. 88-90. Hagerty refers to it as "Fort Rickey", today the site of Fort Rickey Game Farm. Robert Hare identified the ruins at the Canada Creek site as "Fort Richie" when he passed by on May 27, 1774. "Memorandum of a Tour...", *Collections of the Historical Society of Pennsylvania,* I, No. 6 (Nov., 1853), p. 365. The 1776 Sauthier map shows a "Fort Bute" on Wood Creek slightly east of Canada Creek. This might have been a misprint for Fort "Bull."

A "Plan for a military post on Wood Creek at the junction of Canada Creek & Wood Creek" by William Eyre is in Hulbert, I/3, p. 26 (Crown Maps, cxxi 115). Contemporary confusion existed between the Canada Creek post and the former Fort Bull site. In August 1759 Rhode Island Regiment chaplain Henry True described his arrival at "Fort Bull there Canada Creek met with Wood Creek, &c." Henry True (ed.), *Journal and Letters of Rev. Henry True* (Marion, OH, 1900), p. 12.

up to one's knees at the narrowest of fords. The surrounding country would be good for cultivation. **[142]**

Fort Sténix is built at a distance of a musket shot from the river, on the slope of land which goes down in that direction. The slope is very gentle. One notices this fact, because the terrain as one leaves the forest to enter the *désert* slightly overlooks the fort.

This fort is a square measuring 90 *toises* on each exterior side. It is made of earth and faced both internally and externally with large beams in the style of the fort at Chouegen.[1254]

In September 1760, it was not quite finished.[1255] It is the main storage depot of the English for all goods passing from their colonies along the lakes. It is there too that they normally assemble their army and all the bateaux used for navigating the waterways of this country.

It is necessary to haul them for half a league on drays in order to change river. These drays **[143]** are two very light limbers joined together by a trail adjusted to the length of the bateau. It requires eight men only to load them, or even less. The bow of the bateau is first placed on the forward limber and then the rest is hauled up on the rear. Two nags drag this carriage along with ease, at a fast trot. From this one can judge the lightness of these bateaux, which are made of pinewood and can carry 25 men. As a result they rarely last a whole campaign. The English always maintained vehicles for the portages in this fort.[1256]

After the fort, the river is about the same width as the Seine at Paris. It has a uniform current, which is quite fast, especially on bends. Its meanders can extend over three quarters of a league. It flows through a fine flat **[144]** country for 18 to 20 leagues. On the bends there are a number of reefs, but they are not too dangerous. There are also

---

1254 Construction of Fort Stanwix began in 1758 to replace the posts destroyed in 1756. Luzader, pp. 7-17. Rocque includes a "Plan of Fort Stanwix, Built at Oneida Station, By Provincial Troops, in 1758." The nearby commanding height was also noted by General Amherst in 1760 who thought that the "Ground Fort Stanwix is built on might have been better chose." Webster, *Jeffery Amherst,* p. 216.

1255 Pouchot observed Fort Stanwix as a prisoner of war on his way to New York in 1760. On his previous crossing of the Oneida Carry, in 1759, his captors had taken pains to prevent him from seeing the new fortification (note 706).

1256 Lord Adam Gordon observed, in 1765, that the "Batteaus are Slung on a long Carriage with four Wheels" for the portage. Newton D. Mereness (ed.), *Travels in the American Colonies* (New York, 1916), p. 419. Both Gordon and Pouchot also mention the use of vehicles of this sort at the Little Falls portage farther down the Mohawk.

quantities of trees which tend to hamper navigation if one is not careful. The banks of the river are quite high and of solid earth.

The English have built a little fort of stakes half way between the homesteads and Fort Sténix in order to provide their convoys with cover from our raiding parties when they stop. The fort is of no consequence and is called "Schüller."[1257]

Four or 5 leagues above the homesteads which have been abandoned one begins to see banks of little mountains running E. & W. and about half a league distant from one another. They end at the river & form reefs there. The first **[145]** homesteads one comes across are ten or twelve houses destroyed by the raiding party of M. Bellestre.[1258]

The right bank of the river is covered by Fort Harkman, which can hold 200 men & which is always garrisoned. It is a star-shaped redoubt made of earth and faced with saucissons, with a ditch 15 to 18 feet wide. It is palisaded at the far end & on the berm and has three or four poor cannon covering the river. It is within musket range of quite a tall mountain, which overlooks it.[1259]

The valley is no more than a mile wide. Its inhabitants do not appear to be very wealthy. Opposite the fort is the mouth of a river which has its source far away in the mountains. It is quite fast-flowing at its confluence, where it forms a reef **[146]** which compels both small & large bateaux to pass beneath the fort.[1260] The homesteads in this region are so widely separated that they cannot afford one another any protection.

After the fort, for a distance of two leagues, the river has quite a strong current with reefs at all the ridges of the meanders that are very

---

1257 Fort "Schüller" stood on the site of Utica, New York. It was brand new when Pouchot saw it in September 1760, construction having begun only on July 1. The work was done by New Jersey troops under Colonel Peter Schuyler (note 1244) for whom the post was named. Webster, *William Amherst,* p. 58; Webster, *Jeffery Amherst,* p. 215. The fort was simple and hastily built, containing a blockhouse in addition to the stockade. By 1765 all that remained was a "little block-house ... not capable of containing above six or eight people." Izard, p. 7. An account of 1767 also refers to a lone blockhouse in "100 acres of clear land." Francis Grant, p. 188.

1258 Lieutenant François-Marie Picoté de Belestre's November 1757 raid against the Mohawk Valley (note 385).

1259 Located east of modern Mohawk, New York across the river from the city of Herkimer, Fort Herkimer had been established in 1756 around the home of Johan Jost Herkimer. Loudon to Johnson, Sept. 16, 1756, *SWJP,* II, p. 560. Rocque includes a "Plan and Profile of Retrenched Work round Harkemeis house at ye German Flats 1756."

...(continued next page)...

difficult to get round. It is easy to run aground there, which causes damage to freight.

The chain of low mountains becomes higher in this region & one enters a kind of gorge the sides of which are fallen rock interspersed with scrub.

Two miles this side of the falls, the English have a large shed where they keep the vehicles for freight and bateaux. The portage road is in the gorge between the rocks, on a **[147]** marshy base. It is covered with billets.[1261]

The c.1757 French account of the "Topography of the Country between Oswego and Albany" has a detailed description of Fort Herkimer, referred to as Fort "Kouari." It is of interest as an example of a frontier fortification converted from a private home and for what it says of the concerns for attacking such a place with irregular forces. Fort Kouari was "a large three-story stone house with port-holes at each story and likewise in the basement, for the purpose of cross-firing. There are some small pieces above. The house is covered with plank and shingles. It was built as a store and depot for Chouëgen. It is surrounded with a ditch of about 30 feet from it. This ditch is 6 feet deep and seven wide. The crown of the ditch inside is planted with palisades in an oblique form; they are all well jointed to one another. Behind these is a parapet of earth to fire from over the palisades. The four angles of this parapet, which is at the back of the ditch, form, as it were, four little bastions that reciprocally flank each other. On the West side is a house apart from the large one. It backs against the parapet of the palisades, and serves as a barrack and guard-house. There are two doors to the large building; the one at the North is a small swing door. It is used only in going to the river for water. At this side of the house there is no ditch; only palisades fixed in planks set in the scarp of the right bank of the river, to support the earth. The large gate of the house is on the South side; it is folding but not ironed. To go outside the palisades and ditch through this large door, you must leave the house to the left and turn to the Eastward where there is a passage. The ditch has not been excavated at that point. The earth serves as a bridge and road. There are palisades to the right and left, on both sides of the way, the whole width of the ditch. Outside the ditch is a folding gate. There is no other barrier nor chevaux-de-frise in front." *DRCHSNY,* X, p. 676.

When Pouchot passed Fort Herkimer in September 1760 it had a garrison of only one sergeant and 15 invalids of the 77th Regiment of Foot. Webster, *Jeffery Amherst,* p. 214. During the winter of 1759-1760, however, 101 officers and men had been stationed there. "Disposition of his Majesty's Forces ....", Nov. 17, 1759, Gage Papers/4.

1260 West Canada Creek joins the Mohawk at Herkimer, New York.

1261 The portage was at Little Falls, New York. The "billets" formed a corduroy road. Although there was a road on each side of the river, that on the north bank was preferred. *DRCHSNY,* X, p. 678.

Pouchot notes no fortifications at this important point although, on May 24, 1759 General Amherst had ordered a "proper post" constructed at the "Carrying place at Canajoharie." He sent Lieutenant Diedrich Brehm (fl.1756-after 1786), a lieutenant (1756) in the 60th Regiment of Foot with a detachment of 50 provincials to erect a fort for 25 men. Webster, *Jeffery Amherst,* p. 112. Brehm returned to Albany on June 3, and it is unlikely that he accomplished much in so short a time. Scull, p. 75.

...(continued next page)...

The river flows for a good mile between these rocks & at the bottom forms a small waterfall, at the foot of which one can easily embark. Since the rocks of the falls are not very high, the water at their base is extremely calm. It forms a very fine basin, between towering rocks running sheer down into the water and heavily wooded. The basin extends possibly for three hundred *toises* before one leaves the rocks behind.

This is the best post available along this entire route. Just a few men stationed on these rocks could hold up a considerable army. This place seems entirely suitable as a natural border.

The landscape changes entirely here, as does the nature of the terrain. On emerging from this mountain, **[148]** the countryside broadens out. The valley is about a league wide. The soil is very good & well-cultivated. The inhabitants are comfortably off and live in spacious houses. The latter are separated from one another by a quarter of a league, along the river, on the cultivated land & in the mountains.

The landscape here, just like the preceding one, is intersected with low mountains running E. & W. They end at the river and form reefs and small rapids that are more frequent than on the land higher up the same river, which is uninhabited.

The river makes a substantial bend in the region called "Conyoxery." At this point a little river flows down from the mountains and winds across the countryside. It is barely wider than a large ditch.[1262]

The Agniers River is consistently wide **[149]** throughout its course. Its waters are clear except at the reefs that are encountered every league or so. Near the first Agniers village, there is another rather fine river

---

Franklin Hough's translation of Pouchot places the location of the shed "below" the falls. Pouchot's "this side" seems more likely to mean above them. The description probably pertains to his September 1760 crossing of the portage, and the shed was a recent addition. On July 1 of that year Captain William Hervey had complained of Little Falls: "No store houses built; roads on both sides river infamously bad .... Required at this place sheds at both landings; the road to be bridged the whole way .... where the road cant be made broad enough for two carriages another to be struck out; the rocks at the east landing to be blown." Hervey, p. 54.

1262 The bend is identified as "Anthony's Nose" on the map in Mante, p. 61, although this is the name of a mountain at the tip of the bend. Modern Canajoharie, New York is at the southern extremity of Pouchot's bend.

The little river flowing from the mountains is probably East Canada Creek at the western end of the bend and opposite the eighteenth century Iroquois village of Canajoharie. East Canada Creek joins the Mohawk from the north at St. Johnsville, New York after falling from the Adirondacks.

which flows from the W. It does not appear to be navigable as it does not extend over any distance.[1263]

There is an old & poor wooden fort on the bank at the confluence of the two rivers. A number of English homesteads are to be found in this village mingling with those of the Indians.[1264]

Two leagues lower down is the main village of the Agniers, or Mohacks. There is quite a large horizontal log fort called Fort "Hunter." It is built on a fairly large river which skirts the fort. It flows down from the mountains behind the sources of the Delaware. There could be 150 to 200 warriors in these two villages. **[150]** They are the Indians who are the most closely associated with the English. They are of the Protestant religion.[1265]

After this village, the mountains begin to be closer to one another. A league below it, there is a kind of gorge which is nonetheless inhabited along its reverse slopes. The house of Colonel Johnson, who is in charge of all matters concerning the Indians, is on the left two leagues below the second Indian village. It stands at the far end of a

---

1263 The first Agnier or Mohawk village was the "Upper Castle" located at modern Indian Castle, New York where Nowadaga Creek joins the Mohawk from the south. This was the Canajoharie of Pouchot's time.

1264 Sir William Johnson, at the request of the Mohawk, had ordered a fort constructed at Canajoharie in 1755. It was completed in August and came to be known as Fort Hendrick after the Mohawk leader killed at the Battle of Lake George. Johnson to DeLancey, Aug. 8, 1755, *SWJP,* I, p. 842. The c.1757 French account of the Mohawk describes the Canajoharie fort as "a square of four bastions of upright pickets (joined together with lintels), fifteen feet high, about one foot square, with port-holes inserted from distance to distance, with a stage all round to fire from." It was further reported to be "one hundred paces on each side. It is not surrounded by a ditch. There are some small pieces of cannon at each of its bastions, and a house at each curtain to serve as a store and barrack." *DRCHSNY,* X, p. 677.

Descriptions of the fort proper do not mention blockhouses. There were at least three standing in the vicinity of the fort, but it is unclear whether they were incorporated into the walls. For a discussion see George A. Bray, III, "More on Fort Hendrick", unpublished manuscript, Old Fort Niagara Association Library. When Francis Grant visited in 1767 he found "Fort Hendrick, a Mohawk Indian Settlement, at which there are three blockhouses which served as a post during the war." Grant, p. 185.

1265 The primary Mohawk village was the "Lower Castle" or Fort Hunter at modern Fort Hunter, New York. It was situated on the east side of the junction of Schoharie Creek with the Mohawk. The structure seen by Pouchot had been erected in 1755-1756. See Kevin Moody and Charles L. Fisher, "Archaeological Evidence of the Colonial Occupation at Schoharie Crossing State Historic Site, Montgomery County, New York", *The Bulletin: Journal of the New York State Archaeological Association,* No. 99 (Fall 1989), pp. 1-13.

...(continued next page)...

small area of pasturage that extends down to the river's edge. To the right of the house, there is a little stream that flows down from the mountains; it is quite deep. The house is a kind of chateau with the main entrance projecting from the facade; the latter has a soffit as coping. It has a fairly large forecourt, with a wall all the way round it and two quite tall turrets **[151]** on either side of the main door facing the pasturage. The back of the house abuts two hillocks of the mountain. On the hillock on the right there is a blockhouse to provide covering fire for the house, which is overlooked by the heights at pistol range from it. The house is remote & very vulnerable to attack. If we had known of this, our raiding parties could easily have abducted Colonel Johnson. Almost opposite the house, on the mountain facing it, there is a path that runs down into the valley of the Susquehanna.[1266]

From that point on, the river consistently flows down into a gorge. Its meanders are shorter, the extent of a musket shot, or a quarter of a league at the most. At its ridges there is very little water. Passage past

---

A French account of c.1757 described Fort Hunter as "the same form as that of Cannatchocari, with the exception that it is twice as large" and noted that a church and thirty Mohawk cabins were inside the walls. *DRCHSNY,* X, p. 677. Amherst, in 1760, noted that the fort consisted of "four blockhouses with loopholes & the Curtains furnish Storehouses boarded over, which serves for a Platform for men to stand on with small Arms." Webster, *Jeffery Amherst,* p. 213.

In 1711 Queen Anne had ordered construction of a chapel at Fort Hunter for the Mohawks. Mark M. Boatner, III, *Landmarks of the American Revolution* (New York, 1975), p. 243.

1266 Unfortunately, Sir William Johnson was elsewhere on both occasions when Pouchot passed his home, so he was unable to entertain his former adversary. Fort Johnson, located a short distance west of Amsterdam, New York where Kayaderosseras Creek meets the Mohawk, was Sir William's residence from 1749 to 1763. A detailed history and structure report of the property, which is today preserved by the Montgomery County Historical Society, is Mendel, Mesick, Cohen, architects, *Fort Johnson: A Historic Structure Report* (Albany, 1977). A very detailed contemporary view of the house and outbuildings was published in the *London Magazine* in 1759 and is reproduced on p. 5 of the report. It corresponds very closely with Pouchot's description.

Johnson's residence was indeed exposed to attack. General Amherst, who dined with him there on June 24, 1760, worried that it was "very badly situated, commanded entirely by hills." Webster, *Jeffery Amherst,* p. 213. The very detailed c.1757 French account of the route from Oswego to Albany shows that the house had been carefully reconnoitered by Johnson's enemies. It notes "a rise of ground on which is a small house with port-holes, where, ordinarily, is kept a guard of honor of some twenty men, which serves, also, as an advanced post." *DRCHSNY,* X, p. 679.

Late in 1758 Johnson erected two blockhouses "to serve as Outworks, & cover my house." Johnson to Stanwix, *SWJP,* X, p. 77. There was always a small garrison. In the winter of 1759-1760, for instance, one officer and 22 men of the 46th Regiment of Foot kept watch. Disposition of his Majesty's Forces ....", Nov. 17, 1759, Gage Papers/4.

the reefs is extremely difficult. The country is by no means beautiful. [152]

A league from Schenectady, or Corlack, the country opens out considerably into high terrain studded with hills but with no mountains. It is fine country & the soil appears to be fertile.

The river as far as Corlack is shallow and full of reefs. In front of Corlack is a very large grassy island formed by the Mohawk River & another river which runs round almost the entire town.[1267]

Schenectady, or Corlack, is a well-constructed town. The streets are well designed & built in the Flemish style. It could have a population of 3000 souls. Its position would be an admirable one were it not for a stretch of rising ground in front of the Porte d'Orange a musket shot away from it. It makes for a steep climb as one leaves the outskirts of the town. The remainder of the town's contour is a peninsula on top of [153] an earth escarpment 40 feet high. A river which is not fordable encircles the whole town except for the Orange side, which is narrow. On the banks of the river there are a large number of very pretty gardens. The town is only surrounded by cedar stakes without supports and has no defense against a raiding party of any size.[1268]

One can only navigate on the Agniers River from Schenectady to its falls.[1269] It is deeply embanked along all this stretch. There are five leagues from Schenectady to Albany and they are always covered overland. The countryside is hilly & uninhabited. There are only two or three taverns half way between the two towns.

The high ground consists of pine-covered sand dunes. As far as Albany, all the land is on a downward slope. [154]

At Schenectady all the bateaux for the use of the armies sailing on Lake Ontario were constructed. If any were laid down at Orange, they were transported on drays to Corlack.

1267 There are five islands in the Mohawk River at Schenectady. Pouchot probably describes the one known today as Hog Island or Isle of the Cayugas. The others, in addition to modern names, are each named for one of the original five nations of the Iroquois.

1268 Schenectady, founded in 1661, was an important English base for Mohawk River operations. It was the head of the portage between the Hudson and Mohawk rivers. Rocque shows the town, its defenses and the commanding ground on the side toward Albany, in a "Plan of Schenectady." The other features noted by Pouchot, including the creeks and island, are also shown.

1269 Cohoes Falls at Cohoes, New York on the Mohawk River below Schenectady (note 1048).

• CHAPTER IV •

On communication from Lake Ontario to the English frontiers along the River Casconchiagon

The Baye de Casconchiagon,[1270] as has been noted before, would be excellent for the anchorage of ships, but entering it is difficult because of the reef. If the region were inhabited, however, it would be possible to make a convenient channel.

One normally enters the **[155]** Baye du Fort des Sables[1271] in order to negotiate the portage at the far end of it. You sail along its shores before entering the river. Up to now, this navigation has only been undertaken in bark canoes. It would be necessary to have bateaux in reserve above the falls[1272] where the water is quite deep & the current gentle enough for navigating bateaux. The river has only the one portage marked on the map. It crosses all the region of the Five Nations & links up with the Belle-Riviere by way of a small lake, the waters of which flow partly **[156]** into the Casconchiagon, partly into the Ohio.[1273] This is doubtless one of the highest points in America,

1270 The broad bay on Lake Ontario between the mouth of the Genesee River and Irondequoit Bay. Pouchot refers to the bar at the mouth of the Genesee (note 1207).

1271 Irondequoit Bay, four miles east of the mouth of the Genesee River. See note 1207.

1272 Note in original, p. 156 (a): *There are three of them. The first is 60 feet high & two arpents across. The third is 100 feet high & three arpents across. The second is much less extensive.* Journal du P. Charlevoix, *vol. V, p. 330.*

These are the falls of the Genesee River at the site of Rochester, New York. The citation is to Pierre-François Xavier de Charlevoix, *Histoire et description générale de la Nouvelle France, avec le Journal historique d'un voyage fait par ordre du roi dans l'Amérique septentrionale* (6 vols.; Paris: 1744). His description of the Genesee Falls is in Charlevoix, 1761, I, pp. 339-40.

The widths of the falls are given in the original note in linear arpents, each equivalent to 191.835 English feet. Reasonover, p. 27.

The appearance of the falls in Pouchot's time was captured in a series of watercolors done in 1761 by Lieutenant Thomas Davies. Hubbard, pp. 88, 96, 140, 142.

1273 The portage is not clearly identified on Pouchot's map, but it is probably the dotted line running from a pair of villages ("Connectsio" and "Kanonshegon") in the region of modern Geneseo, New York along the upper Genesee River and along its west bank to the headwaters of the Allegheny River. This portage was necessary to circumvent the gorge and falls of the Genesee in modern Letchworth State Park, Livingston County, New York. Charlevoix noted that, having passed the three falls at Rochester, "after sailing fifty leagues higher, you discover a fourth fall, nothing inferior to the third." Charlevoix, 1761, I, p. 340.

...(continued next page)...

since its waters flow on the one side into the Gulf of St. Lawrence & on the other into the Gulf of Mexico. Close by this lake there is a very considerable source of bitumen oil.[1274]

The abundance of lakes, the ease of navigation & the small number of portages provide a good indication that the land consists of very high plains. In fact, one only encounters the great mountains when at a considerable distance from the sources of the rivers.

Traffic along this river will be very substantial indeed if this **[157]** territory were ever to be inhabited by Europeans. As we have just noted, one of its tributaries can link up with the Ohio River while another communicates with the Canestio by way of a portage a league in length. This latter river joins up with the Susquehanna, of which it is a tributary.[1275]

Pouchot shows the lake with two outlets as "Lac mer veilleusse." It does not exist. The Genesee River rises in northern Pennsylvania near the village of Ulysses. The source of the Allegheny, which joins the Ohio at Pittsburgh, is only about five miles to the southwest. Père Charlevoix was informed that after traveling 60 leagues up the Genesee "you have no more than ten to make over land, turning towards the right, to arrive at the Ohio, otherwise, *la Belle Riviere.*" *Ibid.,* p. 340.

1274 Note in original, p. 157 (a): *According to M. de Joncaire, there are two of these springs. The Indians use the liquid to alleviate all kinds of aches and pains.* Journ. cit. *p. 331.*

The citation is to Charlevoix. Although the priest does not name the "officer worthy of credit, and from whom I have received all I am relating to you," this footnote identifies him as Louis-Thomas Chabert de Joncaire (c.1670-1739), a lieutenant in the Marine but most important because of his close relationship with the Iroquois. Chabert had been captured and adopted by the Seneca, probably in the 1680s. He gained great influence over them, and used it to French advantage. *DCB,* II, pp. 125-27. His sons, Philippe-Thomas (note 517) and Daniel (note 331) were well-known to Pouchot and figured prominently in his service at Niagara.

The elder Chabert assured Charlevoix that he had seen "a fountain, the water of which resembles oil, and has the taste of iron. He added, that a little farther there is another exactly of the same kind, and that the Indians make use of its water to mitigate all kinds of pains." Charlevoix, 1761, I, p. 340. Chabert had observed the seeps of crude oil in the hills of northern Pennsylvania in the region where nineteenth century Americans would establish the first economical oil wells. Peter Kalm was shown petroleum obtained by English traders from the Iroquois and describes their method for gathering the crude oil from the surface of lakes. Kalm was told that "nothing was better for wounds than this." Kalm, p. 608.

1275 The upper reaches of the Canestio River come within two miles of Black Creek, a tributary of the Genesee, in Allegany County, New York. Pouchot's map indicates a trail or portage from the village of "Kanonshegon."

1276 Note in original, p. 158 (a): *It is 100 leagues long, according to the P. Charlevoix,* Journ. cit. *p. 330.*

See Charlevoix, 1761, I, p. 340.

The course of the Casconchiagon[1276] & of the Canestio is the region mainly inhabited by the Sonnontoins, who are the most numerous of the Five Nations. All the land around these rivers is beautiful & fertile, as in general all the territory inhabited by the Iroquois. Their villages are located around the lakes.[1277] There are open grasslands, which make for very agreeable landscapes, & soils which **[158]** would be admirable for cultivation. It is in the territory of the Five Nations that the "gin-seng" plant is most commonly to be found.[1278] **[159]**

The Iroquois people, who are made up of six nations, have possibly two thousand warriors according to the status which they hold among themselves namely:

Onontagues ...	300
Sonnontoins ...	700
Goyogoins ...	350
Onoyotes ...	250 **[160]**
Angiers or Mohacks ...	150
Tascarorins ...	100[1279]

This list is a good basis for estimating the population of the nation. Is it conceivable that they have become enfeebled to this extent since they have been in contact with Europeans?[1280] Our historians blithely send armies of 10 or 20 thousand men to subjugate part of the other nations of America. Since that time, we are not aware of any special

1277 The Finger Lakes of western New York. Pouchot's map indicates the locations of the principal Seneca villages.

1278 Note in original, p. 159 (a): *We are indebted to P. Lafitau for discovering this plant. This missionary became convinced that it was to be found in Canada. After quite a long search, he found it in the country. He learned without excessive surprise that the Chinese word "gin-seng" means "likeness of man", or, as the translator of Father Kircher explains, "thigh of man", because the Iroquois word* garent-oguen *had the same meaning:* orenta *in Iroquois means "thighs & legs", and* ouen *means "two things separated." He published this discovery in 1718, in a pamphlet dedicated to the Regent. He called the plant* Aureliana Canadensis, Sinensibus gin-seng, Iroquœis garentoguen. *M. Sarrasin, a Québec physician, had sent back in 1704 some of these plants for the king's garden but no one in Paris at that time could identify them. They are to be found in several regions of North America which are on the same latitude as Korea, where the type that the Chinese prize most highly comes from. Gin-seng is as common in the territory of the Illinois as it is in Iroquoia. Some plants have also been seen in Maryland &c. As soon as people were certain that* gartentoguen *was "gin-seng", there was a great rush to pick it. The Compagnie des Indies sent a cargo of it to China and initially paid the Canadians as much as 96* livres *a pound for it. It subsequently fetched only 4* livres *and fell into discredit for the reasons given by Abbé Raynal in his* Histoire philosophique et politique des éstablisssemens des Européens.

...(continued next page)...

pestilence which may have destroyed them.[1281] There could thus have been [161] exaggeration in the historical accounts.[1282]

The banks of the Canestio River are also inhabited by the Abenakis; we call them "Loups" and the English "Mohaigans."[1283] There is also a village of Renards or Outugamis who have fled since the war which their nation had with the French.[1284]

The Loups, who inhabit the valleys of the Susquehana can put 15 to 1800 warriors in the field. The village of Theaogen alone has 600. The small village of Tateyonons, which cannot supply sixty warriors, is allied to the Iroquois.[1285]

The Susquehana River is navigable right up to the area of its sourcc.[1286] It flows into a beautiful valley [162] of fine free-standing

---

Joseph-François Lafitau (1681-1746) was a Jesuit priest and missionary. His discovery was made in 1716. *DCB*, III, pp. 334-38. Lafitau's pamphlet was entitled *Mémoire présenté à S.A.R. Mgr le duc d'Orléans concernant la précieuse plante du ginseng, découverte en Canada par le P. Joseph-François Lafitau* (Paris, 1718).

Athanase Kircher (1602-1680) was a German Jesuit. *Biographie universelle,* XXI, pp. 642-46.

Michel Sarrazin or Sarrasin (1659-1734) sent Canadian ginseng to France in 1704 under the name *Aralia humilis fructu majore. DCB,* II, pp. 593-600.

1279 Onondagas, Senecas, Cayugas, Oneidas, Mohawks, and Tuscaroras. Here Pouchot acknowledges that the people whom he consistently calls the "Five Nations" actually comprised six.

The total population of the Iroquois country in 1768 has been estimated at 7,500. Assuming a warrior to total population ratio of 1:4, this suggests a potential fighting force of 1,875 men, very close to the 1,850 given by Pouchot. Tanner, p. 66.

1280 Note in original, p. 161 (a): *It cannot be doubted that the population of the Indian nations has enormously diminished since that time. See the end of this work.*

1281 Note in original, p. 161 (b): *Have not smallpox & brandy been two very great pestilences for all the Indians of the continent?*

1282 Note in original, p. 162 (a): *That may well be, but even if the drop in population were less great, there has nonetheless been a steady decline.*

1283 These are the Loups of Theogen or the Munsees described in note 341. Tanner, p. 42. The English equivalent is Mohicans.

1284 Pouchot again mentions the Renards (Mesquakie) discussed in note 343. These refugee people resided with a group of Delaware on the upper Allegheny in Pennsylvania. Tanner, p. 42.

1285 The village of Tateyonons does not appear, at least not in a recognizable form, on Pouchot's map. The closest possibilities are "Halbanyane" on the West Branch of the Susquehanna and "Thayengoa" on the upper reaches of the Allegheny. Both are located in logical areas for a Loups village allied with the Iroquois.

forest. It has good depth along its entire course and sufficient to sail bateaux as far as Fort Schamokin.[1287]

The W. tributary of the river is more of a torrent than a river. Since it is surrounded by rugged mountains, it flows very rapidly. However, the Indians negotiate it by canoe at high water.[1288]

From Schamokin as far as Chesaspeheack Bay, the Susquehana has rapids which criss-cross the chains of mountains running E. & W. through the English colonies. The worst one is at Canowega.[1289] As a result of these rapids, the English make little use of the river for internal navigation between their colonies.

After Schamokin is the most convenient route **[163]** the English have for getting to the Five Nations & the lakes. But the presence in between of the Loups & Iroquois nations has up to the present time prevented them from establishing settlements in those areas.

Before the late war, they had settled as far as the area of Theaogen, which the Indians forced them to abandon and also up to just below the Juniata Valley, which is beautiful and very fertile. There too they have been compelled to withdraw.[1290] **[164]**

1286 The Susquehanna has its source in Otsego Lake at Cooperstown, New York.

1287 Fort Augusta at Sunbury, Pennsylvania.

1288 Pouchot refers to the West Branch of the Susquehanna that joins the main branch just above Sunbury.

1289 The Susquehanna flows into Chesapeake Bay. Canewago was on the Susquehanna near Bainbridge, Pennsylvania.

1290 Note in original, p. 164 (a): *Since the period discussed by M. Pouchot, the English have returned in force to the banks of the Ohio & the rivers that flow into it & have forced the Indians to leave them in peace. Just a few years before the present war, the court at London was elaborating a plan to send out a substantial number of colonists to this area. The famous economist, Young, at that time wrote against the plan, which the troubles in America have aborted.*

Pouchot's editor digresses with his comment about the English return to the Ohio River, a reference either to Henry Bouquet's 1764 expedition to the region or to Lord Dunmore's War of 1774 against the Shawnee and other tribes of the Ohio. See Boatner, *Encyclopedia*, pp. 102, 341-42. Pouchot is discussing events on the Susquehanna.

"Theaogen" was the Delaware/Munsee village of Tioga at modern Athens, Pennsylvania. As early as 1753, settlers from Connecticut had moved into the "Wyoming" region farther down river where the Lackawanna River meets the upper Susquehanna in the vicinity of modern Wilkes-Barre, Pennsylvania. The Wyoming settlements grew in the years following the war and were the scene of bitter fighting during the American Revolution. *Ibid.*, pp. 1221-28.

...(continued next page)...

• Chapter V •

On communication between Niagara and the Belle-Riviere, or Ohio River, in English Alligeny, & between the Ohio and Pennsylvania & Virginia

Fort Niagara is situated at the E. point of the river of the same name, which is still the River St. Lawrence.[1291] The point ends in a triangle. Its base is the head of a hornwork measuring 114 **[165]** *toises* on its exterior side. It is entirely constructed of earth, sodded on both thc inside & the outside, with a ditch eleven *toises* wide and nine feet deep, a demilune & two little lunettes, or retrenched places of arms, a covered way & a glacis built in proportion to the earthworks. The ditches have no revetment.[1292]

The place[1293] & the demilune have a stockade on the berm. The two other sides consist of a simple earth retrenchment, grassed over inside & out. It is seven feet high on its internal side & six feet thick on the top of the parapet, with a fraise on the berm. These two retrenched sides are built on earth cut 40 feet high. The side overlooking the river could be assaulted, although with difficulty. The side overlooking the

---

The Juniata River flows from the west meeting the Susquehanna near Duncannon, Pennsylvania about 15 miles above Harrisburg.

Arthur Young (1741-1820), English agriculturalist and prolific author, published *Political Essays Concerning the Present State of the British Empire* (London, 1772). *DNB*, XXI, pp. 1272-78.

1291 Note in original, p. 165 (a): *Properly speaking, this river is no more than the discharge of the Great Lakes into the ocean, & the Niagara River the discharge of Lake Erie into Lake Ontario. Thus it is pointless to go seeking the sources of this great river in territory situated to the N. or the N.W. of Lake Superior.*

1292 Pouchot completed at least three plans of Fort Niagara in the course of his work there. The earliest was that entitled "Plan Du Fort de Niagara, construit et défendu par M. De Pouchot", later engraved and printed in his *Mémoires*. The second, completed in 1756, was a "Plan de Niagara et des Fortifications faites en 1755 et 1756." The original is in the French archives with copies at the National Archives of Canada, NMC-26647. The third, drawn late in 1757, was entitled "Niagara Cette Place à èté Commencée le 14 Janvier 1756 et Finie le 12 Octobre 1757 ...." It is in the British Library, Add. MSS, 15332.H with a copy in Hulbert. For details on the development of Fort Niagara under Pouchot, see Dunnigan and Scott, pp. 22-27.

1293 The word "place" here refers to the main body of the fort. The more general contemporary definition is: "PLACE, in fortification, signifies in general terms a fortified town, a fortress; hence we say it is a strong place." Smith, *Dictionary*, p. 205.

lake is more steep.[1294] There are no **[166]** stones around the site of Niagara. They have to be brought from the foot of the Côtes, or the Platon.[1295] There are deposits of separate sandstone blocks which are very suitable for all kinds of construction but there is no freestone. Before 1759, it had always been necessary to transport lime for the use of the fort from Frontenac. M. Pouchot, when in command at Niagara, discovered some excellent lime at the top of the escarpment. It is doubtful whether the English know about it. They have to bring lime from Chouegen. One could build a city with the deposits.[1296]

There is a sand bank in front of the fort which extends for a good quarter of a league into the lake. Only bateaux can pass over it. The entrance to the river is difficult when one does not know it because of the sand bank & a strong current **[167]** flowing out of the river. The latter can force you into an eddy which makes you run aground on the sand bank. This passage is well defended by the artillery at the point of the fort, because vessels can only fight against the current, which is right under the fort's batteries, with great difficulty. Boats often have

---

1294 Since they were much less exposed to assault, the lake and river sides of Fort Niagara were more lightly defended. Wall sections of all three sides of the fort were drawn soon after its capture by the English in 1759. These are shown on a plan entitled "Niagara", probably drawn by Lieutenant Thomas Sowers. Public Record Office, MPG 342. It is printed in Dunnigan, *Siege,* p. 65.

1295 The French referred to the Niagara Escarpment as the *Côtes* as in "brow" or "declivity of a hill." The Platon was their name for the flat area at the top of the riverbank but below the escarpment at the site of Artpark in Lewiston, New York. The bank there is cut by a gulley that provided access to the river and served as the lower landing of the Niagara Portage. For a history of the site see Stuart D. Scott, Patricia Kay Scott, et. al., *An Archaeological Survey of Artpark and the Lower Landing, Lewiston, New York* (Lewiston, 1993), pp. 79-101.

1296 It is ironic that, with vast amounts of limestone forming the Niagara Escarpment, the French had usually transported both stone and lime the length of the lake from Fort Frontenac. Limestone for the "machicolated house" or "Castle" of 1726 and the 1757 powder magazine came from that source. Dunnigan, *Glorious Old Relic,* p. 16. When it was decided to revet the Fort Niagara earthworks with masonry, in 1758, the stone was also to have come from Frontenac. The project was aborted by Bradstreet's capture of the latter post (note 514).

The English soon discovered Pouchot's source of limestone. While Niagara's English commandant thought, as late as February 1760, that it would be necessary to obtain stone from Frontenac, he happily reported in May that "After a great deal of difficulty in searching where the French got their Lime Stone, have found the Place to be about ten Miles Up this River, a little more than half Way to the Falls." He had made the discovery only two days before while scouting in the direction of the former Fort Little Niagara. Eyre to Gage, Feb. 24, 1760, Gage Papers/5; Eyre to Gage, May 8, 1760, Gage Papers/6.

to cast a hawser on shore and be hauled into the anchorage which is located at a flat sandy space beneath the center of the fort. Ships anchor there right up against the shore & there is enough deep water for a warship.[1297]

The passage via Niagara is the most frequented of the American continent because this tongue of land links up with three great lakes & because the ease of the journey brings all the Indians that way, wherever they want to go. Niagara **[168]** is in some sense the center for trade between Indians and Europeans and as a result they gladly make their way there from all parts of the continent.

Vessels cannot spend the winter in the Niagara River because it constantly ferries down pieces of ice out of Lake Erie from December to the beginning of March. However, a port or shelter could be constructed for them towards the west at Pointe à Mascoutin.[1298]

The river, from its mouth to three leagues above the place called the "Platon", has a channel about 400 *toises* wide, quite a gentle current & a depth of water sufficient for frigates which could sail up to the Platon & anchor at any point **[169]** along that section of the river. There are three meanders in this stretch, each of them a league long, which offers a fine view of Niagara. For three leagues, the river flows between two almost sheer cliffs 2 to 300 feet high and so swiftly that it is not navigable from the Platon up to the basin beneath the falls.

Half a league above the falls, the river, which is nearly half a league wide, is no more than a swift-moving current. It surges along as far as the falls and hurtles over a sheer drop of 140 feet onto a cluster of very hard rocks below. Its width is about 900 *toises.* The top of this mass of water forms a very wide arc. Two thirds of the way round it there is a little wooded island which always seems as though it must shortly be engulfed with water.[1299] **[170]**

1297 The landing place for Fort Niagara is today the site of a United States Coast Guard station. La Broquerie's 1757 "Carte Du Lac Ontario" depicts the extensive *Bature* [sand bank] *de Niagara* running parallel to the shore to the east of the river mouth.

1298 Pointe à Mascoutin is so marked on the 1757 "Carte Du Lac Ontario" by La Broquerie. It is shown as "Pt. Amascou" on the c.1760 chart "Lake Ontario." This is the prominent headland on the Canadian side of the Niagara River about 1300 yards above Fort Niagara.

The English soon settled upon Pouchot's solution for wintering vessels in the Niagara River. In November 1765 they took advantage of the favorable situation of Pointe à Mascoutin to establish a dock and barracks which soon became known as Navy Hall. Arnot to Gage, Nov. 20, 1765, Gage Papers/45. The point is today occupied by Navy Hall and Fort George in Niagara-on-the-Lake, Ontario.

At the bottom of the falls, the river forms a great basin between the rocks where the water is so calm that it would be possible to sail there in a bateau. From the foot of the falls, the water rebounds nearly 40 feet into the air, which gives the impression of a facing of ice.

On the banks of the basin, one often finds fish, bears, deer, geese, ducks and other birds which have been killed by going over the falls, or have been dragged down by the waters or the draught of air. The Indians collect up these creatures.[1300]

There is a cart track from **[171]** Niagara to the Platon but in the summer the journey is usually made by water. In winter, it is always necessary to take the land route because of the ice. The track from the Platon to the portage fort is also three leagues long; it takes three hours to make the journey. Since it traverses a forest, the going is sometimes muddy. If it were well maintained, it would be a fine road.[1301]

1299 Note in original, p. 170 (a): *P. Charlevoix asserts that this islet is very narrow & about an eighth of a league long. He adds that a number of reefs dotting the river in various places, beside it & higher up the river, slow down the current quite considerably above the falls. Islets of a similar sort, or rocks covered with trees, are to be seen at the falls of the Rhine at Lauffen.*

Pierre-François-Xavier de Charlevoix (1682-1761) was a Jesuit priest who visited Niagara Falls and the Great Lakes in 1721 and descended the Mississippi in 1722. His *Histoire et description générale de la Nouvelle France avec le Journal historique d'un voyage fait par ordre du roi dans l'Amérique septentrionale* was published in Paris in 1744 and has been cited previously. *DCB,* III, pp. 103-10.

Goat Island, New York separates the American and Canadian or Horseshoe Falls. Pouchot describes Niagara Falls as a single cataract. Charlevoix's 1721 description of Goat Island is in Charlevoix, 1761, I, p. 354.

Pouchot's editor cites the islands of the "Rheinfall", a cataract on the upper Rhine River at Laufen, Switzerland. John Gibson, *Great Waterfalls, Cataracts, and Geysers* (London, 1887), pp. 204-08.

1300 Many early descriptions of Niagara Falls mention the harvesting of fish and other animals swept over the falls. Charlevoix "saw nothing of this sort" in 1721 and noted that, "quite the contrary ... I saw small birds flying very low, and exactly over the Fall, which yet cleared their passage very well." Charlevoix, 1761, I, p. 355. Peter Kalm, however, was told by Fort Niagara's officers when he viewed the cataract in 1750 that in "the months of September and October, such abundant quantities of dead waterfowl are found every morning below the Fall, on the shore, that the garrison of the fort for a long time live chiefly upon them; besides the fowl, they find also several sorts of fish, also deer, bears, and other animals which have tried to cross the water above the fall." Letter of Sept. 2, 1750 in the *Gentleman's Magazine* of January 1751 printed in Charles Mason Dow, *Anthology and Bibliography of Niagara Falls* (2 vols.; Albany, 1921), I, p. 58. See also Kalm, pp. 701-02 which gives much the same information.

1301 This track, the portage road, was situated on the east or New York bank of the river and ran from modern Youngstown to Niagara Falls, New York.

At the bottom of the Côtes, or the Platon, there are three sheds which act as storehouses for goods that have arrived. The bank where they are unloaded is a good 60 feet high, which creates great difficulties, as nobody has ever thought of building a facility to ease the problems of unloading.[1302]

The Côtes consist of three curtains of land the height of which, measured from the Platon above them is at least as great as that of Meudon[1303] but is no more steep. To ascend, there are two tracks. [172] One of them, for vehicles, extends over a quarter of a league. There are two ramps, which are fairly gentle. The other is a footpath, which runs straight down the hill. It is very steep. Travellers, & others who are heavily laden, always use this path. However, it never takes more than half an hour to walk up it. At the top of the hill, there is a storage shed.[1304]

The memoir of M. Belin implies that this place resembles one of the difficult passes of the Alps, whereas the top part & the bottom of this hill are immense open spaces.[1305]

The portage fort was nothing more than an enclosure of standing stakes. Sheds had been built there for freight & boats to link up with the fort. It is at this post that shipments for Lake Erie are made up. Beyond this [173] point the river is not navigable for more than a

---

1302 The English solved this problem soon after their capture of Fort Niagara by constructing, in 1762-63, a wharf and a series of machines for lifting boats and cargo up the Niagara Escarpment. See Brian Leigh Dunnigan, "Portaging Niagara", *Inland Seas,* Vol. 42, #3 (Fall 1986), pp. 177-83, 216-23.

1303 Meudon is a suburb of Paris, situated on a height above the left bank of the Seine River, seven miles from Notre Dame cathedral.

1304 Ensign Daniel Chabert de Joncaire (note 331) had been given command of the portage around Niagara Falls in 1751. He introduced horses and vehicles where, previously, only men could "carry the canoes and the goods by narrow, sloping paths over three steep hills" and heavy bales could be "manhandled only with the aid of cables." Chabert, II, p. 2. A list of the animals and vehicles in use on the portage in 1759 is in *Ibid.,* III, p. 5.

By 1758 there were at least two storage sheds at the lower landing. One was on the "Platon", the area atop the riverbank. The other was on the summit of the Niagara Escarpment. Chabert had also been ordered to build a house on the Platon for the clerk of stores. Vaudreuil to Chabert, April 12, 1758, *Ibid.,* III, p. 19. These buildings were all plundered and burned by the Iroquois during the siege of Fort Niagara. Johnson to Amherst, July 31, 1759, *SWJP,* III, p.115.

1305 The work cited is Jacques-Nicolas Bellin, *Remarques sur la carte de l'Amérique septentrionale, comprise entre le 28e et le 72e degré de latitude, avec une description géographique de ces parties .... Paris, 1755.*

For Bellin, see note 973.

quarter of a league and even then one has to be very careful not to get caught by the current leading to the falls. The ground around the fort is flat & very good. Any earthwork one might wish to make could be created on this site.[1306]

On the W. side of the river and opposite the fort, there is a pleasant little river called "Chenondac", on the banks of which there are some fine woods.[1307] It is from this forest that wood was taken for building the bateaux for navigating in the area as well as boards & joists for the use of the fort.

One must be very careful navigating in & out of the Chenondac. After sailing upstream for a league to cross its outflow, one then sails back along the bank as far as its mouth. Similarly when leaving the river, **[174]** one has to proceed upriver & then sail back down to the fort, passing between the islands situated in front of it.[1308]

The river channel is full of islands right up as far as the small rapids, as can be seen from the map. Its current is quite gentle and one can navigate either by oar or sail. A number of these islands are fine open grassland.[1309]

On the E. side of the river, three leagues from the portage fort, is the Riviere aux Bois Blancs.[1310] The Five Nations use this waterway to

---

1306 The Portage Fort or Fort Little Niagara had been constructed in 1751 by Daniel Chabert de Joncaire. It was located at the upper landing above Niagara Falls in what is today the city of Niagara Falls, New York. See Severance, I, pp. 374-79. Severance quotes a description by the Abbé Picquet who said that Little Niagara was a stockade, "badly made ... with two kinds of bastions at the two angles of the side towards the roads which lead to [Fort] Niagara." The gate was on the water side, and Picquet considered all aspects of the design "contrary to rules of fortification." *Ibid.*, p. 379.

Peter Kalm visited the upper landing in 1750. He noted that "the boat landing above Niagara [Falls], when one wished to sail up Lake Erie was said to be about 15 arpents [about a third of a mile above the falls]. I thought it was but a little less than 1/8 of a Swedish mile [3/4 of an English mile]. It is not considered safe to approach the falls nearer with a boat." Kalm, p. 708.

A lineal arpent, the measure used by the French officers who guided Kalm, was equivalent to 191.835 English feet. Reasonover, p. 27.

1307 Chippawa Creek or the Welland River joins the Niagara just above the falls in Niagara Falls, Ontario.

1308 That is, using the channel between Navy Island, Ontario and Grand Island, New York.

1309 These small islands are located in the Niagara River above Grand Island.

1310 Tonawanda Creek, later the western end of the Erie Canal, enters the Niagara River between Tonawanda and North Tonawanda, New York.

join up with the main river. Its current is extremely gentle and several sites along it are cultivated by the Indians. The surrounding land is very fine and the river is full of fish.

The small rapids are the outflow of Lake Erie. They consist of reefs over which the current flows uniformly but with force for half a league. The river is **[175]** a quarter of a league in length and has a rocky bottom. It is not very deep. There are channels where, if special vessels were built, they could sail against the current given a brisk wind. Bateaux can be poled or hauled up it.[1311]

The shores of the lake on the E. side are higher than those to the W. They all seem to be very firm.

Lake Erie has never been explored by anyone capable of giving with any accuracy the lie of the shores, the depth of the bays & the anchorages to be found there, as well as the harbors that could be created to take advantage of its geography.[1312] The shape given to it on the map coincides with the best known accounts of it, especially with respect to **[176]** its S. shore.[1313]

1311 These rapids are today below the Peace Bridge connecting Buffalo, New York with Fort Erie, Ontario.

1312 Pouchot gives no indication that he was aware of the diary, maps and sketches of Lieutenant Joseph-Gaspard Chaussegros de Léry who voyaged to Detroit along the south shore of Lake Erie in 1754. Although Pouchot worked briefly with de Léry at Niagara in 1755, he had little regard for his engineering abilities (note 172). De Léry's sketches of river mouths and his description of the south shore of Lake Erie have been published in *Rapport de l'Archiviste de la Province de Québec pour 1927-1928* (Québec, 1928), pp. 355-408.

The shape of Lake Erie is also well-defined in Nicolas Bellin's "Carte du Cours de l'Ohio ou la Belle Riviere" published in 1755. See Kent, p. 42.

1313 Note in original, p. 177 (a): *We learn from a letter of the Marshal de Bellisle, dated 3 July 1758, that M. Pouchot had handed over his personal map of the lake to M. de Montcalm, who was supposed to send it to the minister. We have found no copy of it in the private papers of M. Pouchot, which is doubtless a great loss.*

The 1781 editor seems to have been under the misapprehension that Pouchot had provided Montcalm with a separate map of Lake Erie. He would have searched Pouchot's papers in vain for a copy because the author was referring to the representation of the eastern three-quarters of Lake Erie on his own general map of the French and English frontiers. This was published as "Carte des Frontieres Françoises et Angloises dans le Canada depuis Montreal jusques au Fort du Quesne" in Pouchot's *Mémoires,* I, pl. 3. The printer of the *Mémoires* presumably worked from a copy in Pouchot's papers. This is implied in the introduction to the third volume. *Ibid.,* III, p. 3.

The development of Pouchot's general map is well-documented. Montcalm, in a letter to Minister of War de Paulmy of February 23, 1758, reported that Pouchot had submitted it to Governor Vaudreuil with some observations on another sent to Vaudreuil by the Chevalier de Lévis. The governor had forwarded Pouchot's map to M. de Moras, Minister

...(continued next page)...

The entrance to the lake as far as the Riviere aux Chevaux forms a large bay entirely covered with pebbles, where there could not be any anchorage. If the mouth of the river could be kept clear, there would be an anchorage for ships.[1314]

The shore from there up to Presqu' Isle[1315] has no known shelter. At Presqu' Isle there is a good bay, but the water is only 7 to 8 feet deep.

Ships could enter the **[177]** river at Seguin[1316] & a good harbor could be made there as well as at Sandoské.[1317] It is said that, generally speaking, the bottom of the lake is very level & navigation on it dangerous. What is true is that storms blow up almost at a moment's notice, that the swell is dangerous & that in heavy weather it often kills fish, which can be found scattered over the shores. But it should be

---

of Marine. It appears from the letter that Montcalm also received a copy as he intended to add some notes to it for the Minister of War. *DRCHSNY,* X, p. 690.

On April 14, Pouchot wrote to Marshal Louis-Charles-Auguste Fouquet, Comte de Belle-Isle (1684-1761; *Biographie universelle,* III, pp. 565-67), that he had given Vaudreuil a map and a memoir regarding the French and English frontiers, prepared from information gathered while at Niagara. He indicated the lack of information on Lake Erie and made no mention of a separate map of it. Lake Erie, he wrote, "is entirely unknown; it is, perhaps, [as] navigable for large vessels as Lake Ontario. The resources of those countries, once known, would furnish opportunities of avoiding long routes and expenses, exclusive of enabling us to occupy more decided points for the security of the country." He appended to the letter some "General Observations on the Frontiers of Canada", probably the same notes previously submitted to Vaudreuil. *DRCHSNY,* X, pp. 694-95. The letter and observations are printed in Appendix A of this volume.

On April 18, Montcalm again wrote to de Paulmy informing him that Lévis was sending "a very large and very handsome map" (see note 88). He added: "Captain de Pouchot of the Bearn Regiment, has drawn another in which the environs of Niagara, the country of the Five Nations and of the Loups will possibly be laid down with still greater exactitude than in that of Chevalier de Lévis. I ought to send it to you from him, but defer doing so until the last ship, because if the map which Chevalier de Lévis sends you should happen to be intercepted, that of Sieur Pouchot would serve as a duplicate." *Ibid.,* p. 698. Neither Montcalm nor Pouchot were yet aware that Belle-Isle had replaced de Paulmy as minister of war on March 3. Belle-Isle was still awaiting a copy of Pouchot's map in July. Belle-Isle to Montcalm, July 4, 1758, Casgrain, *Collection,* III, p. 117. This might be the letter referred to in the 1781 footnote above.

1314 Here Pouchot describes Buffalo Creek, later the harbor of Buffalo, New York.

1315 Erie, Pennsylvania.

1316 Pouchot seems to use Seguin as a place rather than as a name for the river itself which is shown on his map as the "gayougé." This is the Cuyahoga which enters Lake Erie at Cleveland, Ohio.

1317 Sandusky Bay at Sandusky, Ohio.

noted that the lake has only been navigated in a bark canoe & hardly ever in a bateau except from the Niagara River up to Presqu' Isle.

Exploration has been confined to the shoreline, which is very flat. A little farther out in the lake, however, the water may be quite deep. It would have been useful to have had a skiff built and, from May until around the [178] end of September when the weather is always fine, to take soundings and reconnoitre all the natural harbors which may be located around the lake. The ships suitable for sailing there could have been built and they would have spared many special missions & much expense.

The Ohatacoin River is the first river that joins Lake Erie to the Ohio. It was along its course that expeditions in that region were made in the early days of exploration.[1318] It was always negotiated by canoe because of the shallow depth of the water. Indeed, unless the river was in flood, it was very difficult to pass along it. As a result, navigation along the Riviere aux Boeufs was preferred. The fort of Presqu' Isle is the warehouse on that river.[1319]

This fort is quite large and built of [179] horizontal logs, with sheds for storing the trade goods. It is situated on a flat open space in the shape of a peninsula, which gives it its name.[1320] The surrounding land is fertile & pleasant. Carts are maintained there for the portage, which is six leagues long. Although over flat country, the track is not very good as far as the fort on the Riviere aux Boeufs, which is built in a square smaller than that of the Presqu' Isle but, like it, is of horizontal logs.[1321]

1318 Pouchot's map illustrates the series of portages from Lake Erie to the Ohio by way of Chautauqua Lake and its connecting creeks. This route was used by Céloron in 1749 and had been intended for the 1753 expedition before Presqu' Isle was selected as an alternative. The route began where Little Chautauqua Creek enters Lake Erie at Barcelona, New York. It was necessary to portage into Chautauqua Lake and then follow its outlet, the Chadakoin River (shown as "Schatacoin R." on Pouchot's map) and Conewango Creek to the Allegheny River.

The Chautauqua and Presqu' Isle portages are also distinctly shown on Bellin's "Carte du Cours de l'Ohio" in Kent, p. 42.

1319 Presqu' Isle replaced the Chautauqua route in 1753. See note 158.

1320 Pouchot never saw Presqu' Isle. The fort was not located on the peninsula separating Erie Bay from Lake Erie but rather on the mainland opposite. Stotz, pp. 72-73.

1321 Although Pouchot gives the impression that the portage was an easy one, the track from Presqu' Isle to the Rivière aux Boeufs traversed the 400 foot-high Lake Erie

...(continued next page)...

The Riviere aux Boeufs is very winding and quite shallow both when the water is low & in rainy weather. Its waters become greatly swollen and its current is extremely swift. It cuts through a valley which deepens as it approaches the Belle-Riviere.

At its mouth, called in **[180]** English "Vaningo", the French had a very poor, tiny fort called "Fort Machault", which is also a storehouse for goods in transit to Fort du Quesne.[1322]

The two rivers marked on the map beyond Presqu' Isle, which flow into the lake, also link up with rivers which flow into the Ohio, like the Riviere au Castor. But the latter is not deep. It is even obstructed by a number of rapids.[1323]

The river at Séguin provides a much finer link with the Belle-Riviere. Ships could sail up almost to within three leagues of its source & with bateaux it is possible to arrive as far as a portage no more than a mile long. From there one can enter a very good river which the English call "Muskingann."[1324] According to all accounts, it is **[181]** the most attractive place to inhabit in the whole of America. Here are to be found the finest forests, suitable for different uses, & the best land set in excellent plains.

Escarpment and then glacial bogs that made the road extremely difficult for carts except in very dry weather. The longer land carriage also made it necessary to depend on draft animals. When these were not available, as in 1753, the portage was particularly arduous. Captain Péan wrote that "The labor of the troops was excessive. The soldiers, sunk half-leg deep in the mud and weakened by the recent fatigues of the first portage [Niagara], succumbed under their burdens. It was impossible to use the few horses that remained. It was an afflicting spectacle to behold these debilitated men, struggling at the same time against the bad season and the difficulties of the road, broken down by their weapons and the loads which they had to carry." Quoted in Kent, pp. 58-59.

The post at the south end of the portage was that at the Rivière aux Boeufs. Constructed in 1753, it stood on French Creek (Waterford, Pennsylvania) where water transport to Fort Duquesne could resume, at least when the creek was high. The fort was burned by the French in July 1759. Stotz, pp. 74-75.

1322 Fort Machault was the post at Venango, today Franklin, Pennsylvania. Stotz, pp. 76-79.

1323 The two rivers are not named on Pouchot's map but correspond to Conneaut Creek at Conneaut, Ohio, and the Grand River, the mouth of which is at Fairport Harbor, Ohio. A number of streams in northeastern Ohio allow portages to the Shenango and Beaver (Rivière au Castor). The latter joins the Ohio below Pittsburgh near Rochester, Pennsylvania. The Shenango (Chinigués) portage is shown on Bellin's "Carte Du Cours de l'Ohio" in Kent, p. 42.

1324 The river at Séguin is the Cuyahoga. Its upper reaches, at Akron, Ohio, are within portaging distance of the Tuscarawas that joins the Muskingum near Coshocton, Ohio. The Muskingum joins the Ohio at Marietta. Tanner, p. 41, shows the portage.

Sandoské also communicates with the Sonhioto River & with the Riviere à la Roche; the latter joins up with the Ohio by means of extremely short portages. This is the great waterway used by the Indians to reach the Belle-Riviere.[1325]

If we had originally got established at the two latter positions described above, instead of going to settle on the Belle-Riviere, we would have intercepted all contact between the Indians and the English, thus avoiding giving the latter umbrage until we had built up our strength sufficiently to establish ourselves where we wanted. The trade of the Belle-Riviere **[182]** was of no account at all for the French,[1326] because that region is only inhabited by Loups & Iroquois who have fled from their own territory and come to settle there.

The Ohio is navigable by canoe almost from its source and has no rapids. From Kanoagon, the water is constantly excellent for transporting medium-sized bateaux.[1327] Its course is winding and cuts through a valley which deepens & widens out as one navigates down it. It has no rapids but it does have a strong current, especially during the spring run-off.

Navigation of the waterway, however, **[183]** requires some care on the way downriver, because the bends in the river are very frequent & often carry the vessel over tree-trunks with which the water never ceases to be encumbered. Downriver of Fort du Quesne, navigation becomes better, the river bed wider & the water deeper.

The valley is not more than a quarter or a half a league in width as far as Fort du Quesne. The land on the north bank is high without being mountainous. On the south bank are the reverse slopes of the Apalaches or Alligeny mountains. There are no navigable rivers that flow out of these mountains to join up with the Belle-Riviere. The majority of them are torrents or streams rather than rivers.

1325 It was possible to portage from the Sandusky River to the Sciota north of Marion, Ohio. The Sciota joins the Ohio at Portsmouth, Ohio. One could portage from the headwaters of the Sciota to the Riviere à la Roche (Great Miami) near Indian Lake in Logan County, Ohio. There were two other portages from Lake Erie into the Great Miami in northwestern Ohio. The first followed the Maumee and Auglaize and the other the Maumee to the site of Fort Wayne, Indiana and then up the St. Mary's River. The Great Miami reaches the Ohio below Cincinnati at the Ohio-Indiana boundary. Tanner, p. 41.

1326 Note in original, p. 183 (a): *But control of the river banks was essential in order to maintain communications between Canada and Louisiana.*

1327 Pouchot here means the Allegheny River. "Kanoagon" on his map was the Seneca village of La Paille Coupée at the mouth of Conewango Creek on the upper Allegheny at Warren, Pennsylvania. Tanner, p. 41. It was strategically sited on the Chautauqua portage route.

The Manenguelée will carry bateaux **[184]** as far as its confluence with the Oxiogani at the foot of the Mont du Laurier, or Laurel-Hill; in consequence, the English have never attempted to cover these routes other than by land.[1328]

The mountains in the area of the source of the Belle-Riviere are rocks covered with box-trees, like the Cevennes.[1329] I have marked on the map the trails made by the traders. They lead horses laden with goods, like our peddlers.

In his march to the Belle-Riviere, Braddock blazed a trail every day as he advanced. But the English remade it in 1758 and improved it in 1759, as is shown on the map.[1330] **[185]**

Fort du Quesne was situated on a low-lying point near the river & subject to flooding. The English have built their new fort, called "Pittsbourg", on the curtain in front of the former fort. It is a pentagon measuring about 80 *toises* on each exterior side, built of earth and faced inside & out with large pieces of wood in the style of the fort at Chouegen. It can hold 7 to 800 men.[1331]

They had built at Loyal-Anon a fort of upright stakes to hold 200 men. They had also built at that place a retrenched camp of earth, ten feet thick at the top, faced with logs, and with a ditch 12 to 15 feet wide. This camp was situated right up against a mountain at the bottom of a valley & was overlooked from all sides.[1332] **[186]**

1328 The Monongahela River is joined by the Youghiogheny at McKeesport, Pennsylvania. Laurel Hill and Chestnut Ridge are major heights of land running southwest to northeast through southwestern Pennsylvania.

1329 A mountain range to the west of the lower Rhône in the south of France.

1330 Note in original, p. 185 (a): *See also the map by Thomas Hutchins of Colonel Bouquet's march across Indian country in 1764.*

Braddock's route of 1755 was replaced by a new road blazed by the Forbes expedition in 1758. See note 465.

Thomas Hutchins (1730-1789) was a military engineer and geographer. His map was drawn to illustrate William Smith, *An Historical Account of the Expedition Against the Ohio Indians in the Year 1764* (Philadelphia, 1765). A biography of Hutchins is in *DAB*, V, pp. 435-36.

1331 Pittsburgh or Fort Pitt was begun in 1759. Although of similar shape to Fort Ontario at Oswego, the fortress at the Forks of the Ohio was faced with brick on the land fronts and sod on the sides facing the water. Stotz, pp. 126-40.

1332 Fort Ligonier was located on Loyalhanna Creek just west of the Laurel Ridge. It was established during Forbes' 1758 march to Fort Duquesne. The fort itself was constructed

...(continued next page)...

The other forts on this route, as far as Virginia, are palisaded enclosures which function as storehouses. They used to hold a garrison of 25 to 50 men.[1333]

The mountains & tracks along this route are rather difficult to traverse. When the English formed convoys, they had to increase the number of horses by one third in order to carry the oats needed to feed the pack horses.

The towns with warehouses for these expeditions were Lancaster,[1334] & Schippenbourg, where they assembled all the food & ammunition which were to be transported to the Belle-Riviere.

There can be no doubt that if the French had had any substantial force in [187] that region, they would have prevented the enemy from settling there because of the potential for ambush characteristic of this mountainous area. [188]

Observations on the Mountains of North America

One can only form a proper idea of the theory of the earth through a profound knowledge of the structure & orientation of mountains. The highest mountain chains of our continent[1335] run mainly from west to east. Those of the New World, the Cordelieres & the Apalaches,[1336] are by contrast oriented north and south. The learned academicians sent to Peru to measure the earth provided M. Buffon with interesting details about the Cordelieres, which he used as the basis of his system. [189] He did not have available to him the same

with a combination of logwork and stockade walls while the retrenched camp was fortified with a horizontal log wall. Stotz, pp. 118-20. Restoration of the original subsurface features and reconstruction of the above-ground features of much of the post was completed in 1969. The site is preserved and maintained by the Fort Ligonier Association in Ligonier, Pennsylvania. See *Ibid.,* pp. 146-89 for a detailed account of the reconstruction.

1333 Descriptions, plans and accounts of a number of these posts are in Stotz. See also William A. Hunter, *Forts on the Pennsylvania Frontier, 1753-1758* (Harrisburg, 1960).

1334 Note in original, p. 187 (a): *The English reckon the distance from Pittsburgh to Lancaster as 238 miles & from Lancaster to Philadelphia 66 miles.*

1335 Europe.

1336 The Andes and the Appalachians. Pouchot and contemporary Europeans knew little about the Rocky Mountains.

resource in respect to the Apalaches which he, in some sense, forgot. It is for that reason that we propose to transcribe here the judicious observations on the latter mountain chain which we found in the papers of M. Pouchot, who has greatly benefited from the work of M. Evans, without ever acknowledging it.[1337]

The Notre Dame Mountains form a kind of angle to the entrance of the St. Lawrence River & can be envisaged as a continuation, or more precisely as the beginning, of the Apalaches chain.[1338] These mountains are higher in the region of the river mouth & as one travels into the continent **[190]** they seem to become lower & the continent to become higher until one reaches the lakes, where very high plains are to be found. The latter come to an end in the east at the summit of the Apalaches, which they even seem to form.

In the territory occupied by the English colonies, the structure of the mountains varies & they are separated by the Hudson River into two chains which in general run parallel to the ocean. From the eastern region as far as Massachusetts Bay, they make up a single range running almost due north & then tending to veer E. to follow the contour of the sea coast. This stretch of territory can be divided into two belts that run in a westerly direction from Boston.

The first begins in the area of **[191]** Water-Town and forms very steep hills or small mountains until the other side of Vester[1339] & from there for around 20 miles towards the Hudson River. The second belt covers a wider area, full of low mountains in a long chain extending to the south as far as the Sound, separating Long Island from the mainland & causing the slopes, ridges, heights, & landscape full of boulders which can be seen when one travels along the seacoast of Connecticut & which form an obstacle to the creation of a good road into the interior.

1337 Note in original, p. 190 (a): *He often does no more than translate the analytical material of the general map of the British colonies, an English work published in 1755,* in-4, *by M. Evans.*

Pouchot had consulted Lewis Evans' "A General Map of the Middle British Colonies in America" as well as his *Geographical, Historical, Political, Philosophical and Mechanical Essays. The First, Containing an Analysis of a General Map of the Middle British Colonies in America, and of the Country of the Confederate Indians* (Philadelphia, 1755). See note 972.

Much of Evans' information on the mountains of North America was also used by Thomas Pownall in *A Topographical Description of the Dominions of the United States of America,* the first edition of which appeared in 1776.

1338 The Notre Dame Mountains or Monts Chic-Chocs form the spine of the Gaspé Peninsula.

1339 Watertown and Worcester, Massachusetts.

Although the greater part of Connecticut is included in this kind of belt, one nonetheless finds large valleys and fine, fertile country. Between the two chains, the largest valleys extend [192] along the Connecticut River. Some of these valleys are 20 miles wide. The direction of these chains of hills & mountains indicates the direction of the rivers & streams of the country.

To the east of the first belt, on the ocean side, the land is made up of a mass of sand mixed with debris washed up by the N.E. and S.W. tides which have created almost the entire Cape Cod area as far as the E. end of Massachusetts Bay. Longue-Isle, or Long Island, also seems to have been formed by ocean sands mingled with earth washed up from the mainland. The area towards the W. is of the same type. But the mountains are higher as one approaches the frontiers of Canada. [193]

The territory to the S.W. of the Hudson River is divided in a more regular manner by a greater number of belts than has been mentioned above. The first landmark to be found in this area is that curtain of rocks, a kind of soap-stone, two, three & up to six miles wide; its highest points dominate the surrounding country. It extends from the city of New-York towards the south west, and incorporates the lowest falls of the Delaware, Schuylkill, Susquehana, Gun-Powder, Patapsco, Potomack, Rapahannock, James & Ronoack rivers. This chain of rocks, which appears as a regular curve, once formed the sea coast of this part of America.

From the ocean as far as this chain, [194] & from the hills of Navesink[1340] towards the south west as far as the extreme limits of Georgia, the entire territory is included in the first belt. It can be identified as "low plains" which are formed by soil washed down from the highlands that has mixed with ocean sand. These plains are not watered by a single river. The sand is white, about 20 feet deep and completely sterile; there is no vegetation-producing soil to improve it. But the areas in which the rivers are located have been fertilized by the soil washed down by them; it has mixed with the sand, as the salt marshes, shellfish & other strange organisms found there indicate.

The soil is the same over a stretch of territory 40 to 50 miles [195] wide. On the coast from Navesink as far as the Cape of Florida the land is sterile everywhere. Even the areas around rivers have not been fertilized by soil brought down from higher ground. Only a number

1340 Navesink, New Jersey.

of marshes or swamps are to be found and they can barely sustain a few white cedars. One also quite commonly finds beds of clay detached by the ocean from the soapstone hills; some of them are three to four miles wide.

From that chain of rocks, where all the rivers form falls, as far as the chain of broken, low mountains called the "Montages du sud",[1341] there is a territory extending for 50, 60 or 70 miles, which is extremely uneven and which gets higher as one penetrates into the interior. This second belt can be called the Pays **[196]** Supérieur.[1342] It consists of a number of strands of different types of soil & rock debris. It is a few miles long & in a few places is studded with little curtains of land & low mountain ranges. Their steep slope causes the waters of torrents & ravines to flow very swiftly, which washes soil down into the rivers and makes the low plains fertile. The steepness of the slopes & the ravines makes half the territory largely unsuited to cultivation.

The Southern Mountains have no high peaks, like the Endless Mountains.[1343] Rather they are low rocky outcrops, broken at irregular intervals in a few places & remote. Some of them are a few miles long, others are several miles wide. Between the Southern Mountains & the tall Endless Mountains, which **[197]** to distinguish them from those of the north, are called in some places Kittatini, & in others Pequilin,[1344] there are very fine, fertile valleys 8, 10, & 20 miles wide. It is there that the majority of the best cultivated land owned by the English is to be found. This band stretches across New Gersey, Pennsylvania, Maryland and Virginia. No general name has yet been given to this territory but it could be called "Piedmont", because of its similarity to that region of Europe in respect of the richness of the soil. It is the third belt of the northern continent of the Americas.[1345]

1341 Thomas Pownall describes "The South Mountain" crossing New Jersey, Pennsylvania, Maryland, and Virginia. Its southern extension was the Blue Ridge. Pownall, pp. 95-97.

1342 The Pays Supérieur or "Upland" is described by Pownall on pp. 93-94.

1343 The Appalachian Mountains. See Pownall, pp. 110-13.

1344 The Kittatinny Mountains of northwestern New Jersey. The Pequest River, a tributary of the Delaware, flows a short distance to the south. See Pownall, pp. 95-97, who called this Pequa Creek.

1345 Pouchot compares the region to Piedmont in northwestern Italy. He had campaigned there during the closing stages of the War of the Austrian Succession.

The "Sans fin", or Endless Mountains, called thus after the Indian name translated into English, form a long and fairly uniform range. They stand 5 to 600 *toises* [198] above the valleys that intersect them. Their name aptly expresses their extent.

In some places, for example towards the Kaatikill Mountains[1346] & around the sources of the Ronoack, it would be possible to suppose that one was confronted with the far end of the Endless Mountains. But closer examination of their sides reveals that they extend into new branches that are no less extensive. Their final chain, which is that of the Alligeny, or the Belle-Riviere, runs parallel to that first chain of soapstone rock which ends the first belt. This chain ends in large earth mounds & isolated boulders, in the area of the sources of the Ronoack & the New River.[1347]

The most E. chains, which seem to run towards the S., turn imperceptibly towards the W. and [199] as a result the valleys of the belt in the Pays Supérieur & the Piedmont, as we have just called it, are wilder in Virginia than in the N. regions. The chains in the S.W. seem to want to link up with those of the Alligheny. In a number of places, they divide & form new mountain chains, like those of Ouasioto.[1348]

Everywhere, these chains penetrate one another, so to speak, forming counterforts, or spurs which emerge from the larger mountain chain and disperse into separate ranges of low mountains. This would seem to suggest that there are good routes into the interior but when travelling in the region, one never finds a way out. It is safer to journey over the rocks than through areas of mingled rock & earth because the latter lead to [200] ravines which form precipices. Scarcely a tenth of the land in these mountains is suitable for cultivation. This is the fourth belt, which ends at the territories of the Iroquois & the country which becomes the plains of the Ohio.

Let us conclude from these remarks of M. Pouchot, 1st, that all these belts of which he speaks are no more than branches of the

---

1346 Catskill Mountains.

1347 The Roanoke (or Staunton) River rises in the mountains behind Roanoke, Virginia and flows through southern Virginia and North Carolina to the Atlantic. The New River has its sources farther southwest near the Virginia-North Carolina line. It flows generally north through West Virginia to enter the Ohio as the Kanawha at Pt. Pleasant, West Virginia.

1348 The Cumberland Mountains of the Virginia-Kentucky border region. See Pownall, pp. 110-11.

Apalaches, or rather the different areas in which that mountain chain is to be found, whether in its length or its width; 2nd, that the entire territory situated to the east of the Apalaches has quite obviously been covered by the waters of the ocean and that the numerous and unmistakable traces of this development prove that it happened quite recently.

We may also be permitted to add here that the Apalaches chain of mountains & the high **[201]** belt of land to the west which it seems to buttress while constituting its former limit, are a part of a major mountain range which extends from the south east to the north west, from the mouth of the Rio de la Plata[1349] to beyond the Great Lakes of North America. **[202]**

Remarks on Niagara Falls

Since the most northerly part of America is very high, the rivers that flow from it, before discharging into the lakes & rivers, must necessarily follow the slope of the land and create waterfalls of varying sizes. The most famous of these are obviously those at Niagara. The Indians living close to Québec considered them to be situated at the far western end of the continent when the French first went to settle there. They assured the latter that "at the far end of Lake Ontario there are falls that could well be **[203]** a league across, over which a very great current descends into said lake; that beyond these falls, land can no longer be seen on either side, only a sea so great that they had never seen its far end, nor heard of anyone else who had done so; that the sun sets on the right hand of said lake, &c."[1350]

The voyages undertaken shortly thereafter by the French into the interior of America provided them with rather less vague knowledge of this famous cataract. However, this knowledge was at first inaccurate & one can scarcely rely on the details about it supplied by the Baron de La Hontan[1351] & Pére Hennepin.[1352] The description of it that we owe to Pére Charlevoix **[204]** deserves to be given greater

---

1349 The river and estuary between Argentina and Uruguay. With imprecise knowledge of the Rocky Mountains, Pouchot's editor assumed that the Appalachian Chain connected with the mountains of South America.

1350 Note in original, p. 204 (a): *Marc l'Escarbot,* Histoire de la Nouvelle France, *p. 352.* This work was published in Paris in 1609. See *DCB,* I, pp. 469-72.

credence.[1353] M. de Buffon has not disdained to include it in his immortal work. In addition to what M. Pouchot has recorded about these falls, in the observations we have just read, others have been found in his papers which will be used here.

The Portage, or Niagara, River is properly speaking no more than the drainage channel from Lake Erie which flows along this waterway into Lake Ontario, six leagues from the Falls. Since it is not easy to measure the height of the falls with instruments, travellers, who could only in any case view them in profile, have varied very greatly in their accounts. The Baron de La Hontan suggests that they are seven to eight hundred feet high[1354] & the Chevalier de Tonti one hundred *toises.*[1355] The estimate of P. Charlevoix **[205]** is more reliable; he puts the height of Niagara Falls at one hundred forty to one hundred fifty feet.[1356]

1351 Louis-Armand de Lom d'Arce de Lahontan (1666-before 1716) had served in New France as an officer of the Marine and probably first saw Niagara Falls in 1687. In 1703 he published three volumes, *Nouveaux voyages dans l'Amérique septentrionale, Mémoires de l'Amérique septentrionale* and *Supplément aux voyages ....* in which he commented on many aspects of life in New France. *DCB,* II, pp. 439-45.

1352 Louis Hennepin (1626-c.1705) was a Recollet missionary who accompanied René-Robert Cavelier de La Salle (1643-1687) on his 1678-1680 explorations and first saw Niagara Falls on December 8, 1678. In 1683 he published *Description de la Louisiane, nouvellement découverte au sud-ouest de la Nouvelle France* and in 1697 *Nouvelle découverte d'un trés grand pays situé dans l'Amérique entre le Nouveau Mexique et la mer Glaciale ....* with descriptions of the Niagara cataracts. Hennepin's veracity has been, to say the least, questioned by many historians. *DCB,* II, pp. 277-82; Dow, pp. 23-24. For information on La Salle see *DCB,* I, pp. 172-84.

1353 Charlevoix and his *Histoire et description générale de la Nouvelle France avec le Journal historique d'un voyage fait par ordre du roi dans l'Amerique septentrionale* (Paris, 1744) have been cited in note 1299.

1354 Note in original, p. 205 (a): Voyages, *p. 107.*
La Hontan, *Nouveaux voyages.* The passage is translated in Dow, p. 31.

1355 Note in original, p. 205 (b): Dern. dec. de l' Amér. *p. 30. P. Hennepin judges the falls to be 100 fathoms, that is to say 600 feet.*

Henri de Tonty (1649/50-1704), a Neapolitan by birth, served as lieutenant to La Salle. He first saw Niagara Falls in December 1678 as a member of the same expedition as Hennepin. Tonty (or Tonti) later published *Dernières découvertes dans l'Amérique septentrionale de M. de La Sale, mises au jour par M. le chevalier Tonti* (Paris, 1697). *DCB,* II, pp. 633-36. In a *relation* of November 14, 1684, Tonty estimated the height of the falls as 500 feet. Dow, p. 30.

Hennepin claimed, in his edition of 1683, that the fall was "more than five hundred feet." His 1697 work says "above 600 Foot in depth." Dow, pp. 23, 27.

M. de Buffon initially thought that the cataract was the finest in the whole world & that its preeminence was due to its height; but recently he seems to have retracted this view in order to give the preference to the Terni Falls in Italy. Although the majority of travellers only estimate the height of the latter at two hundred feet, the celebrated naturalist assumes that they are three hundred feet.[1357] Without seeking here to refute his testimony, we shall merely note that the Marmora Mountain has an opening of only twenty feet through which the [206] Velino rushes, the perpendicular falls constituting its cataract.[1358]

It is not the height, but the width of a cataract that makes it large. Since Niagara Falls are nine hundred feet across, they are therefore quite obviously superior to all the others. They cannot be compared to the Terni Falls which, in respect of height are lower than a number we know of in the Pays des Grisions, the Valais and Switzerland.[1359] We are astounded that M. de Buffon has not cited as examples of perpendicular falls those to be seen in the famous Vallée de Lauterbrun,

---

1356 In 1721 Charlevoix wrote: "I presently found the baron de la Hontan had committed such a mistake with respect to its height and figure, as to give grounds to believe he had never seen it. It is certain, that if you measure its height by that of the three mountains you are obliged to climb to get at it [the escarpment at Lewiston, New York], it does not come much short of what the map of M. Deslisle makes it; that is, six hundred feet, having certainly gone into this paradox, either, on the faith of the baron de la Hontan or father Hennepin .... it is no easy matter to measure its height with instruments. It has, however, been attempted by means of a pole tied to a long line, and after many repeated trials, it has been found only one hundred and fifteen, or one hundred and twenty feet high. But it is impossible to be sure that the pole has not been stopt by some projecting rock .... For my own part, after having examined it on all sides, where it could be viewed to the greatest advantage, I am inclined to think we cannot allow it less than a hundred and forty, or fifty feet." Charlevoix, 1761, I, pp. 353-54.

The actual free-fall height of the Canadian or Horseshoe Falls is today about 170 feet. The crest of the American Falls is 185 feet above the surface of the river and 70 to 120 above the rocky talus that has accumulated at its base. It should be remembered that the cataracts, particularly the Canadian falls, have retreated significantly since Charlevoix's time.

1357 Note in original, p. 206 (a): Supplément à l'Historie naturelle, *vol. I, p. 469.*

The reference is to Georges-Louis Leclerc, Comte de Buffon, *Supplément à l'histoire naturelle* (7 vols.; Paris, 1774-1789).

1358 Terni is on the Velino River about 45 miles north and east of Rome. The Velino rises in the Appennino Umbro-Marchigiano, flows past Terni and joins the Tiber at Orte. Just above Terni is the Cascata di Marmore, the falls described here.

1359 The "Pays de Grisions" was Grisons (today Graübunden) Canton in the east of Switzerland. The Valais is a canton of Switzerland bordering Italy.

where nature has exposed to view her most awesome beauties.[1360] From the summit of two mountains that end at the glacier, leaving between them a dark & narrow valley, several streams hurtle down to [207] form perhaps the highest cataracts in the universe. The Staubbach Falls have been accurately measured & their perpendicular height is no less than 816 King's feet, or 1,100 Berne feet.[1361] In truth, their width is not very great. This can be judged from the stream they form as they fall, which is barely more than eight or nine feet wide at its greatest extent. We shall not discuss the Myrrebach Cascade[1362] & a number of others of which the mass of water is also quite small & the height a little less.

Niagara Falls are also as remarkable for the phenomena they produce as for their width. When the weather is fine, several rainbows can be seen one above the other. It is not difficult to guess the cause. Sometimes [208] a light fog rises like smoke above the cataract, rather like that of a forest fire. It can be seen from Lake Ontario fifteen leagues beyond Fort Niagara. It is an unambiguous sign of rain or snow & a sure means of reconnoitering this fort or the mouth of the Portage River.[1363]

1360 The Lauterbrunnen Valley, south of Interlaken, Switzerland.

1361 Staubbach Falls are near the village of Lauterbrunnen in the valley of the same name. The falls are formed by the Lütschune River. See Gibson, pp. 200-02. Modern measurements give them a height of 984 feet.

1362 Mürrenbach Falls, reputed to be Europe's highest, are at Stechelberg, Switzerland. This cascade is also in the Lauterbrunnen Valley at the intersection of the Lütschune and Sefinen rivers.

1363 Rainbows are still a staple of a visit to Niagara Falls. Peter Kalm recorded, in 1750, that "On a certain time of day when the sun is shining, we can always see a rainbow against the mist. I arrived a little after eleven in the forenoon, and we had clear, sunshiny weather all day, so that we could see almost constantly a beautiful rainbow on the east side of the river, near the bottom of the cataract, about six fathoms below it. When we stood on the edge or bank of the river the rainbow appeared beneath us." Kalm, p. 707.

Throughout the eighteenth century, the rising mist of Niagara Falls served as a landmark for voyageurs and sailors on both Lake Ontario and Lake Erie. Kalm was one of many visitors to note this phenomenon: "Everyone contended that from the other, east side of Lake Ontario, where only a small portion of this section can be seen, one on a still morning could behold something like a heavy mist rising above the woods in the vicinity of the falls, and that it looked like the smoke from a forest fire, although it was but the vapor from Niagara. The same phenomenon is noticed several miles from the falls, on Lake Erie, if one looks in the right direction." *Ibid.*, p. 700. In 1768 an English traveller reported that "in a fair, calm day, there appears a pyramidal cloud, very high, arising like the steam of a

...(continued next page)...

The noise made by the cataract, augmented by the echoes of the surrounding rocks, travels a greater or lesser distance according to the direction of the wind. It is not rare to hear it from ten or twelve leagues away, but rather like the muted rumble of distant thunder. This fact caused P. Charlevoix to speculate that a cave must have been formed over time beneath the falls. According to him, another reason is that nothing swept over **[209]** the falls has ever reappeared.[1364] This is the result of the whirlpools that are always found either at the bottom of great waterfalls or at places where a river current is forcibly counteracted or too compressed.

The urge to criticize the Baron de la Hontan led P. Charlevoix to deny that fish caught in the current above the falls are dead when they reach the bottom. The Jesuit adds: "I have also been assured that birds which flew over the falls were sometimes enveloped in the swirling draught which the violence of these rapids created in the air. I personally noted quite the contrary. I have **[210]** seen small birds flying about quite low directly above the falls without coming to any harm at all."[1365] I have myself seen birds diving to the foot of the cascades of the Rhine which, on the Chateau de Lauffen side, are eighty feet high,[1366] & subsequently flying away without peril to

---

mighty furnace, from the violence of the falls forcing the spray so high that, becoming lighter than the air, is suspended and said to serve as a mark, in the navigation of the lakes above and below, for fifty miles." Letter from a Gentleman to the *Antigua Gazette,* Aug. 21, 1768, quoted in Dow, I, p. 67.

Kalm was informed that it was the noise level of the cataract that served as a harbinger of bad weather: "M. Joncaire and all others declared that the falls occasionally made a louder racket than at other times. When its roar was unusually strong in any direction, it was said infallibly to indicate rain and bad weather. This was the reason, it was claimed, why the Indians of the neighborhood were able to prophecy the weather so accurately." Kalm, p. 699.

By Portage River (*riviere du Portage* in the original), Pouchot means the Niagara.

1364 Note in original, p. 210 (a): Journal historique du voyage de l'Amérique Septentrionale, *vol. V of the* Histoire de la Nouvelle France, *p. 346.*

The theory of the cavern beneath the falls may be found in Charlevoix, 1761, I, p. 355.

Peter Kalm, like most other travellers of the time, was told by French officers that, in still weather, the roar of Niagara Falls could be heard at Fort Niagara, 14 miles to the north, as well as at the mouth of Buffalo Creek on Lake Erie. Kalm, p. 699.

1365 Note in original, p. 211 (a): Ibid, *pp. 346-347.*

The passage is in Charlevoix, 1761, I, p. 355.

1366 Note in original, p. 211 (b): *On the opposite side, near the ironworks of Neuhauffen, these falls appear to be lower.*

...(continued next page)...

themselves. Birds of prey may have emerged just as unscathed from Niagara in calm weather, but not when the winds blow hard from the southern sector. Then, as M. Pouchot observed several times, the waterfowl which follow the course of the river & fly up to the height of the rocks are compelled to fly close to the surface of the water **[211]** in order to find shelter from the wind. But since in this position they can no longer struggle against the air flow, they are blown down into the basin. More or less the same thing happens to fish swept down by the rapids upriver from the cataract, which are in evidence quite far into Lake Erie. A larger number of animals must also perish in the swirling waters. They are so dreadful above the falls that it is impossible to navigate in them.[1367] Ten or a dozen Outaouais Indians tried to cross the river in their canoes at that point in order to escape a party of pursuing Iroquois. They struggled in vain to resist the fierce currents and were promptly swallowed up in the waters of the cataract.[1368] **[212]**

Although the main body of water falls perpendicularly onto sharp rocks, it nonetheless forms, as a result of the force & mass of water, a fairly considerable ramp. The Baron de la Hontan claims that underneath there is a track along which three men could easily pass from one side to the other without getting soaked, indeed without being wetted by a single drop of water.[1369] Neither P. Charlevoix nor M. Pouchot mention this track, which in all likelihood no one has much felt like exploring.

Around the falls there are curtains of land eighty feet high. They indicate quite obviously that the channel or river that forms the falls

---

The "Rheinfall", located on the upper Rhine River near Schaffhausen. Neuhausen and Laufen stand on opposite banks of the river. The falls are described and pictured in Gibson, pp. 204-08.

"I" seems here to refer to the 1781 editor. In the first two volumes of the *Mémoires,* Pouchot invariably wrote in the third person. The chapters on the mountains of North America and Niagara Falls appear to have been especially incomplete at the time of Pouchot's death, and both exhibit a great deal of contemporary editing.

1367 Note in original, p. 212 (a): Trans. Philos, *vol. VI, part 2, p. 119.*
The reference is to Raynal's work.

1368 Note in original, p. 212 (b): *Charlevoix,* Journal, *p. 345.*
See Charlevoix, 1761, I, p. 354.

1369 Note in original, p. 213 (a): *See p. 107.*
"Between the surface of the water that shelves off prodigiously, and the foot of the Precipice, three Men may cross in a breast without any other dammage, than a sprinkling of some few drops of water." From the English edition of 1703 in Dow, p. 31.

was once almost the same level as Lake Erie. The falls at Niagara must therefore **[213]** have been much higher than they are now & the rock bed which causes them must have been gradually worn away before reaching its present state.

On reaching the top of the mountains in the neighborhood of the falls, one perceives a plain three or four leagues wide which stretches away around Lake Ontario in the direction of Toronto & is oriented, according to the lie of the escarpment, to the north-east & south-west. This curtain, or range of hills begins in the mountains of the north & extends in an easterly direction as far as the territory of the Five Nations. It cannot be doubted that at one time these hills constituted the shore of the lake, of which the water level fell, leaving behind the plain that runs round it.[1370]

The area of all the Great Lakes, & in particular that of Lake Erie, **[214]** which lies above Niagara Falls, has undergone the same change. The banks of the St. Lawrence, into which they drain, have not been immune from the process. The island of Mont-Réal, formed of two branches of the river, provides the proof of this. Its hills rise a certain distance from its banks & hence show that all the land area from the foot of the hills as far as the river bank was once occupied by its waters, which withdrew as the waters of the lakes diminished, by the progressive decline in height of Niagara Falls & the other rapids or cataracts which punctuate the course of the river above Mont-Réal.

Another proof of the change just mentioned should be noted. It is to be sought in the highest mountains of Canada. In them, as well as in the former plains covered with calcinable, sulphurous soil, or soil made up of soapstone & sandstone, one constantly finds **[215]** shellfish of all kinds. The more recent plains, by contrast, are covered with fossils of wood, fruit, snakes, snails & other freshwater shellfish. **[216]**

1370 The Niagara Escarpment once marked the southern margin of glacial Lake Iroquois. See Albert H. Tiplin and George A. Seibel, *Our Romantic Niagara: A Geological History of the River and the Falls* (Niagara Falls, Ont., 1988), pp. 27-32. The geology of the escarpment and Niagara Falls is treated in the above work and in great detail in Irving I. Tesmer (ed.), *Colossal Cataract: The Geologic History of Niagara Falls* (Albany, 1981).

1371 Père Joseph-François Lafitau (note 1278). His account was published as *Moeurs des sauvages ameriquaines, comparés aux moeurs des premiers temps* (2 vols.; Paris, 1724). An English translation, edited by William N. Fenton and Elizabeth L. Moore, was published by the Champlain Society in 1974 and 1977.

On The Manners And Customs Of The Indians Of North America [217]

• Introduction •

It is to the missionaries that we owe precious details about the manners & customs of the Indian nations of North America. The accounts of those apostles of the New World would doubtless be blameless had they not allowed themselves to be swayed by national prejudices which too often impelled them, according to their private interests, either to exaggerate the barbarism of the Indians or to disguise their faults. One of the missionaries, P. Lafitau,[1371] *did not baulk from comparing the Indians to the first peoples of antiquity. His imagination caused him to discover many links between their religion, customs, traditions, &c., although few persons today would recognize any truth in them* **[219]** *since such parallels are no longer much appreciated in our time.*

The first travellers, especially Champlain, the founder & father of the French colony in Canada,[1372] *composed their accounts with that simplicity & truth that will forever render them useful, although their style is almost unintelligible. Those who followed them, instead of rectifying their mistakes, merely multiplied them or misrepresented their accounts. Certain persons have even ventured to exalt the Indians in a manner as ludicrous as it is exaggerated. If their intention had been nothing more than to write an epigram against civilized nations, then they might have been forgiven. But they have sought to lead their contemporaries astray & to deceive posterity. The Baron de la Hontan, in particular, merits this reproach. He wished to change* **[220]** *all the natives of America into great philosophers & unfortunately his work at one time enjoyed a dangerous renown. Jean-Jacques Rousseau even derived from it many ideas that are as false as they are outlandish.*[1373]

1372 Samuel de Champlain (c.1570-1635), founder of Québec. *DCB*, I, pp. 186-99. His work, *Les Voyages de la Nouvelle France occidentale,* was published in Paris in 1632.

1373 Louis-Armand de Lom d'Arce de Lahontan (1666-before 1716) served as a Marine officer in Canada and the Great Lakes during the 1680s and '90s. *DCB*, II, pp. 439-45. His comments are to be found in *Nouveaux voyages de Mr. Baron de Lahontan dans l'Amérique septentrionale* (2 vols.; La Haye, 1703). This was translated into English and published in

...(continued next page)...

A minister from Cleves, to whom paradoxes cost nothing & who always arrives at erroneous conclusions, even when he ceases to reason in an illogical manner, believed himself to be compelled, although he had never been out of Germany, to reject all the testimony of the missionaries & voyageurs in order to promote credence of his own daydream. Since a number of readers were seduced by these reveries, which might generate unfair prejudices concerning M. Pouchot, we think it proper to transcribe here the judgment that M. Buffon passed on the system of the minister in question, M. Paw.[1374] *It is an excellent antidote against the work* [221] *of that rash detractor of the human species.*

"I admit, says the illustrious naturalist, that I do not have sufficient knowledge[1375] *to confirm these details, which I would doubt less if the author had not advanced a very large number of others that are belied by, or directly opposed to, facts which are both widely known & perfectly well established. I shall not bother to cite anything here except the monuments of the Mexicans & Peruvians, whose existence he denies and the traces of which nonetheless still exist & demonstrate* [222] *the greatness & genius of these peoples, whom he treats as stupid beings, degenerate examples of the human species both in body and mind. It seems that M. P. wished to make all the facts conform to this opinion and he chooses them with that end in view. It concerns me that a man of merit, one who seems at all events to be educated, should have stooped to such extreme partiality in his judgments & that he bases the latter on ambiguous data. Is he not very wrong indeed to blame so acrimoniously the travellers & naturalists who may have suggested a number of dubious facts when he himself proposes many which are more than dubious? He embraces & proposes these facts the moment they seem to favor his opinion; he wants us to take him at his word that they are true without citing any authority to vouch for them. For example, when he talks of those frogs* [223] *which moo, so he says, like calves; or of the meat of the*

London in the same year. German and Dutch editions followed in 1709 and 1710, respectively, so Lahontan's work had wide distribution.

Jean-Jacques Rousseau (1712-1778), French writer and philosopher. Kunitz and Colby, pp. 799-802.

1374 Cornelius de Pauw (1739-1799). *Biographie universelle,* XXXII, pp. 321-22. His *Recherches philosophiques sur les Américains, ou Mémoires intéressants pour servir à l'histoire de l'espèce humaine* was published in Berlin in 1768-1769.

1375 Note in original, p. 222 (a): *How different is this language from that of M. Paw! Should a foreign author who writes in our language believe himself to be exempt from adopting a gentlemanly & modest tone, when the first writer of the nation has always conceived such a tone as a duty?*

iguana which communicates venereal diseases to those who eat it; or of the glacial cold of the earth at a depth of one or two feet, &c. He claims that Americans in general are degenerate human beings, that it is not easy to conceive how created beings can be in a state of decrepitude & decline, but that this is the state of the Americans; that there are no shells or other debris from the sea on tall mountains or even on those of average height; that there were no cattle in America before it was discovered; that only those who have not reflected sufficiently on the nature of the climate of America have believed that one might consider the peoples [224] *of the continent as very new; that beyond the eightieth degree of latitude, beings constituted as we are could not breathe for twelve months of the year because of the density of the atmosphere; that the Patagonians are of the same stature as Europeans, &c. But it is pointless to draw up a longer list of all the false and dubious facts that this author has made bold to advance with a confidence which will embarrass any reader who is a friend of the truth."* Supplement à l'histoire naturelle, vol. VIII, ed. in-12, pp. 326-329. [225]

On The Manners And Customs *of the Indians of North America*

The race of men who people this great continent is more or less the same everywhere. Their skin is normally copper-colored though they usually seem darker than that because they are brought up naked & because of their habit of rubbing grease, clay or brown dye into their skin. Combined with their uncleanliness, this makes their [227] skin appear darker than it really is. One of their distinctive features is that they have neither beard nor eyebrows. It is true that they are careful to remove any hair that grows there, but they are individual hairs. If some Indians are currently to be found who have some kind of beard, it is because they are métis.[1376]

They are normally tall. Their height is five feet, 4, 5 or 6 inches & up to nearly six feet. They are very nimble. A number of them are good-looking. In some of the nations they seem rather affected. They are bright-eyed. Generally speaking, they are not so strong as Europeans. The women are less attractive than the men. They become very

1376 People of mixed European and Indian blood.

fat & lose their bloom early. There are certain nations in the area of the Chaouanons **[228]** who have a lighter skin; some of them are even as fair as Germans. That is very rare, however.

Those who have a different type of face are different because of their habits. The Têtes Plates & all the Caräibes[1377] have a flat forehead and a high crown to their heads because in their childhood their heads are compressed between two pieces of wood. Those who are called Têtes de Boule have a round head, which is peculiar to several nations in the N.W. of America.[1378] It is said that in that region, some bearded Indians have been found, which seems very doubtful. Only Indians themselves have attested to this and they may have taken Spaniards occupying that region for natives of the country.[1379] One striking thing is that those who are in the habit of seeing **[229]** Indians can, without speaking to them, judge from their features as well as from the way they dress which nation they belong to.

Each nation can be considered as a family assembled in the same area. There is very little intermarriage between the different nations. Each one inhabits a separate area of this great continent. Unless the interests of the nation demand it or the warriors they have engaged request it, they frequent one another very little & remain isolated in their own districts. Each nation is divided into villages, which in no way resemble those of Europe. The huts of an Indian village are scattered along a river or a lake, sometimes over a distance of one or

---

1377 The Carib peoples of the Caribbean.

1378 Bougainville saw warriors he described by this name at Montréal in June of 1758: "[Some] *Têtes de Boules* brought here by the Nipissings. These Indians, also called *'Gens des Terres'*, live deep in the woods, are great hunters, mediocre warriors, have neither government nor policy, trade more with the English of Hudson's Bay than with us. Their dialect is a corruption of Algonquian. They have few ideas and few words, no foresight for the future. In case of necessity when the hunt is unproductive, they eat their old people, who very peacefully allow themselves to be killed." Hamilton, *Adventure*, pp. 212-13.

The Têtes de Boule were inhabitants of the basin of the St. Maurice River in Québec. The name literally means "round heads", and they were so called because they cut their hair short. They referred to themselves as the "Whitefish People." Adney and Chapelle, p. 107. This would seem to relate them to the Attikamégues or "Whitefish Nation" (note 1442). "Glossary of Indian Tribal Names", *DCB*, I, p. 13.

1379 Some of the northwestern seacoast peoples did, in fact, wear beards. Pouchot presumably heard Indian descriptions of these people. The practice was common among the Kwakiutl and Salish who inhabited the Pacific coast in Washington and British Columbia. Wayne Suttles (ed.), *Handbook of North American Indians, Volume 7: Northwest Coast* (Washington, DC, 1990), pp. 366, 463, 493.

two leagues. Each hut contains the family head, his sons, grandsons and often his brothers and sisters. As [230] a result, some of them hold as many as 60 persons. The hut is generally a large, long rectangle, the sides of which are no higher than 5 to 6 feet. It is made of the bark of young elm trees, as is the roof, with openings in the ridge to let the smoke out and an aperture at each end, but no doors. From this it is easy to guess that they are always filled with smoke. The fire is placed under the opening in the ridge & there are as many fires as there are families.[1380] The cooking-pot is suspended from two forked sticks with a cross piece along the top; there is a large cooking spoon, called a *Mikoine,* beside it. The beds are planks on the ground, or merely animal skins, which they call *Appichimon,* placed along the partitions. They sleep on the skin wrapped in their blankets which, in the daytime, they use [231] as clothing. Each person has his own spot. A man and his wife sleep in a squatting pose in such a way that the backside of the woman is up against the tummy of the man. Their blankets are pulled over their heads & under their feet. It looks rather like a duck paté.[1381] The huts of the Scioux, on the great Mississippi plains, are cone-shaped & made of poles covered with the tanned hides of Illinois cattle.[1382] This presents a pretty picture.

Although the resident, or Christian, Indians[1383] have lost none of their customs, they are nonetheless more decently housed than the others, at the king's expense. Some of them even have furnished rooms where they can receive Europeans when visited by them. Their furniture consists of cooking pots of different sizes [232] according to their needs, & their garments of shirts which are cut as for men. The women wear them too. They like to have them trimmed; the affected young folk & women are quite happy to wear them with embroidered or lace cuffs. They keep them on until they are either worn out or

---

1380 Here Pouchot seems to be describing primarily the Iroquoian nations, the people with whom he had the most contact during his time at Niagara and along the St. Lawrence. See Trigger, p. 303.

1381 The expression used by Pouchot is *un pâté de canards,* and is roughly akin to a modern expression such as "a dog's dinner" or, simply, "a mess."

1382 Plains Indian tepees covered with the skins of bison. Peter Kalm commented at length on these animals which were particularly plentiful "in those parts where the Illinois Indians lived." Kalm, pp. 405-06.

1383 Those people, primarily Iroquois, Abenaki or Nipissing, living in the mission villages along the St. Lawrence River and its tributaries.

rotten. When they first acquire them, they wear them white, after which they rub them with vermillion. They are then red for some time until they become black from use. One can judge from that that they go through considerable quantities of them, since they never wash them. They normally take off their shirts to go to bed. Men sleep naked while women just keep on their *machicotés* for the sake of decency.

Their footwear consists of a [233] kind of gaiter made of crimped flannel and colored red, white or blue. This gaiter is sewn lengthwise to follow the contour of the leg & has four fingers of cloth overlapping the seam. These four fingers of cloth are embroidered with ribbons of different colors mingled with a wide variety of designs. All this looks very pretty, especially when the leg is not too short or too fleshy, which is rare among them. Apart from that they wear embroidered garters of velvet or porcupine skin, four fingers wide & tied into a knot at the side of the leg. The ties of the gaiters fall loosely in front of the leg in order to protect their shins from the undergrowth. Their shoes are a sort of slipper made of deer or caribou skin, tanned like kid gloves & as soft. [234] The top part of the shoe is laced & decorated with a tuck two fingers wide, at ankle height, also embroidered with porcupine skin dyed different colors & with little copper laces decorated with dyed hair and with little bells that tinkle when they walk. This custom may have been conceived to scare off serpents & snakes, which are very numerous. There are special winter shoes, in the form of ankle boots, which are very well adapted to the season. Some of them cost as much as a *louis* and the least well made sell for from 40 *sous* to 3 *livres*. There are pairs of gaiters that cost up to 30 *livres*. The women wear a petticoat, called a *machicoté*, made from a yard of blue or red cloth, of the same quality as material from Berri or Carcassone.[1384] The hem is decorated with different rows of ribbons, [235] yellow, blue & red, or with English lace. This garment resembles a woodsman's skirt. It is held up by nothing more than a strap around the waist. The chemise goes over the petticoat & covers it entirely. The women are laden with necklaces, like maidens in special finery. They are strings of porcelain beads at the ends of which are attached Calatrava crosses,[1385] thimbles,

1384 Berri or Berry is a province of France situated between Tours and Bourges. Carcassone is a city in the south of France located between Toulouse and the Gulf of Lion.

1385 The reference is to the cross adopted by the Christian military Order of Calatrava and displayed on the tunics of its members. This was a scarlet "cross *fleurdelisée*," on which each

...(continued next page)...

silver coins, falling to below their bosom, which is almost entirely covered with them.

They do not pierce their ears, like the men, but wear pendants made of little chains of brass or beads, which fall well down onto the shoulders. They wear their hair parted in the middle & arranged in such a way that it covers a [236] part of their ears & is tied at the back in a tail which reaches down to their waist. This tail, which is shaped like that of a crayfish, is four fingers wide at the top & about three at the bottom & is rather flat. It is tied up with tanned eel skin painted red. A number of women attach a silver plaque, of two or three fingers breadth, to the top part of their hair, & to the bottom part little triangles, also of silver or some other metal. All this is really rather attractive. A woman whose tail of hair is cut off considers herself dishonored & does not dare to show herself in public until it has grown back. The women use their hair to wipe from their hands the grease which they are constantly touching. Their hair is very black, very long, smooth & thick. They sometimes put vermillion [237] in the parting & behind their ears. The Abenaki women paint their whole face when they dress up, the upper part reddish-brown & the lower part vermillion. The Outaouais women often wear, instead of a chemise, a kind of brassiere of blue or red cloth, cut into two pieces in such a way that with four or six tapes they cover half their bodies & arms.

The men, instead of a *machicoté,* wear a breechcloth[1386] which consist of a quarter of a yard of cloth that passes under the thighs and is doubled behind and in front over a tape tied around their waist. Sometimes the breechcloth is embroidered. When they travel and are afraid that the wool will make them too hot, they simply arrange the breechcloth in front of them like an apron. Around their neck they wear a [238] pendant collar, like our knights who belong to an order. At the end of it is a silver plaque as big as a saucer, or a shell of the same

---

end terminated in a *fleur de lis.* The order was formed in Spain to fight the Moors and remained in existence in Pouchot's time. Calatrava is near Ciudad Real in Castile. Some trade silver crosses distributed or traded to Native Americans approximated this shape.

1386 The word used in the original is *braguet.*

1387 Note in original, p. 239 (a): *The porcelains of Canada are shellfish which the English call "clams." They are found on the coasts of New England & Virginia. They are grooved, long, rather sharp & quite thick. See the journal of P. Charlevoix, vol. 5 of the* Histoire de la Nouvelle France, *p. 308, and Calm,* dans l'Amér. Sept., *vol. 2, pp. 385 et seq.*

Charlevoix includes his discussion of wampum in 1761, I, pp. 318-20. Peter Kalm's description of wampum is on p. 129.

size, or a porcelain hoop.[1387] On the forearm they wear silver bracelets of three to four fingers in width and on the upper arm a kind of muff made of wampum or porcupine skin colored with leather laces which form a fringe covering the whole arm.

The Indians gladly wear [239] rings on all their fingers.[1388] Men make up their faces more than women. They can sometimes spend three or four hours on their makeup. It can be said that they are more concerned about it than any affected young woman in France. The care they devote to prettying themselves, or coloring their faces artistically red, black & green with symmetrical patterns that they sometimes change two or three times a day, leaves nothing natural exposed except for their eyes & teeth. The latter are very small but very white & their lips are bright red. Their hair is no wider on their head than a priest's skullcap. It is cut to the length of a finger, swept up into a topknot with grease & powdered with vermillion in the middle. They leave two locks of hair which they enclose in two silver tubes the length of a finger or arrange in a tail with a woggle [240] of porcupine skin. They also stick several birds' feathers in their hair, which forms a kind of crest. When a young man has made war, he cuts away the hair from around his ear & attaches a piece of lead to it in such a way that the weight stretches this cartilage, thus forming an opening in which a rolled gaiter could be placed. They wind around it a brass wire and in the middle they place tufts of feathers or colored hair. These ears come down to their shoulders & flap when they walk. When they travel through the forest, they put a strap around their foreheads to hold in their ears so they are not torn in the bush. They only keep their ears for as long as they are sober for as soon as they fight when they are drunk, they tear them off one another. As a result, there are few [241] men of an advanced age who have whole ears. They pierce the tendon of the nose & insert in the hole a small ring with a little silver triangle that dangles over the mouth.

Men & women wear over their shoulders a blanket, either of wool, which they buy from the Europeans, or of cloth or of tanned skins. It is almost exclusively those from the interior of the country who use the

---

1388 Pouchot probably refers to what are commonly called "Jesuit rings." See Judith Ann Hauser, *Jesuit Rings from Fort Michilimackinac and Other European Contact Sites* (Mackinac Island, MI, 1982). Although the purpose of these rings is poorly documented in the historical record, Hauser (pp. 59-61) concludes from the archaeological evidence that, although they originally served as religious devices, they became primarily trade items after 1700. Pouchot's comment would seem to support this. Peter Kalm, however, does not list rings among the most popular Indian trade goods of 1749 Canada. Kalm, pp. 519-21.

latter. The woolen blankets are made in Normandy of fairly fine wool, better than those supplied by the English, which are coarser. For the children, they are in one point or a point & a half. For men, in two or three point.[1389] After wearing them white for two or three days, they smear them with vermillion, in the first place with a red cross. Some days after that, they cover [242] themselves with the dye, which makes their skin even redder. When the girls intend to seduce a young man, they paint their blankets again. Those of cloth are red or blue and made of a yard & a quarter of material, of the same quality as their *machicotés*. They sew onto the lower part of the garment a dozen strips of yellow, red & blue ribbon & English lace, leaving a space the width of the ribbon between each strip. At the end of each strip they leave five or six fingers length of ribbon dangling. They cover the top part with round silver plaques three quarters of an inch in diameter. The details above all pertain to the attire of the affected young men and women. The men gladly wear greatcoats or a kind of dress-coat trimmed with braid, hats with false brims, turned up [243] on the button side and the remainder turned down and trimmed with blue, red or yellow plumes. Even those who are Christians have always refused to wear breeches, despite the earnest entreaties of the missionaries.

A chemise that is almost black and daubed with red, a braided or glossy jacket, an unbuttoned suit trimmed with braid, an untucked hat, sometimes a wig put on the wrong way round; add to that a face more bizarre than a Venetian mask. Such is an idea of the general appearance of an Indian. The men wear a woolen belt about six inches wide in different colors, which the women make very skillfully with a flame-shaped design on it. They hang from this belt their mirrors and their tobacco pouch, which is made of the flayed and tanned skin of an otter, a beaver, a [244] wildcat or a bird. In this they keep their pipe, their tobacco & their tinderbox. They also have an ammunition bag, which is made like a small scrip, where they keep their ball & shot for war or hunting. They carry their mirror & their tomahawk against their backside. They also wear slung across their shoulders a bull's horn in which they keep their powder. Their knife, in a sheath suspended around their neck, falls over their chest. They also have a hooked knife, which is a knife or sword blade that has been bent into

1389 Points were generally lines woven into a blanket. The number indicated the weight or quality of the blanket.

a curve. They put this implement to many uses. The above constitutes the equipment and wealth of the Indians. They consider the possession of these things as sacred as their children.

Women & girls remain outside their huts when they are menstruating & [245] do not return to them until their period is over & they have cleansed themselves. Men do not sleep with women during this time and very rarely during pregnancy or when the women are nursing their babies. A woman usually gives birth alone. She leaves her hut & clings to the branches of a tree. They do not complain and find it strange that European women cry out. They carry their babies to the nearest water to wash them & return to their hut. Despite that, difficult births are rare.

Both men & women are deeply attached to their children. The women are particularly attentive to them & keep them very clean. Their cradle is a plank of wood on which they place the child and wrap the softest furs around it. They [246] place over the plank some down extracted from a kind of reed, so that the baby does not become soiled by its own excrement. They take care to leave a small opening at the top of the bundle, so that the baby always urinates outside the covers. If it is a girl they place there a little tube made of bark. The plank has holes at the sides for the tapes which bind the child in place. There is a block as a footrest & above the head there is a hoop about three fingers in circumference to which they attach a rug of calico or cloth, as clean as can possibly be, in order to cover the child.

At the top of the plank, they attach leading strings for carrying the child. They place them around their forehead & the cradle hangs down over their shoulders. If the child cries, they sooth it by rocking it. When the women take it off, they hang it [247] from a branch in such a way that the child is always in an upright position. It sleeps in this position, its head lolling over its shoulder. When the child is sick, the mother places it in a horizontal position, never leaves its presence & feeds it little potions, which are excellent. If she gives the child enemas, there is a bladder on which a little tube acts as a nozzle. The Indian women nurse their infants for two or three years & more; they leave the breast of their own accord. They go around entirely naked until the age of four or five. At that age the girls wear only a *machicoté*. Children of both sexes have their own little blanket. They shout, weep & play among themselves without their parents bothering about it. It is very rare that they beat them. When they are obliged to carry them, the child puts its arms around its mother's neck & sits astride [248] her shoulders. He is wrapped in a blanket, which soothes him. Men, women, & the older children all carry the infants thus. When they travel overland, each person carries his own little parcel on his shoulders suspended from his forehead by a belt. Their belongings are all in a blanket folded at each end and tied up with the straps of the belt in a very artistic way. This closes off the two ends like a pouch.[1390]

Up to the age of thirteen or fourteen, the children have nothing else to do except play. The boys make little bows, & use as arrows pieces of wood with a musket ball at one end. They amuse themselves by firing at small birds & become so skilled at it **[249]** that they often kill them. Their favorite game is lacrosse. The adults join in and from 20 up to 50 people have fun playing it. They even gamble high stakes on the game & lose all their possessions. However badly hurt they may be - and injuries are frequent, despite their skill, because of the intensity with which they play - they never lose their temper.

The girls start by making dolls, then they take up sewing or tanning skins. This idle existence makes them knowing at a young age & sometimes they have lost their virginity by the age of 6 or 8 in the course of playing with other children. The parents find nothing wrong in this and say that every person's body is his own. Nonetheless, the girls always maintain an air of decency in their manner of speaking & their general deportment. They would not allow **[250]** anyone to touch their bosom or to kiss them, especially during the daytime. In public they are always wrapped up in their blankets & take very small steps when they walk. They are pigeon-toed in their gait & walk in a mincing manner. The men are also very pigeon-toed. This perhaps arises from going in the forest, which makes them wary about stumbling over roots. Those girls who have a personality of their own follow the men there, while others, equally spirited, remain very virtuous.

At the age of 14, the boys begin to hunt & even go to war. The major pastime of the young folk in the villages is dancing, which they often continue doing up to two or three in the morning. The girl who **[251]** fancies a young man, places herself behind him as they dance and follows him around throughout the evening. These dances are very active round dances. The steps of the men are almost those of an

---

1390 Note in original, p. 249 (a): *This would be an excellent thing for our soldiers to do. The English had adopted the practice.*

Pouchot describes a tumpline. The English adopted a similar practice for their light infantry units during the 1759 Québec campaign. Wolfe ordered the knapsack of the light infantryman "carried very high between his shoulders, and ... fastened with a strap of web over his shoulder, as the Indians carry their pack." Orders, May 31, 1759 in Knox, I, p. 352.

allemande;[1391] those of the women a kind of quick shuffle. Their songs generally have no set words. The leader sings and the others do no more than respond *hé é* at the appropriate cadence. After each stanza there is a general shout followed by a short pause, then they launch into a new verse. The women do not participate. They all dance with so much fervor that they become bathed in sweat. These dances seem more designed to promote good health than European dances, which are not so tiring. There are other ceremonial dances, only performed by men, which will be discussed elsewhere. [252]

Once this dancing is over, those who have no special purpose in being there return to their huts and go to bed. The others do more than that. A girl who fancies a boy follows him without a word to the place where he intends to sleep. When he is in bed, he says to her: "Come to bed." Then she gets undressed and slides under her blanket, which she arranges with his, as we have already explained above. They often remain in bed until 9 or 10 o'clock in the morning. After that, they do not exchange a single word throughout the day. On other occasions the young man who fancies a girl waits for her to retire to her hut. When everything is quiet, he enters, approaches the fire and takes from it a small flaming twig which he holds over the face of the girl. If she hides her head under the blanket, he withdraws without [253] a word. If, on the other hand, she blows out the flame, he throws it away & gets into bed beside her.[1392]

The boys are generally more restrained than the girls. Many of them wish to have no contact with girls before the age of twenty-two or twenty-three. They say that they do not want their stamina to be sapped. It is in fact improper for a young man to show himself to be in love with a girl. Were he not sought after, he would feel that his virility was diminished. Womanizers are not highly regarded among them. Although some of the girls are virtuous, few of them can resist their own feelings or gifts they are offered. They are vain about the price attached to their attractions & make a point of boasting immediately afterwards about their [254] conquest & what they have been given, especially if they have seduced chiefs or Europeans of some rank, whom they have difficulty resisting. They prefer an Indian to a European & normally only succumb to the latter because of

---

1391 A German dance style of the sixteenth through eighteenth centuries performed in 2/2 time.

1392 Note in original, p. 254 (a): *It is what is called "blowing out the match."*

material gain or out of vanity. If they have slept with someone of rank, they tell their family who come and thank you for the honor you have done them. There are some young women who fall genuinely in love & become very jealous; this can have baleful consequences. If they really love their suitors, they bring up the fruit of their love proudly & with maternal attention; otherwise, they have an abortion. A few have even poisoned themselves. As soon as a man and a woman are recognized as a couple, the other girls are careful not to go after the man; they send him back to his mistress. If they **[255]** steal one another's lovers, there are inevitably fights between them.

Three types of love can be identified among the Indians: 1st, a casual relationship arising out of a dance, a present, &c. 2nd, a genuine love match which leads to a sort of temporary marriage; 3rd, love which compels them to contract a legitimate marriage. The first & second are of no consequence among them & do not prevent them in any way from thinking of the third. A number of girls prefer to remain generally available. Those who live so licentiously are very likely to have abortions. In spite of this debauched life, which Indians find as natural as drinking or eating, there is great mutual respect between brothers & sisters. The Iroquois even consider first cousins as brothers & prohibit sexual **[256]** relationships between close relatives. If you ask them why, they reply that it is their custom. There are, however, nations in the area of the Sauteurs, or Ochibois,[1393] who, once they marry a girl from a certain family, have all the other women as wives. The Outaouais & Missisakes take two or three wives if they believe themselves capable of feeding them through their hunting, though that is not common. When they are asked why they only have one wife, they reply that it is to maintain domestic harmony because, if one wife is preferred, the jealousy of the others causes disputes that the husband has to stop with a stick. If the girl fancies a young man strongly, she tries to be in the same place as him; if he is travelling, she takes his parcel of possessions & carries it for him and if the young man finds her to his taste, **[257]** he takes her with him on the hunt. She acts as his wife all the time they are away. He looks after her & on their return they either break up or remain married. The women think in the same way as Turkish women. They believe they are in this world to serve men & ease their domestic burden. The Indians sometimes marry for love but almost always to serve family interests, to secure an

---

1393 Ojibwa.

ally or acquire a hunter in the family, as the husband goes to live in his wife's hut. Thus it is advantageous to have daughters in a household since, when they marry, they bring hunters into it for the welfare of fathers & mothers. Many young people refrain from marriage in order to look after parents who are old.[1394] **[258]**

The marriage ceremony is extremely brief. If it is a marriage agreed upon by the parents, the latter come to a mutual understanding between themselves & inform their children. If the children have a liking for one another, the boy goes to live in the girl's hut & gives her a full set of domestic items. When he is in bed, he suggests to the girl that she should sleep with him. She is standing beside the bed. She allows herself to be asked for quite some time, then takes off her clothes and modestly gets into bed. Among the Iroquois, it is a sign of respect if the man does not touch his wife. Some of them remain in this situation for up to three months to show more clearly **[259]** how highly they respect them.[1395] The husband brings all the furs acquired through hunting to his wife, who prepares the skins for their common use. It is normally she who sells them. She barters them for what is necessary for the family needs while the surplus goes to pay for jewels of the type described above & brandy which they resell in their community or which they drink on ceremonial occasions. The husband looks after the purchase of arms & ammunition. Thus the wives become very sensible women and accompany their husbands everywhere together with her family, except to war. It is their job to grow and prepare the corn, to see to **[260]** the cooking, and to go and collect the firewood & the wild animals killed by their husbands in the vicinity of the hut. Often the husband comes back without a word & lights his pipe. After a little while, he says to his wife: "I have killed such and such an animal more or less at such and such a place in the forest." As he has blazed a trail on trees along his route, the woman sets out, brings the animal back on her shoulders & skins it.[1396] The life of a woman is full of hardship. If she and her husband are not

---

1394 Note in original, p. 258 (a): *Among the Iroquois, families are matrilineal. They talk of the son of such and such a woman, referring to the family by the name of the mother.*

See Trigger, pp. 309-14, for a summary of Iroquois social organization.

1395 Note in original, p. 260 (a): *In what country, or nation, does not common opinion outrage nature?*

1396 Note in original, p. 261 (a): *A European would be very hard pressed to find the place in question.*

temperamentally suited, they separate & share their children between them. The mother prefers to take the daughters. If she is profoundly dissatisfied with her husband, she keeps all the children, who constitute their sole wealth. **[261]** They often remarry immediately.

The long separations of husband & wife, especially when they are young, caused by pregnancy & the nursing period, trigger divorces because the men weary of being without a woman. This is normally the time when they go to war. In this period they find another woman as their new wife. It is not rare to find Indians who have had five or six wives. However, a good many of them are perfectly satisfied to have one wife in a lifetime. Jealousy also often causes divorces. If they believe that their wives have been unfaithful, they bite off their noses & repudiate them. Such instances are quite rare. The Scioux have an even more extraordinary punishment. When they wish to punish adultery, they **[262]** assemble as many young men as they can, sometimes up to 30 or 40. After a great banquet, they hand the woman over to them and the young men all enjoy her at their leisure, after which they abandon her. Some women die as a result of this disgrace. This ceremony is called "dragging over the prairie."[1397] Other tribes simply kill unfaithful wives. It is certain that infidelity among the women is rare & that in this area far fewer men have anything to complain about than in Europe.

When Indian women become old, which happens quite early, around the age of forty, they no longer have any pretensions. It is then that they acquire more esteem & are consulted over delicate matters, especially among the Iroquois, who call them the "Ladies of the Council."[1398] Indeed they enter the great councils of the nation & exercise considerable influence there. The Iroquois never **[263]** declare war without consulting them & it is their advice that clinches the

---

1397 Pouchot seems to imply that this custom was widespread, but he is unlikely to have observed it among the Iroquois, and it appears to have been more common among peoples living farther to the west. Charlevoix heard of a variation of this custom in 1721: "The Sioux have a plurality of wives, and severely punish such as are wanting in conjugal fidelity. They cut off the tip of their noses, and make a circle in the skin on the top of their heads, and afterwards tear it off." Charlevoix, 1761, I, p. 281.

During the nineteenth century, female infidelity among the Sioux or Lakota could be punished by death or mutilation at the hands of the husband. He might also "throw" or give her away, thereby relinquishing marital authority over her. James R. Walker, *Lakota Society* (Lincoln, NE, 1982), p. 42.

1398 The matrons or the eldest females of a family lineage of an Iroquois village. Trigger, p. 312.

decision. When this happens, they exhort the warriors to conduct themselves with courage & to show to the whole world that their menfolk are fit to protect them. Above all, they impress upon them that they must not abandon their wounded.

When they are in their villages, the Indians are not primarily concerned with hunting. When there they only hunt or fish in order to survive. While at home, they get together in huts, almost always the one belonging to the chief. There, smoking their pipes, they discuss political matters & the history of their nation. They speak of the treaties and common interests they have with other nations & of the journeys they made for the purposes of war. The young but experienced [264] warriors listen in order to become better informed, & acquire in these assemblies that spirit of emulation that constitutes the most central aim of their existence. The oldest men are the council chiefs; it is they who preside over the war councils. The Indians between thirty and forty years of age are in charge of the young warriors. For lack of food supplies, the Indians never remain in their villages all at the same time. They do not grow enough corn to feed themselves for more than two or three months. As soon as they find themselves short of provisions, the entire family leaves and settles in a distant place, especially if they intend to stay there for some considerable time. It is in winter that the villages are most deserted, mainly among nations deeply engaged in the beaver hunt. They spread throughout the interior of the country which they [265] consider as belonging to their nation. They live there, separated from one another, along the shores of lakes, ponds or rivers where, in their view, they will find the most game. Upon arrival at their destination, they construct a hut which is always located in some thicket or little valley, so as to be more sheltered from the wind. They gather a woodpile, because of the harsh winter days. The husbands & the young men spread out in all directions around the hut in order to hunt, sometimes to a distance of ten leagues. They place nets underneath the ice to catch beavers or, if they find any out in the open, they shoot them. They hunt bears which they find in the hollows of trees; they can recognize their presence from the state of the bark. If there is a bear inside, they can hear it grunting as it licks its paws. They cast fire into the hollow to make the bear come out or [266] light a fire at the foot of the tree, thus smoking it out. Harried by the fire & the smoke, the bear emerges from its trunk, where it had been quite upright, & as soon as they see it on the tree, they fire at it. Sometimes they have to cut down the tree in order to secure it. They set traps or snares to catch foxes, otters and marten. They also kill *loups-cerviers, pichous,*[1399] mink, wildcats, muskrats, woodrats, caribou, moose,[1400] roedeer, which are mainly hunted in summer, stags, hedgehogs, partridges, which are the grouse of Europe, and turkeys, which are very abundant in certain regions. They **[267]** eat the meat of all of these animals, except that of the fox, the otter and the mink. They also place nets underneath the ice to catch fish. They cut the animals into large pieces after skinning them in expert fashion and then place the quarters of meat onto a sort of grill, which they erect above their fire in order to dry & cure it. They save this meat for the days when hunting is poor or when severe weather compels them to remain in their huts. It might be thought, from the above list of animals, that the Indians enjoy a happy life. But their idleness, bad weather & the scarcity of game in certain areas sometimes reduce them to the most extreme expedients & force them to seek out a few roots in order to sustain **[268]** life; even then, they often suffer want. Some Indians have been reduced to eating their slaves & they have occasionally practiced cannibalism among themselves. Their remoteness from any help, severe weather and frozen rivers make them stay, despite themselves, in the areas where they are. They often change abode, in order to get the best results from the hunt.

When the worst of the cold is over & the ice begins to melt, nature has already begun to renew itself & the trees, which were frozen, contain a fluid between the new bark & the wood which is not yet sap but which precedes it by about a month. When an incision is made, slightly crosswise, in the trees & a knife blade or a little piece of bark is fitted into it, there flows out of this cut a fluid which, on being boiled, leaves a **[269]** residue of crystals which is bitter or sweet according to the type of tree. The crystals of the walnut tree or cherry tree are bitter.[1401] Almost all trees give off this fluid, which could be

---

1399 Note in original, p. 267 (a): *The* pichou *is the* chat-cervier.

The *loup-cervier* is the lynx. Charlevoix, however, describes the *pitoi* (*pichou*) as a relative of the weasel, similar to the European polecat but differing "only in that its fur is longer, blacker, and thicker." The polecat and the *pitoi,* he wrote, "make war on the birds, even of the largest sort." The skin of the *pitoi,* along with that of the mink, weasel, wood rat, ermine, and marten, were considered "lesser peltry" of the fur trade. Charlevoix, 1761, I, pp. 208-09.

1400 Note in original, p. 267 (b): *These two last types of animals are rather rare.*

1401 Franklin Hough pointed out, in his edition of Pouchot, that the sap of the walnut tree, like that of the maple, is sweet. Pouchot, *Memoirs,* II, p. 205n.

of some use, even for medicinal purposes. The maple & plane trees provide a fluid that is so sweet that it makes excellent sugar. It is sweet, refreshing & also very good for the chest. When boiled, it forms crystals or cakes of reddish-brown sugar, which tastes rather like manna.[1402] It is very agreeable & one can eat as much of it as one likes without fear of adverse effects, just like cane sugar. The Indians, who in that season can neither hunt nor fish because of the thaw & because the fish are not yet swimming up the rivers, live off this natural food [270] for fifteen days of a month.

These trees render an abundant quantity of the fluid, which only flows if there has been frost during the night & sunshine the next day. If the sky is overcast or there is rain, the fluid does not flow from the trees. Naturalists would find this a curious fact. The sap is collected in a copper pot or a small wooden trough once or twice a day. It can be kept for some time. It is boiled up in large copper pots and the crystals so produced are the sugar. It is excellent for colds. Very good syrup is made from ferns, although it has a taste of burnt paper.[1403] It is also good for all sorts of jams, gives an added flavor to chocolate & blends well with either milk or coffee which, however, it causes [271] to taste like unpleasant medicine. There is no doubt that the same sort of sugar could be found in Europe, especially after cold winters, if it were sought in trees of some size.

Once the ice has melted, the Indians find a great many swans, geese, bustards, ducks, teal, plovers, woodcock & snipe, which come back from the south of the continent to repopulate the land. It would be impossible to express the countless numbers of birds to be found at that season until they have settled on large ponds & marshes in order to build their nests.

At the same time, the fish begin to leave the large lakes and swim up the rivers, which almost all have only a kind of narrow channel [272] at their mouths.[1404] Normally, at these places, the water is no more than 2, 3, or 4 feet deep and the Indians wait for the fish there

---

1402 Pouchot refers to a contemporary substance and not to the Biblical food of the Israelites in the desert. A sweet, gummy juice can be extracted from the flowering ash (*Fraxinus ornus*) of southern Europe. These trees are also called "manna ash." The substance is a mild laxative.

1403 Hough identified the plant as *Adiantem pedatem* or maidenhair fern. Pouchot, *Memoirs,* II, p. 206n. Peter Kalm recorded its use in Canada as well as a trade in the plant with the English colonies. The leaves "were very much used instead of tea, in consumption, cough, and all kinds of pectoral diseases." Kalm, pp. 437-38.

and spear them, at which they are very skilled. The fish swim up in incredible numbers, especially on certain days. The carp appear first. There are two species, one that resembles the European carp, but less good to eat, another which has wens on its head. They are called *galeuses*. They are fat & very good eating. Six to twelve-pound fish are quite commonly to be seen.[1405] Next comes the brill, which is a fish with a flat head and four large barbels beside its mouth.[1406] It has the taste & color of tench & weighs from two to seven pounds. Sturgeon are from five to seven feet long. Around the months of May & June, one finds pike from [273] seven to fifteen pounds, mullet, salmon trout from 15 to 18 pounds, and golden & green bass.[1407] The latter fish is short, flat & the most delicate eating of all. The *mastilongé*,[1408] which weighs from 10 to 25 pounds, is a type of pike-trout and is

1404 Pouchot refers particularly to the creek mouths along the shores of Lake Ontario which are normally blocked by gravel bars. A narrow channel usually opens only at times of heavy runoff.

1405 The two species compared by Pouchot to European carp (*Cyprinus carpio*) are probably redhorse suckers (genus *Moxostoma*) which spawn in tributaries of the lake each spring. At least six species inhabit the Niagara River. The second type of Pouchot's *galeuses* may be one of the smaller redhorses (*M. erythrurum* or *M. macrolepidotum*). "Wens" are nuptial tubercules, visible during the spawning season.

Pouchot's descriptions of aquatic life are vague and often difficult to identify. For assistance with the following notes I am indebted to Dr. Robert A. Daniels of the New York State Education Department's Biological Survey Laboratory and to Al Clifton of the Aquarium of Niagara Falls. See especially Daniels to Clifton, Feb. 4, 1994, copy in Old Fort Niagara collections.

1406 A catfish, most likely the brown bullhead (*Ameiurus nebulosus*). Eight barbels should have been visible on this fish although Pouchot mentions only four.

1407 Lake sturgeon (*Acipenser fulvescens*) in this size range were common during the eighteenth century, and the Niagara River was later renowned for its sturgeon fishery.

Pouchot's pike were probably walleye (*Stizostedion vitreum*) although, in the size range given, walleye look less like European pike (*Esox lucius*) than perch.

The "mullet" may be the white sucker or the northern hog sucker (*Hypentelium nigricans*) which is still commonly known in New York by this name. There is no association with the saltwater mullet (genus *Mugil*).

Salmon trout may be the lake trout (*Salvelinus namaycush*) or the Atlantic salmon (*Salmo salar*).

The two varieties of bass are probably the largemouth (*Micropterus salmoides*) and smallmouth (*M. dolomieu*). There is also the possibility that the "green" bass is actually the black crappie (*Poxomis nigromaculatus*) sometimes described by modern fishermen as the tastiest and even most "delicate" of lake fish.

1408 Muskellunge (*Esox masquinongy*), a large North American pike.

excellent eating, as is the golden fish which looks like a whiting.[1409] It is very good eating & weighs from 5 to 12 pounds. All species of European fish are to be found, such as perch weighing from 3 to 4 pounds, burbot of the same size, & eels which are both very large & excellent to eat.

In the lakes above Niagara Falls, one no longer finds any sturgeon. However, it has been replaced by whitefish, which are very abundant & excellent to eat, & by a type of herring that is more delicate than the saltwater variety.[1410] When this land has a much larger population of Europeans, its [274] fisheries will become an important economic activity. For all their fishing the Indians use a spear which is made up of two pieces of iron, 10 to 12 inches long & very sharp, and two upturned hooks, like our fishing hooks but proportionately not as large. They attach these two pieces of iron, about a quarter of an inch apart, to the end of a pole 10 to 12 feet long or more. At fords and rapids they wait for the fish to swim up & then spear them. They very rarely miss. They also fish at night in their canoes. Inside the canoe, they light shavings of cedar wood. One man stands at the bow with his spear while another, in the stern with an oar, guides along the shore the fish which comes to frolic [275] in the light. They spear it as much as ten feet beneath the surface of the water; a fish no thicker than an arm does not escape them.

In the summer the Indians concentrate above all on deer hunting. Since that animal is tormented by little flies - mosquitoes, which we call *cousins*, & *brûlots*, almost imperceptible gnats in which the forests abound - it frequents rivers looking for places where there is a muddy bottom. It wallows in the water in order to give itself a coating of mud that will make it impervious to stings. The Indians know of these places and lie in wait for the deer. They kill quite a number of them in a single day. If they give chase in the forest, they do not need dogs, as they can see the tracks of the animals on the snow. In the other seasons, the most favorable time is when the leaves are rather [276]

1409 Although they bear little resemblance to a whiting, the most likely candidates for this fish are the golden shiner (*Notemigonus crysoleucus*) or one of the smaller redhorse suckers (genus *Moxostoma*), most likely the "golden."

1410 Whitefish (genus *Coregonus*) are common to the upper Great Lakes. Lake herring (*Coregonus artedii*), another form of whitefish known popularly as a cisco, was an important part of the diet of both the French and Indians of the upper Great Lakes. See Elizabeth M. Scott, *French Subsistence at Fort Michilimackinac, 1715-1781* (Mackinac Island, MI, 1985), esp. pp. 10, 76.

wet & do not make any noise when a man treads on them. An Indian can initially recognize the animal by earth that has been stamped down or by bent leaves. He can even judge whether it is far away or close by. He follows the tracks quietly, looking all the time to right & left in the hope of seeing the animal. Sometimes he imitates the cry of the fawn. As soon as he espies the deer, he stops & makes no further movement until the animal starts feeding again. If it lifts its head, the hunter freezes on the spot. He creeps up to it until he is within range. If the animal is wounded, the hunter shows surprising skill in following the trail of blood. It is very rare that he returns without his prey.

When the Indians live in the vicinity of Europeans, they trade with them anything that is surplus to their [277] needs. To preserve the meat of the deer which they decide to keep, they cut away strips from the sides, which they cure thoroughly. Then they roll them up like leather & cut pieces off as they need them. When they have no fresh meat, it is perfectly edible. They always keep the brains of the deer to tan the hides, as they can soften them as perfectly as the processes used by our tanners. To prevent the skins from becoming stiff after they have got wet, they cure them. To do this they collect a pile of rotten wood and place sticks around it in the shape of a cone, over which they drape the skins. They set fire to the wood, which gives off a lot of smoke that thoroughly permeates the skins. To remove the smell & filth of the smoke, they then wash the skins, which turn very [278] white & supple again. They do not become stiff any more than do our skins tanned with oil. They keep the lard of bears in bladders because this grease does not coagulate unless it is mixed with that of the deer. It is yet more delicate than goose grease. One can even use it in salad, which tastes better that way than with butter.

In the forest, in the months of May & June, one can find chervil, very good small onions and wild garlic that is milder & larger than ours. The cloves are pear-shaped, & the Europeans use them successfully as a cure for scurvy, which the Indians[1411] never suffer from, [279] gout or rheumatism, although they are constantly lying down on the ground in the rain & damp.

In the fall, the Indians eat nuts, including chestnuts, but as the trunks of the trees which bear them are generally 60 to 90 feet tall and without branches, & as it would be very difficult to climb them, they

---

1411 Note in original, p. 279 (a): *They do not eat them raw, as we do. They cook all their vegetables.*

cut them down to harvest the fruit. They boil up the kernels in their copper pots and extract oil for their needs. They are quite happy to throw into these pots all kinds of meat mixed with mashed corn. They eat directly from the pot without taking it off the fire. When they are thirsty, they swallow a *mikoine* of bouillon. They rarely drink fresh water. They have no set time for their meals. Day or night [280] they eat when they are hungry. They rarely use salt, although they find it very good.

We have not yet spoken of the most abundant form of game in America, the pigeon to which the French have given the name *tourtes*. There are almost fabulous numbers of them between the months of May and September. Flocks of them fly past for two or three hours on end & they are so thick in the sky that it could be said they form a shadow. This lasts the whole day long. The Indians do not bother to shoot them. They are killed with a long pole with foliage left on the end of it. Some individuals have killed hundreds of them by this method. They nest in forests which are thick with them over an area four leagues long and half a league wide. [281] As soon as an Indian announces to his village that he has found a flock of them, he is presented with a set of gear as a reward for the good news. The whole village goes to the forest. Men, women, & children settle down to eat the eggs & the young pigeons for as long as the birds are roosting. This happens twice a year, & there has never been any perceptible diminution in their numbers.[1412]

The Indians travel on foot or by canoe. In summer, journeys on foot are always short. The Iroquois & those who live along the Ohio have horses which they have stolen from the English, who leave them to graze in the forest. They have considerable numbers of them, but they do not raise them themselves. In the course of journeys on foot, each carries his bundle, which contains all the household utensils. Its strap [282] goes over the shoulders &, in the case of women, around their forehead because of their bosom. They set up camp early. The women

---

1412 Pouchot describes the extinct wild pigeon (*Ectopistes migratorius*). When Franklin Hough annotated his translation, in the early 1860s, he noted that passenger pigeon hunts were still being conducted in northern New York. Pouchot, *Memoirs*, II, p. 212n. Uncontrolled hunting by Indians and European settlers eventually had its effect. The last known passenger pigeon died at the Cincinnati Zoological Gardens in 1914.

Charlevoix recorded that these birds, whose "plumage is a dark brown, excepting their wings, in which there are some feathers of a very fine blue," were commonly called "turtles." Charlevoix, 1761, I, p. 262. The traditional French Canadian *tourtière* or meat pie, today usually made with pork, was originally a passenger pigeon dish.

& children make a hut out of branches & light the fire. The men go hunting to find something for supper. If the hunt is not good, they stay there in order to have at least a small quantity of provisions & live thus from day to day. An Indian often goes off alone, travelling 60 to 100 leagues into the forest with only his musket, powder and shot, a tinderbox, a knife, his tomahawk & a little pot. When the Indians have several rivers to cross, they make little rafts out of light pieces of wood, lashed together with cords. They have a little oar, which they make, or a long stick. They stand at one end and cross in this way rivers as wide as the Rhône & the Rhine. **[283]**

Journeys on foot in the summer are more tiring because of the marshes, hollows full of water which are always clogged with fallen pines and cedars. Some of them are a quarter or a half a league in width. As a result, they hardly ever travel in this fashion except to make war. Although the weather is harsher in the winter, nonetheless there is much advantage in the rivers being frozen & the forest covered in snow, which covers over these obstacles. By means of their snowshoes, however awkward they may appear at first when one is not used to them, the Indians overcome all these difficulties. The snowshoes are 4 to 5 feet long & about two feet wide at the bulges, which are strung with animal gut. They place the front of the foot in a slot which is **[284]** about two thirds of the way along the snowshoe, made up of straps passed behind the heel and over the foot. They adjust it so that the heel can be raised. You have to walk with a swinging motion & with the foot pointing well inwards, otherwise you would get tangled up. If you fall, you have great difficulty standing up again. The Indians do not have to fear that. The elasticity of the snowshoe carries you forward & makes movement easier, which compensates for its awkwardness. It does not sink more than 4 or 5 inches into the finest snow. They also make toboggans, very practical ones to carry their equipment. They consist of two runners, made of hard but flexible wood, 10 to 12 feet long. They are used as the framework of a type of sledge, between a foot and a foot & a half in width. The base is made up of birch bark or elm wood & the front curves up **[285]** in a semi-circle to pass easily over the snow. They tie their gear onto it. With straps over both their shoulders, they pull it or have it pulled by a dog. This sledge can carry 80 pounds.

They set up camp early in thickets, building a shelter against the wind in the shape of a half-roof, with two forked sticks holding up little poles which they cover with branches of conifer, flattened foliage or reeds gathered from marshes. In front of this shelter they light a good fire. This arrangement, although simple, is preferable to a square tent or a bell tent where you would freeze because you are more or less cut off from the heat of the fire. In their travels the Indians take precautions against the cold. Their shoes, [286] although of simple tanned leather, are very warm because the snow is so dry that it does not give off any humidity. They wrap pieces of blanket around their feet and the sides of the shoe form a tight fit with the ankle that prevents the snow from entering. Their feet would freeze in European shoes, an experience some of them have unfortunately suffered. The Indians put their belt around the lower part of their blanket and adjust it to go over their head in the form of a hood. They arrange it so well that one can only see their nose & hands. They have mittens of leather or flannel hanging by a string around their neck. They are better than gloves because fingers that are separated are more likely to freeze. They make bonnets out of a square piece of material, sewing one side in such a way as to cover the neck & [287] ears. We have gone into this matter in detail because similar clothing would be very good for troops that one might want to lead on a winter campaign & would spare the soldiers much discomfort. If they notice that any part of the neck or body has frozen, which is immediately obvious from its whiteness, they pick up some snow & rub the place in question until the blood has started to circulate again. They take care not to get too near the fire because if that part of the body was thawed out by the heat, it would become gangrenous. The greatest inconvenience during these spring journeys is the glare of the sun's rays on snow or ice. It can make you lose your sight for a number of days and experience very great pain owing to inflammation of the eyeballs. There can be no other remedy than [288] the use of tinted glasses.[1413]

The canoes of elm bark are not suitable for long journeys, as they are too frail. When the Indians want to make a canoe of elm, they choose the most regular tree trunk at any time when the sap is flowing. They mark it all around its circumference at both top and bottom of

---

1413 French soldiers experienced snow blindness during the war, notably on the return journey from Rigaud de Vaudreuil's expedition against Fort William Henry in March 1757 (note 337). Montcalm reported the phenomenon to the Minister of War: "Among the various sufferings experienced by this detachment, there has occurred a singular accident, namely, total loss of sight from the reflection of the sun on the ice. One-third of the detachment has returned blind. Canadians, Indians and our men, to the number of fifteen score, had to be led by their comrades, but at the end of twice twenty-four hours, sight is restored with simple remedies." Montcalm to Argenson, April 24, 1757, *DRCHSNY,* X, p. 549.

its entire length of 10, 12 or 15 feet, according to the number of persons it is to carry. After cutting it down in one piece, they remove the roughest part of the external bark, which will make up the interior of the canoe. They make ribbands the thickness of a finger & of the length which the canoe is to be & for this they use young oaks or other soft & sturdy wood. They slide the two long sides of the bark between two ribbands. Hoops, also made of pieces of wood **[289]** fit into the ribbands which are about two feet distant one from the other. They sew the two ends of the bark with strips made from the second layer of elm bark, taking care to make the two ends, which are called *pinces,* curve upwards slightly and to make the middle wider than the rest & the two sides shaped in a curve to give the boat protection against the wind. If there are any splits, they sew them up with these strips and cover them over with resins which they mash up. They smear this over the splits and heat it with a burning brand from the fire. The bark as well as the ribbands are held in place by these straps. They then add a mast cut from a piece of wood & a spar to act as a yard. They use their blankets as sails. These canoes can carry from three to nine people and all their gear. With these flimsy boats they sometimes undertake **[290]** journeys of a dozen leagues along the lakes. They as well as their children squat down on their heels without changing position when they are in the canoe. For if one loses one's balance, even slightly, the whole canoe capsizes. This happens very rarely, however, except as a result of a sudden gust of wind. The oars are 4 to 5 feet long. The very sight of such a vessel, which is barely three inches out of the water, is quite sufficient to scare any European out of his wits. If the canoe capsizes, they jump in the water & right it, then get back in over one of the raised ends. When they approach land, they are very careful not to let the canoe run aground, because it would break up. They carry it ashore & place it in the water again to relaunch it. They use these vessels especially for their war parties. They make them **[291]** wherever they have rivers to negotiate.[1414]

The canoes made of birch bark are far more solid & more artistically constructed. The shell of these canoes is made with strips of cedar wood, which is very soft & which they make as thin as the straps of a scabbard, three to four fingers wide. They all abut one another and join up in a point between the two ribbands. This shell is covered with pieces of birch bark sewn together like furs, fastened between the two

---

1414 Pouchot's detailed discussion of Iroquois elm bark canoes is expanded in Adney and Chapelle, pp. 212-19.

ribbands & tied up along the ribbands with the second layer of bark from the root of the cedar tree, in the same way as we twist wicker around the hoops of barrels. All the stitches & holes are covered over with a mash of resin, just as they do with canoes made of elm wood. **[292]** They fit cross pieces into them to retain their shape and to act as benches & also, in heavy weather, a long pole running from bow to stern to stop the vessel being broken up by shuddering when the vessel pitches. There are canoes of 3, of 6, of 12, and even of 24 places, indicated by the number of benches. The French are almost the only people to use these canoes for long journeys. They can carry up to three thousand pounds. Four men can take them over the portages and two are sufficient for other occasions. These little vessels can survive squalls that would cause real difficulties for many ships. One only has to make sure that they do not run aground. If they are holed, spare pieces of bark are carried to effect the repairs. The vessel is also used as a hut. A side of it is lifted and rested on one or **[293]** two oars and one then stretches out beneath it out of the wind. This is the normal form of shelter during voyages & when one is hunting.[1415]

If man was created to be the master of the earth, it can be said that he only seems to be really its master in this part of the world. Since he is subject only to his own desires, without tiresome restrictions, capable of satisfying all his needs, which are few, he lives a truly free life. To the misfortune of the Indians, we arrived in their land & taught them to use our materials for their garments. They could not now dispense with gunpowder & brandy without the majority of them perishing. Such necessities make them keep the peace with regard to the English, whom they despise because English merchants do all they can to cheat them. **[294]** The Indians used to have a way of life & utensils to which they could re-accustom themselves today only with the greatest difficulty. They made pottery, they made fire from wood, & they supplied their food by using bow and arrow. They made needles & fishhooks from fish bones. Animal sinew is used as thread; they dry it, pound it &, on dividing it, they extract a thread as fine as they wish. Their womenfolk are skillful and industrious in making their clothes attractive.

When the Indian has sufficient to eat, his needs are satisfied. He then thinks only of amusing himself, smoking, or sleeping, without worrying about what the next day will bring. Unless some urgent

---

1415 For further details on the construction of birch bark canoes, see Adney and Chapelle, particularly pp. 113-31.

matter crosses his mind, he thinks of nothing. He is extremely calm & [295] patient, which makes him appear melancholy. The fact that he is solitary & isolated may contribute to this appearance, but it is so deeply ingrained in his nature that the Indian who was treated & housed better than any other would die of boredom before the month was out if he was unable to spend time in the forest & lead his normal life. This has been proved by experience. He thinks only of hunting, of his nation's enemies & of means to remain in comfort on his mat, that is to say in his own area. He has no conception of what we commonly term "ambition:" he never covets what belongs to others. His sole ambition is to be recognized as a great hunter & a formidable warrior who has killed many men. If a European decides to tell him about the power of the King of France or the King [296] of England, he listens very carefully to what is said, then asks very cooly: "Is he a good hunter? Has he killed many enemies?" When someone assures him that the King has been to war & that he is a good shot, he exclaims: "Oh, then he is a real man!" That is the greatest praise that an Indian can bestow. They are extremely hospitable. If you enter their hut, you can take anything from their cooking pot & eat it without them protesting. They offer you the best of what they have & even deprive themselves of necessities in order to give them to a stranger. This is reciprocal; they imagine that you can refuse them nothing. In some nations, they even offer you women so that you do not become bored.

They believe that the master of life caused them to be born in the territory [297] they inhabit and that no one has the right to interfere with what they possess. Since they do not recognize land ownership, they consider that the whole country is common property & that any territory they inhabit where the bones & spirit of their ancestors reside is sacred & inviolable. They believe that they cannot leave it without going to take the land of some other nation, which would have the right to chase them off it. This instinctive feeling makes them very touchy over the matter and it is always a cause for war when one nation comes to hunt on territories of another. Indian travellers even take good care to leave the skins of the animals they have killed on the lands of another nation hanging from trees so it can get the benefit of them. It is thus entirely inappropriate for the English to say that they have bought a number of territories from [298] some of them. Europeans were only tolerated at first because they originally represented themselves as beneficent beings who possessed everything they desired and thus came to offer things to them to drag them out of poverty &

minister to their needs. They were welcomed in that capacity alone. The Europeans subsequently volunteered to support them against nations with which they were at war and were regarded as benefactors & friends. But when the Europeans came in force, they compelled the Indians to cede the lands they needed to them. The Indians found their hunting territory too restricted and withdrew into the interior, where they found themselves obliged to seek asylum among other nations, which welcomed them out of charity and absorbed them. Since that **[299]** time, they have been hostile, especially the Loups of the Belle-Riviere,[1416] a hostility which they express towards the English on every possible occasion.

Since the French only occupied the banks of the St. Lawrence, they have not worried the Indians up to now, as the latter have retained all the interior. On the contrary, the French policy was to keep the Indians as close to them as possible by establishing Christian villages. Apart from the advantage of spreading Christianity, this was a very good idea because the well-being which the Indians enjoyed as a result of the King's bounty, attracted their esteem & friendship to our nation. The English **[300]** are now very troubled because they were unable to conciliate them when their colonies were first established.

The Indians consider as their chiefs the eldest of the first branch of their nations. Some nations allow them to have a little more authority than others. This authority extends to members of such a nation allowing themselves to be beaten by the chiefs without seeking to avenge themselves, which is very rare. The Loups of Theaogen are such a nation. All rights of preeminence come down to that. In any case, they only have the right of suggestion or exhortation. If someone refuses to do as he is told, they have no means to compel him. The chief is only useful as a meeting point, for councils or deliberations, and it is in his name that the nation speaks on public matters. The **[301]** young men all have a most exemplary respect & deference for

---

1416 Note in original, p. 300 (a): *It was they who, in 1763 & 1764, sparked an uprising of almost all the peoples beyond the Appalachians against the English, with whom they became involved in a bitter war.*

The Loups of the Ohio, or the Delaware, were a refugee people by this time, having been pushed west by English expansion in Pennsylvania. The nation was prominent in the uprising of 1763-1764. Neolin, known to the English and Americans as the "Delaware prophet," advocated a return to more traditional lifestyles and was instrumental in inciting them to war. His teachings also influenced the Ottawa chief Pontiac and other tribes of the Great Lakes and Ohio region. See Peckham, *Pontiac and the Indian Uprising,* pp. 98-101, 113-16, 187-88. A new treatment of the subject is Gregory Evans Dowd, *War Under Heaven: Pontiac, the Indian Nations, and the British Empire* (Baltimore, 2002).

their parents & elders, which induces them to comply quite willingly with what they are told or what is suggested to them. They obey them without grumbling & perform all chores without complaint. When they are travelling, the young men, without a word, undertake the task of building the huts & collecting the firewood, while the old men smoke in peace.

No metaphysical or moral idea enters the heads of Indians. They believe what they may be told about these questions without having any deep personal conviction. They say that they are not sufficiently bright to understand things amenable to simple reason. In consequence, it may well be imagined that they are not very good Christians. When a missionary speaks to them **[302]** of the Holy Trinity or the Incarnation of the Word, they calmly reply: "Such matters are well and good for you who are endowed with intelligence, but we do not have enough to be persuaded. We believe, since you tell us." They compare the Trinity to a piece of bacon in which are to be found meat, fat & rind, three distinct things which combine to make up the one piece. The missionaries have convinced[1417] them that Judas had red hair & that the English, who generally have hair of the same color, belong to his race, which makes yet another reason to hate them.

Since the King sent out with the first colonists, priests, **[303]** foreign missions, Sulpicians, Recollects & Jesuits, these missionaries, who involved themselves as much as possible in all forms of trade, invited the Indians to come and settle at a given spot, where they would be housed, fed & clothed and all their needs would be satisfied. A number of them, out of sloth, allowed themselves to be won over, came to take advantage of this easy life & willingly consented to instruction. If they had been given nothing & had derived no clear advantage from taking this decision, very little would have come of the initiative. This can be gauged from the very small number of them who were induced, despite the easy circumstances arranged for them. Their indifference as to whether they should believe or not believe made them make the decision, as did the promises of an afterlife. Children reared in the Christian **[304]** religion follow the example of their fathers &, unless their passions compel them to abandon the missions, which continually happens, those who remain in them follow the religious ceremonies with great reverence. It is very edifying to see them in church, men on one side, women on the other, always

1417 Note in original, p. 303 (a): *The author is quite wrong in attributing to all missionaries in general views which can only have been expressed by a few of them, or by some woodsmen.*

on their knees with a more contemplative air than a Capuchin friar, never talking among themselves, never turning their head. The women, their bodies very well covered, sing, in a double choir with the men, the prayers the missionaries have translated for them & which they have learned by heart. Their singing is sweet & harmonious; one does not hear voices off key, as in our villages. As for the sacraments, they take them as often as their missionaries want them to, [305] & submit themselves to the private & public penances imposed upon them, for the priests have to some extent invoked the rules of the early church. The Indians pay punctiliously the tithe on their corn & furs which the missionaries have levied on them.

Nonetheless, drunkenness is still the dominant vice of these newly converted souls. The obligation of keeping to one woman, which they do not like, is the one thing that greatly bothers them. As a result, many of them abandon the missions and return to their villages in order to live in freedom. In general, it can be said that they have become rather more human & have developed a special attachment to the French as persons of the same religion or communion, because they make a distinction between Catholicism [306] and the religion of the English. Our priests have taken pains to convince them that the latter is barely Christian. If the Indians who have embraced our religion have become more human, it must also be admitted that, having had from that point on more contact with us, they have also been more infected with our vices than others less exposed to the contagion.

All these facts are the exact truth; it is distressing that they scarcely coincide with the exaggerated accounts of the missionaries. The small number of Indians who have been converted over the past 180 years or so that we have occupied this huge continent is a proof of their indifference towards religion. However much they are preached at, they listen very calmly & without ill-will, but they always return to their usual refrain, that they are not [307] sufficiently intelligent to believe & follow what they are told, that their forefathers lived like them & that they adopt their way of life. When they killed a number of priests, it was not because they hated their dogmas, but because they considered them as an enemy nation.

It has already been noted that the King covered all the expenses of the missions. When the Indians considered themselves to be ill at ease there, because of too much clearing of the ground, they asked to be transferred to more remote spots. The King paid for these new establishments & the missionaries ceded their former villages to

Europeans, although they retained seigniory over them. As a result they acquired title to eight villages out of ten in Canada; all those on the island of Mont-Réal & the city of that [308] name thus came into the possession of the Sulpicians.[1418]

There are no known nations which have not heard of our religion & a good number of them are well situated to witness its practice. In their own way, they all respect it. The Indians have a kind of veneration for our priests, whom they call the "Fathers of prayer" because they have a high regard for their morals & because of their sermons. They consider them as men solely engaged in praying to the Supreme Being.

These peoples of America have no very clear idea of that Infinite Being. They do not honor him by any special observance & they refer to him as the "Master of Life." They believe that it is upon him that all events depend & that they cannot succeed in any enterprise without him willing [309] it. From the way they talk, one might suppose that they are fatalists. They believe that there exist evil spirits which cause sinister events & everything that appears to them to be out of the ordinary. They give these spirits the name *Manitous.* The ocean, the lakes, the rivers, all things created, have their Manitou. It is to these maleficent spirits that they make offerings or sacrifices. If they are assailed by a squall on a lake & are in peril, they throw away tobacco, or a piece of equipment or a tool in order to appease the Manitou of the lake.[1419] They act in the same way when they are crossing dangerous rapids. When they go hunting or on the warpath, in order to be successful, they make a kind of sacrifice. It consists of erecting

---

1418 Charlevoix also commented on the Sulpician influence over Montréal: "This city was first called *Ville Marie* by its founders, but this name has never obtained the sanction of custom in conversation, and holds place only in the public acts, and amongst the lords proprietaries, who are exceedingly jealous of it. These lords, who are not only lords of the city, but also of the whole island, are the governors of the seminary of St. Sulpicius; and as almost all the lands on it are excellent, and well cultivated, and the city as populous as Quebec, we may venture to say, this lordship [seigneury] is well worth half a score the best in all Canada." Charlevoix, 1761, I, p. 213. In 1749 Peter Kalm recorded that the Sulpicians' owned the "ground on which the town of Montreal is built, together with the whole island of that name." Kalm, p. 531. See also p. 452.

The role of religion and the clergy, particularly the Sulpicians, in the founding of Montréal is discussed in Eccles, pp. 46-48.

1419 Pouchot recorded a particularly dramatic example of this practice in June of 1759. Caught in a violent Lake Ontario storm while aboard the brig *Outaouaise,* a party of Mississaugas first tossed their personal possessions into the water. They then attempted to throw a Canadian dwarf overboard but were restrained by the crew. See note 564.

a stake from which they hang a dog that has been killed, or another animal, feathers, tobacco, **[310]** or something else that crosses their mind. This is the sum total of all their religious ceremonies. They are very superstitious. Everything that seems to them to be an ill omen can make them cancel a war party, even if they had already travelled several hundred leagues & were on the point of attacking. It is the same for a hunting expedition or anything else. They call prayer "making medicine."

The habit of being alone must make them prone to dreaming, which they call *jongler.* Men or women in such a frame of mind wrap themselves up in their blanket & remain there for a very long time. Their imagination becomes excited &, in the grip of inspiration, they believe they can see future events & predict them with confidence. In fact their prophecy is often not **[311]** fulfilled, but it also does happen that they make good guesses. The Europeans who have been among them recount surprising things on this subject. We shall report two facts which are well known in the French army. In 1756, M. Duplessis, the commandant at Niagara,[1420] sent a party of 25 Missisakes to Chouegen. The womenfolk remained beneath the fort. They gathered every evening to make medicine. An old woman sang and the others responded in chorus. The officers used to go and watch this ceremony. After six or seven days, the women were asked why they were no longer making medicine. The old woman replied that their men had been involved in a battle, that she had had a vision of it & that they had killed some of the enemy. An officer who knew about these visions immediately wrote down the day she had indicated. When the party **[312]** had returned, he questioned the Indians & their prisoners. Their reply was exactly as the old woman had said. Niagara is at least 60 leagues from Chouegen & no one had arrived from there either by water or land. The Indians had lost two men, taken twelve scalps & brought back as prisoners three ship carpenters.[1421] In 1758, around the month of March, M. de Vaudreuil sent a large party of resident Indians to Carillon. Once arrived at the fort, they went to see the commandant & asked him to give them some food, because they wanted to rest for a few days before setting out for Fort George. They returned to their campsite and one of them began to prophesy. He said

---

1420 Captain François Lefebvre Duplessis Faber (note 280).

1421 Pouchot describes this raid in Volume I. It was composed of two parties, the first of 25 Chippewas and the second of 21 Mississaugas. They left Niagara on May 7, 1756 and returned on the 15th. See notes 297-298.

that the English were very close to Fort Carillon and that they should go out immediately and attack them. The other **[313]** Indians were convinced by this example of inspiration & went back to the commandant's quarters to inform him that they wanted to set out the next day as they were sure they would come across an English raiding party. The commandant, although astounded by their idea, was very pleased with it as a means of getting rid of them. A number of officers & soldiers wanted to participate in the expedition. They had covered three leagues when they came upon the tracks of the English on the lake, which was still frozen over, heading in the direction of Carillon. They followed the trail and their scouts saw from on top of a hillock the English, numbering 250 men, who were going down into a hollow that lay between them. The scouts warned their men who soon caught up with the enemy and attacked at the moment when they were beginning to emerge from the valley. Scarcely eight of them escaped. They were all volunteers **[314]** with regular officers under the command of Rober Roger [sic]. They were all killed & very few prisoners were taken.[1422]

In almost every village there are acknowledged shamans. They are also doctors, or rather, authentic charlatans. They shut themselves up by themselves in a hut, where they fling themselves around like men possessed. When they emerge, they spout their prophecies, which normally concern the fate of a sick person or the outcome of a war or hunting party. With regard to the first case, the shaman will say that the Manitou of the illness demands a feast, which will enable him to get drunk. Then he provides a remedy. If the latter has an adverse effect, the so-called doctor is sometimes beaten or killed, because the Indians are angry at having been duped. Each family has its own special remedies, but these shamans also have very good ones. They know all about excellent remedial plants, **[315]** especially for wounds. It is certain that, unless there are broken bones, no surgeon could treat them with such success, so swiftly and with less fuss. They know of a large number of plants, roots & barks which they use to great effect. Since the settlement of Europeans in America, they have not taken sufficient trouble to be taught about these remedies, which would be extremely useful to our sort of medicine. It is true that the Indians jealously guard their recipes for themselves, but with time & through

---

1422 The commandant was Captain Louis-Philippe Le Dossu d'Hébécourt (note 275). The "Battle on Snowshoes" against Robert Rogers and his rangers was fought March 13, 1758. See notes 393-397.

gifts, it would have ultimately been possible to extract their secrets from them. They cure themselves of the most chronic venereal diseases without the use of mercury. They have a root which grows like a very large turnip & which causes deep abcesses to come to a head and burst. [316] Cuttings can be taken of the plant, which grows in damp or slightly swampy spots.

Other plants cure cuts[1423] as quickly as the best balsam. The one called anise, which tastes like aniseed, is excellent. The diseases to which they are the most subject are pleurisy, pneumonia & various forms of consumption caused by the quantity of brandy they drink. Although they like self-preservation, one can say that there exists no type of man who has less fear of death, which they regard as [317] a crossing to another life. As soon as they feel they are about to die, they sing their death chant until they can no longer speak. This chant is a kind of lament which each person makes up for himself, although they all have the same tone.

When an Indian dies, one hears neither cries nor lamentations in the hut, but the people come to pay a farewell visit. The man is buried with his finest clothes, his arms & a cask of brandy to help him make the journey. Above his grave a kind of hut of stakes is erected in the shape of a mausoleum & beside it there is another large stake with the arms of his family. Figures are cut into the stake to represent the number of scalps & prisoners he has taken. Some nations have the custom of sending women for a week or so [318] to light a little fire beside this mausoleum. Squatting down on their heels, they remain there motionless for a quarter or a half an hour. If a man dies while on the hunt, even if it were three or four months before, they exhume him & bring him back in their canoe for burial in the village. This practice even extends to children.

After a year, they commemorate the anniversary by having a feast near the grave. Then they dig up the cask of brandy and drink some of it. They believe that after death, their souls go up to a large country beyond the sea which they call by the name of "great lake." They imagine that there they will find all the animals they could wish for hunting & that they will spend their time dancing. [319]

1423 Note in original, p. 317 (a): *The plant that cures them is called by P. Charlevoix* plant universelle. *Its ground-up leaves close all sorts of wounds. These leaves are the size of a man's hand & are shaped like a fleur-de-lis.*

This might have been the arbor vitae (genus *Thuya*), the medicinal qualities of which are discissed at length in Kalm, pp. 468-70.

They have no tradition which has preserved even the slightest notion about their origin. Everything they say about it has been suggested to them by the Europeans. Some of them believe that a woman came down from Heaven onto water & that when she placed her foot onto a turtle in order to rest, the earth formed around the turtle & the woman gave birth to the world.[1424] As they have adopted as their clan a bear or a turtle, or a wolf or a fox, &c.,[1425] & designate their families by the names of these animals, [320] they think that they are descended from them. Some of them are convinced that they emerged from the land they inhabit. Since they cannot write, they identify themselves by the figures of the animals mentioned, which they carve on pieces of bark or on tree trunks. Some of these sculpted figures are quite up to the level of those done by our average artisans. To make the figures, which are very expressive, they use a hooked knife of a sort employed to make their wooden utensils.

When a family goes into mourning for one of its members, they take off all their jewelry, dress as simply as they can & smear only black dye on themselves. The period of mourning lasts for about a year. Relatives, friends & those who are interested in the family present it with gifts to bury the dead man, which consist [321] of a wampum belt and a suit of clothing. They substitute in the hut a scalp or a prisoner for the dead man. There are some families which do not relinquish mourning until this last ceremony is accomplished; this is one of the principal causes of their wars, as they are constantly obliged to have another nation from which they can acquire prisoners or scalps to replace their dead.

All the nations of the eastern part of America had more or less the same language but the bonds between them were not sufficiently close to avoid wars, a list of which can be seen in the works of English & French authors. The English found them in a warlike disposition when they first began to settle the land, which favored the establishment of their colonies. However, common misfortunes [322] united the Indians. These nations also had to sustain long wars with the

1424 Note in original, p. 320 (a): *This notion seems to have come to them originally from the Indians of Asia, who communicated it to the inhabitants of Japan, and from there the fable crossed to America.*

Pouchot recounts the basic Iroquois and Huron creation legend. Trigger, p. 319. This tradition long predated the arrival of Europeans in North America.

1425 Note in original, p. 320 (b): *The Baron de La Hontan found in this custom the childish rules of our vain science of heraldry.*

Iroquois, who had made them wear the *machicoté*, that is to say, in their terms, had forbidden them to bear arms, thus treating them as women.[1426]

The Iroquois are six nations united together and include the Tascarorins, whom they had almost destroyed & whom they absorbed, as they did the Erié or Cat nation, which was reduced to just a few persons who were adopted by the Sonnontoins.[1427] The Iroquois nation is the most intimately linked of all those in America & forms a true federated republic. This union has given them a marked superiority over all the other peoples who, being less numerous, were crushed by them without being able [323] to escape. The Iroquois sought out their enemies with thousands of warriors, going as far as the rivers that flow into the Mississippi & Lake Superior. All these invasions, however, never had any other purpose than to kill or abduct men & to destroy these nations, or what the Indians call "eating them." They never had in mind to extend their territory in order to enjoy larger hunting grounds, or to render any nation subject to them. The proof is that in spite of all their advantages over the different peoples whom they nearly destroyed, like the Huron, the Nepicins & the Algonquins, who were formerly very numerous, the Iroquois have nonetheless never sought to annex the lands of these nations & to reduce the peoples to slavery. The claims of the English are thus very frivolous [324] when they assume that, by virtue of their supposed alliance with the Iroquois, they have rights over the territories of all the nations with whom that people has been at war and has subjugated. These nations may have declined but they have never changed their status in the various areas they inhabit.[1428]

1426 Specifically, the Delaware. See note 342.

1427 The Tuscarora had not been destroyed by the Iroquois but, rather, had been forced from their original homeland in North Carolina about 1713 by encroaching English settlement. The nation moved north to the country of the Five Nations of New York and was eventually accepted as the sixth nation in 1722-1723. See David Landy, "Tuscarora Among the Iroquois", in Trigger, pp. 518-24.

The Erie or Cat, an Iroquoian people, occupied the region to the south of Lake Erie between Buffalo, New York and northeastern Ohio. The nation was destroyed by the Iroquois in 1654-1656. Tanner, p 33; Marian E. White, "Erie", in Trigger, pp. 412-17. For details of the war see George T. Hunt, *The Wars of the Iroquois: A Study in Intertribal Trade Relations* (Madison, 1967), pp. 100-04.

1428 The Iroquois rivalry with the Huron and the victorious Iroquois campaigns of 1648-1649 are discussed in Hunt, pp. 66-96. See also Conrad E. Heidenreich, "Huron", in Trigger, pp. 368-88.

The Outaouais, the Sauteurs or Ochibois, & the Missisakes, who share more or less the same language & are closely related to one another, have always supported one another against them because of their unity. Other nations which are more remotely situated & less united have nonetheless suffered little. They have even formed alliances in order to expel them. One might even express considerable scepticism about those huge Iroquois armies which, so it was said, sometimes totalled 20000 men. If they had existed, it is certain **[325]** that they would have destroyed all the nations in their path.[1429]

When the French entered America, they arrived on the territories of the Algonquins, one of the oldest nations of the continent, as can be gauged from the area over which their language is spoken, & on those of the Népicins & Hurons, who were at war with the Iroquois. The French sided with the latter nations and, as a result, the Iroquois became natural enemies of the French settlements. We dislodged them from the island of Mont-Réal & the eastern plains of the St. Lawrence, where they used to come and hunt. In consequence, they did us considerable harm in the early period of the colony.[1430] **[326]** They were constantly encouraged in this first by the Swedes & Dutch, who were founding settlements on the Hudson River, & subsequently by the English who replaced them. This gave us the opportunity to forge links more especially with all the nations of the continent, because they all dreaded the Iroquois, and we were in consequence always prepared to support them or encourage alliances between them in order to repel the common enemy. In any case, since the French normally met the Indians only at trading posts where they supplied them with goods to satisfy their needs, they became necessary to the native peoples. When they had a falling out with one or another of these nations, they were always careful to unite with the others against

1429 Estimates of Iroquois military strength vary, but the largest reliable number for their seventeenth century armies is about 2,000 fighting men. Hunt, p. 66. A chart in Trigger, p. 421, estimates warriors available for a number of periods between 1660 and 1779. The most dependable of these estimates range from a high of 2,340 warriors (1665) to a low of 1,600 (1779) during the American Revolution. The Iroquois population was much larger before the onset of devastating epidemics in 1634, and before that time they might have fielded as many as 4,000 to 5,000 fighting men.

1430 Note in original, p. 326 (a): *On several occasions they brought it to within a hair's breadth of destruction. See the* Histoire de la Nouvelle-France *by P. Charlevoix.*

New France was plagued by war with the Iroquois throughout much of the seventeenth century. In 1650 the situation was so bad that it was feared that the colony might have to be abandoned. Eccles, p. 51.

the nation in question & soon compelled it to live in peace, because the [327] Indians are far more fearful of one another than apprehensive at having to deal with Europeans.

Since the Indian nations sometimes meet together, either when they return from the hunt or when they send embassies to one another, they pass the time in feasting, or rather in debauches of brandy. They often attempt to take precautions against the occurrence of some mishap. The women seek to hide the arms of their husbands as far as possible, because there are always some who refrain from drinking in order to serve the others. There are some feasts where everything is eaten, even if they were to burst as a result. Usually, their feasts consist of drinking bouts at which no food is eaten. Ten to twelve Indians will drink on average 15 to 20 jars of brandy, & more. The youngest man has the job [328] of pouring the drinks. Each man drinks the same quantity in turn. The waiter draws the measures so accurately that they are all equal to one another. Those women who do not care to drink take their share, put it in their mouth & spit it out into their small cooking pot. They later sell it when the guests have run out of brandy. The Indians have such a great passion for this drink that once they are well into drinking it, they generally dispose of everything they have in order to swallow one more mouthful. It can easily be guessed what a confused din is occasioned by all this. They start by singing, which is followed by frightful yells. They wait to get drunk before levelling reproaches and insults at one another. They always quarrel about lack of courage. [329] Someone[1431] almost invariably dies as a consequence of one of these brawls; at the very least the result is a few torn ears. If they attack a man who is not drunk & he is not smart enough to beat a hasty retreat, he often becomes a victim, because it would be dishonorable on his part to strike a man in a drunken state, their view being that drunken men have lost their wits. If he withdraws he is afraid that the other man will boast that he put him to flight. Faced with that alternative, he shows his courage by telling the other man to strike him & is killed by him. Accidents of this kind have caused the death of very good and brave Indians.

1431 Note in original, p. 330 (a): *A man wounded with a firearm or a sharp blade is considered to be dead & demands vengeance.*

It was this practice that caused Pouchot to order the men of his garrison to resist with their fists rather than their weapons if the Iroquois molested them after the surrender of Fort Niagara. Pouchot, *Mémoires,* II, p. 118 and note 695.

On the effect of drink see Peter C. Mancall, *Deadly Medicine: Indians and Alcohol in Early America* (Ithaca, NY, 1995).

If the dead man is from the same nation [330] as the man who killed him, nothing is said at first, but another bout of drinking will be organized. If the killer is present, which almost always happens, as they are unable to resist their craving for drink, one of the dead man's relatives will kill him. Some of them pretend to be more drunk than they really are in order to carry out their plan. If they were sober, they would not dare to try and exact revenge, because they would not have the excuse of saying that they were out of their wits. In order to put an end to these vendettas, the relatives of the murderer bury the dead man, which causes another war party to be formed to capture someone to replace him. This process does not always prevent the mother or the wife of the dead man, unable to accept his loss, from hiring someone [331] to kill his assassin. The number of men destroyed by these incidents can easily be imagined. If the assassin has killed several men, the nation as a whole agrees to execute him at a feast for that purpose, which even his father attends. When Europeans try to exhort them to desist from such practices, they reply calmly: "He wants to die. It is of no consequence."

If a similar incident occurs between different nations, it is much more serious. The whole nation considers itself offended & is obliged to avenge the death. If the murderer's nation wants to avoid war, it has to hand him over & bury the dead man, and even then that does not always give satisfaction. The injured parties want to take their revenge, even much later, & these disputes often only end with the destruction of [332] one group or the other, or when other nations interpose their authority. The more harm done to an Indian nation, the more intractable it becomes & one can only manage to recover its friendship through a gentle approach.

In diplomatic missions from one nation to another to discuss peace or an alliance or some other political matter, the delegation is always a large one. It is made up of chiefs or elders of the nation, war chiefs & young warriors. They are accompanied by an orator. A number of the latter are found in almost every village. Generally, they are the best speakers. They bear belts of white wampum on which the subject of the mission is expressed. If it is a question of alliance, they design huts to represent their villages, the course [333] of the trail from one village to another & Indians grasping one another's hands. The meaning naturally has to be guessed. The more important the matter at hand, the larger must be the belts. They carry a pipe of peace with them; it is a pipe made of marble, or of very soft stone, reddish brown in color, or black. The grain of the stone is very fine & the pipe can be shaped with a knife. They insert into it a wooden tube two or three feet long. The wood, which is very hard, has a thick pith which they extract with a brass wire heated over the fire. It is painted yellow & black or covered with a twine made of porcupine skin colored white, yellow, red & black, & with a wing of eagle's feathers attached to it by porcupine twine and dangling, multi-colored ribbons. [334] All this makes it look very attractive. The group of men make their way to the chief's hut and all those in the party sit down on the ground without a word & then light up the pipe. The chief takes the first puff, then he hands it to the chief of the nation with which they are negotiating. The pipe passes around the group. Each man takes a puff. This is the greatest sign of peace & friendship they can make to one another. If they were not in agreement, they would refuse to smoke. After the ceremony, the orator gets up, gives an account of their journey, says that they are very tired & concludes by indicating the day of their departure. If they are on friendly terms, food is brought in & then they all retire to their huts without saying a word. On the agreed day, they attend the council. The orator sets out everything they have to say, presenting [335] wampum strings or belts according to the importance of the proposal. During all this you could hear the wings of a fly beating. They all puff the pipe, including, quite frequently, the orator. At important points in the oration, the delegates express approval with a *ho ho.* The other party does not respond the same day. They simply indicate the day upon which, with the same ceremony, their nation's orator will respond on behalf of all, taking up each item in turn & presenting belts or strings for each topic. When the council is finished, they break up & it is then that the joint dances & feasting begin; they last until the leave-taking.

If they want to persuade a nation to join them in making war, they wear belts of black wampum, with axes painted on them and painted vermilion. The larger it is, [336] the more pressing is the invitation. It is their way of offering the axe.

The pipe dance, which is one of their great ceremonies, is really rather attractive. All the Indians are seated in a circle with their axes or tomahawks & their knives in their hands. Some of them only have *chichiquois,* which are little gourds filled with small pebbles, or spurs of a deer's foot tied together at the end of a handle. Some of them make a kettle drum out of a cooking pot over which a skin is stretched. The most nimble man is chosen as a dancer and another man sings a melody appropriate to the dance. All the rudimentary instruments beat time & the dancer, with his pipe in one hand & a *chichiquoi* in the other, adapts his dance to the melody by making rhythmic movements with every part [337] of his body. He stoops right down to the ground & after many contortions he stands suddenly upright again. He allows a few moments to elapse, then starts the dance again. To do it, one has to be both lithe and strong. The dance has a number of pleasing features & would not be disagreeable if performed in a theater.

From what has been said above, it can easily be guessed that the Indians often have opportunities to make war and, in consequence, they have to have some other nation to "eat", because they cannot be happy without war. An Indian who remained three years without going to war would not be considered a real man & would suffer reprimands during their feasts.

When war against another nation has been decided, there is no lack of war chiefs seeking to form raiding parties. The chief [338] who has the highest reputation is chosen. He acquires a supply of brandy & some sets of gear. He invites the young men, especially his relatives, to feast, which consists of eating a dog that has been killed and flayed. Once the meal is over, they begin to drink brandy. The war chief stands up and sings his war chant; each of them has his own special song. He recounts all his exploits, strikes his fist against a post & throws down a pledge to guarantee what he has said. He speaks of all the nations against whom he has fought, pointing with his hatchet in the direction where they are situated. He announces his plan with as threatening an air as he can manage & invites those who are courageous to follow him. He ends by throwing disdainfully to the ground a belt of black wampum with red paint on it, inviting those men with pluck to pick it up & announcing that he will award it to the man who [339] shows himself to be the bravest. All the young men sitting in a circle respond with a rhythmic *hé hé* which acts as a refrain to the song.

When the chief has finished, the first man to make up his mind stands and goes through the same ceremony. He sings his war song, enumerates his exploits as he strikes the post, throws down pledges as a guarantee, still singing picks up the belt &, brandishing it, claims that he will be worthy of winning it. He then throws it down again. As they make up their minds, all the others stand up & go through the same ceremony. Those who do not want to join the war party drink but do not dance. Once the ceremony is over, all those who are to participate go and dress themselves as warriors. They strip naked, down to their moth-eaten breechcloth. They smear their entire body in a reddish brown color, with stripes that they make with their fingers. They [340] cut their hair, or rather they remove it except for the topknot. They decorate their faces, covering it with vermillion over which they throw white down, which is the sign that they are going off to make war.

As soon as they are all assembled, they march out together, their instruments in their hands. They dance in a circle and go from hut to hut singing a tune which always ends in *ha ha, he he, heu, hi hi,* &c. They turn circles around one another &, with their tomahawks or knives in their hands, they gesticulate a great deal. People give them little presents. During this time, the girls follow them everywhere, dancing with them in a loose-limbed fashion throughout the evenings, as has been mentioned above. Everybody tries to treat them as well as possible. All this lasts for several days, right up to the time they set out. They then stand in single file, [341] with the chief at the head preceded by the youngest warrior carrying the medicine mat in which there is either a bird or an animal that each war chief selects as his Manitou. In the mat there are also a few medicinal herbs for wounds or sickness. They have a great deal of respect for this mat which always precedes them when they are on the march. The war chief sings the parting song, which is a prayer to the master of life to look kindly upon their projects. A number of girls follow the party carrying the gear of the young warriors. They sometimes accompany them for three or four days, after which they return to the village. They travel always by canoe, because the rivers are the only highways of the country.

As has already been noted, the young men have to perform [342] all the chores, which they share among themselves without grumbling. If someone is idle and decides to do nothing, he suffers no reprimand. If someone is inspired to make predictions & desires to go back to the village, he leaves unopposed. They set up camp early in order to hunt, since they bring with them no other provisions than their ammunition. Nonetheless, they sometimes have a small bag of *sagamité,* which is mashed corn, grilled & cooked in a copper pot with fat & maple sugar. They save this type of flour for the time when they are near their enemies, or for when they are short of food. Simply by mixing it with water, they can make a meal which is healthy, nourishing & pleasant to the taste. Two handfuls of this food a day are sufficient to sustain life if they fear that food may be in short supply. When they are in [343] enemy territory, they do not fire their guns. If they do not have a bow and arrow, they live on fish, a few roots and their *sagamité.* But as soon as they are on the point of attacking, or when they are retreating, or after an operation, they will go three or four days without food. As they approach the frontier, where they might encounter the enemy, they take care to seek out the thickest part of the bush and to cover over all the tracks around the place where they entered the territory. They also hide their canoes, together with all their packs & ornaments. They smear black paint over the entire body & only take with them their arms & their Manitou, without forgetting their mirrors. They frequently hold council to determine how they will scatter after the attack, where their rendezvous point will be, &c. They never march without sending on [344] scouts to reconnoitre two or three leagues all around. They take their decisions on the basis of the scouts' reports. Their skill in discovering all kinds of tracks is extraordinary: marks in the earth, bent leaves, dew disturbed, none of this can prevent them from recognizing the trail of wild animals. The footprints of Indians are identified by the type of footwear but above all by the manner in which the foot is placed or turned in the ground. From the footprint and the sole of the shoe, they can judge if it is Europeans even more easily. They can even distinguish between an Englishman and a Frenchman & can make quite an accurate estimate of the number of people either by their tracks or by the fires they have lit or by the signs around the place they slept, if the tracks belong to a raiding party. The party which picks up the trail first is almost certain to beat the others. They will track them for several [345] days until they have caught them in a position which gives them the advantage, as for example in a hut where the enemy has spread out to sleep or in a stretch of country where they have become separated from one another. They hide near the place where they intend to strike. Each man has his position decided by the war chief & lies low until the latter gives the signal, whooping by striking his hand against his mouth. The cry is taken up by all the assailants who have already drawn a bead on their target. At the very instant they surprise the enemy, they fire & very rarely fail to shoot their man down. They immediately rush forward, axe in hand, throw themselves on their adversary & do not desist until he is finished off. If they think he is not fatally wounded or can still defend himself, they [346] give him an axe blow to the head. If he flees, they throw it at him so that it sticks in his back, at which they are very skilled. As soon as the man is felled, they run up to him, thrust their knee in between his shoulder blades, seize a tuft of hair in one hand &, with their knife in the other, cut around the skin of the head & pull the whole piece away. The whole thing is done very expeditiously. Then, brandishing the scalp, they utter a whoop which they call the "death whoop." When fighting, they utter the most frightful yells they can manage, in order to give themselves courage &

to intimidate the enemy. If they are not under pressure & the victory has cost them lives, they behave in an extremely cruel manner towards those they kill or the dead bodies. They disembowel them & smear their blood all over themselves. [347]

Although they find these atrocities repugnant, they nonetheless commit them in order to steel themselves to the slaughter & induce in themselves a kind of rage. This makes them appear insensitive to danger & more courageous to their fellows. They bind up all the prisoners they can take with the straps they use to carry their packs, which they always take with them. They tie them up so tightly around the neck, the arms & the waist that it is impossible for them to free themselves. If they fear an attack, they scatter at a run towards an agreed-upon rendezvous point, which is sometimes nine or ten leagues away, depending on the territory & circumstances in which they find themselves. They sometimes go in pairs in order to help the prisoners to march, taking them by the hand. If, in spite of that, the prisoners cannot keep up with them, they take their scalps. [348]

When they are concealed near some village or fort, since there are usually scouting parties in the vicinity, they try to take up their hiding places at night. If it is daytime, they lie down on their belly and cover their back & head with grasses, leaves or straw, according to the coloring of the surrounding land. They creep forward on their hands, pushing their muskets ahead of them, until they are within comfortable range. They make an estimate as to whether there are enough of them to attack the position or not. This decision is always a cautious one for, however little the danger may appear to them to be, they forego the opportunity & wait for the favorable moment and occasion with very great patience, especially when there are only two or three of them. When their mission is completed, they utter a cry as they run away, [349] & leave behind some very conspicuous signs to indicate which nation was responsible for the attack.

They hardly ever indulge in pillaging.[1432] If they have time, they kill animals to eat. They very rarely burn houses, because they do not wish to reveal their position. Their principal aim is to return with prisoners or to take scalps. When they have no hope of achieving this, they do not open fire. They rarely bother to search or strip the dead. It is essential to note at this point that if they have the opportunity to take

---

1432 Pouchot presumably means while engaged in raids into enemy territory. In Volume II of his *Mémoires* (pp. 118-19), he describes Iroquois pillaging following the surrender of Fort Niagara (see note 696).

a large number of prisoners or to lift a lot of scalps, they do it until the operation is completed. But even if there are three hundred of them & they were to take only one or two scalps, they would not begin another operation, even were they capable **[350]** of devastating an entire territory and killing other men. They say that if they were not content, the master of life would be angry with them & that they would run the risk of failure and of losing men. Thus they return home to report to base, so to speak, even if it is two or three hundred leagues away, embark on other raids & then go back.[1433]

As soon as they are at their rendezvous point, they deck out the prisoners according to the taste of their nation.[1434] If it is a European, they cut his hair in their style & dress him as an Indian. In the evening, they put paint on him & tie him to the branches of a tree by an arm and a leg, so that he cannot extricate himself. Around his neck, they place a necklace of wampum, of the sort worn by our ladies. It is painted red and signifies his status as a slave. They ensure that he is **[351]** fed everything they eat themselves, so that he does not suffer during the journey. They spread out their scalps, with the hair hanging down, around a small hoop in the shape of a drum skin. They grease them and powder them with vermillion, the inside of the skin as well as the outside.

The war chief is careful to give the promised belt to the man who was the first to attack the enemy, or the one who performed the most admirable exploit. The judgment is made fairly among themselves, without any grumbling. If the chief has spare gear & one of them is lacking some, he deprives himself of it and gives it to the man. It is through exemplary conduct, bravery, good fortune & generosity that

---

1433 The traditional Indian manner of making war was a constant source of frustration for French officers from Europe, accustomed to military forces disciplined in the manner of that continent. One of the most outspoken was Bougainville. Among the disadvantages he saw in massing large numbers of these irregular troops was that "once they have struck, have they taken only a single scalp or one prisoner, back they come and are off again for their villages. Then for a considerable time the army is without Indians. Each one does well for himself, but the operation of the war suffers." Bougainville concluded that "in the end they are a necessary evil. It would be better to have on hand only a specified number of these mosquitoes, who would be relieved by others, so that we would always have some on hand." Hamilton, *Adventure,* p. 60.

1434 A particularly detailed account of the treatment of war prisoners is to be found in the narrative of Ensign Thomas Gist, captured by Hurons or Wyandots from Detroit in 1758. Their practices were similar to those of the Iroquois, to whom they were related. See Peckham, "Thomas Gist's Indian Captivity", pp. 293-301.

he acquires the reputation as a good war chief. Even were he to succeed beyond his expectations in the attack, if he had the misfortune to [352] lose one of his men, great sadness prevails & the glory with which he has covered himself is counted as nothing. They compel him to return to battle in order to avenge the blood of the dead man & to provide a replacement to his family. The Iroquois are especially punctilious in bringing back all the wounded, even if they are from another nation. It is one of their first duties. They make a stretcher or they pass under the thighs of the wounded man a strap which they attach to their forehead & they carry him for hundreds of leagues in this fashion if they do not have any canoes. Other nations abandon their prisoners in the forest, leaving them what they can to live on; these poor wretches have to find the remainder of their subsistence in the forest, as well as something to bandage themselves. After spending months there, a number of them make their way back; the others [353] perish of hunger. As a result, they consider both a wounded man & a prisoner as dead men.

Until they arrive back at the village, the prisoners are treated well & without ill-will. On the return of the party, those men who participated march in single file, just as when they departed. The man holding the scalps heads up the procession. They are dangling all the way along a pole. Then come the prisoners holding a *chichiquoi* and singing, although they do not feel like doing so. All the warriors are silent. The man holding the pole with the scalps first utters several shouts to indicate the number of men they have lost. It is a lugubrious cry which gradually dies away. After these, he utters a number of more lively shouts to express the number of scalps or prisoners taken. They all shout together [354] to finish their report. They continue shouting until they arrive at the chief's hut. It can just be imagined how the adolescents, women and children rush up to greet them. The nimblest of them takes the pole to carry it to the chief, as if to announce good news. Others seize the prisoners, whom they are all eager to lead to the chief's hut. They are lucky if they are steady on their legs for they are then assailed with a hail of stones & blows with sticks; the village folk compete in striking them the most. They all join in, except for the warriors who calmly let them do it and continue their march as in a procession. When one of these unfortunates falls, he is ill-treated even more harshly, especially if he screams, because that [355] amuses them. It is rare for them not to arrive covered in bruises at the hut, where the chiefs & elders are assembled. The leader of the war party recounts the details of the journey & the operation, renders justice to each man & praises his warriors, mentioning their exploits. After that, they present to the gathering the prisoners, each of whom is compelled to dance in turn. It may well be imagined that they have no desire to do this, especially if they are Europeans. The Indians, however, do not need to be asked twice, as it gives them the opportunity of recounting their bravery. Once the ceremony is over, the war chief disposes of the scalps & the prisoners, according to decisions he has previously made. Normally, among the Iroquois, the prisoner destined to replace a dead man occupies the latter's place in the family. The entire nation **[356]** regards him as one of its members & his new relatives remove the belt of slavery from his neck. If they do not want to adopt him & say that they are too upset to think of replacing the dead man, the prisoner is handed over to the young men so that they may amuse themselves with him. This verdict is irrevocable and the unfortunate man is burned. We shall not enter into details about this horrible ceremony, which can be found in all authors on the subject. Fortunately, these events are becoming more and more rare.[1435] In other nations, prisoners are even more to be pitied, because they are treated like their dogs. They kill them with impunity when in a drunken state & in periods of food shortage they do not hesitate to eat them any more than they might an animal. If a prisoner is lucky enough to find a wife **[357]** among these nations, his family enjoys no respect & is subjected to all the chores that they can dream up. Some prisoners are fortunate enough to find good-hearted Indians with whom they lead a life that is neither hard nor perilous, especially if they are careful to avoid the orgies of drinking by using the time to go hunting.

Often when the victors have lost a number of highly reputed chiefs, it is almost impossible to stop them sacrificing some of their prisoners to the shades of the dead men. It is on those occasions that, to satisfy such a desire, they ceremonially eat a prisoner. It must, however, be

---

1435 Among the authors was Charlevoix. Although he also omits the worst details of the sufferings inflicted on a prisoner, he discusses burning by the Iroquois and Huron at much greater length. Charlevoix, 1761, I, pp. 373-80.

Montcalm, fresh from Europe in 1756, claimed that the burning of prisoners as described by Charlevoix "has so to say gone out of fashion this year," although he added that they "did indeed, burn one, in the vicinity of La Belle Rivière so as not to lose the habit." Montcalm noted that the French "ransom prisoners from them, from time to time, who, when they come into our hands, are treated according to the laws of war." Montcalm to La Marquise de St. Véran, June 16, 1756, *Report on the Public Archives for 1929,* pp. 44-45. Burnings did continue, but adoption, as described by Pouchot below, was very common. Thomas Gist was so treated upon his arrival at Detroit. Peckham, "Thomas Gist's Indian Captivity", pp. 298-302.

stated firmly that they taste human flesh only with repugnance. More than once, young men have been seen to vomit it up. It is solely **[358]** through bravado & to harden their hearts that they sometimes consume such food.

It is certain that the best recourse Europeans have in fighting the Indians is to corner them in some place from which there is no exit & to march at them with fixed bayonets. A company that tried to shoot it out would very soon be beaten because of the accuracy of their musket fire. If through misfortune one's troops lose contact with one another, destruction is certain because of their skill in attacking with hatchet and spear.[1436] **[359]**

Although the Indian has little knowledge, some of them have nonetheless a lively & well-informed mind & considerable finesse. Some of them are very stupid. But is that not also the case among our peasants? The Indians have a good memory. When dealing with them, one must take good care not to tell lies. It is always wise to hold in reserve some means of extricating oneself from a difficult situation in order to show them that they have not been deceived. They are by nature so even-tempered that they cannot understand our urge to walk for exercise and are always surprised when they see us raising our voices in argument. On those occasions they say that we have lost our wits.

We have already noted that, in order to attract their friendship - and some of them are indeed very ready to have a close **[360]** friendship with us - one must be generous. However, it is not through the quantity given that one acquires the reputation of generosity. One is considered generous or miserly according to the way one acts. For example, if you give them some brandy and present them with a large goblet half full, you will be regarded as stingy. They will be perfectly satisfied, though, if you give them a small one, provided it is full. If you

---

1436 Note in original, p. 359 (a): *The* Reflexions sur la guerre avec les Sauvages, *&c., to be found at the end of the account of General Bouquet's expedition against the Indians of the Ohio in 1764, are well worth consulting. The principles of its author do not differ from those of M. Pouchot but have the advantage of being more fully developed.*

The work referred to is William Smith, *A Historical Account of the Expedition Against the Ohio Indians in the Year 1764,* which includes, on pp. 37-59, "Reflections on the War with the Savages of North-America." Bouquet had perfected and successfully employed tactics for fighting against the Indians in 1763 and during his punitive march across Ohio in 1764.

Henry Bouquet (1719-1765) was lieutenant colonel (1756) in the 60th Regiment of Foot. *DNB,* II, p. 917 and *DAB,* I, pp. 480-81. There is an eighteenth century sketch of his career before entering the British service in Stevens, *The Papers of Henry Bouquet,* I, pp. xvi-xxvii. The originals of Bouquet's papers are in the British Library, Add MSS 21631-21660.

offer them some bread, let it be a whole loaf. Half a loaf, even if it is much larger, would compel them to say that you want to make them die of hunger, which would be quite enough to render an entire nation hostile toward a commandant. When giving them presents, you must hand them less than you desire to give them, because if they ask for more, you are still able to oblige, which [361] will lead them to extol your generosity.

Without any doubt at all, brandy is what the Indians like most in the world. Everything depends on the appropriate manner of dispensing it to them, whether for drinking or trading. It is a way of enlisting the support of all the nations. There has been far too much trade in it & on occasions a better use could have been made of it. Although, at the moment, they have not suffered from being deprived of that pernicious liquor, they are very angry because they have become addicted to it & consider its use as the principal cause of their ruin.

Apart from one or two minor differences, in customs or in character, the Indians of this continent are all the same. It may simply [362] be noted that those of them who have frequent contacts with the European colonies are the most docile & enlightened. They refer to their relationships with each other & to their respective status by familial terms - uncles, nephews, cousins and so on. The Outaouais & the Abenakis call the Iroquois their uncles and the latter treat them as nephews, which indicates the considerable seniority of the Iroquois. Another proof could be found in a number of place names, like "Ohio", which means beautiful river. "Theaogen" the confluence of two rivers, "Schenutar",[1437] "Niagara", & other place names in very remote areas, all of which [363] are words in the Iroquoian language. The Indians address the English as brothers when speaking to them & refer to the French as fathers in order to show that their first contacts were with the latter. As a consequence, they believe that the French should not allow their children to go without anything.

When they want to refer to each nation in its entirety, they call the English *Saganach,* the French *Mistigouch* & the Indians *Michinabé.* They identify French officers by the name *Onontio,* which means mountain, because one of the first they got to know was called "Mont-Magny." [1438] By extension, they [364] call the King of France the "Great Mountain," *Onontio-Goa,* & the king of England the "Great Belt." [365]

1437 Note in original, p. 363 (a): *This is the name of the town of Orange, situated on the Hudson River in the province of New York.*

The reference seems more likely to be to Schenectady than to Orange (Albany).

❀ SUPPLEMENT ❀
concerning the Indian population of North America

If the first explorers had provided a detailed account of the population of North America, & if those who followed them had also taken care over the matter, we would be able to estimate the gradual decrease it has suffered. But we lack these data & are reduced to imperfect notions and vague accounts. Their conclusions are none the less distressing for humanity. **[367]**

When Champlain laid the first foundations of the French colony in Canada, a number of sizeable nations whose names today are scarcely known occupied the country at that time. The language of the Algonquins, which is still used by several Indian tribes, is the only thing to have preserved the memory of that great nation.[1439] The Hurons no longer constitute a nation. Those faithful & powerful allies of the French, after their dispersal, took refuge in two villages very far from one another, the first near Québec & the second at the far end of the Great Lakes.[1440] The Outaouais, formerly so numerous, now occupy only three villages & the Poutéoutamis two.[1441] No trace can be found of the Bersiamiamites, the Papinachois, the Montagnais, the

1438 Note in original, p. 364 (a): *The Chevalier de Mont-Magny succeeded Champlain in 1636. He was the second governor of New France & the first to enjoy the prestige of pacifying the Indian nations of the country & of negotiating a treaty with the Iroquois.*

Governor Charles Huault de Montmagny (note 800).

1439 See Ives Goddard, "Eastern Algonquian Languages" in Trigger, pp. 70-77.

1440 Following the dispersal of the Huron peoples of southern Ontario by the Iroquois in 1648-1649, one group of refugees settled at Lorette near Québec while another was driven west to Michigan. From 1661 to 1671 many of the latter lived at the mission of La Pointe on Chequamegon Bay at the southwestern end of Lake Superior. They then moved to the Straits of Mackinac. In 1701 these Hurons began to move to Detroit where they came to be known as the Wyandot. Their name for themselves was "Wendat." Tanner, pp. 30-35; Christian Morissonneau, "Huron of Lorette", in Trigger, pp. 389-93; Elisabeth Tooker, "Wyandot", in Trigger, pp. 398-406.

1441 At the period discussed by Pouchot, the late 1750s, there were three major concentrations of Ottawa: L'Arbre Croche (Cross Village, Michigan), the mouth of Michigan's Au Sable River on Lake Huron and opposite the village of Detroit on the Detroit River. Major Potawatomi villages were located at Detroit and near Fort St. Joseph (Niles, Michigan) on the St. Joseph River. Tanner, pp. 40-41.

Amikoués, the Attikamégues, &c. The latter tribes were surrounded by a [368] number of other peoples whose territory extended to the Lac St. Jean area & as far as lakes Mistassins and Némiscau.[1442] They were all destroyed, mainly in the Iroquois wars. The latter, who were so formidable & who, at the end of the last century, could field 7500 warriors, could barely assemble fifteen hundred at the present time.[1443]

The nations to the east have suffered an even more perceptible decrease. Formerly they constituted, so to speak, only one people, known to the French by the general name of Abenakis. Their homesteads were spread out over the vast territory extending from the banks of the St. Lawrence down the Appalachian mountain chain as far as the southern tip of Carolina. Although separated into a number of [369] tribes, this people spoke the same language. Those of them who had settled near the sea coast were occupied solely with fishing, while the remainder concentrated on hunting.

As the English colonies expanded, these Indians withdrew into the interior without, however, ceding, as the English have claimed, the territory they had been forced to abandon, since they have never had any notion of what we call "sale by contract, or cession by treaty." The different groups of this nation who had settled in Acadia were differentiated by names like "Abenakis", "Etchemins", "Souriquois",

---

1442 Charlevoix discusses these early groups. He places the "Bersiamites" on the north side of the lower St. Lawrence in the vicinity of the Papinachois and Montagnais. Charlevoix, 1761, I, p. 284-85.

The Papinachois were identified by early missionaries in the highlands between the Mistassini region and Labrador. They were ancestors of the Montagnais-Naskapis. "Glossary of Indian Tribal Names", *DCB*, II, p. xxxvii.

Montagnais were the northern Algonquians of the region to the north of the lower St. Lawrence. *Ibid.*, p. xxxv.

The Amikoués were an Algonkian people identified by French traders in the seventeenth and eighteenth centuries. They lived on the north shore of Lake Huron and were related to or part of the Chippewa with whom they had become consolidated by Pouchot's time. *Ibid.*, p. xxvii.

Attikamégues or "the Whitefish Nation" lived in the upper St. Maurice Valley. That river joins the St. Lawrence at Trois Rivières, Québec. They were a hunting people who, ravaged by European diseases, had been absorbed by the Montagnais by 1670. "Glossary of Indian Tribal Names", *DCB*, I, p. 13.

Lac St. Jean, located about 150 miles north of Québec, is drained into the St. Lawrence by the Saguenay River. Lac Mistassini is also in Québec, nearly 200 miles northwest of Lac St. Jean. Lac Némiscau is west of Mistassini. The two latter bodies of water drain into James Bay.

1443 In 1685 the Iroquois could reportedly field 2,050 warriors. The number of men had indeed declined by Pouchot's time, but was still believed to be 1,810 in 1763. By the time Pouchot's *Mémoires* was printed in 1781, their number had decreased to 1,600. Trigger, p. 421.

"Micmack" &c.[1444] After the foundation of New Halifax,[1445] they made war on the English who destroyed almost all of them. Some of them escaped destruction, but only just enough to form a small number of villages which **[370]** might be able to field a thousand warriors.

The most important of the Abenaki tribes that remained beyond the Appalachians is that of the Loups, whom the English call "Delawares." The tribe lives on the banks of the Ohio. At the end of the late war, they could muster eighteen hundred fighting men, but that number must by now have greatly decreased because of the losses sustained by the tribe in 1763 & 1764, when it attempted to organize an uprising of all the Indians of the continent against the English.[1446]

The other nations which participated in this conspiracy lived mainly around the Great Lakes of Canada & close to the rivers that flow into them. Their losses were just as considerable. With the exception of the Outagumis & the Missisakes, they currently have only a very small number of men at their disposal. **[371]**

The Indians of the north & the north-west have very little contact with Europeans and in consequence are very little known to them. We only know that the Scioux, the Cristinaux, & the Assiniboels are still quite numerous. Although the Eskimaux[1447] & the other peoples of Labrador have a few commercial links with the English, we are nonetheless ignorant of the extent of their population. According to English reports, about a thousand persons, men and women, arrive every year in 500 canoes at Fort Nelson or York[1448] to trade their furs.

The names of a considerable number of the peoples of Louisiana are known but that is almost all that remains of them. They have not

---

1444 Etchemins were also known as Malecites and originally inhabited the Saint John River area of New Brunswick. "Glossary of Indian Tribal Names", *DCB,* II, p. xxxii. Their language was extinct by the end of the seventeenth century. Trigger, p. 71.

Souriquois and Micmac were branches of the same people, who were, in turn, part of the Abenaki Confederacy of Nova Scotia, Prince Edward Island and northern New Brunswick. *DCB,* p. xxxiii.

1445 Halifax, Nova Scotia, founded in 1749.

1446 By 1768 the total population of the Loups or Delaware numbered only about 3,500 individuals. Tanner, p. 66. In 1779 there were only 3,200. Trigger, p. 214. Pouchot's estimate of 1,800 fighting men must therefore be considered inflated. His identification of the Delaware with the Abenaki is due to their use of the same Algonquian language.

1447 The Inuit or Eskimo had regular contact with Europeans along the Labrador coast. "Glossary of Native Peoples", *DCB,* IV, p. liii.

stopped disappearing, so to speak, since the French settled among them. In the space of only **[372]** twenty-five years, their losses have been immense. We can judge this with some accuracy thanks to an excellent memoir which M. de Bienville, the governor of the French colony, wrote while he was there.[1449]

He mentions more than fifty nations who, before the year 1700, could field 54550 men. Twenty-five years later, they were down to 24260. A number of nations who formerly had 400, 500 & as many as 600 fighting men, subsequently had no more than 30, 40 & 50. Since that last period, very powerful nations have continued to suffer a great **[373]** decrease in population. The Chactas, who before the French settlement had 20000 warriors & at the time of M. de Bienville's census 8000, would barely field 4000 today.[1450] The Chicachas, the most dreaded enemies of the French colony, were considered to be as numerous as the Chactas but could not currently be compared to them.[1451]

From these facts, the conclusion of M. de Buffon is unavoidable: that the most populous nations of America amount to no more than three or four thousand persons. He is rightly convinced that "it could be suggested, without fear of error, that in a single city, like Paris,[1452] there are more **[374]** people than there are Indians in the whole of that

---

1448 Fort Nelson was a post of the Hudson Bay Company at York Factory, Manitoba at the mouth of the Nelson River. The river drains Lake Winnipeg into Hudson Bay.

1449 Note in original, p. 373 (a): *The title of this memoir is* Le Cours du Mississipi, ou Les Sauvages de la Louisiane, leur nombre & le commerce qu'on peut faire avec eux. *It was published in the* Journal de Trevoux *of March 1727.*

Jean-Baptiste Le Moyne de Bienville (1680-1767) served as military commander of Louisiana from 1701 to 1717 and commandant general (the position of governor having been abolished) from 1717 to 1725. He was finally appointed governor in 1732 and held the position until 1743. *DCB,* III, pp. 379-84. Bienville's article in the *Journal de Trévoux* was published while he was in France.

1450 The Choctaw, a Muskhogean people of middle and southern Mississippi, numbered as many as 15,000 to 20,000 people in 1700. Hodges, I, pp. 288-90.

1451 Note in original, p. 374 (a): *A modern author estimates the strength of the Chicachas at 750 warriors.*

The Chickasaw were a Muskhogean people related to the Choctaw. They inhabited northern Mississippi and parts of Tennessee. In 1744 an English trader estimated that one of the two divisions of this group could field 450 fighting men. The population of the entire nation was probably 3,000-4,000 individuals. Hodges, I, pp. 260-62.

1452 Note in original, p. 374 (b): *On the assumption that this city has a population of no more than six or seven hundred thousand souls.*

area of North America between the northern sea & the southern sea, from the Gulf of Mexico to the north coast, although this expanse of territory is far larger than the whole of Europe."[1453]

This rapid decrease in Indian population can be ascribed to the following causes: 1st, the excessive use of brandy; 2nd, outbreaks of smallpox;[1454] 3rd, the wars triggered by the arrival of the Europeans; 4th, the custom of replacing the dead with prisoners, which causes **[375]** a state of perpetual war among these peoples. Man was thus born there with the nature the famous writer Hobbes[1455] claimed he had. Did not their conduct also justify his ideas when he defined man as a robust child well aware of its strength? Indeed, the Indian abuses his strength, because he is too conscious of it. He succumbs too easily to the impulse for revenge & he is awakened by the sound of arms, which he always takes up in order to destroy, and never to acquire or preserve. He is a slave to his tastes & needs, which have both multiplied since the discovery of the New World. In order to satisfy them, he has forgotten those interests that are the most dear to him & has become the instrument of the hatred of two rival powers as well as that of his own destruction.

END [376]

❁ SUPPLEMENT ❁
to the preface to the topographical observations

Since the time of M. de La Verandiere,[1456] the regions he explored have been visited by M. Carver.[1457] After spending the winter on the banks of the Mississippi, at 44 degrees latitude, he headed towards the

---

1453 Note in original, p. 375 (a): Histoire naturelle, *V, p. 176, ed. in-12.*

1454 Note in original, p. 375 (b): *To this epidemic can be added the most terrible of all, the plague, that one of the King's ships brought in 1704 to Mobile, where it destroyed two populous nations called the large & small "Thomés", and others.*

This small nation was a Muskhogean group of the Gulf Coast. They spoke a dialect of Choctaw. In 1702 they inhabited a village eight leagues north of the French settlement at Mobile and had a reported population of 350. Hodges, II, p. 771. A contemporary French account of these people has been published by Vernon J. Knight, Jr. and Sherée L. Adams (eds.), "A Voyage to the Mobile and Tomeh in 1700, with Notes on the Interior of Alabama", *Journal of Alabama Archaeology*, XXVII, No. 1 (June 1981), pp. 32-56.

1455 Thomas Hobbes (1588-1679), English philosopher. *DNB*, IX, pp. 931-39.

north of the American continent & traversed its highest area which is the source of the various rivers that water it, some of them flowing into the Atlantic and others towards the west of the Straits of Anian.[1458] This English explorer visited the territory of the Sakis & Nadouessis[1459] &c. He also reached the banks of the Bourbon River, Lake Ouinipigon, &c. Their existence is thus no longer problematic. A French translation of the account M. Carver published in 1778 of his explorations in the years 1766, 1767, 1768 would be very desirable. It provides us with fresh information & extends our geographical knowledge.[1460] **[377]**

❁ Table Of The Chapters ❁ Contained In The Third Volume

[Not translated]
[378-379]

❁ Approbation ❁

I have read the work entitled *Mémoires sur la derniere guerre de l'Amerique Septentrionale &c.* & have found nothing in it that should prevent its publication. Yverdon, 20 December 1780.

É. Bertrand[1461]
Censor **[380]**

1456 Pierre Gaultier de La Vérendrye de Boumois (note 968).

1457 Jonathan Carver (1732-1780), former officer of rangers and explorer. *DNB,* III, pp. 1146-47. Carver's *Travels Through the Interior of North America in the Years 1766, 1767 and 1768* was published in London in 1778. For further information on Carver and his travels see John Parker (ed.), *The Journals of Jonathan Carver and Related Documents, 1766-1770* (St. Paul, MN, 1976).

1458 The Bering Strait.

1459 The Sauk and the Sioux or Dakota. "Glossary of Indian Tribal Names", *DCB,* II, p. xxxix.

1460 Carver's travels are discussed in Parker, *The Journals of Jonathan Carver.* Carver crossed Nipigon Bay on the northern side of Lake Superior on August 13, 1766. He noted the river flowing from it but he did not proceed to the lake itself. *Ibid.,* p. 174.

1461 Élie Bertrand (1712-1797), protestant clergyman, natural scientist, writer, and member of a distinguished Swiss scientific family. *Biographie universelle,* IV, p. 183.

Pierre Pouchot's value as an engineer was greatly enhanced by the British capture of the 64-gun ship of the line *Alcide* on June 8, 1755. Lost with her were all three engineers of the French force in transit to Canada. *Alcide* was taken into the Royal Navy and is seen here at Havana in 1762, her stern gallery still adorned with the French coat of arms. *Courtesy, William L. Clements, University of Michigan.*

Corsica, the scene of Pouchot's final fight. General Vaux's French army was between Lento (rendered here as "Honto") and Canavaggio ("Canavagia") facing the crossing of the Golo River at "Pte Novo." Pouchot was reconnoitering the road from Lento to Borgo when he met his end near the village of Vignale ("Vignole"). Detail from Nicolas Bellin, "Carte de L'Isle de Corse," Paris, 1768. *Courtesy, William L. Clements Library, University of Michigan.*

Appendix A: Other Writings By Captain Pouchot

While Captain Pierre Pouchot's published *Mémoires* represent the bulk of his known writings, ten other letters or brief *mémoires* have been identified and are printed here. Six of the letters were published in the nineteenth century, either in the original French or in English translation. The other four letters, all relating to Pouchot's two periods as a prisoner of war, are in manuscript form among the papers of Major General Jeffery Amherst. Additional correspondence with officers of the Department of War or of the Marine undoubtedly exists in French archives. One additional item, a journal from the summer of 1757, unsigned but almost certainly kept by Pouchot, will be found as "Appendix D."

In addition to the letters printed in this appendix, fragments of three more survive in other works. In Volume I of the *Mémoires* Pouchot includes a partial text of a dispatch from Niagara to Governor Vaudreuil dated May 30, 1756. Volume II contains part of a long letter of instruction to Captain de Lignery dated June 1, 1759. Another communication of July 11, 1759, directed to the captain of the schooner *Iroquoise,* was transcribed into René La Force's journal on July 12. The journal is printed as "Appendix C" of this volume.

Another and much longer document is an unsigned journal recording the first weeks of the siege of Fort Niagara. It spans the period July 6, 1759, through the night of the 18th-19th. Pouchot's authorship of this piece is less certain than for the 1757 journal. The text differs enough from the narrative in his *Mémoires* to suggest that it might have been kept by another officer, although almost certainly under Pouchot's direction as an "official" journal of events. This account of the siege is incomplete because it was probably among the dispatches for Montcalm and Vaudreuil sent by Pouchot to the schooner *Iroquoise* on the evening of July 20 (see note 650). The journal is in the La Pause Papers in *RAPQ,* 1933-34, pp. 92-94. It probably found its way into Captain La Pause's possession because he held a position on Montcalm's staff.

Annotation of the documents below follows the same sequence and form as that for the *Mémoires.*

Geographical Observations Taken From the Letter of M. Pouchot of 11 April [1757][1462]

The Kanestio River or the western branch of the Susquehanna, as the English call it, has its source in the mountains, flows through the territory of the Loups and the Renards who live in the highlands, communicates by way of a short portage with the Kanouchiagon River above which stands the Tsonnonthouans village called Sonnechio.[1463] That is all that needs to be said as to the area of the Renard country in the highlands. From Kanestio to the Kanoagon River, there is only one portage, six leagues in length, a very good trail according to the Iroquois. The Kanoagan River flows into the Ohio in the area of the Rivière au Boeuf, where Fort Machault is located. It is an area of mountains, passes and dry watercourses.[1464]

An Englishman resident among the Loups maintains that this river is broad but shallow, full of beaches covered with rocks, doubtless like the torrents of the Dauphiné;[1465] that it was full of rapids; that at the very most only very small bateaux could be hauled up that far; and that he did not believe it could be of much use to them. He also said that the territory was full of passes, whether going up to the source of the Kanoagon or coming down it.

1462 This fragment of Pouchot's geographical observations was taken from one of his letters to the Marquis de Montcalm. It was transcribed into the general's journal and published in Casgrain, *Collection,* VII, p. 197. This translation is by Michael Cardy.

1463 Pouchot describes a number of features found on his map of the French and English frontiers. The contents of these observations seem to have provided at least part of the basis for his discussion of the route between the Genesee River and the Susquehanna in Volume III of the *Mémoires.* The Canestio River and its portage to the Genesee (here rendered Kanouchiagon instead of "Casconchiagon" as in the *Mémoires*) are discussed in note 1275.

In other deviations from the forms used in his printed work, Pouchot begins the spelling of his name for Seneca with a "Ts." Their principal village was at the site of Geneseo (Sonnechio), New York.

1464 The Kanoagaon is the upper Allegheny River and here takes its name from a Seneca village also known as La Paille Coupée. See note 1327. Fort Machault was at Venango (Franklin, Pennsylvania) where French Creek enters the Allegheny. Stotz, pp. 76-79.

1465 The mountainous region of southeastern France bordering Italy and the Maritime Alps.

To M. de Paulmy[1466]

Montreal, this 3rd 9ber, 1757

My Lord,

I arrive from Niagara where I have been relieved by the Marquis de Vaudreuil. I may say I have finished that place, according to the plan I had the honor of transmitting last fall, on the 18th 8ber.[1467] There remain to be sodded only a small portion of the interior parapet; the epaulement from the gate to the covert-way along the river, and both the traverses of the place of arms, salient from the half moon.

It will be essential, my Lord, to sod the scarps and counterscarps of the ditches, in consequence of the nature of the soil which, notwithstanding its hardness, wastes extremely the moment 'tis exposed to the air.

I would have wished much, my Lord, to be able to transmit to you a plan of the present situation of that place; I could never find a moment to draw it, in consequence either of the details attached to the command for the Indians, or the assiduity those works required, as I was alone to superintend them.

I am astonished, my Lord, at being able to bring them to their present perfection, with the few men I have had for so many different operations as were necessary to be executed. I have had them done with all the economy of a gentleman cultivating his farm, and I dare assert that I have saved more than fifty thousand *écus,* according to the estimate of connoisseurs.

I have made some changes in the locations of the buildings on the plan I had the honor to transmit you last year. There are now two large barracks, one church, one powder magazine, one store for provisions and merchandise. Barracks will have to be built in the course of the winter to lodge one hundred men additional. I can assert, my Lord,

---

1466 Antoine-René de Voyer d'Argenson, Marquis de Paulmy (note 380) served as Minister of War from February 1, 1757 until February 25, 1758. *DRCHSNY,* X, p. vii.

This letter was copied from the original in the Départment de la Guerre in Paris and published in *DRCHSNY,* X, pp. 667-68. The translation was done in the nineteenth century under the direction of John Romeyn Brodhead. The form and spelling of the translation have been retained. The date is rendered in French form for November 3.

1467 October 18, 1756. The work referred to is almost certainly Pouchot's "Plan de Niagara et des Fortifications faites en 1755 at 1756." The original is in the French Archives with a copy in the National Archives of Canada (NMC-26647).

from all I have seen, that that fort influences extremely all the bordering Nations. 'Tis not for me, my Lord, to praise my work. The Marquis de Vaudreuil and the M[arq]is de Montcalm will have the goodness to render you an account of it, which, I dare believe, will be of advantage to me. They have not allowed me to remain ignorant of the favor the King has been pleased to grant me.[1468] Might I hope from your goodness, my Lord, that you will be pleased to procure me an augmentation of it. I am fortunate enough to be in a peculiar position, which may procure for me some exception to what his majesty has proposed to do, and when I desire the commission of Lieutenant-Colonel, 'tis to procure for myself the advantage of serving his Majesty still more usefully, and my time is passing, my Lord, for I am 46 years of age. You have had the goodness to promise me in Paris that you would be favorable to me; I dare hope for your confidence.[1469]

My Lord, when I assumed the command at Niagara, the Marquis de Vaudreuil wished me to engage the Five Nations to observe neutrality. I dare flatter myself that I have, by my proposals to, and treatment of, them, contributed not a little to embroil them with the English, and to make them decide on waging war against the enemy. I do not pretend, my Lord, to detract from the merit therein of M. Chabert Joncaire, a Colonial officer, who is much beloved by the Senecas.[1470]

1468 On March 20, 1757 de Paulmy had written to Governor Vaudreuil that "I have not forgotten the favorable testimony you bear of Captain Pouchot ... to whom you have entrusted the direction of the fortifications of Fort Niagara, and he will find some marks of his Majesty's satisfaction in the list of favors which will reach M. de Montcalm." *DRCHSNY*, X, p. 535. On August 8 the army on Lake George received a list of *grâces* for officers of the army. Pouchot was awarded a pension of 200 *livres*. Rélation de Poulariès, *RAPQ*, 1931-32, pp. 59-60. This was presumably in response to a request for recognition of Pouchot's work at Niagara and as an engineer at the siege of Oswego. A list of proposed *grâces*, dated August 30, 1756, had recommended him for a lieutenant colonel's commission. *RAPQ*, 1944-45, pp. 82-83.

1469 Captain Pouchot's hope was in vain. By the time this letter reached Paris, Paulmy had been replaced as Minister of War, and whatever influence Pouchot had was gone. He was clearly disappointed to have received only a small pension instead of promotion. His conversation with Paulmy on the matter had taken place sometime before April 1755 when Pouchot went to Brest to embark for New France. Paulmy was then working in the office of his uncle and predecessor as Minister of War whom he served as General Secretary of the War Department from 1751 to 1757. *DRCHSNY*, X, p. 535n. His uncle, Marc-Pierre de Voyer, Comte d'Argenson (1696-1764) was Minister of War from 1743 to 1757. *Biographie universelle*, XLIV, pp. 149-52.

1470 Lieutenant Daniel Chabert de Joncaire (note 331).

I have been in the same position in regard to the Loups who were become very cold since last year. I dare say that I had gained their confidence as well as that of all the Nations with which I have had to treat. The number of parties, prisoners, and scalps I have had are solid proofs thereof. They have principally infested New Jersey and Pennsylvania which have been laid waste by them.

I applied myself, my Lord, throughout the whole of my mission, to be in a position to render an account of the state of the country as regards the frontier of the English; of the dispositions of the Indians, and of the advantages that can be derived from them, as well as of the arrangements for the safety of those parts.

I ardently desire, my Lord, that the knowledge I have acquired may merit your approbation and the honor of your confidence.

I am, with the most profound respect, my Lord,

Your most humble and
Most obedient servant,
POUCHOT
Captain in the Bearn.

To Marshal de Belle-Isle[1471]
Montreal, 14th April, 1758

My Lord,

I have handed to the Marquis de Vaudreuil a map and Memoir on the subject of the French and English frontiers, which I have drawn up on the best information I have been able to procure during my sojourn at Niagara. I have laid down, in my voyages, the course of the river from Montreal, Lake Ontario as far as Lake Erie; therefore it has

1471 Although this letter was addressed to de Paulmy's successor as Minister of War, Louis-Charles-Auguste Fouquet de Belle-Isle (note 1313), his appointment was not yet known to the officers in New France. Belle-Isle's term of office had begun only on March 3, 1758 and would continue until January 26, 1761. *DRCHSNY*, X, p. vii. Montcalm recorded news of the change of ministers only on May 22. Casgrain, *Collection*, VII, p. 354. Belle-Isle was a prominent military figure and a Marshal of France. The correspondence of Pouchot, Montcalm and other officers with him prior to his ministry was due to his status in the army.

This letter was transcribed from the original in the Départment de la Guerre in Paris. The translation was done in the nineteenth century under the direction of John Romeyn Brodhead, and published in *DRCHSNY*, X, pp. 694-95.

an appearance of truth which correctly represents the country and is not contained in any other maps. As I have not been on the English frontier, I laid that down according to their best maps, which I again corrected on such reliable information as I have obtained. It is in sufficient detail, however, to show the interest we ought to take to prevent the English interposing obstacles in the way of the Iroquois and Loups who form a barrier. That country, my Lord, would be well worth being seen by experienced eyes, which has not as yet been the case; the well known Carrying place of Niagara is an evident proof. The most recent accounts thereof describe it as the most rugged of Alps, whilst 'tis only a rise of ground a little more elevated than that of Bellevue;[1472] below and above are very fine plains, as can be seen on my map.

The detail of Lake Erie, which is entirely unknown; it is, perhaps, [as] navigable for large vessels as Lake Ontario. The resources of those countries, once known, would furnish opportunities of avoiding long routes and expenses, exclusive of enabling us to occupy more decided points for the security of the country.

If you have the goodness, my Lord, to signify to me that this essay might please you, I shall set about perfecting that work in order to accomplish all the objects which will possibly be required of it.

I hope, by my zeal, to deserve the kindness you have promised to honor me with, my Lord, on the recommendation of Mde de Meillian.[1473] Since our arrival in this country I have had the good fortune to be always pretty usefully employed. I constructed the Frontenac intrenchments, completed Fort Niagara and the siege of Chouaguen. I dare hope, my Lord, that you will be graciously pleased to give attention to the good reports of me which our Generals are so good as to render.

I am, with the most profound respect,

My Lord,
Your most humble and
Most obedient servant,
POUCHOT,
Captn in the Bearn regiment

1472 In his *Mémoires,* (III, p. 172) Pouchot compares the height of the land at Niagara to Meudon, near Paris (note 1303). Bellevue is perhaps an alternate name for the place.

1473 Pouchot's patroness, Madame "de Meillian", has not been identified. The Marquis de Montcalm corresponded with a Madame de Massillan, and she is perhaps the same person. Montcalm to La Marquise de St. Véran, June 2, 1758, *Report of the Public Archives,* 1929, p. 69.

Permit me to annex hereunto some Observations adapted to the positions of the affairs of this country.

General Observations on the Frontier of Canada

Acadia being, so to speak, a separate portion of the main-land, events may decide its limits or its recovery *in toto.*

The English must have little inclination to settle the district North of their Colonies only so far as they are convenient to the sea. They would not settle in the interior of those countries, which are mountainous, cold, consequently of little value unless they could thereby excite our jealousy, which they will not fail doing as much as they will be able; moreover, there are very few Indians in those parts with whom to trade.

The English, once masters of the heights of land, and with posts on Lake St. Sacrament,[1474] are in a position to excite continual alarms in the Colony, by being conveniently situated to repair thither in force. But they will attempt to establish in that quarter only posts of war, the country not being fit for cultivation.

The Provinces in their rear, belonging to the Five Nations and Delawares,[1475] are very populous.

The outlets they possess by the River des Armes[1476] and the Canestio[1477] will tempt them always to come and settle on the South shore of Lake Ontario; the country being very fine and fit for cultivation they will make large settlements in it.

Placing the Five Nations and Delawares in their rear, they would be masters of them and would attract the other Nations by their commerce.

We should soon be driven from our isolated posts.

1474 Lake George.

1475 Here the original French text undoubtedly used the name "Loups."

1476 The translator here wrote *sic* and indicated that the word was intended to be "Anniés" for the Mohawk River. Pouchot, in his *Mémoires,* usually spells this *agniers* or *âniers.* See note 345.

1477 The translator identifies this with the Canestio River flowing from Allegany County, New York to join the Tioga River in the town of Erwin, four miles southwest of Corning. On Pouchot's map, however, the name "Kanestio" is applied to the entire Chemung River that, fed by the Tioga and Canisteo, takes its name at Corning and joins the Susquehanna at Athens, Pennsylvania. Pouchot therefore considered the Susquehanna and its north-western tributaries as the second potential route by which the English might reach Lake Ontario.

The outlets of the English on the Beautiful River[1478] furnish nearly the same reasons. 'Tis only to be observed that fewer Indians are among them than among us, and that their road is long and difficult, but they would compensate themselves in return by profitable settlements.

The lower parts[1479] are not so dangerously menaced at present; being at a great distance from their settlements, the English can throw only war posts in advance there. We have equal advantage to dislodge them, either from Canada or from Louisiana.

To Chevalier de Lévis[1480]
Niagara, 5 May 1759

We arrived here as comfortably as one could wish on 30 April, followed by Montigny on 2 May.[1481] From that time onwards, there have been storms and high winds from the south-west. You have learned the news from the Chouaguen River following M. Villejoin's arrival;[1482] they more or less confirm what I am being told by the Indians coming from Johnson's camp. The Tsonnonthouan and Goyogouin[1483] chiefs have sent messages to let me know that they, as well as the Loups, were coming to see me.

To date I have received news almost every day from M. Des Ligneris[1484] who is still at his post and I hope it will be he who will detect the enemy. You will see, sir, from the statements of prisoners whom I am sending to M. de Vaudreuil, that I have good reason to

---

1478 La Belle-Rivière or the Ohio.

1479 The lands south of the Ohio River.

1480 Soon after his return to Niagara Pouchot addressed this letter to François de Lévis, Duc de Lévis (note 284), second-in-command to Montcalm. It was printed in Casgrain, *Collection,* X, pp. 119-21. This translation is by Michael Cardy.

1481 Captain Jean-Baptiste Testard de Montigny (note 546).

1482 Ensign Michel Rousseau de Villejouin (note 507).

1483 Seneca and Cayuga.

1484 Captain François-Marie Le Marchand de Lignery (note 328), commanding French forces on the Ohio from Fort Machault at Venango.

hope so.[1485] I am only waiting for the return of a scouting party that I have in the Chouaguen River area - it should arrive very shortly - before sending out M. Des Ligneris well equipped to act before the enemy receive reinforcements. I now dare believe that this decision bodes well for us. May God grant that things go as well in the lower country.[1486] When you receive my letter, you will doubtless have had news from France. It seems, from everything that I can deduce, either that the enemy has serious designs on Québec or Carillon or that they have some major problems in their own camp that we are not yet aware of. They will shortly begin their campaigning season in your area and are currently marching towards their rendezvous positions. You are very well acquainted with the region around Carillon, hence I am hoping for victory or at least that they will not inflict any great damage on us.

I am extremely busy here. Quite a lot has been done, but as though by someone who knows nothing about military matters. I shall have to do a number of the jobs over again.[1487] However, I have said that things have been done well and it's now up to me to find solutions. I trust that before long I shall be able to give you news of New England; all the Indians seem well disposed, even the Têtes Plates who have sent wampum collars asking that our nations should make peace with them and who are volunteering to wage war against the Englishman with whom they seem to be displeased.[1488]

But if we are not sent any paper, we shall have to stop writing; there's not a quire[1489] of it left here.

I beg you, sir, to continue your kindnesses towards me. I shall always endeavor to deserve them through my attachment to everything that concerns you.

I am, sir, with respect, etc.

1485 Pouchot, in his *Mémoires*, (II, p. 7) recorded questioning prisoners taken on the Belle-Rivière on May 2 and 3.

1486 The St. Lawrence Valley and Québec.

1487 Pouchot complains in his *Mémoires* about the poor state of the earthworks of Fort Niagara. See note 514.

1488 Pouchot had news of the inclinations of the Têtes Plates from some Huron and Iroquois who had been in the vicinity of Pittsburgh. See note 538.

1489 The twentieth part of a ream of writing paper or 25 sheets.

To Chevalier de Lévis[1490]
Niagara, 27 June 1759

I am sending you, General, statements by the chiefs of the village of Sonnechio.[1491] They seem to be upset and apologize because they did not win a victory. La Militiere has been taken prisoner despite their opposition and they had a great deal of difficulty in saving Joncaire, who is not yet back. But he will not be lost. One of Chabert's enlisted men was killed. They looted his possessions and burned his hut. This has been a worrying uprising on the part of some of the Iroquois and it has frightened those on our side. I await your decision, General, as to how you wish me to deal with them. They are also dissuading as many of our nations as they can from making war. They told our Mississaugués, who were going to Fort Bull, that the English had hired them to stop others from passing.[1492]

M. Des Ligneris informs me that he has been told about six thousand English bound for the Belle-Rivière and that there are currently six hundred at Fort Pétersbourg,[1493] where they are holding council. He adds that men from all our nations have gone there. They are saying that if we wish to withdraw, they will go back over the mountains, if not, they will attack Fort Machault.[1494]

M. Des Ligneris will send out some raiding parties against the supply convoys but I am afraid that I may be obliged to bring all men

---

1490 Published in Casgrain, *Collection,* X, pp. 121-22. This translation is by Michael Cardy.

1491 The Seneca village at Geneseo on the Genesee River. Tanner, p. 41. The chiefs had sent wampum collars to Pouchot apologizing for the capture of Lieutenant Honoré Dubois de La Milletière on their territory. See note 561.

1492 The Iroquois attack on Joncaire and La Milletière had occurred sometime before June 17. See note 556. Lieutenant Honoré Dubois de La Milletière of the Languedoc Regiment (notes 555-556) was captured. Captain Philippe-Thomas Chabert de Joncaire (note 517) escaped and eventually made his way to Fort Niagara. "Chabert" was his brother, Lieutenant (1757) Daniel-Marie Chabert de Joncaire de Clausonne (note 331). The unfortunate soldier was presumably from Chabert's detachment at Fort Little Niagara.

1493 Pittsburgh.

1494 The French post at Venango (Franklin, Pennsylvania) where de Lignery and his troops were concentrated.

and equipment back here before everyone has joined up with him. I have heard from Indian intelligence that they are currently coming down the river. I cannot, though, order any operation before knowing how large this body of men is.[1495] I have told M. de Portneuf[1496] to stop at Presqu' Isle all Frenchmen and Indians who reach him there in order that I may be able to order them back as soon as I receive any definite news.

Our greatest problem is supplies, of which we are very short. If some do not get through to us very soon, our army will be forced to disband and throw itself on God's protection. I must advise you of this, General. If only you could get some to us.

I am, sir, with the deepest respect, etc.

To Major General Jeffery Amherst[1497]

Albany, 14 August 1759

Sir,

The Niagara garrison has arrived here in accordance with your orders.[1498] Allow me, Sir, to request your kind attentions on its behalf. I venture to flatter myself that you will not refuse them to us as well as to the officers who were members of it. I beg you to recommend us to the Governor of New York.[1499] We have nothing but praise for the attentions and courtesies of Mr. Dufes, Captain in the Royal Americans,[1500] and Mr. Grent, Captain in the Royal Scottish, and of all the other officers under whose conduct we have been.[1501]

1495 Indian reports of increasing enemy activity had caused Pouchot to investigate Oswego more carefully. A party of Mississaugas, sent to obtain information about English movements on the Oswego River, returned to Fort Niagara on June 27, the day Pouchot wrote this letter. They reported no sign of the enemy, but must, as Pouchot later claimed, have only narrowly missed the advance elements of John Prideaux's army destined to besiege Fort Niagara. See note 566.

1496 Captain Pierre Robinau de Portneuf (note 549), commandant of Fort Presqu' Isle.

1497 Major General Jeffery Amherst (note 533) commanded English forces in North America.

The original of this letter is in the Amherst Papers, Public Record Office, WO 34/10, ff. 85-86. This translation is by Michael Cardy.

1498 The prisoners had arrived in Albany by August 12. See note 746.

1499 Lieutenant Governor James De Lancey (1703-1760). *DAB,* III, pp. 212-13.

Near the Portage, a lieutenant in the Marines named Larminas made his escape. He is an officer new to this territory and a rash young man. As soon as I was made aware of it, I informed the officers of our escort in order to express to them how much this event had affected us. We have in consequence lost our freedom that I hope you will be kind enough to have restored to us, as always on our word of honor.[1502]

I am, Sir, with deep respect your very humble and obedient servant

Pouchot de Maupas
Captain in the Bearn Regiment,
Commandant[1503]

1500 Captain George De Fez of the 60th (Royal American) Regiment of Foot commanded the prisoner escort from Oswego to Fort Stanwix (note 707).

1501 William Grant (fl.1745-1778), captain (1758) in the 42nd (Royal Highland) Regiment of Foot. Richards, p. 69; *Army List,* 1759. Grant was yet another officer encountered by Pouchot who had been wounded at Carillon in 1758. *DRCHSNY,* X, p. 729.

Captain Grant was in charge of the prisoners on the march from Fort Stanwix to Albany during which time Lieutenant Larminat made his escape. Webster, *Jeffery Amherst,* p. 162.

1502 The escape of Lieutenant Jean-François de Larminat (note 581), who had arrived in New France only in 1757, took place on the Mohawk River. Amherst believed that Larminat had slipped away at Fort Hunter. Webster, *Jeffery Amherst,* p. 162. Another account says near Fort Herkimer, and Pouchot would seem to confirm this by his reference to the Little Falls portage. The lieutenant was accompanied by a commissary and two soldiers. *New York Mercury,* Aug. 20, 1759. On August 21 a teamster from the English post at Lake George discovered the starving fugitives along the upper Hudson River near the modern site of Glens Falls, New York. A detachment was sent out from Fort Edward, and the Frenchmen were apprehended the following day. *Boston Gazette,* Sept. 10, 1759; Montresor to Amherst, Aug. 22, 1759, Amherst Papers/77; Scull, p. 91.

After some confusion, it was determined that the men, whom Amherst listed as a lieutenant, a commissary, a sergeant, and two privates, had escaped from Captain Grant's detachment. They were confined at Fort Edward. Webster, *Jeffery Amherst,* p. 162. Larminat was in a dangerous situation. "If this Lieut. Signed the Capitulation he deserves to be hanged," the general wrote, "or if he should turn out to be the Lieut. that deserted on the march, he deserves the Same fate." Amherst to Montresor, Amherst Papers/80.

On September 10 Amherst received a letter from Larminat stating his reasons for violating the capitulation of Niagara and attempting to escape. The lieutenant requested clemency. Larminat to Amherst, n.d. (received Sept. 10, 1759), Amherst Papers/78. Amherst relented. On November 29, as Pouchot and the other Niagara prisoners were moving north to be exchanged, he ordered Larminat and the eight French then held at Fort Edward to join them and return to Canada. Webster, *Jeffery Amherst,* p. 194.

1503 General Amherst received Pouchot's letter at Crown Point on August 18th. He noted that the French captain "disapproved much" of Larminat's violation of the articles of capitulation. Amherst Journal, Knox, III, p. 53; Webster, *Jeffery Amherst,* p. 158. The

...(continued next page)...

To Major General Jeffery Amherst[1504]
Jamaica, 29 August 1759

Sir,

I have the honor to send you some letters which the officers who are here and I, Sir, beg you to forward to Canada. The desire we have, Sir, to pay back the money so kindly advanced to us to find a means of covering our subsistence compels us to ask this favor of you.[1505] We venture to flatter ourselves that you will not refuse us.

I am, Sir, with deep respect your very humble and very obedient servant,

Pouchot
Captain in the Béarn
Regiment[1506]

English general responded from Crown Point on September 7. He wrote that the prisoners had been remanded to the Governor of New York who would show every courtesy permissible under the circumstances. Anthony Wheelock, a former officer of the 27th Regiment of Foot, was to serve as commissioner of prisoners. He would be responsible for their good treatment and would provide advances for subsistence. Amherst made no mention of Larminat or the impact of his escape on the prisoners, probably because the lieutenant had, by this time, been apprehended. Amherst Papers/81.

Anthony Wheelock (fl.1745-1760) had been a captain (1747) in the 27th Regiment of Foot. He resigned his commission, and his company was obtained by William Stewart on April 17, 1759. Notation on the copy of the 1759 *Army List* in the William L. Clements Library.

1504 The original letter is in the Amherst Papers, Public Record Office, WO 34/10, f. 87. This translation is by Michael Cardy.

1505 Pouchot refers to the official subsistence money noted in Amherst's letters of September 7 and 11. Pouchot, and presumably some of his officers, had also received personal loans from English officers. *Mémoires,* II, p. 127. This money was necessary because the officers had to obtain their own food and lodgings. In 1760 "J.C.B." recorded French prisoners, apparently enlisted men, boarding at an inn on Broadway at their own expense. Stevens, *Travels,* p. 122.

1506 Amherst replied from Crown Point on September 11 that he had forwarded the letters to Montréal the day before. On the matter of subsistence, the general informed Pouchot that he had ordered Captain Wheelock to reimburse Lieutenant Governor De Lancey the 50 *piastres* advanced to each French officer. Amherst Papers/81.

On September 10 Amherst sent "a large Packet from the prisoner officers to their Relations &c." with a French officer who had approached Crown Point under a flag of truce. Webster, *Jeffery Amherst,* p. 167. The officer was Bernard La Tour d'Asserat (fl.1744-1760), captain (1746) in the La Reine Regiment. "Officiers du régiment de la Reine", p. 385.

To Major General Jeffery Amherst[1507]

Jamaica, 30 October 1759

Sir

If the length of the campaign were to prevent the planned exchange, could I ask you, Sir, without indiscretion for an authorization for a French officer who would travel on parole to Canada under whatever conditions it might please you, Sir, to grant him in order there to settle some minor matters we have concerning our respective armies. This, Sir, will add further to the obligations we shall have towards you for the kindnesses that you are good enough to show us. We are deeply sensible of them through the good offices of Mr. Willyocks[1508] who, in following your intentions, fulfills them admirably.

I am, Sir, with profound respect, your most humble and obedient servantPouchot

Pouchot
Captain in the Béarn
Regiment[1509]

To Major General Jeffery Amherst[1510]

Fort Levis, 25 August 1760

Sir

I have just this very moment realized that the fort where I have the honor to command on behalf of the King my master is under fire with red-hot shot. I did not think I deserved from your Excellency unjust treatment.

I venture, Sir, to entrust myself to your power in questions of war and I do not believe I have placed myself in the position of a

1507 The original letter is in the Amherst Papers, Public Record Office, WO 34/78, f. 110. This translation is by Michael Cardy.

1508 Wheelock.

1509 No response from Amherst has been found, probably because one was no longer necessary. On November 2 a French flag of truce had arrived at Crown Point bearing satisfactory terms for an exchange of prisoners.

[illegible].[1511] If necessity placed me in the position of surrendering the fort I hope to deserve favorable treatment if you wish me to request conditions [of surrender] from you.

I am with very deep respect, Sir, your very humble and very obedient servant,

Pouchot[1512]

1510 The original letter is in the Amherst Papers, Public Record Office, WO 34/101. This translation is by Michael Cardy.

This letter and the circumstances surrounding it are described in the *Mémoires*, II, pp. 280-81 (note 936). Pouchot requests terms in a very oblique manner, and his writing seems to reflect the stress and fatigue of the bombardment sustained over the preceding several days.

1511 The word is illegible. Based on Pouchot's published account, it is probably "rebel" or a similar term. The captain reported that he had complained to Amherst "about this manner of making war, which was only used against rebels and not against a brave garrison." Pouchot, *Mémoires*, II, p. 280.

1512 Amherst was not inclined to dicker about Pouchot's intent. He reported that the French captain beat a parley and sent this letter which was immediately answered. Webster, *Jeffery Amherst*, p. 239. The general demanded surrender and enclosed three abrupt conditions (note 936). Amherst to Pouchot, Aug. 25, 1760, Amherst Papers/101. Although Pouchot claims in his *Mémoires* (II, p. 280) that he was allowed half an hour to consider what he called "a kind of surrender document," the general's letter offers only ten minutes. The terms were accepted.

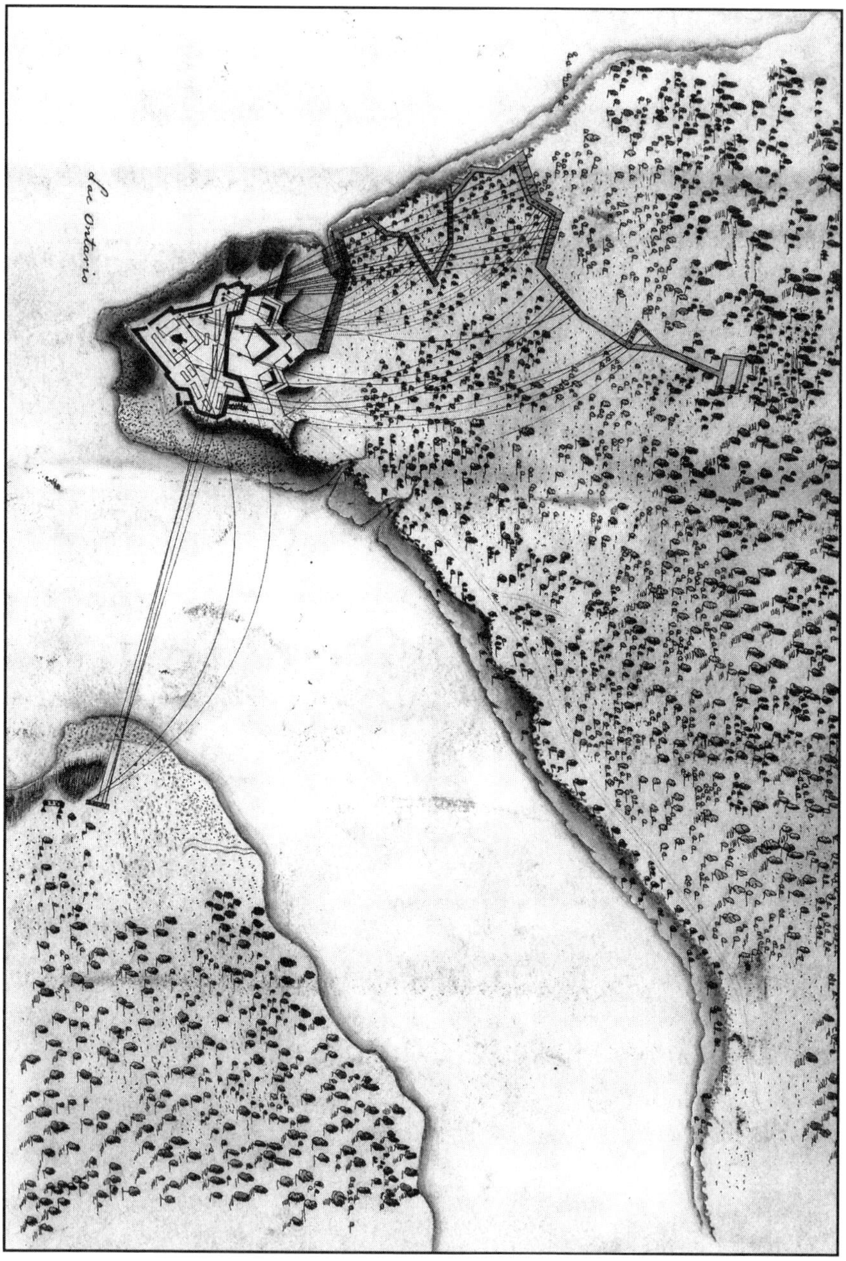

This untitled plan provides a graphic impression of the intense bombardment endured by Fort Niagara's garrison in the final days of the 1759 siege. Charles Baudouin made this copy from a French original in 1933. *Courtesy, National Archives of Canada (C8679).*

Appendix B: Anonymous Journal Of The Siege Of Niagara

On August 20, 1759, at the height of excitement about the conquest of the important French posts of Ticonderoga, Crown Point and Niagara, a translation of an anonymous journal appeared in the pages of the *New York Mercury*. It had been kept by a member of the Niagara garrison and recovered by the English following the capitulation. The editor of the *Mercury* informed his readers that "The above with some Letters were found in an Embrasure, after we were in Possession of the Fort, since which translated, and the Original given to SIR WILLIAM JOHNSON."

The reported circumstances of the journal's recovery might very well have been accurate, and the document could have been lost, discarded or hidden by its author during the final hours of the siege. But it is just as likely that the writer's baggage was not secured in the fort's powder magazine, and that his journal was part of the plunder carried off by the victors. "Everything that had not been locked away there was taken, either by the English officers or by soldiers designated for the purpose," Captain Pouchot reported of the aftermath of the surrender.[1513]

Regardless of the manner by which it was obtained, this journal provided the first general knowledge in the English colonies of occurrences in the besieged garrison. It accordingly received wide attention. Following its publication in the *Mercury,* the translation was soon repeated by a number of colonial newspapers including the *Pennsylvania Gazette* of August 23, the *Boston Evening Post* of August 27, the *Maryland Gazette* of August 30, and, belatedly, the *Boston Gazette* of September 10. It is likely that it was printed in others as well.

Years after it had ceased to be news, the anonymous journal of the siege of Niagara became an early historical source documenting one of the significant events of the French and Indian War. It first reappeared as an appendix to Volume I of William Smith's *History of*

1513 Pouchot, *Mémoires,* II, p. 119.

Canada.[1514] In 1850 the journal, along with several colonial newspaper accounts describing the siege of Fort Niagara, was printed in Oliver Turner's *Pioneer History of the Holland Purchase.*[1515] It surfaced again in 1869 in the *Historical Magazine.*[1516]

The anonymous Fort Niagara journal was thus readily available to English-speaking American historians long before Pouchot's account, the first published translation of which did not appear until 1858 when that part of the *Mémoires* covering July 6-25, 1759 was included in Volume X of *Documents Relative to the Colonial History of the State of New-York.*[1517] Franklin Hough's complete translation was published eight years later. Until 1858, therefore, the anonymous journal provided the only first-hand account in English of events within the besieged fortress.

The fate of the original manuscript journal is unknown other than that it was presented to Sir William Johnson in 1759. There is no mention of it in his published papers nor in the list of documents destroyed in the New York State Library fire of 1911. The only known version is the translation first printed in the *New York Mercury.*

The translated and published journal does not identify its author. Nor does it indicate whether it was written by an officer or an enlisted man, although the nature of the document and the writer's obvious access to current official information in the garrison strongly imply the former. The only clue is provided by an incident on the night of July 12-13. The journalist recorded for that period that "I went with 30 Men to observe where the Enemy wrought." The implication is that he, an officer, led this party on its reconnaissance of the English siege lines. Pouchot's *Mémoires* state that, on the same night, he "sent eight volunteers, under the orders of M. Cornoyer, who approached right up close to the battery."[1518] It seems unlikely that the journal was kept by anyone other than this officer.

Lieutenant Jacques-Ange Hertel de Cournoyer was born on Isle Royale in 1727. He followed a typical career path for a colonial officer,

---

1514 William Smith, *History of Canada* (2 vols.; Québec, 1815), I, Appendix 8, pp. 20-23.

1515 Oliver H. Turner, *Pioneer History of the Holland Purchase* (Buffalo, 1850), pp. 209-13.

1516 *Historical Magazine,* V (2nd Series), No. 3 (March 1869), pp. 197-99.

1517 "Journal of the Siege of Fort Niagara", *DRCHSNY,* X, pp. 977-92.

1518 Pouchot, *Mémoires,* II, p. 72.

serving first on the Isle Royale establishment where he was appointed a cadet at the age of 16 in 1743. Cournoyer was promoted ensign *en second* on April 15, 1750 and ensign *en pied* April 1, 1754. He transferred to the establishment of New France in 1757 and received his lieutenancy on January 1, 1759.[1519]

Cournoyer's wartime activities are not particularly well documented. He appears to have served primarily on the Lake Ontario and Lake Champlain frontiers and to have been in Canada prior to his transfer to the New France establishment. The ensign was assigned to Lieutenant Gaspard-Joseph Chaussegros de Léry's expedition against Fort Bull in 1756, although illness forced him to withdraw before the attack.[1520] He appears to have marched with Captain Rigaud de Vaudreuil on his winter raid against Fort William Henry in 1757.[1521] Cournoyer probably fought with Montcalm's army in the Lake Champlain campaigns of 1757 and 1758 before going to Niagara with Pouchot in 1759. He was not among the Niagara officers exchanged in December 1759 and remained a prisoner in New York.[1522] The lieutenant appears to have been mistakenly listed as killed in action in 1760.[1523]

Cournoyer was returned to France in 1761, probably aboard the same vessel as Pouchot. There, in 1766, he obtained a commission as captain in the troops of Saint-Domingue. Cournoyer was made *aide-major* (assistant adjutant) of the Regiment Du Cap in 1772 and obtained the coveted Cross of St. Louis in 1773. He was promoted battalion commander (*chef de battalion*) in his regiment in 1784 and later returned to Paris where he died in 1789. During his colonial service Cournoyer had married Monique Amidieu du Clos in 1774.[1524]

Annotation of the anonymous journal follows the same form as that for Pouchot's *Mémoires.* Otherwise, the text, including spelling and punctuation, conforms to that printed in the *New York Mercury* of August 20, 1759.

1519 Fauteux, *Les Chevaliers,* p. 201.

1520 Composition of de Léry's Detachment, Casgrain, *Collection,* VIII, p. 114; Journal of the de Léry Campaign, *Ibid.,* XI, pp. 58, 63; Hagerty, p. 37.

1521 Relation of the Expedition to Fort George, Casgrain, *Collection,* XI, p. 77.

1522 "List of the French Officers Prisoners of War", Amherst Papers/8; "List of the French Prisoners Sent to Canada", Dec. 14, 1759, Amherst Papers/8.

1523 "Tableau ... des Officiers de la Marine", Brymner, 1886, p. clxxxiv.

1524 Fauteux, *Les Chevaliers,* p. 201.

JOURNAL
Of the Siege of Niagara, translated from the French

6th July 1759.

About 7 at Night a Soldier, who was hunting, came with all Diligence to acquaint Monsieur Pouchot, that he had discovered at the Entrance of the Wood, a Party of Savages, and that they had even fired on some other Hunters. Monsieur Pouchot immediately sent Mr. Selviert, Captain in the Regiment of Roussillon,[1525] at the Head of one Piquet, a dozen Canadian volunteers preceded them, and on their coming to the Edge of the Woods, a Number of Indians fired upon them, which they returned, and were obliged to retire: They took Messieurs Furnace and Aloque, Interpreters of the Iroquois, two Canadians and two other Gentlemen.[1526] They made another Discharge and retired. Monsieur Pouchot fired some Cannon upon them. Monsieur Selviert lay all Night, with 100 Men, in the Demylune, and the rest of the Garrison was under Arms on the Ramparts 'till Midnight.

Saturday 7th July.

We perceived 7 Barges on the Lake, a League and a Half Distance from the Fort; we judged by that it was the English come to besiege us: Monsieur Pouchot ordered the General[1527] to be beat, and employed all Hands to Work on the Batteries, to erect Embrasures, all being En Barbet before. He immediately dispatched a Courier to Monsieur Chevert,[1528] to give him Notice of what happened; he also sent out Monsieur Leforce, Captain of the Schooner Iroquoise,[1529] to destroy the English Barges where he cou'd find them. All that Day

---

1525 Captain Serviès of the Royal Roussillon Regiment (note 511).

1526 Regarding Furnace and Aloque, see note 574. The "gentlemen" were probably the "Sieur Saumandre" and "Blois" named in La Force's journal (Appendix C). The former was a Marine cadet, recorded as "Soumandre" in a list of prisoners in Sir William Johnson's journal. *SWJP,* XIII, p. 115.

Pouchot says five men were captured. An English writer, possibly Chaplain John Ogilvie, also reported five prisoners. Letter from an officer, July 14, 1759, Severance, II, p. 291.

1527 This was a particular drum beating, here used to signal the alarm. "The general, is to give notice to the troops that they are to march." Smith, *Dictionary,* p. 79.

1528 Lieutenant Daniel-Marie Chabert de Joncaire de Clausonne (note 331).

1529 René-Hippolyte La Force (note 573).

several Savages shewed themselves on the Edge of the Deserts.[1530] Monsieur Leforce fired several Cannon Shot at them; and perceived they were working at an Entrenchment at the Little Swamp, which is a League and a Half from the Fort. The Guards this Night as the Night before.

Sunday 8th July.

The Schooner continued to cruize and fire on the English Camp. About 9 in the Morning an English Officer brought a Letter from Brigadier Prideaux, to Monsieur Pouchot, to summons him, proposing him all Advantages and good Treatment; all which he very politely refused, and even seemed to be unwilling to receive the English General's Letter.[1531] The Remainder of this Day the English made no Motions.

Tuesday 10th.

At 2 o'Clock all our Men were on the Ramparts, and at Day-break we perceived they had opened their Trenches, at the Entrance of the Wilderness, at about 300 Toises from the Fort; we made a very hot fire upon them all Day. M. Chabourt[1532] arrived with the Garrison of the little Fort and seven or eight Savage Iroquoise and Missagoes. Monsieur Pouchot went to pallisade the Ditches: The Service as usual, only the Addition of two Officers to lie in the covered Way. About 11 o'Clock at Night Orders were given to make all the Piquets fire from the covered Way, to hinder the Workmen of the Enemy. M. Laforce sent his Boat on Shore for Monsieur Pouchot's Orders.

Wednesday 11th July.

The Works continue on both Sides. At Noon a Party of about 15 Men, Soldiers and Militia, went very nigh the Trenches of the Enemy, and perceived them sally out between 4 and 500, who came towards them at a quick Pace, but they were stop'ed by our Cannon. They began on the other Side of the Swamp, which is to the Left of their

---

1530 The cleared ground or *désert* around the fort.

1531 This was the summons carried by Captain Walter Rutherford (note 603).

Neither Pouchot nor the anonymous journalist has an entry for Monday, July 9. See note 599.

1532 Chabert.

1533 The French journalist surely used the *toise* as a measure of distance. The eighteenth century English translator rendered it here, and subsequently, as the more familiar "yard."

Trench, another about 20 Yards;[1533] and at 5 o'Clock they began to play two Granadoe Royal Mortars.[1534] At six o'Clock two Savages of the Five Nations, who were invited by one Cayendesse[1535] of their Nation, came to speak to M. Pouchot; the Firing ceased on both Sides during this Parly. At 10 O'Clock we began to Fire again, and then we found the English had eight Mortars.

Night between the 11th and 12th.

The Enemy ran their Paralel from their first Trench to the Lake Side, where it seemed they intended to establish a Battery. At 2 in the Afternoon four Chiefs of the Five Nations, came to us on Parole, and said they were going to retire to Belle-Famille. The Enemy wrought the rest of that Day, and perfected their Night's Work. M. Laforce had Orders to proceed to Frontenac, and to return immediately: In the Night between the 12th and 13th they fired many Bombs: I went with 30 Men to observe where the Enemy wrought.[1536]

Friday 13th July.

A Canoe arrived from Monsieur DeVille,[1537] to hear how we stood at this Post (or rather for the Canada Post.)[1538]

The Enemy threw a great many Bombs all this Day, and continued to Work to perfect their Trenches: We fired a great many Cannon Shot. Many of their Savages crossed the River, and desired to speak with us; there was but two of those Nations with us. I went out with five Volunteers to Act as the Night before. The Enemy fired no Bombs 'til about Midnight.

Saturday 14th July.

At Day-break we found they had prolong'd their Trenches to the Lake Side, in spite of the great Fire from our Cannon and Musquetry, during the Night, and perfected it during the Day Time; they have

---

1534 "Royals" were mortars or howitzers of 5 1/2-inch caliber (note 916). Prideaux's army had two Royal howitzers but no mortars of that size. Dunnigan, *Siege,* pp. 136-37.

1535 Pouchot renders his name as Kaendaé (note 615).

1536 This incident identifies Lieutenant Cournoyer as the author of the journal. "M. Pouchot immediately sent eight volunteers, under the orders of M. Cornoyer, who approached right up close to the battery." Pouchot, *Mémoires,* II, p. 72.

1537 Captain Alexander Dagneau Douville (note 714), commandant of Toronto.

1538 Presumably he means to take mail and dispatches to Canada.

placed four Mortars, and threw many Bombs. All our Garrison lay in the covered Way and on the Ramparts.

Sunday 15th July.

In the Morning we perceived they had finish'd their Works began the Night before: During the Night they threw about 30 Bombs; the rest of the Day and Night they threw a great many, but did not incommode us in any Shape.

Monday 16th July.

At Dawn of Day we spied, about Half a League off, two Barges, at which we discharged some Cannon, on which they retired: They have already disabled us about 20 Men. All our Men lye on Beaver or in their Cloths and armed. We do what we can to incommode them with our Cannon.

Tuesday 17th July.

Until Six this Morning we had a thick Fog, so that we could not discern the Works of the Enemy; but it clearing a little up, we saw they had raised a Battery of three Pieces of Cannon and four Mortars on the other Side of the River; they began to fire about 7 A.M. and M. Pouchot placed all the Guns he could against them: The Fire was brisk on both Sides all Day, they seemed most inclined to batter the House where the Commandant lodges.[1539] The Service as usual for the Night.

Wednesday 18th July.

There was a great firing on the preceding Day: we had one Soldier dismembered and 4 wounded by their Bombs.

Thursday 19th July.

At the Dawn of the Day we found the Enemy had began a Parrallel about 80 Yards long in Front of the Fort. The Fire was very great on both Sides. At 2 P.M. arrived the Schooner Iroquoise, from Frontenack, and laid a-breast of the Fort, waiting for a Calm, not being able to get in the Enemy having a Battery on the other Side of the River: Monsieur Pouchot will have the Boat on Shore as soon as the Wind falls.

Friday 20th.

The English have made a third Parrallel, toward the Lake; they are to-day about 160 Yards from the Fort: They cannot have worked

---

1539 Fort Niagara's "French Castle."

quietly at the Sappe, having had a great Fire of Musquetry all Night long, which they were obliged to bear. During the Day they made a great Firing with their Mortars, and they perfected their Works began the Night of the 19th to the 20th. We had one Man killed and four wounded. The Fire of the Musquetry was very hot on both Sides, 'til eleven at Night, when the Enemy left off, and we continued ours all Night. Two Canoes were sent on board the Schooner, who are to go to Montreal and Tironto.

Saturday 21st.

During the Night the Enemy made a fourth Parrallel, which is about 100 Yards from the Fort, in which it appears they will erect a Battery for a Breach in the Flag Bastion. They have hardly fired any Cannon or Bombs in the Day, which gives Room to think they are transporting their Cannon and Artillery, from their old Battery to their new one. The Service as usual.

Their Battery on the other Side, fired but little in the Day. The Schooner went off to see her two Canoes over to Tironto, one of which is to post to Montreal, and from thence she is to cruize off Oswego, to try to stop the Enemy's Convoys when on their Way. The Company of Voluntiers[1540] are always to pass the Night in the covered Way.

Sunday 22nd.

All the Night was a strong Conflict on both Sides. We had one Man killed by them and by our own Cannon. We fired almost all our Cannon with Cartridges.[1541] They worked in the Night to perfect all their Works began the Night before. The Enemy began to fire red hot Balls in the Night; they also fired Fire-Poles.[1542] All Day they continued at Work to establish their Batteries. They fired as usual, Bombs and Cannon. The Service as usual for the Night of the 22d

---

1540 This "Company of Volunteers" might have been the group formed for Cournoyer's scouts on the nights of July 12-13 and 13-14 or a unit organized to help bolster flagging morale in the defenses.

1541 The English translator here probably rendered *cartouche* or "case shot" as "cartridge." See note 656. Pouchot noted for this night that three cannon placed in the covered way each fired 50 rounds of grapeshot at the English.

1542 The translator's "Fire-Poles" were probably *pots de feu* or "fire pots" in the original. See note 935 for a description of these incendiaries.

and 23d. They worked hard to perfect their Batteries, being ardently sustained by their Musquetry.

Monday 23d.

We added two Pieces of Cannon to the Bastion of the Lake, to oppose those of the Enemy's Side. At 8 A.M. four Savages brought a Letter from Monsieur Aubry[1543] to Monsieur Pouchot, by which we learn, that he was arrived at the great Island, before the little Fort,[1544] at the head of 2500 Men, half French and half Savages: Monsieur Pouchot immediately sent back four Savages with the Answer to M. Aubry's Letter, informing him of the Enemy's Situation. The Savage, before he came in, spoke to the Five Nations, and gave them five Belts to engage them to retire from the Enemy: They saw Part of the Enemy's Camp, and told us the first or second in Command was killed by one of our Bullets, and two of their Guns broken, and one Mortar:[1545] We have Room to hope, that with such Succours we may oblige the Enemy to raise the Siege, with the Loss of Men; and as they take up much Ground, they must be beat, not being able to rally quickly enough.

At [blank] P.M. they unmasked another Battery of — Pieces of Cannon, 3 of which were eighteen Pounders, the others 12 and 6. They began with a brisk Fire, which continued two Hours, then slackened: About five P.M. we saw a Barge go over to Belle Famille, on the other Side of the River, and some Motions made there: One of the four Savages which went off this Morning, returned, (his Porcelaine, i.e. Wampum) he had nothing new. The Service of the Night as usual. We worked hard to place two Pieces, 12 Pounders, on the Middle of the Curtains, to bear upon their Battery.

1543 Captain Charles-Philippe Aubry (note 468).

1544 Grand Island, New York in the Niagara RIver above Niagara Falls and opposite Fort Little Niagara at the head of the portage.

1545 This was the garrison's first intelligence of the deaths of Colonel John Johnston and Brigadier John Prideaux on the evening of July 20. See note 644. Prideaux (note 125) commanded the besiegers. John Johnston (note 125) was colonel of the New York Regiment and Prideaux's second in command. Command of the army devolved to Sir William Johnson. Dunnigan, *Siege,* pp. 71-75.

The rest of the Indian's information was quite accurate. After the siege the English reported that a number of bronze six-pounders were "quite spoiled" by overheating from rapid firing and that one of the valuable iron 18-pounders was "broke at the Muzle." Return of Ordnance, Aug. 4, 1759, Gage Papers/3; Gage to Amherst, Aug. 21, 1759, Gage Papers/ "Oswego Letterbook."

Tuesday 24th July.

The Enemy began their Fire about 4 o'Clock this Morning, and continued to fire with the same Vivacity for the rest of the Day. At 8 A.M. we perceived our Army was approaching, having made several discharges of Musquetry at Belle Famille.

At 9 the Fire began on both Sides, and lasted half an Hour: We wait to know who has the Advantage of those two: At 2 P.M. we heard, by a Savage, that our Army was routed, and almost all made Prisoners, by the Treachery of our Savages: When immediately the English Army had the Pleasure to inform us of it by summoning us to surrender.

The above with some Letters were found in an Embrasure, after we were in Possession of the Fort, since which translated, and the Original given to SIR WILLIAM JOHNSON.

Appendix C: Captain La Force's Journal Of The Siege Of Niagara

While Pouchot's *Mémoires* and the anonymous, or Cournoyer, journal detail events within Fort Niagara during the siege of July 6-25, 1759, a third work presents an altogether different perspective. The journal of Captain René-Hipployte La Force, master of the armed schooner *Iroquoise,* is a French account from outside the walls of the besieged fortress.

The French possessed two especially powerful weapons for the defense of Fort Niagara: a pair of new, well-armed and fully crewed warships. These could cruise at will on Lake Ontario, and the English had nothing with which to oppose them. Nor could they until great amounts of time, labor and material had been expended to construct vessels at Oswego of equal or greater strength. The existence of *Iroquoise* and her near-sister, the brig *Outaouaise,* was of great concern to Brigadier John Prideaux as he advanced on Oswego and planned his water-borne strike against Fort Niagara. Information from prisoners and deserters repeatedly confirmed the existence and the readiness of a pair of armed "sloops" on Lake Ontario.[1546]

Despite the threat presented by these vessels, Prideaux plunged forward from Oswego on July 1. Luck was with him, and, five days later, his army landed safely at Four Mile Creek east of Fort Niagara. The large and nearly defenseless flotilla of whaleboats and bateaux had completely escaped discovery by the French warships.[1547]

If Prideaux was favored by fortune, then Pouchot suffered miserable luck. His significant naval advantage was neutralized by circumstances and by what can only be considered a lack of initiative on the part of his naval officers. In June, a scouting party of Mississaugas sent aboard *Outaouaise* to reconnoitre the Oswego River missed the vanguard of Prideaux's army by, at most, a few days. Then the brig was

---

1546 See "Examination of a French Prisoner", May 29, 1759 and a similar document with an account by a captured Indian, Elias Pegan, of June 27. Both are in Gage Papers/2.

1547 Dunnigan, *Siege,* pp. 35-36.

caught in a squall off Oswego, losing her mainmast and bowsprit and necessitating a return to Oswegatchie.[1548] Repairs do not appear to have been completed until after Niagara had fallen.

That left only the schooner *Iroquoise,* commanded by La Force. Admittedly, the logistical needs of the Niagara garrison and French forces on the Ohio required the use of this vessel as a transport as well as a warship. But, when La Force anchored in the Niagara River on the afternoon of July 6, he reported no sign of the enemy at Oswego. How he had missed a huge fleet of English boats that had spent six days coasting along the exposed south shore of Lake Ontario in generally good weather is difficult to understand.

Pouchot, in his *Mémoires,* is implicitly critical of La Force, writing that "this was a misfortune, since these ships were armed for no other purpose." He nonetheless offers a plausible reason for La Force's lack of vigilance: when both ships were in service, one was to be on patrol off Oswego while the other transported men and supplies up and down the lake.[1549] It is possible that La Force had had little or no contact with Captain La Broquerie following the storm that damaged *Outaouaise* and might even have been unaware that the other warship was off station. It is more likely, however, that the demands of supplying Niagara prevented him from taking up the Oswego patrol. Wind and weather conditions might also have made it imprudent to sail *Iroquoise* too close to the dangerous south shore of the lake.

Once the English had appeared at Niagara, Pouchot attempted to utilize La Force's powerful warship in the defense of his post. Despite the schooner's mobility and firepower, however, her guns had little effect on the enemy once they were safely ashore. *Iroquoise* presented a greater threat to the besiegers by her potential for disrupting their communications with Oswego. Pouchot eventually admitted this and dispatched La Force on that mission, but the schooner made no contact with English boat convoys, and supplies moved steadily from Oswego to the siege lines.[1550] Once Niagara had surrendered, the French warships did little more than guard the head of the St. Lawrence and periodically reconnoitre the enemy at Oswego and Niagara.[1551]

1548 Pouchot, *Mémoires,* II, p. 38.

1549 *Ibid.,* pp. 40-41.

1550 See note 627.

1551 Vaudreuil to Lévis, Sept. 15, 1759, Casgrain, *Collection,* X, pp. 49-51.

Although the English were cautious because of their presence, the vessels had no real impact on efforts to supply Fort Niagara's garrison for the winter.

Iroquoise and *Outaouaise* were employed just as timidly during 1760. On August 1 *Iroquoise* was badly damaged by grounding, never to sail again under French colors. Her captain, crew and armament were brought ashore to supplement the defenders of Fort Lévis.[1552] *Outaouaise* remained in service but, in spite of a spirited defense by Captain La Broquerie, she was trapped in the river, captured and incorporated into the English squadron.[1553]

Much of the responsibility for the lackluster performance of the French warships must be placed with René-Hippolyte Pépin *dit* La Force, master of *Iroquoise,* commander of the naval squadron on Lake Ontario and author of the journal translated below. La Force spent most of his life as a sailor and served as a naval officer in two wars. Biographical sketches of this officer must, however, be used with great care.[1554] These contain much confusion between René and his brothers Michel Pépin (b.1721), Pierre Pépin (1725-1760) and François Pépin (b.1727). All were called La Force. They had active careers on land and participated in widely separated events of the war.

Michel, eldest of the three La Forces, was involved in the earliest clash with the English. He was captured in 1754 during the Jumonville affair and remained in enemy hands, his exchange frustrated by the dispute over the violation of the Fort William Henry capitulation.[1555] Michel was finally on his way to Canada with Pouchot's party in 1759 when, near Saratoga on November 29, General Amherst "stopd Mons La Force whom Capt Wheelock had sent until I can learn whether Mr Van Braham may be exchanged for him."[1556] No arrangements had

---

1552 See notes 884 and 900.

1553 See note 905.

1554 See Yves Desloges, "La Force, René-Hippolyte", *DCB,* V, pp. 470-71; Robinson, pp. 177-81; Severance, II, *passim.* The latest effort to unravel and identify members of the complex La Force clan of the French and Indian War is Paul W. Truax, *La Force Descendants in North America* (Hinesburg, VT, 2001), pp. 12-38.

1555 Alberts, pp. 45-46; Montcalm Journal, Nov. 22, 1757, Casgrain, *Collection,* VII, p. 319.

1556 Webster, *Jeffery Amherst,* pp. 194-95. That the name of the officer captured in 1754 was Michel is confirmed in a list of receipts for money advanced to Canadian prisoners by Anthony Wheelock in which Michel La Force is described as a captain of militia. List in Wheelock to Gage, Aug. 22, 1763, Gage Papers/9.

...(continued next page)...

been made by the time of the capitulation of Montréal, and Michel probably returned home only in the autumn of 1760.

Another La Force, probably François, was *garde-magasin* at Carillon from as early as July 1756 until at least March 1757. It is unlikely that he still held the position in the spring of 1758 when the storekeepers at Chambly, St. Jean, St. Frédéric, and Carillon were dismissed, according to Bougainville, because their "thefts too glaring" and "rules to prevent them useless."[1557] La Force had probably left his storekeeper's post as early as March 1757 to take up more active duties as an Indian interpreter on Rigaud's expedition against Fort William Henry. During that summer's campaign a La Force was an interpreter for the mission Iroquois, including those of La Présentation, and he later led a party of Mississauga on Belestre's raid against German Flats in November. During the latter campaign he was identified as the interpreter from Fort Frontenac.[1558] When Frontenac was lost to the English the following summer, the list of employees taken there included one François "Papin", clerk to storekeeper Joseph Papin.[1559] The latter was almost certainly a relative of François. Both were immediately paroled.

Throughout this time, René La Force was on Lake Ontario. He had been born in 1728 and spent much of his youth at Fort Niagara where his father, Pierre Pépin *dit* La Force, was then storekeeper. By 1751 he was engaged in the merchant trade between Québec and Louisbourg. He probably gave up that occupation to command the naval vessels on Lake Ontario when appointed by Governor Vaudreuil in April 1756.[1560] During that summer's campaign he commanded the schooner *Marquise de Vaudreuil* and successfully led the French

---

Captain Jacob Van Braam was one of two English hostages given up to the French by the capitulation of Fort Necessity in 1754. He remained in Canada until 1760 (note 170).

1557 Instructions to Lévis, July 1756, Casgrain, *Collection,* VI, p. 18; Montcalm to Lévis, Oct. 26, 1756, *Ibid.,* p. 40; Montcalm Journal, March 4, 1757, *Ibid.,* VII, p. 166; Hamilton, *Adventure,* p. 203.

1558 Rélation, March 1757, Casgrain, *Collection,* XI, p. 78; Hamilton, *Adventure,* p. 150; Journal, 1757, Casgrain, *Collection,* XI, p. 130.

1559 "List of the King's Employees at Fort Frontenac", Aug. 27, 1758, Preston and Lamontagne, p. 260.

1560 Vaudreuil to Commandant of Frontenac, April 12, 1756, cited in Robinson, p. 179; *DCB,* V, p. 470.

squadron in battle and in support of Montcalm's army.[1561] He retained his position in 1757 and 1758.[1562] La Force narrowly escaped capture at Frontenac when, as captain of one of a pair of vessels, probably *Marquise de Vaudreuil,* he was trapped in the harbor by an English battery. The sailors abandoned ship and escaped in small boats, thereby avoiding inclusion in the capitulation.[1563] La Force soon thereafter found himself appointed to command *Iroquoise,* one of two new vessels constructed in the winter of 1758-1759. In this position he served during 1759.

The campaign of 1760 found La Force still in command of *Iroquoise.* His activities are detailed in Pouchot's *Mémoires.* He was taken prisoner by the surrender of Fort Lévis and went to New York with Pouchot and the rest of the garrison.[1564] René was among those Canadians sent home on October 3 after having taken the oath of allegiance to King George II.[1565]

René La Force was described by at least one contemporary as a "brave man."[1566] He was a competent sailor and personally courageous in the three actions in which he fought against the English between 1756 and 1760. His sailors earned Pouchot's esteem for their skill and spirited service on the Fort Lévis batteries.[1567] But La Force was not the aggressive naval officer needed for the defense of Lake Ontario from 1758 through 1760, and he failed to exploit the superiority of his powerful ships during all three campaigns. It is difficult to avoid the conclusion that, with greater vigilance and energy, he might have prevented the capture of Frontenac and Niagara and at least delayed

---

1561 "State of Our Naval Forces on Lake Ontario", 1756, Preston and LaMontagne, p. 254.

1562 Bigot to Lévis, Sept. 11, 1758, Casgrain, *Collection,* IX, p. 34.

1563 Vaudreuil to Massiac, Sept. 2, 1758, *DRCHSNY,* X, p. 823.

1564 "A Return of the State of the Garrison", Aug. 25, 1760, Amherst Papers/8. La Force gave his parole at Fort Lévis on August 25. This was necessary because he had not signed the articles of capitulation. Parole, Aug. 25, 1760, Amherst Papers/101.

1565 *New York Mercury,* Oct. 6, 1760. The Canadians were sent from New York to Montréal in the care of La Force and his fellow sailor, Joseph Boucher de La Broquerie. Wheelock to Gage, Nov. 5, 1761, Gage Papers/8.

1566 Courville, p. 76.

1567 Pouchot, *Mémoires,* II, pp. 274-75.

the fall of Fort Lévis. Although Pouchot's published criticism of La Force's 1759 performance was implicit, rumors of explicit dissatisfaction even reached the enemy. In the spring of 1760, Lieutenant Colonel Haldimand at Oswego heard "several times" from his Indian informants, that "Monsr Pouchaud and the officers of the garrison of Niagara had carried complaints against M. La Force who commanded the vessels last year & that the M de Vaudreuil had replaced him with another."[1568] The latter news was incorrect, and La Force, perhaps through influence, retained the senior naval appointment on Lake Ontario throughout the war.

La Force's life after 1760 is described in the *Dictionary of Canadian Biography.*[1569] He lived in Kamouraska, Québec from 1762 until 1766 and returned to Québec City in 1767 to engage in the West Indian trade. This occupied him until the outbreak of the American Revolution in 1775. He was then appointed captain of the town's militia artillery company but was soon commanding the vessel *Providence* in the St. Lawrence River. She was probably a craft of the "Naval Department" operated under the direction of the Governor of Canada.[1570] From 1778 until 1784 La Force once again served on Lake Ontario as master of the Naval Department schooner *Seneca* and at the Carleton Island shipyard. In 1780 he briefly commanded the Lake Ontario squadron once again, this time for the King of England. At the conclusion of the war, René La Force returned to the West Indian trade until at least 1788. He continued to hold a commission in the Québec militia and died at Québec on February 3, 1802.

The following journal, recording La Force's activities during the first week of the siege of Niagara, was originally published in French as an appendix to the Champlain Society's 1914-1916 edition of John Knox's *An Historical Journal of the Campaigns in North America for the Years 1757, 1758, 1759, and 1760.*[1571] The text was obtained from "a copy of the original Journal" then in the hands of M. René de Kerallain of Quimper, France. M. Kerallain's marginal notes were published with the journal. In 1917 Frank Severance included translations of a

---

1568 Haldimand to Amherst, June 1, 1760, Amherst Papers/9.

1569 *DCB,* V, pp. 470-71.

1570 For information on the British Naval Department see Blanco, I, pp. 179-80.

1571 Knox, III, pp. 233-41.

few passages in *An Old Frontier of France.*[1572] Severance worked from a copy in the Public Archives of Canada, probably the one used in the preparation of Knox.

In 1935 the Public Archives of Canada (now the National Archives of Canada) obtained a transcript directly from M. de Kerallain.[1573] No further information on the whereabouts of the original document had been discovered by the time the first Old Fort Niagara edition of Pouchot's *Mémoires* went to press in 1994. It was subsequently learned that La Force's Fort Niagara journal was sold at Sotheby's, London,on June 24, 1957. Less than four years later it was again offered for sale by the London firm of Henry Stevens, Son & Stiles. Purchased by the Burton Historical Collection of the Detroit Public Library on May 14, 1962, it is now a part of that institution's Bougainville Papers.[1574]

The manuscript La Force journal covers seven and one-half pages of 12 1/2 by 8-inch paper. It is probably a contemporary copy of La Force's original document. Aside from some very minor errors in transcription, the text of the journal as published in the 1914-1916 Champlain Society edition of Knox is identical. The translation that follows (pages 550-558) is by Michael Cardy. Annotation is in the same form as that for Pouchot's *Mémoires.*

1572 Severance, II, pp. 300-02.

1573 Information from Timothy Dubé, National Archives of Canada. The reference number of this transcript is MG 18 – N20.

1574 Correspondence files, Henry Stevens, Son & Stiles, Burton Historical Collection.

Log Written by the Sieur LaForce, Commander of the Vessels on Lake Ontario, Concerning the Siege of Niagara. Beginning 6 July 1759 Until the 14th of the Same Month

In the name of God and the Blessed Virgin Mary, etc.

On Friday 6 July, I arrived at Niagara at 6 o'clock in the evening, with a light wind from the North-East.

At 7:30 on the same evening, there arrived a soldier who reported that the Indians had taken 4 or 5 prisoners on the edge of the *désert* and that he alone had escaped. Immediately 18 or 20 volunteers from the fort went out to recover the dead. M. de Serviez,[1575] a captain in the Royal Roussillon Regiment, also sortied with a *picquet* of 50 regular troops in support of the first detachments. When these volunteers were three quarters way around the *désert,* they received a volley of musket fire from about 60 to 80 men. None of our men was killed. Our young soldiers returned fire but all in vain as they were surrounded by the troop of Indians and so they escaped towards the *piquet* of M. Serviez, who held his ground for awhile and then fell back on the fort to conform with an order received from M. Pouchot.

Killed or taken prisoner in this skirmish were the Sieur Saumandre, Blois the Iroquois interpreter and the son of the elder Forge. We also suffered 2 wounded, the Sieurs St. Martin and Pramont, who managed to get back to the fort.[1576]

It was our view that our adversaries were a party from the 5 nations who had come on a raid that had been threatened on Niagara for more than a month.

That morning, at daybreak on the 7th, the English appeared in the *désert* and talked a good deal among themselves. A number of them speak good French and they hurled a good many taunts at us. We then saw 6 barges which were leaving the Petit Marais a league to the east

---

1575 Captain Serviès of the Royal Roussillon Regiment (note 511).

1576 Saumandre was a cadet of the Marine and is so recorded in Sir William Johnson's list of prisoners. *SWJP,* XIII, p. 115. The "son of the elder Forge" might be the "Furnace" named in the anonymous journal (Appendix B). Pramont, a commissary, is discussed in note 574. St. Martin has not been further identified but might also have been an interpreter and the same man who translated for the Hurons of Detroit and Lorette under the Abbé Picquet at the siege of Fort William Henry in 1757. Hamilton, *Adventure,* p. 150.

of Niagara.[1577] They sailed out to examine the fort and I twice sent my longboat on reconnaissance on the orders of M. Pouchot. It reported that they had seen many barges and tents in the Petit Marais and many men at work in the area between the camp and the edge of the *désert.* I immediately received the order to set sail, wind permitting, to discover if the enemy was really there in force and if they had artillery; if they did not, I was to try and destroy their barges and to cruise up and down in front of their camp.

At noon, I sailed out of the River with the winds in the N.W. and arrived in front of the English camp. They saluted me with 2 cannonballs to which I responded with 4 shots. They continued their fire and so did I but their balls did not reach the ship since my orders are that I should not engage fire beneath their batteries. However, many of my balls fell in their camp.

At 7:30 in the evening, I sailed close to their advanced guard and fired 3 shots at them, though I do not know what effect they had. At sunset, I noticed that they had taken down their tents at the Petit Marais in order to get closer to the fort, leaving very few of them at the reserve camp. As far as I could see, they undertook no operation near the fort.

Sunday the 8th, from last night until this morning I continued to cruise up and down in front of the fort watching to see if the English were crossing to the other side of the river.[1578] I saw nothing new. At 6 o'clock in the morning I approached the English camp and fired in 3 separate volleys 15 cannon shots at them.[1579] They responded with more or less the same number. At 9 o'clock, M. Pouchot signalled to me to set sail for Chouaguin. I sent my longboat ashore to get his orders, which I received at noon. I was to remain here and continue doing what I have done since yesterday. At 1 o'clock in the afternoon I approached to within a good range of the camp and fired on their barges. I was not able to see what effect my fire had on them. They responded with vigorous fire. I was pleased that we were not hit by

1577 Four Mile Creek.

1578 The English had no way to cross the river other than to proceed by boat along the lake or to carry their small craft overland and launch them in the Niagara River. Pouchot expressed concern in his *Mémoires* (II, pp. 49-50) on this date that a crossing of the river by the English would threaten the retreat to Fort Niagara of the garrison of Fort Little Niagara above the falls.

1579 These would appear to have been full broadsides of five guns each.

their balls although two shots just missed. In this time I fired 11 shots with my cannon and their batteries responded with 6. Up till this point, the fort only fired a few times at a small number of men who appeared in the *désert* but from 2 o'clock until 8 they fired a very great deal. We must assume that the English are opening up a trench into the *désert*. We saw them working all day at the edge of the forest where they made their own *désert*.[1580]

Their barks must have been hauled up into the forest as I have only seen 60 to 80 at the most, together with a longboat which seems to be large and flat and has a mast.[1581] There is also a bark canoe.

Since this morning, I have seen a stretch of the coast on fire. I think it must be the soldiers from the small fort who have set fire to it and withdrawn to the large fort.[1582]

At 7 in the evening we saw evidence of a great deal of musket fire close to the fort. We think it was a sortie made by our troops.[1583]

Monday 9th. From last night until this morning I patrolled between the fort and the camp. I saw nothing new. Last night the fort only fired its cannon a few times, 7 or 8 shots perhaps. There was no response from the enemy. At 6 o'clock this morning I sailed close to the Petit Marais where their camp was. I saw neither tents nor barges and very few men. I think that we discomfited them yesterday and that they have hauled their boats up into the forest. I fired a few cannon shots at them, to which they made no response. I have not seen any artillery, although it may be concealed. From 4 o'clock until 6 this morning, there was a good deal of firing from the fort's batteries and

---

1580 The English were indeed creating their own *désert* for the security of their encampment. The besieging army received orders on the evening of July 7 that "The front of the camp be cleared of all brush for the space of 200 yards." In addition, a work party of eight officers and 516 enlisted men was assigned to turn the brush into fascines and gabions for use in siege operations. Orders, July 7, 1759, *SWJP*, III, pp. 63-64.

1581 By "barks" La Force probably means barges or bateaux.

1582 La Forces uses the word *costes*, translated here as "coast." He refers to the Niagara Escarpment which was known to the French as the *Côtes* (Pouchot, *Mémoires*, III, p. 167. See note 1295). On July 7 Lieutenant Chabert had been ordered to destroy Little Niagara and withdraw his garrison to the large fort. It appears from his narrative that the place was set afire on the 8th. Chabert, pp. 18-19. Chabert informed Pouchot of the destruction of his post on the 8th. Pouchot, *Mémoires*, II, pp. 51-52.

1583 On the evening of July 8 Pouchot recorded the appearance of "several hundred Indians" who began to fire on the fort and were dispersed by cannon firing grapeshot. Pouchot, *Mémoires*, II, p. 51.

since then they have been fairly quiet. At noon I sent my longboat ashore to fetch some gun carriage wheels as all mine were broken, also some pump spikes as our vessel has been taking on water since yesterday. The cannon have shaken her and opened up one end of the starboard gangway. I have been forced to put 2 guns in the hold because they or their carriage wheels were faulty.[1584] At 3 o'clock in the afternoon my longboat returned with the information that an English officer, the Sieur Blain, had come to the fort this morning on behalf of his General to demand M. Pouchot's surrender, a proposal that was at once rejected since there has been no firing from either side.[1585] At 3:30 I went to a station in front of where the English fort had been,[1586] having received an order from M. Pouchot to go and see if there still seemed to be many barges in the Marais. I saw no more than were there this morning. I fired 3 shots with my cannon and they responded likewise one after the other. I think there was only one gun there and that the others are being transported for service around the fort. We do not know how to interpret their very slow maneuvers.

The skiff spent last night anchored over the sandbar with two cannon on board; it was charged with observing whether any barges crossed to the other side while I patrolled over the open lake.[1587] The musket fire we heard yesterday evening was directed at Indians who were approaching the fort. They were fired at from the demilune.[1588]

1584 Evidence of the hasty construction of *Iroquoise* had begun to appear following two days of firing.

1585 Curiously, neither Pouchot's *Mémoires* nor the anonymous journal (Appendix B) have a clear entry for July 9. Both seem to place the English surrender summons on the morning of the 8th. A careful reading of Pouchot, however, makes it clear that the date was simply omitted in his narrative. La Force and the "Siège de Niagara" (*RAPQ*, 1933-34, p. 93) record the summons on the morning of the 9th.

La Force's account is the source of the statement, perpetuated in a number of histories, that Prideaux's letter was carried by an English officer named "Blain" or "Blane." In fact the bearer was Captain Walter Rutherford. See note 603.

1586 Presumably La Force means the battery defending the English boats.

1587 This sandbar was off the lake side of Fort Niagara and extended from the mouth of the river to a point some distance east of the walls. The French were concerned that the English would carry boats overland in order to cross to the west side of the Niagara River. This they eventually did and subsequently constructed a battery near Montréal Point. It opened fire on July 17. Dunnigan, *Siege*, pp. 61-62.

1588 La Force confirms Pouchot's report discussed in note 1582.

Tuesday, 10th. From last night until this morning the winds have been variable from the W. to the S.W. and very brisk. We tacked all night in front of the fort. At daybreak we had rain and thunder. About 2 hours later the weather cleared up. There was not much firing from the fort last night but since this morning the artillery has been continuously in action. I think that the enemy is opening up the trench.[1589] At 11 o'clock this morning I was in range of the English camp into which I fired 2 volleys without response from them. I think they wanted to allow me to commit myself further in order to disable me. I then tacked for 3 leagues along the south coast to see if another fleet of barges was advancing along it. I would have attempted to destroy them if there had been. I saw nothing except a French bateau floating along the coast; it is most certainly a bateau which has drifted from Niagara. I noted signs of wind from the N.W. I went about into the open lake since I was only a musket shot's distance from shore. At 3 p.m. a storm blew up in the N.W. with appalling claps of thunder, lightning, heavy rain and only a very light wind blowing from all points of the compass. I stayed well out in the lake in order not to get too close inshore. At 5 in the evening, the weather calmed down with a fairly brisk wind from the N.W. I sailed back to Niagara where heavy artillery fire was continuing.

Wednesday the 11th. Around 9 last evening the French in the fort kept up a heavy artillery and musket fire that lasted for 1 hour and 1/2. I was not able to see to what extent the English were returning fire as I was well out in the lake and becalmed. Then the winds began to blow briskly from the North with every appearance of becoming violent. This morning I was stationed between the high bluffs and Niagara, about 2 leagues lower down.[1590] The winds shifted to the N.E., fairly brisk. I sailed hard for the fort and then they changed to the south. Throughout the whole day I was unable to get close to Niagara. They did not fire as much today as yesterday although we quite frequently saw volleys of 5 or 6 cannon shots at a time. It seems that the enemy have not yet got their batteries established, as they are not returning artillery fire.

1589 The besiegers had indeed opened their trenches. Daybreak had revealed to Pouchot the beginnings of what he termed a "parallel." Despite the rain, he bombarded it heavily. See note 613.

1590 About six miles east of Niagara near the mouth of Six Mile Creek.

Thursday the 12th. At 9 o'clock yesterday evening the fort began heavy artillery and musket fire which continued right up to this morning.[1591] It is surprising that the fire is so heavy given the relatively small garrison of the fort. The enemy also unleashed a number of volleys of musketry. I do not know if they fired their cannon during the night as I was unable to determine exactly where the artillery fire came from. This morning the enemy fired several bombs at the fort which did not appear to do any serious damage since they exploded in mid-air.[1592] They continued firing them throughout the day, though very slowly. At noon, M. Pouchot signalled to me to send my longboat ashore. I immediately despatched it and brought the ship up to the entrance to the river where I observed a trench that the English have opened up half way across the *Désert.* The trench did not seem very substantial. At 4 in the afternoon my longboat returned with a wampum collar that 3 Indians of the 5 Nations currently with the English brought to M. Pouchot. This collar is of white porcelain and was brought by an old chief to ask for peace with Onontio.[1593] He said that the English, by means of presents, trading goods and brandy, have forced them to take up the hatchet, having given them to understand that upon their arrival at Niagara, the fort would immediately be available to them. They now acknowledged their fault and were cross that they had struck out at Onontio. They begged him to forget everything. M. Pouchot replied that the intention of their father had never been to cause grief to his children, that he had had a heavy heart when he had seen them coming to strike their brothers, but that, since he acknowledged their fault, they only had to withdraw freely and he would forget the past. Their response to him was that

---

1591 Pouchot's council with the Seneca chief Kanendaé (note 615) and two representatives of the Iroquois allied to the English had concluded at 9:00 p.m. on July 11 when the envoys were led blindfolded back to the English lines. The truce concluded, and firing recommenced. The anonymous journal says the batteries reopened fire at 10:00 p.m.

1592 The English mortars were firing at extreme range, and it is possible that this caused them to be elevated at an unfavorable angle to properly allow the bombs to burst at ground level without burying themselves. "In a siege, shells should never be thrown with an angle of 45 degrees, excepting one case only; that is, when the battery is so far off that they cannot otherwise reach the works: for when shells are thrown out of the trenches into the works of a fortification ... they should have as little elevation as possible, in order to roll along and not bury themselves .... when shells are thrown upon magazines, or any other buildings with an intention to destroy them, the mortars should be elevated as high as possible, that the shells may acquire a greater force in their fall, and consequently do more execution." Smith, *Dictionary,* p. 186.

they would abandon the English that very night; they also begged their father to protect and take care of any of their young men remaining with the English, who might fall into the hands of the French. He so promised.

They presented a second collar for the following. Yesterday afternoon, there were 2 men at the fort on parole for the subject treated above. There was a cease-fire while the men were at the fort and during that time the English worked hard to open a trench 200 *toises* away from the fort. The second collar is by way of an apology to Onontio for not preventing the English from working during that time as it had not occurred to them that the English would seize the opportunity to work on it.[1594]

They presented a 3rd collar to some Mississaugués and Outa8ois[1595] who were in the fort. This collar is sent by the Loups from over the mountains who are also in the English camp. This nation invites the Outa8ois to keep calm and not to get involved in the disputes between the French and the English; they should go to an agreed place some distance from the fort to hold council, as they had a great many things to tell them. The Outa8ois responded through the same collar that they could not accept, that they did not know them just because they had taken the side of the English. For their part, they said, they had come to help their father and wished to die with him. M. Pouchot responded to the 2 collars with 2 others and gave them a 3rd to invite them to leave as soon as they could and make their way to their home village. If they did that, everything he had done in the past would be buried and he would think of it no more. This they promised to do.[1596]

M. Pouchot wrote me a letter, of which what follows is a copy. "My dear Laforce: we are holding off the Enemy quite well, but he is continuing his advance. He is now 200 *toises* from the glacis. This is

---

1593 The chief was Collier Pendu or Kaghswagtaniunt (note 623). Pouchot records the council.

1594 Pouchot makes it clear in his *Mémoires* (II, p. 65) that he was greatly annoyed at the English violation of the cease-fire of July 11. He does not, however, record the Iroquois apology for their allies' poor manners.

1595 La Force employs the numeral "8" (*huit*) in an attempt to duplicate the pronunciation of Ottawa. This form was frequently used at the time. Pouchot spells the word "Outaouais."

1596 Pouchot records most of these details in *Mémoires*, II, pp. 66-72.

when we shall begin to give him a really hot time of it. Stay within range if you are able to do so in order to bombard the plain. You will damage the reverse side of his trenches. If you cannot manage that, go out into the lake to see what is going on if the wind allows. Come and report at the first favorable wind. That is the only useful service you can render us. Otherwise wait in the vicinity for the outcome. We shall let you know of what we need. Come in close. If we require you ashore, I shall fly a white flag from the back window of the house.[1597] If it is urgent, we shall fly the main flag and the small one together. 11 July 1759. Signed: Pouchot."

"M. Corbin will give you details about all the rest that I cannot tell you. I have still no news from Presqu' Isle."

"The Iroquois inform me that there is no one at La Presentation. If you, Laforce, can do it, will you please hand over this collar. Signed: Pouchot."

The M. Corbin mentioned informs me through the Sieur Corbin, my second in command,[1598] to send the said Sieur Corbin with the wampum collar to the General and that he has not time to write to him as he is still trying to hold council with the Iroquois. He also tells me to inform the General about what is happening here since he cannot do it himself. He adds that I should sail along the south coast of the lake to see if I cannot encounter a convoy of 500 men bringing supplies to the enemy and try to destroy them. I should then go and communicate all this news. The Iroquois have told him that only 3700 English were involved in the siege and that there were 3000 of them at Chouaguin. Since 5 o'clock this evening I have been sailing there to find out. Around 7 I approached the English entrenchments and fired 10 cannon shots at them. I cannot have done them much damage because of the coast, which is very high, and the river battery which prevented me from getting in closer.[1599] I then sailed to the Petit

---

1597 The "French Castle."

1598 The Sieur Corbin, La Force's lieutenant, was probably Étienne Corbin (note 585). The messenger, also named Corbin, has not been further identified, but was very likely a son of the second in command of *Iroquoise.* It is possible that these men were related to La Force by marriage. On January 10, 1757, the captain had wed Madeleine Corbin, daughter of Gaspard Richard Corbin. *DCB,* V, p. 470; Robinson, p. 178.

1599 The nature and location of La Force's "river battery" is unclear. It could not have been the siege battery on the west side of the river since that position did not open fire until July 17. *Iroquoise* was in the lake east of Fort Niagara, so La Force presumably means that the English placed some guns in their approaches to fire at him.

Marais where I fired 12 shots at their camp. They responded in kind but I was not hit. They have mounted 3 cannon at that point.

The fort kept up a continuous fire all day and greatly increased it from sunset until 11 at night. The cannon and muskets were firing so much that it sounded like constant thunder. As I am sailing for Frontenac and Chouaguin, I can no longer hear the firing. At midnight, the winds are to the south and quite brisk.

Friday, 13 July. Since last night the winds have been in the south and blowing hard. Today they have veered to the S.W. and are just as brisk. At 6 o'clock in the morning I was at Kaskouchagon[1600] about a musket shot away from the coast. I proceeded to Chouaguin in the hope of discovering the convoy that is to carry supplies to Niagara but was not able to see it. It cannot have set sail yet. At 2 o'clock in the afternoon, I arrived in front of Chouaguin where I found the 3 thousand English troops the 5 Nations had spoken about at Niagara. Their camp is situated to the East of the River where Fort Ontario used to be. There is a large retrenched space beginning at the edge of the lake and extending as far as the river. Inside there are about 250 tents, well arranged and providing a fine spectacle. About 400 *toises* upriver from this retrenchment there is a guard post that is also retrenched. I saw a barge emerging from the river and coming to reconnoitre us. They certainly thought that Niagara was captured and that their men had sent our vessel to Chouaguin. I allowed the barge to get close. They knew we were not English and avoided passing under our guns. I hoisted the French flag and as I passed the camp I fired my cannon 5 times with shot, to which there was no response. I continued my observations of the position and then set sail for Frontenac where I very much hope I shall not find the enemy.

Saturday the 14th. I arrived at Frontenac at 11 o'clock in the morning after sending at daybreak a longboat to spy out the land there. I remained 3 leagues offshore with the vessel. They returned to tell me that M. Corbière was there.[1601]

1600 The Genesee River.

1601 Lieutenant Corbiere of the Marine (note 475).

Appendix D:
Journal Of Indian Affairs, 1757

One of the most important tasks of Fort Niagara's commandant was the maintenance of good relations with Native American groups of the region. This was crucial for trade and alliances, and the duty became even more critical after France and Britain commenced open hostilities in America. The location of Fort Niagara was particularly strategic. It stood at the western end of the country of the Iroquois, who maintained an official policy of neutrality until late in 1758, but whose individual nations were influenced to varying degrees by the European combatants. To the north lived the Mississauga and related nations, who generally favored the French. To the south were the Loups or Delaware of Pennsylvania and splinter groups of Iroquoian peoples. Many of nations of the upper Great Lakes also visited Niagara because of its commanding position on the primary east-west water route.

Few documents provide a better sense of the demands placed on Niagara's commandant than the journal of Indian affairs kept at the post during the summer of 1757. This was a critical time. The French, flushed with success after the destruction of the British post of Oswego the preceding summer, enjoyed unprecedented influence among the Six Nations. Warriors from Pennsylvania and the Great Lakes passed through the post in a steady stream on their way to support Montcalm's offensive on Lake Champlain. In spite of these successes, however, the effect of British naval power was making itself felt by a shortage of the trade goods and presents that were so important to retaining French prestige and influence.

The journal of Indian affairs that follows was kept for two months during the summer of 1757. Unsigned, it was most likely the work of Captain Pierre Pouchot. Many accounts speak very highly of Pouchot's abilities as a diplomat to the Native American allies and enemies of New France. This journal provides a revealing, day-to-day picture of how he went about that part of his duties.

NIAGARA JOURNAL from the month of June to the month of August 1757 [1602]

News of the Indians

6 June - Arrival of four Goyogouins who brought me[1603] a wampum belt on behalf of the Onontagués, as being more faithful than they are to the French, to know if it was I who had sent some Ottawas[1604] to their territory to do battle with them. They had said that they came on a warlike footing to avenge the death of a chief who had been killed. Through the other half of the belt, they besought me not to send out raiding parties across their lands. I responded through a belt that Johnson[1605] had told them that he would return this summer to rebuild Chouaguen,[1606] even if he were to lose many men. I added that I wished to know from reliable men if there were any new developments along the river[1607] and that I wanted them to push ahead as far as the English homesteads in an attempt to capture some of them or to kill them if they could not capture any; but that I had not told the Mississagués to do battle except with the English. On the contrary, I remarked, I was very pleased with the Five Nations and, for fear of making them anxious, I had recommended them [87] to take

1602 This journal was first published in Casgrain, *Collection*, XI, pp. 87-116. The translation is by Michael Cardy. Bracketed numbers reference the end of that page in Casgrain.

1603 The opening statement would seem to confirm that Pouchot kept this journal. A wampum belt was unlikely to have been intended for anyone other than the commandant of Fort Niagara. Pouchot might suggest his authorship in *Mémoires*, I, p. 97, when he states that it "would be too long & boring to give the details of all the raids which were made" during the summer of 1757. His text, on pp. 97-100, then recounts several events and anecdotes recorded in this journal.

1604 Throughout this journal the writer uses the English spelling for "Ottawa." By contrast, in his *Mémoires*, Pouchot consistently employs the more common French spelling "Outaouais." It is likely that the change to "Ottawa" was made when the original manuscript was transcribed for publication in Casgrain, *Collection*.

1605 Sir William Johnson (note 256).

1606 The British post at Oswego had been besieged, captured and destroyed by the French in August 1756 (notes 301-21).

1607 The Oswego River.

the left bank when coming back up the river.[1608] I noted that I had told them throughout the winter and repeated now that the intention of their father was that they should be without worry on their own lands and that we had no reason to complain about their conduct. I finally said that a number of them had taken up the axe of their father and that I urged them to follow that example and to be united in their feelings.

Through branches[1609] that I rejected, they told me: "The Frenchman and the Englishman are very large. We are afraid that they may crush us. We can see that we shall be dead men. Nothing is sold to us at Frontenac and on the Belle-Rivière, and, ever since we have taken up the axe, we are given no ammunition, or very little worth speaking of. We are thus giving these branches back to you to stop our warriors and to prevent them from going to do battle." I replied that, although we were very large, they could be at ease on their mats and that I had always assured them, on behalf of their father, that we would not trample on them when they were on their mats. I said that I had had no reason to be astonished that the Onontagués, whom I did not know, were complaining because they had no ammunition since they had taken up the axe, because I was unaware of the fact. They only had to come and see me and they would judge by the manner I treated them and through others who had come here that the intention of their father is that they should lack nothing they needed. I said I was astonished that they should present these branches to me in order to stop their warriors, and that I rejected them because I had always assured them that their father did not wish to force them to take up the axe, but that he would distinguish through **[88]** his kind deeds those who, of their own free will, sided with him from those who remained on their mats. I thus wanted them to be responsible for doing what they wanted to do. I was very grateful to those who were making war and urged all of them to be of the same mind and to be united among themselves.

This man [sic] told me that he himself had been to Orange[1610] to fetch a blacksmith some twenty days before and that the man had

---

1608 The reference may be to the left (south or west) bank of the Oswego River. However, Pouchot might also have meant the Mohawk River. Both were in the territory of the Five (actually six) Iroquois nations.

1609 Strings of wampum.

1610 Albany.

come with him. Not having any tools, though, he had gone back after four or five days.[1611] He asked us not to be angry about it, but he had had no tools with which to plant their cereal crop. I offered to have them repaired for them and to have those they lacked made for them *gratis*.

He said to me that Johnson had told them that the governor of Canada had told them that he had written to them to arrange a suspension of hostilities for three years, but that nonetheless during that time he would give ammunition and presents to the Indians and that after that period there would be an agreement to abandon and even destroy them. He also claimed that he had captured a vessel from us, I believe last fall. I assured them in the strongest terms that all this was false. Following that, I took the opportunity to tell them what Loudon was saying about them in Philadelphia and to recount the adventure of Fort George, which they knew from their own people who spent the winter in Montréal.[1612]

7 [June] - Some Ottawas arrived from Michilimackinac to trade, who told me that some of their people had gone down to Montréal and others to the Belle-Rivière. **[89]**

In the afternoon, arrival of two missing Mississagués from the raiding party on the loose who had been on the Chouaguen River. They brought with them an English scalp. They admitted that they had met an Onontagué by himself who was stripping bark, that the man had seen them before they saw him and that he had addressed them in Ottawa first. I asked them if they had aimed their guns at him. They said no, that they had told him what they intended to do and that the man had shown them the path they could take to some homesteads where they would find some English.[1613] They had continued their

---

1611 The Cayuga had applied to Sir William Johnson for a smith to mend their tools and arms. On May 2 Johnson issued instructions to one Ryar Bowen to go to their villages and remain until the necessary work had been completed. Johnson to Bowen, May 2, 1757, *SWJP*, IX, pp. 713-14. Bowen was thought to be near Wood Creek when the two Mississauga spies mentioned below were discovered there, and it was feared that he had been killed. Butler to Johnson, May 17, 1757, *Ibid.*, p. 724. The smith apparently arrived safely, however, and worked in later years for the Indian Department.

1612 Major General John Campbell, Earl of Loudon (note 347), had been in Philadelphia in March 1757 for a meeting with the governors of the southern provinces. Minutes of a Meeting with the Southern Governors, [March 15, 1757], *BP*, I, pp. 91-93. Pouchot might have derived his information from the Philadelphia newspapers.

The "adventure of Fort George" was Rigaud de Vaudreuil's raid on Fort William Henry in March 1757 (notes 337-38).

journey and had come across a fork or crook in the river where there were two villages, one on their side, one on the other, not very far from a fort of stakes.[1614] They saw two men on horseback pass by, carrying no other arms but an axe. They fired on them, one of the men fell and they took his scalp. They found a black man and according to them they very much wanted to capture him but he was too close to the houses. They said that they did not want to kill him as they were hoping to be able to take him alive. I asked them if they had not taken any papers from the dead man and they replied that, as they had seen men on alert in the villages, they had fled.[1615] They have a canteen, which they have broken. With the bottom they have made a little neckband and with the sides two bracelets. I read on it H/44; at this, I recognized it as belonging to the fourth company of Halcet's Regiment, no. 28, and Keniskent.[1616] They told me that they had seen no troop movements and that they had found the river that flows down to Orange all blocked up. The piles of wood the English had thrown into it [90] last year had been heaped together when the river was in spate.[1617] I took care to explain in detail all the above to the Goyogouins, whom I summoned in order to give them presents, as they requested. They told me that they had been at Orange only twenty-two days before, that those men had apparently gone to do battle in the Palatine village and that the river was indeed blocked up.

1613 On May 17 John Butler reported to Sir William Johnson that he had heard that French Indians had been seen on Oneida Lake. An Indian sent by Butler to investigate encountered two Mississaugas. Butler's agent, "not understanding them," took them to a village, where they were interrogated by three other Iroquois. The Mississaugas admitted that they were going to spy on German Flats. Butler to Johnson, May 17, 1757, *SWJP*, IX, pp. 723-24.

1614 The warriors described Fort Herkimer and the settlement known as German Flats (note 1259).

1615 The incident has not been identified. The Mississaugas appear to have attacked a work detail or wood-cutting party, probably of soldiers of the 44th Regiment of Foot stationed at Fort Herkimer.

1616 Their trophy was a kidney-shaped, tin canteen of the British 44th Regiment of Foot (Sir Peter Halkett's regiment until his death at Braddock's Defeat; James Abercrombie was colonel of the 44th from March 13, 1756.). The markings indicated the regiment and the name and number of the soldier to whom the canteen had once belonged. Excavated examples are described and illustrated in Jacob L. Grimm, *Archaeological Investigation of Fort Ligonier, 1960-1965* (Pittsburgh, 1970), pp. 76-79, 166-67.

1617 The Indians described Wood Creek and the obstructions placed in it by the British following the fall of Oswego and the abandonment of the Oneida Carry (note 323).

They said that there was certainly a small, poorly constructed fort there, the houses of which were so low that you could hardly stand upright inside and that the stakes were very low. They added that the fort was of little importance and that what they had taken for a village were nothing more than separate houses scattered here and there.[1618] I again took the opportunity to speak to them about the English and that they themselves ought to be assured of the affection their father has for them, because they still seemed to me afraid that an army corps would be sent to their lands to march up the Chouaguen River. I assured them that, provided the English did not proceed in force to that area, we would not march over their mats and that, were that to happen, they should be confident that we would do them no harm, as we were sure that they will always follow the will of their father. They told me that they had firmly resolved no longer to allow the English to set foot in their country because, if that were to happen, they saw themselves as killed by both the French and the English. They claimed that Johnson was a liar and that he was adept at deceiving them. He seduced them to the English side by special marks of friendship, for example, if he was wearing some fine clothes, he would take them off to give to them. Their chiefs, they claimed, had expressly forbidden any of their **[91]** warriors to take up arms on behalf of the English and, if that happened, the nation would be ordered to kill them because the only reason they could have decided to take up arms on behalf of the English was the hope of booty. They had sent messages to the Agniers[1619] summoning them to visit their lands, while those at the Sault and La Présentation[1620] had also sent them messages, which had reached them. We should therefore be convinced that they were doing all they could to dissuade them from such action. All this, they claimed, was the absolute truth they were telling me.

A Chaouénon arrived and told me that he had had a brother and a nephew killed in the winter, in the course of which they had lost five men in all. He said that his brother, who was a war chief, had gone on a scouting expedition in the area of Fort Cumberland[1621] with ten

---

1618 The Cayugas' description of Fort Herkimer and the settlement at German Flats agrees more closely than that of the Mississaugas with Pouchot's own observations in *Mémoires*, III, pp. 146-47.

1619 Mohawks.

1620 The mission villages at Caughnawaga (note 396) and La Présentation or Oswegatchie (note 242).

men. It had been their intention to prowl around other forts in the region, and they had promised not to return empty-handed. He said that he was giving me this message on his own behalf so that I would give him some tobacco and brandy to commune with his brother and lay their dead to rest.[1622] He also said that some of the Catabas who were around Fort Cumberland had withdrawn, repelled by the search parties sent out against them.

He confirmed that this Loup had said that the Chéroquis had, through their allies the Têtes-Plates, requested to enter into alliance with Onontio and to settle on the Belle-Rivière, and that half of the Têtes-Plates were to come this summer, if they had not already arrived.[1623]

A Loup who had come to work at the portage[1624] told me they were [92] sowing seeds at Théoga,[1625] and that the chief who had been to Philadelphia was very content there.

8 [June] - The representatives of the Onontagués told me that twenty-one days ago they had been at Orange which was still knee-deep in snow and that in the winter it had come up to people's bellies. They said that foodstuffs were being transported to Lac Saint-Sacrement[1626] but that there was still black ice in the area. They did not perceive any settlers being recruited into the army; there were only soldiers there and they had only identified Scotsmen.[1627] They added

---

1621 Fort Cumberland in Maryland (note 246).

1622 The ceremony of "covering the dead" is described by Pouchot at the siege of Fort Niagara in *Mémoires*, II, pp. 68-69 and in greater detail in III, pp. 321-22.

1623 Pouchot discusses this development in *Mémoires*, I, p. 97. A delegation of the Têtes-Plates or "Flatheads" had met with the commandant of Detroit on January 10. The events of their council are described in some detail in Vaudreuil to Machault, April 19, 1757, *DRCHSNY*, X, pp. 539-40. Bougainville recorded in his journal on March 5 that the Cherokees had expressed their intention of coming to Detroit in the spring to ally themselves with the French. It was not until the first week of September that he acknowledged news of their desire to settle on the Belle-Rivière. Hamilton, *Adventure*, pp. 89, 180.

1624 The Niagara portage.

1625 The village of Tioga (note 1290).

1626 Lake George.

1627 The Onondagas refer to New York provincial troops being recruited from among the "settlers." The Scots were soldiers of the 42nd Regiment of Foot.

that Johnson had told them that he would not go to Chouaguen this year, because the governor of Canada had asked for a truce.[1628] They assured me that they had sent for the Agniers. I also asked them if there were any troops marching towards Boston. They said that there were not, that everything was heading in the direction of La Pointe[1629] and that Johnson had told them that he would lose a lot of men this year, but also that they would see troops circulating in that region throughout the summer.

In the evening, arrival of the raiding party of Pakens, a Mississagué who is going to war in Pennsylvania.[1630]

9 [June] - Arrival of several couriers announcing the visit of three chiefs who were to give me an account of what has happened at Onontagué[1631] and of a Tsonnonthouan chief, a friend of the English who is said to be a spy.[1632]

10 [June] - The Tsonnonthouan friend of the English, in order to prove that he has abandoned them, handed over to me two medals which he had from them and said that he had sent word to the village that a packet of wampum given to him by the English to work in their cause should be burned. **[93]**

This man told me that there were only Scottish highlanders at Orange and in the area of La Pointe, and that he did not know where they had marched the two regiments.[1633] He said that Johnson was a liar who had claimed to them that we had lost a thousand men last

---

1628 Sir William Johnson's papers make no mention of a request for a truce from the Marquis de Vaudreuil. Nor do French sources identify a specific offer of a truce aside from constant efforts to keep the Iroquois neutral. Bougainville acknowledged the difficult position of the Five Nations, "undecided between Onontio, their father [the Governor of New France] and Corlar, their brother [the New York Dutch trading interests in Schenectady]." Hamilton, *Adventure*, pp. 106-07.

1629 This place name occurs several times and apparently refers to the French post at Crown Point on Lake Champlain or, more broadly, to the Lake Champlain military frontier.

1630 The Mississauga chief Pakens (Poton?) was active in the French interest as late as May 1759, when he formed a war party at Fort Niagara (note 519).

1631 The chief town (Onondaga) of the Iroquois and not the nation of that name.

1632 Throughout this journal the writer has used the spelling "Tsonnonthouan" to identify the Seneca. The form "Sonnontoin" is generally employed in Pouchot's published *Mémoires*.

1633 The missing regiments were probably those that had left Albany to join Loudon's expedition against Louisbourg. Among them were the 44th and 48th of Foot, perhaps the two units referred to here. Knox, I, p. 32.

winter at Fort Georges.[1634] He told them nothing of what we had burned there nor that, after the capture of Chouaguen, an order had arrived to destroy the three forts which the English had had built on that road; nor that this had been announced by a proclamation read out loud and that the Five Nations had been informed that, since the forts had been built without orders, they were to be demolished;[1635] nor that, the governor of Canada having submitted a request for a truce, they were not operating in that area.

On the same day, departure of Pakens and ten men for Pennsylvania.

Arrival of three Tsonnonthouan chiefs, representatives of the Five Nations.

11 [June] - These chiefs did not approve the deputation of the Goyogouins, who ought to have waited for them because they had only good things to say.

12 [June] - I had a meeting with the representatives of the Five Nations. [They told me] that, as a result of their deliberations, the Onontagués are determined to detach the Agniers from the English. A chief from each village went to their territory with that purpose, at the same time as the others set out to come here. They asked me for a blacksmith. They also requested that their father should not send troops **[94]** along the Chouaguen River and that our Indian raiding parties should take care not to confuse them with the English. Arrival of Poutéotamis chiefs who are going down to Montréal.[1635] I learned from M. Dumuys,[1637] who has it from the Illinois,[1638] that the

---

1634 During Rigaud de Vaudreuil's raid on Fort William Henry (notes 337-38).

1635 The forts guarding the Oneida Carry had been ordered abandoned and destroyed by Brigadier Daniel Webb following the loss of Oswego in August 1756 (note 323).

1636 The Potawatomi were represented among the large numbers of western Indians then gathering at Montréal to participate in Montcalm's summer campaign on Lake Champlain. Bougainville reported the arrival of canoe loads of warriors at Montréal every day during the week of June 20. Those from the "Far West," he claimed, already numbered more than one thousand, and some were "from a nation so far away that no Canadian interpreter understands their language." Hamilton, *Adventure*, pp. 117-18.

1637 Jacques-Pierre Daneau de Muy (1695-1758), captain (1748) in the Marine and commandant of Detroit from 1754 until his death there on May 18, 1758. *DCB*, III, pp. 161-63.

1638 The French posts in the Illinois country.

Chéroquis have made peace with the French as well as with all the nations of this continent except for the Chicasas. They are all to do battle with the English.

Arrival of the great Loup chief from Théoga and twenty-seven warriors. I introduced them to the Poutéotamis and the Iroquois. They greeted one another very warmly.

13 [June] - I was not able to speak to the Loups, who had been drinking. They postponed the meeting until the next day.

14 [June] - I held council with the Loups, in the presence of the Ottawas, the Poutéotamis and the representatives of the Five Nations. They thanked those nations for having received their belts and for helping them, and urged them always to remain united. They told me that they had seen themselves abandoned on all sides, that they had come to have recourse to their father and to ask him to fill their needs as they were deprived of everything.

A number of Indians have come to trade, Ottawas and Sioux.[1639]

15 [June] - I responded to the Tsonnonthouans, assuring them that their father was grateful for the initiative they had taken with the Agniers and that, although he was forcing no one to take up his axe, they should not be surprised if he distributed his largesse to those who took his side.

The same morning, I responded to the Loups and invited them to go down to Montréal. **[95]**

16 [June] - The Loup chief and four of his lieutenants decided to go down to Montréal on our boats.[1640]

17 [June] - I took my leave of the Loups, who are to return to their village, and gave them presents, with which they seemed happy. In the afternoon, arrival of M. La Moëlle.[1641] No news from Fort Duquesne. From the Presqu' île I am informed that the English have abandoned their homesteads well away from the frontier.

1639 Although Sioux from the Mississippi River might well have come as far east as Niagara to trade, Bougainville does not record any of their warriors among the "Indians of the Far West" with Montcalm's army on Lake George. Hamilton, *Adventure*, pp. 150-51.

1640 The sailing vessels maintained by the French on Lake Ontario.

1641 Pierre Raimbault de La Moëlle (1723-after 1761), ensign *en pied* (1755) in the Marine. Fernand Grenier (ed.), *Papiers Contrecoeur* (Québec, 1952), I, p. 10n; "Tableau ... des Officiers de La Marine" in Brymner, 1886, p. clxxxiv. The "Tableau" renders his name as "La Moette." He was among those officers captured at La Belle Famille in July 1759. Dunnigan, *Siege*, p. 131.

From the Rivière aux Boeufs, the return of the cadet Montcourt. After eighty-five days on short rations, he has brought back two scalps. He had been in Pennsylvania, where different raiding parties had been formed. He says that five prisoners have been taken, and four scalps.[1642]

From Fort Machault I am informed that a raiding party has returned from Pennsylvania with seven prisoners and three scalps. They say that they scouted around Schamoukin, that they saw a number of people there, and animals too, but that they were all so much on their guard that there was no opportunity to attack. They were obliged to advance well into the interior of the territory before they found inhabited land.

It must be supposed that these are settlers and animals that have withdrawn to Schamoukin, for they said that they had come across about forty abandoned houses.[1643]

Arrival of two Loups who had travelled from Fort Duquesne in eleven days. They told me that two hundred and fifty Frenchmen had arrived from the Illinois, together with Indians from that nation, Sakis, Sioux, Renards, Mascoutins and Kikapous. They had seen them.[1644]

18 [June] - I took my leave of the Tsonnonthouan delegation **[96]** with the customary presents. They have left very happy.

19 [June] - The Loups came to find me with branches to tell me that they had been reflecting, that the Iroquois had given them messages not to side either with the English or the French and that

---

1642 Cadet Pierre Hertel de Montcour (note 700). News of his raid, and of the taking of the other prisoners and scalps, was duly reported to France on July 12. Vaudreuil to Moras, July 12, 1757, *DRCHSNY*, X, p. 581.

1643 The party of seven warriors had been sent out by Michel Marest de La Chauvignerie (1704-after 1761), lieutenant (1755) in the Marine and commandant of Fort Machault at Venango. Vaudreuil to Moras, July 13, 1757, *DRCHSNY*, X, p. 589. For La Chauvignerie see Fauteux, *Les Chevaliers*, p. 184. The post at Schamokin was Fort Augusta on the Susquehanna (note 344). Pouchot reported, on June 21, that the people and animals observed at Schamokin comprised refugees from the frontier settlements and not a concentration of enemy troops and boats. Hamilton, *Adventure*, p. 119.

1644 The French attracted a host of Native American warriors from a wide variety of nations during the campaign of 1757. Pouchot recorded the arrival of Kickapoos and Mascoutins at Fort Niagara about July 2. *Mémoires*, I, pp. 100-01. Some went on to join Montcalm's army. Others remained at the Belle-Rivière. Bougainville had news of the arrival there of many war parties late in May as well as a large convoy with provisions from the Illinois settlements. Hamilton, *Adventure*, p. 114.

they could not go down [to Montréal]-. I rejected them. The Iroquois have withdrawn that message and given branches encouraging them to go. They replied giving their consent.[1645] In the evening arrival from Détroit of a party of Ottawas on its way to Montréal. I am told that many more are to arrive.

20 [June] - I took leave of the Loups who are going back to their villages.

21 [June] - Departure of the Loups for their villages. In the evening, the Ottawas spoke to me. They were very angry at not having had any brandy. I urged them to leave the next day. There are a hundred of them here, men, women and children.

22 [June] - Departure of La Broquerie with fourteen Loups and thirty-six Ottawas.[1646] The mother of Techicabaoui told me that her son had been in action because for ten days she felt her breasts quivering and since then she has felt nothing, which she has always found to be a sure sign. Her reaction turned out to be quite accurate; her son was involved in action at Fort George at around that time.[1647]

23 [June] - I made the Hurons whose arrival had been announced to me wait in the vessels.

1645 The Loups, or Delaware, had a tributary relationship to the Iroquois that had required their permission for important decisions. Although the Loups had formally rejected this status during a council at Fort Niagara in January 1757 (note 342), it appears here that the Iroquois continued to exert some of their former influence. Bougainville recorded the arrival at Montréal on June 29 of "a band of Delawares from Tioga." "This nation has never before come to Montreal," he added. Hamilton, *Adventure*, p. 120.

The delegation was led by the "great Loup chief of Théoga" who had arrived at Niagara on June 12. While at Montréal he was entertained by Governor Vaudreuil, who reported that he had "sent him home in such a manner that he and all his Nation were attaching themselves warmly to the French and waging war on the English." Some of the Loup warriors joined the army on Lake Champlain. Vaudreuil to Moras, July 13, 1757, *DRCHSNY*, X, p. 590.

1646 Joseph Boucher de La Broquerie (note 790) commanded one of the French sailing vessels operating on Lake Ontario. His passengers were destined for Fort Frontenac from which point they could descend the St. Lawrence to Montréal. The warriors were apparently those Ottawas and Loups who reached Montréal on June 29. Hamilton, *Adventure*, p. 120.

1647 This is one of the few incidents recorded in this journal that was included in Pouchot's *Mémoires* (I, p. 98). See note 357. Pouchot did not learn of the accuracy of the woman's dream until the warrior returned to Niagara on July 5.

1648 Armand-François (?) Mézières (Mésières) de Maisoncelles (fl. 1741-1760), captain (1754) in the Marine. "Tableau ... des Officiers de La Marine" in Brymner, 1886, p. clxxii. This officer had come from the country of the Ouias or Weas in the upper Wabash Valley of Indiana. He arrived at Montréal with his party on July 10. Bougainville identified him as a captain. Hamilton, *Adventure*, p. 125.

24 [June] - Arrival of M. de Mézière from the Ouillas,[1648] with four Acadians who had deserted from the English of Carolina. **[97]** They have ended up near the mouth of the Ohio. They are some of the most important persons of that colony.[1649]

25 [June] - A Mississagué came to me to announce that he was setting out to fetch some men and would leave on the next boats.

26 [June] - The Acadians told me that the English of Carolina had in their capital only forty men, garrison militia, and that this territory would be easy to ravage, as the homesteads are extremely scattered. The English, he said, had built a fort close to the lands of the Chérakis along the waters of the river Mississippi.[1650]

27 [June] - Some Iroquois have sent me a message asking for ammunition and vermilion.

28 [June] - Some Tsonnontouans came to me to ask me to have pity on them and said that on their return they would make war.

In the evening, I learned that the Hurons - thirty of them - were on the Platon.[1651] We have also had an alert among the Ottawas, who said they had met three tattooed Indians whom they could not understand. We sent a detachment of sixteen troops,[1652] an officer and as many Indians - Loups, Ottawas and Iroquois - who made a

---

1649 For the dispersal of the Acadians in 1755 see note 142. Bougainville deferred mention of the refugees, who had made their way from South Carolina to Niagara, until September 1-10. He does not identify them by name, but his journal entry implies that they travelled to Montréal with the Dunkers discussed by Pouchot in his *Mémoires* (see notes 378-79). Hamilton, *Adventure*, pp. 180-81.

1650 The information provided by the Acadians had recently become outdated with the transfer of five companies of the 60th (Royal American) Regiment of Foot under Lieutenant Colonel Henry Bouquet. The troops reached Charleston on June 15 and immediately began work on fortifications and barracks. Bouquet to Loudon, June 23, 1757, *BP*, I, pp. 124-26. By late June the five companies of the 60th in South Carolina mustered over 400 rank and file. They were supported by three Independent Companies and two from the Virginia Regiment for an additional 500 soldiers plus officers. Returns, June 24, 1757, *Ibid.*, pp. 127-29. Most of the men of the Independent Companies had been assigned to frontier posts with only about fifty remaining in the capital. They were probably the "garrison militia" mentioned by the Acadians.

In September 1756 the South Carolina government began construction of Fort Loudon on the Tennessee River. The post and its garrison were intended to influence the Overhill Cherokees and to protect them and the South Carolina frontier from the French and their allies. Gipson, VII, p. 30. Fort Loudon was located in modern Monroe County, Tennessee.

1651 The lower landing of the Niagara Portage. See note 1295.

1652 Army or Marine troops from the garrison of Fort Niagara.

thorough search of the area around the landing and found nothing suspicious.

29 [June] - I learned that the Têtes-Plates had burned two sheds in Carolina and killed those who were guarding them, having pretended to be friends who had come there to offer their services against the French. This news comes from the Iroquois who have returned from war on the Belle-Rivière. Arrival of ... Hurons going to war. In the evening, arrival [98] of M. de Bellestre,[1653] with some Ouillas and Miamis, seventeen in number.

30 [June] - I learned through some Indians that M. de la Saussaye and M. de Bellestre, the son, and another officer whose name was unknown, having gone on a raiding party with a hundred Indians and seven other Frenchmen, found themselves abandoned by the Indians who were annoyed at not finding anything and went off to make war on their own. The French, who had become wearied in their turn, resolved to turn back. The three officers and three French soldiers who were leading the way fell into an ambush and the four who were following fled upon hearing the musket fire and later said that they believed their comrades to have been killed or captured with the officers.[1654] They also said that two Indians had been killed on the *déserts*[1655] near Fort Duquesne.

1 July - I held a council in which the Iroquois confirmed through a wampum belt to the Hurons, Ouillas, Miamis, and Ottawas that they had taken up the axe of their father and would not put it down again.[1656] I had them smoke and drink to each other's health. They were all pleased to make acquaintance.

In the evening, a Tsonnontouan from Canestio[1657] arrived and told me that the governor of Philadelphia had sent them messages

---

1653 Lieutenant François-Marie Picoté de Belestre (note 385). Belestre had served in the western posts and among the Miamis since 1749. *DCB*, IV, p. 633. He and his party of warriors were passing through Niagara on their way to join Montcalm's army.

1654 Pouchot discusses this incident in *Mémoires*, I, p. 97. The three unfortunate officers are identified in note 356.

1655 The cleared ground around the fortifications.

1656 The conference held at Fort Niagara on July 1-2, which Pouchot calls "a great council," is discussed in his *Mémoires*, I, pp. 99-100.

1657 This Seneca resided along the northwestern branch of the Susquehanna. On his map Pouchot labels the entire length of the Chemung River as the "Kanisteo", and he identifies a village of the same name.

asking them to attend a council with him and that the messages had been rejected. The governor had persisted and sent, through an Iroquois, medals, neckbands and a beautiful pipe with silver hoops, warning those who came to wear them around their neck and that these presents should reach no one else but the Tsonnontouans. **[99]** They had been taken to the villages but had not had the desired effect.[1658]

In the evening, arrival of a courier from Fort Duquesne. M. Des Ligneris[1659] informs me of the adventure of the three officers and the three Canadians.

2 [July] - In the afternoon, the Ouillas replied to the Iroquois through the same wampum belt painted vermilion. They encouraged them to follow the will of their father and not to listen to evil messages. They said that they were very upset that they had followed the advice of the Hurons, who, as much as they themselves, had always made fine promises without acting on them. Through the belt, they were inviting them to bring me meat.[1660] They told the Hurons that they were intelligent, that they said prayers and went to Mass, and that they advised them to take advantage of the fact by following the will of their father, as they themselves did, although they were not intelligent.[1661]

3 [July] - Departure of the Hurons with my despatches for Montréal.

4 [July] - An Iroquois chief told me that, although well disposed to us, he had withdrawn his support because he, together with his party of eighteen men, had been presented with two medals and a silver

---

1658 The governor of Pennsylvania was then William Denny (1709-1765) who was organizing a grand council at Easton to convene later in the month. John W. Raimo, *Biographical Directory of American Colonial and Revolutionary Governors, 1607-1789* (Westport, CT, 1980), p. 336. Pouchot passed on this news, which reached Montcalm's army on July 22. Hamilton, *Adventure*, p. 138.

1659 Captain François-Marie Le Marchand de Lignery (note 328), commandant of Fort Duquesne.

1660 The red-painted belt was an invitation to join in the war as allies of the French and to bring in prisoners ("meat") to Fort Niagara. The presentation is recorded in Pouchot, *Mémoires*, I, p. 100.

1661 The Huron of Detroit were the most Christianized of the western nations, having a missionary and a church in their village. The use of the expression "not intelligent" by Native Americans is discussed in Pouchot, *Mémoires*, III, pp. 302-03.

neckband.[1662] As a result, they had hung up their equipment. He communicated to me **[100]** a message asking for six of his men to be equipped, the other twelve being equipped already, and requested a supply of food for his family.

Two Loups from Théoga, who arrived this evening, told me that everything was quiet on their territory and that two raiding parties were to arrive with prisoners.

5 [July] - Arrival of Techicabaoui, who has had a man wounded.[1663]

6 [July] - I took my leave of the Mississagués and invited Techicabaoui to come to the action at the Pointe.[1664] He promised to think about it.

The Miamis arrived and strongly assured me of their zeal and invited me to allow them to trade for brandy and cloth.

The Iroquois of Kanoagon[1665] also requested to trade for brandy. I told them of the latest news.

I learned through the Loups and the people from Tateyonon[1666] that, not long before, they had been to Johnson's estate and that there might have been around his village five or six hundred men in blue uniforms and some Scotsmen.[1667] They told me that Johnson had told them that he wanted to withdraw for fear of losing his scalp. He had refused them rum, either in return for payment or as a present, telling them to request it from the French, for whom they were making war. They stated that they had been given little to eat and had had to pay for it.[1668] They had heard it said by some Indians that I

1662 These presents had been given to the Iroquois warriors by the British.

1663 This is the earliest date at which Pouchot could have confirmed the premonition expressed to him by Techicabaoui's mother on June 22. The incident is described in *Mémoires*, I, p. 98.

1664 Crown Point or Lake Champlain.

1665 A Seneca town on the Genesee River (note 636).

1666 A Loup village in western Pennsylvania (see note 1285).

1667 Sir William Johnson's estate was then Fort Johnson on the Mohawk River near modern Amsterdam, New York (note 1266). The Scottish soldiers encamped there were from the 42nd Regiment of Foot, and the blue-coated troops were New Jersey provincials. Detachments of both units had passed through Fort Johnson in April on their way to German Flats. Johnson to Abercromby, April 6, 1757, *SWJP*, IX, p. 671. A guard was maintained at Fort Johnson as well.

treated them well and they had come to see me, certain that I would have pity on them.[1669]

In the evening, arrival of the Kikapous, who are going to Montréal.

The Tateyonons told me that some Chaouénons and Loups would soon arrive and would bring me prisoners. **[101]**

7 [July] - The Kikapous and Mascoutins came to tell me that they had arrived on their way down to Montréal. They gave me branches to show that the route that had been traced for them was still quite open. They asked me to favor them in their trading.

I equipped six Tsonnontouans of a raiding party of eighteen men, who promised to make war on the other side of Orange and confirmed that another party of twenty men had already set out.

8 [July] - Arrival of some Iroquois and Loups from the Belle-Rivière to trade, and some Ottawas from Détroit.

9 [July] - Departure of M. de Bellestre and seven canoe-loads of Indians.[1670] Arrival of forty Iroquois to tell me news by means of branches. They told me that the governor of Philadelphia had summoned them by sending three medals, two neckbands, a silver shell, and a pipe, which were to be worn by their representatives.[1671] He had said that they should come within the week and that they would find a supply of food. They refused. They told me that the Loups and Chaouénons had gone with horses to fetch food.

10 [July] - The Ottawas asked me in a council to let them know when they would have brandy and trading goods. I replied that

---

1668 On the evening of April 14, 1757, Sir William Johnson received a message at Fort Johnson from the "Kings and Chiefs of the Shawanese and Delaware Nations, living on the West Branch of the Susquehana River" announcing that they, with 165 men and some women, would arrive that day to "smoke a friendly Pipe" with him. The timing was bad, for the Indian superintendent was then "destitute of Goods" to give as presents. Johnson to Loudon, April 15, 1757, *SWJP*, IX, pp. 677-79. Sir William's visitors were clearly aware of his lack of largesse. Johnson, nonetheless, made a speech to them on April 19 to which, he reported, they gave a "verry favourable" answer on the 21st. Johnson to Loudon, [April 21, 1757], *Ibid.*, p. 685.

1669 Their confidence reinforces Pouchot's and others' opinions of the captain's good reputation among the Indians.

1670 Lieutenant Belestre and his party joined Montcalm's army on August 2. Hamilton, *Adventure*, p. 158. The departure of the Kickapoos and Mascoutens is noted in Pouchot, *Mémoires*, I, p. 100.

1671 The invitation received by these Iroquois was the same as that reported by the Kanisteo Seneca who spoke with Pouchot on July 1.

nothing would please me more than to be able to send them some, but that I could not tell them when that would be and hoped it would be when the ships returned.[1672]

Council with the Iroquois. They gave me branches to open the teat of my milk, as they had realized for some time that it was blocked up.[1673] Through other branches, the matrons[1674] and warriors begged me **[102]** to give them something, together with vermilion to please the warriors. They said that they did not have time to go hunting to procure their needs, which they could no longer obtain from the English. Through other branches, they asked me to urge the quarter-master to have pity on them, to treat them well and always to refill the barrel when they had no more brandy. Through yet other branches, they begged me to do what they were asking and to pay good attention.

The warriors, through branches, asked me for powder, shot and vermilion. They told me that M. Des Ligneris must have written to me, that he was very pleased with them, that they in turn had been very pleased with him, that he had given them everything and that his breasts on both sides had been open for them. He had told them that he was not well placed to give them anything, because his post was so remote, but that they would find everything at Niagara as well as at Montréal.

I replied to them, as I had to the Ottawas, that I would satisfy them as soon as the milk arrived. I said I was surprised that their warriors were not pleased with me and that I had given something to all those who had come and told me that they were departing for or returning from war. I noted that I had believed everything they said, even though they had given me no signs, and that I knew some of them who had had up to three sets of new equipment. I said that I was speaking to war chiefs and that I was asking them, when men went to war, to come here together, or the chief by himself. If they did, I would equip him and those men he designated, either then or on their return. I

1672 Pouchot notes in his *Mémoires*, I, p. 101, that hardly any trade goods or brandy arrived at Fort Niagara during the summer of 1757 and that the resources of the French posts had been exhausted by the need to outfit large parties of warriors. This, he reported, "depressed the Indians."

1673 The comparison of presents to "milk" was a common Native American figure of speech in their dealings with the French. A delegation of Iroquois employed a similar analogy during a council with Pouchot at Niagara on May 28, 1759. *Mémoires*, II, p. 23.

1674 The Iroquois clan mothers.

requested that, if they came by themselves, they should bring me a few proofs that they had indeed been to war. **[103]**

They brought me an old woman, captured 1 May by one of my raiding parties of forty men. She told me that she was the sixth to be captured, that they had had people killed, that there were twenty-four of these Indians, that she did not know how many of their people had perished and that her house and barn had been burned as well as several others. She said her name was Muller. She lives on the eastern section of the Susquehanna.[1675] There are two forts in the area of her homestead but she does not know anything about the people who man them.[1676] I believe her. She is German.

Some Tateyanons asked me through branches given by the Iroquois to have pity on them. They wished to open up a path in order to come here as they had learned that I treated Indians well.

11 [July] - Arrival of a canoe from Toronto and twelve Indians from the rivière de Saint-Joseph.[1677]

12 [July] - Arrival of the Tsonnonthouan chief who has replaced Caitoton, followed by three others who had never been here before.[1678] They told me that their chiefs, who had visited the Agniers, had not yet returned. They said that they were afraid to leave the territory of the English for fear that some trick might be played on them. They had come by a new path along a hill, a fine path without

---

1675 A Philadelphia newspaper reported that, on Monday, May 2, Indian raiders struck Northampton County. They killed fourteen people and burned the house and barn of Abraham Miller. *Pennsylvania Gazette*, May 12, 1757. Miller himself turned up at Carlisle on June 17, having made his escape. He said that he had been captured on May 1, along with his mother and one Hannah Snell, by a party of twenty Mingoes (western Iroquois), two Shawnees and two Frenchmen. The trio were carried to "Diahogo" [Tioga] where, Miller believed, his mother and Snell had been killed. Pouchot's account suggests that Miller's mother survived to be taken to Niagara as a captive. *Pennsylvania Gazette*, June 30, 1757.

1676 Fort Allen and Fort Penn are shown in central Northampton County on William Scull's "Map of the Province of Pennsylvania" of 1770. However, both are located well below and some distance from the east branch of the Susquehanna and may not be the forts to which the woman referred.

1677 These men were probably Potawatomi from the St. Joseph River in southwestern Michigan. See Tanner, pp. 170-71.

1678 The new Seneca leader has not been further identified. He might have been Kaendaé (note 615) who remained an ally of the French throughout the 1759 siege of Fort Niagara. Nor is there good information on Caitoton, who was presumably a leader of the Genesee Seneca. It would appear from the entry for August 1, below, that Caitoton had been killed during a military action.

marshes, which cut out thirty leagues of travelling.[1679] They added that the English were constantly saying that they would have to come to Fort Duquesne, to Fort Machault and here, along the Canestio River with a hundred bateaux. They said that a party of six Loups intent on war had come across an Englishman alone in a canoe and that when he had gone to warn the fort that he had discovered some Indians, the commandant **[104]** had sent him back to scout around the vicinity while he had them surrounded. The man had returned and had been killed, along with five Loups. They told me that they had four raiding parties in the field who would give us news of the English.

13 [July] - The main Tsonnonthouan chiefs came to assure me that they would follow in the footsteps of Caitoton and the Petit-Chief[1680] and that they wanted to remain ever united with their Father, and Chabert, their son.[1681] A Loup arrived from the village to confirm to me their adventure, recounted above. He said that a detachment had set out to avenge the death.

14 [July] - I took leave of the Iroquois ladies of the council with presents. The chiefs stressed to them how happy they should be with the way I had treated them. The council paid me many compliments and thanked me because Chouaguen was razed to the ground.[1682] They thus retained the means of obtaining brandy. M. de Joncaire left to form raiding parties for Orange with his brother at the little fort.[1683]

1679 The exact location of this path is uncertain, although Pouchot's map shows a route from the Genesee River villages to Fort Niagara. This description suggests that it might have followed the high ground of the Niagara Escarpment, perhaps in part along the route later chosen for the Ridge Road (New York Route 104).

1680 The Petit-Chief has not been further identified.

1681 The Seneca referred to the Governor of New France as their father. Lieutenant (1757) Daniel-Marie Chabert de Joncaire (note 331) had spent much of his youth among the Iroquois and had been adopted by them. Thus, the Seneca chiefs naturally referred to him as their son. See *DCB*, IV, p. 137.

1682 The British post at Oswego had been captured and destroyed by the French in August 1756, and the Senecas' nearest source of British trade goods had thus been eradicated.

1683 "Joncaire" is Captain Philippe-Thomas Chabert de Joncaire (note 517). His younger brother, Lieutenant Daniel-Marie Chabert de Joncaire was commandant of Fort Little Niagara at the head of the portage. Daniel's own *Mémoire* is vague about his activities during 1757. However, he mentions working with his brother to encourage the Iroquois to form war parties in support of the French. The Chabert brothers were hamstrung in their efforts by the same shortage of goods to give as presents that impeded Pouchot. Chabert, Part 1, pp. 13-14. See also Part 2, pp. 5-6.

15 [July] - The Ottawas spoke to me about leaving.

In the evening arrival of Baby[1684] from Fort Duquesne. M. Des Ligneris informs me that the prisoners were constantly claiming that the English intended to attack him this year but our scouting parties have seen no troop movements. Currently he has four hundred men scouting over his sector of the frontier.

16 [July] - A council of Tsonnonthouan chiefs was held at the little fort to form raiding parties. They accepted the wampum belts. **[105]**

17 [July] - The Ottawas departed, since they did not have time to wait for the vessels.

18 [July] - A Chaouénon who accompanied Baby told me that the Master of Life had determined that he should go to Montréal. His chiefs, he said, had not been able to agree to go down, but in the spring I would see many of them. Although by himself, he was determined to go down to Montréal. He told me that four Chaouénons had gone as a delegation, two to the Chactas, two to the Chicasas, who were at war. Together they had arranged for the two nations to make peace and had offered them their axes, which were only of wood, against the English, but they joined to them those of their ancestors, which were of iron. The two nations accepted this with good grace, saying that they did not know the French but that they believed them to be good, since the Chaouénons loved them so much. From there they travelled with them to all the nations in the South, on the frontiers of Louisiana and the English, and they all agreed to join and went to Louisiana to receive arms, which were given to them in abundance, together with presents. They left again to make war against the English. A number of these nations have promised to come and settle on the Belle-Rivière. I learned that the governor of Pennsylvania had sent branches to the Five Nations to tell them that, since they were absolutely determined to make war, they should go to Virginia and not cross his territory.

20 [July] - M. de Montisambert[1685] arrived with eleven Loups and six Iroquois who are going down to Montréal. They report that the

---

1684 Probably Jacques Baby, *dit* Dupéront (1731-1789), who had been a trader and French agent since 1753 at the Indian village of Logstown, near Fort Duquesne. His brothers Louis and Antoine were also active in the war in the Ohio Valley. *DCB*, IV, pp. 38-40.

1685 Pierre-Louis Boucher de Niverville de Montizambert (1723-1803), ensign *en pied* (1756) in the Marine. "Tableau ... des Officiers de La Marine", Brymner, 1886, p. clxxiv; Grenier, I, p. 39. Grenier says that he was a lieutenant, and Montizambert was identified as such when he was captured at La Belle Famille in July 1759. Dunnigan, *Siege*, p. 131. The promotion, however, probably came after 1757.

prisoners were still anticipating movements on the part of the enemy, but that our scouting parties [106] had seen nothing but devastation. They all brought scalps and prisoners. There are currently more than three hundred men in the field. They announced to me the return of a number of my own men[1686] with prisoners and scalps.

In the afternoon there arrived a Goyogouin, a nephew of Toniaac,[1687] who told me that he had set out for war, but that their party had met another, which had killed three men. He had surrendered a scalp to them, which had led to his release, but he did not consider himself on that account to have discharged his debt. They wanted to go and look for some themselves. Their chiefs had told them not to set out yet, until their chiefs, who had gone to seek the Agniers, had returned. They recommended them to grip their axes tightly and, if something happened to them, to rise up and strike at those who had done them harm. These views had prevented all the Goyogouins from marching, although they were all ready and well disposed. They were waiting to discover what Johnson would say to them. At the instant they were setting out, a man and a woman had arrived who had been to Philadelphia for a council meeting following the message sent by the governor. They were not certain of what they were reporting, but they had heard it said that more than a hundred and forty Indians of different nations were there. They believed themselves lucky to escape as threats had been made to do them harm.[1688] The Indians were so dissatisfied that, on their return, they had split up into small groups to launch attacks on the English. As soon as their chiefs returned from [107] Johnson and the Agniers, they would come here to report on what had happened and others

1686 Native American raiding parties based at Niagara, some of which might have included soldiers or militia from the garrison.

1687 Neither Toniaac nor his nephew have been further identified. Pouchot, in his *Mémoires*, uses the name "Tonniac" to refer to a group or village band of Iroquois.

1688 These people had probably witnessed the preliminaries of Governor William Denny's council held at Easton from July 21 to August 8. Numbers of Iroquois had already reported to Pouchot (July 1 and 9) that they had rejected Denny's invitation. By July 14 at least 130 Delawares had arrived at Easton, led by Teedyuscung (c.1700-1763; *DAB*, XVIII, pp. 360-61). *Pennsylvania Gazette*, July 14, 1757. Although the Iroquois and western Indians did not attend, the 1757 Easton meeting was successful in removing the threat posed to the province by the eastern Delaware. See Anderson, pp. 205-07. The man and woman who had been to Easton had presumably encountered some of the deep hostility harbored by Pennsylvanians against the Indians, who had devastated the frontier settlements.

would go to Montréal. His uncle, who had known that he was coming to see me, had invited him to tell me all this.

21 [July] - Arrival of four Loups who announced that a raiding party with a prisoner would soon be here. There also arrived the senior chief from the small village of the Tsonnonthouans.[1689]

22 [July] - The vessels arrived. Since I received no brandy or trading goods, which greatly saddened the Indians who have been waiting for a long time, I announced the fact to them in such a way as to encourage them to have patience and resolve. Quite a few chiefs decided to go down to Montréal. An exceptionally large number of Tsonnonthouans came to see me. They announced to me that a party of four men that had been in the field for twenty days was bringing a scalp back for me. The great war chief[1690] assured me of his attachment to the French and said that he would be flattered to receive a medal from the governor. To prove his services, he noted the war parties he had formed in the previous three months, which had captured or killed twenty men.

I equipped a party of seven men, two of them nephews and one a cousin of this chief, who are setting out immediately for war. Sayoons and Lèvre-Coupée[1691] have returned with a scalp given to them by a party of Iroquois who had two prisoners and five scalps. The Mascoutins and Kikapous gave me a large wampum belt as a trade for brandy.

23 [July] - The Ottawa chiefs told me that they were going down to Montréal and that their womenfolk would return to the village. The Mascoutins want to wait here. The chief of the **[108]** little village told me by means of branches: "Father, since you have thrust our brother the Englishman under your foot and ruined our house at Chouaguen, we realize that we are to be pitied. We lack absolutely everything. You have only to look at our equipment. You have told us that we would find everything here and we find nothing. We have no clothes, no brandy. I have come to ask you what your intentions are with regard to us."

Through other branches: "Father, we left our village dying of hunger. Since the time we took up your axe, we have been dying of hunger in our villages. We have come to die here, unless you have pity on us until our crops are ripe. There are four huts of us."

1689 This village was probably Kanoagon or Kunoagon (note 636).

1690 Kaendaé?

1691 Sayoons and Lèvre-Coupée have not been further identified. The latter man's French nickname ("Cut Lip") might refer to a hare lip.

Through a third lot of branches: "Father, I have come to throw myself in your arms. Have pity on me. You can see how naked we all are. Listen to what we are telling you and pay attention. Since you have destroyed our house, we have no other recourse."

24 [July] - I held a great council with some Iroquois, Loups and Chaouénons from the Belle-Rivière who are going down to Montréal with M. de Montisambert. They wanted to turn back but I finally convinced them to continue their journey.

25 [July] - Arrival of M. La Moëlle[1692] with a number of Loups and Chaouénons.

Departure of a number of Ottawas.

26 [July] - The Loups and Chaouénons from the Belle-Rivière came to me to say that, since M. de Montisambert **[109]** and their comrades had left, they were turning back. I had to go through various ceremonies in order to persuade them to continue their journey, which they finally promised to do. They told me like the others that what had impelled them to withdraw their services was that they had received a poor welcome in the various posts and that I had offered them inadequate sleeping arrangements.[1693] A bottle of brandy resolved the whole matter. It got the Chaouénon chief so worked up that he sold all his wampum to some Iroquois in order to have some more, and he got so drunk that he was ill.

27 [July] - Arrival of a party of Loups, thirty in number, with a scalp and a female prisoner. Her name is Marie-Blanche, a native of

---

1692 Ensign *en pied* Pierre Raimbault de La Moëlle (note 1641).

1693 Accommodation for visiting Indians is not mentioned in the sources on Fort Niagara. Pouchot's own "Plan Du Fort de Niagara....", published in his *Mémoires*, identifies a pair of "cabins" for the Indians situated on the low ground between the fort and the river. The plan represents Fort Niagara as it was during the winter of 1755-56. These or similar structures were probably maintained to shelter visiting Indians.

1694 This prisoner has not been positively identified, nor have the exact circumstances of her capture been determined. It is possible that Pouchot recorded the French equivalent of her English name (i.e. Mary White), but it seems more likely that she was one of the numerous Blanchan family of Ulster County, New York. Paul Huey to the editor, May 4, 2002. "Schangen" is one of several variant renderings of "Shawangunk", the long mountain ridges running from Kingston and nearby Esopus on the Hudson to the Delaware River. The range is identified as the Shawangoak Mountains on John Montresor's "A Map of the Province of New York" published in 1775. The valley to the west of Montresor's Shawangoak Mountains bears the additional label "Cut off and [blank] by ye Indians in 1756." This isolated frontier area would have been easily accessible to the Loups in 1757.

Schangen in the Soupusa region, New York province, ten leagues from the Hudson River, four days journey from New York and slightly fewer from Albany.[1694] She has been living there for three years. Her husband was killed. There is no fort near her home. The homesteads are a league distant from one another. It was the first time there has been an attack in that area, and it occurred about forty days ago. She said that no order has been issued to recruit troops this year, that this year they levied a hundred and forty pounds in her village, which is a new settlement. Last year, they recruited men and requisitioned horses and carts. They were on the alert for more than four months, doing nothing and receiving no payment. They had become very thin and sickly. She said that Johnson was still the one who was in charge of their operations, but that he did not bestir himself and thought only of drinking. It was being said that if he did nothing this year, they would kill him and **[110]** surrender to the French.[1695] She had heard that for three months there had been a convoy at New York ready to sail but that it had not departed at the time she was captured. She does not know if there are troops on board, nor if there are any in the vicinity who are to embark. It was said that the convoy was to sail to Cape Breton.[1696]

The Loup chief told me that he had respected my wishes and gone to find me some fresh meat, as he had promised. He had been where a number of paths crossed and had seen a great many tracks. He had even seen about twenty men together, but they themselves could not attack because there were only five of them. When they had set out together there had been many of them, but they had become dispersed. He had then crossed a great mountain and, going down the other side, he had come across the trail of a man, which he followed for three days as far as some homesteads. He had hidden in some

1695 Although it is not certain to whom she referred, it was probably Sir William Johnson (note 256) who held a commission in the New York militia in addition to his position as superintendent of the northern Indians. Marie-Blanche's story is probably an example of the wild rumors that circulated among the threatened frontier settlements.

1696 The Earl of Loudon embarked with his army from New York, bound for Louisbourg, on June 20 after a long delay awaiting an adequate naval escort. Loudon took with him a large part of the British regular troop strength in America. The expedition was abortive, and it weakened the defenses of the upper Hudson River, thereby contributing to the disaster at Fort William Henry. See Anderson, pp. 184, 208-09. Loudon's inability to invest Louisbourg was due, in part, to the late arrival of British naval strength and the presence of a powerful French squadron at the fortress port (note 386).

ruined houses and had seen people passing by without being able to attack them. When night fell he had struck tinder in order to set fire to a homestead, which had given the alarm. An old woman came out, but he did not wish to kill her. A man appeared at the door with a musket. He shot him down and when the woman he had brought to me came out, he captured her.[1697] The fire destroyed the house and the other bodies. He told me that all the nations among whom he had travelled, even those on English territory, were going to make war on them. He had seen nothing but darkness in that region and could only see clearly here. He begged me to have pity on them. For two years they **[111]** had been making war without respite. They were destitute and had no powder, no shot, no clubs and no spears. As soon as they had rested a little here, they would go back to the village but would remain there for one night only. They felt hostile in their hearts towards an Irish village near the one they had attacked and wished to ravage it.[1698]

A man named Charles Peller, a Bavarian, who was taken prisoner in Pennsylvania by an Iroquois party that left here around three months ago, says that he has been in this country for nine years. He is a carpenter by trade and belongs to an independent company whose captain is called Harmestran. His company numbers sixty men but he does not know how many companies there are. They are garrisoned in little forts of which there are a number sited at mountain passes as a defense against Indians. Each fort may contain a hundred men and they are all built of standing stakes. The one where he was captured is called Kanegouhi after the river on which it stands. He says that he has heard about Fort Schamoken, that it was not extensive and could not hold three hundred men. The fort at Kanegouhi is a hundred and sixty miles away from Philadelphia.[1699] He says that there are only five hundred men garrisoned at Philadelphia.[1700]

28 [July] - The Tateyonons have left. The son of the great chief of the village was among them. He promised to form a raiding party on

---

1697 The Loup chief describes the capture of "Marie-Blanche" and the death of her husband.

1698 The name and location of the "Irish village" have not been identified. It was in the general vicinity of the "Schangen" settlements where "Marie-Blanche" was captured. The inhabitants were probably Ulster Scots, and they might have originated in the families brought to New York in 1738 by Alexander Montgomery and his partners. Many of these people, dissatisfied with the arrangement, settled in Orange County. Paul Huey, unpublished history of Charlotte County, New York. Huey to the editor, May 6, 2002.

his arrival and to bring me a prisoner, if God wishes to protect him. Two Tsonnonthouan and Goyogouin parties left to attack in the area of Orange and several others are being formed. **[112]**

29 [July] - Some Indians have returned from Toronto, where they have been trading. Immediately there departed eight canoes of different nations in search of skirmishing. Arrival of thirty Loups, almost all of them old folk, women and children. They said that the men were all making war, they were hungry and begged me to take pity on them. A Goyogouin arrived carrying two English scalps taken to replace the Goyogouin chief from the Belle-Rivière.[1701]

30 [July] - Arrival of one of my parties of resident Iroquois, thirteen of them.[1702] They took four scalps quite close to Schamoken. They said that the fort was almost as big as this one, terraced to cover the houses, with stakes inside. The fort was very much dominated by two mountains. There were hardly any *déserts* and it seems to them to be

---

1699 Peller has not been identified, but he must have been a soldier of the Pennsylvania Regiment. The provincial government had begun constructing small forts along the frontier in 1755 and continued this effort into 1756-57. Most were concentrated in a band running northeastward from the Susquehanna River to shield the most populous part of the province. Stotz, p. 101. No place called Kanegouhi has been identified, but the name probably refers to Fort Loudon, which was located on Conococheague Creek, well to the westward of the Susquehanna and about 145 miles from Philadelphia. *Ibid.*, p. 110.

Peller was probably taken on June 12, when Lieutenant Holliday led 25 men out of Fort Loudon to march to the Great Cove. The party was attacked, and the bodies of Holliday and three others were found the next day. Four men of the detachment were missing. *Pennsylvania Gazette*, June 23, 1757.

Peller's Captain "Harmestran" might have been John Armstrong (1717-1795), who, though promoted lieutenant colonel in May 1756, was still a company commander at the end of 1757. His company was then assigned "westward of the Susquehanna." "Officers and Soldiers in the Service of the Province of Pennsylvania, 1744-1764," *Pennsylvania Archives, Second Series, II* (Harrisburg, 1876), p. 543. There was also a Captain George Armstrong in the regiment in 1756. *Ibid.*, p. 536. John Armstrong led a successful attack on the Delaware town of Kittanning in September 1756 and was a Continental Army general during the American Revolution. *DAB*, I, pp. 353-54.

1700 Peller refers to regular soldiers of the 60th (Royal American) Regiment of Foot. Several battalions of the newly formed unit were assembling at Philadelphia in December 1756. The 1st Battalion spent the winter there before five of its companies embarked for South Carolina on May 16. The other five remained in Pennsylvania for frontier defense. See *BP*, I, pp. 10, 76, 104. Peller's figure of 500 men represents roughly half of the 1st Battalion at its authorized strength. Return of First Battalion, Dec. 24, 1756, *Ibid.*, p. 40.

1701 The lost Cayuga chief might have been the man mentioned in the entry of June 6.

1702 Mission Iroquois from the St. Lawrence Valley villages.

easy to capture. Along the river, there is a little fort, then a larger one and even farther on another one. It appeared to them that there were a lot of people. Those they had killed were leading cattle to grazing. They believed that they had killed a chief as they had only brought back the skin of his head, which was shaved.[1703]

31 [July] - Arrival of thirty-eight Loups, almost all of them old folks, women and children, who asked me to give them food and shelter. All their men are making war.

1 August - I equipped thirty-three Loups of the party who brought me the woman prisoner. They promised me to make up another party as soon as they arrived at their villages. I persuaded them to accept a French officer to go with them. **[113]**

I also equipped four Tsonnonthouans who are forming a war party. The cousin of Caitoton also told me that he was going to form a party to avenge the death of that chief.

I equipped the Iroquois party who brought me two scalps. The chief gave me a message that he would form another one straight away.

One of the four Tsonnonthouans whom I equipped had just returned from war during which he was wounded in the arm and the back by musket fire. He says that they killed fifteen men in this skirmish and that, for his own part, what consoled him for his wound was the fact that he had killed the chief.[1704]

2 [August] - Arrival of forty Tsonnonthouans who have been making war. The principal chiefs are involved. They told me about their expedition.

In the evening, arrival of forty Goyogouins and a canoe from Detroit, who informed us that many Indian parties were returning, all of them with prisoners and scalps.

3 [August] - The Tsonnonthouan war chiefs held councils, then summoned me to tell me how ready they were to respect the will of the

---

1703 The Iroquois had struck around 10 a.m. on June 23, when they attacked the cattle guard outside Fort Augusta. Several of the Pennsylvania soldiers were wounded and four were killed. At least three of the bodies were scalped, but relief parties dispatched from the garrison drove off the attackers and claimed to have prevented the fourth scalping. The dead men were identified as Corporal Robert Parker and Privates Matinas Coal, George Kelly, and James Kelly. Colonel James Burd estimated the number of attackers at 40. "Journal of Col. James Burd: While Building Fort Augusta at Shamokin, 1756-7," *Pennsylvania Archives, Second Series, II*, p. 798.

1704 Presumably in the June 23 attack on the cattle guard at Fort Augusta (Schamokin).

late Caitoton by retaining the axe of Ononthio, which they had tested and found good. I gave them a wampum belt in order to hone it. They showed how pleased they were with it and wanted me to tell them whom I wished them to replace.[1705] I left them entirely free to choose. They told me that they did not wish to put the axe down again, and the more the English wept, the more they would be delighted to attack them. In the evening, **[114]** they came to me to say that they intended the belt, since I had left them free to choose, to replace the late Blanche du Sault-Saint-Louis whom they very much missed.[1706] They were going to trade a few furs and would then return to the village where they would determine with Jonquière[1707] the date of their departure, which would be soon. They received substantial presents.

The Goyogouins came to tell me by means of branches that I should have pity on them and that I should give their womenfolk vermilion so that they could arrange their *couettes*,[1708] and ammunition so that they could go hunting and provide themselves with clothing. They said that we were now the only ones to supply them with their needs and that we did not get angry over their indiscreet requests. The warriors made the same request. They told me it was being said that we intended to march an army by way of the Pointe, one along the Chouaguen River and another which was to leave Niagara and cross their territory to proceed beyond Orange. I explained the true situation to them.[1709] They said that they were unaware of any troop movements since the operations in the spring and that they did not know of any army at the Pointe. They told me that their chiefs had left for Montréal twenty-one days before and that those who had been to the Agniers had been back for just a few days. Tomorrow or later I would, they said, receive their news. **[115]**

1705 Pouchot's belt was to symbolically replace a Seneca who had been killed.

1706 The otherwise unidentified Blanche du Sault Saint-Louis was probably an Iroquois resident of the mission village of St. François Xavier or Caughnawaga (note 396).

1707 Captain Philippe-Thomas Chabert de Joncaire (note 517).

1708 The tresses of their hair, according to the editor's note in Casgrain, *Collection*, XI, p. 114.

1709 Presumably Pouchot explained that French forces under the Marquis de Montcalm were to move south from Crown Point and Ticonderoga against Fort William Henry. Montcalm's army was even then beginning its siege of the British post, which surrendered on August 10 (see notes 363-72).

Statement of Marie-Catherine Heilerin, captured by the Loups and adopted as a replacement for Caitoton, a Tsonnonthouan chief

Her father and her mother are still alive. She was captured in a homestead isolated in some woods, two days march from Philadelphia, on the road to Virginia and Lancaster. She was captured in May.[1710] She says that the governor of Virginia had with him four hundred Indians, and others who came in small numbers to make up raiding parties. She says that there is a fort four leagues from her homestead and another fifteen leagues away. There are some troops in her area, but very few. The forts are of stakes and have been built quite recently.[1711] She says that she has not seen any troops marching in the area of Fort Cumberland but many had passed who were proceeding towards the Pointe. **[116]**

1710 The date of this statement is not given, although it was presumably taken about August 3. The *Pennsylvania Gazette* carried many notices of Indian attacks along the frontiers of the province in 1757. Names of victims are often given, but that of Marie-Catherine Heilerin does not appear. Nor can the location she gives be associated with a particular incident. The *Pennsylvania Gazette* of May 26 reported numerous attacks and deaths in the Lancaster area around May 17-20 but admitted that those who had been "carried off" could not be accounted for.

1711 These forts have not been individually identified, but they were probably some of those erected by the province of Pennsylvania during 1756-57. See Stotz, p. 101.

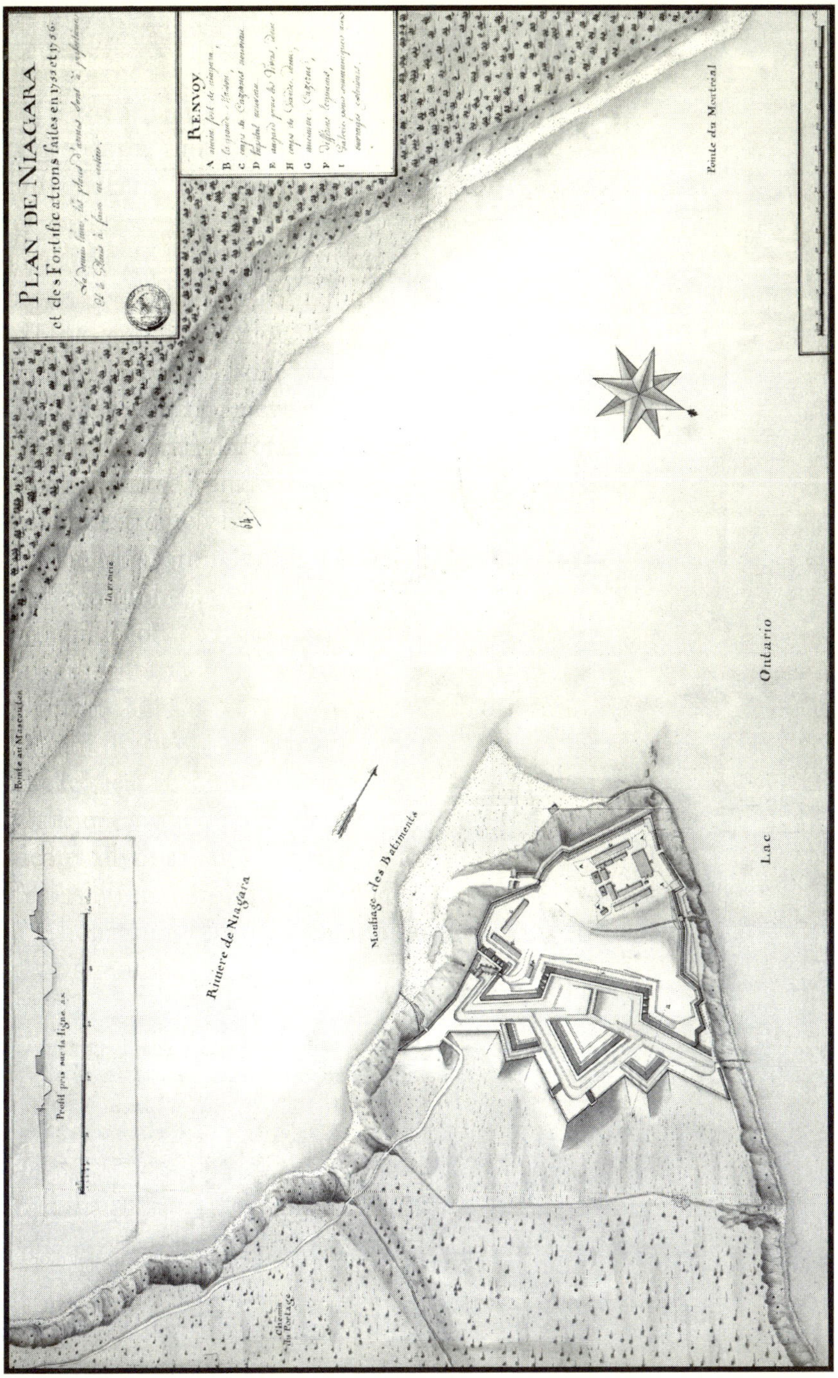

Pouchot probably drew this plan of Fort Niagara during 1756 and submitted it to the Marquis de Paulmy on October 18th. It shows the first buildings constructed between the area of the original stockade and the new walls. Pouchot included a section through the earthworks. This plan was not included in the *Mémoires*. *Courtesy, National Archives of Canada, NMC 0026647.*

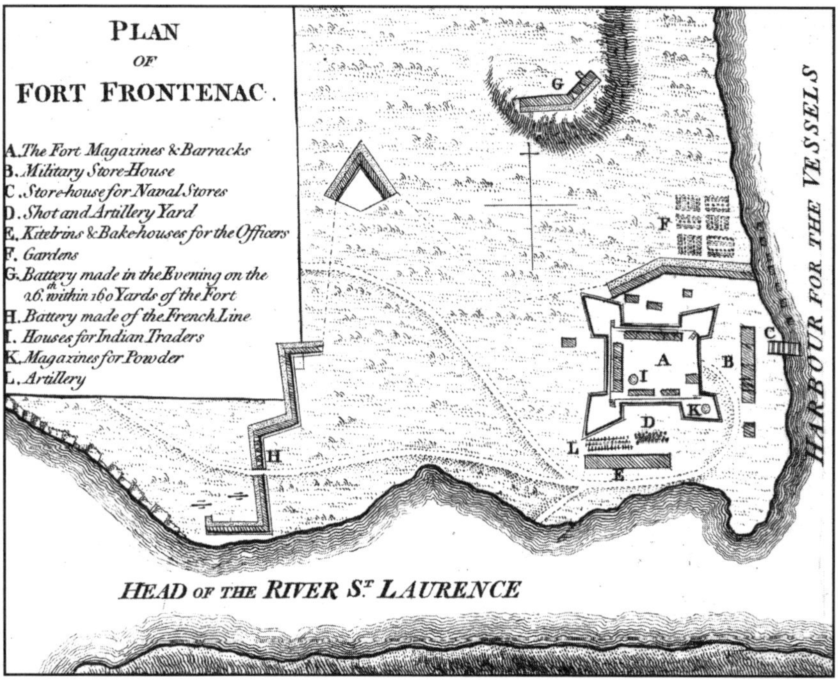

Pouchot's first engineering task in America was to fortify the French camp at Fort Frontenac on Lake Ontario in 1755. Three years later the English used the remains of his defensive lines (shown as "H" and by a dotted line) as part of their siege approaches. *Courtesy, William L. Clements Library, University of Michigan.*

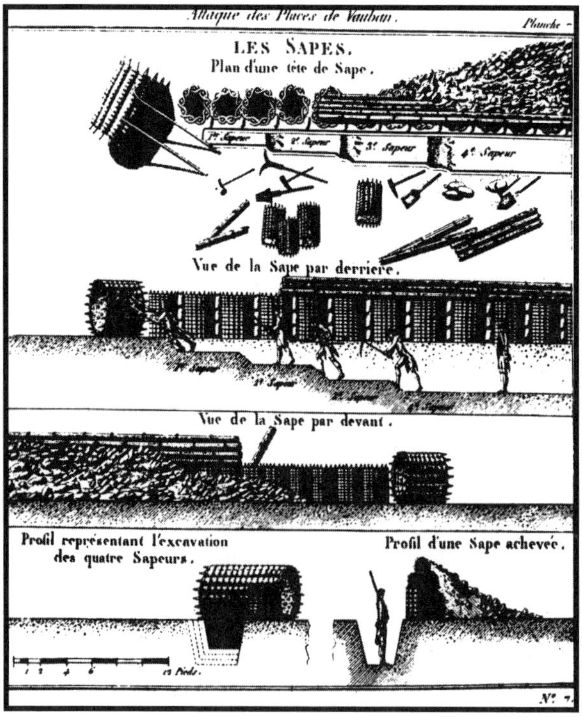

Pierre Pouchot's military engineering skills were honed by practical experience in both the attack and the defense of fortified places according to the system of Sébastien Le Prestre de Vauban. This plate illustrates Vauban's method for digging "saps", the trenches by which besiegers approached an eighteenth century fortress. French troops, under Pouchot's direction, used this technique at Oswego in 1756. Three years later, the British employed the same methods against him at Niagara.

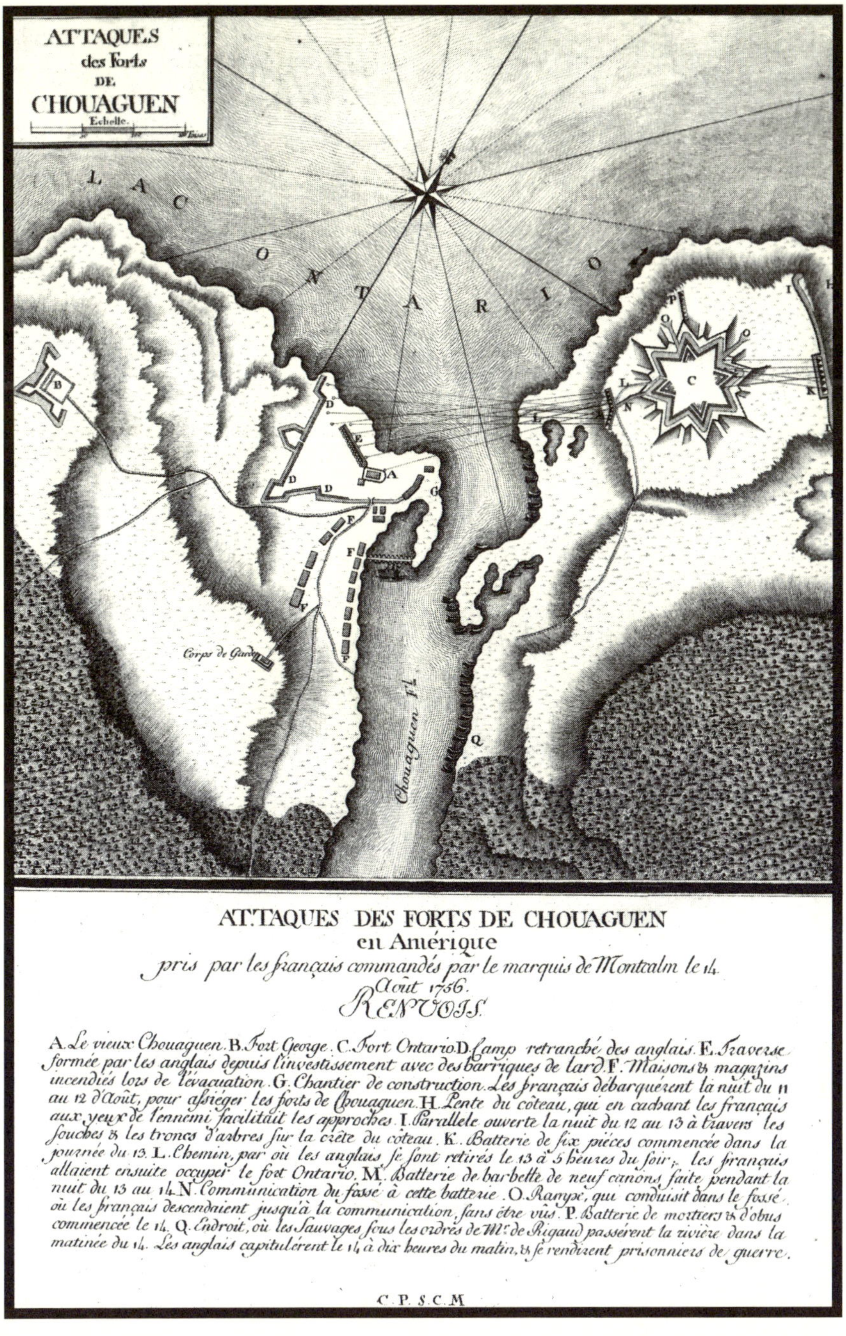

A German engraving of about 1792 details the 1756 siege of Oswego. Pouchot assumed functional control of operations after the French army's engineer was killed. He laid out the siege batteries identified here as "K" and "N". *Courtesy, William L. Clements Library, University of Michigan.*

Lieutenant Therbu's "Plan du Fort Carillon" shows both the unsuccessful English attack of July 8, 1758 and the additional fortifications (identified as "M") constructed by the French following the battle. The "Redan de Pouchot" is the fortification closest to the lakeshore at top. *Courtesy, William L. Clements Library, University of Michigan.*

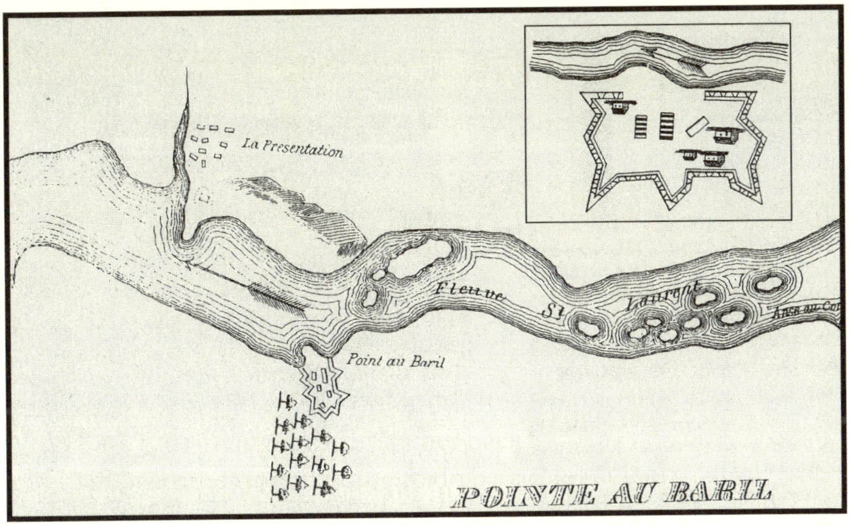

Pouchot designed the temporary defenses of the shipyard at Pointe au Baril. This crude engraving (re-oriented segment), from the 1873 edition of Courville's *Mémoires sur le Canada,* gives a rough idea of its appearance (arrows indicate flow of river).

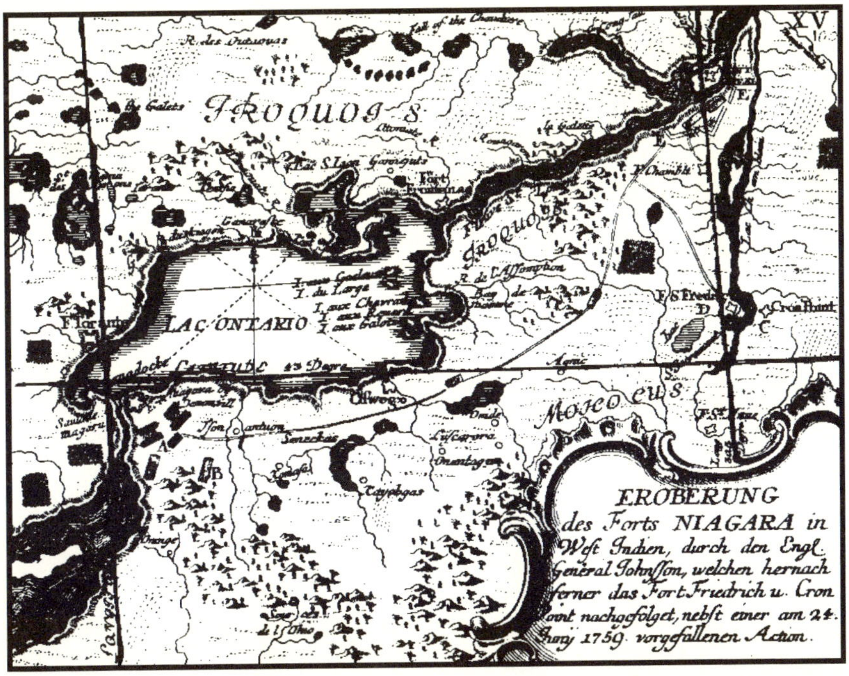

A contemporary German engraving of the 1759 conquest (*Eroberung*) of Fort Niagara is crude and inacccurate compared to Pierre Pouchot's carefully compiled representation of the same area. Pouchot's map is illustrated on page 41. *Courtesy, William L. Clements Library, University of Michigan.*

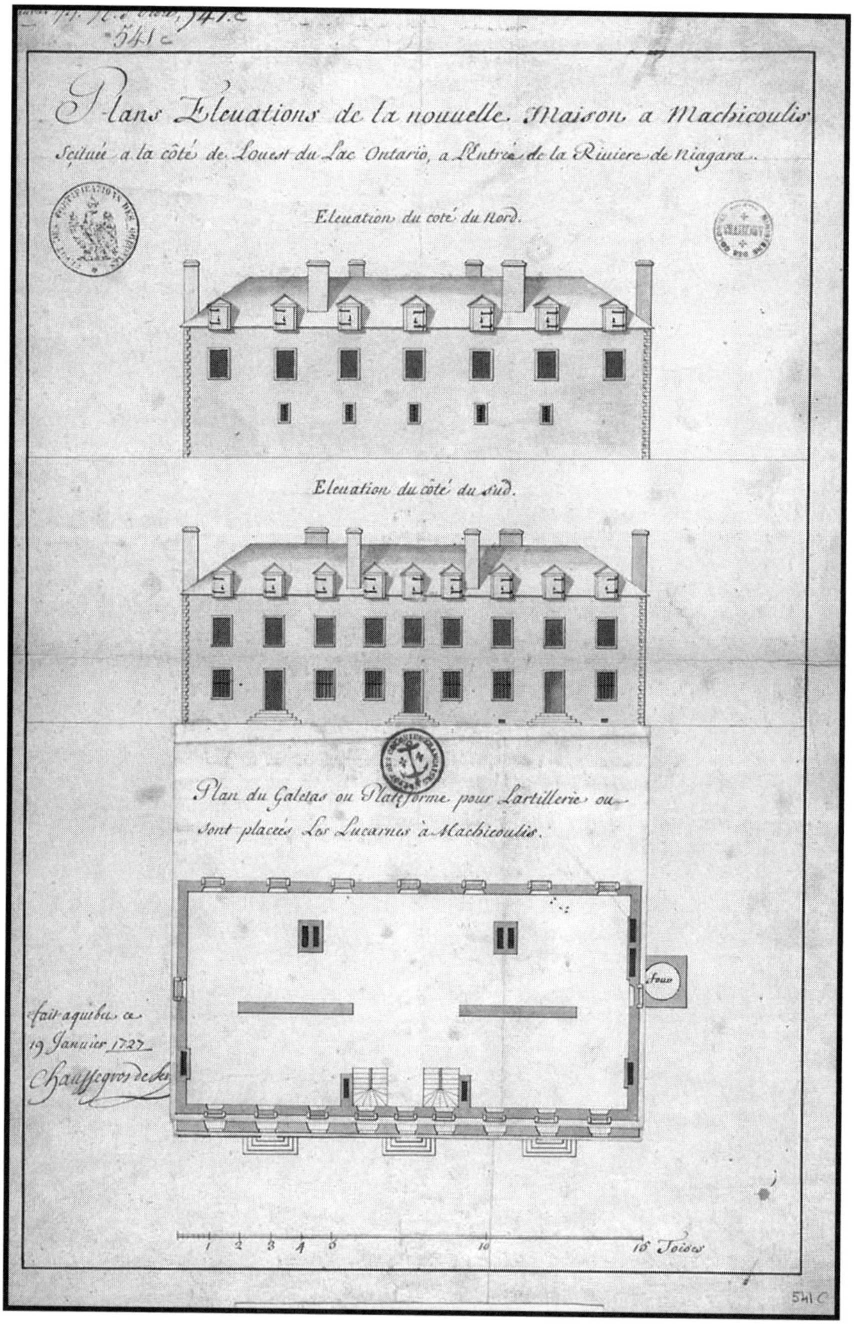

The edifice known today as the "French Castle" was the original building of Fort Niagara when it was constructed in 1726-1727. It was described by its designer, Captain Gaspard-Joseph Chaussegros de Léry, as a "machicolated house" because of the overhanging dormers that allowed defenders to fire on enemies below. The building was later converted to officers' quarters, probably under the direction of Pierre Pouchot in 1755-1756. This is Chaussegros de Léry's drawing of January 19, 1727, completed at Quebec. All exterior and interior walls are highlighted in red. The ground-floor plan was drawn directly on the master sheet, along with the exterior views. The upper two stories were rendered separately, attached on top, and turned up to be viewed. Floor plans appear on Page 595. An earlier plan of October 11, 1726, indicates in yellow the parts not completed that year. A later drawing of December 9, 1738, by de Léry's son, uses red highlights on structural elements. *Courtesy, Centre des archives d'outre-mer (CAOM), France. CAOM 03DFC 541bsC01; Plans Elevations de la nouvelle Maison a Machicoulis Scituée a la côté de L'ouest du Lac Ontario a L'Entrée de la Riviére de Niagara.*

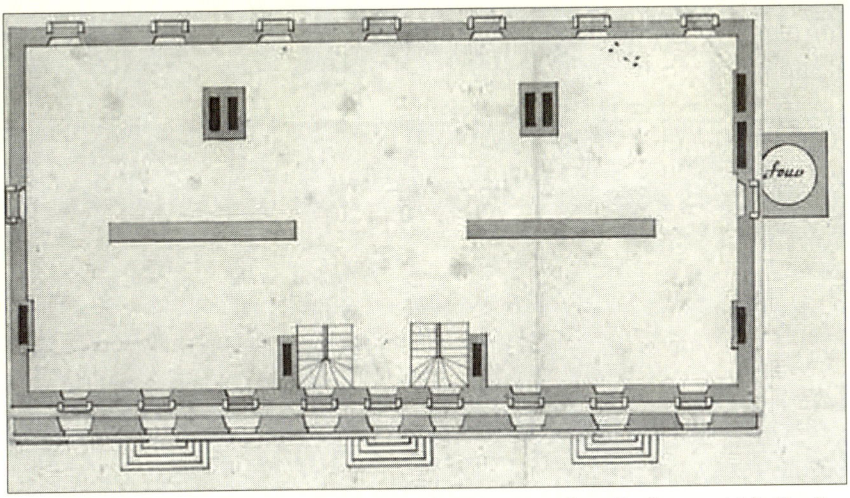

Third floor: *Plan du Galetas ou Platforme pour L'artillerie ou sont placeés Les Lucarnes à Machicoulis.* (CAOM 03DFC 541bsC01)

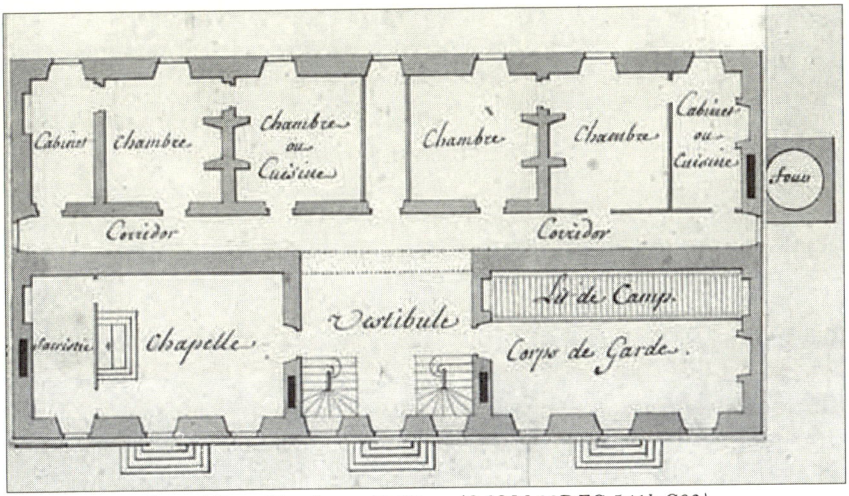

Second floor: *Plan du premièr Etage.* (CAOM 03DFC 541bsC02)

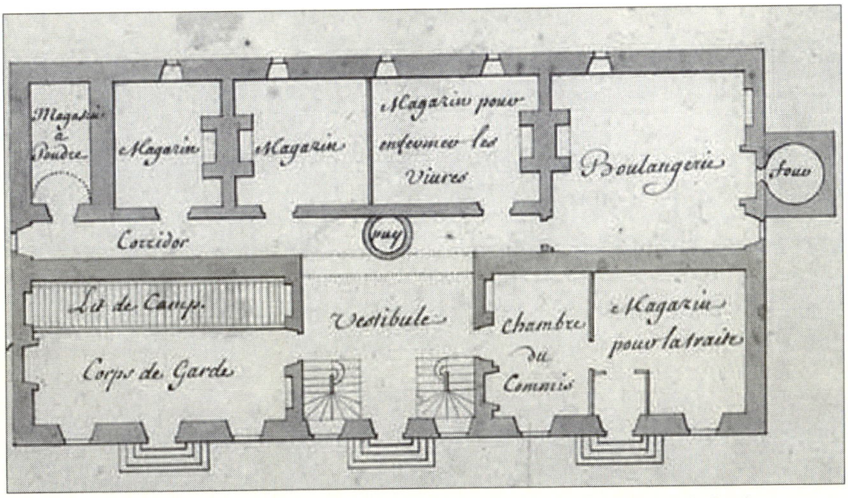

First floor: *Plan de L'Etage au Rez de Chaussée.* (CAOM 03DFC 541bsC03)

Bibliography

Abbreviations

The following abbreviations have been used to identify frequently cited sources.

DAB .... Johnson and Malone, *Dictionary of American Biography.*
DCB .... *Dictionary of Canadian Biography.*
DHNY....................... O'Callaghan, *Documentary History of New York.*
DNB .... *Dictionary of National Biography.*
DRCHSNY .... O'Callaghan, *Documents Relative to the Colonial History of the State of New York.*
RAPQ .... *Rapport de l'Archiviste de la Province de Québec.*
SHA .... Service historique de l'Armée.
SWJP......................... Sullivan and Hamilton, *The Papers of Sir William Johnson.*

• • • • • • • • • • • •

Because of the great variety of sources used in the annotation of Pouchot's *Mémoires* and related documents, it has seemed more convenient to simply list them alphabetically by author or collection.

Adney, Edwin Tappan and Chapelle, Howard I. *The Bark Canoes and Skin Boats of North America.* Washington: Smithsonian Institution Press, 1983.

Alaire, Abbé J.B.A. *Dictionnaire biographique du clergé Canadien-Français.* 2 vols.; Montréal: Imprimerie de l'Ecole Catholique des Sourds-Muets, 1910.

Alberts, Robert C. *The Most Extraordinary Adventures of Major Robert Stobo.* Boston: Houghton Mifflin Co., 1965.

Aman, Jacques. *Une campagne navale méconnue à la veille de la guerre de sept ans: L'escadre de Brest en 1755.* Vincennes: Service historique de la Marine, 1986.

Amherst, Jeffery. Jeffery Amherst Papers. Great Britain, Public Record Office, WO 34.

Anderson, Fred. *Crucible of War: The Seven Years' War and the Fate of Empire in British North America, 1754-1766.* New York: Alfred A. Knopf, 2000.

Antoine, Michel. *Louis XV.* Fayard, France: Librairie Arthème Fayard, 1989.

Aquila, Richard. *The Iroquois Restoration: Iroquois Diplomacy on the Colonial Frontier, 1701-1754.* Detroit: Wayne State University Press, 1983.

Archives Nationales, France, MSS C13A. Copies are in the National Archives of Canada.

Army List. London, 1755-1800.

Back, Francis and Chartrand, René, "Canadian Militia, 1750-1760", *Military Collector & Historian,* XXXVI, No. 1 (Spring, 1984), pp. 18-21.

Back, Francis and Chartrand, René. "French Engineers, New France, 1750-1763," *Military Collector & Historian,* XXXVIII, No. 1 (Spring 1986), pp. 26-27.

Baker, Charles E. "The William Starling Powder Horn and the Niagara Campaign of 1759", *New-York Historical Society Quarterly,* XXXII, No. 2 (April 1948).

Bass, George F. (ed.). *Ships and Shipwrecks of the Americas: A History Based on Underwater Archaeology.* London: Thames and Hudson, 1988.

Bellico, Russell P. *Sails and Steam in the Mountains: A Maritime History of Lake George and Lake Champlain.* Fleischmanns, NY: Purple Mountain Press, 1992.

Bellin, Jacques-Nicolas. *Remarques sur la carte de l'Amérique septentrionale, comprise entre le 28e et le 72e degré de latitude, avec une description géographique de ces parties ....* Paris, 1755.

Bénézit, E. *Dictionaire critique et documentaire des peintres, sculpteurs, dessinateurs et graveurs.* 8 vols.; Paris: Librairie Gründ, 1966.

Bernier, Olivier. *Louis the Beloved: The Life of Louis XV.* Garden City, NY: Doubleday and Co., 1984.

Bertsch, Maj. W.H. "The Defenses of Oswego", *Proceedings of the New York State Historical Association,* XIII, 1914, pp. 108-27.

Bienville, Chevalier de. "Le Cours du Mississipi, ou Les Sauvages de la Louisiane, leur nombre et le commerce qu'on peut avoir avec eux", *Journal de Trévoux,* (mars 1727).

Biographie universelle (Michaud) ancienne et moderne ..., Nouvelle edition. 45 vols.; Paris: Madame C. Desplaces, 1854-1865.

Bittinger, Emmert F. *Allegheny Passage: Churches and Families, West Marva District, Church of the Brethren, 1752-1990.* Camden, ME: Penobscot Press, 1990.

Blanchard, Anne. *Dictionnaire des ingénieurs militaires.* Montpellier, France, 1981.

Blanco, Richard L. (ed.). *The American Revolution, 1775-1783: An Encyclopedia.* 2 vols.; New York: Garland Publishing, Inc., 1993.

Boatner, Mark M. III. *Encyclopedia of the American Revolution.* New York: David McKay Co., Inc., 1966.

Boatner, Mark M. III. *Landmarks of the American Revolution.* New York: Hawthorn Books, Inc., 1975.

Boston Evening Post, Boston, 1758.

Boston Gazette. Boston, 1759.

Boston News-Letter. Boston, 1758-1760.

Bowman, Rufus D. *The Church of the Brethren and War, 1708-1941.* Elgin, IL: Brethren Publishing House, 1944.

Boyesen, Persis Yates. "The French and Indian Settlement at Ogdensburg." *The Quarterly* (St. Lawrence County Historical Association), XXXV, No. 1 (January 1990), pp. 4-10.

Boyesen, Persis Yates. "French Personnel at La Presentation." *The Quarterly* (St. Lawrence County Historical Association), XXXV, No. 1 (January 1990), 11-15.

Bray George A., III. "Leather Caps During the French & Indian War," *Military Collector & Historian,* XXXVIII, No. 2 (Summer 1966), p. 66.

Bray, George A., III. "More on Fort Hendrick," unpublished manuscript. Copy in Old Fort Niagara Association Library.

Bremner, John. Journal, 1756-1764. New-York Historical Society, New York.

Brumwell, Stephen. *Redcoats: The British Soldier and War in the Americas, 1755-1763.* Cambridge: Cambridge University Press, 2002.

Brymner, Douglas (ed.). *Report on Canadian Archives, 1886.* Ottawa: MacLean, Roger & Co., 1887.

Brymner, Douglas (ed.). *Report on Canadian Archives, 1888.* Ottawa: A. Senécal, 1889.

Buffon, Georges-Louis Leclerc, comte de. *Histoire naturelle générale et particulière.* 15 vols.; Paris, 1749-1767.

Buffon, Georges-Louis Leclerc, Comte de. *Supplément à l'histoire naturelle.* 7 vols.; Paris, 1774-1789.

Canada Gazetteer Atlas. Ottawa: Minister of Supply and Services Canada, 1980.

Cardy, Michael. "The Memoirs of Pierre Pouchot: A Soldier's View of a Doomed Campaign", *War, Literature, and the Arts,* IV (Spring 1992), pp. 1-23.

Carter, Clarence Edwin (ed). *The Correspondence of General Thomas Gage With the Secretaries of State, and With the War Office and Treasury.* 2 vols.; New Haven: Yale University Press, 1931.

Carver, Jonathan. *Travels Through the Interior of North America in the Years 1766, 1767 and 1768.* London: J. Waller, 1778.

Casgrain, H.R. (ed.). *Collection des manuscrits du Maréchal de Lévis.* 12 vols.; Montréal and Québec: C.O. Beauchemin & Fils and Demers & Frère, 1889-1895.

Casgrain, H.R. (ed.). *Extraits des archives des ministères de la Marine et de la Guerre à Paris.* Québec: L.-J. Demers & Frère, 1890.

Chabert de Joncaire, Daniel. *Mémoire pour Daniel de Joncaire-Chabert, ci-devant commandant au petit fort de Niagara.* Paris, [1763].

Champlain, Samuel. *Les voyages de la Nouvelle France occidentale.* Paris, 1632.

Chapin, Howard M. *A List of Rhode Island Soldiers & Sailors in the Old French & Indian War, 1755-1762.* Providence: Rhode Island Historical Society, 1918.

Charbonneau, André. *The Fortifications of Île aux Noix.* Ottawa: Parks Canada, 1994.

Charlevoix, Pierre-François-Xavier de. *Histoire et description générale de la Nouvelle France, avec le Journal historique d'un voyage fait par ordre du roi dans l'Amérique septentrionale.* 6 vols.; Paris: 1744.

Charlevoix, Pierre-François-Xavier de. *Journal of a Voyage to North America. Undertaken by Order of the French King. Containing the Geographical Description and Natural History of that Country, particularly Canada. Together with An Account of the Customs, Characters, Religion, Manners and Traditions of the original Inhabitants.* 2 vols.; London: R. and J. Dodsley, 1761.

Chartrand, René. *Canadian Military Heritage: Volume I, 1000-1754.* Montréal: Art Global, Inc., 1993.

Chartrand, René. "Les drapeaux en Nouvelle-France," *Conservation Canada,* I, No. 1 (1974), pp. 24-26.

Chartrand, René. *The French Soldier in Colonial America.* Ottawa: Museum Restoration Service, 1984.

Chartrand, René. *Louisbourg, 1758: Wolfe's First Siege.* Oxford: Osprey Publishing, 2000.

Chartrand, René. *Quebec, 1759.* Oxford: Osprey Publishing, 1999.

Chartrand, René. *Ticonderoga, 1758: Montcalm's Victory Against All Odds.* Oxford: Osprey Publishing, 2000.

Chartrand, René. "The Troops of French Louisiana, 1699-1769," *Military Collector & Historian,* XXV, No. 2 (Summer 1973), pp. 58-65.

Chatham, Earl of (Pitt, William), Chatham Papers, Great Britain, Public Record Office.

Collection de manuscrits contenant lettres, mémoires, et autres documents historiques relatifs à la Nouvelle-France, recueillis aux Archives de la Province de Québec, ou copiés à l'étranger. 4 vols.; Québec, 1883-1885.

Collections of the New-York Historical Society for the Year 1871. New York: The New-York Historical Society, 1872. Papers of Charles Lee.

Collections of the New-York Historical Society for the Year 1891. New York: New-York Historical Society, 1892.

Conyngham, Redmond. "An Account of the Settlement of the Dunkers at Ephrata, in Lancaster County, Pennsylvania ... to Which is Added a Short History of That Religious Society by the Late Rev. Christian Endress of Lancaster," *Memoirs of the Historical Society of Pennsylvania, Vol. II.* Philadelphia: Carey, Lea & Carey, 1827, pp. 135-53.

Coolidge, Guy. *The French Occupation of the Champlain Valley from 1609 to 1759.* Harrison, NY: Harbor Hill Books, 1979.

Coti, Rinatu (ed.). *Ponte Novu: Journal d'un officier français an 1769.* [Corsica], 1986. This reproduces an earlier edition of *Journal des campagnes de 1768 & 1769 en Corse par le chevalier de Lenchères.*

Courcelles, Jean B.D., Chevalier de. *Dictionnaire historique et biographique des généraux français depuis le onzième siècle jusqu'en 1823.* 8 vols.; Paris: Imprimerie de Plassan, 1823.

Courville, Louis de. *Mémoires sur le Canada depuis 1749 jusqu'à 1760.* Québec: T. Cary & Cie., 1838.

Cuneo, John R. *Robert Rogers of the Rangers.* New York: Oxford University Press, 1959.

Dahlinger, Charles W. "The Marquis Duquesne, Sieur de Menneville, Founder of the City of Pittsburgh", *Western Pennsylvania Historical Magazine,* XV (1932), pp. 2-33, 121-51, 219-62.

Dictionary of Canadian Biography. 12 vols. and index; Toronto, 1966-1991.

Dictionary of National Biography. 22 vols.; London: Smith, Elder & Co., 1908.

Dobson, John. *Chronological Annals of the War; From Its Beginning to the Present Time.* Oxford: Clarendon Press, 1763.

Doughty, A.G. and Parmalee, G.W. *The Siege of Quebec and the Battle of the Plains of Abraham.* 6 vols.; Québec: Dussault & Proulx, 1901.

Dow, Charles Mason. *Anthology and Bibliography of Niagara Falls.* 2 vols.; Albany: State of New York, 1921.

Dowd, Gregory Evans. *War Under Heaven: Pontiac, the Indian Uprising, and the British Empire.* Baltimore: Johns Hopkins University Press, 2002.

Downey, Fairfax. *Louisbourg: Key to a Continent.* Englewood Cliffs, NJ: Prentice-Hall, Inc., 1965.

Duffy, Christopher. *The Fortress in the Age of Vauban and Frederick the Great, 1660-1789.* London: Routledge & Kegan Paul, 1985.

Dunnigan, Brian Leigh. *Glorious Old Relic: The French Castle and Old Fort Niagara.* Youngstown, NY: Old Fort Niagara Association, 1987.

Dunnigan, Brian Leigh. "Portaging Niagara," *Inland Seas,* XLII, #3 (Fall 1986), pp. 177-83, 216-23.

Dunnigan, Brian Leigh. *Siege 1759: The Campaign Against Niagara.* Revised Edition. Youngstown, NY: Old Fort Niagara Association, 1996.

Dunnigan, Brian Leigh and Scott, Patricia Kay. *Old Fort Niagara in Four Centuries: A History of Its Development.* Youngstown, NY: Old Fort Niagara Association, 1991.

Dupuy, R. Ernest and Dupuy, Trevor N. *The Encyclopedia of Military History.* New York: Harper & Row, 1970.

Duverger de Saint-Blin, [Paul-François]. *Mémoire pour le sieur Duverger de Saint-Blin, lieutenant d'infanterie dans les troupes étant ci-devant en Canada.* Paris, Moreau, 1763.

Eccles, W.J. *France in America.* Toronto: Fitzhenry & Whiteside, Ltd., 1972.

Edmunds, R. David and Peyser, Joseph L. *The Fox Wars: The Mesquakie Challenge to New France.* Norman, OK: University of Oklahoma Press, 1993.

Elting, John (ed.). *Military Uniforms in America: The Era of the American Revolution, 1755-1795.* San Rafael, CA: Presidio Press, 1974.

Entick, John. *The General History of the Late War Containing It's Rise, Progress, and Event in Europe, Asia, Africa, and America.* 5 vols.; London: Edward Dilly and John Millan, 1763.

Ephémérides du citoyen, ou Chronique de l'esprit national. Paris, 1765-1772 and 1774-1776.

Erffa, Helmut von and Staley, Allen. *The Paintings of Benjamin West.* New Haven: Yale University Press, 1986.

Evans, Lewis. *Geographical, Historical, Political, Philosophical and Mechanical Essays. The First, Containing an Analysis of a General Map of the Middle British Colonies in America, and of the Country of the Confederate Indians.* Philadelphia, 1755.

Fauteux, Aegidius. *Les Chevaliers de Saint-Louis en Canada.* Montréal: Les Éditions Dix, 1940.

[Fauteux, Aegidius] "Officiers de Montcalm," *Revue d'Histoire de l'Amérique Française,* III, No. 3 (December 1949), pp. 367-82.

Fauteux, Aegidius. "Quelques officiers de Montcalm," *Revue d'Histoire de l'Amérique Française,* V, No. 3 (December 1951), pp. 404-15.

Ford, Worthington Chauncey, ed. *The Writings of George Washington.* 14 vols.; New York: G.P. Putnam's Sons, 1889-93.

Fortier, Paul. "La Galette versus Cataraqui: The Forgotten Rivalry for Dominance on the Upper St. Lawrence, 1673-1796," unpublished paper, 1979. Copy in the Old Fort Niagara Association library.

Frégault, Guy. *Canada: The War of the Conquest.* Toronto: Oxford University Press, 1969. Translated by Margaret M. Cameron.

Fry, Bruce W. *"An Appearance of Strength:" The Fortifications of Louisbourg.* 2 vols.; Ottawa: Parks Canada, 1984.

Fuller, Nathaniel. Manuscript journal, March 13-October 28, 1760, William L. Clements Library, Ann Arbor, Michigan.

Gabriel, Abbé Charles-Nicolas. *Le Maréchal de camp Desandrouins, 1729-1792.* Verdun: Renvé-Lallement, 1887.

Gage, Thomas. Papers of Thomas Gage. William L. Clements Library, Ann Arbor, Michigan.

Gallay, Alan, ed. *Colonial Wars of North America, 1512-1763: An Encyclopedia.* New York: Garland Publishing, Inc., 1996.

Gélinas, Cyrille. *The Role of Fort Chambly in the Development of New France, 1665-1760.* Ottawa: Parks Canada, 1983.

Gentleman's Magazine. London, 1756-1760.

Gibson, John. *Great Waterfalls, Cataracts, and Geysers.* London: Thomas Nelson and Sons, 1887.

Gipson, Lawrence Henry. *The British Empire Before the American Revolution, Volume IV: Zones of International Friction: North America South of the Great Lakes Region, 1748-1754.* New York: Alfred A. Knopf, 1939.

Gipson, Lawrence Henry. *The British Empire Before the American Revolution, Volume V: Zones of International Friction: The Great Lakes Frontier, Canada, the West Indies, India, 1748-1754.* New York: Alfred A. Knopf, 1942.

Gipson, Lawrence Henry. *The British Empire Before the American Revolution, Volume VI: The Great War for the Empire, The Years of Defeat, 1754-1757.* New York: Alfred A. Knopf, 1949.

Gipson, Lawrence Henry. *The British Empire Before the American Revolution, Volume VII: The Great War for the Empire, The Victorious Years, 1758-1760.* New York: Alfred A. Knopf, 1946.

Gipson, Lawrence Henry. *The British Empire Before the American Revolution, Volume VIII: The Great War for the Empire, The Culmination, 1760-1763.* New York: Alfred A. Knopf, 1954.

The Grand Canal Cruise 'n Chart. Albany: New York State Office of Parks and Recreation, 1974.

[Grant, Francis]. "Journal from New York to Canada, 1767," *Proceedings of the New York State Historical Association,* XXX, 1932, pp. 181-96, 305-22.

Grant, W.L. "The Capture of Oswego in 1756," *Proceedings of the New York State Historical Association,* XIII, 1914, pp. 338-67.

Graves, Donald E., ed. *Fighting for Canada: Seven Battles, 1758-1945.* Toronto: Robin Brass Studio, 2000.

Graves, Donald E. *French Military Terminology, 1670-1815.* Saint John, New Brunswick: The New Brunswick Museum, 1979.

Green, Ernest. "Corvettes of New France," *Ontario Historical Society Papers and Records,* XXXV (1943), pp. 29-38.

Green, John. *Remarks in Support of the New Chart of North and South America, in Six Sheets.* London, 1753.

Grenier, Fernand, ed. *Papiers Contrecoeur et autres documents concernant le conflit Anglo-Français sur l'Ohio de 1745 à 1756.* 3 vols.; Québec: Les Presses Universitaires Laval, 1952.

Grimm, Jacob L. *Archaeological Investigation of Fort Ligonier, 1960-1965.* Pittsburgh: Carnegie Museum, 1970.

Hager, Robert E. *Mohawk River Boats and Navigation Before 1820.* Syracuse: Canal Society of New York State, 1987.

Hagerty, Gilbert. *Massacre at Fort Bull: The de Lery Expedition Against the Oneida Carry, 1756.* Providence, RI: Mowbray Company Publishers, 1971.

Haldimand, Frederick. Frederick Haldimand Papers. Great Britain, British Library, Add. MSS 21661-21892.

Hall, Thadd E. *France and the Eighteenth Century Corsican Question.* New York: New York University Press, 1971.

Hamilton, Edward P. (trans. and ed.). *Adventure in the Wilderness: The American Journals of Louis Antoine de Bougainville, 1756-1760.* Norman, OK: University of Oklahoma Press, 1964 (1990 edition).

Hamilton, Edward P. *The French Army in America.* Ottawa: Museum Restoration Service, 1967.

Hamilton, Stanislaus Murray, ed. *Letters to Washington and Accompanying Papers.* 5 vols.; Boston: Houghton Mifflin, 1898-1902.

Hammersley, Sydney Ernest. *The History of Waterford, New York.* Waterford: Privately printed, 1957.

Hare, Robert. "Memoranda on a Tour Through a Part of North America in Company with Mr. Wm. Allen," *Collections of the Historical Society of Pennsylvania,* I, No. 6 (November 1853), pp. 363-76.

Harrington, J.C. *New Light on Washington's Fort Necessity: A Report on the Archaeological Explorations at Fort Necessity National Battlefield Site.* Richmond, VA: Eastern National Park and Monument Association, 1957.

Harris, R. Cole (ed.).*Historical Atlas of Canada, Vol. I: From the Beginning to 1800.* Toronto: University of Toronto, 1987.

Hastings, Hugh (ed.). *Orderly Book and Journal of Major John Hawkes.* New York: Society of Colonial Wars in the State of New York, 1911.

Hauser, Judith Ann. *Jesuit Rings from Fort Michilimackinac and Other European Contact Sites.* Mackinac Island, MI: Mackinac Island State Park Commission, 1982.

Heldman Donald P. and Grange, Roger T., Jr. *Excavations at Fort Michilimackinac: 1978-1979: The Rue de la Babillarde.* Mackinac Island, MI: Mackinac Island State Park Commission, 1981.

Hennepin, Père Louis. *Description de la Louisiane, nouvellement découverte au sud-ouest de la Nouvelle France.* Paris: Sébastien Huré, 1683.

Hennepin, Père Louis. *Nouvelle découverte d'un trés grand pays Situé dans l'Amérique entre le Nouveau Mexique et la mer Glaciale avec les cartes & les figures necessaires, & de l'Histoire naturelle & Morale, & les avantages, qu'on en peut tirer par l'etablissement des colonies.* Utrecht: Guillaume Broedelet, 1697.

Hervey, William. *Journals of the Hon. William Hervey.* Bury St. Edmunds, England: Paul & Mathew, 1906.

Hill, William H. *Old Fort Edward Before 1800.* Fort Edward, NY: Privately printed, 1929.

Hodges, Frederick Webb (ed.). *Handbook of American Indians North of Mexico.* 2 vols.; New York: Pageant Books, Inc., 1959.

Holden, James A. and Inglis, John A. "The Lineage of Colonel George Monro," *Proceedings of the New York State Historical Association,* XIII, 1914, pp. 389-403.

Houlding, J.A. *French Arms Drill of the 18th Century.* Bloomfield, Ont.: Museum Restoration Service, 1988.

Hubbard, R.H., ed. *Thomas Davies, c.1737-1812: An Exhibition Organized by the National Gallery of Canada.* Ottawa: National Gallery of Canada, 1972.

Hulbert, Archer Butler (ed.). *The Crown Collection of Photographs of American Maps.* 3 series, 12 vols.; Cleveland: Arthur H. Clarke Co., 1907.

Hunt, George T. *The Wars of the Iroquois: A Study in Intertribal Trade Relations.* Madison: The University of Wisconsin Press, 1967.

Hunter, William A. *Forts on the Pennsylvania Frontier, 1753-1758.* Harrisburg: Pennsylvania Historical and Museum Commission, 1960.

[Izard, Ralph]. *An Account of a Journey to Niagara, Montreal and Quebec in 1765; or "'Tis Eighty Years Since."* New York: William Osborn, 1846.

James, Alfred Procter. "Decision at the Forks," *The Western Pennsylvania Historical Magazine,* XLI, Nos. 3 and 4 (Autumn, 1958), pp. 1-56.

James, Alfred Procter (ed.). *Writings of General John Forbes Relating to his Service in North America.* Menasha, WI: Collegiate Press, 1938.

Johnson, Allen and Malone, Dumas (eds.). *Dictionary of American Biography.* 10 vols. and supplements; New York: Charles Scribner's Sons, 1927-1936 and subsequent editions.

The Journal of Major George Washington. Williamsburg, VA: Colonial Williamsburg, 1959.

"Journal of the Siege of Niagara, Translated from the French," *New York Mercury,* August 20, 1759.

Kalm, Peter. *Travels into North America; Containing Its Natural History, and A circumstantial Account of its Plantations and Agriculture in general, with the Civil, Ecclesiastical and Commercial State of the Country, The Manners of the Inhabitants, and several curious and Important Remarks on various Subjects.* Warrington: William Eyres, 1770. Reprinted as *Peter Kalm's Travels in North America; The English Version of 1770.* Edited by Adolph B. Benson. New York: Dover Publications, 1987.

Karpinski, Louis C. "Manuscript Maps Prior to 1800, Relating to America; Photographic Facsimiles Made from Originals in Various Libraries and Archives in Paris, Spain and Portugal." Copy in the William L. Clements Library.

Kennett, Lee. *The French Armies in the Seven Years' War: A Study in Military Organization and Administration.* Durham, NC: Duke University Press, 1967.

Kent, Donald H. *The French Invasion of Western Pennsylvania, 1753.* Harrisburg: Pennsylvania Historical and Museum Commission, 1954.

Kimball, Gertrude Selwyn (ed.). *Correspondence of William Pitt.* 2 vols.; New York: The Macmillan Company, 1906.

Klein, Walter C. *Johann Conrad Beissel, Mystic & Martinet, 1690-1798.* Philadelphia: University of Pennsylvania Press, 1942.

Knight, Vernon J., Jr. and Adams, Sherée L. (eds.). "A Voyage to the Mobile and Tomeh in 1700, with Notes on the Interior of Alabama," *Journal of Alabama Archaeology,* XXVII, No. 1 (June 1981), pp. 32-56.

Knox, John. *An Historical Journal of the Campaigns in North America for the Years 1757, 1758, 1759, and 1760.* 2 vols.; London, 1764. Edited by Arthur G. Doughty and reprinted in 3 vols.; Toronto: The Champlain Society, 1914-1916.

Kopperman, Paul E. *Braddock at the Monongahela.* Pittsburgh: University of Pittsburgh Press, 1977.

Kunitz, Stanley J. and Colby, Vineta. *European Authors, 1000 to 1900.* New York: The H.W. Wilson Co., 1967.

Lacour-Gayet, G. *La Marine militaire de la France sous le règne de Louis XV.* Paris: Librairie ancienne, 1910.

Lafitau, Joseph-François. *Customs of the American Indians Compared with the Customs of Primitive Times.* Trans. and edited by William N. Fenton and Elizabeth L. Moore. 2 vols.: Champlain Society, 1974 and 1977.

Lafitau, Joseph-François. *Moeurs des sauvages ameriquaines, comparées aux moeurs des premiers temps.* 2 vols.; Paris: Saugrain, 1724.

Lafitau, père Joseph-François. *Mémoire présenté à S.A.R. Mgr le duc d'Orléans concernant la précieuse plante du ginseng, découverte en Canada par le P. Joseph-François Lafitau.* Paris, 1718.

Lahontan, Louis-Armand de Lom d'Arce. *Nouveaux voyages de Mr. Baron de Lahontan dans l'Amérique septentrionale.* 2 vols.; La Haye: les Fréres l'Honoré, 1703.

Lawson, C.C.P. *A History of the Uniforms of the British Army,* Volume II. London: Norman Military Publications, 1941 (1963 edition).

Leach, Douglas Edward. *Roots of Conflict: British Armed Forces and Colonial Americans, 1677-1763.* Chapel Hill, NC: University of North Carolina Press, 1986.

Lelièpvre, Eugène and Chartrand, René. "Béarn Regiment, New France, 1755-1757," *Military Collector & Historian,* XXXIX, No. 3 (Fall 1987), pp. 126-27.

Lelièpvre, Eugène and Chartrand René, "Corps of Cavalry, Canada, 1759-1760," *Military Collector & Historian,* XXVIII, No. 3 (Fall 1976), pp. 131-32.

Lescarbot, Marc. *Histoire de la Nouvelle France.* Paris: 1609.

Le Sueur, William D. *Count Frontenac.* Toronto: University of Toronto Press, 1964 edition.

Lincoln, Charles Henry (ed.). *Correspondence of William Shirley, Governor of Massachusetts and Military Commander in America, 1731-1760.* 2 vols.; New York: The Macmillan Company, 1912.

Lincoln, Charles Henry (ed.). "Manuscript Records of the French and Indian War in the Library of the Society," *Transactions and Collections of the American Antiquarian Society, Vol. XI.* Worcester, MA: American Antiquarian Society, 1909.

Lintvelt, Jaap and Paré, François, eds. *Shifting Boundaries: Place and Space in the Francophone Cultures of Canada.* Amsterdam: Rodopi, 2001.

Loescher, Burt Garfield. *The History of Rogers Rangers.* 2 vols.; San Francisco, 1946 and San Mateo, CA 1969.

Loescher, Burt Garfield. *The History of Rogers Rangers, Volume III: Officers and Non-Commissioned Officers.* Burlingame, CA: Privately printed, 1957.

Loudon, Earl of. Loudon Papers, Huntington Library, San Marino, California.

Luzader, John F. "Fort Stanwix: Construction and Military History, 1758 to 1777," *Fort Stanwix: History, Historic Furnishing, and Historic Structures Report.* Washington: National Park Service, 1976.

MacKenzie, Frederick. Papers of Frederick MacKenzie. William L. Clements Library, Ann Arbor, Michigan.

MacLeod, Malcolm. "Fight at the West Gate, 1760," *Ontario History*, LVIII, No. 3 (September 1966), pp. 172-94.

Malcomson, Robert. *Warships of the Great Lakes, 1754-1834.* London: Chatham Publishing, 2001.

Mancall, Peter C. *Deadly Medicine: Indians and Alcohol in Early America.* Ithaca: Cornell University Press, 1995.

Master Drawings from the National Gallery of Canada. Washington: National Gallery of Art, 1988.

McCardell, Lee. *Ill-Starred General: Braddock of the Coldstream Guards.* Pittsburgh: University of Pittsburgh Press, 1958.

McCulloch, Ian. "*Believe Us, Sir, This Will Impress Few People!:* Spin Doctoring – 18th Century Style," *The Bulletin of the Fort Ticonderoga Museum*, XVI, No. 1 (1998), pp. 92-107.

McIlwain, Charles Howard (ed.). *An Abridgement of the Indian Affairs ... Transacted in the Colony of New York from the Year 1678 to the Year 1751 by Peter Wraxall.* New York: Benjamin Blom, 1968.

McLennan, John Stuart. *Louisbourg from its Foundation to its Fall, 1713-1758.* London: Macmillan and Co., Ltd., 1918.

Malartic, Gabriel de Maurès de and Gaffarel, Paul (eds.). *Journal des campagnes au Canada de 1755 à 1760 par le comte de Maurès de Malartic.* Paris: Librairie Plon, 1890.

Manders, Eric, Dunnigan, Brian Leigh and Elting, John. "80th Regiment of Foot, 1757-1764," *Military Collector & Historian,* XXXVI, #4 (Winter, 1987), pp. 172-73.

Mante, Thomas. *The History of the Late War in North America and the Islands of the West-Indies.* London, W. Strahan & T. Cadell, 1772.

Maps, Reports, Estimates, &c. Relative to Improvements of the Navigation of the River St. Lawrence and a Proposed Canal Connecting the River St. Lawrence and Lake Champlain. Toronto: John Lovell, 1856.

Mémoires des Commissaires du roi et de ceux de Sa Majesté britannique, sur les possessions & les droits respectifs des deux couronnes en Amérique; avec les actes publics & piéces justificatives .... 4 vols.; Paris: Imprimerie royale, 1755-1757.

Mémoires des Commissaires du roi et de ceux de Sa Majesté britannique, sur les possessions & les droits respectifs des deux couronnes en Amérique; avec les actes publics & piéces justificatives .... 6 vols.; Paris: Imprimerie royale, 1756.

Mendel, Mesick, Cohen, Architects. *Fort Johnson: A Historic Structure Report.* Albany: Mendel, Mesick, Cohen, 1977.

Mennonite Encyclopedia. Scottdale, PA: Mennonite Publishing House, 1959.

Mereness, Newton D. (ed.). *Travels in the American Colonies.* New York: The Macmillan Company, 1916.

Millar, John F. *American Ships of the Colonial & Revolutionary Periods.* New York: W.W. Norton & Company, Inc., 1978.

Mitchell, John. *The Contest in America Between Great Britain and France.* London, 1757.

Mohar, Ronald E. "Oswego's First Permanent Settlement: The Connection Between Oswego & Niagara," unpublished manuscript, 1993. Copy in the Old Fort Niagara Association library.

Monypenney, Alexander. "The Monypenny Orderly Book," *The Bulletin of the Fort Ticonderoga Museum,* XIII, No. 2 (June 1971), pp. 151-84.

Moody, Kevin and Fisher, Charles L. "Archaeological Evidence of the Colonial Occupation at Schoharie Crossing State Historic Site, Montgomery County, New York," *The Bulletin: Journal of the New York State Archaeological Association,* No. 99 (Fall 1989), pp. 1-13.

[Moreau, Jacob Nicolas], *Mémoire contenant le précis des faits avec leurs piéces justificatives, pour servir de résponse aux Observations envoyées par les ministres d'Angleterre dans les cours de l'Europe.* Paris: Imprimerie royale, 1756.

Mouillard, Lucien. *Les Régiments sous Louis XV.* Paris: Librairie Militaire de J. Dumaine, 1882.

Muller, John. *The Attack and Defence of Fortified Places.* Third edition, London: J. Millan, 1765.

Muller, John. *A Treatise Containing the Elementary Part of Fortification, Regular and Irregular.* London: J. Nourse, 1746; reprinted Ottawa: Museum Restoration Service, 1968.

Mulligan, Robert E. Jr. (ed.). "Colonel Charles Clinton's Journal of His Campaign in New York July to October, 1758 During the French War," *The Bulletin of the Fort Ticonderoga Museum,* XV, No. 4 (1992), pp. 292-315.

Munsell, Joel. *Collections on the History of Albany.* Albany: J. Munsell, 1867.

"Murray Atlas", 1761-1765, particularly "Plan of the Part of Canada and the River St. Lawrence Which Lies Between the Coteau des Cedres." William L. Clements Library, Ann Arbor, Michigan.

Nebenzahl, Kenneth and Higginbotham, Don. *Atlas of the American Revolution.* New York: Rand McNally, 1974.

New Hampshire Gazette, Portsmouth, 1758.

New York Mercury. New York City, 1755-1761.

New York State Atlas & Gazetteer. Third Edition; Freeport, ME: DeLorme Mapping, 1991.

Nicolai, Martin L. "A Different Kind of Courage: The French Military and the Canadian Irregular Soldier During the Seven Years' War," *Canadian Historical Review,* LXX, No. 1 (1989), pp. 53-75.

The Northwest Passage Cruise 'n Chart. Albany: New York State Office of Parks and Recreation, 1974.

O'Callaghan, E.B. (ed.). *Documentary History of New York.* 4 vols.; Albany: Weed, Parsons & Co., 1849-51.

O'Callaghan, E.B. (ed.). *Documents Relative to the Colonial History of the State of New York.* 15 vols.; Albany: Weed, Parsons & Co., 1856-77.

"Officiers du régiment de Béarn," *Le bulletin des recherches historiques,* LI, No. 10 (October 1945), pp. 352-58.

"Les Officiers du régiment de Guyenne," *Le bulletin des recherches historiques,* LI, No. 5 (May 1945), pp. 189-94.

"Officiers du régiment de Languedoc," *Le bulletin des recherches historiques,* LI, No. 8 (August 1945), pp. 283-89.

"Officiers du régiment de la Reine," *Le bulletin des recherches historiques,* LI, No. 11 (November 1945), pp. 385-91.

"Officiers du régiment de la Sarre," *Le bulletin des recherches historiques,* LI, No. 7 (July 1945), pp. 251-60.

"Officiers du régiment de Royal-Roussilon," *Le bulletin des recherches historiques,* LI, No. 12 (December 1945), pp. 425-32.

O'Meara, Walter. *Guns at the Forks.* Englewood Cliffs, NJ: Prentice-Hall, Inc., 1965.

Ordonnance du Roy sur le Maniement des armes de l'Infanterie Françoise & Étranger Du 7 May 1750. Arras: Imprimerie de la Veuve Duchamp, 1750.

Pace, Antonio (trans. and ed.). *Luigi Castiglioni's Viaggio.* Syracuse: Syracuse University Press, 1983.

Pargellis, Stanley M. *Lord Loudon in North America.* New Haven, CT: Yale Historical Publications, 1933.

Pargellis, Stanley M. (ed.). *Military Affairs in North America, 1748-1765.* New York: D. Appleton-Century Company, Inc., 1936.

Parkman, Francis. *A Half-Century of Conflict.* New York: Collier Books, 1962.

Parker, John (ed.). *The Journals of Jonathan Carver and Related Documents, 1766-1770.* St. Paul, MN: Minnesota Historical Society Press, 1976.

Pauw, Cornelius de. *Recherches philosophiques sur les Américains, ou Mémoires intéressants pour servir à l'histoire de l'espèce humaine.* Berlin, 1768-1769.

Pease, Theodore Calvin and Jenison, Ernestine (eds.). "Illinois on the Eve of the Seven Years' War, 1747-1755," *Collections of the Illinois State Historical Library, XXIX.* Springfield: The Illinois State Historical Library, 1940.

Peckham, Howard H. *Pontiac and the Indian Uprising.* New York: Russel & Russel, 1947 (1970 edition).

Peckham, Howard H. (ed.). "Thomas Gist's Indian Captivity, 1758-1759," *The Pennsylvania Magazine of History and Biography,* LXXX (July 1956), pp. 285-311.

Pell, S.H.P. *Fort Ticonderoga: A Short History.* Ticonderoga, NY: Fort Ticonderoga Museum, 1966.

Pennsylvania Archives, Second Series, Vol. II. Harrisburg: Pennsylvania Archives, 1876.

Pennsylvania Archives, Second Series, Vol. VI. Harrisburg: Pennsylvania Archives, 1876.

Pennsylvania Gazette. Philadelphia, 1755-1760.

Peyser, Joseph L., trans. and ed. *Jacques Legardeur de Saint-Pierre: Officer, Gentleman, Entrepreneur.* East Lansing: Michigan State University Press, 1996.

Peyser, Joseph L., trans. and ed. *On the Eve of the Conquest: The Chevalier de Raymond's Critique of New France in 1754.* East Lansing: Michigan State University Press, 1997.

[Pichon, Thomas]. *Lettres et mémoires pour servir à l'histoire naturelle, civile et politique du Cap Breton ... jusqu'à la reprise de cette isle par les Anglois en 1758.* La Haye: P. Gosse, 1760.

Pond, Mrs. Nathan Gillett (ed.). "Experiences in Early Wars in America .... Original Journal of Peter Pond, Born in 1740," *The Journal of American History,* I, No. 1 (First Quarter 1907), pp. 89-93.

Pouchot, M. *Memoir Upon the Late War in North America Between the French and English, 1755-60.* Translated and edited by Franklin B. Hough. 2 vols.; Roxbury, MA: W. Elliot Woodward, 1866.

Pouchot, Pierre. *Mémoires sur la dernière guerre de l'Amérique septentrionale entre la France et l'Angleterre. Suivis d'Observations, dont plusieurs sont relatives au théatre actuel de la guerre, & de nouveaux détails sur les mœurs & les usages des Sauvages, avec des cartes topographiques.* 3 vols.; Yverdon, Switzerland, 1781.

Pownall, Thomas. *A Topographical Description of the Dominions of the United States of America.* Pittsburgh: University of Pittsburgh Press, 1949. Edited by Lois Mulkearn.

Preston, Richard A. and LaMontagne, Leopold. *Royal Fort Frontenac.* Toronto: The Champlain Society, 1958.

Pritchard, James. *Louis XV's Navy, 1748-1762: A Study of Organization and Administration.* Kingston, Ont.: McGill-Queen's University Press, 1987.

Proulx, Gilles. *The Garrison of Québec from 1748 to 1759.* Ottawa: National Historic Sites Parks Service, 1991.

Public Archives of Canada. *The Northcliffe Collection.* Ottawa: F.A. Acland, 1926.

Rahmer, Frederick A. *Dash to Frontenac: An Account of Lt. Col. John Bradstreet's Expedition to and Capture of Fort Frontenac.* Rome, NY: Frederick A. Rahmer, 1973.

Raimo, John W. *Biographical Directory of American Colonial and Revolutionary Governors, 1607-1789.* Westport, CT: Meckler Books, 1980.

Rapport de l'Archiviste de la Province de Québec pour 1923-1924. Québec: L. A. Proulx, 1924. "La Mission de M. de Bougainville en France en 1758-1759"; "Mémoire sur l'état de la Nouvelle-France" (1757); "Le Journal de M. de Bougainville."

Rapport de l'Archiviste de la Province de Québec pour 1924-1925. Québec: L. A. Proulx, 1925. "Mémoire du Canada."

Rapport de l'Archiviste de la Province de Québec pour 1926-1927. Québec: L. Amable Proulx, 1927. Journals of Joseph-Gaspard Chaussegros de Léry for 1749 and 1754-1755 and for the attack on Fort Bull and the siege of Oswego in 1756.

Rapport de l'Archiviste de la Province de Québec pour 1927-1928. Québec: L. Amable Proulx, 1928. "Mémoire sur les postes de Canada" by Chevalier de Raymond (1754) and Joseph-Gaspard Chaussegros de Léry's journal for 1754-1755.

Rapport de l'Archiviste de la Province de Québec pour 1931-1932. Québec: Rédempti Paradis, 1932. La Pause, "Mémoire et observations sur mon voyage en Canada" and "Relation de Poulariès Envoyé à Mr le Marquis de Montcalm."

Rapport de l'Archiviste de la Province de Québec pour 1932-1933. Québec: Rédempti Paradis, 1933. "Les 'mémoires' du chevalier de La Pause."

Rapport de l'Archiviste de la Province de Québec pour 1933-1934. Québec: Rédempti Paradis, 1934. Antoine Roy (ed.), "Les 'papiers' La Pause."

Rapport de l'Archiviste de la Province de Québec pour 1944-1945 Québec: Rédempti Paradis, 1945. "Les Lettres de Doreil."

Ravaisson-Mollien, F. et L. (eds.). *Archives de la Bastille: documents inedits.* 19 vols.; Paris, 1866-1904.

Raynal, Abbé Guillaume-Thomas-François. *Histoire philosophique et politique des établissemens et du commerce des Européens dans les deux Indes.* 6 vols.; Amsterdam, 1770.

Reasonover, J. Roy. *Land Measures: French, Spanish and English Land Measures of the United States with Units, English Equivalents, Reduction and Conversion Factors.* Houston, TX: privately printed, 1946.

Report on the Public Archives for 1929. Ottawa: F.A. Acland, 1930. "Montcalm's Correspondence."

Richards, Frederick B. *The Black Watch at Ticonderoga and Major Duncan Campbell of Inverawe.* Ticonderoga: Fort Ticonderoga Museum, 1926.

Roberts, Robert B. *Encyclopedia of Historic Forts: The Military, Pioneer, and Trading Posts of the United States.* New York: Macmillan Publishing Co., 1988.

Robinson, Percy J. *Toronto During the French Regime.* Toronto: The Ryerson Press, 1933.

Robison, Kenneth G. "John Peter, Jr. of Early Albemarle County, Virginia." Unpublished manuscript. Copy in the Old Fort Niagara Association Library.

Rocque, Mary Ann. *A Set of Plans and Forts in North America, Reduced from Actual Surveys.* London, M.A. Rocque, 1763.

Rogers, Robert. *Journals of Major Robert Rogers.* London, 1765; reprint, New York: Corinth Books, 1961.

Rolt, Richard. *An Impartial Representation of the Conduct of the Several Powers of Europe Engaged in the Late General War: Including Particular Account of All the Military and Naval Operations; from the Commencement of Hostilities Between the Crowns of Great Britain and Spain in 1739 to the Conclusion of the General Treaty of Pacification at Aix La Chapelle, 1748.* 4 vols.; London, 1749.

Roy, Pierre-Georges. *Bigot et sa bande et l'affaire du Canada.* Lévis, Québec, 1950.

Roy, Pierre-Georges. "Les Commandants du Fort Niagara," *Le bulletin des recherches historiques,* LIV, Nos. 5-7 (May-July 1948), pp. 131-40, 163-77, 195-99.

Roy, Pierre-Georges. "Les Commandants du Fort Saint Frédéric," *Le bulletin des recherches historiques,* LI, No. 9 (September 1945), pp. 317-32.

Roy, Pierre-George. *Hommes et choses du Fort Saint-Frédéric.* Montréal: Les Editions Dix, 1946.

Roy, Pierre-Georges (ed.). *Inventaires des papiers de Léry.* 3 vols.; Québec: Archives de la Province de Québec, 1939-40.

Saint-Croix, Guillaume-Emmanuel-Joseph Guilhem de Clermont-Lodève, Baron de. *Observations sur le traité de paix conclu à Paris le 10 février 1763 entre la France, l'Espagne & l'Angleterre: relativement aux intérêts de ces puissances dans la guerre présente.* Yverdon, 1780.

Sautai, Maurice. *Les débuts de la guerre de la succession d'Austriche. Tome premier: L'entrée en Bavière et en Haute-Austriche.* Paris: Librairie militaire R. Chapelot et Cie., 1909.

Sautai, Maurice. *Montcalm at the Battle of Carillon.* Ticonderoga, NY: Fort Ticonderoga Museum, 1941.

Saxe, Maurice, Comte de. *Les Rêveries ou mémoires sur l'art de la guerre de Maurice, Comte de Saxe.* Hague, 1756.

Schoenfeld, Max. *Fort de la Presqu'Ile.* Erie, PA: Erie County Historical Society, 1989.

Scott, Elizabeth M. *French Subsistence at Fort Michilimackinac, 1715-1781.* Mackinac Island, MI: Mackinac Island State Park Commission, 1985.

Scott, Stuart D., Scott, Patricia Kay, Nasca, Paul, Hughes, Christopher J., and Mauzy, David. *An Archaeological Survey of Artpark and the Lower Landing, Lewiston, New York.* Lewiston: Edwin Mellen Press, 1993.

Scull, G.D. (ed.). "The Montresor Journals," *Collections of the New-York Historical Society for the Year 1881.* New York: The New-York Historical Society, 1882.

Service historique de l'Armée, Paris. Mss A^1 and Y^b. Copies in the National Archives of Canada.

Severance, Frank H. *An Old Frontier of France.* 2 vols.; New York: Dodd, Mead & Co., 1917.

Skrine, Francis Henry. *Fontenoy and Great Britain's Share in the War of the Austrian Succession, 1741-1748.* Edinburgh: William Blackwood and Sons, 1906.

Smethurst, Colin, ed. *Romantic Geographies.* Glasgow: University of Glasgow French and German Publications, 1996.

Smith, Capt. George. *An Universal Military Dictionary.* London: J. Millan, 1779; reprinted Ottawa: Museum Restoration Service, 1979.

Smith, William. *A Historical Account of the Expedition Against the Ohio Indians in the Year 1764 ...* Philadelphia, 1765.

Smith, William. *History of Canada.* 2 vols.; Québec: John Neilson, 1815.

Stacey, C.P. *Quebec, 1759: The Siege and the Battle.* Toronto: The Macmillan Co., 1959.

Stacey, C.P. *Quebec, 1759: The Siege and the Battle.* Revised edition, edited and with new materials by Donald E. Graves. Toronto: Robin Brass Studio, 2002.

Stacey, Kim R. "Major General John Small, 1726-1796," *Journal of the Society for Army Historical Research*, LXXIV, No. 298 (Summer, 1996), pp. 102-04.

Stanley, George F.G. *New France: The Last Phase, 1744-1760.* Toronto: McClelland and Stewart, Ltd., 1968.

Steele, Ian K. *Betrayals: Fort William Henry &the "Massacre."* New York: Oxford University Press, 1990.

Steele, Ian and Rhoden, Nancy L., eds. *The Human Tradition in Colonial History.* Wilmington, DE: SR Books, 1999.

Stevens, Sylvester K. et. al. (eds.). *The Papers of Henry Bouquet.* 5 vols.; Harrisburg: The Pennsylvania Historical and Museum Commission, 1972-84.

Stevens, Sylvester K. and Kent, Donald H. (eds.). *Wilderness Chronicles of Northwestern Pennsylvania.* Harrisburg: Pennsylvania Historical Commission, 1941.

Stevens, Sylvester K., Kent, Donald H. and Woods, Emma Edith (eds.). *Travels in New France by J.C.B.* Harrisburg: Pennsylvania Historical Commission, 1941.

Stotz, Charles Morse. *Outposts of the War for Empire: The French and English in Western Pennsylvania: Their Armies, Their Forts, Their People, 1749-1764.* Pittsburgh: Historical Society of Western Pennsylvania, 1985.

Sullivan, James and Hamilton, Milton W. (eds.). *The Papers of Sir William Johnson.* 14 vols.; Albany: University of the State of New York, 1921-65.

Summers, Jack L. and Chartrand, René. *Military Uniforms in Canada, 1665-1970.* Ottawa: Canadian War Museum, 1981.

Susane, Louis. *Histoire de l'infanterie française.* 5 Vols.; Paris, 1876.

Suttles, Wayne (ed.). *Handbook of North American Indians, Volume 7: Northwest Coast.* Washington: Smithsonian Institution, 1990.

Sutton, John L. *The King's Honor and the King's Cardinal: The War of the Polish Succession.* Lexington, KY: The University Press of Kentucky, 1980.

Syrett, David and Dinardo, R.L. *The Commissioned Sea Officers of the Royal Navy, 1660-1815.* Aldershot: Scolar Press, 1994.

Taillemite, Étienne. *Dictionnaire des marins français.* n.p.: Editions maritimes & d'outre mer, 1982.

Tanner, Helen Hornbeck (ed.). *Atlas of Great Lakes Indian History.* Norman: University of Oklahoma Press, 1987.

Tesmer, Irving I., ed. *Colossal Cataract: The Geologic History of Niagara Falls.* Albany: State University of New York Press, 1981.

Thatcher, Joseph. *Cast for War: A History of Muzzle-Loading Artillery.* Albany: New York State Bureau of Historic Sites, 1983.

Thwaites, R.G. (ed.). *The Jesuit Relations and Allied Documents.* 73 vols.; Cleveland: The Burrows Bros. Co., 1896-1901.

Tiplin, Albert H. and Seibel, George A. *Our Romantic Niagara: A Geological History of the River and the Falls.* Niagara Falls, Ont.: The Niagara Falls Heritage Foundation, 1988.

Tonti, Chevalier Henri de. *Dernières découvertes dans l'Amérique septentrionale de M. de La Sale, mises au jour par M. le chevalier Tonti.* Paris, 1697.

Tooley, Ronald Vere. *Tooley's Dictionary of Mapmakers.* Tring, Hertfordshire, England: Map Collector Publications, Ltd., 1979.

Trigger, Bruce G. (ed.). *Handbook of North American Indians, Volume 15: Northeast.* Washington: Smithsonian Institution, 1978.

Truax, Paul W. *LaForce Descendants in North America.* Hinesburg, VT: Vermont Built, Inc. Publications, 2001.

True, Henry (ed.). *Journal and Letters of Rev. Henry True.* Marion, OH: Privately printed, 1900.

Turner, Oliver H. *Pioneer History of the Holland Purchase.* Buffalo: George H. Derby and Co., 1850.

Universal Magazine. London, September 1759.

Van Kley, Dale K. *The Damiens Affair and the Unraveling of the Ancien Régime, 1750-1770.* Princeton: Princeton University Press, 1984.

Vauban, Sébastien Le Prestre de. *A Manual of Siegecraft and Fortification.* Ann Arbor: The University of Michigan Press, 1968. Translated and edited from the edition of 1740 by George A. Rothrock.

Voltaire. *Candide and Other Writings.* New York: The Modern Library, 1956.

Walker, James R. *Lakota Society.* Lincoln: University of Nebraska Press, 1982.

Way, Peter. "Soldiers of Misfortune: New England Regulars and the Fall of Oswego, 1755-1756," *The Massachusetts Historical Review,* III (2001), pp. 50-88.

Webster, J. Clarence (ed.). *Journals of Beauséjour. Diary of John Thomas; Journal of Louis de Courville.* Sackville, New Brunswick: The Tribune Press, 1937.

Webster, J. Clarence (ed.). *The Journals of Jeffery Amherst.* Toronto: The Ryerson Press, 1931.

Webster, J. Clarence (ed.). *The Journal of William Amherst in America, 1758-1760.* London: Butler & Tanner, Ltd., 1927.

Webster, J. Clarence (ed.) and Webster, Alice (trans.). *The Siege of Beauséjour in 1755: A Journal of the Attack on Beauséjour Written by Jacau de Fiedmont.* St. John, New Brunswick: New Brunswick Museum, 1936.

Welles, Lemuel Aiken. "Letters of Col. Nathan Whiting, Written From Camp During the French and Indian War," *Papers of the New Haven Colony Historical Society, Vol. VI.* New Haven: New Haven Colony Historical Society, 1900.

West, J. Martin (ed.). *War for Empire in Western Pennsylvania.* Ligonier, PA: Fort Ligonier Association, 1993.

Westbrook, Nicholas. "*Like Roaring Lions Breaking From Their Chains:* The Highland Regiment at Ticonderoga." *The Bulletin of the Fort Ticonderoga Museum,* XVI, No. 1 (1998), pp. 17-91.

Wilkinson, Spencer. *The Defence of Piedmont, 1742-1748.* Oxford: Clarendon Press, 1927.

Willcox, Cornélius De Witt. *A French-English Military Technical Dictionary.* Washington: Government Printing Office, 1917.

Williamson, George C. (ed.). *Bryan's Dictionary of Painters and Engravers.* 5 vols.; Port Washington, NY: Kennikat Press, 1964. Originally published in 1816.

Wisconsin Historical Collections. 31 vols.; Madison: State Historical Society of Wisconsin, 1855-1931.

Woodhull, Nathaniel. "A Journal Kept by General Nathaniel Woodhull," *The Historical Magazine and Notes and Queries Concerning the Antiquities, History and Biography of America,* V, No. 9 (September 1861), pp. 257-60.

Yates, Austin A. *Schenectady County New York: Its History to the Close of the Nineteenth Century.* N.p.: The New York History Company, 1902. Contains the 1759 journal of Lieutenant Christopher Yates.

Young, Arthur. *Political Essays Concerning the Present State of the British Empire.* London, 1772.

Zaboly, Gary S. (ed.) "A Royal Artillery Officer with Amherst: The Journal of Captain-Lieutenant Henry Skinner, 1 May-28 July 1759," *The Bulletin of the Fort Ticonderoga Museum,* XV, No. 5 (1993), pp. 363-87.

Zacharchuk, Walter and Waddell, P.J.A. *The Excavation of the Machault.* Ottawa: Parks Canada, 1984.

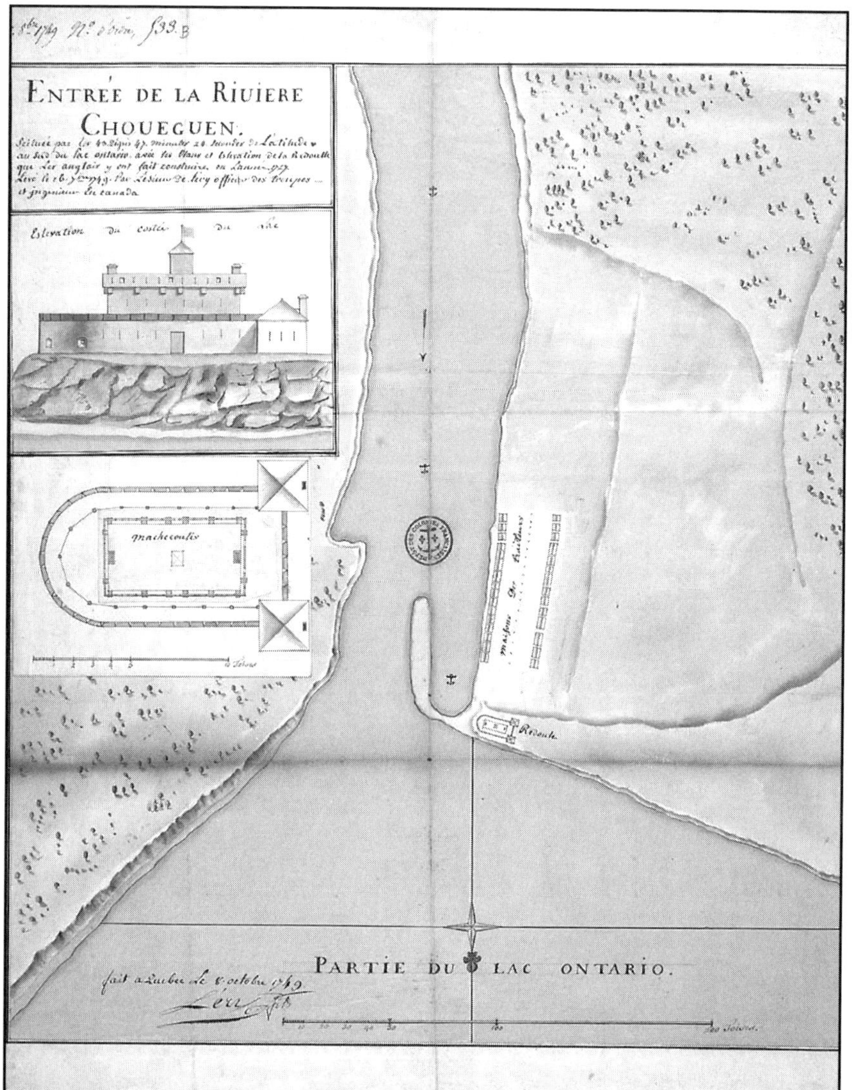

Like Fort Niagara, the post established by the British at Choueguen or Oswego in 1727 was built around a defensible stone house. The old building stood on the west bank of the Oswego River and formed part of the defenses besieged by the French in August 1756. Following the British surrender, Captain Pierre Pouchot, the army's most capable engineer, probably helped direct the demolition of this substantial building. It is shown here in a plan prepared at Quebec, October 8, 1749, by Ensign Gaspard-Joseph Chaussegros de Léry, the younger. The British would later retake Oswego; it was from there that they embarked their assault force for the final attack on Fort Niagara in July of 1759. *Courtesy, Centre des archives d'outre-mer (CAOM), France; CAOM 03DFC 533B.*

INDEX

Note: Square brackets enclose names used by Pouchot when they vary substantially from the usual spellings. Dates for battles and treaties, military units of soldiers, the armament rating or type of naval vessels, alternate names for places and similar information are shown in parentheses. A page number followed by lower case "n" indicates a note on that page. See also "Index Addendum," on Page 631.

• Index Addendum •

Reference Pages 36-37, Pierre Pouchot: a Biographical Addendum

Pouchot's largest known manuscript map is his "Niagara...Finie le 12 Octobre 1757....." It depicts the mouth of the river and the completed Fort Niagara as he left it in the autumn of 1757. This is a detail of the fort. The plan ultimately fell intoEnglish hands and was not included in the *Mémoires*. *Courtesy, British Library, Add MSS 15332-H.*

Editorial Acknowledgements

The preparation of an annotated edition of Pouchot's *Mémoires* has been a project involving many patient and helpful individuals. It is impossible to adequately express my appreciation for the time and effort they spent on generous responses to inquiries. These were provided by the harried staff members of large and small institutions and by individuals who have explored many different aspects of the French-English conflict in America.

Invaluable research assistance was provided by several institutions. *Tim Dubé* of the National Archives of Canada helped obtain much-needed copies of French documents. The staff of the Niagara Falls Public Library, particularly *Don Loker* of its Local History Department, were important local resources. The University of Michigan Libraries made it possible to find rare sources.

As in all my past projects, the William L. Clements Library of the University of Michigan has been both the single most valuable source of eighteenth century documentation and the most congenial place imaginable to conduct research. Special thanks to *David Bosse, Robert Cox,* Director *John Dann, Dick Ryan,* Assistant Director *Arlene Shy,* and *Don Wilcox* for sharing information on their respective specialties and for providing advice. Their colleague and my old friend, *John Harriman,* was, as so many times before, of inestimable assistance during my visits and in long-distance contacts with "the Clements."

Another institution, much closer to home, provided many of the bibliographic resources needed to keep the project moving ahead. Completion of this work would, in fact, have been impossible without access to the fine collections of the library of Brock University in St. Catharines, Ontario. *John Burtniak,* Special Collections Librarian and University Archivist, provided sound advice, morale boosts and great assistance in obtaining rare Canadian publications. *Colleen Beard* of Brock's Geography Department not only made the collections of the Map Library available but identified places I had despaired of locating.

The scope of the project made it necessary to call on friends and colleagues for help in identifying references relating to Pouchot's North American experience. *George Bray* offered his knowledge of journals and newspapers of the period to lead me to obscure accounts of events. *René Chartrand* provided advice based on his unparalleled knowledge of the French military forces in North America. *Al Clifton,* Director of the Niagara Falls Aquarium and *Robert A. Daniels* of the New York State Education Department's Biological Survey Laboratories, helped a historian deal with piscine identifications. *James Corbett* led an "on the ground" research tour of Lake George battlefields and historic sites despite being encumbered with a boisterous train of camp followers. *Dr. Jane Denovchek-Harwell* enthusiastically investigated a not-so-current affair. *Bruce Egli* offered insights on French infantry drill and the *piquets* with which Pouchot spent so much of his time. *Katie Emerson* explained the fine distinctions between Pennsylvania religious sects. *André Gousse* of Parks Canada located appropriate contemporary graphics. *Donald Graves* provided assistance relating to the military history of Canada. *Ned Jenkins,*

Archaeologist/Park Manager of Fort Toulouse, helped with information on the Indians of Alabama. *Patricia Martin* gave botanical advice (most of it in Latin). *Joseph Thatcher* was a resource on period artillery. *Paul Truax* made available fascinating details about his ancestors of the complex La Force clan.

Other friends strengthened the manuscript through their advice and criticism. Special thanks to archaeologist *Paul Huey* of the New York State Office of Parks, Recreation and Historic Preservation, *Dr. Kim McQuaid,* History Department, Lake Erie College, *Timothy Titus,* Interpretive Programs Assistant of Crown Point State Historic Site, *J. Martin West,* Director of the Fort Ligonier Association, and *Nicholas Westbrook,* Director of the Fort Ticonderoga Museum.

Production was as involved as the research. Many members of the Old Fort Niagara staff helped in the conversion of the original translation to a completed work. *Kathryn Packman* and *Brian White* helped ferret out errors in the manuscript. The support of *Ray Wigle, Dennis Farmer* and the other members of the Old Fort Niagara staff was invaluable in permitting my frequent absence from day-to-day operations to enjoy peace and quiet while editing Pouchot's work.

The usual fine support has also been forthcoming from the Publications Committee of the Old Fort Niagara Association. The design and layout of this particularly involved book has once again been skillfully handled by the committee's chair, *Harry M. DeBan.* It was also Harry who originally encouraged the commencement of this long-discussed project.

I have more than special gratitude to *Candice Cain Dunnigan* who moved my part of the project off dead center by transferring Michael Cardy's translation to the computer. Not only was she the first to get to know Pierre Pouchot in English, but she subsequently listened with patience and interest to chatter about problems and discoveries as I learned more about the man whom she soon came to refer to as "your favorite dead person." On days when I worked at home, she also ran interference with *James Cain Dunnigan* and *Claire Beausom Dunnigan* who participated by enhancing several drafts of the manuscript with full-color illustrations of North American fauna.

Finally, I share the gratitude and enthusiasm expressed by translator *Michael Cardy* for having had the wonderful opportunity to enter the front lines of the French and Indian War under the steady guidance of Captain Pierre Pouchot. Michael and I have become fellow travellers in this respect as we approached Pouchot's narrative (and the meanings tucked between his lines) from the differing perspectives of linguist and historian. The time spent on this project with Michael has been stimulating and most educational. He has been a wonderful collaborator.

Brian Leigh Dunnigan
Youngstown, New York

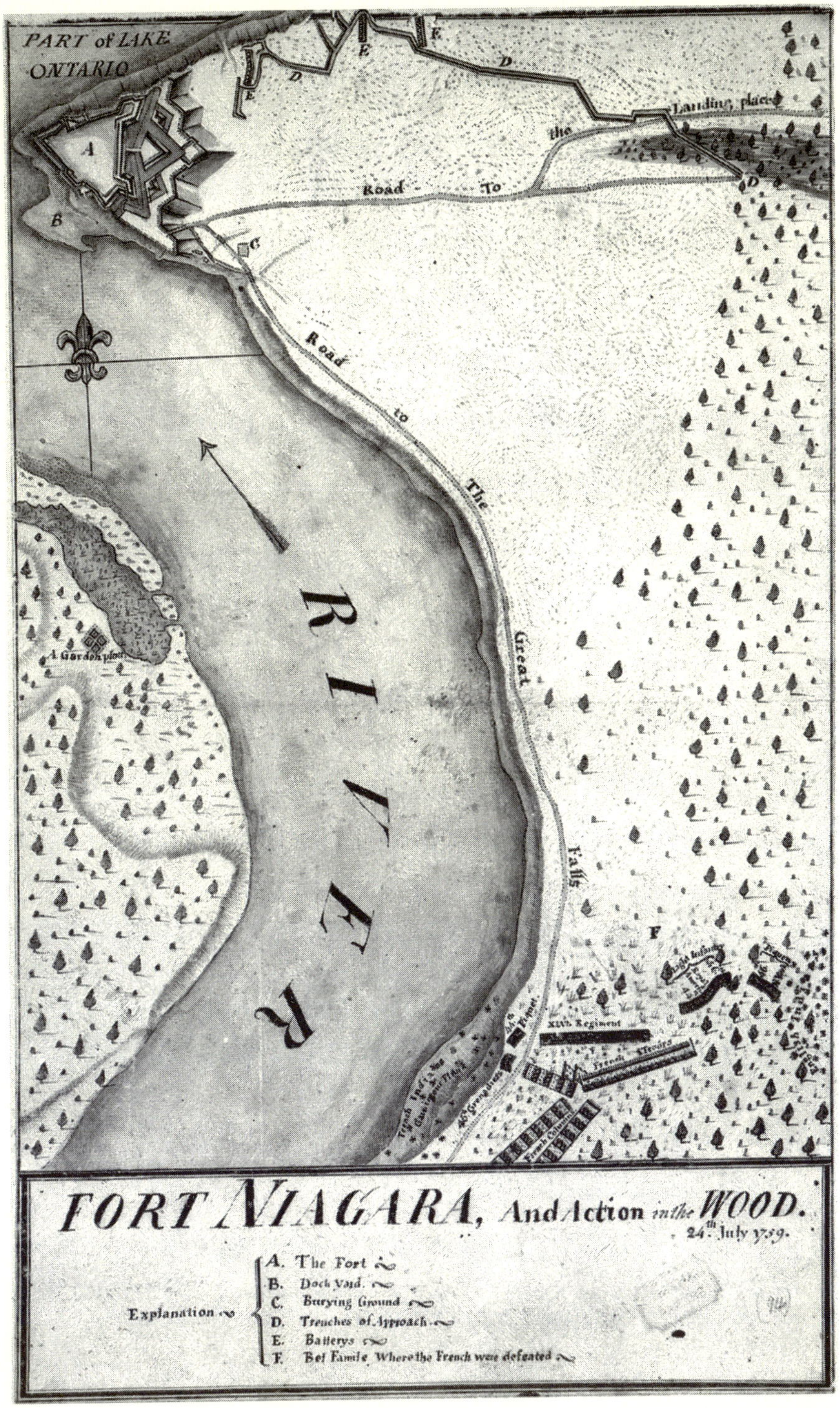

Pierre Pouchot is not known to have drawn a representation of the 1759 siege of Fort Niagara. This plan, prepared by an English officer in 1760, shows both the besiegers' approaches and the battle fought at La Belle Famille on July 24th. *Courtesy, Public Record Office, MPI 277.*

Following the defeat of the French relief force on July 24, 1759, and the surrender of Fort Niagara the following day, Sir William Johnson went to great lengths to keep French officers from captivity among his Iroquois allies. Artist Benjamin West later commemorated Johnson's efforts in a painting entitled "General Johnson Saving a Wounded French Officer from the Tomahawk of a North American Indian." Although West did not specifically identify the scene of his action, numerous elements of the composition suggest that it depicts the siege of Niagara and the aftermath of the battle at *La Belle Famille. Courtesy, Derby Museum and Art Gallery.*

In 1745, Pierre Pouchot was made a *Chevalier de l'ordre royal et Militaire de Saint-Louis* (Knight of the Royal and Military Order of Saint Louis), a coveted honor and distinction aspired to by all French military officers.

The decoration, hung from its red ribbon, is shown in a portrait of Captain Jean-Baptiste-Phillipe Testard de Montigny (1724-1786). Montigny is depicted in upper body armor, a common convention of military portraiture of his time. He served with Pouchot on several occasions and was wounded and captured at the battle of *La Belle Famille. Courtesy, Château Ramezay Museum collection, Montréal*.

The Royal and Military Order of Saint Louis
(l'ordre royal et Militaire de Saint-Louis)

The Order of Saint Louis was the first "Order of Merit" established in France, and was created by Louis XIV in 1693. Dedicated to Saint Louis (of France), it was intended to reward and honor distinguished military service, and was specifically restricted to French Catholic officers who had served a minimum of 10 years in the army. Recipients were not required to be nobles. There were three ranks within the Order: *chevalier* (knight), *commandeur* (commander), and *grand-croix* (grand cross). The king was "Grand Master," and the *Dauphin* (heir apparent to the throne) was automatically a member. The "Cross of St. Louis" was worn by *chevaliers* hanging from a red ribbon, from a red sash for *commandeurs*, and accompanied by a plaque for the *grand-croix*. Receipt, alone, did not confer a title of nobility; however, the Edict of 1750 established that three successive generations of recipients would confer such recognition under certain circumstances. The Order's name was changed to *Décoration militaire* in July 1791, and abolished on October 15, 1792. It was reestablished in 1814, and abolished again in 1830, although recipients were allowed to wear it into the 1860s.

In 1759, a version called the *Mérite militaire*, was instituted and reserved for Swiss, German and Swedish Protestant officers serving in the French army . The *Mérite militaire* did not have a grand-master, but did have the same ranks, and a similar cross, but with a dark blue ribbon. In 1814, it was extended to Muslims and French Protestants. Like the Order of Saint-Louis, it was abolished in 1830.

About The Translator

Michael Cardy, was born in the United Kingdom and holds degrees from the University of Oxford. He served for 24 years on the faculty of Brock University in St. Catharines, Ontario, prior to his appointment as Professor of French and Head of Department at the University of Wales, Swansea, in 1991; there, taking early retirement in 1999. Professor Cardy has published a book and munerous articles on eighteenth-century French literature and ideas, as well as work relating to French discourse about North America in the period of 1530-1800. In the 1990s, he co-edited two books of essays – one on French fiction of the nineteenth and twentieth centuries, the other on twentieth-century theater in French. He recently completed work on a book dealing with two plays by the modern Québec dramatist, Michel Tremblay.

Over the years Professor Cardy has undertaken several translation assignments, though the Pouchot project is by far the largest. He has also published poetry and short fiction in various literary magazines, mainly in Canada. He lives with his wife, Bernice, in Swansea.

About The Editor - Annotator

Brian Leigh Dunnigan is Curator of Maps and Head of Research & Publications at the William L. Clements Library, University of Michigan. He was Executive Director of the Old Fort Niagara Association from 1979 until 1996. A Michigan native, Mr. Dunnigan had previously served as Managing Director of Historic Fort Wayne, Indiana, and in several museum positions for the Mackinac Island State Park Commission. He holds Masters Degrees in History from the University of Michigan and in Museum Studies from the Cooperstown Graduate Program.

The early history of the Great Lakes has long held a particular fascination for Mr. Dunnigan. He has written extensively on the Niagara, Detroit, and Mackinac regions and collaborated in the development of Old Fort Niagara's publication program. He is the author of *Siege, 1759: The Campaign Against Niagara* (1996) and *Frontier Metropolis: Picturing Early Detroit, 1615-1838* (2001).

Mr. Dunnigan lives with his wife, Candice, and children, James and Claire, on a farm in Grass Lake, Michigan.n.

Preserving History At Old Fort Niagara

Since 1927 the preservation and interpretation of Old Fort Niagara have been the goals of the Old Fort Niagara Association, Inc. The Association is a private, not-for-profit organization. Membership is open to anyone with an interest in the Fort and its long history. The Association operates Old Fort Niagara, a State Historic Site, in cooperation with the New York State Office of Parks, Recreation and Historic Preservation.

Old Fort Niagara Association

Old Fort Niagara Publications

Publications are an extension of the Old Fort Niagara Association's educational purpose. Created in 1984, the Publications Committee has been charged with establishing and maintaining an ongoing program of works relevant to the history of Old Fort Niagara. This includes new titles as well as the re-publication of older works, as well as production of other designated audio and visual materials.

Additional information about the Fort's publications, exhibits, programs, or membership in the Old Fort Niagara Association may be obtained from:

Old Fort Niagara National Historic Site
Fort Niagara State Park
P.O. Box 169
Youngstown, New York 14174-0169

Production Notes...The Pouchot Project, Revised Edition, 2004

In Memory of: *Dean Ford Ulrich* **(Old Fort Niagara Publications and Museum Committees);**
Richard Cary, Jr. **(Old Fort Niagara Publications Committee and Director Emeritus, Old Fort Niagara Association);**
Linda Leone **(Fort Niagara Guide amd Volunteer);** ***Sister Mary James Catherine, OSF-Allegany; and Jeannette F. DeBan.***

Type Face: This work is set in ADOBE® CASLON and composed at the Old Fort Niagara Publications office.The type used in the original 1781 edition was various styles of *Caslon*. For ease of reading, this edition does not use all of the *Caslon* ligatures of 1781.

Production Coordinator: Harry M. DeBan

Design & Layout: Harry M. DeBan

Technical Assistance: David J. Bertuca *(State University of New York at Buffalo)*, James F. Egloff *(Rapid Service Engraving, Buffalo, NY)*, Matthew W.T. Jacobs *(PAR Technology Corporation, New Hartford, NY)*

Word Processing: Brian Leigh Dunnigan, Candice Cain Dunnigan, Harry M. DeBan

Additional Research: Harry M. DeBan, Dean F. Ulrich, Harry G. DeBan

Indexing: John Jacob Ulrich Rivardi, Claire Beausom Dunnigan

Proofing: Brian Leigh Dunnigan, Harry M. DeBan, Claire Beausom Dunnigan, Henri-Michel du Lac

Production Assistance: James F. Egloff, John P. Egloff, Joseph Castiglia and Laurie Webb (*Rapid Service Engraving, Buffalo, NY)*

Printing Production: Thomson-Shore, Inc., Dexter, Michigan, USA

Production Support: Marty Abramson; Dennis Griesser; Dr. Keith G. Kozminski *(University of Virginia, Charlottesville);* Laird A. Burkett; Thomas M. O'Donnell; Michelle Sunderlin; Robert A. Marinucci; Craig O. Burt, III; John Burtniak; Douglas W. DeCroix, Cynthia Liddell, Sue Allen, Eric Bloomquist, Ray Wigle, Jerome P. Brubaker and Robert Emerson *(Old Fort Niagara);* André Gousse *(Parks Canada);* Jean François Royal *(Musée du Château Ramezay, Montréal);* Jill Iredale *(Derby Museum and Art Gallery, Derby, England);* François, Sandrine and Henri-Michel du Lac *(du Lac Archives);* James A. Hoffman; George and Carole Bolton; Chuck and Evie Leonard; William Dann, Bruce Strassburg; Rev. Emeric J. Hydo, CM; Rev. John A. Doyle, CM *(Coach !);* Janet Krause *(You are not forgotten !);* and . . . Lt. Col. & Mrs. Harry G. DeBan *(USAF Ret.)* . . . this one's still for you ! — HMD